T0290925

Fundamentals of Groundwater

Fundamentals of Groundwater

Second Edition

Franklin W. Schwartz
Ohio State University
Columbus
USA

Hubao Zhang
Sahuarita
USA

Published by John Wiley & Sons, Inc., Hoboken, New Jersey.
Published simultaneously in Canada.

For general information on our other products and services or for technical support, please contact our Customer Care Department within the United States at (800) 762-2974, outside the United States at (317) 572-3993 or fax (317) 572-4002.

Wiley also publishes its books in a variety of electronic formats. Some content that appears in print may not be available in electronic formats. For more information about Wiley products, visit our website at www.wiley.com.

Library of Congress Cataloging-in-Publication Data applied for:

ISBN: HB: 9781119820130, ePDF: 9781119820147, epub: 9781119820154

Cover Design: Wiley
Cover Image(s): © John Carnemolla/Getty Images

Set in 9.5/12.5pt STIXTwoText by Straive, Pondicherry, India

SKY10061631_120423

This book is dedicated to our families:
(Schwartz) Diane and Cynthia
(Zhang) Ping, Dayu, and Michael

Contents

Preface

After a 20-year hiatus, we are pleased to bring you the 2nd Edition of *Fundamentals of Groundwater*. Like its predecessor, the book is written at an introductory level to facilitate learning and teaching, while maintaining appropriate rigor. Not surprisingly, much has changed with this edition in terms of its look and feel and importantly the content. Most obvious is the use of color in many figures throughout the book. As the book progressed, the dearth in high-quality colored figures and photographs in traditional journal articles became obvious. Fortunately, here in the United States, groundwater-related documents and reports of the U.S. Geological Survey represent a valuable treasure of high-quality graphical material.

This text covers what we consider the basic areas of theory and practice in groundwater. The first half of the book is similar in coverage to the first edition with changes focused on updating methods and concepts. What is new is a "deep dive" into aquifers (Chapter 4) with a focus on large aquifers experiencing problems associated with overpumping. Chapter 5 focuses on basic groundwater flow theory and calculations. Chapter 6 introduces basic concepts in vadose or unsaturated zone. We have also updated and expanded Chapter 7 on geologic and hydrogeologic investigations with the addition of geophysical methods appropriate for larger-scale evaluations of aquifers, and novel approaches for the installation of piezometers. Chapter 8 covers regional groundwater flow and how groundwater systems interact with other components of the hydrologic cycle.

We have maintained our comprehensive treatment of classical well hydraulics and common methods used to interpret results from aquifer tests. Chapter 9 is focused on confined aquifers that are homogeneous, isotropic, and infinite in extent. A new Excel spreadsheet curve fitting method is provided to show students a modern tool to solve aquifer testing problem. Chapter 10 looks at the response of leaky confined aquifers to pumping. Chapter 11 describes the response of an unconfined aquifer to pumping. Slug, step, and intermittent tests are treated in Chapter 12. Chapter 13 introduces how to deal with complex geological settings with analytical and numerical methods.

Subsequent chapters in the back half of the book will return to several at-risk aquifers, considering problems, such as subsidence, streamflow depletion, destruction of riparian ecology, seawater intrusion, and contamination. Chapters 14 and 15 are new with a focus on groundwater sustainability and technical approaches (such as long-term monitoring, water balance, and modeling) to identify, manage, and to treat aquifers at risk. Chapters 16 and 17 address the basic concepts of water chemistry necessary to tackle both natural and contaminated systems. Material added to these chapters reflect a new and emerging interest in evaluation of water quality impacts on human health and novel approaches for health assessments. Studies of groundwater and health in India and Bangladesh emphasize the need to study trace constituents in groundwater, present at μg/L concentrations. Studies by the U.S. Geological Survey, particularly in California have pioneered new strategies in health-risk assessments associated with water quality.

As before, we have maintained our strong coverage of isotopes in groundwater, mass transport, and contaminant hydrogeology in Chapters 18 through 20, respectively. In the case of isotopes, we get beyond basic theory with case studies illustrating how isotopes are used in practice. Our approach to mass transport in Chapter 19 uses transport pathways as an organizing principle along with discussions of physical and chemical processes, which operate to impact concentrations of key constituents. We examine the role of hydrogeologic settings influencing the chemical evolution of natural processes, as well as a focus on redox processes. Chapter 20 introduces the basics of contaminant hydrogeology including dissolved contaminants, LNAPLs and DNAPLs, and strategies for evaluating and interpreting contaminant transport.

This book is designed as an introductory groundwater text for upper-division undergraduate students, or lower-division graduate students, or a refresher for groundwater professionals. Elementary college calculus and chemistry will help to understand the derivation of equations in the book. However, this book does not require readers to have this background because it provides step-by-step solutions for the application of mathematical equations. The book has been improved through the integration of an "aquifer" theme across various chapters to facilitate the teaching of the material and concepts. Students will benefit from a focus on aquifers at risk of depletion in the United States and Asia, and especially the chemical risks to sustainability. As before, we purposely made the text longer than one course can cover to provide instructors choice of materials.

The publisher, John Wiley & Sons, maintains companion websites for the book. On the student site that is open to everyone, readers will find Chapter 21 Modeling Contaminant Transport, software files and spreadsheets. A password-protected instructor site is also available with worked answers to questions found at the end of chapters.

As with most book-writing efforts, we benefited from the support and assistance of our colleagues and, especially, our families. We acknowledge the help of Dr. Sam Lee in producing the book and Dr. Rob Schincariol for his time in effort with suggestions for improving the first edition. Our Wiley Executive Editor, Summers Scholl provided great help in assistance in the books' preparation along with the production team led by Managing Editor Kubra Ameen.

Although it has been more than 20 years since the passing of mentor and colleague, Pat Domenico, his ideas, and influence still live on in this edition. He was a champion of the process-based organization in books, the strong linkages between theory and practice, and the use of quantitative approaches and case studies.

Franklin W. Schwartz
Hubao Zhang

About the Companion Website

This book is accompanied by an instructor and a student companion websites:

www.wiley.com/go/schwartz/fundamentalsofgroundwater2 

The Instructor site is password protected which includes the material shown below:

- Solutions Manual

The Student site is open access which includes the materials shown below:

- Bonus chapter 21
- Analytics spreadsheet
- User guides and confined aquifer spreadsheet
- Software files

1

Introduction to Groundwater

CHAPTER MENU

This book is concerned with the theory and practice of *groundwater hydrology*, the science of water in subsurface environments. It is divided into two main parts; (1) basic concepts dealing with the origin, movement of groundwater, and its recovery from wells, and (2) important areas of practice that these days includes groundwater sustainability, geogenic, and anthropogenic problems of contamination. The book is developed around a process-oriented theme that organizes hydrologic phenomena based on physical and mathematical principles. It emphasizes the application of knowledge to the solution of practical problems focused on aquifers and human impacts, as well as hydraulic testing and groundwater contamination.

One of the important ways to learn about groundwater problems is to experience them firsthand. Thus, the book relies on case studies and demonstrations of techniques through worked problems. Although most of the hydrogeological world is hidden from view, there are exciting things to see in the field. We brought some of these features to life through the colored photographs and illustrations.

1.1 Why Study Groundwater?

There are a variety of reasons why scientists and engineers study groundwater. First and foremost, groundwater is a key source of drinking water that is essential to life on earth, as we know it. The earth has something like 1375 million cubic kilometers with most occurring as non-potable seawater. Groundwater, interestingly, is a tiny fraction, just 0.06% of the earth's available water. However, this relatively small volume is critically important because it represents 98% of the freshwater readily available to humans (Zaporozec and Miller, 2000). Abundant fresh water is tied up in glaciers and icecaps, but essentially unavailable. Of the other available reservoirs, large rivers have been particularly important for their role in sustaining societies for millennia. However, now, the explosive growth of human populations around the world over the last 150 years has required unsustainable development of groundwater supplies, which is among the most serious problems affecting humanity today.

Groundwater is found in aquifers, which have the capability of both storing and transmitting groundwater. An *aquifer* is defined formally as a geologic unit that is sufficiently permeable to supply water to a well. Commonly, the large volumes of water stored in aquifers could be counted on as reliable source during periods of drought lasting months or years. Moreover, hydrogeologists have always been taught that groundwater is a renewable resource, recharged by rain and snow-melt runoff. Indeed, major aquifer systems store impressively large quantities of water. For example, the High Plains Aquifer located in America's Midwest covers an area of 450,000 km^2. It stored an impressive 4000 km^3, roughly equivalent to Lake Huron, the fifth-largest lake in the world. Now, that important function is threatened by groundwater technologies capable of depleting groundwater in just a few generations.

Fundamentals of Groundwater, Second Edition. Franklin W. Schwartz and Hubao Zhang.

Companion website: www.wiley.com/go/schwartz/fundamentalsofgroundwater2

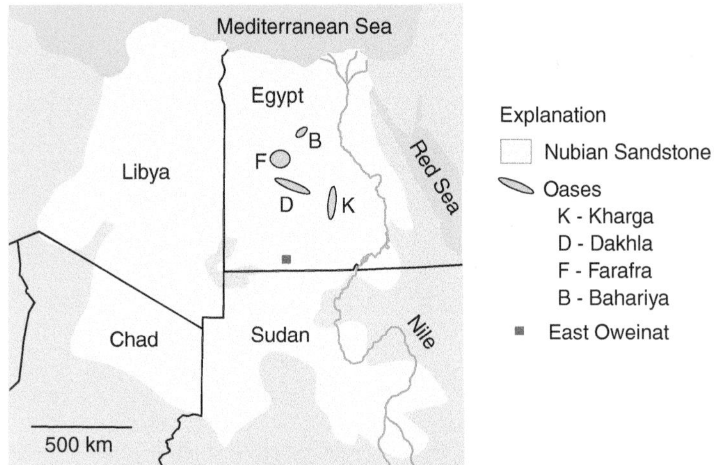

Figure 1.1 The Nubian Sandstone Aquifer System underlies Egypt, Libya, Chad, and Sudan. The oases shown were localized discharge areas for waters leaking upward from the aquifer. East Oweinat is an area where groundwater is mined for irrigation of crops (FWS).

The sad reality is that aquifers can be pumped to the extent that they easily flip to being a nonrenewable resource. Our book is chock full of examples of water supplies in aquifers that are no longer renewable, such as the High Plains aquifer system in the United States. Particularly illustrative in this respect has been the historical utilization of groundwater in the oases of the Western Deserts of Egypt (Figure 1.1) The main oases there occur are associated with erosional depressions within the largely desert landscape. They exist because underlying the Sahara Desert is the Nubian Sandstone Aquifer System. Fracture systems facilitated the natural upward flow of artesian groundwater that discharge into these oases. This water has nurtured human societies for nearly 25,000 years (Caton-Thompson and Gardner, 1932). Moreover, these oases (Figure 1.1) have been centers of trade even for millennia with the first occupation before the Old Kingdom (~2500 BCE) time in Egypt.

The Kharga Oasis (Figure 1.1) was likely a seasonal home for paleolithic humans, who migrated there as early as 25,000 years B.P. Groundwater originally discharged as a series of mound springs in the valley floor. Nearby is the Dakhla Oasis which also provided unique agricultural products during Middle Kingdom of Egypt (Boozer, 2015). Crops in these oases required irrigation from groundwater given that the mean annual rainfall is less than 1 mm year. These oases came under control of the Romans about 30 BCE and with expanded irrigation (Figure 1.2) supplied the empire with olives, dates, and wine (Kaper and Wendrich, 1998).

The Nubian Sandstone aquifer is the largest aquifer in the world with an areal extent of 2.6 million km^2. Rainfall is the original source of water, having infiltrated the subsurface in the south-western corner of Egypt and flowed northeastward hundreds of kilometers to the northeast Studies with krypton isotopes and other isotopes (Sturchio et al., 2004) indicated groundwater ages from 200,000 years (sampled at Dahkla) to 1 million years (at Bahariya).

Since the 1930s, increasing numbers of deep, high-capacity groundwater wells eventually resulted in water-level declines. For example, through the 1950s, deep wells in the Kharga Oasis flowed at the surface. By 1975, all the deep wells had ceased flowing (LaMoreaux et al., 1985). The future now appears to be groundwater mining using deep (800–1000 m) high-capacity wells until storage is depleted.

Groundwater in most places is at risk of depletion because of its essential role in irrigation and food production. Potential solutions to aquifer restoration are technical complicated, expensive, and impractical in developing countries (Schwartz et al., 2020). Thus, although aquifers often facilitate food production in arid lands for a while, the water will eventually be used up when rates of production exceed rates of replenishment. Shown in Figure 1.3 is a satellite view the East Oweinat Project, located in southern Egypt (map Figure 1.1). This is one of several ambitious projects to diversify agriculture geographically within Egypt. In this case, irrigation of desert lands is accomplished by mining groundwater water from the Nubian Sandstone Aquifer System. The circles on the figure are indicative of central pivot irrigation with crops like wheat and potatoes.

Figure 1.2 The photograph shows the remains of a Roman-age water well at the Dakhla, Oasis. Groundwater discharged originally as artesian and mound springs, which were associated with fault and fracture zones (With permission from Phil LaMoreaux).

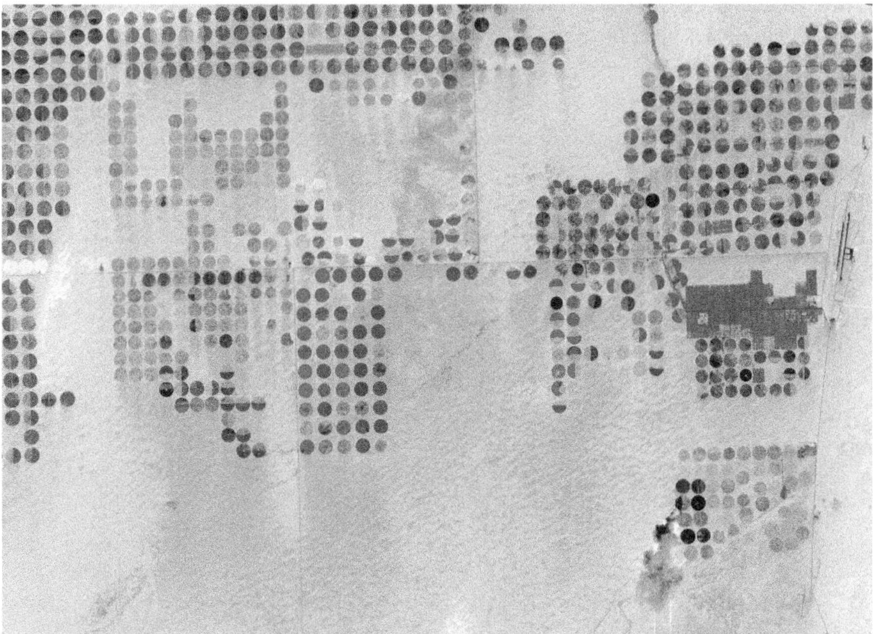

Figure 1.3 This image covers a 33.3 km × 47.2 km area of south-western Egypt and shows irrigated lands of the East Oweinat Project. Crops are watered with pivot irrigation systems using groundwater mined from the Nubian Sandstone Aquifer System (National Aeronautics and Space Administration/https://photojournal.jpl.nasa.gov/targetFamily/Earth/last accessed under 5 May 2023).

One feature of groundwater that makes it valuable as a resource is its physical and chemical quality. Unlike many surface-water supplies, natural groundwater has few suspended solids, small concentrations of bacteria and viruses, and often only minimal concentrations of dissolved mineral salts. These characteristics make groundwater an idea source of water to support human life. The connection of the groundwater pathway in hydrologic cycle to the land surface, unfortunately, provides the opportunity for humans to pollute natural groundwaters and devalue the resource.

Not surprisingly, then, issues of groundwater pollution and the protection of groundwater resources provide another important reason to study groundwater. In the latter part of the 20th century, hydrogeologists became aware of the health threat posed by contamination and the daunting technical challenges in cleaning up contaminated sites.

The human attack on groundwater comes from many different directions. Significant contamination in groundwater has come from the careless disposal of human and animal wastes; haphazard disposal of industrial wastes, and contamination associated with mining and oil operations; leaks from storage tanks, pipelines, or disposal ponds; and everyday activities such as farming and solid waste disposal. Developed countries have moved aggressively to clean up or mitigate historical problems of contamination and to prevent new problems from developing. Other countries will require more time and money to confront the problems of contamination.

Since publication of the first edition of this book in 2002, new problems of contamination have emerged. From a health perspective, the most serious are *geogenic* contaminants (due to geological processes), such as arsenic and uranium with the potential to create huge health risks. On the industrial side, contamination by PFAS (per- and polyfluoroalkyl substances) and PFOA (perfluorooctanoic acid) compounds has created significant problems for countries around the world. The issues around these compounds are discussed in later chapters.

Groundwater also influences the design and construction of engineered facilities. The long-term viability of dams depends upon controlling groundwater flow under or around such structures. The stability of excavations (e.g., trenches or open-pit mines) often requires that groundwater levels in adjacent geologic units be controlled through an engineering dewatering program. Similarly, tunnels in rock often rely on extensive grouting programs to control groundwater inflow, while mining proceeds. These problems rely on an understanding of response of systems due to withdrawals of water with wells—a fundamental area which remains an important area of coverage in this edition.

Groundwater contributes to the development of landslides, rock falls, and stream channel networks. The dissolution and channelization within near-surface carbonate rocks or salt deposits leads to the formation of *karst*, a landscape typified by sinkholes, sinking streams, and subsurface drainage. Karst landscapes occur in many countries around the world and provide the most spectacular evidence of the geologic work of groundwater.

As we make clear in the next section, the focus of hydrogeological practice continually changes as new problems emerge or old problems reemerge with a different focus. The sophistication and complexity of methods have grown to the extent of creating areas of interest that are disciplines in their own right, such as flow and transport modeling, contaminant hydrogeology, and groundwater microbiology.

1.2 Brief History of Groundwater

In his career, our mentor and friend Patrick Domenico authored and coauthored three textbooks (Domenico, 1972; Domenico and Schwartz, 1990; Domenico and Schwartz, 1998). Each of these books began with a backward look at hydrogeology. This section reprises some of the themes that Domenico developed but is embellished by more recent developments and the critical water issues of the day.

1.2.1 On Books

Pat Domenico through his career was a student of hydrogeological science. Concerning the Domenico and Schwartz books (1st ed 1990; 2nd ed 1998), he lamented that the title "Physical and Chemical Hydrogeolgy" came about in part because "most other potential titles have been preempted." The title for this book reflects this same issue.

Domenico appreciated the appeal of a simple title like "Groundwater," which was the title of a book by Tolman (1937) and Freeze and Cherry (1979). However, the books had little in common because in the 42 years separating their publication hydrogeology underwent a "tremendous expansion of knowledge" that was "expertly captured by Freeze and Cherry (1979)," particularly the fundamentals of contaminant hydrogeology. The title "Hydrogeology" was the title of a book by Lamarck (1802) and Davis and DeWiest (1966). Again, the scopes of these books were obviously quite different. Other 1960s-era books included "Geohydrology" (DeWiest, 1965) and "Groundwater Hydrology" (Todd, 1959, 1980). The "*Hydrogeology*" and "*Geohydrology*" titles were designed to represent a focus on geological aspects with the former and "hydraulics and fluid flow" with the latter. However, this subdivision of the field never gained much traction.

Another focus with books was to place groundwater-related ideas within the context of the hydrological cycle. For example, Mead's (1919) book "Hydrology" saw groundwater as a geological agent contributing to the surface-water systems.

But as Domenico pointed out his view of groundwater was narrowly focused on the author's "special interest" in surface water. Similarly, Meinzer published an edited book called "*Hydrology*" in 1942, which set groundwater as a component within the hydrologic cycle. Domenico was satisfied with this definition but considered that "it does not go far enough."

With the boom in contaminants, there has been no shortage of books, but the scope of hydrogeology has been substantially redefined and broadened. The more recent timeline in terms of books reflects this substance with specialized books on modeling, contaminants, microbiology, etc. Fetter's book "Applied Hydrogeology" first published in 1980 and updated through several editions has been especially popular in university courses.

Our book here, nominally a revision of Schwartz and Zhang (2003), is substantially different in look, feel, and most importantly in our mind, it reflects the emerging concerns of the world with substantial overpumping of groundwater.

1.2.2 On the Early Evolution of Hydrogeological Knowledge

The growth of consumer products like typewriters or phones, biological populations, and many other things can be described by an S-shaped curve (Figure 1.4). Schwartz and Ibaraki (2001) explained how a research strand, for example well hydraulics, or regional groundwater flow, behaves in this manner. Research on problem begins with a so-called innovation stage just a few scientists. Interest grows in the strand with the publication of the early pioneering papers, which attract other researchers. Next is a significant growth phase created as many researchers contribute their ideas and papers. Eventually growth in the strand wanes as the strand becomes mature, and researchers find other problems to work on (Figure 1.4). A rich field like hydrogeology has been able to grow and evolve because it involves many research strands with new strands replacing those that have died out.

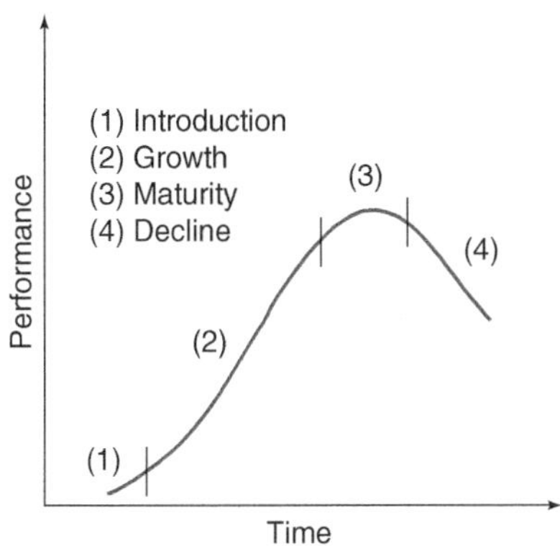

Figure 1.4 Examples of life-cycle curves. The red line represents a simple growth curve with distinct periods that include (1) introduction, (2) growth, (3) maturity, and (4) decline. Adapted from Schwartz et al. 2017. (Reproduced by permission of National Groundwater Association. Copyright © 2017. All rights reserved/John Wiley & Sons).

Schwartz et al. (2017) explained the evolution of hydrogeological science differently within a complexity-understanding space, y and x-axes respectively (Figure 1.5). The axis origin is starting place at time zero (t_0), where the new field begins. The innovation stage ($t_0 \rightarrow t_1$) involves a few researchers and a few strands. The branching (Figure 1.5) during the growth phase ($t_1 \rightarrow t_2$), reflects expansion and diversification of research with increasing activity. The last stage at (t_3) is maturity where growth slows, leading to several outcomes. At this stage, research could decline without much further progress. Another possibility is a renewed growth spurt from a small collection of branches.

Ward (2015) refers to Figure 1.5*b* as the simplicity cycle. It tracks the evolution of innovations or research ideas with axes of understanding and complexity (Figure 1*b*). Progress is evident as ideas move along the pathway $1 \rightarrow 2 \rightarrow 3$. With time ideas become more sophisticated moving from naïve simplicity to simplicity

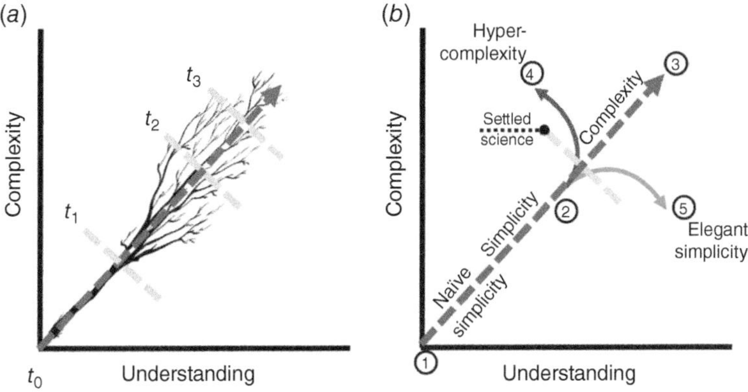

Figure 1.5 Panel (*a*) depicts the evolution of a scientific idea from its inception (t_0 to t_1), through a period of growth (t_1 to t_2) to maturity (t_3) and beyond. The cycle of simplicity in panel (*b*) depicts this same evolution but in a different manner that focuses on the maturity of ideas. Along simplicity curve (1 to 2, and the line of settled science) ideas evolve through naïve simplicity to complexity (Schwartz et al. 2019). (Reproduced by permission of National Groundwater Association. Copyright © 2019. All rights reserved/John Wiley & Sons).

(i.e., Point 1 to 2; Figure 1*b*) to complexity (Point 2 to 3, and beyond). Naïve simplicity is an early time in research where "you don't know what you don't know." First results are "based on assumptions, stacked on top of guesses, and are likely wrong and incomplete" (Kolko, 2016). Eventually, naivety's time may be left behind, but the results are simple with no significant leaps. It can take many decades to move away. Real problems emerge when naïvely simple or simple methods are subsumed into practice.

Pat Domenico's description of the evolution of the field of hydrogeology (Domenico and Schwartz, 1990) can be placed in the framework of simplicity and complexity. He described four different time periods and some of the historically important innovations. The first period is "Physical Hydrogeology Before the 1940s." This period included the experimental investigations of Henry Darcy in 1856 that provided foundational knowledge on the flow of water through porous media. Another important contribution related to work of T.C. Chamberlain in 1885 on aquifers and "artesian" flow. Domenico viewed the culmination of studies during this period as coming with the Meinzer's 1923 book on the occurrence of groundwater in the United States. Creation of a nationwide understanding of key aquifers was the essential first step (exploration, development, inventory, and management) necessary for groundwater-based water supplies. About this time, the water work of the United States Geological Survey turned towards inventory (Domenico and Schwartz, 1990).

Another noteworthy scientific accomplishment was Meinzer and Hard's 1925 work on the Dakota Sandstone. They showed that the elastic behavior of aquifers contributed to the release of groundwater from storage. A few years later came CV Theis' classic work that drew on analogies with heat flow to describe the behavior of water levels due to the withdrawal of groundwater from wells. These works led to a two-decade effort to extend well hydraulics into many different aspects of hydrogeology. This quantitative strand of research eventually led to modern computer-based modeling of flow and transport. Figure 1.5 depicts these early, simple days of hydrogeology.

The end of this early period in 1940 was the setting for two significant contributions, Hubbert's (1940) paper on the theory of groundwater flow, and Jacob's (1940) solution of the groundwater-flow equation. Among the most influential papers in hydrogeology, Hubbert's work explained the theory of groundwater flow in geologic basins. Interestingly, it was another 20 years before the implications of this paper were fully understood and studies of large basinal systems became a mainstream pursuit (Domenico and Schwartz, 1990). Jabob's (1940) paper was titled "*On the flow of water in an elastic artesian aquifer*" and was concerned with the equation of groundwater flow that included Meinzer's concepts of elastic storage. Domenico considered this work to be "the most important of the first half of the twentieth century" (Domenico and Schwartz, 1990). Simply put, a "differential equation is an expression of a law of nature." It explains "the relation between one state of nature and a neighboring state, both in time and space" (Domenico and Schwartz, 1990). The contributions of Theis, Jacobs, and Hantush (a Jacob student) led to the decades of studies, which focused on the flow of water to pumping wells. These quantitative ideas and practical methods in well-design and testing remain an integral part of our discipline some 75 years later.

Domenico also wrote about the development of "Chemical Hydrogeology Before the Early 1960s." This was a new research strand but less ordered than the physical side of hydrogeology. There was evident progress in the graphical representation of water chemistry as plots or on maps, such as Piper (1944) and Stiff (1951). During this period, workers began to understand how the chemistry of groundwater evolved as groundwater moved into the subsurface. Important works in this respect were (1) Foster (1950) that explained processes at work to produce unique sodium-bicarbonate groundwater, (2) Chebotarev's (1955) studies of common evolutionary patterns, and (3) seminal work by Bill Back (1960) on hydrogeochemical facies.

The kinds of processes at work to modify the chemical and isotopic composition of groundwater were well-known by the early 1960s. The work of Garrels, in his book (Garrels, 1960) and coauthored classical textbook (Garrels and Christ, 1965), provided important thermodynamic glue to cement ideas of geochemical processes together. Together with Hubbert's concepts of regional groundwater flow, the newly found understanding around chemical hydrogeology opened the door to the modern geochemical investigations of regional-scale flow systems. This unified understanding of physical and chemical processes was the foundation for literally decades of follow-on studies concerned with the computational modeling of multi-constituent geochemical systems.

1.2.3 1960–2005 Computers and Contaminants

The modern era of hydrogeology began around 1960. The importance of groundwater as a resource naturally led to a growth in practice, research, and education. Everything took off at the same time. Engineering and geology departments began to offer groundwater as a graduate specialty, and academic programs of excellence began to emerge in the United States, e.g., Nevada, California, and Illinois. Important journals with a groundwater focus, e.g., Water Resources Research and Groundwater began publication in the early 1960s.

Early computers of the 1960s opened the door for powerful computer-based solutions to problems of groundwater flow and contaminant transport. It was not long thereafter that pioneering research realized the potential with numerical models for aquifer evaluation (e.g., Pinder and Bredehoeft, 1968) and mass transport (Bredehoeft and Pinder, 1973). The MODFLOW code, created by researchers at the USGS (McDonald and Harbaugh, 1984, 1988) was an important technical milestone, which unified modeling for decades with updates and useful variants. This technical area of groundwater continues to flourish more than 40 years with emphasis on more sophisticated numerical approaches, calibration techniques, and increasing flexibility and power to address realistic problems.

While the undisputable value of this kind of modeling is evident, it did not substantially change the trajectory in the growth of groundwater practice, research, and education. However, our field literally exploded in the mid-1970s with the discovery of pervasive contamination in shallow groundwater across the United States. The sources were many, including leaking underground tanks, landfills, illegal chemical disposal sites, and more. The large numbers of contaminants were poorly understood in terms of potential health impacts, and behavior in the subsurface. The types of wastes included industrial wastes, most commonly organic solvents, leaking fuels of all kinds including oil, gasolines, diesel, and exotic compounds such as PCBs (polychlorinated biphenyls), creosote, coal oil, etc. Defense-related facilities and national laboratories were associated with their own unique assortments of organic, radioactive, and dissolved metals in groundwater. Commonly unlined landfills contributed their own collections of dissolved metals, organic, compounds, nutrients, which were "spiced up" by used industrial chemicals that joined the typical urban refuse. If all these problems weren't bad enough, there was a quite limited capacity in terms of site investigations, because scale-appropriate know-how and methods for sampling and site assessment did not exist. Similarly, there was also limited information as to how to clean up contaminated sites.

These concerns about groundwater safety of America's groundwater created a tsunami of government funding in contaminant-related area and legislation to force cleanups. The research was designed to support research concerned with the assessment and cleanup of industrial sites, military bases, Department of Energy facilities, and landfills. Some of the largest contaminated sites were subject to litigation to sort out companies to ultimately pay for remediation.

This boom ended up creating or redefining careers for young academics in those times, including those in the developed European countries The employment opportunities for geologists and engineers in the groundwater-consulting industry were unprecedented. Memberships in professional societies grew substantially, along with specialty conferences, training courses, and post-graduate education.

The contaminant boom extended from ~1975 to 2005, peaking in the late 1990s, when it began to wane (Schwartz et al., 2019). Most noticeable in this respect was a reduction in research funding and site cleanups, followed by a halt in the U.S. program for nuclear-waste disposal.

Schwartz et al. (2019) discussed some of the reasons why a decline could have been expected. The first reason is obvious—science made progress in establishing the scope of the problem. Indications were that health impacts, although potentially serious, were localized and avoidable through water-quality testing. Costs for site remediation were expensive but manageable over decadal time frames. Experience convinced regulators that remediation was feasible, although often problematic, and problems were manageable.

The second reason is the natural progression in growth cycles from boom to bust. For example, mining boomtowns of the western United States emerged around discoveries of gold or silver. The 1859 discovery of silver of the Comstock Lode in western Nevada brought several tens of thousands of miners and fortune hunters to Virginia City. By the mid-1870s, the ore had been extracted and the city largely abandoned. The growth and decline with a variety of activities, enterprises, or products commonly follow an "S-shaped" growth curve (Figure 1.4). Examples include the growth in populations of animals in an area, industrial enterprises, (e.g., typewriter companies), or sales of consumer products with time (e.g., tape recorders). Studies by Schwartz and Ibaraki (2001) found that such growth curves also describe evolution in research strands, where ideas are created, develop through a period of growth, mature, and then decline. For example, a strand-like analytical well hydraulics grew, prospered in the 1950s and 1960s, and subsequently declined. Addition research might occur, but the numbers of papers will have substantially declined (Schwartz et al., 2019).

A research theme, such as contaminant hydrogeology, is somewhat different because it is comprised of many research strands, each following a growth curve. Thus, the research theme would persist over a longer time frame with the ingrowth of later strands. Because the boom in contaminants led to explosive growth across groundwater, there was a noticeable bust, associated with declining industrial work and research (Schwartz et al., 2019). Nevertheless, work on contaminant-related problems continues to be an important part of groundwater education and practice.

To be clear, contaminant hydrogeology did not run out of fundamental and interesting research problems. However, the simple existence of problems does not automatically guarantee the existence of funding. Our case for funding of issues and problems across hydrogeology is probably less compelling now than in other areas of science and engineering.

1.2.4 2005 and Onward: Research Diversified

This section leaves history behind because our narrative here has finally arrived at the present in terms of this book. The groundwater enterprise has continued to grow mainly because people need abundant and clean water, which is increasingly difficult to sustain. Research trends are no longer defined by what is happening in America and Europe but in response to national needs and problems. Not surprisingly, China has invested heavily in water resources issues, because of their overall needs for water and overpumping of key aquifers, like the North China Plain aquifer system. Global warming is a topic of concern there with respect to rivers rising on the Tibetan Plateau and the loss of permafrost. The boom-and-bust cycle in contaminants was mostly a North American and European phenomenon. Thus, worldwide, there are countries newly invested in problems of contamination, especially in areas of geogenic contaminants like arsenic and uranium and industrial contamination.

Schwartz et al. (2019) have discussed the diversification of research interests in America and Europe, as support for contamination-related studies wound down. A determined group of researchers is continuing work in groundwater contaminants albeit without the financial support of past years. More proactive adjustments tend to fall in one of three broad areas (1) integrated hydrologic investigations involving groundwater, surface water, and climate, (2) water resource sustainability in the face of growing populations, global climate change, and worldwide depletion of aquifers, and (3) water and health broadly defined.

The foundation for integrated studies of the hydrologic cycle—precipitation, runoff, evapotranspiration, surface water, and groundwater—generally lies conceptually with the terrestrial water cycle and associated modeling. The conceptual blueprint for water-cycle modeling was laid out more than 50 years ago in a seminal paper by Freeze and Harlan (1969). However, another 30 or more years would pass before these ideas became mainstream within robust numerical models (Simmons et al., 2019). For hydrogeologists, the important takeaway is the importance of integrating groundwater and surface water resources. This kind of thinking has been a long time in coming. Key concepts are discussed in reports of the United States Geological Society, such as "*Groundwater and surface water: A single resource*" (Winter et al., 1998) and understanding of basic processes "*Streamflow Depletion by Wells—Understanding and Managing the Effects of Groundwater Pumping on Streamflow*" (Barlow and Leake, 2012).

The state-of-the-art in water-cycle modeling is represented by HydroGeoSphere, (Li et al., 2008; Hwang et al., 2023) an amazingly powerful and flexible code that seamlessly integrates surface and subsurface hydrological processes with capabilities for addressing issues like urban and agricultural flooding, climate change impacts, and real-time hydrologic prediction.

Another emerging research strand is water resource sustainability. Technological advances in drilling and well construction, and continuing needs for water for crop irrigation have led to pervasive decimation of aquifers worldwide. The realization of the seriousness of these problems of overpumping has come rather recently mainly through geospatial measurements with GRACE (Gravity Recovery and Climate Experiment). For groundwater applications, variations in the Earth's gravity field can be represented in terms of mass loss from groundwater withdrawals and highlighted surprising storage declines in large aquifers (Rodell et al., 2009; Famiglietti et al., 2015; Scanlon et al., 2015; Velicogna et al., 2015; Rodell et al., 2018; Thomas and Famiglietti, 2019).

In California, recent legislation has required that groundwater basins be operated sustainably. It is likely that similar legislation will develop elsewhere. Yet, there are reasons to be concerned that groundwater resources are simply being mined rather than being managed. For example, Schwartz et al. (2020) provided a pessimistic assessment of prospects for groundwater sustainability in Asia. Groundwater impacts primarily due to irrigation continue to worsen with storage declines and contamination of some aquifers. For some countries, such as Pakistan, India, and China, the governance of groundwater is fragmented within government with little operational capacity to regulate groundwater, in part because of the huge numbers of groundwater users. Groundwater-related problems are mostly invisible (Biswas et al., 2017), subsumed with greater concerns around food production (Schwartz et al., 2020).

Climate change has the potential to aggravate sustainability problems. Chapter 15 explains one business-as-usual approach to groundwater management in the United States—mine groundwater to exhaustion locally (the most resilient water supply) and replace groundwater with surface water (the least resilient supply). The obvious problem is that climate change associated drought has the potential to disrupt food production.

Water and health represent a third broad area of interest with a water aspect that includes groundwater and surface water, with a focus on human or ecosystem health. The scope of potential topics is broad, encompassing problems associated with severe human health problems associated with (1) geogenic contamination in groundwater, for example, arsenic, uranium, or other trace constituents (2) pathogens in surface water and groundwater, and (3) ecological damage from overpumping, and drainage of lakes and wetlands.

Our treatment here of this collection of diversified topics is what sets this book apart from other contemporary groundwater books on groundwater. Our subject area has grown beyond the region below the water table to systems influenced by groundwater and problems influencing groundwater.

References

Back, W. 1960. Origin of hydrochemical facies of groundwater in the Atlantic Coastal Plain, *Proceedings of 21st International Geological Congress, Copenhagen*, p. 87–95.

Barlow, P. M., and S. A. Leake. 2012. Streamflow depletion by wells: understanding and managing the effects of groundwater pumping on streamflow. U.S. Geological Survey Circular 1376, 84 p.

Biswas, A., C. Tortajada, and U. Saklani. 2017 Pumped dry: India's invisible and accelerating groundwater crisis. Asia and the Pacific Policy Society Policy Forum, https://www.policyforum.net/pumped-dry/ (accessed on 1 April 2023).

Boozer, A. L. 2015. Tracing everyday life at Trimithis (Dakhleh Oasis, Egypt). Archeological Papers of the American Anthropological Association, v. 26, no. 1, p. 122–138.

Bredehoeft, J. D., and G. F. Pinder. 1973. Mass transport in flowing groundwater. Water Resources Research, v. 9, no. 1, p. 194–210.

Caton-Thompson, G., and E. W. Gardner. 1932. The prehistoric geography of Kharga Oasis. The Geographical Journal, v. 80, no. 5, p. 369–406.

Chamberlin, T. C. 1885. The requisite and qualifying conditions of artesian wells. U.S. Geological Survey Annual Report 5, p. 131–173.

Chebotarev, I. I. 1955. Metamorphism of natural waters in the crust of weathering-1. Geochimica et Cosmochimica Acta, v. 8, no. 1, p. 22–48.

Davis, S. N., and R. J. M. De Wiest. 1966. Hydrogeology. John Wiley & Sons, New York, 463 p.

De Wiest, R. J. M. 1965. Geohydrology. John Wiley & Sons, New York, 366 p.

Domenico, P. A. 1972. Concepts and Models in Groundwater Hydrology. McGraw-Hill, New York.

Domenico, P. A., and F. W. Schwartz. 1990. Physical and Chemical Hydrogeology. John Wiley & Sons, New York, 824 p.

Domenico, P. A., and F. W. Schwartz. 1998. Physical and Chemical Hydrogeology, 2nd edition. John Wiley & Sons, New York, 506 p.

Famiglietti, J. S., A. Cazenave, A. Eicker et al. 2015. Satellites provide the big picture. Science, v. 349, no. 6249, p. 684–685.

Fetter, C. W. 1980. Applied Hydrogeology. C.E. Merrill, Columbus, 488 p.

Foster, M. D. 1950. The origin of high sodium bicarbonate waters in the Atlantic and Gulf Coastal Plains. Geochimica et Cosmochimica Acta., v. 1, no. 1, p. 33–48.

Freeze, R. A., and J. A. Cherry. 1979. Groundwater. Prentice-Hall, 624 p.

Freeze, R. A., and R. L. Harlan. 1969. Blueprint for a physically-based, digitally-simulated hydrologic response model. Journal of Hydrology, v. 9, no. 3, p. 237–258.

Garrels, R. M. 1960. Mineral Equilibria at Low Temperature and Pressure. Harper & Brothers, New York, 254 p.

Garrels, R. M., and C. L. Christ. 1965. Solutions, Minerals and Equilibria. Harper and Row, New York, 450 p.

Hubbert, M. K. 1940. The theory of groundwater motion. The Journal of Geology, v. 48, no. 8, p. 785–944.

Hwang, H. T., A. R. Erler, O. Khader et al. 2023. Estimation of groundwater contributions to Athabasca River, Alberta, Canada. Journal of Hydrology: Regional Studies, v. 45, 101301.

Jacob, C. E. 1940. On the flow of water in an elastic artesian aquifer. Eos, Transactions American Geophysical Union, v. 21, no. 2, p. 574–586.

Kaper, O. E., and W. Z. Wendrich. 1998. East and west in Roman Egypt: an introduction to Life on the fringe. Life on the Fringe: Living in the Southern Egyptian Deserts During the Roman and Early-Byzantine Periods. Proceedings Netherlands Institute for Archaeology and Arabic Studies in Cairo, December 9–12, 1996, p. 1–4.

Kolko, J. 2016. Simplicity on the other side of complexity. http://www.jonkolko.com/writingSimplicityComplexity.php (accessed on 1 April 2023).

Lamarck, J. B. 1802. Hydrogéologie. Hydrogeology, translated by A.V. Carozzi, 1984. University of Illinois Press, 152 p.

Lamoreaux, P. E., B. A. Memon, and H. Idris. 1985. Groundwater development, Kharga Oases, Western Desert of Egypt: a long-term environmental concern. Environmental Geology and Water Sciences, v. 7, no. 3, p. 129–149.

Li, Q., A. J. A. Unger, E. A. Sudicky et al. 2008. Simulating the multi-seasonal response of a large-scale watershed with a 3D physically-based hydrologic model. Journal of Hydrology, v. 357, no. 3, p. 317–336.

McDonald, M. G., and A.W. Harbaugh. 1984. A modular tree-dimensional finite-difference groundwater flow model. U.S. Geological Survey Open-File Report 83-875, 528 p.

McDonald, M. G., and A. W. Harbaugh. 1988. A modular three-dimensional finite difference groundwater flow model. U.S. Geological Survey, Techniques of Water-Resources Investigations 06-A1, 576 p.

Mead, D. W. 1919. Hydrology. McGraw-Hill, New York, 647 p.

Meinzer, O. E. 1923. The occurrence of groundwater in the United States, with a discussion of principles. U.S. Geological Survey Water Supply Papers 489, 373 p.

Meinzer, O. E., and H. A. Hard. 1925. The artesian water supply of the Dakota sandstone in North Dakota, with special reference to the Edgeley quadrangle. U.S. Geological Survey Water Supply Papers 520-E, p. 73–95.

Meinzer, O. E. (ed). 1942. Hydrology. McGraw-Hill, New York, 712 p.

Pinder, G. F., and J. D. Bredehoeft. 1968. Application of the digital computer for aquifer evaluation. Water Resources Research, v. 4, no. 5, p. 1069–1093.

Piper, A. M. 1944. A graphic procedure in the geochemical interpretation of water-analyses. Eos, Transactions American Geophysical Union, v. 25, no. 6, p. 914–928.

Rodell, M., J. S. Famiglietti, D. N. Wiese et al. 2018. Emerging trends in global freshwater availability. Nature, v. 557, no. 7707, p. 651–659.

Rodell, M., I. Velicogna, and J. S. Famiglietti. 2009. Satellite-based estimates of groundwater depletion in India. Nature, v. 460, no. 7258, p. 999–1002.

Scanlon, B. R., Z. Zhang, and R. C. Reedy. 2015. Hydrologic implications of GRACE satellite data in the Colorado River Basin. Water Resources Research, v. 51, no. 12, p. 9891–9903.

Schwartz, F. W., and M. Ibaraki. 2001. Hydrogeological research: beginning of the end or end of the beginning? Groundwater, v. 39, no. 4, p. 492–498.

Schwartz, F. W., G. Liu, P. Aggarwal et al. 2017. Naïve simplicity: the overlooked piece of the complexity-simplicity paradigm. Groundwater, v. 55, no. 5, p. 703–711.

Schwartz, F. W., Z. Yiding, and M. Ibaraki. 2019. What's next now that the boom in contaminant hydrogeology has busted? Groundwater, v. 57, no. 2, p. 205–215.

Schwartz, F. W., and H. Zhang. 2003. Fundamentals of Groundwater. John Wiley & Sons, New York, 583 p.

Schwartz, F. W., G. Liu, and Z. Yu. 2020. HESS opinions: the myth of groundwater sustainability in Asia. Hydrology and Earth System Sciences, v. 24, no. 1, p. 489–500.

Simmons, J. A., K. D. Splinter, M. D. Harley et al. 2019. Calibration data requirements for modelling subaerial beach storm erosion. Coastal Engineering, v. 152, 103507.

Stiff, H. A. Jr. 1951. The interpretation of chemical water analysis by means of patterns. Journal of Petroleum Technology, v. 3, no. 10, p. 376–379.

Sturchio, N. C., X. Du, R. Purtschert et al. 2004. One million year old groundwater in the Sahara revealed by krypton-81 and chlorine-36. Geophysical Research Letters, v. 31, no. 5.

Thomas, B. F., and J. S. Famiglietti. 2019. Identifying climate-induced groundwater depletion in GRACE Observations. Scientific Reports, v. 9, no. 1, p. 4124.

Todd, D. K. 1980. Groundwater Hydrology, 2nd edition. John Wiley & Sons, New York. 552 p.

Todd, D. K. 1959. Groundwater Hydrology. John Wiley & Sons, New York. 535 p.

Tolman, C. F. 1937. Groundwater. McGraw-Hill, New York, London. 593 p.

Velicogna, I., M. Rodell, J. T. Reager et al. 2015. Satellites provide the big picture. Science, v. 349, no. 6249, p. 684–685.

Ward, D. 2015. The Simplicity Cycle: A Field Guide to Making Things Better Without Making Them Worse. Harper Collins, New York, p. 224.

Winter, T. C., J. W. Harvey, O. L. Franke, et al. 1998. Groundwater and surface water: a single resource. U.S. Geological Survey Circular 1139, 79 p.

Zaporozec, A., and J. C. Miller. 2000. Groundwater pollution. UNESCO International Hydrological Program, 24 p.

2

Hydrologic Processes at the Earth's Surface

CHAPTER MENU

Water circulates on earth from the oceans to the atmosphere to land and back to the oceans in what is known as the hydrologic cycle. The main pathways in the hydrologic cycle are shown schematically in Figure 2.1. Water evaporates from oceans, lakes, and rivers into the atmosphere. This water vapor is transported with the atmospheric circulation and eventually falls as rain or snow onto land, lakes, rivers, and oceans. Of the water falling on land, some proportion quickly evaporates, some flows to streams or lakes as overland flow, and some infiltrates the subsurface. Of the water entering the soil, some is transpired back into the atmosphere by plants. The remaining water (i.e., groundwater) follows a subsurface pathway back to surface (Figure 2.1).

Non-potable ocean water stored within the hydrologic cycle constitutes approximately 96.9% of the water (~1.34 million × 10^3 km^3). Another 1.91% (26,350 × 10^3 km^3) is frozen in icecaps and glaciers (Trenberth et al., 2007). Most of the world's available fresh drinking water is stored as groundwater (1.1%, or 15,300 × 10^3 km^3). Although the quantity of water stored in rivers is relatively small, ~0.0132% or (178 × 10^3 km^3), rivers are important because the flow is localized, and they deliver large quantities of water every day.

2.1 Basin-Scale Hydrologic Cycle

Most hydrogeological investigations are conducted at smaller than global scales. A useful scale is that of a drainage basin or watershed, the fundamental hydrologic unit that organizes the investigation of surface waters. In essence, a drainage basin is topographic feature that functions to move rainfall and associated runoff to the outlet. That outlet might be another stream channel, a lake, or ocean. In other words, some of rain or snow falling in a drainage basin eventually ends up at its outlet. Figure 2.2 is a photograph showing a smaller drainage basin within the larger Agashashok River basin, located in the northwestern Arctic of Alaska. Looking at this photograph, it is not difficult to imagine how snow-melt runoff from the broad slopes to the left and right would flow downhill in small streams, eventually entering larger river in the foreground. In most basins, land surface elevations increase away from the rivers and streams to the basin boundary. This boundary is also called a *drainage divide*, the topographic line that separates one basin from the neighboring basin.

By their very nature, drainage basins are nested, such that each tributary of the largest river in the basin drains smaller basins associated with even smaller collections of subbasins. Figure 2.3 illustrates some of the important features of drainage basins. The Apalachicola–Chattahoochee–Flint River Basin (ACFB) covers an area of ~50,700 km^2 of western Georgia,

Fundamentals of Groundwater, Second Edition. Franklin W. Schwartz and Hubao Zhang.

Companion website: www.wiley.com/go/schwartz/fundamentalsofgroundwater2

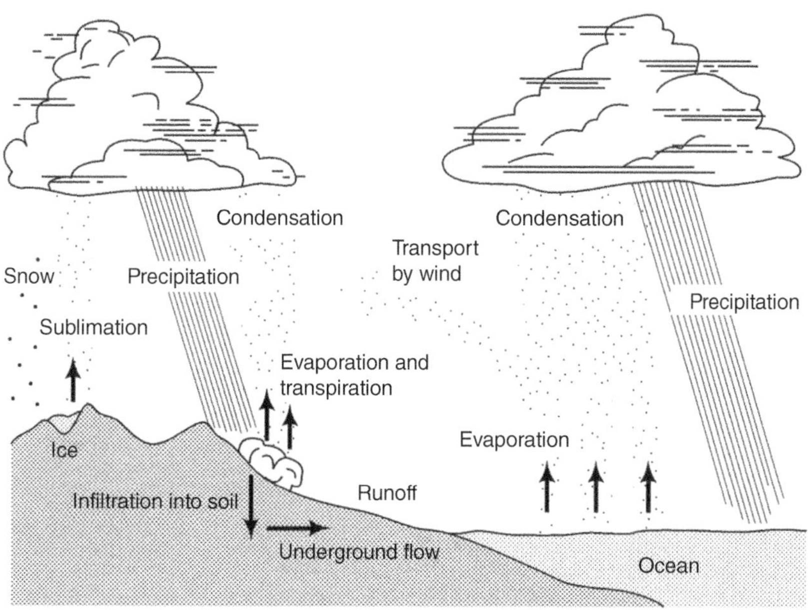

Figure 2.1 Schematic representation of the hydrologic cycle (Reproduced with permission Domenico and Schwartz, 1998/John Wiley & Sons).

Figure 2.2 Example of a topographic basin where runoff due to precipitation collects in streams and flows down the stream network toward the basin outlet in the foreground. This tributary basin is located within the larger Agashashok River Basin in Alaska (U.S. Geological Survey/https://www.usgs.gov/media/images/a-tributary-agashashok-river-watershed/last accessed under on 5 May 2023).

eastern Alabama, and a piece of Florida (LaFontaine et al., 2013). The basin outlet is the Gulf of Mexico (Figure 2.3). As the name implies, ACFB is comprised of three smaller basins, the Chattahoochee River Basin to the west (22,600 km^2), the Flint River Basin to the east (21,900 km^2), and the Apalachicola River Basin to the south (6200 km^2) (Figure 2.3). Each of the three basins includes many smaller basins. For example, the blue circles on Figure 2.3, show the approximate locations of two tributary basins within the Flint River Basin, i.e., Spring Creek Basin, and the Ichawaynochaway Creek Basin. The Chipola

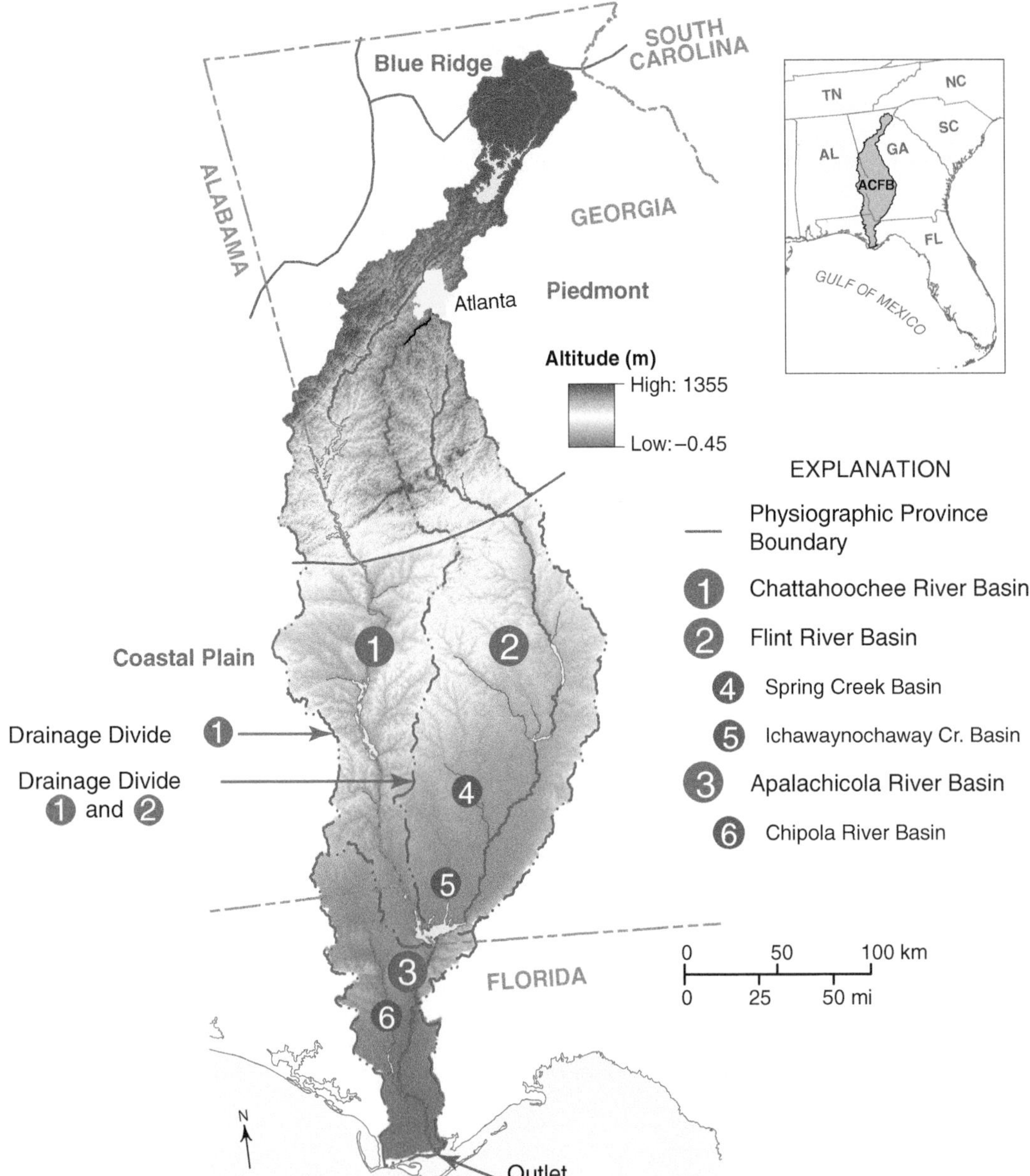

Figure 2.3 Digital elevation map of the Apalachicola–Chattahoochee–Flint River Basins showing drainage divides, the nested character of watersheds, and boundaries of the associated physiographic regions (U.S. Geological Survey LaFontaine et al., 2013/ Public domain).

River Basin is an example of smaller basin within the larger Apalachicola River Basin. The small size of the figure precludes the depiction of even smaller basins.

Inspection of the shaded contour map shows the position of the drainage divides for the Chattahoochee River, Flint River, and Apalachicola River Basins. Figure 2.3 also illustrates how the drainage divide running down the middle of the figure follows the topographic high between the Chattahoochee and Flint Rivers with associated creek valleys running downhill away from the divide.

The contour map for the ACFB also illustrates the influence of regional physiography in determining the character of drainage basins. For example, the northern third of the ACFB lies within the Piedmont Physiographic Province, a zone of rolling hills and narrow valleys. The Piedmont separates the Blue Ridge Mountains to the north from the Coastal Plains, the flat lands present across the southern two-thirds of the ACFB (Figure 2.3). An example of basin-scale hydrologic cycle is depicted in Figure 2.4. This conceptual model emphasizes pathways and processes associated with water within the near-surface portion of the terrestrial water cycle. Notice that the lateral boundaries for this basin are similar to the drainage

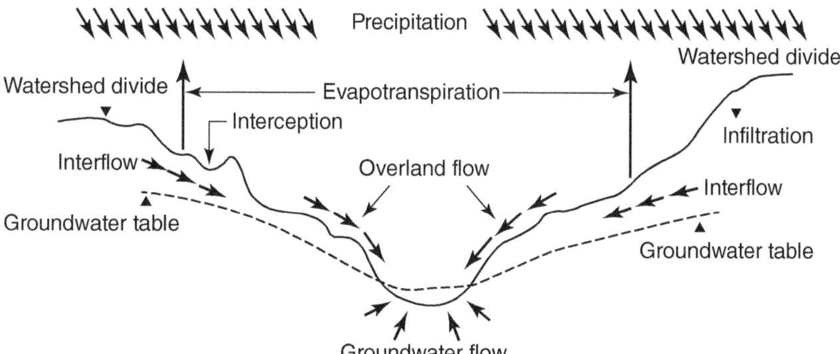

Figure 2.4 Processes and pathways in the basin hydrologic cycle (Reproduced with permission Domenico and Schwartz, 1998, *Physical and Chemical Hydrogeology*. Copyright © 1990, 1998 John Wiley & Sons, Inc. All Rights Reserved/John Wiley & sons).

divides in Figure 2.3. Because of the nested character of drainage basins, the scale of the conceptual model (Figure 2.4) is related to the width of the basin of interest. However, processes like surface runoff typically operate at small scales, making this conceptual model a small-scale depiction (i.e., few 100s of meters).

Water falling as rain on the ground undergoes a variety of different processes that function to move water within the hydrologic cycle (Figure 2.4). For example, water moves toward streams via overland flow, interflow that is flow in the unsaturated zone above the water table, and groundwater flow. It also can return to the atmosphere through evaporation and evapotranspiration. Once in streams or rivers, the channel network conducts water through watersheds and beyond. Upcoming sections in this chapter will discuss these different processes.

Another important feature of the basin hydrologic cycle is the opportunity for water to be stored for periods of time in various parts of the hydrological system. Such storage might range from small quantities of rainwater stored for a few hours on tree leaves, to somewhat larger quantities stored in the soil zone, to the large volumes of water stored for long times in aquifers. Figure 2.5 presents the basin hydrologic cycle in a different way, which emphasizes where water is stored. Storage within the basin cycle is indicated by the rectangles in Figure 2.5, which represent the plant canopy, snowpack, unsaturated zone, and active groundwater system. Stream water is also stored within the channel network. As mentioned, groundwater is by far the largest reservoir for water in the basin hydrologic cycle.

2.2 Precipitation

Precipitation includes the various forms of water, liquid or solid, that fall to the ground from the atmosphere. Precipitation can occur as rain, drizzle, snow, hail, sleet, and ice crystals. It can be measured at a location using a variety of devices. The simplest instruments are calibrated containers that collect precipitation during a storm event. The quantity of precipitation is measured by an observer following the storm. The standard U.S. precipitation gage has a 20.32 cm (8 in.) diameter mouth and height of about 76.2 cm (30 in.) (Figure 2.6*a*). More sophisticated rain gages, such as the tipping bucket rain gage (Figure 2.6*b*), can record the time, duration, and intensity of precipitation. Typically, they store this information or transmit results electronically.

The accuracy of precipitation measurements is affected by the physical setting and by disturbances due to the presence of the gage itself. For example, strong winds during storm events can cause considerable difference between measured and actual precipitation. With snow measurements, all gages typically underestimate the actual precipitation, particularly at high wind speeds.

Precipitation data are often available from networks of government weather stations, such as those operated by the National Weather Service. Historical data can be purchased for specified geographic areas in the United States in computer-compatible formats.

Many applications require that amounts of precipitation be estimated over relatively large areas. A few simple estimation approaches provide areal estimates of precipitation from a distributed network of weather stations. A simple *arithmetic average* of precipitation for an area is given as

$$P_a = \frac{1}{N}\sum_{i=1}^{N} P_i \tag{2.1}$$

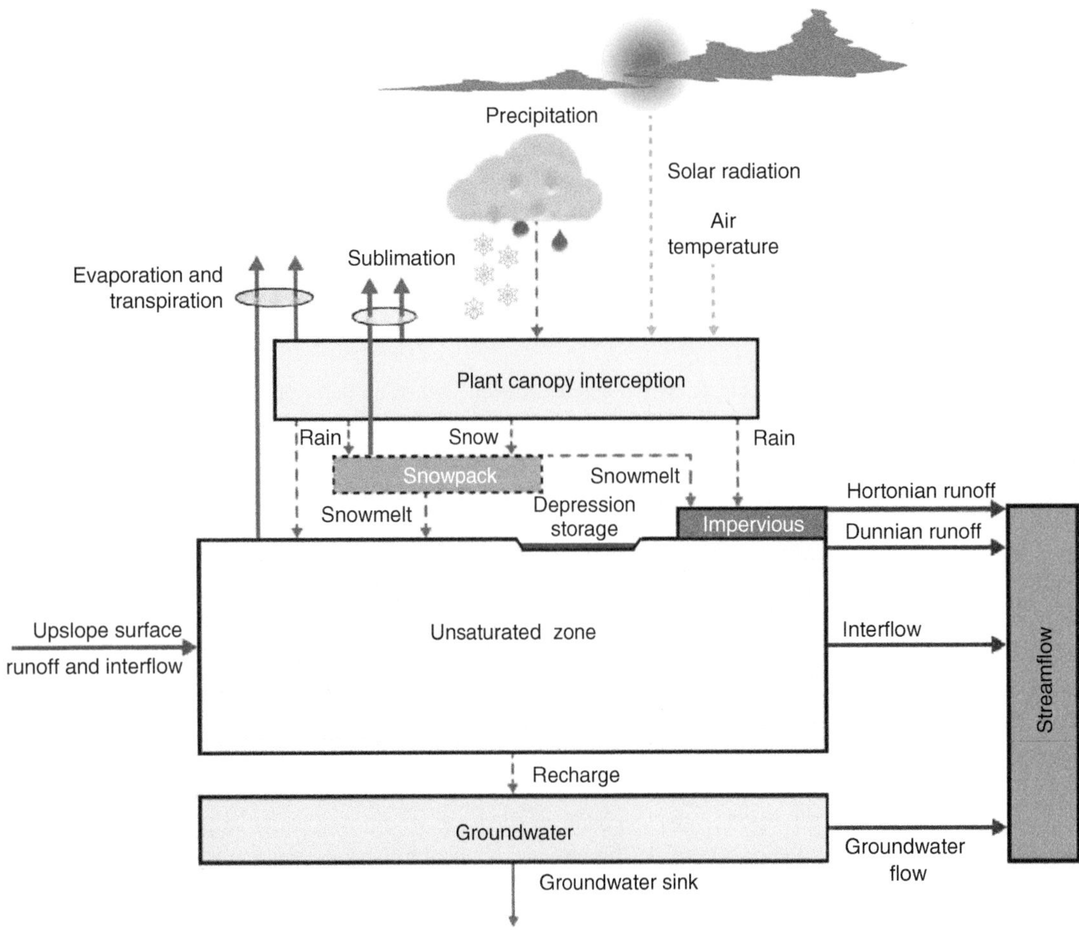

Figure 2.5 The basin hydrologic cycle represented in terms of storage elements (i.e., rectangles) and processes that work to change the quantities of water stored (U.S. Geological Survey Regan and LaFontaine, 2017/Public domain).

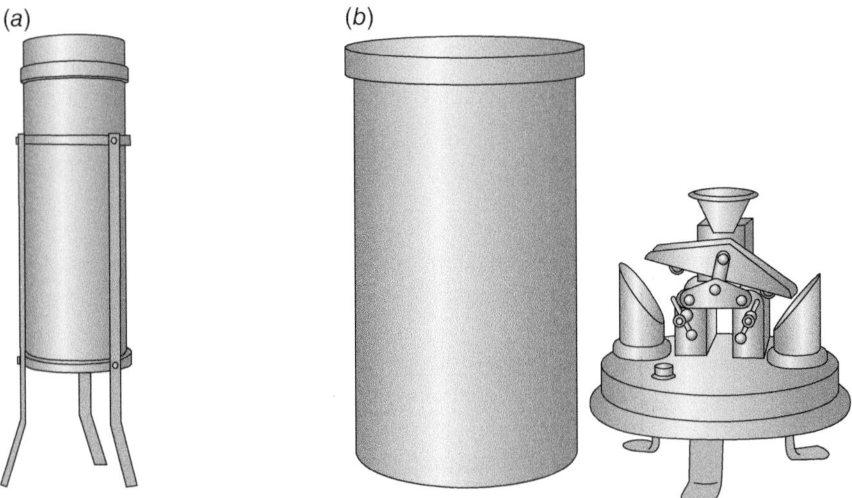

Figure 2.6 Examples of gages used for measuring precipitation. Panel (*a*) shows an 8-in. standard U.S. rain gage. The collection funnel inside directs the rain to a plastic measurement tube. Panel (*b*) shows a tipping bucket recording rain gage. The buckets are calibrated to tip after 0.01 in. of rainfall and are recorded by a data system (Schwartz and Zhang, 2003/John Wiley & Sons).

where P_a is the average precipitation for the area, N is the number of weather stations in the area, and P_i is the precipitation at station i.

The *Thiessen-weighted average* method is a more sophisticated approach. It is based on the idea that precipitation data for a station are most representative of an area immediately surrounding it. A Thiessen network is constructed by plotting the stations on a map and connecting adjacent stations by straight lines and bisecting each connecting line perpendicularly. The two sets of perpendicular lines define a polygon around each station. Precipitation at a station is applied to each polygon closest to it. The area-weighted average precipitation is given by the following equation:

$$P_a = \frac{\sum_{i=1}^{N} P_i a_i}{\sum_{i=1}^{N} a_i} \tag{2.2}$$

where a_i is the area of the polygon around station i.

The *isohyetal method* is based on areas computed from a contoured precipitation map. This average can be calculated by

$$P_a = \frac{\sum_{i=1}^{N_c-1} P_{ai} A_i}{\sum_{i=1}^{N_c-1} A_i} \tag{2.3}$$

where N_c is the number of contour lines, A_i is the area between two contour lines, and P_{ai} is the average precipitation of adjacent two contour values.

Example 2.1 Precipitation (in inches) was recorded at several stations (Figure 2.7*a*). Characterize the precipitation over the area as the arithmetic average, the Thiessen-weighted average, and the isohyetal average.

Solution

Arithmetic Average:

$$p_a = \frac{1}{9}(4.5 + 4.4 + 4.4 + 4.1 + 4.0 + 3.7 + 3.5 + 2.5 + 2.7) = 3.76 \text{ in.}$$

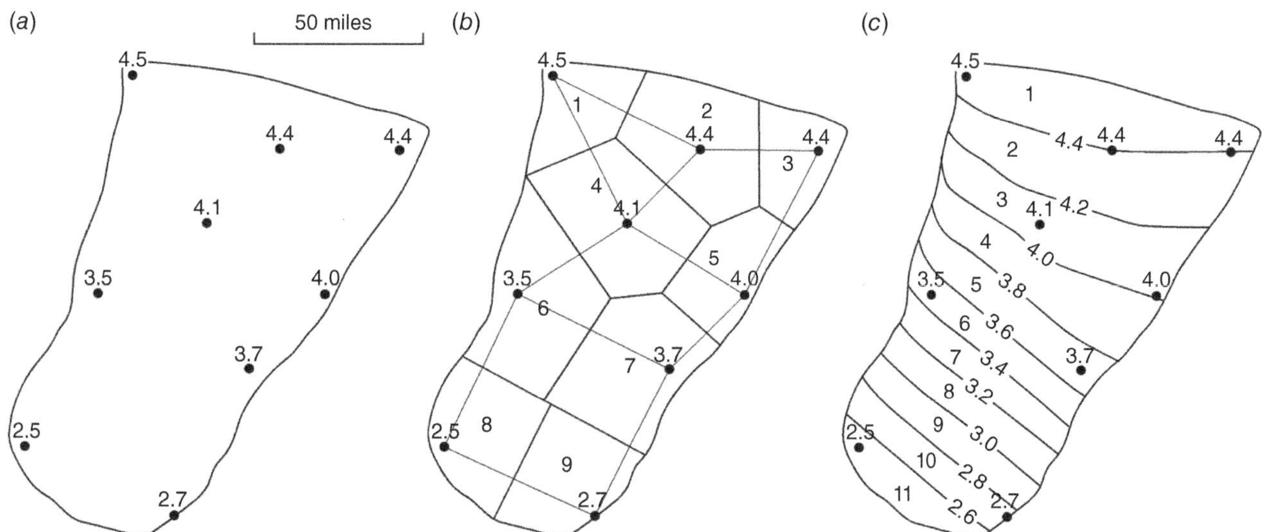

Figure 2.7 Calculation of the average rainfall using (*a*) arithmetic average method; (*b*) Thiessen-weighted average method; (*c*) isohyetal method (Schwartz and Zhang, 2003/John Wiley & Sons).

Table 2.1 Calculation of average precipitation by Thiessen-weighted average method.

Polygon index	a_i (mile^2)	P_i (inches)	a_iP_i
1	781	4.50	3516
2	1262	4.40	5552
3	625	4.40	2750
4	1563	4.10	6406
5	888	4.00	3553
6	1575	3.50	5513
7	1375	3.70	5088
8	1164	2.50	2910
9	952	2.70	2570
Total	10,185		37,857

Source: Schwartz and Zhang (2003)/John Wiley & Sons.

Thiessen-weighted Average:
Adjacent stations are connected by a straight line, and the connecting line is bisected perpendicularly (Figure 2.7*b*). The area of each polygon around each station is measured and tabulated in Table 2.1. The total area and the sum of the product of the polygon area and the precipitation are calculated and listed in Table 2.1. The average precipitation is

$$P_a = \frac{\sum_{i=1}^{N} P_i a_i}{\sum_{i=1}^{N} a_i} = \frac{37{,}857 \text{ mile}^2 \cdot \text{in.}}{10185 \text{ mile}^2} = 3.72 \text{ in.}$$

Isohyetal Average:
Contour lines of precipitation are drawn using an interval of 0.2 in. (Figure 2.7*c*). We measure the area between each pair of contour lines and find the average of the pair of contour values to apply to the area (Table 2.2). Other calculations are shown in the table. The average precipitation by isohyetal method is

Table 2.2 Calculation of precipitation by isohyetal method.

Contour index	A_i (mile^2)	P_i (inches)	A_iP_i
1	1303	4.45	5799
2	1528	4.3	6571
3	1453	4.1	5958
4	967	3.9	3772
5	967	3.7	3578
6	661	3.5	2314
7	661	3.3	2182
8	661	3.1	2049
9	661	2.9	1917
10	661	2.7	1785
11	661	2.55	1686
Total	10,185		37,609

Source: Schwartz and Zhang (2003)/John Wiley & Sons.

$$P_a = \frac{\sum_{i=1}^{N_c-1} P_{ai} A_i}{\sum_{i=1}^{N_c-1} A_i} = \frac{37{,}609 \text{ mile}^2 \cdot \text{in.}}{10{,}185 \text{ mile}^2} = 3.69 \text{ in.}$$

In the S.I. unit system, the precipitation is in centimeter (cm) (1 in. =2.54 cm, or 1 cm = 0.3937 in.).

Weather radar has provided a comprehensive approach to measuring precipitation on an areal basis. Weather radar works by emitting a radio wave away from an antenna. When the radar waves hit raindrops or snowflakes, the signal is scattered, producing waves that are returned to the antenna. Difference in the reflectivity can be interpreted in terms of rainfall types. Newer Doppler weather radars can be processed to yield almost continuous space/time estimates of many meteorological variables, including precipitation types and rainfall amounts. There is nearly complete coverage of the continental U.S. by Doppler weather radar, with individual sites capable of covering thousands of square kilometers on a continuous basis. For a given storm, the strength of the reflected radar signal is proportional to the precipitation intensity.

Commonly, intensities are presented in qualitative terms (i.e., light or heavy). When calibrated, Doppler radar can provide highly resolved estimates of precipitation amounts, especially for complex rainstorms or storms in areas where the coverage by conventional gages is limited.

The coverage possible with radar-based techniques is evident by looking at data for a significant precipitation event (April 29–May 1, 2012), which produced severe weather and heavy rainfall in northeastern Oklahoma and southeastern Kansas. NEXRAD radar reflectivity data provided the basis for mapping the distribution of precipitation. The National Weather Service provided several data products for this weather event, which included 24-hour precipitation for each day of the storm and a total precipitation map for the storm (Figure 2.8). Areas of higher precipitation, i.e., 5–10 in. (~12.5 to 25.0 cm), were indicated by the "hotter" colors (red and pink) (Figure 2.8).

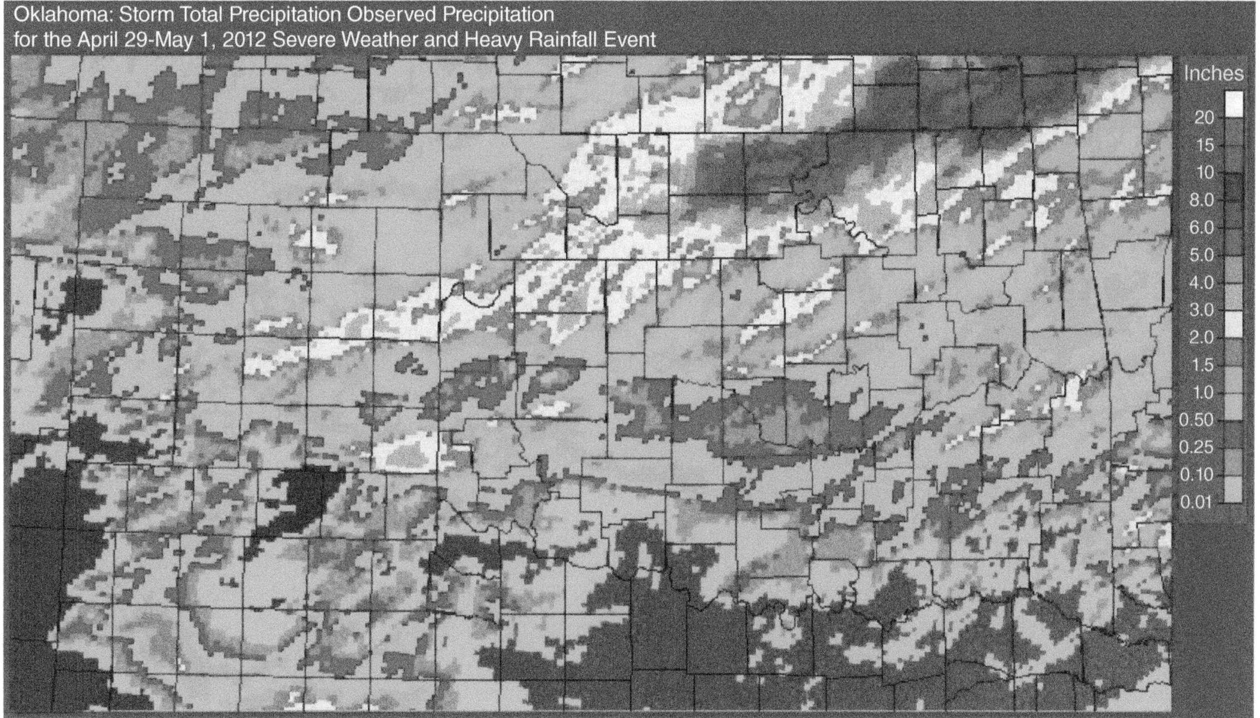

Figure 2.8 Total precipitation map for a severe 3-day weather event in parts of Kansas and Oklahoma, April 29–May 1, 2012. The hot colors indicate the areas of southeastern Kansas and northwestern Oklahoma with the highest rainfalls. The map was prepared using the NEXRAD radar in Norman Oklahoma (National Weather Service-NOAA, https://www.weather.gov/oun/events-20120430/last accessed under 5 May 2023).

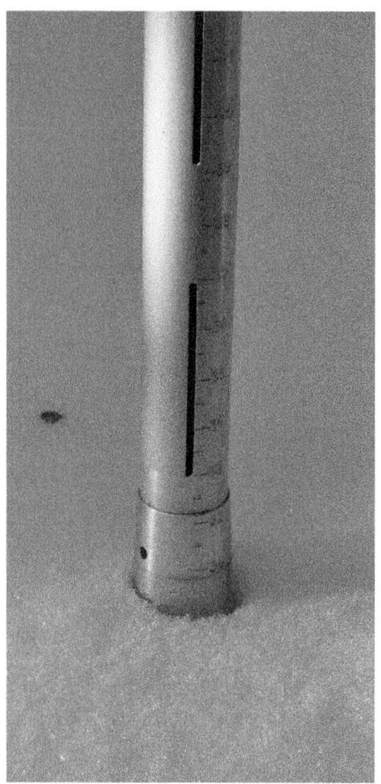

Figure 2.9 Photograph showing a Mt. Rose sampler that provides measurements of snow water equivalent (SWE) of a snowpack. The weight of the snow core captured inside the tube readily yields an estimate of SWE. The scale along the outside of the tube indicates the depth of the snowpack (U.S. Department of Agriculture/https://www.flickr.com/photos/160831427@N06/39059498621/in/photolist-22vy98p/ last accessed under 5 May 2023).

2.2.1 Snowpack Distributions

As is the case with rainfall measurements, various gages are available to measure the quantity of snow from a particular storm or over some fixed time interval (e.g., monthly). However, a snow parameter of much greater interest is snow water equivalent (SWE), which describes the quantity of water snowed within a snowpack and potentially available as runoff in spring or early summer. Unlike rain, snow is stored as a solid on the ground surface. With time, however, a snowpack can decline due to sublimation—the transformation of snow to water vapor and eventually spring warming. Given the importance of snowmelt runoff to irrigated agriculture, information on SWE is a key variable for water-resource planning for irrigation in the western U.S.

High-elevation snow in Sierra Nevadas and other mountain ranges in the western U.S. is an immensely important resource of potential water that is carefully monitored. Historically, monitoring of SWE involved monthly measurements at high-mountain stations, i.e., snow courses. A Mt. Rose sampler (Figure 2.9) provides for the collection of a core through the snowpack from the surface of the snow through to the ground surface. The SWE represented by snow in the tube is determined by weighing. The cutting-edge of the coring device has a specific design size, so that 1 oz of snow in the tube is equivalent to 1-in. SWE. In addition to the measurement of SWE, the device also provides measurements of snow depth in inches and eventually an estimate of the mean snow density. For example, a snowpack with a SWE of 10 in. of water and a snow depth of 36 in. yields a snow density of 10/36 or approximately 28%.

Records of snow surveys for many courses in the western U.S. are longer than 50 years and monitoring efforts continue to this day. However, to an increasing extent, automated systems provide continuous measurements at key stations. The Snow Telemetry (SNOTEL) Network involves climate data collection at more than 900 sites at remote sites in the West (Figure 2.10). SWE is measured using a snow pillow, a thin, extensive bladder, filled with water and antifreeze. As the snowpack accumulates, the pressure of fluid in the bladder increases, which is calibrated to provide an estimated SWE. The thickness of the snowpack on the pillow is measured using an acoustic measuring device (Figure 2.10).

2.3 Evaporation, Evapotranspiration, and Potential Evapotranspiration

Evaporation is the physical process by which a liquid is transformed into a gas. In hydrologic applications, the term evaporation refers to the loss of water from soils, rivers, and lakes because of the phase transfer from water-to-water vapor. A variety of methods can be used to estimate or measure the amount of evaporation. It is beyond the scope of this book to discuss these approaches in detail. Interested readers can refer to standard references such as Gray (1970).

Estimates of the quantity of evaporation of surface water historically were made with an evaporation pan that contained water. The U.S. Weather Service Class A pan (Figure 2.11) is 4 ft (1.22 m) in diameter and 10 in. (25.4 cm) deep. Water depth is measured daily along with maximum and minimum temperatures, and the volume of water added or removed from the pan to adjust for losses from evaporation or gains from precipitation. The measurement of daily evaporation from the pan is straightforward. However, adjustments to the measured values are necessary because the quantity of evaporation from a pan is usually higher than nearby surface waters, such as a lake. Water in an evaporation pan can heat up considerably during the day and such that water can evaporate more rapidly from the pan than a cooler lake. Lake evaporation is determined as the product of the pan value and a *pan coefficient*, an empirical correction factor that varies between about 0.58 and 0.78 depending on the month (Roberts and Stall, 1967).

Plants use water in the process of *transpiration*. The combination of water loss from evaporation and transpiration is *evapotranspiration*. The idea of potential evapotranspiration recognizes that evapotranspiration cannot occur if there is no water to evaporate. Thus, p*otential evapotranspiration* is the quantity of water that would evaporate or transpire if

Figure 2.10 Soil water equivalent is measured automatically at remote sites, especially at high elevations, by a fluid-filled snow pillow, which lays on the ground surface. Typically, an acoustic device measures the depth of the snow accumulated on the pillow. In the photograph, other weather monitoring devices are also present. At this SNOTEL site on Mt. Eyak, Alaska in April 2012, there was approximately 10.5 ft of snow (3.2 m) at 45% density (U.S. Federal Government, 2012).

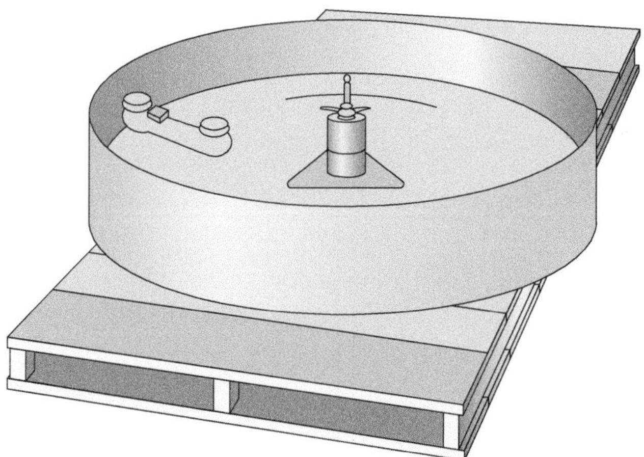

Figure 2.11 Sketch showing the U.S. Weather Service Class A evaporation pan. It has a diameter of 47.5 in. and is 10 in. deep. Mounted on the interior are a stilling basin, a hook gage for accurate measurements, and a min/max thermometer (Schwartz and Zhang, 2003/John Wiley & Sons).

sufficient water were available in the soil to meet the demand. In other words, the potential evapotranspiration is the maximum evapotranspiration, assuming the water is there to evaporate. In arid areas, the actual evapotranspiration would be less than the potential rate. For example, the actual evapotranspiration in a desert is small because there is little water present in the soil.

As was the case with evaporation, there are both measurement and empirical techniques for estimating evapotranspiration. For example, a *lysimeter* is a soil-filled tank, installed in a field setting and populated with growing plants. By carefully accounting for the quantity of water added through precipitation, water lost through surface runoff, outflows through the bottom of the tank, and storage changes in the tank, the rate of evapotranspiration can be calculated. In effect, water balance equations take account of inflows (precipitation), changes in storage (determined by weighing), and known outflows (such as surface runoff and deep percolation), to calculate evapotranspiration as the unknown.

Lysimeter data are typically rare, and one must turn to one of several empirical model approaches. The Thornthwaite (1948) method is among the simplest approaches for estimating the potential evapotranspiration from an area given a few simple environmental variables. The Thornthwaite equation is (Gray et al. 1970)

$$E_T = 1.62\left(\frac{10T_{ai}}{I}\right)^a \tag{2.4}$$

$$I = \sum_{i=1}^{12}\left(\frac{T_{ai}}{5}\right)^{1.5} \tag{2.5}$$

$$a = 0.492 + 0.01179I - 0.0000771I^2 + 0.000000675I^3 \tag{2.6}$$

where E_T is the potential evapotranspiration in cm/month, T_{ai} is the mean monthly air temperature in °C for month I, I is annual heat index, and a is constant. It is assumed that each day has 12 hours of sunshine with 30 days in a month. Table 2.3 lists the correction factors to adjust the hours of sunshine in each day for various latitudes. Temperature may also be in °F (°C = [°F − 32] × 5/9, or °F = °C × 9/5 + 32).

Example 2.2 (from Gray and Others 1970)

The second column of Table 2.4 lists the monthly temperatures between April and October in 1961 in Saskatoon, Canada. Calculate the potential evapotranspiration for an alfalfa crop using the Thornthwaite method.

Solution

Temperature needs to be converted from °F to °C. Next, the heat index is calculated using Eq. (2.5). Each term is calculated and listed in the fourth column of Table 2.4. The heat index is the summation of all the terms.

$$I = \sum_{i=1}^{6}\left(\frac{T_{ai}}{5}\right)^{1.5} = 0.21 + 3.56 + 7.92 + 8.90 + 2.32 + 0.89 = 31.48$$

The exponential constant a is calculated as

$$a = 0.492 + 0.01179 \cdot 31.48 - 0.0000771 \cdot 31.48^2 + 0.000000675 \cdot 31.48^3 = 1.0$$

Calculations of potential evapotranspiration for each month are listed in the fifth column of the table. With Saskatoon located at 50° north latitude, one can determine the daylight correction factors using Table 2.3. The corrected potential

Table 2.3 NRC mean hours of bright sunshine expressed in units of 30 days of 12 hours each day.

North lat.	J	F	M	A	M	J	J	A	S	O	N	D
0E	1.04	0.94	1.04	1.01	1.04	1.01	1.04	1.04	1.01	1.04	1.01	1.04
10E	1.00	0.91	1.03	1.03	1.08	1.06	1.08	1.07	1.02	1.02	0.98	0.99
20E	0.95	0.90	1.03	1.05	1.13	1.11	1.14	1.11	1.02	1.00	0.93	0.94
30E	0.90	0.87	1.03	1.08	1.18	1.17	1.20	1.14	1.03	0.98	0.89	0.88
35E	0.87	0.85	1.03	1.09	1.21	1.21	1.23	1.16	1.03	0.97	0.86	0.85
40E	0.84	0.83	1.03	1.11	1.24	1.24	1.27	1.18	1.04	0.96	0.83	0.81
45E	0.80	0.81	1.02	1.13	1.28	1.28	1.31	1.21	1.04	0.94	0.79	0.75
50E	0.74	0.78	1.02	1.15	1.33	1.33	1.37	1.25	1.06	0.92	0.76	0.70

Source: Gray et al. (1970); Public domain.

Table 2.4 Potential evapotranspiration for an alfalfa crop at Saskatoon, 1961.

Month	T_m (°F)	T (°C)	i	ET (cm)	Daylight factor	Adjusted ET (cm)	Adjusted ET (in.)
Jan.							
Feb.							
Mar.							
Apr.	35.2	1.8	0.21	1.21	1.15	1.39	0.55
May	52.8	11.6	3.56	3.99	1.33	5.30	2.09
June	67.4	19.7	7.92	7.29	1.36	9.91	3.90
July	66.8	19.3	7.68	7.08	1.37	9.70	3.82
Aug.	70.3	21.3	8.90	8.15	1.25	10.18	4.01
Sept.	47.7	8.7	2.32	3.11	1.06	3.30	1.30
Oct.	40.2	4.6	0.89	2.01	0.92	1.85	0.73
Nov.							
Dec.							

Source: Gray et al. (1970); Public domain.

evapotranspiration for each month is listed in the last two columns of the table. To apply the results to an alfalfa field, it is important to include only the growing season for the crop. For example, if the crop only grew from May 16 to September 24, 1961, the total potential evapotranspiration was

$$ET_{\text{alfalfa}} = \frac{16}{31}(3.13) + 5.43 + 5.36 + 5.39 + \frac{24}{31}(1.87) = 19.30 \text{ in.}$$

Evapotranspiration processes are complicated by variability in the origin of water being evaporated. In topographically low settings, even in arid areas, where the water table can be relatively close to the ground surface, significant losses of groundwater can occur because of evapotranspiration. In this setting, two processes could be operative (Moreo et al., 2017). The first is transpiration of water from phreatophytes (Figure 2.12*a*), plants with root systems extending downward into the saturated groundwater system or the capillary fringe (Meinzer, 1927). In the American Southwest, phreatophytes commonly occur in areas adjacent to streams. They are efficient users of groundwater from alluvial aquifers to the extent that streamflow might be reduced by the loss of groundwater that otherwise might discharge into the stream or by induced outflow from the stream. We will return to phreatophytes in later chapters.

Sometimes, a water table will be sufficiently close to the ground surface (e.g., several meters) such that water within capillary fringe is available for evaporation. In this setting, this shallow evaporation produces a "persistent upward flow of water" from the shallow groundwater (Moreo et al., 2017), which is the second process leading to the depletion of shallow groundwater.

In a desert area with a deep water table, the water evaporates from the unsaturated soil zone and comes from occasional rain storms that result in infiltration into the subsurface (Moreo et al., 2017). In this case, soil water evaporates directly from soil with heating or with transpiration associated with dry-region or xerophytes, dry-region plant species that are adapted to live through long periods with no rain (Figure 2.12*b*). In this setting, groundwater losses from the groundwater system would be negligibly small. It is also possible that these processes associated with shallow water tables can be complicated by the addition of precipitation. Thus, the water being evaporated would not only include groundwater but also water added from periodic the infiltration by precipitation.

2.4 Infiltration, Overland Flow, and Interflow

The relationship between precipitation, groundwater flow, and stream flow was introduced in Figure 2.4 as the basin hydrologic cycle. Here, we examine the key features of the basin water cycle in more detail. *Interception* is a process where some precipitation is stored on leaves and branches of plants and quickly evaporates back into the atmosphere. The total energy

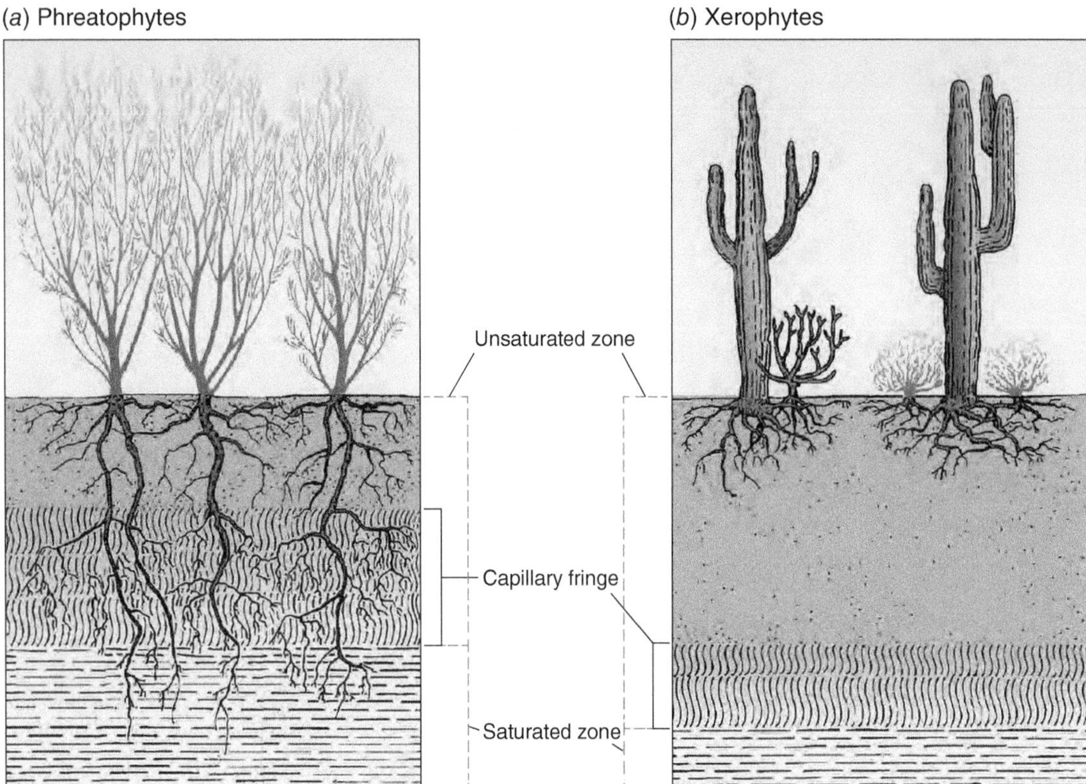

Figure 2.12 Plants growing in arid areas contribute to evapotranspiration. In Panel (*a*), phreatophytes rooting in the capillary fringe and saturated groundwater systems groundwater can deplete shallow groundwater. In Panel (*b*), xerophytes or dry-region plants survive with infiltration provided from rare rainfall events (U.S. Geological Survey Moreo et al., 2017 as modified from Robinson, 1958/Public domain).

available for evaporation of water from leaf surface is the same whether evaporation is supplied from intercepted water or from within the leaves of plants.

Infiltration is the process of downward water entry into the soil. The rate of infiltration is usually sensitive to near-surface conditions as well as the antecedent water content of the soil.

Hence, infiltration rates are subject to significant change with soil use and management, and time. Horton (1933, 1940) pointed out that the maximum permissible infiltration rate decreases with increasing time. When water is initially applied to a soil surface, the infiltration rate is equal to the rate of application. All the water infiltrates, and there is no overland flow. As water continues to be added, a time will be reached where free water accumulates or ponds on the ground surface. Initially, such ponding would be restricted to local depressions and essentially absent from the rest of the surface.

Once ponding begins, control of infiltration shifts from the rate of water addition to the characteristics of the soil. Surface-connected pores and cracks become effective in conducting water downward. Infiltration with free water present on the ground surface is referred to as *ponded infiltration*. With the initiation of ponding, the infiltration rate usually decreases appreciably with time because of the deeper wetting of the soil, which results in a reduction in the energy available for flow, and the closing of cracks and other surface-connected macropores. After long-continued wetting under ponded conditions, the rate of infiltration reaches a constant or steady-state infiltration rate. This stage is referred to as steady, ponded infiltration. Under ideal conditions, the saturated hydraulic conductivity of the soil (this term will be introduced in Chapter 3) within a depth of 0.5–1 m is a useful predictor of the steady ponded infiltration rate. Figure 2.13*a*,*b* illustrates an infiltration rate and accumulated infiltration through time. Total depth of infiltration increases linearly with time although the infiltration rate is reduced to a steady-state infiltration rate.

Once ponding begins, overland flow is possible. *Overland flow* or *surface runoff* is flow across the land surface to a nearby channel. Another pathway for near-surface flow to streams is interflow. *Interflow* is a lateral flow of water above the water table during storms. Interflow can proceed directly to a lake or stream or reemerge at the ground surface. Overland flow and interflow are sometimes grouped together as *direct runoff*.

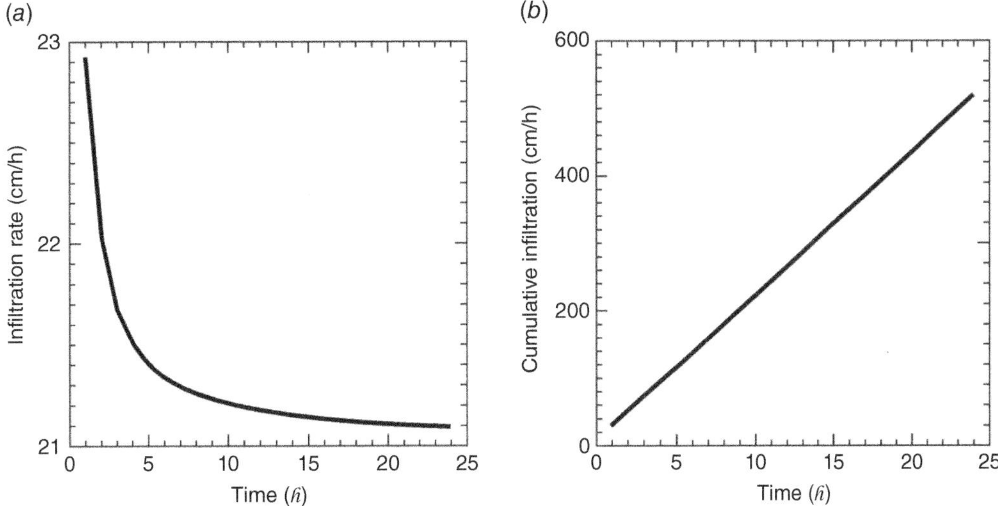

Figure 2.13 Infiltration change with time: (*a*) infiltration rate; (*b*) cumulative infiltration (Schwartz and Zhang, 2003/John Wiley & Sons).

2.5 Simple Approaches to Runoff Estimation

There has been a long and continuing effort to estimate runoff from land surfaces as some fraction of precipitation events. Such calculations are important for the design of infrastructure to control storm runoff. In this section, we briefly examine two simple, empirical schemes, the rational and curve number approaches. However, these days, water cycle models provide a useful approach for large and complex watersheds.

The rational method has been around for more than 100 years in the United States and is applicable to small basins less than 200 acres in size.

The rational method, which applies to small basins of up to 200 acres. The maximum rate of runoff (Q) has units of cubic feet per second and is calculated as

$$Q = CiA \tag{2.7}$$

where C is a dimensionless runoff coefficient, which varies between 0 and 1 depending upon the setting, i is the design rainfall intensity (inches per hour) for a duration equal to T_C (min.), the time of concentration, and A is drainage area. Values of the runoff coefficient are typically determined from table lookups that include both urban (e.g., residential, industrial, and recreational) agricultural and forested land uses, with adjustments in some cases associated with soil and crop type. A quick search of the web usually yields values of the runoff coefficient appropriate for regions of interest. Tables 2.5 and 2.6 provide illustrative values for urban and rural areas. When land uses vary within a watershed, a representative value of the coefficient can be calculated as the area-weighted contributions of the runoff coefficients for individual areas.

The maximum rate of runoff (Q) at some point of interest is achieved when contributions come from entire area of interest. This condition is met when the duration of some constant, steady rain is greater than the time of concentration (T_c). In other words, runoff at some observation point during a steady rain increases through time as more and more of the area contributes discharge. Once the rainfall duration exceeds T_c, the rate of runoff becomes constant, as calculated with Eq. (2.7). More formally, the *time of concentration* is the estimated time required for runoff to flow from the most remote part of the area under consideration to the point under consideration. The time of concentration may include overland flow time and channel flow time. The normal procedure to calculate the time of concentration is to sum the time spent in flowing overland and in channels. The actual procedures for calculating overland flow times can involve the use of nomograph or simple equations. Interested readers can find sample calculations with nomographs in USACE (2011) page 7 and with equations in USDA (1986) Chapter 3. Example 2.3 illustrates the calculation of the maximum rate of discharge for a small area (Virginia Soil and Water Conservation Commission, 1980).

Table 2.5 Illustrative values of runoff coefficient for urban areas.

Urban areas	
Type of drainage area	**Runoff coefficient *C***
Lawns:	
Sandy soil, flat 2%	0.05–0.10
Sandy soil, average, 2–7%	0.10–0.15
Sandy soil, steep, 7%	0.15–0.20
Heavy soil, flat, 2%	0.13–0.17
Heavy soil, average, 2–7%	0.18–0.22
Heavy soil, steep, 7%	0.25–0.35
Business:	
Downtown areas	0.70–0.95
Neighborhood areas	0.50–0.70
Residential:	
Single-family areas	0.30–0.50
Multi-units	0.40–0.60
Detached multi-units	0.60–0.75
Attached suburban	0.25–0.40
Apartment dwelling areas	0.50–0.70
Industrial:	
Light areas	0.50–0.80
Heavy areas	0.60–0.90
Parks, cemeteries	0.10–0.25
Playgrounds	0.20–0.35
Railroad yard areas	0.20–0.40
Unimproved areas	0.10–0.30
Streets:	0.70–0.95
Asphaltic	0.80–0.95
Concrete brick	0.70–0.85

Source: U.S. Department of Agriculture (USDA, 2019); Public domain.

Table 2.6 Illustrative values of runoff coefficients for rural areas.

	Runoff coefficient *C* soil texture		
	Soil texture		
Topography and vegetation	**Open sandy loam**	**Clay and silt loam**	**Tight clay**
Woodland			
Flat 0–5% slope	0.10	0.30	0.40
Rolling 5–10% slope	0.25	0.35	0.50
Hilly 10–30% slope	0.30	0.50	0.60
Pasture			
Flat	0.10	0.30	0.40
Rolling	0.16	0.36	0.55
Hilly	0.22	0.42	0.60
Cultivated			
Flat	0.30	0.50	0.60
Rolling	0.40	0.60	0.70
Hilly	0.52	0.72	0.82

Source: U.S. Department of Agriculture (USDA, 2019); Public domain.

Example 2.3 A small drainage area of 80 acres in Lynchburg, Virginia has four different land use categories, 30% of rooftops (C 0.9), 10% of streets and driveways (C 0.9), 20% of lawn on sandy soil (C 0.15), and 40% of woodland (C 0.10). The height of most remote point is 100 ft above flow outlet with a maximum travel length of 3000 ft. Calculate the maximum rate of runoff from a 10-year frequency storm. Assume that a nomogram estimate gave a time of 15 minutes. This duration when used with a rainfall-frequency-duration chart (10 year curve) gave a rainfall intensity of 4.9 in./h.

Solution

To calculate Q using Eq. (2.7), start by determining runoff the coefficient C.

The runoff coefficient of the basin is the area weighted sum of the C values for the four different land uses.

$$C_{\text{basin}} = (30\% \cdot 0.9) + (10\% \cdot 0.9) + (30\% \cdot 0.15) + (40\% \cdot 0.10) = 0.43$$

Thus, the runoff coefficient is 0.43 for the basin.

Rainfall intensity: The rainfall intensity can be read from a rainfall frequency-intensity-duration chart for a given return period (10 years) and duration (15 minutes) ($I = 4.9$ in./h).

Peak discharge:

$$Q = CiA = 0.43(4.9)(80) = 169\text{ cfs}$$

Another approach to estimate direct runoff due to a precipitation event was developed by the Natural Resources Conservation Service and described in detail in Chapter 10 Part 630 Hydrology National Engineering Handbook (NEH-630, chapter 10). Other background material is provided in other, variously dated chapters of the handbook and USDA (1986).

The simplest application of this approach involves the use of curves on a plot of depth direct runoff (Q) in inches (y-axis) versus depth of rainfall (P) in inches (x-axis). This nomogram provides a graphical solution to an equation that converts precipitation to direct runoff. In other words, assuming a known quantity of rainfall is uniformly distributed across some small watershed, it is a simple matter to use that precipitation value and a curve to estimate Q (Figure 2.14). Given the variability in runoff due to variability in features of the physical setting, associated with soils, plant cover, impervious areas, infiltration capacity, etc. The nomogram needs to contain many individual curves that are identified by Curve Numbers (CN) (Figure 2.14). In practice, features of the physical and biological setting can be evaluated to identify the most appropriate CN. Figure 2.14 illustrates how the nomogram yields an estimate of runoff knowing the quantity of rainfall, for example, 5 in. for a watershed characterized by a CN of 60. Users also have the option of using a table lookup approach to estimate Q (NEH-630, chapter 10).

The simplified development of the basic theory comes from USDA (1986). The governing equation for the calculation of runoff (Q) in inches from some watershed is given as

$$Q = \frac{(P - I_a)^2}{(P - I_a) + S} \tag{2.8}$$

where P is rainfall in inches, S is potential maximum retention after runoff begins in inches, and I_a is initial abstraction in inches after runoff begins. Both S and I_a are related and correlated variables that represent storage in the basin associated with closed depressions and vegetation and water losses caused by infiltration into the soil. The relationship between these variables is given as

$$I_a = 0.2S \tag{2.9}$$

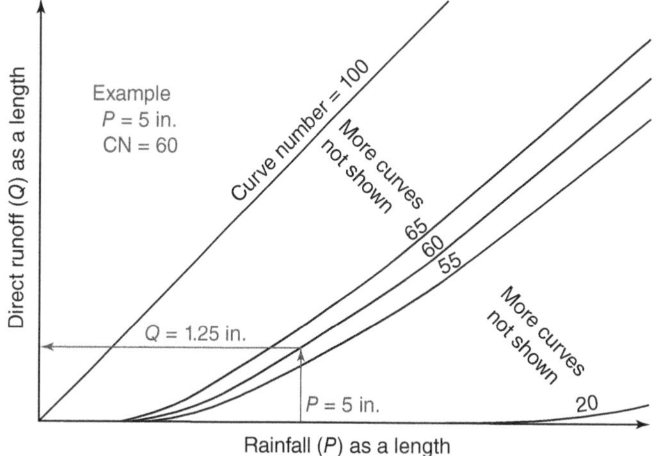

Figure 2.14 A simplified nomogram illustrating how information on the quantity of rainfall in a storm (e.g., 5 in.) and a CN (e.g., 60) for a watershed yield an estimate of direct runoff (Q) of 1.25 in. (U.S. Department of Agriculture USDA, 2019, adapted from NEH-630, chapter 10/Public domain).

Substitution of Eq. (2.9) into (2.8) yields the following equation

$$Q = \frac{(P - 0.2S)^2}{(P + 0.8S)} \tag{2.10}$$

Finally, S is related to CN by the following equation.

$$S = \frac{1000}{\text{CN}} - 10 \tag{2.11}$$

The graphical solution for Q involves the solution of these latter two equations.

Thus, a significant effort in data collection and assessment is necessary to choose the correct curve or interpolate between two adjacent curves among those shown on the nomogram.

As shown in Figure 2.14, each curve carries a number that ranges typically between 20 and 100. These numbers are not arbitrary but a reflection of the tendency for some fraction of the rainwater to be retained in the watershed due to surface depressions, evaporation, and infiltration. The curve number (CN) can be calculated by rearranging Eq. (2.11)

$$\text{CN} = \frac{1000}{10 + S} \tag{2.12}$$

where S is the potential maximum retention after runoff begins (inches) (USDA, 1986). Inspection of Figure 2.14 shows, for example, that a CN of 100 reflects the condition where the watershed is essentially a paved parking lot (i.e., $S = 0$) where all the rainfall ends up as surface runoff (i.e., $Q = P$). With a CN of 20, the watershed functions as a sponge, where there is no surface runoff except in the case of large rainstorms.

Figure 2.15 shows an actual version of a nomogram with CN values ranging from 20 to 100, increasing in steps of 5. In most cases, curve numbers are provided for two significant figures, e.g., 84, so that it will be necessary to interpolate between CNs of 80 and 85.

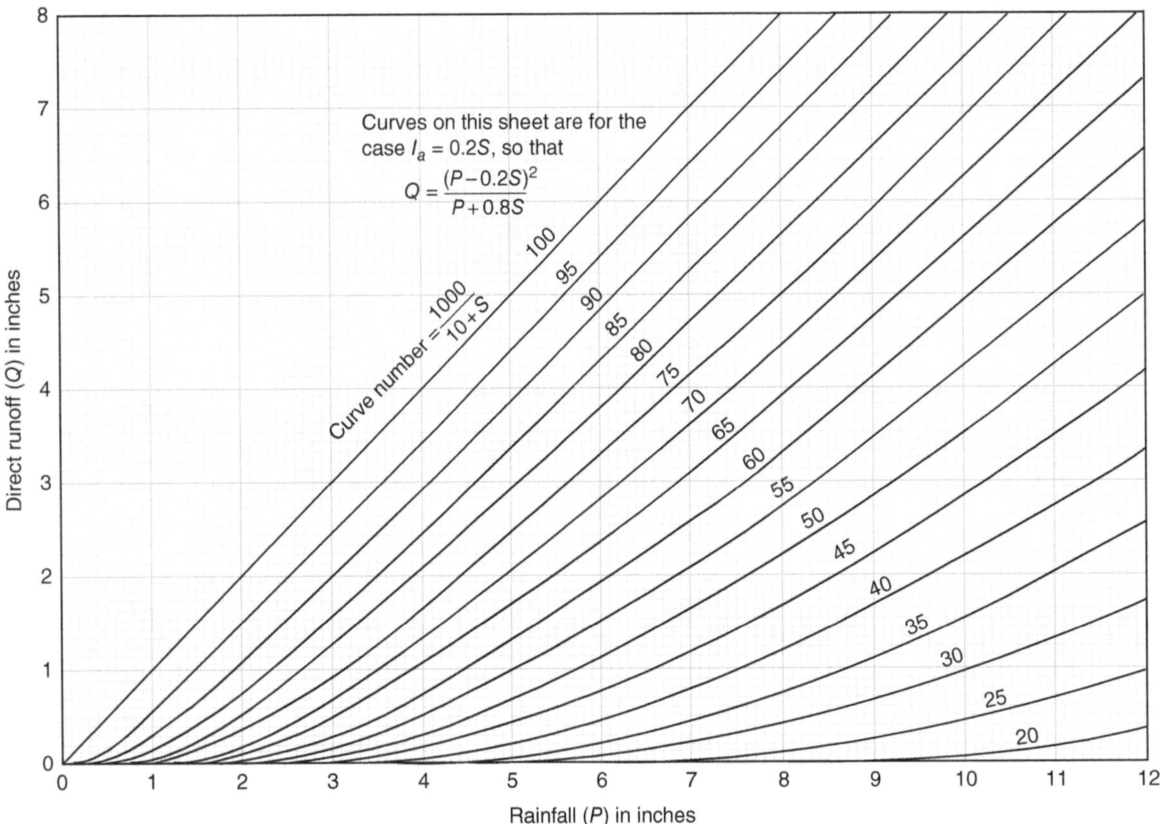

Figure 2.15 Nomogram for the graphical solution for direct runoff (Q) in inches equation given the rainfall in inches and the curve number (CN) for the case of $I_a = 0.2S$ (U.S. Department of Agriculture USDA, 2019, adapted from NEH-630, chapter 10/Public domain).

Table 2.7 Examples of runoff curve numbers for two cover types associated with agricultural lands.

Cover description			CN for hydrologic soil group			
Cover type	**Treatment**	**Hydrologic condition**	**A**	**B**	**C**	**D**
Fallow	Bare soil	—	77	86	91	94
	Crop residue cover (CR)	Poor	76	85	90	93
		Good	74	83	88	90
Row crops	Straight row (SR)	Poor	72	81	88	91
		Good	67	78	85	89
	SR + CR	Poor	71	80	87	90
		Good	64	75	82	85
	Contoured (C)	Poor	70	79	84	88
		Good	65	75	82	86
	C + CR	Poor	69	78	83	87
		Good	64	74	81	85
	Contoured and terraced (C & T)	Poor	66	74	80	82
		Good	62	71	78	81
	C & T + CR	Poor	65	73	79	81
		Good	61	70	77	80
Pasture, grass-land, or range-continuous forage-grazing		Poor	68	79	86	89
		Fair	49	69	79	84
		Good	39	61	74	80

Source: U.S. Department of Agriculture (USDA, 2019), adapted from NEH-630, chapter 10; Public domain.

The process of assigning curve numbers is relatively complicated. Chapter 9 of the National Engineering Handbook provides tables for different settings such as agricultural lands, arid and semiarid rangelands urban areas, etc. Associated with each of these settings are factors to be considered in arriving at a unique CN value, such as land cover description, hydrologic condition, and hydrologic soil group. Table 2.7 illustrates two of nine cover types in the category runoff in agricultural lands. In this illustrative example, selection of the CN for the cover type "Row crops" requires information on what is happening to the land surface because of the farming—termed "treatment" (Table 2.7).

The hydrologic condition is categorized as "poor" and "good." Poor refers to a hydrologic setting where factors tend to promote runoff. Good refers to a setting that promotes infiltration. Note that what constitutes a poor or good hydrologic condition, may change as a function of the cover type. Thus, when attempting to categorize the hydrologic, it will be necessary to consult the appropriate chapters in the Handbook for various cover types.

The last category in Table 2.7 "hydrologic soil group" exerts considerable influence on CN values. Essentially, it represents the influence of water transmitting capabilities within the soil zone. Group A soils are coarse grained and highly capable of transmitting water, which reduces the quantity of runoff from the watershed (relatively smaller values CN). Group D soils are much finer grained and may contain low permeability layers that minimize transfers from water on the land surface into the soil (relatively higher curve numbers). Again, the Handbook provides detailed quantitative guidance to select appropriate values.

Example 2.4 (from NEH-630, chapter 10)
This is a simple example of the application of the curve-number approach for the estimation of direct runoff due to a single storm of 4.3 in. in a watershed with uniform properties. Assume the land use is pasture that is in good condition with soils belonging to hydrologic group C.

Solution

The first step in solving the problem is to take the known information and determine the CN value. On Table 2.7, find the pasture category, with a hydrologic condition of good. Hydrologic group C yields a CN value of 74.

Next, use the rainfall value of 4.3 in. with an interpolated CN of 74 and the nomogram (Figure 2.15) to estimate direct runoff (Q). This value is approximately 1.82 in.

Our goals here with the curve-number approach for the estimation of direct runoff were to introduce what is a rather complicated method and to build an understanding of the basics. With the help of the Handbook, readers should be able to pursue more advanced topics. For example, direct runoff from a given storm also is complicated because inherent variability in curve numbers for a given watershed due to *Antecedent Runoff Conditions* (ARC). The Handbook suggests various causes for this variability but notes that previous work had focused on antecedent soil moisture in the days before the storm of interest. The curves in the nomogram (Figure 2.15) represent ARC conditions for class II. If the antecedent condition is dry, i.e., class I, the selected curve number from the nomogram is adjusted downward because there will be less runoff. Similarly, if the antecedent conditions are wet, i.e., class III, the curve number is adjusted upward. The Handbook (NEH-630, chapter 10) provides a table to do these adjustments.

Another complication arises, for example, when a watershed is comprised of different cover types, such as an area with row crops and an area of meadow. There are different strategies to aggregate these different areas to produce a single curve number for the watershed.

There are other useful applications of the theory developed here that are beyond the scope of this brief overview. For example, values of CN can be used together with information on the average watershed slope (%) and length of the longest flow path to estimate times of concentration. Theory is also available to create discharge hydrographs for streams in a watershed.

2.6 Stream Flow and the Basin Hydrologic Cycle

Streams and rivers do work in the hydrologic cycle by returning surface runoff, interflow, groundwater discharge, and channel precipitation back to the oceans. From a groundwater perspective, understanding features of stream flows directly and indirectly lets us learn something of the subsurface hydrologic properties. *Discharge* of a stream or river is the volume of water flowing per unit of time (Manning, 2016). Thus, discharge has units of $[L]^3\,[T]^{-1}$, for example m^3/sec or ft^3/sec. Commonly, hydrologists present discharge data graphically as a *discharge hydrograph,* which displays the discharge as a function of time (Figure 2.16).

2.6.1 Measuring Stream Discharge

Stream discharge is a basic parameter that needs to be measured to support a variety of hydrologic assessments. Because discharge is so difficult to measure, it is more convenient to provide velocity measurements instead. Discharge is easy calculated as the product of stream flow velocity and area of flow (Figure 2.17). In real streams, this approach for estimating discharge is complicated by the fact that velocity varies within the stream. However, you can get around this complication by subdividing the stream cross-section into smaller vertical sections and estimating the discharge through each piece, as the product of the velocity in that section and its area (Figure 2.18). The total discharge is the sum of discharges through all the segments. The average velocity in each vertical section is determined by averaging values at two-tenths and eight-tenth depths. In shallow water near shore, a single measurement is made at a six-tenth depth. The vertical sections are selected so that no more than 10% of the flow comes through each.

A Price current meter measures the flow velocities (Figure 2.19). The cups on the meter are turned by the flowing water, with the speed of rotation proportional to velocity. Calibration information provided with the meter lets you convert the number of revolutions over some interval of time to velocity. One keeps track of where the measurements are made by using a long tape stretched across the stream and the scale on the wading rod that holds the current meter (Figure 2.19).

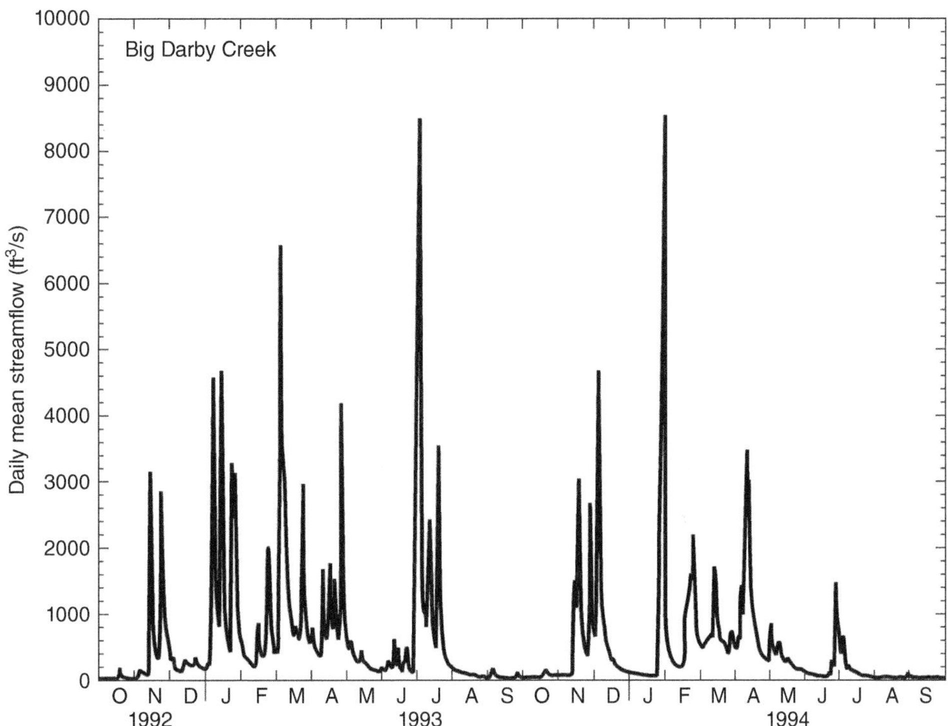

Figure 2.16 A streamflow hydrograph of daily mean discharge from 1992 to 1984 for Big Darby Creek in central Ohio (U.S. Geological Survey, Hambrook et al., 1997/Public domain).

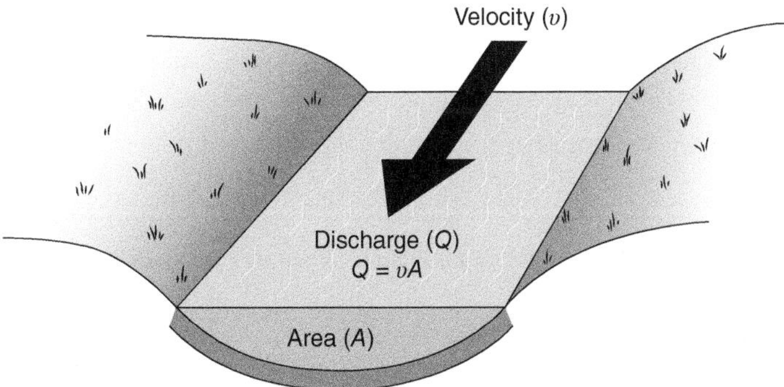

Figure 2.17 Ideally, stream discharge for a stream is a product of the flow velocity and the area of the channel cross section. (Schwartz and Zhang, 2003/John Wiley & Sons).

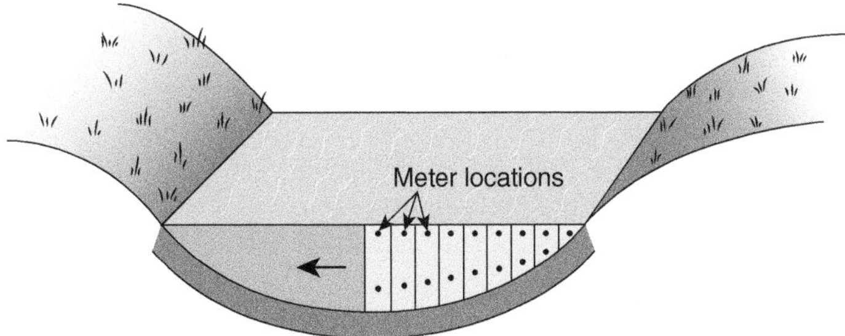

Figure 2.18 The cross section of the stream is divided into vertical segments. The streamflow is commonly measured at two-tenths and eight-tenths depths (Schwartz and Zhang, 2003/John Wiley & Sons).

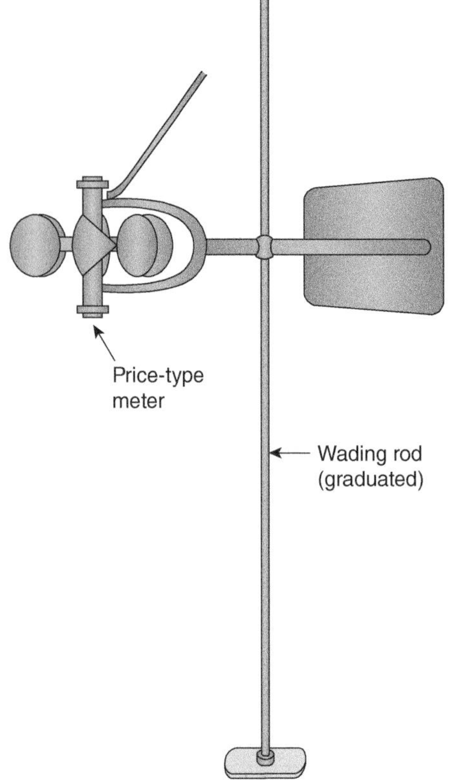

Figure 2.19 The Price current meter attached to a wading rod is one of the most common streamflow meters used for stream gaging (Schwartz and Zhang, 2003/John Wiley & Sons).

Example 2.5 Shown in Table 2.8 are a series of unpublished measurements collected on Big Darby Creek in central Ohio. Estimate the stream discharge.

Solution

Proceed by estimating the area and the mean velocity in each vertical section. Discharge for each vertical section is calculated as the product of velocity and discharge. The summation in Table 2.8 provides a total discharge of 10,317 ft^3/sec. Using Figure 2.16 for comparison, it is obvious that Big Darby Creek was gaged during a major flood event.

Monitoring stream discharge by gaging on a routine basis is not practical because the approach is so labor-intensive. However, methods are available to provide estimates of stream discharge from stream stage. As Figure 2.20 shows, *stream stage* is the elevation of the stream above some datum. During storm flows, the stream stage increases (Figure 2.20).

The simplest way of measuring stream stage is by using a staff gage (2.20). An observer visits the site and records the stream stage using the staff gage. A more automated approach uses a stilling well together with a continuous water-level recorder (Figure 2.21). This latter approach requires that the gaging site be visited much less frequently. With the most modern systems, stage data are collected automatically and sent via satellite to ground stations for further processing. Within a matter of a few hours, discharge data are posted on the Web for interested users. Take a look at the near-real-time data for Big Darby Creek at https://waterdata.usgs.gov/oh/nwis/uv/?site_no=03230500&PARAmeter_cd=00065,00060,00010 and see what the discharge is today.

Turning stage data into discharge estimates requires a *rating curve*. A rating curve relates stream stage to stream discharge at a gaging site (Figure 2.22). A rating curve is created over a number of years by measuring discharge and stage together on periodic visits to the gaging site. It is important in constructing a rating curve to make measurements across a complete spectrum of river conditions, particularly high flows.

Gaging sites are usually established at places where the stream channel is relatively stable. Erosion or deposition in the channel can cause the rating curve to change over time. In small watersheds, structures called *weirs* are sometimes installed to provide stable cross-sections at gaging stations. Specialized weirs, such as triangular weirs, permit discharge estimates from stage measurements alone. In other words, one calculates the discharge from equations rather than empirical relationships with stage.

2.6.2 Hydrograph Shape

A look at a typical hydrograph reveals that the discharge of a stream varies with time. During a storm flow event, the total streamflow increases due mainly to runoff components capable of quickly adding water to the stream. Typical examples include overland flow, interflow, and direct precipitation on the stream (Figure 2.4). The term *quickflow* describes the contribution to stream discharge from the collection of processes (i.e., overland flow, interflow, direct precipitation on the stream, and sometimes enhanced groundwater discharge), which deliver rainfall to a nearby stream in minutes to a few days. Thus, a storm hydrograph has three distinct parts—a rising limb, a peak, and a falling limb. The *rising limb*, at the beginning of a storm, reflects significant new contributions from overland flow and interflow, and perhaps enhanced groundwater discharge. Once the storm ends, the stream discharge will reach a peak and begin to decline—*the falling limb*. The term *recession* is used to describe this behavior in a streamflow hydrograph.

As quickflows finally die out following a rainstorm, streams commonly continue to flow until the next rainstorm. This dry-interval flow is termed baseflow and is classically considered to be produced by groundwater that flow paths through the subsurface, which end with discharge into the streams (Raffensperger et al., 2017). In some settings, there are anthropogenic activities that also sustain streamflows during dry intervals. Examples include releases from wastewater treatment plants, surface-water reservoirs, slow drainage from stormwater abatement ponds (Raffensperger et al., 2017). In natural system,

Table 2.8 Unpublished stream-gaging data for Big Darby Creek at Darbyville, Ohio, April 30, 1996.

Section number	Width (ft)	Stream depth (ft)	Observation depth (ft)	Velocity at point (ft/sec)	Mean velocity (ft/sec)	Area (ft^2)	Discharge (ft^3/sec)
1	13	2.4	0.6	0.242	0.242	31	8
2	20	4.5	0.6	0.402	0.402	90	36
3	20	7.0	0.2	2.24	1.81	140	253
			0.8	1.38			
4	50	7.9	0.2	1.66	0.982	395	388
			0.8	0.303			
5	47.5	10.1	0.2	1.29	1.51	480	653
			0.8	1.73			
6	12.5	11.0	0.2	3.80	3.89	138	537
			0.8	3.98			
7	8.5	11.0	0.2	4.85	4.85	94	456
			0.8	4.85			
8	7.0	11.7	0.2	6.41	6.24	82	512
			0.8	6.06			
9	7.0	12.9	0.2	7.26	7.04	90	634
			0.8	6.81			
10	6.0	14.3	0.2	8.17	7.56	86	650
			0.8	6.96			
11	5.0	14.7	0.2	8.38	7.60	74	562
			0.8	6.81			
12	5.0	15.2	0.2	8.38	7.52	76	572
			0.8	6.67			
13	5.0	15.7	0.2	8.54	7.82	78	610
			0.8	7.11			
14	6.0	15.7	0.2	7.78	6.98	94	656
			0.8	6.17			
15	7.0	15.8	0.2	6.96	6.62	111	735
			0.8	6.29			
16	7.0	16.0	0.2	6.54	6.14	112	688
			0.8	5.74			
17	8.5	15.9	0.2	5.32	5.32	135	718
			0.8	5.32			
18	10	15.3	0.2	4.26	3.84	153	588
			0.8	3.43			
19	10	12.7	0.2	3.37	3.58	127	455
			0.8	3.80			
20	12.5	11.6	0.2	2.10	2.36	145	342
			0.8	2.61			
21	15	10.7	0.2	1.41	1.52	160	243
			0.8	1.62			
22	17	4.0	0.6	0.308	0.308	68	21
						◄ 2,959	◄ 10,317

Source: USGS, Public domain.

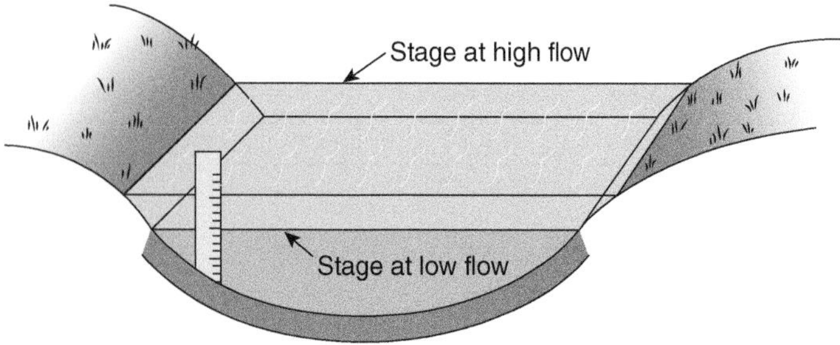

Figure 2.20 Stream stage is the elevation of the water surface above some datum. Stage measurements can be used to estimate discharge with the help of a rating curve (Schwartz and Zhang, 2003/John Wiley & Sons).

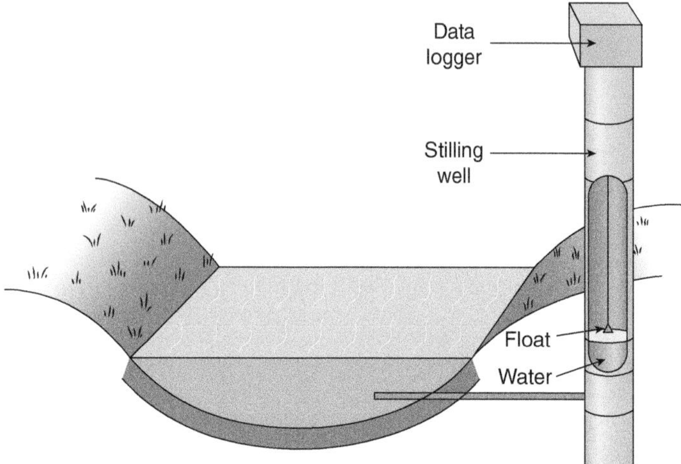

Figure 2.21 A gaging station equipped with a water-level recorder provides a continuous record of stream stage. The stilling well protects the fragile float and damps out waves that might be generated during floods. Data are recorded digitally and downloaded to a PC or sent via satellite (Schwartz and Zhang, 2003/John Wiley & Sons).

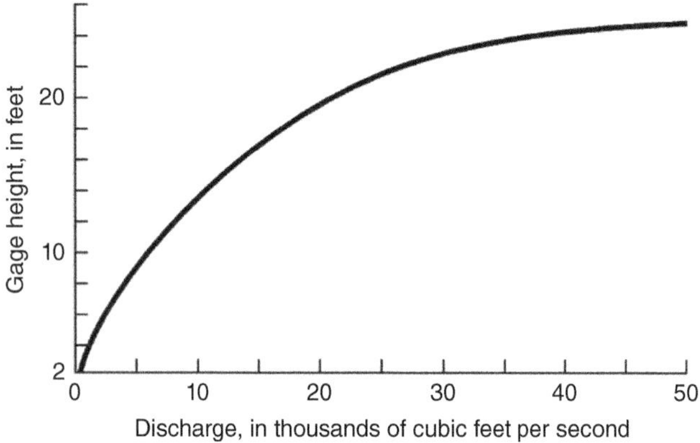

Figure 2.22 A rating curve, as shown, is constructed by measuring stream stage and discharge during a variety of different flow conditions (U.S. Geological Survey Kennedy, 1984/Public domain).

stream flow usually declines from one day to the next during this period of recession until the next rainstorm. If there are contributions from constant releases from reservoirs or wastewater treatment plants, the patterns of stream discharge can be more complicated.

This characteristic behavior in streamflow was evident with a notable storm hydrograph for Packhorse Creek (Figure 2.22). This creek is a small tributary to Wilson Creek, a small stream draining the Manitoba Escarpment in the Province of Manitoba (Schwartz, 1970). The two main tributaries, Packhorse and Bald Hill Creeks rise at an elevation

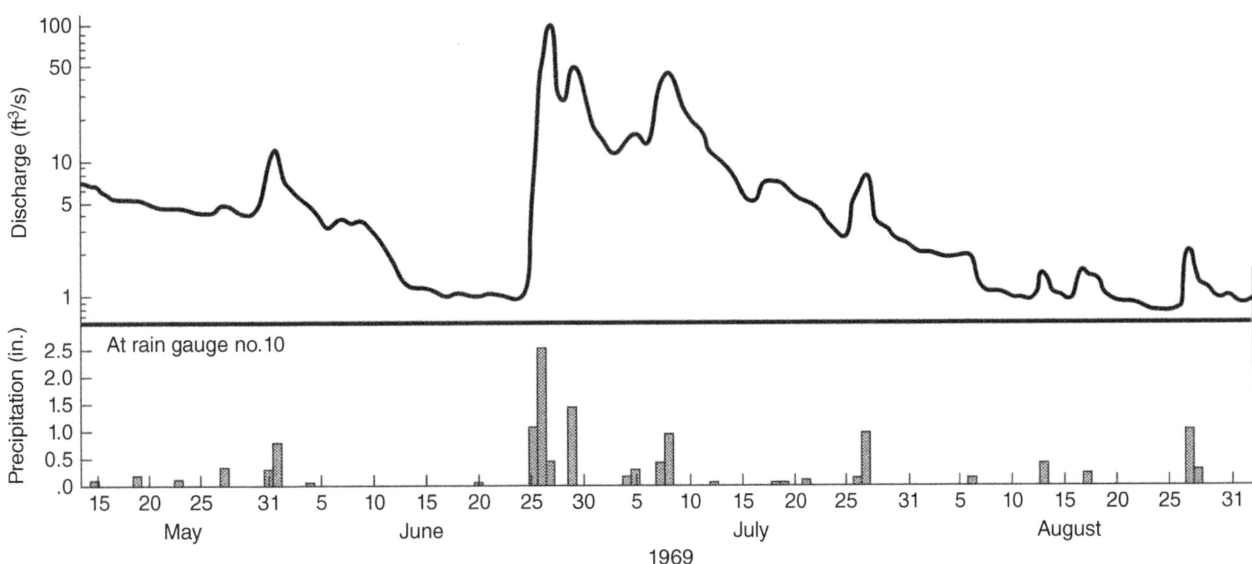

Figure 2.23 Discharge hydrograph for Packhorse Creek from 1969 (Schwartz, 1970 with permission).

of approximately 2450 ft a.s.l. and flow down the escarpment to its base at approximately 1100 ft a.s.l. Thus, although the Wilson Creek watershed is relatively small in area, 8.5 mi^2, it has high relief.

In June 1968, the watershed experienced a rare 100-year rainfall event. Over a three-day period, approximately 10 in. of rain fell on the higher elevation parts of the watershed. At lower elevations, rainfall amounts were about 4.9 in. The hydrograph on Packhorse Creek exhibited a steep rising limb (Figure 2.22) with a peak flow of approximately 100 ft^3/sec. Although not obvious on the hydrograph, the peak flow occurred approximately three to five hours following the mid-point of the precipitation event (Schwartz, 1970).

The relatively large volume of water added to groundwater and upland ponds was dissipated during a long period of recession that lasted approximately two months (Figure 2.23). Typically, baseflow is thought of as groundwater discharge to surface water systems. However, in this example, slowly draining impoundments played a more important role in maintaining stream flows through relatively dry summer months than groundwater (Figure 2.23). Discharge hydrographs due to subsequent rainstorms were obviously complicated by the effects of the lingering recession of the 100-year storm.

In other studies (e.g., Jones et al., 2006), the sources of streamflow during a storm are described in terms of event and pre-event sources. Event sources include water associated with the rainstorm, such as surface runoff and rain directly falling on the stream. With pre-event sources, the water contributed to the stream during the storm event was already present in the saturated and unsaturated zones adjacent to the stream (Jones et al., 2006). Although similar to the baseflow/quickflow terminology, there are differences. For example, interflow is considered as quickflow but is otherwise included with groundwater as pre-event water.

2.6.3 Estimation of Baseflow

In a groundwater context, baseflow characterization has been a fundamental strategy in water-resource investigations for more than half a century. Such studies contribute to knowledge concerning "the dynamics of the groundwater system, groundwater discharge to streams, and the transport of chemicals to streams" (Raffensperger et al., 2017) and are increasingly important in studies concerned with the sustainability of stream ecosystems.

Studies of baseflow can involve different scales. For example, Raffensperger et al. (2017) discuss a *reach scale* and *watershed scale*. The reach scale (Figure 2.24*a*) involves a relatively small steam segment between two measurement points. Surface and subsurface flows of water enter from an adjacent contributing area. When contributing areas are relatively small, i.e., several square kilometers, the key features controlling runoff to the stream are relatively homogeneous (Raffensperger et al., 2017). The watershed scale (Figure 2.24*b*) represents a larger scale that integrates complications in the geometry of the stream network and in the characteristics of the contributing areas with differing land uses and hydrologic features.

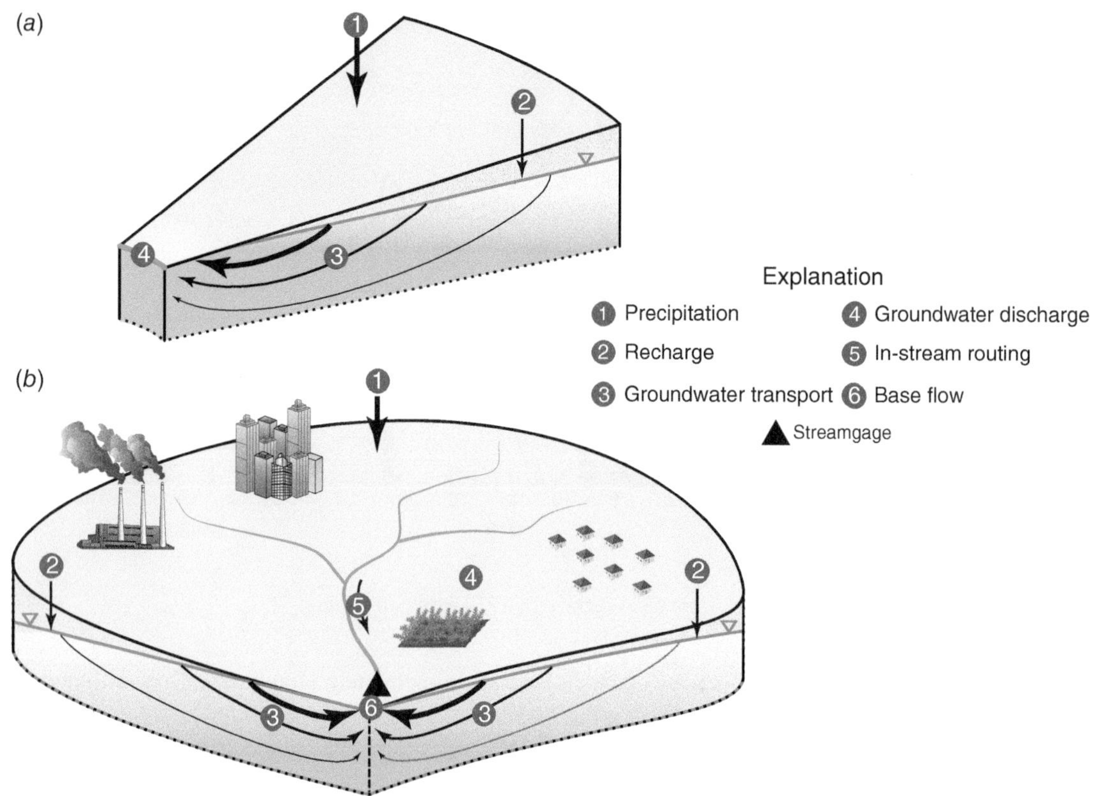

Figure 2.24 Conceptual models of useful scales in the investigation of stream baseflow. Panel (*a*) illustrates a reach-scale of investigation with a relatively small contributing area. Panel (*b*) illustrates the larger watershed scale with inherent variability in land use (Modified from Raffensperger et al., 2017; Public domain).

Long-term interest in baseflow studies has translated into many different approaches for separating a streamflow hydrograph into various components. Typically, quickflow and baseflow are the major focus (Raffensperger et al., 2017). However, in more complicated settings, baseflow might be further subdivided into a groundwater component and releases of surface waters from a variety of different sources.

There are three approaches for determining the baseflow contribution to the total stream discharge with time. The earliest approaches were mechanical, essentially visual techniques that used simple rules to draw lines on a paper copy of the hydrograph. These were empirical because no independent data were available to validate the approach. With time, these approaches were computerized (Raffensperger et al., 2017) in codes like PART (Rutledge, 1998) or HYSEP (Sloto and Crouse, 1996) within the USGS.

The graphical approaches commonly use patterns in the discharge hydrograph to separate baseflow and quickflow. An important step forward has involved the use of chemical and isotopic tracers for direct hydrograph separation or adding constraints to graphical approaches. The approach is based on mass-balance equations for water and mass. For example, consider the separation of a composite stream-discharge hydrograph at any time t during the storm event (Q_t) into three components: Q_r, Q_u, Q_s (i.e., surface runoff, unsaturated zone, and saturated zone, L^3/t, respectively). For example, in many settings, chemical constituents in groundwater have concentrations > interflow > surface runoff. Thus, monitoring the dilution in water chemistry through a storm even provides a chemical signature that can be resolved into various sources. Solving for the three unknown discharge components requires three mass balance equations, a water balance equation (2.13), and two constituents (2.14)

$$Q_t = Q_r + Q_u + Q_s \tag{2.13}$$

$$C_{ti}Q_t = C_{ri}Q_r + C_{ui}Q_u + C_{si}Q_s \tag{2.14}$$

$$i = 1, 2$$

where the concentrations of the chemical tracers from the three sources C_{ri}, C_{ui}, C_{si}, and Q_t are measured and known (Jones et al., 2006).

As noted by Jones et al. (2006), chemical hydrograph separation often found that storm flows were dominated by pre-event water, i.e., groundwater inflows and interflow. This result was noted as counter-intuitive because groundwater flow tends to be slow. Model studies confirmed issues in the application of chemical and isotopic tracers for hydrograph separation associated with dispersion (Jones et al., 2006). Indeed, water-cycle modeling of the type described by Jones et al. (2006) holds promise as the most realistic approach for base-flow modeling.

Techniques continue to evolve with new approaches such as recursive digital filtering constrained by chemical mass balances (Raffensperger et al., 2017). Readers should understand that all methods except for water-cycle modeling provide approximate values at best because the inherent complexities of reaches and basins are unappreciated. Because baseflow is difficult to measure, empirical approaches are rarely validated by independent field measurements.

2.7 Flood Predictions

Historical flood discharge records in a stream can be used to determine the probability or recurrence interval of floods in the future. The *recurrence interval* is defined as (Riggs, 1968)

$$T = \frac{n+1}{m} \tag{2.15}$$

where T is recurrence interval in years, n is number of items in a sample, and m is order number of the individual in the sample array. The probability of an exceedance in any one year is

$$P = \frac{1}{T} \tag{2.16}$$

Following now is a graphical procedure for creating a plot of the annual flood discharge versus recurrence interval:

1) Determine total number of records in the sample (n).
2) Arrange the historical flood data in order of magnitude—beginning with the largest flood first. This rank value is m in Eq. (2.15).
3) Calculate the recurrence interval for each flood using Eq. (2.15).
4) Plot the annual flood on a linear scale against recurrence interval on log scale using semi-log graph paper.

Example 2.6 Annual discharges for years 1915–1950 are listed in Table 2.9. Calculate the recurrence interval and the probability of exceedance in any one year.

Table 2.9 Computation of recurrence interval.

Water year	Q (cfs)	Order number (m)	T	P
1915	264	34	1.09	0.92
1916	374	11	3.36	0.30
1917	332	19	1.95	0.51
1918	346	16	2.31	0.43
1919	359	13	2.85	0.35
1920	333	18	2.06	0.49
1921	483	3	12.33	0.08
1922	417	5	7.40	0.14
1923	346	17	2.18	0.46

(*Continued*)

Table 2.9 (Continued)

Water year	Q (cfs)	Order number (m)	T	P
1924	320	21	1.76	0.57
1925	271	31	1.19	0.84
1926	214	36	1.03	0.97
1927	530	2	18.50	0.05
1928	304	25	1.48	0.68
1929	271	32	1.16	0.86
1930	271	33	1.12	0.89
1931	304	26	1.42	0.70
1932	400	9	4.11	0.24
1933	327	20	1.85	0.54
1934	415	6	6.17	0.16
1935	402	8	4.63	0.22
1936	362	12	3.08	0.32
1937	320	22	1.68	0.59
1938	272	30	1.23	0.81
1939	244	35	1.06	0.95
1940	279	28	1.32	0.76
1941	303	27	1.37	0.73
1942	310	24	1.54	0.65
1943	275	29	1.28	0.78
1944	317	23	1.61	0.62
1945	350	15	2.47	0.41
1946	387	10	3.70	0.27
1947	359	14	2.64	0.38
1948	449	4	9.25	0.11
1949	406	7	5.29	0.19
1950	570	1	37.00	0.03

Source: U.S. Geological Survey Riggs (1968); Public domain.

Solution

The order of data in the sample (m) is numbered in column 3 of Table 2.9. The total number of records in the sample is 36. Thus, $n + 1 = 37$. The recurrence interval (T) is calculated as $(n + 1)/m$ and listed in column 4. The probability (P) of exceedance of flood discharge in any one year is represented as $1/T$ in column 5.

Exercises

2.1 Calculate the average precipitation in the Figure 2.25 using an arithmetic average, Thiessen-weighted average, and isohyetal methods. Note that the precipitation in the figure is given in inches.

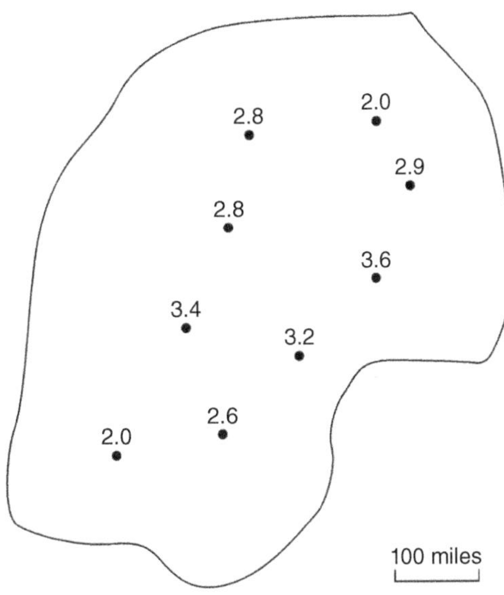

Figure 2.25 Precipitation in inches for a basin (Schwartz and Zhang, 2003/John Wiley & Sons).

2.2 Following here is a summary of the average monthly temperature (°F) at Minneapolis-St. Paul, Minnesota.

YEAR	JAN	FEB	MAR	APR	MAY	JUN	JUL	AUG	SEP	OCT	NOV	DEC
1964	20.0	23.9	25.8	46.8	61.5	68.7	76.0	68.5	58.9	48.2	35.0	14.8
1993	14.6	17.2	29.5	44.2	57.2	64.5	70.3	70.4	55.0	46.5	30.6	22.2

Source: Schwartz and Zhang (2003). *Fundamentals of Groundwater*. Copyright 2003 © John Wiley & Sons, Inc. All Rights Reserved.

Calculate the potential evapotranspiration for each month of the two years using the Thornthwaite method and compare the results.

2.3 The length from a remote point on a strip of land to a stream is 800 ft. The slope of the bare soil land is 10%. Calculate overland flow time using Figure 2.10. If the length of the stream from the point where the overland flow enters the stream to the exit point in the basin is 3000 ft and average stream flow velocity is 10 ft/min, what is the time of concentration at the exit?

2.4 For a drainage area of 150 acres in a single-family residential area, calculate the peak discharge due to surface runoff for a rainfall intensity of 2.5 in./h using the rational method.

2.5 The following table contains annual peak discharges for a gaging station in La Paz County, Arizona, Hydrologic Unit Code 15030104, Latitude 33°59′58″, Longitude 114°26′41″. Calculate the recurrence interval and the probability of exceedance in any one year (USGS Surface Water data for USA: USGS Surface-Water Annual Statistics)

Year	Flow (cfs)	Year	Flow (cfs)	Year	Flow (cfs)
1970	234	1989	145.6	2006	121.3
1974	227.6	1990	147.9	2007	100.4
1975	190.2	1991	142.6	2008	101.8
1976	170.1	1992	136.8	2009	107.4
1977	153.1	1993	134.3	2010	134.6
1978	148.1	1994	145	2011	134.7
1979	132	1995	132.3	2012	155.7
1980	127.2	1996	119.9	2013	150

(Continued)

Year	Flow (cfs)	Year	Flow (cfs)	Year	Flow (cfs)
1981	132	1997	118.2	2014	141
1982	129	1998	118.3	2015	160.3
1983	143.9	1999	106.7	2016	150.1
1984	165.4	2000	98	2017	171.3
1985	139.5	2001	109.1	2018	199.3
1986	126	2002	108.4	2019	159.1
1987	145.6	2003	89	2020	93.3
1988	136.7	2004	103.1	2021	102

2.6 a) Delineate a watershed and calculate its area using a topographic map provided by your instructor.
b) For a storm in the basin, calculate how many days will be needed for all of the direct runoff to reach a stream in the basin.

References

Domenico, P. A., and F. W. Schwartz. 1998. Physical and Chemical Hydrogeology. John Wiley & Sons, New York, 506 p.

Gray, D. M., G. A. McKay, and J. M. Wigham. 1970. Handbook on the Principles of Hydrology. Water Information Center, Inc., Port Washington, NY.

Hambrook, J. A., G. F. Koltun, and B. B. Palcsak, et al. 1997. Hydrologic disturbance and response of aquatic biota in Big Darby Creek Basin, Ohio. U.S Geological Survey Water-Resources Investigations Report 96-4315, 82 p.

Horton, R. E. 1933. The role of infiltration in the hydrologic cycle. Eos, Transactions American Geophysical Union, v. 14, no. 1, p. 446–460.

Horton, R. E. 1940. An approach toward a physical interpretation of infiltration-capacity 1. Soil Science Society of America Journal, v. 5, no. C, p. 399–417.

Jones, J. P., E. A. Sudicky, A. E. Brookfield et al. 2006. An assessment of the tracer-based approach to quantifying groundwater contributions to streamflow. Water Resources Research, v. 42, no. 2, 15 p.

Kennedy, E. J. 1984. Discharge ratings at gaging stations. U.S. Geological Survey Techniques of Water-Resources Investigations, Book 3, Chapter A10, 59 p.

LaFontaine, J. H., L. E. Hay, R. J. Viger, et al. 2013. Application of the Precipitation–Runoff Modeling System (PRMS) in the Apalachicola–Chattahoochee–Flint River Basin in the Southeastern United States. U.S. Geological Survey Scientific Investigations Report 2013-5162, 118 p.

Manning, J. C. 2016. Applied Principles of Hydrology. Waveland Press.

Meinzer, O. 1927. Plants as indicators of groundwater. U.S. Geological Survey Water-Supply Paper 577, 95 p.

Moreo, M. T., B. J. Andraski, and C. A. Garcia. 2017. Groundwater discharge by evapotranspiration, flow of water in unsaturated soil, and stable isotope water sourcing in areas of sparse vegetation, Amargosa Desert, Nye County, Nevada. U.S. Geological Survey Scientific Investigations Report 2017–5079, 55 p.

Raffensperger, J. P., A. C. Baker, J. D. Blomquist, et al. 2017. Optimal hydrograph separation using a recursive digital filter constrained by chemical mass balance, with application to selected Chesapeake Bay watersheds. U.S. Geological Survey Scientific Investigations Report 2017–5034, 51 p.

Regan, R. S., and J. H. LaFontaine. 2017. Documentation of the dynamic parameter, water-use, stream and lake flow routing, and two summary output modules and updates to surface-depression storage simulation and initial conditions specification options with the Precipitation-Runoff Modeling System (PRMS). U.S. Geological Survey Techniques and Methods, Book 6, Chapter B8, 60 p.

Riggs, H. C. 1968. Frequency curves. U.S. Geological Survey Techniques of Water-Resources Investigations, Book 4, Chapter A2, 15 p.

Roberts, W. J., and J. B. Stall. 1967. Lake evaporation in Illinois. Illinois State Water Survey. Report of Investigation, No. 57, 111 p.

Robinson, T. W. 1958. Phreatophytes. U.S. Geological Survey Open-File Report 55-152, 20 p.

Rutledge, A. T. 1998. Computer programs for describing the recession of groundwater discharge and for estimating mean groundwater recharge and discharge from streamflow records: Update. U.S. Geological Survey Water-Resources Investigations Report 98-4148, 43 p.

Schwartz, F. W. 1970. Geohydrology and hydrogeochemistry of groundwater: streamflow systems in the Wilson Creek experimental watershed, Manitoba. Master's Thesis, University of Manitoba, 124 p.

Schwartz, F. W., and H. Zhang. 2003. Fundamentals of Groundwater. John Wiley & Sons, 583 p.

Sloto, R. A., and M. Y. Crouse. 1996. HYSEP: A computer program for streamflow hydrograph separation and analysis. Water-Resources Investigations Report, 96-4040, 46 p.

Thornthwaite, C. W. 1948. An approach toward a rational classification of climate. Geographical Review, v. 38, no. 1, p. 55–94.

Trenberth, K. E., L. Smith, and T. Qian. 2007. Estimates of the global water budget and its annual cycle using observational and model data. Journal of Hydrometeorology, v. 8, no. 4, p. 758–769.

U.S. Federal Government. 2012. U.S. Climate Resilience Toolkit. [Online] Snow Telemetry (SNOTEL) Data Viewer | U.S. Climate Resilience Toolkit (accessed 5 May 2023).

USACE (US Army Corps of Engineers Afghanistan Engineer District). 2011. AED Design Requirements: Hydrology Studies. https://www.tad.usace.army.mil/Portals/53/docs/TAA/AEDDesignRequirements/AED%20Design%20Requirements-%20Hydrology%20Studies_Feb-11.pdf (accessed 6 May 2023).

USDA (U.S. Department of Agriculture). 1986. Urban hydrology for small watersheds. Technical Release, v. 55, p. 2–6.

USDA (U.S. Department of Agriculture), National Resources Conservation Service. 2019. National Engineering Handbook. Part 630: Hydrology, Chapter 4 online at: https://directives.sc.egov.usda.gov/viewerfs.aspx?hid=21422 (accessed 1 April 2023).

Virginia Soil and Water Conservation Commission. 1980. Virginia Erosion and Sediment Control Handbook. Virginia Soil and Water Conservation Commission, Richmond, Virginia, 600 p.

3

Basic Principles of Groundwater Flow

CHAPTER MENU

Water flows in sediments or rocks through open spaces, which range from tiny imperfections along crystal boundaries in igneous rocks to huge caverns in limestone. This chapter will introduce the basic concepts and principles of groundwater flow in sediments and rock.

3.1 Porosity of a Soil or Rock

Groundwater is important to humanity because it provides the most important reservoir of available freshwater on Earth. That water resides in the pores of near-surface rocks and sediments. The volume of open space available for water in the subsurface is impressively large, which accounts for the large volume of groundwater. *Total porosity* of a rock or soil is defined as the ratio of the void volume to the total volume of material.

$$n_T = \frac{V_v}{V_T} = \frac{V_T - V_s}{V_T} \tag{3.1}$$

where n_T is the total porosity, V_v is the volume of voids, V_s is the volume of solids, and V_T is the total volume. Because V_T is always greater than V_v, total porosity is always a fraction less than 1. Sometimes, values of porosity are written as percentages. As an example, values of n_T for sands and gravels typical range from 0.25 to 0.35. Values for rocks like sandstone are lower, 0.05–0.15, because pore volumes are reduced by cements that bind the grains together.

Porosity may also be defined in terms of grain density and bulk density,

$$n_T = 1 - \frac{\rho_b}{\rho_s} \tag{3.2}$$

where ρ_b is the *bulk density* (density of dry soil or rock sample) and ρ_s is the density of solids.

Primary porosity refers to the original interstices (or voids) created when a rock or soil was formed. These interstices include pores in soil or sedimentary rocks, vesicles, lava tubes, and cooling fractures in basalt (Figure 3.1*a*,*b*; Heath, 1998). *Secondary porosity* refers to joints, faults in igneous, metamorphic, and consolidated sedimentary rocks, and solution-enlarged openings in carbonate and other soluble rocks (e.g., Figure 3.1*c*–*f*; Heath, 1998).

Fundamentals of Groundwater, Second Edition. Franklin W. Schwartz and Hubao Zhang.

Companion website: www.wiley.com/go/schwartz/fundamentalsofgroundwater2

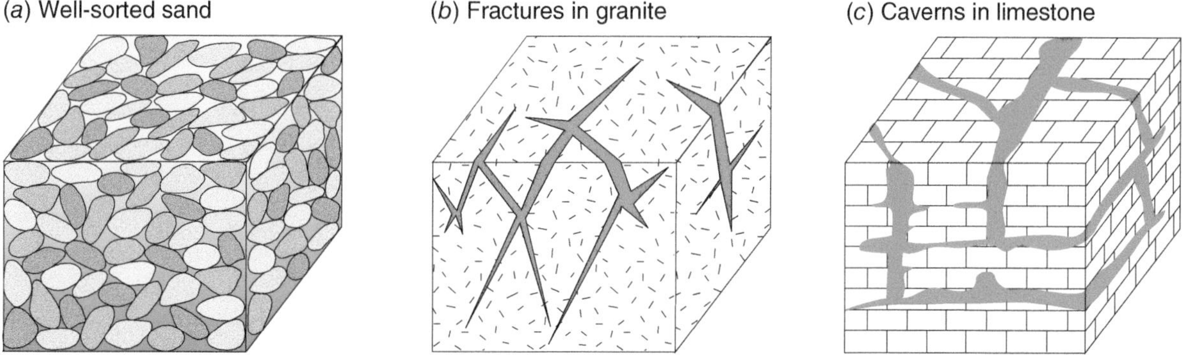

Figure 3.1 Types of openings in selected water-bearing rocks. Block (*a*) is several millimeters wide and shows primary porosity for unlithified sand. Blocks (*b*, *c*) are larger potential meters to 10s of meters in width. For granite (*b*), secondary porosity created by fracturing facilitates flow. Block (*c*) illustrates caverns in limestone produced by solution-enhanced enlargements of fractures (U.S. Geological Survey Barlow, 2003/Public domain).

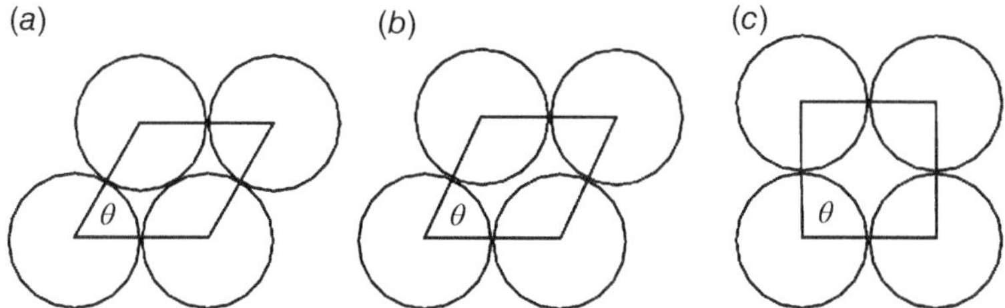

Figure 3.2 Sections of four contiguous spheres of equal size: (*a*) the most compact arrangement, lowest porosity; (*b*) less compact arrangement, higher porosity; (*c*) least compact arrangement, highest porosity (U.S. Geological Survey Slichter, 1899/Public domain).

The magnitude of the primary porosity of a soil or rock depends upon the degree of compaction of grains, the shape of grains, and the particle-size distribution. If a material consists of spheres of equal size, the less the compaction, the greater the porosity (Figure 3.2). The shape of the grains can cause the porosity to be larger or smaller than average, depending on how the grains are arranged and connected. A strongly sorted medium (soil particles of relatively equal sizes) possesses a higher porosity than a poorly sorted medium because particles of a smaller size tend to occupy the void spaces between larger ones.

The porosity of a soil and rock can range from zero to more than 50% (Table 3.1). On average, the porosity of unlithified materials is much higher than that for lithified materials.

For unlithified deposits, the smaller the grain size, the higher the primary porosity. These deposits generally exhibit modest variability. Glacial till is variable because it is a mixture of sand, silt, and clay that occurs in different proportions, e.g., sandy till and clayey till and with different degrees of compaction. The presence of cements in sandstone lowers the porosity as compared to sand. Diagenetic processes tend to be most efficient in reducing porosity in older and deeply buried deposits. Because dolomite and calcite are reactive, porosity is highly variable and often rather small. As Figure 3.1 implies, simply knowing whether an aquifer is sandstone or limestone is not helpful in assigning a porosity value. Finally, unaltered igneous rocks have relatively small porosities.

Fractured rocks are more complicated because they commonly possess two porosities, the primary porosity representing the voids in the rock blocks and a secondary porosity represented by the volume of the fracture apertures. Think of the fracture apertures as small planar volumes between two slabs of rock. Figure 3.3 is a photograph that shows an outcrop of fractured Coconino Sandstone north of Snowflake, Arizona. It is the primary aquifer unit in the C aquifer that occurs across much of northeastern Arizona and northwestern New Mexico in the United States (Leake et al., 2005) The Coconino Sandstone ranges in thickness from approximately 100 to 300 m (Leake et al., 2005). Experience has shown that the

Table 3.1 Illustrative ranges of porosity values including standard deviation (SD).

Material	Porosity			
	Min	Mean	Max	SD
Unlithified deposits[a]				
Sand	28.1	38.9	44.4	4.9
Dune	[c]	42.1	[c]	8.3
Silt	31.5	45.2	50.8	5.6
Clay	40.1	46.1	55	4.5
Glacial till	14.3	23.5	34	[c]
Peat	[c]	92	[c]	[c]
Sedimentary rocks[a]				
Sandstone	4	20.4	8.6	8.6
Limestone	0.44	8.67	34.7	8.6
Dolostone	0.4	7.46	26	[c]
Shale	0.75	10.8	27.2	8.3
Igneous rocks[b]				
Granite	<0.001		0.01	
Basalt	<0.01		0.1	

[a] Data abstracted from Wolf (1982).
[b] Data from INTERA (1983)
[c] Insufficient porosity measurements
Source: (FWS).

Figure 3.3 Outcrop of the Coconino Sandstone along Silver Creek in northeastern Arizona. Notice fracturing here can be both vertical and horizontal (FWS).

fracturing in the unit is variable. How well the C aquifer functions as a groundwater supply depends upon the extent to which fracturing is developed.

The concept of *effective porosity* recognizes that not all the void space participates meaningfully in the flow of water. A case in point is flow through fractured shale. Although the unfractured shale contains water-filled pores, they are poorly connected, so that the flow occurs predominately through the well-connected fractures. In other words, the fracture system provides the effective pathway for flow through the rock. The *effective porosity* of a sediment or rock is the ratio of volume of the fracture or secondary porosity to the total volume of the soil or rock. The concept of effective porosity is different than total porosity because there is an added inference about connectivity. Typically, the effective porosity of a fractured unit is much smaller than the total porosity because the volume of scattered, open fractures is small. For example, because the aperture of a fracture away from outcrops is small (10s to 100s of microns) the fracture porosity $[n_f]$ for a shale might be 5×10^{-2} or even smaller.

Water-filled porosity, as a foundational concept in understanding a groundwater resource, needs further elaboration. The concept of porosity provides a relative measure of void space in a porous medium, but no information on the relative ease with which water can move from pore to pore. The ease of flow is represented by another porous medium property—*hydraulic conductivity, or informally permeability*—which is coming up later in this chapter. Thus, a well completed in a coarse-grained and permeable (easy for flow) unit, such as a gravel, with $n_T = 0.30$ might yield large quantities of water to a well, whereas a lower permeability, fine-grained silt unit with $n_T = 0.30$ would yield only small quantities of water. What is important here is understanding that because of flow constraints, it is the permeable aquifer units that provide the most significant sources of groundwater.

3.2 Occurrence and Flow of Groundwater

The last section discussed pores and voids in groundwater. But what is contained in those pores? In the shallow subsurface, water occurs in the unsaturated zone (Figure 3.4), and the pores of granular materials contain both water and air. In the soil zone, most plants utilize this water. The quantity of water (i.e., water content) in pores of the unsaturated zone fluctuates, increasing when the water from a rainstorm infiltrates downward through the soil zone and decreasing as water is lost by evapotranspiration and downward percolation (Figure 3.4). Deeper in the subsurface is the so-called saturated zone, where the pores contain only water and no air. The water there is known as "groundwater."

The *water table* separates the unsaturated and saturated zones and is in a general sense the top of the groundwater system. The pathway for precipitation to replenish the groundwater now should be evident. Some fraction of the rain falling on the

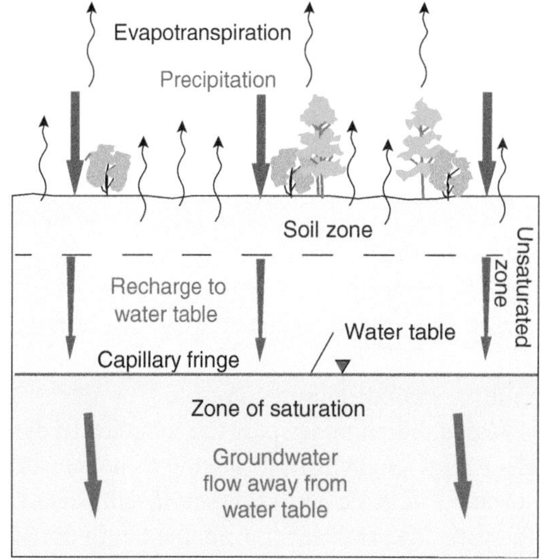

Figure 3.4 Conceptualization of the occurrence of subsurface water within the hydrologic cycle. Water falling as precipitation infiltrates the soil zone and percolates through the unsaturated zone. Water is lost from the soil zone as evapotranspiration. Water joins the groundwater system at the water table and begins to flow downward. The *capillary fringe* is a zone of saturation created due to forces of surface tension due to the presence of small pores (U.S. Geological Survey Alley et al., 1999/Public domain).

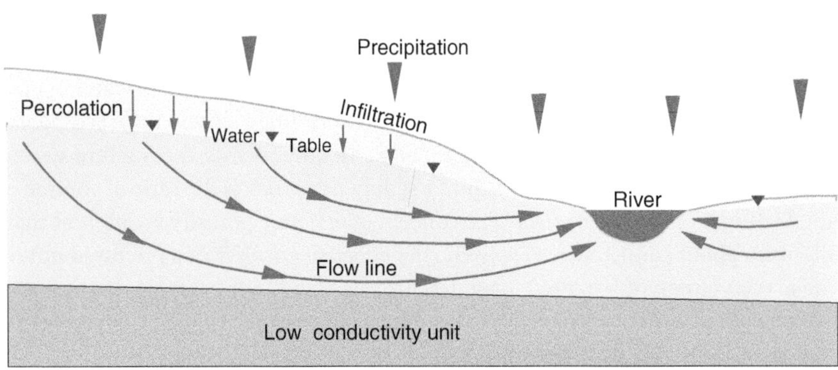

Figure 3.5 Example of a groundwater flow system. Some precipitation ends up as recharge. The low in the groundwater system is illustrated by the red arrows. Groundwater eventually discharges in the nearby river (U.S. Geological Survey, https://www.usgs.gov/media/images/groundwater-flow-showing-natural-conditions/Public domain).

ground surface infiltrates into the subsurface. A smaller fraction of this water within the unsaturated zone eventually percolates to the water table and deeper (Figure 3.4). Water added to the groundwater system in this manner is recharged. A later section will provide a more formal definition.

What happens then to water in the groundwater system? Essentially, the groundwater shown in Figure 3.4 slowly moves away from the water table with the continuing addition of "new" recharge. The simple flow system in Figure 3.5 illustrates how groundwater moves downward and sideways to a nearby stream following the flow paths shown on the figure. Reaching the streambed, the groundwater leaves the system as discharge into the stream. Thus, in this example, surface runoff flowing over the ground surface and subsurface runoff moving along the groundwater system both end up in the nearby stream. The main difference in surface and subsurface pathways is the time required for water to reach the stream. Surface runoff from a rainstorm might get there in a matter of hours, while groundwater could take several years.

3.3 Darcy's Experimental Law

Our exploration of groundwater has gotten to the point of requiring an understanding of basic theory of groundwater flow represented by the Darcy equation. Henry Darcy was a civil engineer in the mid-1800s concerned with the public water supply of Dijon, France. He was interested in acquiring data that would improve the design of filter sands for water purification.

Gravitational forces cause the movement of groundwater. A drop of water present in a flow system at a higher elevation has gravitational energy available to move along a flow path to some place where the energy is lower. Consider the following thought experiment that involves a drop of water at the top of a children's slide. By virtue of its higher elevation, a drop of water at the top of the slide has more gravitational energy $[e_{\text{top}}]$ than at the bottom $[e_{\text{bot}}]$. Turn that drop loose on its available pathway, the slide, it moves from a place of relatively high gravity potential (top of the slide) to a place with a lower gravity potential (bottom of the slide).

These ideas are directly transferable to our example of the simple flow system in Figure 3.5. A water drop joining the groundwater system at the water table beneath the upland has more gravitational potential energy than at the stream where it leaves the groundwater system. Thus, water moves along the flow path because of gravitational forces. Because it keeps raining, the groundwater system is dynamic, providing a continuous parade of water drops through the system. The quantity of flow is represented by the volume of water exiting into the river per unit time (a second, minute, day, etc.).

Most children understand the complicated dynamics of sliding. Firstly, the sliding velocity down a slide is dependent on the energy gradient from the top to bottom of the slide, i.e., the steepness of the slide. The technical term describing steepness in this context is *gradient*, defined as how fast the gravitational potential energy changes per unit distance along the path. The gradient (i) from the top to the bottom of the slide is given as:

$$i = \frac{(e_{\text{top}} - e_{\text{bot}})}{\Delta l} \tag{3.3}$$

where e_{top} and e_{bot} are the energy at the top and bottom, respectively, and Δl is the distance (e.g., meters) along the flow path (i.e., slide) between the two measurement points. We have been deliberately vague about units of measurement for energy to keep the discussion here simple for the moment. The sliding analogy illustrates the first of Darcy's important principles—namely, that the velocity groundwater moving along a flow system is proportional to the hydraulic gradient [i] or the change hydraulic potential along a ground flow path.

A second feature of sliding dynamics is represented by the situation where two slides have identical gradients, but children move more quickly down one of the slides than the other. Personal playground experience again provides the explanation. The friction between a person and the slide slows the descent. Thus, a shiny metal slide yields a greater "ease of sliding" compared to an old rusty surface.

In a groundwater context, the theoretical velocity of water moving along a flow system also depends upon the ease with which water moves through a porous medium. For example, water can through coarse gravel much more easily than a finer-grained porous medium like silt because of the smaller frictional resistance to flow associated with the gravel. As the pores in sediments or rocks become smaller and smaller, it becomes increasingly difficult for water to flow. The parameter, *hydraulic conductivity*, with units of length per unit time (e.g., m/day or m/sec), provides a quantitative measure of the ease with which groundwater flows through a porous medium. Values can vary over an amazingly broad range. For example, well sorted gravels might have a hydraulic conductivity of 1×10^2 m/day, whereas unfractured igneous or metamorphic rocks could have a value of 10^{-7} m/day.

Example 3.1 Shown in Figure 3.6 is a simple flow system with a dipping aquifer. Hydraulic head in the aquifer is measured at the two points shown (1 and 2), located at the top and bottom of the hill, respectively. Assume that volumetric flow rate of water [Q] coming out of the aquifer at the bottom is the same as the same going in at the top. In other words, there is no recharge on the hill slope. Answer these questions.

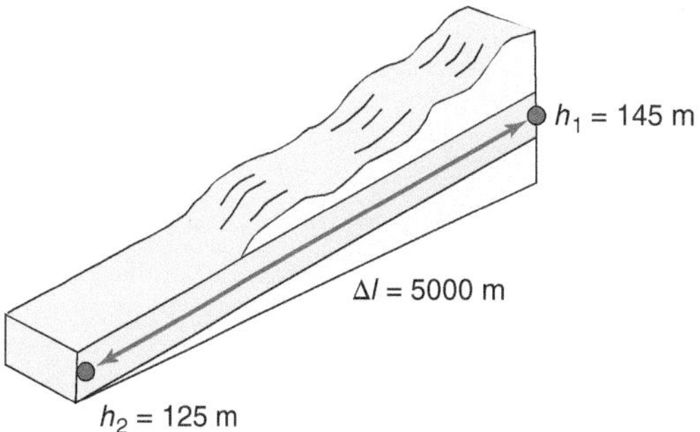

Figure 3.6 Simple flow system (FWS).

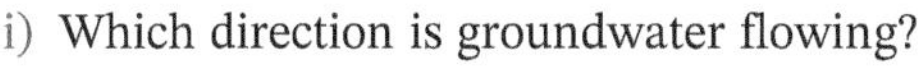

i) Which direction is groundwater flowing?
   Hydraulic head at $h_1 = 145$ m
   Hydraulic head at $h_2 = 125$ m
   Thus, flow downhill is from high head at h_1 directly to the lower head at h_2

ii) Given the information on the figure calculate the gradient (i)

   $i = (h_1 - h_2)/\Delta l = (145 - 125)/5000 = 0.004$ m/m or dimensionless

iii) Where does the groundwater flow leave the system? Flow leaves through the downhill face of the aquifer (blue).

3.3.1 Darcy Column Experiments

Darcy's testing system consisted of a cylinder having a known cross-section area A (L^2), which was filled with various filter sands (Figure 3.7). He provided plumbing to flow water through the column. The cylinder contained two manometers, whose intakes were separated by a distance Δl (L). The *manometers* are nothing more than small open tubes that provide measurements of the energy available for flow at their open ends in the medium (analogous to e_{top} and e_{bot}).

Water was flowed into (and out of) the cylinder at a known rate Q (L^3/T) and the elevation of water levels in the manometers, h_1 and h_2 (L), were measured relative to a local datum to provide the measures of energy, hydraulic head. He derived the following relationship, known as *Darcy's equation*

$$\frac{Q}{A} = -K\frac{(h_1 - h_2)}{\Delta l} \tag{3.4}$$

where K is a constant of proportionality termed *hydraulic conductivity* that is the ease with which water moves through the medium and h_1 and h_2 values of hydraulic head. Shortly, we will discuss concept in greater detail. The term $(h_1 - h_2)/\Delta l$ is known as the *gradient in hydraulic head.* So, just like slide, the velocity of flow through the Darcy column depends upon the hydraulic gradient (i.e., the energy gradient) and hydraulic conductivity (i.e., the ease of motion along the flow path).

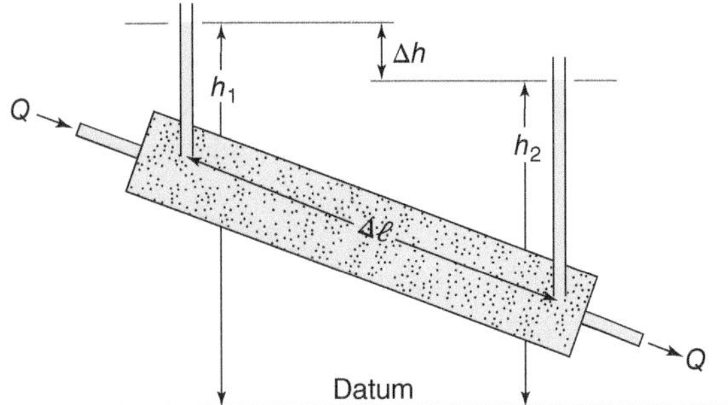

Figure 3.7 Laboratory apparatus to demonstrate Darcy's equation (Reproduced with permission, Domenico and Schwartz, 1998, Physical and Chemical Hydrogeology. Copyright © 1990, 1998/John Wiley & Sons).

The term Q/A representing the volumetric flow rate per unit cross-section area of the cylinder is the *Darcy velocity* (q), or *specific discharge*. In words, Eq. (3.4) states that the velocity of flow is proportional to the hydraulic gradient. If the hydraulic gradient is denoted as i, then

$$i = \frac{(h_1 - h_2)}{\Delta l} = \frac{dh}{dl} \tag{3.5}$$

and Darcy's equation is written as

$$q = -Ki \tag{3.6}$$

or

$$Q = -KiA \tag{3.7}$$

Darcy's equation is valid for flow through most granular material if the flow is laminar. Under condition of turbulent flow, the water particles take more tortuous paths. At the other extreme, in very low-permeability materials, a minimum threshold gradient could be required before flow takes place (Bolt and Groenevelt, 1969).

3.3.2 Linear Groundwater Velocity or Pore Velocity

Built into the Darcy velocity is an assumption that flow occurs over the entire surface area of the soil column. Because water only flows in the pore space, the *actual flow velocity* or *pore velocity* is greater than the Darcy velocity. The pore velocity (v), also termed the *linear velocity*, is defined as the volumetric flow rate per unit of interconnected pore space.

$$v = \frac{q}{n_e} \tag{3.8}$$

where q is the Darcy velocity and n_e is the effective porosity. For a granular medium, like sand and gravel, the effective porosity is approximately the same as the total porosity (i.e., $n_e = n_T$). The pore velocity or true velocity of water flow in a porous medium is always higher than the Darcy velocity because, in Eq. (3.8), n_e is less than 1.

To be clear, the Darcy velocity is a hypothetical velocity and does not describe the "real" velocity of water flow through the porous medium. In a porous medium, the area for groundwater flow $[A]$ needs to be reduced to reflect the fact the presence of the porous medium reduces the area available for flow. This modification involves multiplying the area for flow $[A]$ by the effective porosity $[n_e]$, which is always a number less than 1. Equation (3.9) is the form of Darcy equation for calculating the pore velocity $[v]$ or linear groundwater velocity

$$\frac{Q}{An_e} = \frac{q}{n_e} = v = -\frac{K}{n_e}\frac{(h_1 - h_2)}{\Delta l} \tag{3.9}$$

Example 3.2 reinforces these concepts about velocity and illustrates how effective porosity contributes to large linear groundwater velocities in fractured media.

Example 3.2 Calculation of the Linear Groundwater Velocity

This example uses the information from the previous example (Figure 3.6), which yielded a hydraulic gradient of 0.004. The hydraulic conductivity for this aquifer is 1×10^{-4} m/sec.

a) Assume first that the aquifer is granular porous medium with $n_T = 0.3$. Use Darcy's equation to calculate the linear groundwater velocity.

b) Now for the second calculation, assume the aquifer is a fractured shale with the same hydraulic conductivity, 1×10^{-4} m/sec, and a total porosity, $n_T = 0.1$, and a fracture porosity, $n_f = 3 \times 10^{-3}$. Again, calculate the linear groundwater velocity.

Solution

a) Because this is a granular medium, $n_e \approx n_T = 0.3$. Substituting appropriate values into Eq. (3.9) gives:

$$v = -\frac{K}{n_e}\frac{(h_1 - h_2)}{\Delta l} = \frac{1 \times 10^{-4}}{0.3} 0.004 = 1.33 \times 10^{-4} \text{m/sec or } 0.152 \text{ m/day}$$

b) Because this is a fractured medium, the fractures are effective in conducting the flow, so $n_e = n_f = 3 \times 10^{-3}$.

$$v = -\frac{K}{n_e}\frac{(h_1 - h_2)}{\Delta l} = \frac{1 \times 10^{-4}}{3 \times 10^{-3}} 0.004 = 1.33 \times 10^{-4} \text{m/sec or } 11.5 \text{ m/day}$$

If this groundwater was contaminated, this linear groundwater velocity through the fracture network would be 100 times faster than if the medium was porous.

3.3.3 Hydraulic Head

In Darcy-type experiments (Figure 3.7), the fact that water moves implies the existence of an energy gradient. In effect, water at one point in the column has more energy than at another point. The measure of energy available for flow is reflected by the height of water in the manometer above some datum—the higher the water level, the greater the energy available for flow. Thus, in Figure 3.7, the fact that the water level is higher at h_1 than at h_2 implies that flow is moving from left to right in the column.

In field settings, the direction of groundwater flow is determined by looking at how water levels change from place to place. The measuring device in theory is an open standpipe installed in a borehole in the ground. The *hydraulic head* is the elevation to which the column of water in the standpipe rises above a datum. Imagine there to be an elevation scale beside the well. In the example (Figure 3.8*a*), the elevation of the water column is 110 m above sea level (asl), which is hydraulic head at the measurement point (red circle) at the bottom of the standpipe. Notice then that the unit of measurement for hydraulic head has the units of length [*L*], for example, meters or feet above some datum. Usually, we use sea level as the datum for elevation.

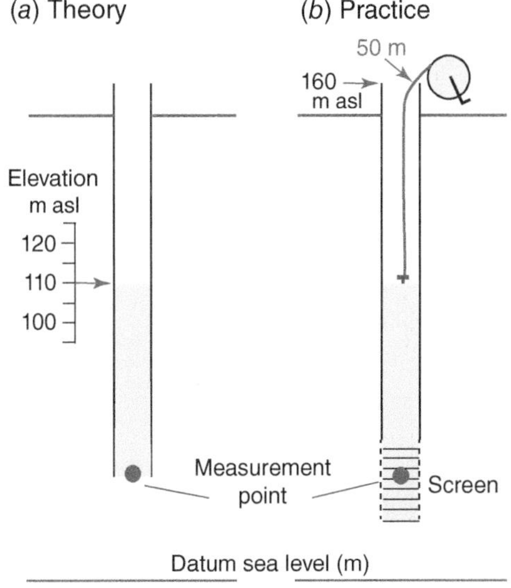

Figure 3.8 Hydraulic head is defined as the elevation to which water in a standpipe open at the bottom. In Panel (*a*), that elevation is 110 m asl. The actual measurement applies to the red dot, called the measurement point. Panel (*b*) shows how a hydraulic head measurement is made in practice. The depth to water is subtracted from the elevation of the top of the casing. The screen lets the device respond more quickly when hydraulic head changes (FWS).

In practice (Figure 3.8*b*), the approach to measuring hydraulic head recognizes that the pipe in the ground does not come with an elevation scale. Thus, the approach is different. First, a high-precision GPS measurement or surveying provides the elevation of the top of the casing (EL_{TOC}), which is 160 m in the example. Next, the depth to water in the well (DTW) is measured (50 m in Figure 3.8*b*) as the distance from the top of the casing to the water surface in the well.

The measurement device is water-level tape (e.g., electric tape). The weighted tape is lowered down the borehole until a buzzer indicates that the probe has hit water. The depth to water is read from markings on the tape, 50 m. The elevation of the water surface (EL_{ws}) in the well is the hydraulic head [*h*] and is calculated as:

$$h = EL_{ws} = EL_{TOC} - DTW \tag{3.10}$$

$$h = EL_{ws} = 160 - 50 \text{ or } 110 \text{ m}$$

A comparison of the two figures shows that the head in Figure 3.8*a*, defined by theory (i.e., 110) is indeed the value obtained using the approach depicted in Figure 3.8*b*.

In practice, standpipes have a screen on the bottom of the casing. A screen is constructed by cutting slots in a casing, providing a much larger area for water to enter the standpipe. In units with a small hydraulic conductivity, adding the screen speeds up water-level adjustments. The measuring point is taken to be the midpoint of screen (Figure 3.8*b*).

3.3.4 Components of Hydraulic Head

A deeper dive into the concept of hydraulic shows that the head measured in the field (e.g., Figure 3.9), as the elevation of the water surface in a well is the sum of three component heads, elevation, pressure, and velocity. This relationship among total head (h) and the elevation, pressure, and velocity head components is described by the Bernoulli equation.

$$h = z + \frac{P}{\sigma_w g} + \frac{v^2}{2g} \tag{3.11}$$

where h is the hydraulic head $[L]$, z is the elevation of the measurement point above the datum $[L]$, P is the pressure exerted by water column $[M/LT^2]$, ρ_w is the fluid density $[M/L]$, g is the gravitation acceleration $[L/T]$, and v is the velocity $[L/T]$. In groundwater settings, the flow velocity is so low that the velocity head component is zero. Thus, the hydraulic head is written as

$$h = z + \frac{P}{\rho_w g} \tag{3.12}$$

This relationship is illustrated in Figure 3.9. The hydraulic head is the sum of elevation head and pressured head. In SI units, h is in meters (m) and z is in meters (m) above the datum (usually sea level), P is in Pascal (Pa), ρ_W is in kg/m^3, and g is in m/sec^2. The density ρ_w is a function of temperature and chemical composition, with fresh water at 15.5°C having a density of 1000 kg/m^3. The gravitational constant, g, is 9.81 m/sec^2. The Pascal is defined as

$$1 \text{ Pascal} = 1 \text{ kg/m/sec}^2 \tag{3.13}$$

In English engineering units, the pressure is in psi (pounds per square inches). A variety of websites are available to convert between different units of pressure.

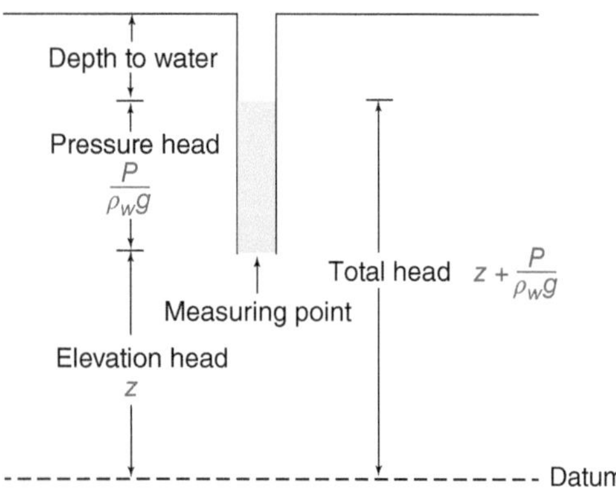

Figure 3.9 Diagram show elevation, pressure, and total head for a point in the flow field (Reproduced with permission, Domenico and Schwartz, 1998, Physical and Chemical Hydrogeology. Copyright © 1990, 1998/John Wiley & Sons).

Example 3.3 With reference to Figure 3.9, assume that the elevation of the ground surface is 1000 m above sea level, the depth to water is 25 m, the total length of the piezometer is 50 m, and the water has density (ρ_w) of 1000 kg/m^3. The gravitational constant is 9.81 m/sec^2. What are (a) the total hydraulic head at the measurement point, (b) the pressure head, and (c) the pressure?

Solution

a) Total hydraulic head at the measurement point at the bottom of the piezometer is

$$h = 1000 - 25 = 975 \text{ m}$$

b) Pressure head can be calculated by rearranging Eq. (3.12)

$$\frac{P}{\rho_w g} = h - z = 975 - 950 = 25 \text{ m}$$

c) Pressure

$$P = \rho_w g(h - z) = (1000\text{kg/m}^3)\,(9.81\text{m/sec}^2)(25\text{ m}) = 2.45 \times 10^5 \text{kg/m/sec}^2 = 0.245 \text{ MPa}$$

We can also use online calculators to convert pressure meters of water or by simple calculation knowing that 1 m of water is 9806 Pa.

$$P = 25 \text{ m of water} = 25 \times 9806 \text{ Pa} = 0.245 \times 10^5 \text{ Pa}$$

3.4 Hydraulic Conductivity and Intrinsic Permeability

Hydraulic conductivity was introduced in the Darcy equation as a constant of proportionality relating specific discharge to the hydraulic gradient. Qualitatively, *hydraulic conductivity* is a parameter describing the ease with which flow takes place through a porous medium. Values are large for permeable units like sand and gravel, and relatively small for poorly permeable materials like clay or shale. Hydraulic conductivity has units of velocity $[L/T]$. Thus, for problems with hydraulic head and lengths in meters, the hydraulic conductivity has units of m/day.

In the English Engineering Unit System with flow rate in gallons per day, hydraulic conductivity has units of gallons per day per foot squared (gpd/ft^2). The following equations describe conversions among these units.

$$1\text{m/day} = 3.28\ \text{ft/day} = 21\ \text{gpd/ft}^2 \tag{3.14}$$

$$1\ \text{ft/day} = 0.3048\ \text{m/day} = 7.48\ \text{gpd/ft}^2 \tag{3.15}$$

$$1\ \text{gpd/ft}^2 = 4.075 \times 10^{-2}\text{m/day} = 0.1337\text{ft/day} \tag{3.16}$$

Example 3.5 that follows illustrates the concept of hydraulic conductivity as applied to an aquifer.

Example 3.4 Groundwater flows through a confined slab aquifer with a cross-section area (A) of 9.3×10^4 m^2 and a length of 6000 m. Hydraulic heads at the groundwater entry and exit points in the aquifer are 300 and 290 m, respectively. At the downstream end of the aquifer, groundwater discharges into a stream at a rate (Q) of 1.0×10^5 ft^3/day. What is the hydraulic conductivity of the aquifer in m/day? A representation of this setting is shown in Figure 3.10

Solution

Begin with Eq. (3.8), a simple form of the Darcy equation written as

$$Q = -KiA$$

where Q is discharge through the aquifer, A is the area of flow, i is the gradient, and K is the unknown hydraulic conductivity. For this example, Q and A are known values and only the gradient i needs to be calculated.

i) Calculate the hydraulic gradient given the known values of hydraulic head

$$i = \frac{(300\ \text{m}) - (290\ \text{m})}{6000\ \text{m}} = 1.7 \times 10^{-3}$$

ii) Calculate the hydraulic conductivity

$$K = \frac{Q}{A \cdot i} = \frac{2850}{9.3 \times 10^4 \cdot 1.7 \times 10^{-3}} = 18\ \text{m/day}$$

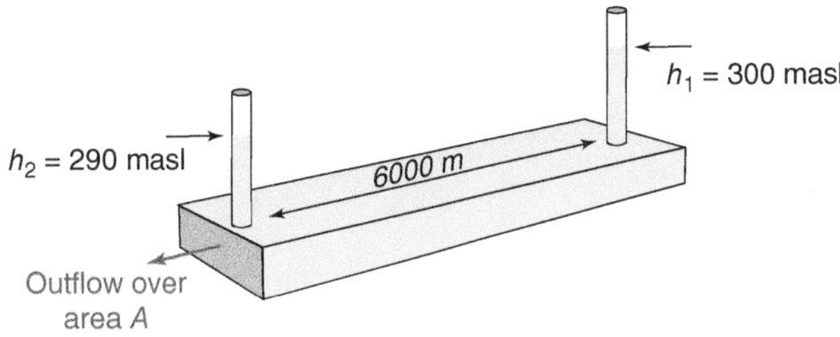

Figure 3.10 Sketch showing horizontal flow through a slab aquifer with hydraulic heads measured with manometers (FWS).

3.4.1 Intrinsic Permeability

Experiments have shown that hydraulic conductivity depends upon both properties of the porous medium and the fluid (e.g., density and viscosity). For most groundwater studies, water is the fluid of interest, providing more-or-less constant values of density and viscosity (neglecting temperature dependencies). Thus, measurements of hydraulic conductivity are useful in comparing differences in hydraulic character of aquifer materials. In looking more generally at systems where fluids other than water are present (such as air, oil, and gasoline), hydraulic conductivity becomes an awkward parameter because the density and the viscosity of the fluid vary together with the medium properties.

A convenient alternative is to write Darcy's equation in a form where the properties of the medium and the fluid are represented explicitly

$$q = -\frac{k\rho_w g}{\mu}\frac{dh}{dl} \tag{3.17}$$

where q is the rate of flow per unit area, k is the intrinsic permeability, ρ_W is the density of water, g is the acceleration due to gravity, μ is the dynamic viscosity of water, and dh/dl is the unit change in hydraulic head per unit length of flow. The *intrinsic permeability* of a rock or soil is a measure of its ability to transmit fluid as the fluid moves through it. The permeability is independent of the fluid moving through the medium. If q (Eq. (3.17)) is measured in m/sec, μ is in kg/(m · sec), ρ_W is in kg/m^3, g is m/sec^2, and dh/dl in m/m, the unit for k is m^2.

$$[k] = \frac{(\text{m/sec})(\text{kg/m}\cdot\text{sec})}{(\text{kg/m}^3)(\text{m/sec}^2)(\text{m/m})} = \text{m}^2$$

Intrinsic permeability also has units like cm^2 or darcy. Equations are available for conversion among units.

$$1\,\text{m}^2 = 10^4\,\text{cm}^2 = 1.013\times10^{12}\,\text{darcy} \tag{3.18}$$

$$1\,\text{cm}^2 = 10^{-4}\,\text{m}^2 = 1.013\times10^{8}\,\text{darcy} \tag{3.19}$$

$$1\,\text{darcy} = 9.87\times10^{-13}\,\text{m}^2 = 9.87\times10^{-9}\,\text{cm}^2 \tag{3.20}$$

Comparison of Eqs. (3.4) and (3.17) show that hydraulic conductivity is a function of both the porous medium (permeability) and the fluid (density and viscosity) such that

$$K = \frac{k\rho_w g}{\mu} \tag{3.21}$$

For water, viscosity and density are functions of pressure and temperature. At 20°C and 1 atm. pressure, the density and dynamic viscosity of water are 998.2 kg/m^3 and 1.002 × 10^{-3} kg/(m·sec), respectively. Under these conditions, Eq. (3.21) can be simplified to

$$K = \frac{k\rho_w g}{\mu} = \frac{(998.2\,\text{kg/m}^3)(9.81\,\text{m/sec}^2)}{1.002\times10^{-3}\,\text{kg/(m}\cdot\text{sec)}}k = \left(9.77\times10^6\frac{1}{\text{m}\cdot\text{sec}}\right)\cdot k \tag{3.22}$$

where K is in m/sec, and k is in m^2. To convert from hydraulic conductivity to intrinsic permeability, Eq. (3.22) can be written as

$$k = \left(1.023\times10^{-7}\,\text{m}\cdot\text{sec}\right)\cdot K \tag{3.23}$$

Example 3.5 What is the intrinsic permeability of a water-saturated medium that has a hydraulic conductivity of 15.24 m/day? Assume the groundwater is at 20°C and 1 atm pressure.

Solution

Convert the hydraulic conductivity in m/day to m/sec as

$$K = 15.24\,\text{m/day} = (15.24\,\text{m/day})\left(\frac{1\,\text{day}}{24\times60\times60\,\text{sec}}\right) = 1.76\times10^{-4}\,\text{m/sec}$$

Intrinsic permeability is calculated from Eq. (3.23) as:

$$k = \left(1.023 \times 10^{-7}\,\text{m} \cdot \text{sec}\right)\left(1.76 \times 10^{-4}\,\text{m/sec}\right) = 1.8 \times 10^{-11}\,\text{m}^2$$

There are several ways to determine the hydraulic conductivity of geological materials. These methods can encompass various field hydraulic tests, methods based on associations with rock type, correlation with grain sizes, and laboratory measurements on core samples. Usually, field hydraulic tests or aquifer tests are considered to provide the most reliable estimates for they facilitate the testing of a large volume of rock with one pumping well and one or more observation wells. Methods for the hydraulic testing of aquifers are coming up.

3.4.2 Hydraulic Conductivity Estimated from Association with Rock Type

Many tens-of-thousands hydraulic conductivity values have been measured for geological materials of all kinds. Subsequent chapters will describe the values for a variety of geological materials and different structural settings. However, in the absence of data, hydraulic conductivity estimates can come from knowledge of the rocks or sediments. This method is only slightly better than an educated guess, but occasionally that is all there is. Hydraulic conductivity for various materials is listed in Figure 3.11.

The table indicates the potential indicates the potential of rocks or sediments as an aquifer. As expected, coarse-grained sediments (clean sand and gravel) and fractured rocks provide the best potential as groundwater aquifers.

3.4.3 Empirical Approaches for Estimation

The intrinsic permeability of a rock or soil is related to the diameter of the grains and porosity. Several empirical equations have been derived to estimate the intrinsic permeability and hydraulic conductivity from grain-size properties (Table 3.2). A grain-size analysis of the material is necessary to use these equations. The *effective grain diameter* is the grain size of the smallest 10% of the grains (90% larger). The grain size distribution of coarse-grained soils is determined directly by sieve analysis. A sieve analysis involves passing a sample through a set of sieves and weighing the amount of material retained in each sieve. Sieves are constructed of wire screens with square openings of standard sizes from 3 in. to 0.074 mm (No. 200).

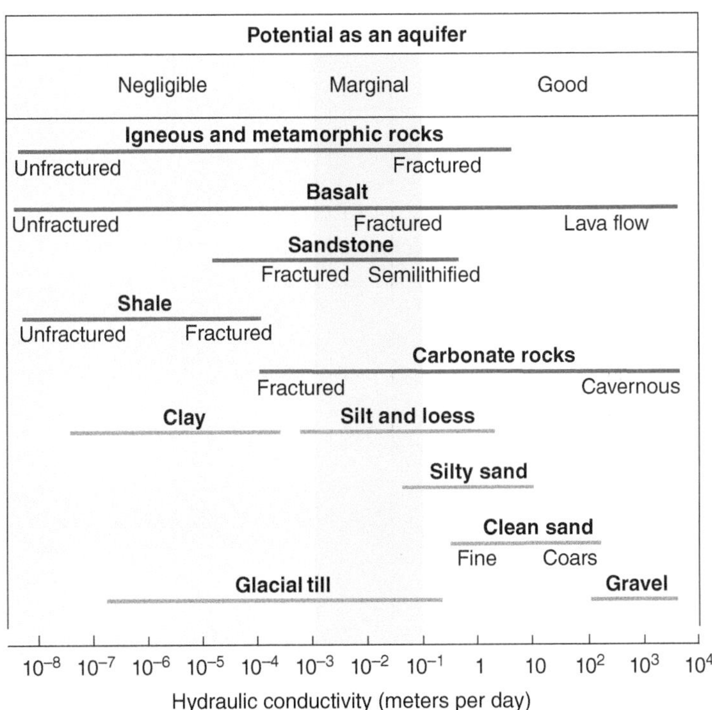

Figure 3.11 Hydraulic conductivity for selected aquifer materials (U.S. Geological Survey, adapted from Heath, 1998/Public domain).

Table 3.2 Examples of empirical relationships for estimating hydraulic conductivity or permeability values.

Source	Equation	Parameters
Hazen (1911)	$K = Cd_{10}^2$	K = hydraulic conductivity (cm/sec) C = constant 100 to 150 (cm/sec) for loose sand d_{10} = effective grain size cm (10% particles are finer, 90% are coarser)
Harleman et al. (1963)	$k = (6.54 \times 10^{-4})d_{10}^2$	k = permeability (cm^2)
Krumbein and Monk (1943)	$k = 760d^2e^{-1.31\sigma}$	k = permeability (darcys) d = geometric mean grain diameter (mm) σ = log standard deviation of the size distribution
Kozeny (1927)	$k = Cn^3/S^{*2}$	C = dimensionless constant: 0.5, 0.562, and 0.597 for circular, square, and equilateral triangle pore openings k = permeability (L^2) n = porosity S^* = specific surface-interstitial surface areas of pores per unit bulk volume of the medium
Kozeny-Carmen Bear (1972)	$K = \left(\frac{\rho_w g}{\mu}\right)\frac{n^3}{(1-n)^2}\left(\frac{d_m^2}{180}\right)$	K = hydraulic conductivity ρ_w = fluid density υ = fluid viscosity g = gravitational constant d_m = any representative grain size n = porosity

Source: Reproduced with permission, Domenico and Schwartz (1998), Physical and Chemical Hydrogeology. Copyright © 1990, 1998/John Wiley & Sons.

The particle distribution for fine-grained soils (or the fine fraction of a coarse soil) is indirectly determined by hydrometer analysis of the size fraction passing the No. 200 sieve.

Example 3.6 Figure 3.12 shows the results of a grain size analysis for a sample from a study site in eastern Idaho (Twining and Rattray, 2016). Estimate the permeability and hydraulic conductivity.

Solution

The upper horizontal axis shows grain size while the vertical axis is the percent finer or percent passing by weight. The effective grain diameter is 0.095 mm or 0.0095 cm. The permeability can be estimated from Harleman's equation in Table 3.2.

$$k = (6.54 \times 10^{-4})d_{10}^2 = (6.54 \times 10^{-4}) \cdot 0.0095^2 = 5.9 \times 10^{-8}\ \text{cm}^2$$

The hydraulic conductivity is calculated knowing the permeability from Eq. (3.22)

$$K = \left(9.77 \times 10^6 \frac{1}{\text{m} \cdot \text{sec}}\right) \cdot k = \left(9.77 \times 10^6 \frac{1}{\text{m} \cdot \text{sec}}\right) \cdot \frac{1\ \text{m}}{100\ \text{cm}} \cdot 5.9 \times 10^{-8}\ \text{cm}^2 = 5.8 \times 10^{-3}\ \text{cm/sec}$$

Or the hydraulic conductivity may be estimated from the Hazen's equation

$$K = Cd_{10}^2 = \left(100 \frac{1}{\text{cm} \cdot \text{sec}}\right)(0.0095\ \text{cm})^2 = 9.0 \times 10^{-3}\ \text{cm/sec}$$

The two results are in same order of magnitude.

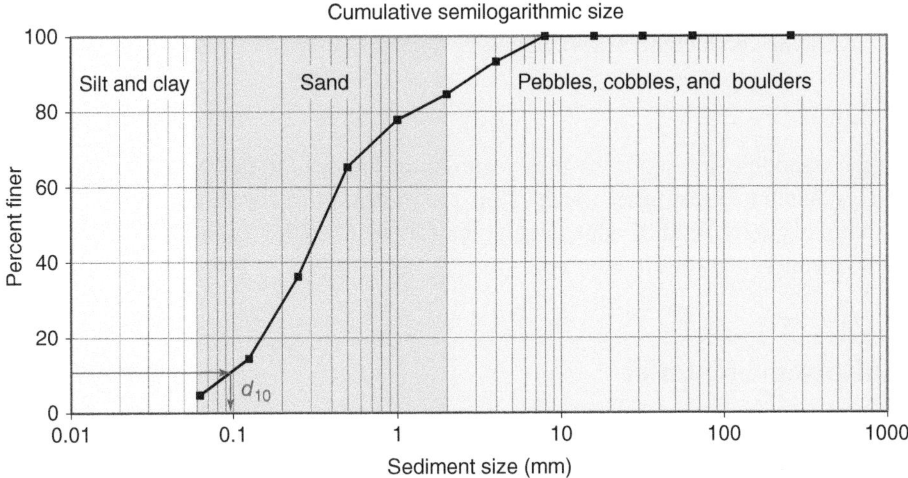

Figure 3.12 Cumulative grain-size distribution for a subsurface sample (BP2_R1). Most of the sample is comprised of sand ranging from very fine to very coarse. The indicated d_{10} value for this sample is 0.095 (U.S. Geological Survey Twining and Rattray, 2016/Public domain).

3.4.4 Laboratory Measurement of Hydraulic Conductivity

This section introduces two laboratory methods for testing cores: the constant head and the falling head tests (Figure 3.13). In the constant head test, a valve at the base of the sample is opened and water starts to flow through the core. Once a steady flow has developed, the volumetric flow rate is determined using data on hydraulic heads and the volume of outflow as a function of time. Hydraulic conductivity is determined with the Darcy's equation written in the following form

$$K = \frac{QL}{Ah} \tag{3.24}$$

where L is the length of the sample, A is the cross-sectional area of the sample, and h is the constant head shown in Figure 3.13a.

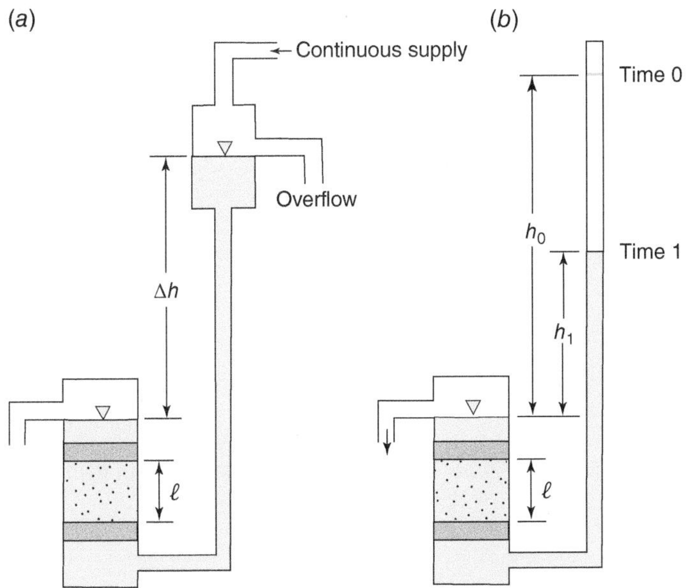

Figure 3.13 Laboratory equipment for measuring hydraulic conductivity of porous media. Panel (*a*) shows a constant-head permeameter. Panel (*b*) shows a falling-head permeameter (Reproduced with permission, Domenico and Schwartz, 1998, Physical and Chemical Hydrogeology. Copyright © 1990, 1998/John Wiley & Sons).

In the falling head test, hydraulic head is measured in the column (Figure 3.13*b*) as a function of time. For a sample of length L and a cross-sectional area A, hydraulic conductivity is given by

$$K = 2.3\frac{aL}{A(t_1 - t_0)}\log_{10}\frac{h_0}{h_1} \tag{3.25}$$

where a is the cross-sectional area of the standpipe, $(t_1 - t_0)$ is the time required for the head to fall from h_0 to h_1.

For fine-grained soils, Q is small and is difficult to measure accurately. Thus, the constant-head test is used principally for testing coarse-grained media (clean sands and gravels) with K values greater than about 10×10^{-4} cm/sec. The falling-head test is generally used for less permeable media (fine sands to fat clays).

3.5 Darcy's Equation for Anisotropic Material

In this section, we examine cases where the hydraulic conductivity changes as a function of the direction of flow. The term *anisotropic* is used to describe the permeability or hydraulic conductivity of materials, which at a point exhibit directional dependency. When the permeability is the same in all directions, the material at that point is *isotropic*. Davis (1969) cites several cases of bedded sediments where permeability is greater in the direction of stratification and smaller perpendicular to the stratification. The Darcy's law in anisotropic materials is expressed as

$$q = -K\nabla h \tag{3.26}$$

where q is the Darcy velocity vector, K is the hydraulic conductivity matrix tensor, and ∇h is the gradient of hydraulic head. In Cartesian coordinates, q, K, and ∇h are written as

$$q = q_x i + q_y j + q_z k \tag{3.27}$$

$$-\nabla h = \frac{\partial h}{\partial x} i + \frac{\partial h}{\partial y} j + \frac{\partial h}{\partial z} k \tag{3.28}$$

$$K = \begin{bmatrix} K_{xx} & K_{xy} & K_{xz} \\ K_{yx} & K_{yy} & K_{yz} \\ K_{zx} & K_{zy} & K_{zz} \end{bmatrix} \tag{3.29}$$

where i, j, and k are unit vectors along the x, y, and z directions, respectively. Equation (3.24) is Darcy's equation for fluid flow in anisotropic media. Combining Eq. (3.24) through (3.27), the Darcy velocity vector may be broken down into components in x, y, and z directions,

$$q_x = -K_{xx}\frac{\partial h}{\partial x} - K_{xy}\frac{\partial h}{\partial y} - K_{xz}\frac{\partial h}{\partial z} \tag{3.30}$$

$$q_y = -K_{yx}\frac{\partial h}{\partial x} - K_{yy}\frac{\partial h}{\partial y} - K_{yz}\frac{\partial h}{\partial z} \tag{3.31}$$

$$q_z = -K_{zx}\frac{\partial h}{\partial x} - K_{zy}\frac{\partial h}{\partial y} - K_{zz}\frac{\partial h}{\partial z} \tag{3.32}$$

where

$$K_{xy} = K_{yx}, K_{xz} = K_{zx}, K_{yz} = K_{zy} \tag{3.33}$$

If the principal directions of anisotropy coincide with the x, y, and z directions, the non-diagonal components of hydraulic conductivity tensor in Eq. (3.27) are zero.

$$K_{xy} = K_{yx} = K_{xz} = K_{zx} = K_{yz} = K_{zy} = 0 \tag{3.34}$$

And the Darcy velocities in x, y, and z directions are simplified as

$$q_x = -K_{xx}\frac{\partial h}{\partial x} \tag{3.35}$$

Table 3.3 The anisotropic character of some rocks.

Material	Hydraulic conductivity (m/sec)	Vertical conductivity (m/sec)
Anhydrite	10^{-14} to 10^{-12}	10^{-15} to 10^{-13}
Chalk	10^{-10} to 10^{-8}	5×10^{-11} to 5×10^{-9}
Limestone, dolomite	10^{-9} to 10^{-7}	5×10^{-10} to 5×10^{-8}
Sandstone	5×10^{-13} to 10^{-10}	2.5×10^{-13} to 5×10^{-11}
Shale	10^{-14} to 10^{-12}	10^{-15} to 10^{-13}
Salt	10^{-14}	10^{-14}

Source: Reproduced with permission, Domenico and Schwartz (1998), Physical and Chemical Hydrogeology. Copyright © 1990, 1998/John Wiley & Sons.

$$q_y = -K_{yy}\frac{\partial h}{\partial y} \tag{3.36}$$

$$q_z = -K_{zz}\frac{\partial h}{\partial z} \tag{3.37}$$

Table 3.3 provides a summary of information on the anisotropic nature of some sedimentary materials as determined from core samples. The ratio between the horizontal and vertical hydraulic conductivities for these rocks is between one and several thousands.

3.6 Hydraulic Conductivity in Heterogeneous Media

The classical definition considers a unit to be *homogeneous* if the permeability in a given direction is the same from point to point in a geological unit. Materials that do not conform with this condition are *heterogeneous.* A simple example of heterogeneity in permeability is represented by a layered geological unit with variable hydraulic conductivities (Figure 3.14).

According to Leonards (1962), an equivalent horizontal hydraulic conductivity for the four units together in the horizontal direction is

$$K_h = \frac{\sum_{i=1}^{M} b_i K_{hi}}{\sum_{i=1}^{M} b_i} \tag{3.38}$$

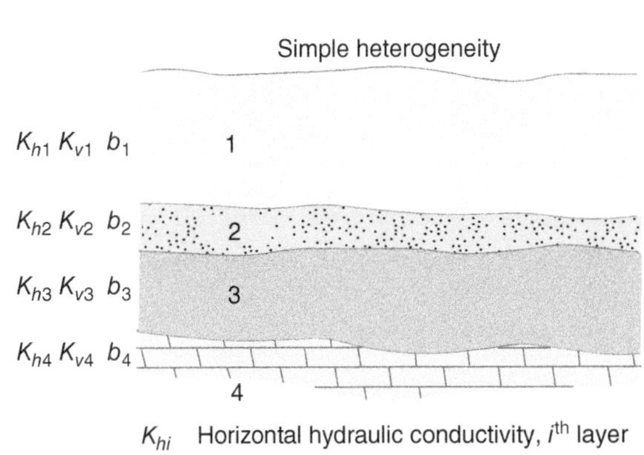

Figure 3.14 An example of simple heterogeneity where the vertical and horizontal hydraulic conductivities within each of the four depositional units are the same (Adapted from Domenico and Schwartz, 1998, Physical and Chemical Hydrogeology. Copyright © 1990, 1998/John Wiley & Sons, Inc.).

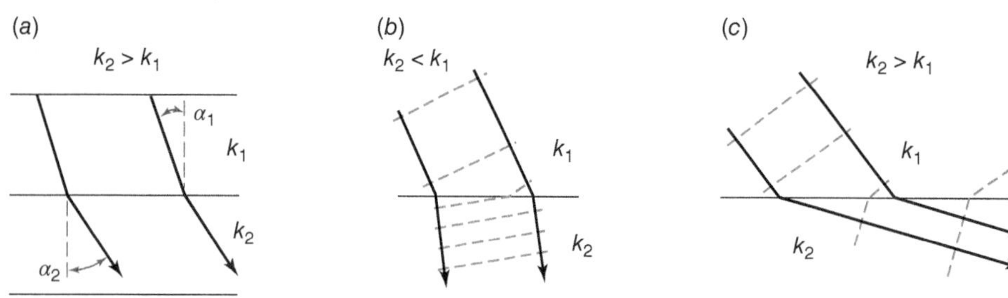

Figure 3.15 Diagrams of flow line refraction and conditions at the boundaries between materials of differing permeability. With (*a*) $k_2 > k_1$, with (*b*) $k_2 < k_1$ and with (*c*) $k_2 >> k_1$ (Reproduced with permission, Domenico and Schwartz, 1998, Physical and Chemical Hydrogeology. Copyright © 1990, 1998/John Wiley & Sons).

where K_h is the equivalent horizontal hydraulic conductivity, K_{hi} is the horizontal hydraulic conductivity of layer i, and b_i is the thickness of layer i. An equivalent vertical hydraulic conductivity is

$$K_v = \frac{\sum_{i=1}^{M} b_i}{\sum_{i=1}^{M} b_i / K_{vi}} \tag{3.39}$$

where K_v is equivalent to vertical hydraulic conductivity, and K_{vi} is the hydraulic conductivity of layer i.

In cases where flow is at some angle to a geological boundary between layers, groundwater flow obeys the tangent refraction law (Figure 3.15*a*). The extent of refraction of the flow lines will depend on the hydraulic conductivity ratio of the two layers. Mathematically, the ratio of the tangents of the angles that the flow lines make with the normal to the boundary is equal to the ratio K_1/K_2, or

$$\frac{K_1}{K_2} = \frac{\tan(\alpha_1)}{\tan(\alpha_2)} \tag{3.40}$$

where groundwater flows from unit 1 into unit 2, K is hydraulic conductivity, and α is the refraction angle. In Figure 3.15*b*, the hydraulic gradient in the lower hydrological unit is steepened to accommodate the flow crossing the boundary from the unit of higher hydraulic conductivity; in Figure 3.15*c*, both the hydraulic gradient and the cross-sectional flow area in the high-conductivity unit are decreased to accommodate the flow crossing the boundary from the low-conductivity unit.

Example 3.7 Consider a 300-m sequence of interbedded sandstone and shale that has 75% sandstone. The sandstone has a horizontal and vertical hydraulic conductivity of 10^{-5} m/sec, and shale has a horizontal and vertical hydraulic conductivity of 1.92×10^{-12} m/sec. Calculate the equivalent conductivities for system of layers. If the groundwater flows in the sandstone with an angle of 45° away from the vertical direction toward the sandstone-shale boundary, what is the flow direction in the shale?

Solution

The equivalent horizontal hydraulic conductivity is

$$K_h = \frac{(225\,\text{m})(1 \times 10^{-5}\text{m/sec}) + (75\,\text{m})(1.92 \times 10^{-12}\text{m/sec})}{(300\,\text{m})} = 7.5 \times 10^{-6}\text{m/sec}$$

The equivalent vertical hydraulic conductivity is

$$K_v = \frac{(300\,\text{m})}{\dfrac{(225\,\text{m})}{1 \times 10^{-5}\text{m/sec}} + \dfrac{(75\,\text{m})}{1.92 \times 10^{-12}\text{m/sec}}} = 7.7 \times 10^{-12}\text{m/sec}$$

Thus, for horizontal flow the most permeable units dominate the system; for vertical flow, the least permeable units dominate the system. The flow direction in the shale is calculated by Eq. (3.40).

$$\alpha_2 = \tan^{-1}\left[\frac{K_2}{K_1}\tan(\alpha_1)\right] = \tan^{-1}\left[\frac{(1.92 \times 10^{-12}\ \mathrm{m/sec})}{(10^{-5}\ \mathrm{m/sec})}(1)\right] = 0°$$

The result shows the flow in the shale is normal to the boundary.

For a uniform flow, arithmetic, harmonic, and geometric means are commonly used to scale measured hydraulic conductivities in heterogeneous fields. The arithmetic mean (η) is the summation of measurements divided by the number of measurements.

$$\eta = \frac{1}{N}\sum_{i=1}^{N} X_i \tag{3.41}$$

where N is the number of measurements in the sample, and X_i is an individual measurement. The harmonic mean (H) of a set of numbers is given as

$$H = \frac{N}{\sum_{i=1}^{N}\frac{1}{X_i}} \tag{3.42}$$

Geometric mean (G) is the Nth root of a product of N numbers.

$$G = (X_1 X_2 3 \ldots X_N)^{1/N} \tag{3.43}$$

Example 3.8 The second column of Table 3.4 lists hydraulic conductivity values measured for a geological unit. Calculate arithmetic, geometric, and harmonic means.

Solution

It is easy to use a spreadsheet program to calculate the mean values as shown in Table 3.4.

Columns 3 and 4 are log(K) and $1/K$ for measured values, respectively. The last row in the table is the summation of K, log(K), and $1/K$. The arithmetic mean is the sum in column 2 divided by the number of measurements.

$$\eta = \frac{8.45 \times 10^{-6}}{32} = 2.64 \times 10^{-7}(\mathrm{m/sec})$$

The logarithm of geometric mean is the sum in column 3 divided by the number of measurements. Thus,

$$G = 10^{-224.1/32} = 9.93 \times 10^{-8}(\mathrm{cm/sec})$$

The harmonic mean is the number of measurements divided by the sum in column 4 of the table.

$$H = \frac{32}{5.59 \times 10^{9}} = 5.72 \times 10^{-9}(\mathrm{cm/sec})$$

Matheron (1967) shows that for a uniform flow in a statistically isotropic hydraulic conductivity field, the average hydraulic conductivity for a system of any dimension is between the harmonic and the arithmetic means.

In many studies, hydraulic conductivity has been shown to be a lognormally distributed parameter. Like a typical normal distribution, this distribution is characterized by an arithmetic mean (or mean) and standard deviation. The standard deviation (σ) is defined as

$$\sigma = \left[\frac{1}{N-1}\sum_{i=1}^{N}(X_i - \eta)^{\frac{1}{2}}\right] \tag{3.44}$$

The equation for a normal (Gaussian) probability density curve is

$$f(X) = \frac{1}{\sigma\sqrt{2\pi}} e^{\frac{(X-\eta)^2}{2\sigma^2}} \tag{3.45}$$

where $f(X)$ is the normal probability density function.

A plot for the frequency of observed values versus the observed values is called a *frequency histogram*. In constructing the histogram, a frequency table records how often observed values fall within certain intervals. The frequency histogram is often constructed to see how well its shape matches certain distributions, like a normal distribution.

Table 3.4 Sample data for the calculation of arithmetic, harmonic, and geometric means.

No.	K	log(K)	1/K	[log(K) − η]^2
1	4.00×10^{-8}	−7.40	2.50×10^{7}	0.16
2	1.00×10^{-7}	−7.00	1.00×10^{7}	0.00
3	2.00×10^{-8}	−7.70	5.00×10^{7}	0.49
4	4.00×10^{-7}	−6.40	2.50×10^{6}	0.36
5	1.00×10^{-6}	−6.00	1.00×10^{6}	1.00
6	1.50×10^{-7}	−6.82	6.67×10^{6}	0.03
7	6.00×10^{-8}	−7.22	1.67×10^{7}	0.05
8	1.50×10^{-6}	−5.82	6.67×10^{5}	1.38
9	8.00×10^{-9}	−8.10	1.25×10^{8}	1.20
10	1.00×10^{-7}	−7.00	1.00×10^{7}	0.00
11	1.00×10^{-7}	−7.00	1.00×10^{7}	0.00
12	1.00×10^{-8}	−8.00	1.00×10^{8}	1.00
13	8.00×10^{-7}	−6.10	1.25×10^{6}	0.82
14	5.00×10^{-8}	−7.30	2.00×10^{7}	0.09
15	4.00×10^{-7}	−6.40	2.50×10^{6}	0.36
16	5.00×10^{-7}	−6.30	2.00×10^{6}	0.49
17	2.50×10^{-8}	−7.60	4.00×10^{7}	0.36
18	2.00×10^{-7}	−6.70	5.00×10^{6}	0.09
19	2.00×10^{-7}	−6.70	5.00×10^{6}	0.09
20	5.00×10^{-8}	−7.30	2.00×10^{7}	0.09
21	2.00×10^{-7}	−6.70	5.00×10^{6}	0.09
22	2.50×10^{-8}	−7.60	4.00×10^{7}	0.36
23	1.00×10^{-7}	−7.00	1.00×10^{7}	0.00
24	1.00×10^{-6}	−6.00	1.00×10^{6}	1.00
25	2.00×10^{-10}	−9.70	5.00×10^{9}	7.28
26	1.00×10^{-7}	−7.00	1.00×10^{7}	0.00
27	8.00×10^{-8}	−7.10	1.25×10^{7}	0.01
28	4.00×10^{-8}	−7.40	2.50×10^{7}	0.16
29	3.00×10^{-7}	−6.52	3.33×10^{6}	0.23
30	6.00×10^{-7}	−6.22	1.67×10^{6}	0.61
31	2.50×10^{-7}	−6.60	4.00×10^{6}	0.16
32	4.00×10^{-8}	−7.40	2.50×10^{7}	0.16
Sum	8.45×10^{-6}	−224.10	5.59×10^{9}	18.12

Source: Schwartz and Zhang (2003), © John Wiley & Sons/John Wiley & Sons.

Example 3.9 Prepare a frequency distribution histogram for each of the two columns of hydraulic conductivity data {K, log(K)} in Table 3.4. Determine by inspection whether one or both distributions are normally distributed. If the hydraulic conductivity appears to be normal, calculate the mean and the standard deviation.

Solution

The histograms for hydraulic conductivity and log-transformed hydraulic conductivity data are shown in Figures 3.16*a* and *b*, respectively. Notice how the histogram of the log-transformed hydraulic conductivity data is similar to a normal

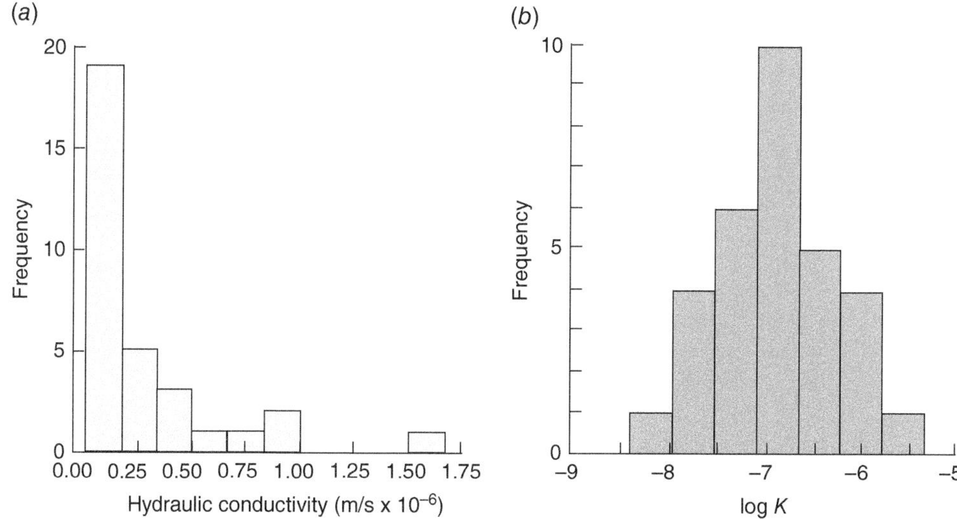

Figure 3.16 (*a*) Frequency distribution histogram of the 32 values of hydraulic conductivity. (*b*) Histogram of the log-transformed hydraulic-conductivity data (Reproduced with permission, Domenico and Schwartz, 1998, Physical and Chemical Hydrogeology. Copyright © 1990, 1998/John Wiley & Sons).

distribution. The mean for the transformed hydraulic conductivity is the summation of log(*K*) listed in column 3 of Table 3.4 divided by the number of measurements.

$$\eta = \frac{-224.1}{32} = -7.0$$

The standard deviation is the square root of summation of $[\log(K) - \eta]^2/(N - 1)$ listed in column 5 of the Table 3.4.

$$\sigma = \sqrt{\frac{1}{32-1} 18.12} = 0.76$$

So far, this chapter has presented a rather simple model of hydraulic conductivity for individual geological units. Hydraulic conductivity values vary from point to point in a systematic manner. Hydraulic conductivity in a geological unit is a correlated random variable. The extent of correlation changes as a function of direction.

3.7 Investigating Groundwater Flow

3.7.1 Water Wells, Piezometers, and Water Table Observation Wells

There are three types of devices for making hydraulic head measurements: (1) conventional water wells, (2) piezometers, and (3) water table observation wells. They are unique in how they are used and what is measured. In Figure 3.17, we illustrate these devices all installed together at a hypothetical site. The hydrogeological system consists of deep aquifer overlain by a shallow, less permeable unit. Flow is mainly downward in the shallow unit and horizontal in the aquifer.

Hydraulic head measurements involving conventional water wells provide the only feasible way to conduct large-scale water resources investigations. This approach eliminates the difficulty in finding sites for drilling, and the costs of drilling holes and installing wells. By their nature, water wells are constructed with relatively large diameter casings and long screened sections. They are best suited for making measurements in well-defined aquifers with predominantly lateral flow. Figure 3.17, for example, illustrates a well with a long screen in the deep aquifer. At the measurement point, the hydraulic head is 73 m asl. In an aquifer like this where flow is horizontal, equipotential lines are vertical (e.g., the 75 contour). Thus, it does not matter where the water well is screened in the aquifer, near the top or near the bottom, the head will be the same.

Piezometers are typically wells with smaller casing diameters and much shorter screens. These design features make them most useful for precision monitoring applications involving problems of contamination or construction dewatering. Typically, piezometers are installed at smaller study sites at places and depths that are carefully selected. The shorter screens

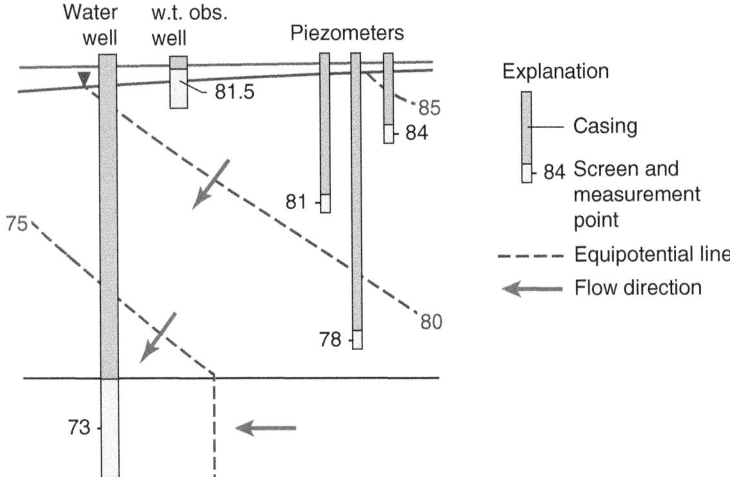

Figure 3.17 The devices for measuring water levels include water wells completed in aquifers, water table observation wells, and piezometers. Water levels measured in these wells indicate hydraulic heads at the measurement points (FWS).

help to assure that hydraulic head measurements and water samples are representative of a "point" in the aquifer. Bentonite in the borehole above the screened interval also aids in this respect.

Often piezometers are installed in clusters called nests. For example, Figure 3.17 features a nest of three piezometers. In the shallow, lower conductivity zone, groundwater flow is mainly vertical with a high hydraulic gradient. The nest of piezometers in this example provides the scale of sampling required to quantify the vertical decline in hydraulic head with depth. Thus, the hydraulic head at the measurement point for each of the piezometers provides an accurate measurement of head that compares favorably to the actual conditions described by the equipotential lines.

Water table observation wells are fundamentally different devices. Their job is to assist in monitoring the elevation of the water table, which can fluctuate significantly at a site. To work properly, the screen should be long enough to accommodate the range of fluctuations in water levels. In other words, the water table should always lie somewhere in the screened section. This explains why in Figure 3.17 the water table observation well is shown with a relatively long screen.

Another unique feature of this well is that the measurement point is not fixed because it is the water table, which moves up and down. Remember, the hydraulic head at any point along the water table is equal to its elevation. Thus, in Figure 3.17, the measurement (81.5 m) is at the water table and not in the middle of screen.

3.7.2 Potentiometric Surface Maps

Mapping the spatial distribution of hydraulic heads in an aquifer is an essential approach to aquifer evaluation. For example, this information is helpful in identifying recharge and discharge areas, areas impacted by pumping. In problems of contamination, understanding patterns of groundwater to anticipate directions of contaminant migration. Such a map of hydraulic-head measurements is a *potentiometric surface* map. Meinzer (1923) defined the potentiometric surface as an imaginary surface that everywhere coincides with level of groundwater measured in an aquifer.

There are three essential points to understand in creating and in interpreting potentiometric surface maps

- a potentiometric map must be related to a single aquifer. Other aquifers deeper or shallower sections can have substantially different potentiometric surfaces with hydraulic heads that are higher or lower than the aquifer of concern. Thus, it is essential to know in which aquifer a well or piezometer is completed,
- flow in an aquifer is assumed to be horizontal, that is, parallel to upper and lower confining layers. Thus, the hydraulic head is presumed not to change as a function of depth in the aquifer given that equipotential lines are vertical. Hence, the potentiometric surface is a projection of vertical equipotential lines into a horizontal (2D) plane, the potentiometric surface map. This detail is important because within the aquifer it does not matter where within the aquifer a well is completed or whether the well has a long or short screen,

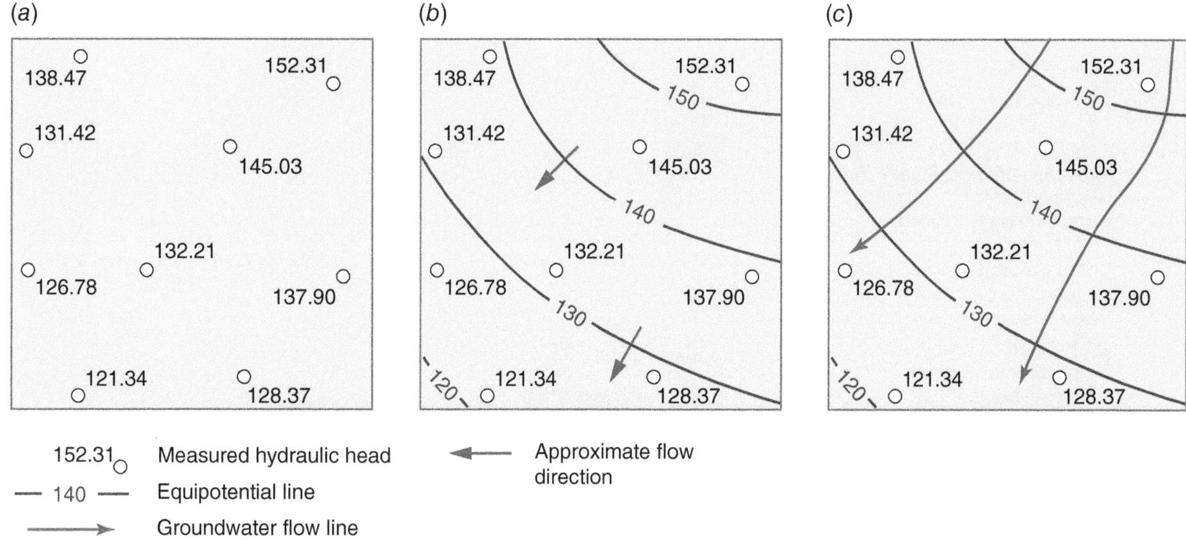

Figure 3.18 Creating a map of hydraulic head involves three steps (*a*) plotting the location of piezometers or monitoring wells on a map together with associated hydraulic head values; (*b*) contouring the data points to create equipotential lines and adding approximate flow directions (red arrows) or (*c*) as flow lines (U.S. Geological Survey, adapted from Winter et al., 1998/Public domain).

- if head losses between adjacent pairs of equipotential lines are equal, the hydraulic gradient varies inversely with distance between lines of equal head. In other words, the further apart the equipotential lines, the smaller the hydraulic gradient.

  The process of making a potentiometric surface map involves three steps (Figure 3.18). The first step is to plot the location of the data points on a map of the aquifer together with the associated value of hydraulic head (Figure 3.18*a*). Next, contouring the collection of data points provides lines along which the hydraulic head is the same (i.e., equipotential lines). The final step involves adding arrows to indicate the approximate direction of groundwater flow (Figure 3.18*b*). The arrows are oriented perpendicular to the equipotential lines with arrowhead pointing in the direction of decreasing hydraulic head. On occasion, more rigorous interpretations might involve the addition of flow lines to create a flow net (Figure 3.18*c*).

An example of a potentiometric surface map is associated with the Memphis aquifer in Tennessee (Kingsbury, 1996; Taylor and Alley, 2001). The Memphis aquifer, also known as the Memphis Sand, is part of the larger Sparta aquifer system that underlies seven states, mainly Mississippi, Arkansas, and Louisiana. The map of the potentiometric surface map in and around Memphis (Figure 3.19) was prepared in 1995 by contouring hydraulic head data for wells (red dots) completed in this single unit. At this time, there were significant drawdowns associated with several pumping centers. The local cones of depression related to these well fields coalesced to form a dominant potentiometric low in the center of the map. The equipotential lines and red flow arrows indicate flow from the surrounding areas into this large potentiometric low (Figure 3.19). In recent years, reductions in groundwater utilization have facilitated some recovery of water levels in the aquifer.

3.7.3 Water-Level Hydrograph

A water-level hydrograph illustrates the variability in water levels with time at one or more monitoring wells. In the past, water-level measurements necessary to create the hydrograph involved regular visits to the wells (e.g., monthly) and measuring the depth to water in the well. These days important sites are monitored remotely through satellite uplinks or wireless networks.

Water-level hydrographs are often used in conjunction with a potentiometric maps to provide a historical perspective on the pattern of water level changes with time. Potentiometric maps represent a snapshot of conditions at some point in time using contemporaneous measurements. In most cases, it is difficult to create a potentiometric map back in time because the necessary water-level measurements are not available. What may be available at a few wells are historical data appropriate to create a water-level hydrograph. Shown in Figure 3.20 is an example of a hydrograph for Sh:P–76, a well completed

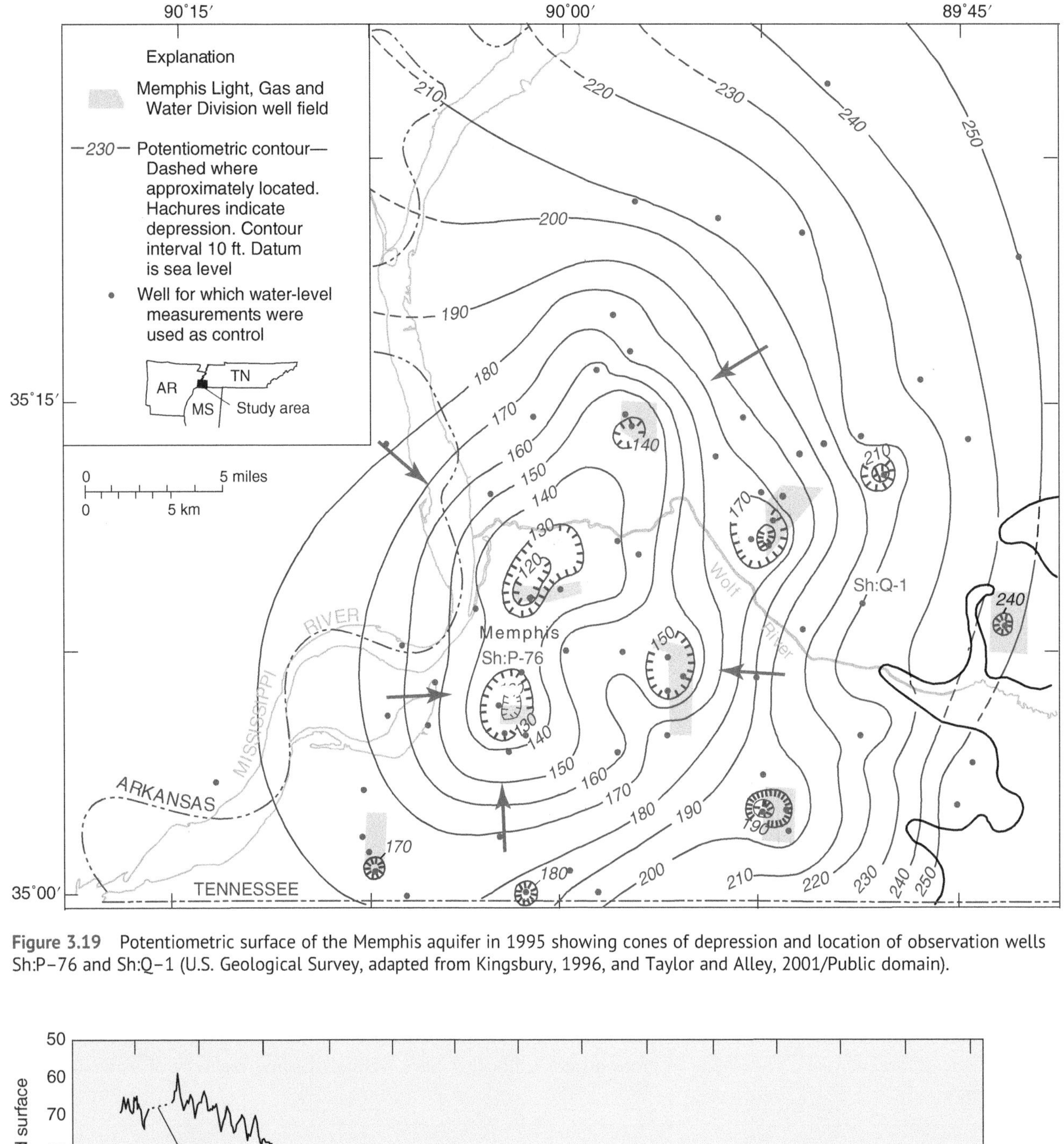

Figure 3.19 Potentiometric surface of the Memphis aquifer in 1995 showing cones of depression and location of observation wells Sh:P-76 and Sh:Q-1 (U.S. Geological Survey, adapted from Kingsbury, 1996, and Taylor and Alley, 2001/Public domain).

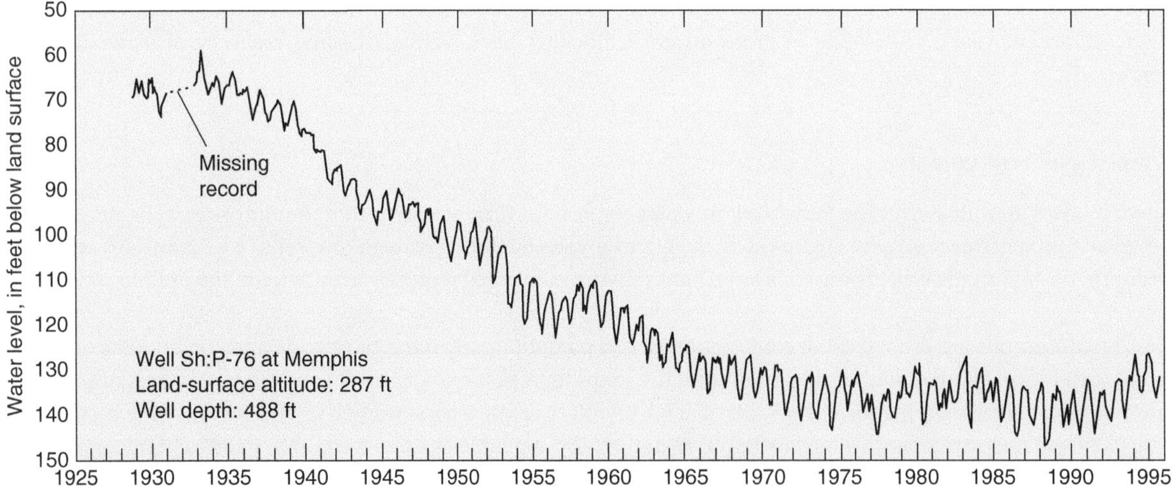

Figure 3.20 Long-term hydrograph of a well completed in the Memphis aquifer (U.S. Geological Survey Taylor and Alley, 2001/Public domain).

Memphis aquifer and monitored from 1925 to 1995 (Taylor and Alley, 2001). The hydrograph indicates a steady and substantial decline in water levels for the first 40 years, which began to level off, beginning in approximately 1970. This well is unique both in terms of the length of the record and its association with the regional potentiometric low (Taylor and Alley, 2001).

3.7.4 Hydrogeological Cross Sections

Besides piezometric surface maps, it is common to find distributions of hydraulic head depicted in vertical cross sections. *Hydrogeological cross sections,* thus, are vertical sections through a three-dimensional flow region. By aligning the section parallel to the direction of mean groundwater flow, flow conditions can be represented accurately in terms of a two-dimensional cross section. Hydrogeological cross sections are more complicated than piezometric surface maps because several units are likely present, each with their own hydraulic conductivity. This vertical pattern of layering often causes flow lines to change direction abruptly, such as when flow enters a more conductive unit from a less conductive unit. These plots are also complicated by different length scales in the vertical and horizontal directions. Exaggeration in the vertical direction increases the vertical height of the section and make it easier to see the different hydrogeologic units. On exaggerated cross section, flow lines need not cross the equipotential lines at right angles, although they are often drawn that way.

Figure 3.21 illustrates a hydrogeological cross section drawn parallel to the direction of mean flow from an upland area to an adjacent valley bottom. Normally, such a plot includes information about the stratigraphy and variations in hydraulic conductivity, as well as hydraulic-head data from nests of piezometers located along the section. The example (Figure 3.21) represents a two-layer system with a less permeable unit (K_1) overlying an aquifer (K_2). The hydraulic head measurements come from four nests that include piezometers and water table observation wells. The measurement points for the piezometers are indicated with circles and with squares for the water table observation wells. Three equipotential lines define the spatial variation in hydraulic head. For this simple example, groundwater flow is depicted with arrows, perpendicular to the equipotential lines. In the upper, low K unit, flow is near vertical downward with a relatively high hydraulic gradient, indicated by the relatively small spacing between equipotential lines. In the more permeable lower unit, the equipotential lines are vertical, with flow in the aquifer is essentially horizontal. As discussed in the previous section equipotential lines in aquifers are vertical as shown here. The horizontal gradient in this more permeable unit is much smaller than in the upper layer. Finally, in the center of the section notice how flow is refracted across the boundary of the two units.

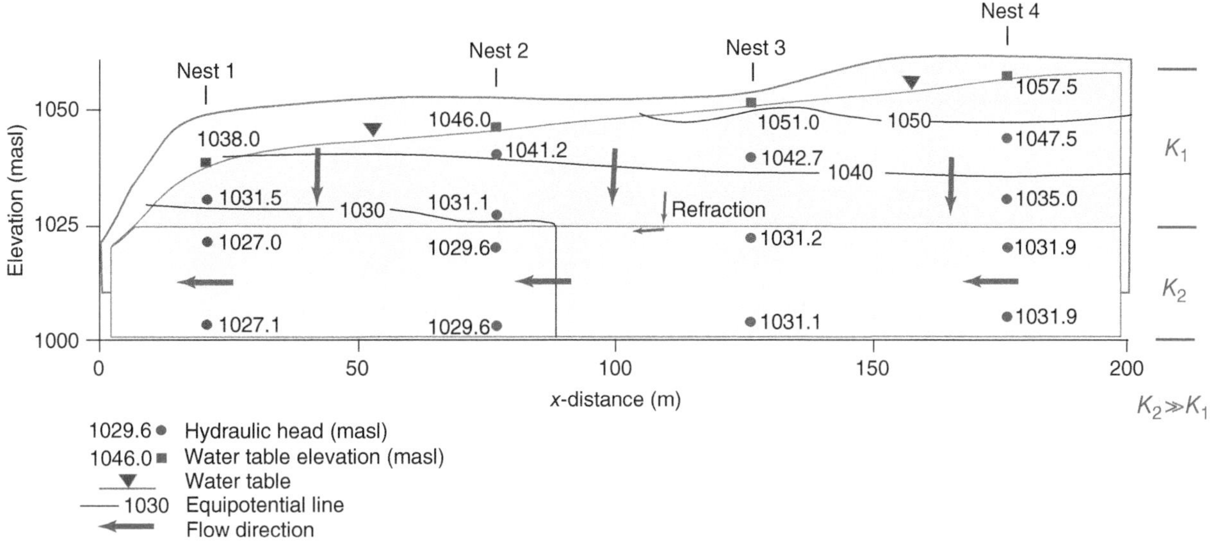

Figure 3.21 Example of a hydrogeologic cross section illustrating how the distribution in hydraulic head depends upon the water table configuration and pattern of layering. The equipotential lines describe downward flow through the lower permeability upper layer and lateral flow in the aquifer (Reproduced with permission, Domenico and Schwartz, 1998, Physical and Chemical Hydrogeology. Copyright © 1990, 1998/John Wiley & Sons).

Exercises

3.1 With reference to Figure 3.21, calculate the average horizontal and vertical hydraulic gradients in the two hydrogeological units (the upper unit having a lower hydraulic conductivity than the lower unit). Explain why the vertical hydraulic gradient is higher in the lower-conductivity unit than the horizontal hydraulic gradient in the higher-conductivity unit using Darcy's law.

3.2 Discuss the similarity and difference of hydraulic conductivity and intrinsic permeability. What is the fundamental difference between them?

3.3 In an experiment, groundwater flows through a sediment column (43.5 cm in length and 5 cm in diameter) at a rate of 4.4 mL/min. The porosity of the sediment is 0.4. Calculate the Darcy velocity and linear groundwater velocity. If the hydraulic conductivity is 0.11 cm/sec, calculate the hydraulic gradient.

3.4 A grain-size analysis yields a d_{10} value of 0.023 mm. Estimate the hydraulic conductivity and intrinsic permeability using the equations in Table 3.2.

3.5 In a constant-head test, the Darcy flux is 5 cm^3/sec for a column of 5 cm in diameter and 50 cm in length. The constant head is 60 cm. Calculate the hydraulic conductivity.

3.6 In a falling-head test, the initial head at $t = 0$ is 60 cm. At $t = 30$ min, the head is 57 cm. The diameters of the standpipe and the specimen are 1 and 20 cm, respectively. The length of the specimen is 20 cm. Calculate the hydraulic conductivity and intrinsic permeability of the specimen.

3.7 A hydrological system consists of five horizontal formations. The hydraulic conductivities of the formations are 20, 10, 15, 50, and 1 m/day, respectively. Calculate equivalent horizontal and vertical hydraulic conductivities. If the flow in the uppermost layer is at an angle of 30 away from the normal direction relative to the boundary, calculate flow directions in all of the formations.

3.8 Table 3.5 lists measured hydraulic conductivity values for a formation. (a) Calculate the arithmetic, geometric, and harmonic means of K; (b) Construct histograms of K and log(k) and determine the probability distribution of the sample. If the distribution is normal, or log-normal, calculate the standard deviation of the sample.

Table 3.5 Measured hydraulic conductivities in a formation (cm/sec).

4.17×10^{-7}	7.59×10^{-3}	2.63×10^{-7}	1.78×10^{-8}	2.29×10^{-6}
1.00×10^{-4}	5.25×10^{-4}	1.62×10^{-6}	6.17×10^{-10}	2.04×10^{-7}
3.16×10^{-16}	3.16×10^{-13}	3.31×10^{-6}	2.09×10^{-2}	2.42×10^{-1}
6.03×10^{-9}	5.50×10^{-9}	8.71×10^{-7}	5.01×10^{-12}	5.01×10^{-5}
4.57×10^{-8}	2.14×10^{-5}	4.90×10^{-7}	8.32×10^{-8}	7.41×10^{-8}
4.27×10^{-10}	1.12×10^{-3}	2.24×10^{-3}	1.02×10^{-6}	1.91×10^{-8}
1.74×10^{-5}	2.40×10^{-4}	2.82×10^{-5}	2.51×10^{-9}	1.74×10^{-9}
1.26×10^{-11}	1.26×10^{-9}	5.01×10^{-11}	2.14×10^{-10}	9.12×10^{-4}
2.69×10^{-8}	4.79×10^{-6}	1.00×10^{-5}	3.98×10^{-11}	9.77×10^{-9}
1.26×10^{-4}	1.26×10^{-7}	7.41×10^{-1}	1.02×10^{-10}	6.17×10^{-6}

Source: Schwartz and Zhang (2003), © John Wiley & Sons/John Wiley & Sons.

3.9 Figure 3.22 is a diagram showing measured heads for a confined aquifer. Hydraulic tests determined the transmissivity of the aquifer is 2000 m/day. Draw contour lines of measured heads, and calculate the average hydraulic gradient in the aquifer and water discharge rate from the aquifer to the stream.

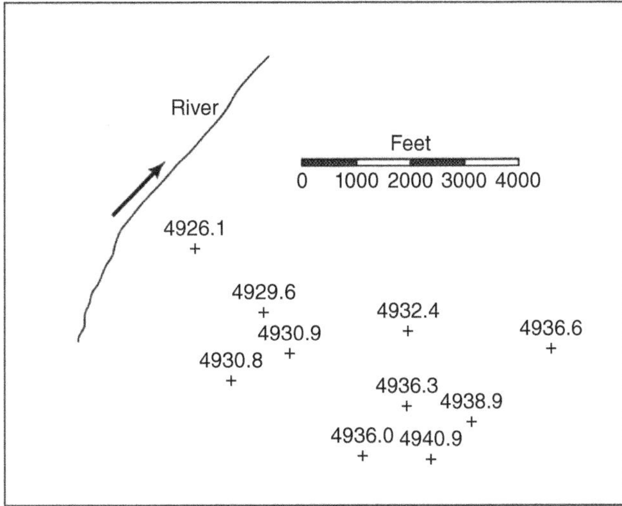

Figure 3.22 Measured hydraulic heads for a confined aquifer (Schwartz and Zhang, 2003, © John Wiley & Sons/John Wiley & Sons).

References

Alley, W. M., T. E. Reilly, and O. L. Franke. 1999. Sustainability of groundwater resources. U.S. Geological Survey Circular 1186, 79 p.

Barlow, P. M. 2003. Groundwater in freshwater-saltwater environments of the Atlantic coast. U.S. Geological Survey Circular 1262, 113 p.

Bear, J. 1972. Dynamics of Fluids in Porous Media. American Elsevier Publishing Company, Inc., New York, 764 p.

Bolt, G. H., and P. H. Groenevelt. 1969. Coupling phenomena as a possible cause for non-Darcian behavior of water in soil. Bulletin of the International Association of Scientific Hydrology, v. 14, no. 2, p. 17–26.

Davis, S. N. 1969. Porosity and permeability of natural materials, In R. J. M. DeWiest (ed.), Flow through Porous Materials. Academic Press, New York, p. 54–89.

Domenico, P. A., and F. W. Schwartz. 1998. Physical and Chemical Hydrogeology, 2nd edition. John Wiley & Sons, Inc., New York, 505 p.

Harleman, D. R. E., P. F. Mehlhorn, and R. R. Rumer. 1963. Dispersion-permeability correlation in porous media. Journal of the Hydraulics Division American Society of Civil Engineers HY2, v. 89, no. 2, p. 67–85.

Hazen, A. 1911. Discussion: dams on sand foundations. Transactions, American Society of Civil Engineers, v. 73, no. 3, p. 199.

Heath, R. C. 1998. Basic groundwater hydrology. U.S. Geological Survey Water-Supply Paper 2220 (eighth printing), 86 p.

INTERA Environmental Consultants, Inc. 1983. Porosity, permeability, and their relationship in granite, basalt, and tuff. Office of Nuclear Waste isolation, Technical Report, ONWI-458, 90 p.

Leake, S. A., J. P. Hoffmann, and J. E. Dickinson. 2005. Numerical groundwater change model of the C aquifer and effects of groundwater withdrawals on stream depletion in selected reaches of Clear Creek, Chevelon Creek, and the Little Colorado River, Northeastern Arizona. U.S. Geological Survey Scientific Investigations Report 2005-5277, 29 p.

Leonards, G. A. 1962. Engineering properties of soil, In G. A. Leonards (ed.), Foundation Engineering. McGraw-Hill, New York, p. 66–240.

Kingsbury, J. A. 1996. Altitude of the potentiometric surfaces, September 1995, and historical waterlevel changes in the Memphis and Fort Pillow aquifers in the Memphis area, Tennessee. U.S. Geological Survey Water-Resources Investigations Report 96–4278, 1 pl.

Kozeny, M. 1927. Uber kapillare Leitung des Wassers im Boden. Sitzungsber, Akademie der Wissenschaften in Wien, v. 136, p. 271–306.

Krumbein, W. C., and G. D. Monk. 1943. Permeability as a function of the size parameters of unconsolidated sand. Transactions of the American Institute of Mining Engineers, v. 151, no. 01, p. 153–163.

Matheron, G. 1967. Eléments Pour une Théorie des Milieux Poreux. Masson et Cie, Paris, 166 p.

Meinzer, O. E. 1923. Outlines of groundwater in hydrology with definitions. U.S. Geological Survey Water-Supply Paper 494, 71 p.

Schwartz, F. W., and H. Zhang. 2003. Fundamentals of Groundwater. John Wiley & Sons, New York, p. 583p.

Slichter, C. S. 1899. Theoretical investigation of the motion of groundwaters. The 19th Annual Report of the U.S. Geophysical Survey, p. 304–319.

Taylor, C. J., and W. M. Alley. 2001. Groundwater-level monitoring and the importance of long-term water-level data. U.S Geological Survey Circular 1217, 68 p.

Twining, B. V., and G. W. Rattray. 2016. Characterization of sediment and measurement of groundwater levels and temperatures, Camas National Wildlife Refuge, eastern Idaho. U.S. Geological Survey Data Series 1024, 23 p.

Winter, T. C., J. W. Harvey, O. L. Franke, et al. 1998. Groundwater and surface water, a single resource. U.S. Geological Survey Circular 1139, 79 p.

Wolf, R. G. 1982. Physical properties of rocks: Porosity, permeability, distribution coefficients. U.S. Geological Survey Water-Resources Investigations, Open-File Report 82-166, 118 p.

4

Aquifers

CHAPTER MENU

Life as a hydrogeologist would be downright boring if field settings looked anything like the soil-filled pipes or simple media that we have considered so far. The reality is that one of the major areas of historical interest in hydrogeology, aquifers, is likely more important now than in the past. In little more than a Century, aquifers have been transformed from the Earth's reliable storehouse for clean, potable water to a resource, suffering from overdevelopment and depletion on a global scale.

Any discussion of groundwater resources begins with aquifers and confining beds. Understanding the role of these bodies in a water supply context is complicated because they are complex manifestations of the geological setting. This chapter provides an understanding of how water is transmitted and stored in aquifer in terms of hydraulic variables like transmissivity and storativity, and the variety of units that function in these respects.

4.1 Aquifers and Confining Beds

From a resource perspective, the primary unit in groundwater investigations is the *aquifer*, a lithologic unit or combination of lithologic units capable of yielding water to pumped wells or springs (Domenico, 1972). An aquifer can be coextensive with geologic formations, a group of formations, or a part of a formation. It may cut across formations in a way that makes it independent of any geologic unit. Units of low permeability that bound an aquifer are called *confining beds*.

Linking the definition of an aquifer to features of water supply can create confusion. In areas with prolific aquifers, a low permeability unit might be considered a confining bed. However, in groundwater-poor regions, the same deposit could be considered an aquifer. In actual field studies, this ambiguity in the definition of an aquifer turns out not to be much of a problem because hydraulic conductivity or porosity values explicitly define the hydraulic character of the unit.

Aquifers and confining beds come in flavors. The terms *water table or unconfined* are applied to aquifers where the water table forms the upper boundary (Figure 4.1). When shallow wells or piezometers are installed into such an aquifer, the water levels in these wells approximately define the position of the water table.

A *confined* (or *artesian*) *aquifer* has its upper and lower boundaries marked by confining beds (Figure 4.1). Stated another way, an aquifer is confined by overlying and underlying low permeability beds. The water level of a well or piezometer

Fundamentals of Groundwater, Second Edition. Franklin W. Schwartz and Hubao Zhang.

Companion website: www.wiley.com/go/schwartz/fundamentalsofgroundwater2

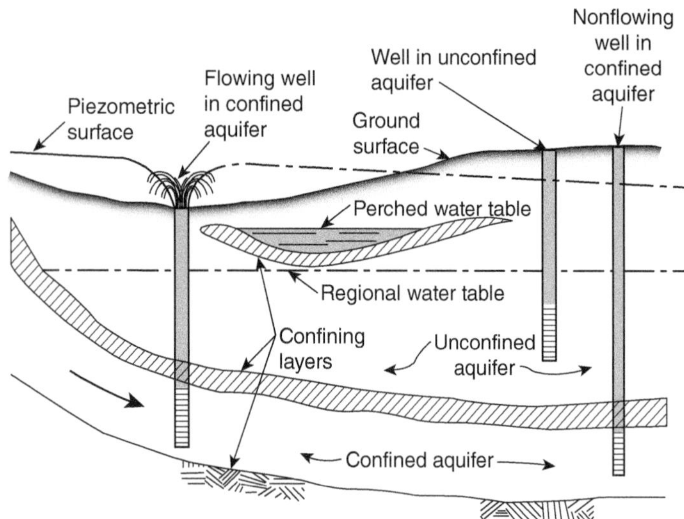

Figure 4.1 Conceptual model of aquifers developed in a field setting (U.S. Bureau of Reclamation, 1995/U.S. Government Printing Office/Public domain).

installed in a confined aquifer occurs somewhere above its upper boundary. Occasionally, the water level of a well occurs above the ground surface. This condition can produce a flowing artesian well (Figure 4.1). As noted in Section 3.7, a contoured map of hydraulic heads from measurements wells installed in the same aquifer is a potentiometric surface.

A *perched aquifer* is an unconfined aquifer that develops above the regional water table. In effect, there is an unsaturated zone below the low hydraulic conductivity layer on which the perched zone develops.

Occasionally, the terms aquifuge, aquiclude, and aquitard are applied to various types confining beds. An *aquifuge* is the ultimate low hydraulic conductivity unit, which is an extremely poor conductor of groundwater and without storage capacity. An aquiclude has similarly low permeability but is able to store water. An *aquitard* is a low permeability unit (i.e., leaky confining bed) that can store and transmit water between adjacent aquifers. This stored and transmitted water is available to wells being pumped in nearby aquifers.

Besides influencing rates of recharge, aquitards can be capable of protecting underlying aquifers from contaminants migrating downward from the ground surface (Runkel et al., 2018). Yet, detailed characterization of sites in the midcontinent region of North America shows that this simple concept of a aquitard does not capture what in reality is more complex behaviors. Runkel et al. (2018) use the term "aquitardifers" to describe confining beds that behave both as an aquitard and aquifer at the same place. Typically, such units possess extreme anisotropy in hydraulic conductivity in vertical and horizontal directions. For example, permeable partings parallel to bedding could transmit water horizontally as an aquifer might. Yet, layering of low permeability units would produce sufficiently low permeabilities in a vertical direction to inhibit vertical transport of contaminants downward. In some shallow settings, these same rocks might develop vertical fractures that create fast-flow pathways over relatively large areas (Runkel et al., 2018).

4.2 Transmissive and Storage Properties of Aquifers

Aquifers play a key role in supplying water to wells. When a pump is turned on in a well, the water level in the well casing (and the hydraulic head) is reduced, causing groundwater to flow from the aquifer into the well. Of the water that is pumped from the well, much of it initially comes from "storage" in the aquifer. Thus, aquifers have at least two important characteristics—some ability to store groundwater and to transmit this water to a nearby well. These properties depend to an important extent on the geologic setting.

4.2.1 Transmissivity

The term *transmissivity* describes the ease with which water can move through an aquifer. More explicitly, it is the rate at which water of prevailing kinematic viscosity is transmitted through a unit width of the aquifer under a unit hydraulic

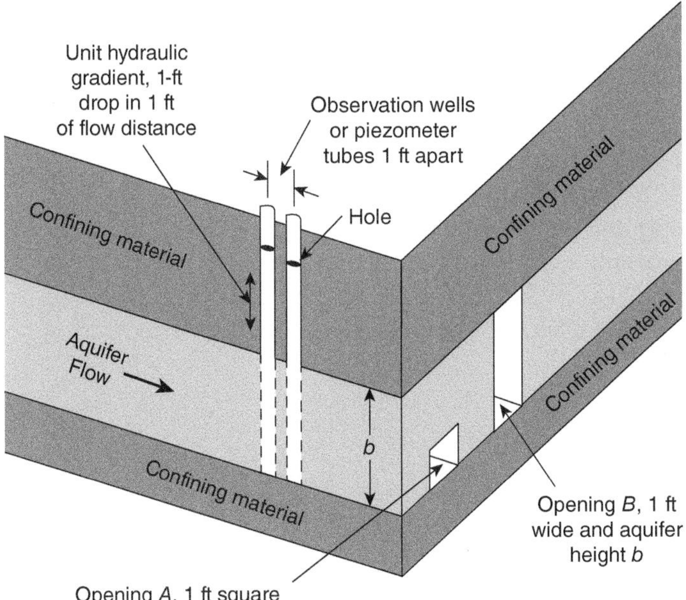

Figure 4.2 The concept of hydraulic conductivity is associated with the unit hydraulic gradient associated with flow through an opening of unit area ("A" 1-foot square). With transmissivity is associated with flow through opening "B" (1 foot wide and b high) (U.S. Geological Survey Ferris et al., 1962/Public domain).

gradient (Figure 4.2). The concept of transmissivity is similar to hydraulic conductivity. The main difference is that transmissivity is a measurement that applies across the vertical thickness of an aquifer. If the thickness of the aquifer is b, the transmissivity (T) is

$$T = bK \tag{4.1}$$

where K is the hydraulic conductivity of the aquifer. Transmissivity has units of $[L^2/T]$ (for example ft^2/day, m^2/day). In English engineering units, transmissivity has units of gpd/ft. The following set of factors provide a basis for unit conversions

$$1\ \text{m}^2/\text{day} = 10.76\ \text{ft}^2/\text{day} = 80.52\ \text{gpd/ft} \tag{4.2}$$

$$1\ \text{ft}^2/\text{day} = 0.0029\ \text{m}^2/\text{day} = 7.48\ \text{gpd/ft} \tag{4.3}$$

$$1\ \text{gpd/ft} = 0.01242\ \text{m}^2/\text{day} = 0.1337\ \text{ft}^2/\text{day} \tag{4.4}$$

Not surprisingly, one can develop a form of the Darcy equation that applies to an aquifer. This equation, which applies to a homogeneous confined aquifer (Figure 4.2), is written as

$$Q = -WT\frac{dh}{dl} \tag{4.5}$$

where W is the width of the aquifer $[L]$, T is the transmissivity of the aquifer $[L^2/T]$, and Q is discharge rate $[L^3/T]$.

Example 4.1 The hydraulic conductivity of a confined aquifer with a thickness of 10 ft is 374 gpd/ft^2. Calculate the transmissivity of the aquifer in gpd/ft, ft^2/day, and m^2/day.

Solution

$$T = Kb = (374\ \text{gpd/ft}^2)(10\ \text{ft}) = 3740\ \text{gpd/ft}$$

Conversion to ft^2/day and m^2/day

$$T = 46.45\ \text{m}^2/\text{day} = 500.0\ \text{ft}^2/\text{day}$$

4.2.2 Storativity (or Coefficient of Storage) and Specific Storage

Aquifers can store water. How this storage is accomplished differs depending upon whether the aquifer is confined or unconfined. When a well is pumped in a confined aquifer, the declining hydraulic head in the vicinity of the well enables the pressurized water to expand slightly, adding a small volume of additional water. In addition, the decline in hydraulic head lets the aquifer collapse slightly, thereby compensating for the volume of water that flowed to the well.

In an unconfined aquifer, the main source of water is the drainage of water from pores as the water table declines in response to pumping. For a comparable unit decline in hydraulic head, an unconfined aquifer releases much more water from storage than a confined aquifer (Figure 4.3). *Storativity* of an aquifer is defined as the volume of water that an aquifer releases from or taken into storage per unit surface area of the aquifer per unit change in head (Figure 4.3).

$$S = \frac{\text{volume of water}}{(\text{unit area})(\text{unit head change})} = \frac{\text{m}^3}{\text{m}^3} \tag{4.6}$$

where S is storativity (dimensionless). With a confined aquifer, values of storativity range from 10^{-3} to 10^{-5}. A related measure of the water stored in an aquifer is specific storage. *Specific storage* is defined as the volume of water that an aquifer releases from or taken into storage per unit surface area of the aquifer per unit aquifer thickness per unit change in head.

$$S_s = \frac{\text{volume of water}}{(\text{unit area})(\text{unit aquifer thickness})(\text{unit head change})} = \frac{1}{\text{m}} \tag{4.7}$$

where S_S is the specific storage of an aquifer $[1/L]$. Equation (4.7) indicates that the unit of specific storage in the SI unit system is 1/m. Specific storage is related to storativity by

$$S = S_s b \tag{4.8}$$

where b is the thickness of an aquifer.

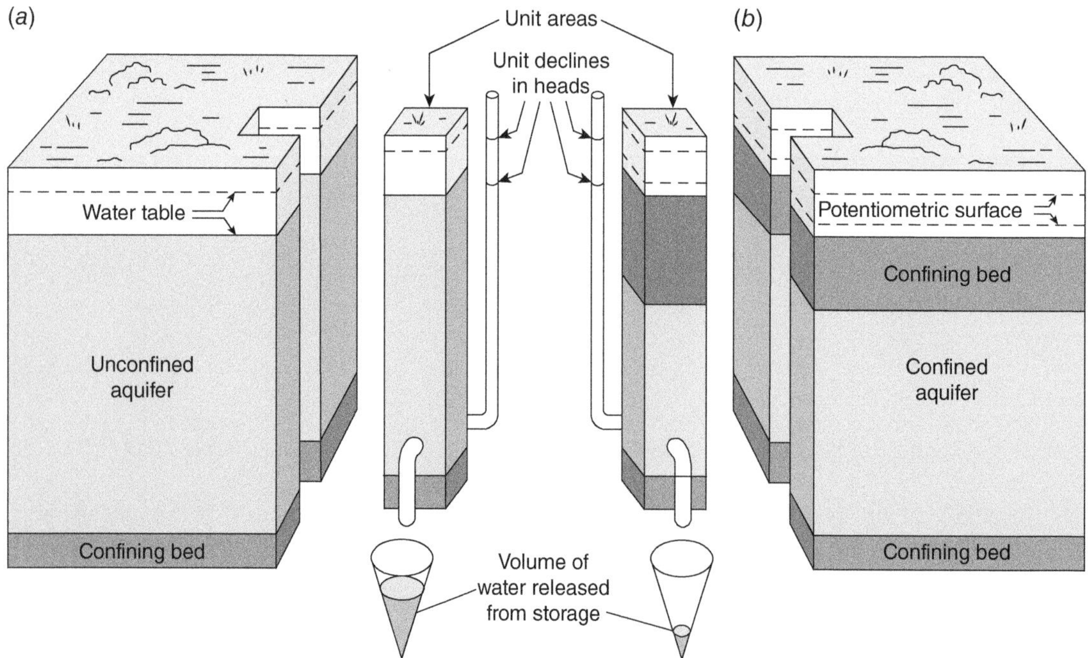

Figure 4.3 Diagrams illustrating the concept of storativity in (*a*) an unconfined aquifer and (*b*) a confined aquifer (U.S. Geological Survey Heath, 1989/Public domain).

4.2.3 Storage in Confined Aquifers

For a confined aquifer, the mathematical definition of specific storage reflects the storage coming from compression of the granular matrix and the expansion of water. The specific storage in a confined aquifer is given by Domenico and Schwartz (1998) as

$$S_s = \rho_w g\left(\beta_p + n\beta_w\right) \tag{4.9}$$

where ρ_w is the density of water $[ML^{-3}]$, g is gravitational constant (9.81 m/sec^2) $[LT^{-2}]$, n is porosity of the aquifer, β_p is the vertical compressibility of rock matrix, and β_w is the compressibility of water. The unit for compressibility is the inverse of pressure. The compressibility of groundwater (β_w) is 4.8×10^{-10} m^2/N (or 2.3×10^{-8} ft^2/lb) at 25 C. To use the compressibility, it is important to review the units of force.

$$1\ \text{Newton(N)} = 1\ \text{kg} \cdot \text{m/sec}^2 \tag{4.10}$$

$$1\ \text{kg of force} = 9.80665\ \text{N} = 2.02046\ \text{lb of force} \tag{4.11}$$

Equation (4.9) can be used to estimate the range of specific storage and storativity of some aquifer types. Values for vertical compressibility of sediments and rocks are presented in Domenico and Schwartz (1998, Table 4.1).

Example 4.2 A confined aquifer is comprised of dense, sandy gravel with a thickness of 100 m and a porosity of 20%. Estimate the likely range for specific storage and storativity. For a total head drop of 100 m in an area of 1×10^9 m^2, how much water is released from the storage?

Solution

From Domenico and Schwartz (1998, Table 4.1), the vertical compressibility of water is 4.8×10^{-10} m^2/N with dense, sandy gravel ranging from 5.2×10^{-9} to 1.0×10^{-8} m^2/N. To calculate the specific storage, it is computationally convenient to replace unit N with kg·m/sec^2.

The specific storage due to the compressibility of water is simplified from Eq. (4.9) as

$$\begin{aligned} S_s^w &= \rho_w g n \beta_w \\ &= \left(1000\ \text{kg/m}^3\right)\left(9.81\ \text{m/sec}^2\right)(0.2)\left(4.8 \times 10^{-10} \frac{\text{m}^2}{\text{kg·m/sec}^2}\right) \\ &= 9.4 \times 10^{-7}\ 1/\text{m} \end{aligned}$$

The specific storage due to the range in compressibility of granular matrix is similarly

$$S_s^M = \rho_w g \beta_p = (1000)(9.81)\left[(0.52 - 1.0) \times 10^{-8}\right] = (5.1 - 9.81) \times 10^{-5} 1/\text{m}$$

Table 4.1 Selected values of porosity, specific yield, and specific retention. Values are in percent by volume.

Material	Porosity	Specific yield	Specific retention
Soil	55	40	15
Clay	50	2	48
Sand	25	22	3
Gravel	20	19	1
Limestone	20	18	2
Sandstone (semi-consolidated)	11	6	5
Granite	0.1	0.09	0.01
Basalt (young)	11	8	3

Source: U.S. Geological Survey, Heath (1989)/Public domain.

The specific storage for the aquifer combines these

$$S_s = \rho_w g\left(\beta_p + n\beta_w\right) = 9.4 \times 10^{-7} + (5.1 - 9.81) \times 10^{-5} = (5.2 - 9.9) \times 10^{-5} 1/\text{m}$$

To convert specific storage values to storativity values simply multiply values by aquifer thickness values to make S^W equal to 9.4×10^{-5} and S^M range from 5.1 to 9.81×10^{-3}. The overall storativity (S) of the aquifer is the sum of S^W and S^M or 5.2 to 9.9×10^{-3}. The volume of water, which is withdrawn from the storage due to a drop in hydraulic head of 100 m in an area of 10^9 m^2 is

$$V = S\Delta hA = (5.2 - 9.9) \times 10^{-3}(100\,\text{m})\left(1 \times 10^9\,\text{m}^2\right) = (5.2 - 9.9) \times 10^8\,\text{m}^3$$

In this example, most of the water comes from the compression of the matrix.

4.2.4 Storage in Unconfined Aquifers

In an unconfined aquifer, the groundwater response to pumping is different from a confined aquifer. At early time, when there is no significant change in water level, water comes from expansion of the water and compression of the grains. Later on, water mainly comes from the gravity drainage of pores in the aquifer through which the water table is falling. The storativity of an unconfined aquifer is expressed as

$$S = S_y + bS_s \tag{4.12}$$

where S_y is the specific yield of the aquifer. The specific yield ranges from 0.1 to 0.3, while the product of aquifer thickness and specific storage is in the range of 10^{-3} to 10^{-5}. Thus, specific yield is the storage term for an unconfined aquifer. In some cases, an aquifer may be confined at early stage of pumping, only to become unconfined at late time. Water levels that started out initially above the aquifer end up falling below the top of the aquifer as it dewaters. As the aquifer changes from a confined to unconfined aquifer, storativity values change accordingly.

4.2.5 Specific Yield and Specific Retention

Specific yield is the water released from a water-bearing material by gravity drainage. The specific yield is expressed as the ratio of the volume of water yielded from a soil or rock by gravity drainage, after being saturated, to the total volume of the soil or rock (Meinzer, 1923).

$$S_y = \frac{V_d}{V_T} \tag{4.13}$$

where S_y is the specific yield, V_d is the volume of water that drains from a total volume of V_T.

Not all, the water present initially in the rock or sediment is released from storage. The term *specific retention* describes the water that is retained as a film on the surface of grains or held in small openings by molecular attraction. The specific retention is expressed as the ratio of volume of water that is retained, after being saturated, to the total volume of the soil or rock (Meinzer, 1923).

$$S_r = \frac{V_r}{V_T} \tag{4.14}$$

where S_r is the specific retention and V_r is the volume water retained against gravity. The porosity defined in Section 3.1 is related to specific yield and specific retention by

$$n = S_y + S_r \tag{4.15}$$

That is, the sum of specific yield and specific retention equals porosity. The specific retention increases with decrease of grain size and pore size of a soil or rock (Table 4.1).

Example 4.3 After a soil sample is drained by gravity, the weight of the soil sample is 85 g. After the sample is oven-dried, the sample weighs 80 g. The bulk density of the wet soil is 1.65 g/cm^3 and density of water is 1 g/cm^3. Calculate the specific yield, specific retention, and porosity of the sample. Assume water that was drained by gravity is 20 g.

Solution

The total volume of the sample is

$$V_T = \frac{(85\,\text{g} + 20\,\text{g})}{1.65\,\text{g/cm}^3} = 63.6\,\text{cm}^3$$

The volume of water retained in the sample after it was drained by gravity is

$$V_T = \frac{(85\,\text{g} - 20\,\text{g})}{1\,\text{g/cm}^3} = 5\,\text{cm}^3$$

The volume of water that was drained by gravity is

$$V_d = \frac{(20\,\text{g})}{1\,\text{g/cm}^3} = 20\,\text{cm}^3$$

The specific retention is

$$S_r = \frac{V_r}{V_T} = \frac{(5\,\text{cm}^3)}{(63.636\,\text{cm}^3)} = 0.079 = 7.9\%$$

while the specific yield is

$$S_y = \frac{V_d}{V_T} = \frac{(20\,\text{cm}^3)}{(63.636\,\text{cm}^3)} = 0.314 = 31.4\%$$

The porosity of the sample is now calculated as

$$n = S_y + S_r = 7.9\% + 31.4\% = 39.3\%$$

4.3 Principal Types of Aquifers

The next several sections examine the principal types of aquifers. Discussions include some of the most important aquifers worldwide and examples of important systems in the United States for which information is readily available. Organizationally, we followed the USGS system for classification of aquifers in America, which is systematic and well-developed (Figure 4.4). There is a broad variety of aquifer types capable of being developed, which range from basaltic and other volcanic aquifers to unconsolidated sand and gravel aquifers. Within each category, the principal aquifers are listed, several of which we discuss in more detail in this chapter. In the central and western U.S., and other countries, unconsolidated sand and gravel aquifer systems represent an important class of aquifer and where we begin our overview.

4.4 Aquifers in Unconsolidated Sediments

The terms *fluvial deposit* or *alluvial aquifer* describe deposits formed by the action of running water. In the United States and worldwide, most of the highest-yielding aquifers are deposits of unconsolidated alluvial materials. They are formed mainly by alluvial fans and within channels and flood plains of large rivers. Formation of alluvial aquifers requires a reliable source of eroding sediments and flows of water capable of transporting these sediments downstream.

4.4.1 Alluvial Fans and Basin Fill Aquifers

Alluvial fans form at the base of mountains where erosion provides a supply of sediment and snowmelt runoff provides a source of water. As illustrated by Figure 4.5, alluvial fans develop due to sediments deposited as mountain streams flow into adjacent valleys. The fan shape develops as the rapid deposition of sediments causes flows of surface water to sweep back and forth,

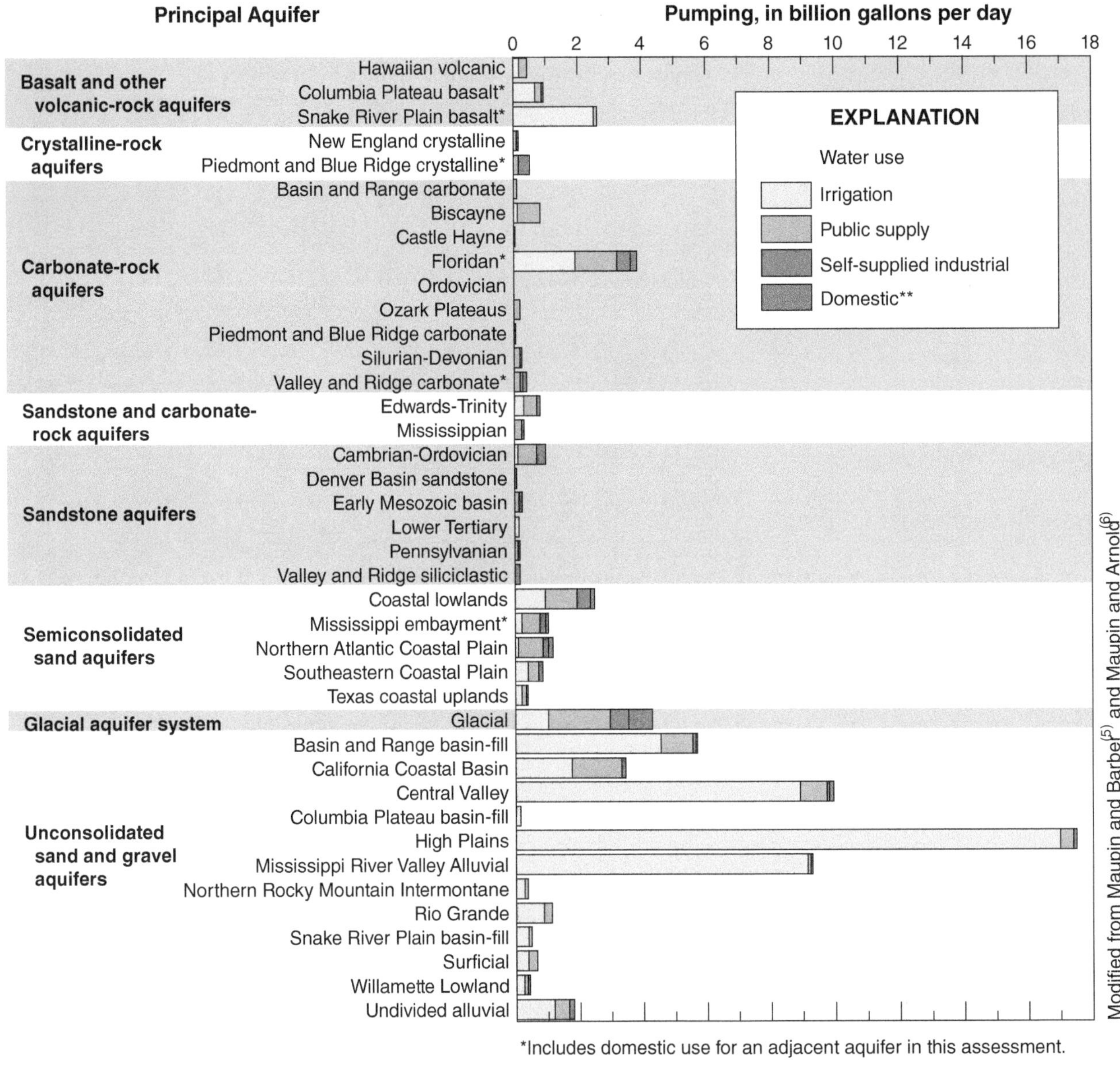

Figure 4.4 Listed here are the Principal Aquifers in the United States. The list is organized around common aquifer types. This collection of aquifers accounts for about 90% of daily groundwater production. Unconsolidated sand and gravel aquifers are pumped extensively, for example, providing much of groundwater needed for irrigation in the western U.S. (U.S. Geological Survey DeSimone et al., 2014, as modified from Maupin and Barber, 2005 and Maupin and Arnold, 2010/Public domain).

moving deposition across 180°. Farmers in arid regions take advantage of the availability of water to grow crops (Figure 4.5). Along a mountain front, alluvial fans, associated with a series of streams can coalesce into a continuous fan apron.

Fans can occur in arid settings, such as Death Valley, California, or in humid areas, such as the front side of the Himalayan Mountains. In arid settings, water delivered to the valley bottom evaporates from temporary playa lakes. In humid settings, sediments associated with alluvial fans can be further distributed by rivers as is the case with the Indo-Gangetic aquifer system in India.

Because of the way alluvial fans develop, the character of the sediment and hydraulic conductivities within the alluvial fan varies as a function of elevation. The upper part of a fan is characterized by coarse gravel-sized sediments because rocks are delivered by physical erosion to the top of the fan. The toe of the fan has the finest sediment, fine sand, or muds, because coarser materials will have been deposited further upslope. Because of fluctuations in flow, the mid-slope region of an alluvial fan contains an interbedded sequence of coarse- and fine-grained units (Figure 4.6).

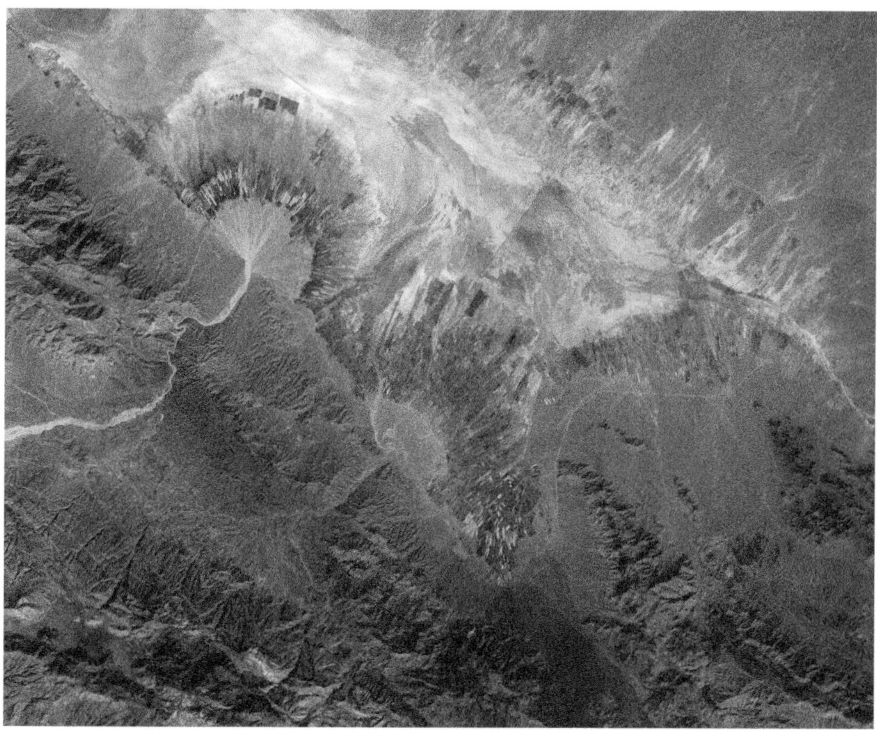

Figure 4.5 This "simulated natural image" shows a pair of alluvial fans developed as mountain streams of the Zagros Mountains in southern Iran flow into an adjacent valley bottom. Evaporation created surface salinization along the valley bottom (National Aeronautics and Space Administration/https://earthobservatory.nasa.gov/images/36041/alluvial-fan-in-southern-iran/last accessed 5 May 2023).

Figure 4.6 Section of an alluvial fan photographed in a gravel pit in the Jordan River Valley, Jordan. Notice how coarse and fine-grained units are interbedded (United States Geological Survey/https://www.usgs.gov/media/images/alluvial-fan-deposits-gravel-pit-jordan-river-valley/last accessed under 5 May 2023).

An alluvial fan by itself can be an important local water supply. However, sometimes structural basins or troughs associated with mountain ranges amplify the quantity of sediments deposited. The resulting aquifers can be hundreds even thousands of meters thick and are known as *basin-fill* aquifers. Among the list of principal aquifers (Figure 4.4), the productivity of basin-fill aquifer systems within category of unconsolidated aquifers ranks them among the most important aquifer systems in the United States. Examples include basin-fill aquifer systems of the western U.S., such as the Basin and Range, California Coastal region, Central Valley, and Rio Grande (Figure 4.4).

By far, the largest number of basin-fill aquifers occurs in the Basin and Range physiographic province of California, Nevada, Utah, and Arizona (Figure 4.7). The area underlain by these aquifers is large ~383,000 km^2. The associated

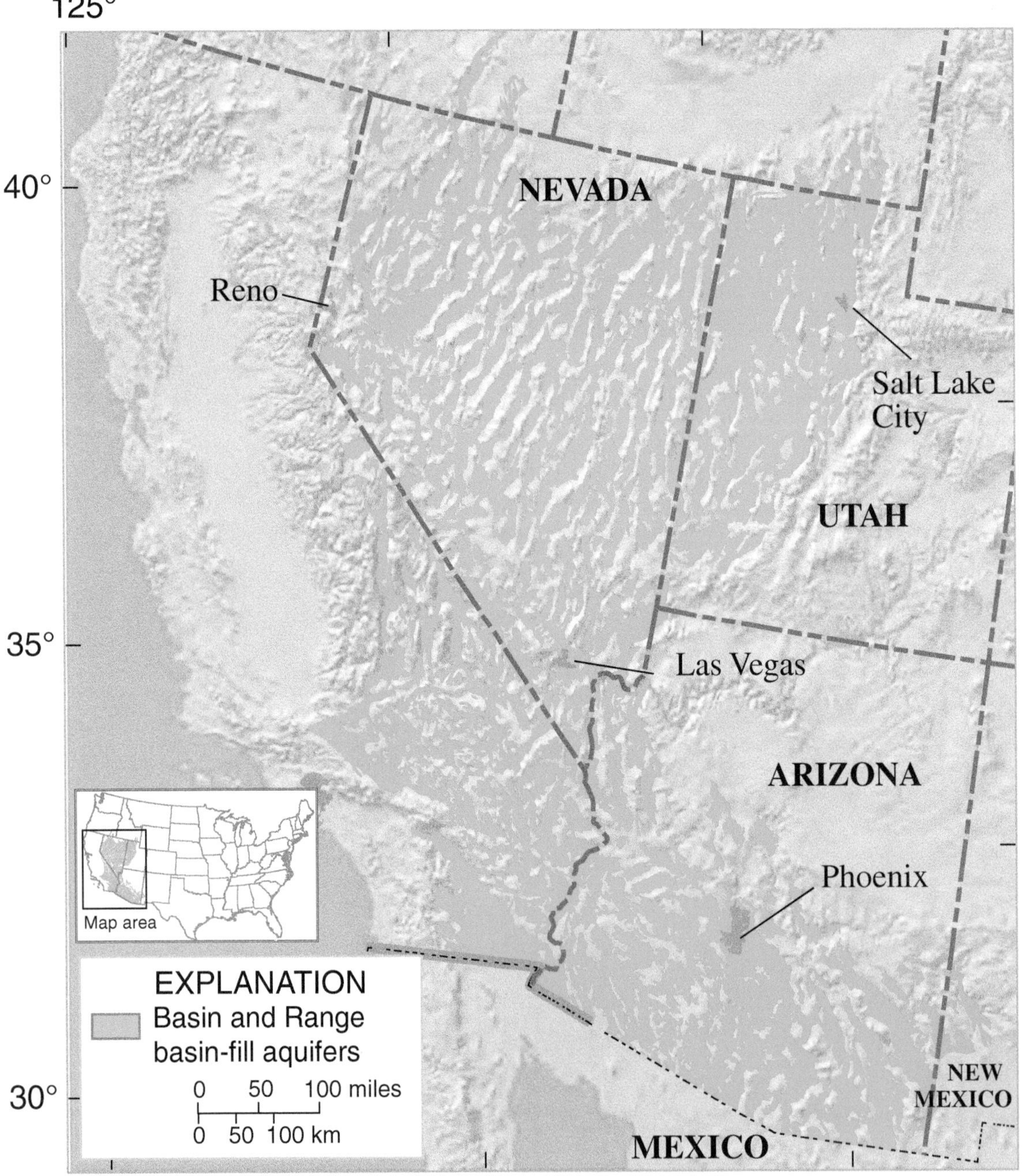

Figure 4.7 The many basin-fill aquifers of the Basin and Range physiographic province provide groundwater for major western cities, such as Phoenix, Las Vegas, Reno, and Salt Lake City (U.S. Geological Survey, adapted from Thiros et al., 2010/Public domain).

production for all uses places this aquifer system rank 4th among the Principal Aquifers (Figure 4.4). It provides water to cities such as Las Vegas, Phoenix, Reno, and Salt Lake City (Musgrove and Belitz, 2016).

The U.S. Geological Survey investigated several Basin and Range basin-fill aquifers as part of their larger Southwest Principal Aquifer Study (Thiros et al., 2010). Basin-fill aquifers in the Basin and Range tend to be long and narrow bodies comprised of sand and gravel interbedded with finer sediments, such as silt and clay. Extensional tectonics led to the formation of grabens that created the down-faulted basins among northerly-trending ranges of mountains (Musgrove and Belitz, 2016). Some basins are drained by streams (Thiros et al., 2010), while others are topographically closed basins.

Other important basin-fill aquifers include the Central Valley aquifer system of California and Indo-Gangetic aquifer, Pakistan, India, and Bangladesh, among the top 10 of the world's most important aquifers.

4.4.2 Fluvial Aquifers

Fluvial processes are also important in creating aquifers as well. Sediments carried by rivers end up deposited in rivers. Fluvial deposits take on the shape of the river valleys in which they form and typically produce aquifers that are long, narrow, and thin. Boggs (1987) recognizes two main environmental settings with fluvial systems rivers, meandering and braided rivers. Meandering rivers are comprised of a looping, sinuous channel that occupies some portion of the flood plain. Meandering streams produce linear shoestring sand bodies that are aligned parallel to the river course, which are bounded below and on both sides by finer materials (Cant, 1982). The shoestring sands are many times wider than they are thick and are relatively discontinuous laterally.

Braided rivers represent higher energy environments with comparatively larger flows and sediment loads. They create coarser sheet-like sands that contain finer materials enclosed within them. The more extensive deposition of sand is a function of rapidly changing channel geometries as a function of constant sediment deposition. Figure 4.8 shows a braided stream located in Alaska (O'Donnell, 2021). Under low flow conditions, the systems of channels across the riverbed are evident. With rapid sedimentation, flow is redistributed by a constantly evolving channel network with sands distributed broadly across the stream valley.

Meandering streams are confined more rigorously in narrow channels as compared to braided streams with many channels, separated by islands or bars. Meandering streams have a greater sinuosity and finer sediment load with the highest permeability sediments associated with point bars and natural levees. Overbank settings or back swamps commonly have much finer silt and clay deposits. River systems in North America developed at the margins of glaciers (e.g., Ohio River)

Figure 4.8 This braided stream is in Alaska. The large sediment load is manifested under low flow conditions by expansive areas of sand between the active flow channels (National Parks Service O'Donnell, 2021/https://www.nps.gov/articles/000/stream-communities-ecosystems-resource-brief-for-the-arctic-network.htm/last accessed under 5 May 2023).

often start out as a braided stream with large quantities of sediments and meltwater from continental glaciers. With the retreat of glaciers, streams change to become meandering. Sediments associated with braided streams typically create the most productive alluvial aquifers.

4.5 Examples Alluvial Aquifer Systems

Alluvial aquifers are among the most productive and important aquifer systems in the United States and the world. As a group, they are also the most-impacted aquifer system due to overpumping. This section describes world-class examples of these aquifers. First is the Central Valley aquifer system of California.

4.5.1 Central Valley Alluvial Aquifer System

The shaded relief map (Figure 4.9*a*) shows the Central Valley of California (Galloway et al., 1999). It is a long (650 km) and narrow (80 km) structural trough that is filled with old marine deposits and more recent terrestrial sediments. Of the surrounding ranges of mountains, the Sierra Nevada have been historically important as the source of surface waters and sediments that comprise this aquifer system. Subsidence of the Central Valley has facilitated the accumulation of a huge

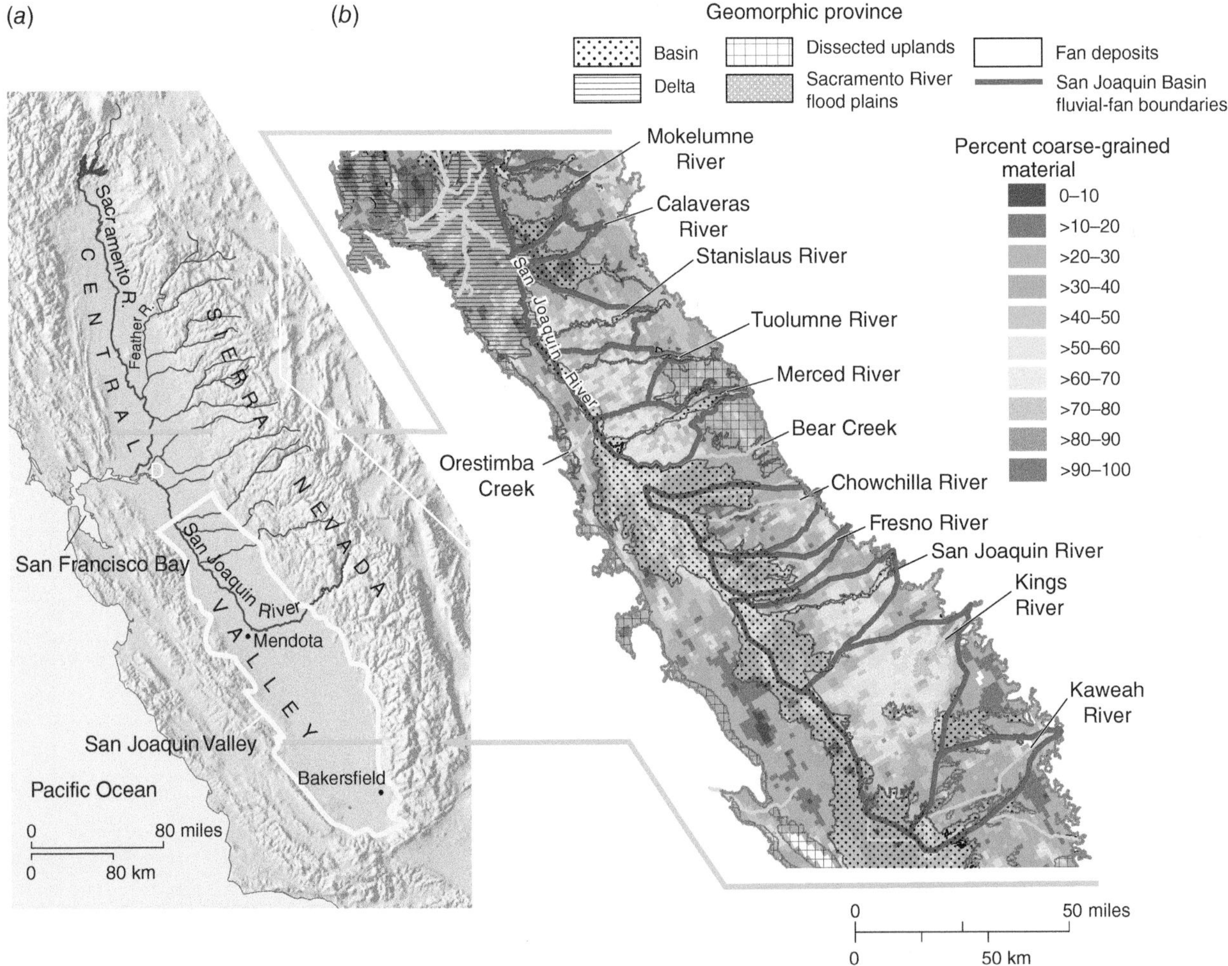

Figure 4.9 The Central Valley of California (*a*) lies between the Sierra Nevada and Coast Range Mountains. In the San Joaquin Valley [yellow outline], sediment distribution is associated with alluvial fans (*b*) created by westward-flowing mountain streams. (U.S. Geological Survey (*a*) modified from Galloway et al. 1999 and (*b*) modified from Faunt ed. 2009/Public domain).

thickness of sediments. This type of aquifer is also an example of a basin-fill aquifer. Within the Valley, the surface water system is unique. Most rivers flow westward from the Sierra Nevada mountains into the Central Valley. Tributary flows are accumulated in the Sacramento River flowing south and the San Joaquin River flowing northward to the Delta area and then San Francisco Bay (Figure 4.9*a*). Irrigated agriculture and cities depend on the Central Valley aquifer system to provide water. High-capacity wells can be counted on to provide >1,000 gal/min (~225 m^3/h) "nearly everywhere" (Bertoldi, 1989). In the 1940s and 1950s, production from the aquifer totaled ~14.2 km^3/yr, an unstainable rate.

Given the importance of this aquifer system and associated problems, there is a large and diverse history of previous investigations, led by the United States Geological Survey (USGS). The present status of knowledge on the Central Valley aquifer system and recent model-based investigations are reported in USGS Professional Paper 1766 (Faunt, 2009).

Here, and in later chapters, the San Joaquin Valley (Figure 4.9*a*) is of particular interest. This area of the Central Valley extends from the Delta southward past the city of Bakersfield (Figure 4.9*a*). The Delta is a lowland area of stream channels and islands protected by levees that collects drainage from the Valley on its way to San Francisco Bay.

Alluvial processes associated with the Sierra Nevada Mountains have been immensely important in shaping the aquifer system. Figure 4.9*b* shows distribution of coarse-grained deposits for the upper 50 ft (15.2 m) (Faunt, 2009). The complexity of the setting is evident with both coarse- and fine-grained clastic materials. The coarsest materials are expressed in alluvial fans developed by westward-flowing rivers off the Sierra Nevada Mountains. These rivers are shown in Figure 4.9*b* with their associated alluvial fans (Faunt, 2009). The internally draining, southern end of the San Joaquin Valley receives streamflow from the Kings, Kaweah, and Kern Rivers. Historical terminal lakes (not shown) included Tulare, Kern, and Buena Vista Lakes. Alluvial deposits associated with the Coast Range along the western edge of the Valley (Figure 4.9*b*) are less well developed and much finer-grained. The causes include finer-grained source rock and absent snowmelt runoff.

Professional Paper 1766 discusses how certain of the alluvial fans of the Sierra Nevada mountains vary in grain texture (Faunt, 2009). For example, the coarse sediment of the Kings River fan (Figure 4.9*b*) comes from an association with glaciated portions of the Sierra Nevada. The more northerly Fresno River and Chowchilla fans are finer-grained because their watersheds are associated with non-glaciated areas. The valley bottom sediments more distal from the fans tend to be finer-grained except for channels of the San Joaquin River, where permeable aquifer units are associated with San Joaquin and other tributaries. This same complexity in sediment variability is present with depth with unconfined aquifer conditions present at shallow depth and confined units at depth.

Although fine-grained units (clay, sandy clay, sandy silt, and silt) are abundant within the aquifer in the San Joaquin valley, there are coarser units as well. Model fits found values of hydraulic conductivities for coarse- and fine-grained end members to be 3.3×10^3 ft/day (1000 m/day) and 2.4×10^{-1} ft/day (0.07 m/day), respectively (Faunt, 2009). Hydraulic testing of wells in the San Joaquin Valley determined yields ranging from 500 to 1,500 gal/min (113–340 m^3/h) (Davis et al., 1964).

4.5.2 High Plains Aquifer System

The High Plains aquifer system is the largest aquifer in the United States. It underlies an area of about ~450,000 km^2 (174,000 mi^2) in parts of Colorado, Kansas, Nebraska, New Mexico, Oklahoma, South Dakota, Texas, and Wyoming (Figure 4.10). The land surface slopes from west to east. In 2000, ~30% of groundwater pumped for irrigation came from this aquifer (Dennehy, 2000). A variety of problems are associated with overpumping in parts of the High Plains aquifer systems, which we will take up in Chapter 15.

Much of the aquifer system is comprised of the Ogallala Group, late Miocene in age approximately 5 to 16 million years old (Smith et al., 2017). The High Plains aquifer also includes several older rock units and much younger units such as the Nebraska Sand Hills (Figure 4.10) and alluvial deposits (Dennehy, 2000). The Ogallala Group is comprised of an assortment of unconsolidated gravel, sand, silt, and clay. It was deposited by an extensive eastward-flowing system of braided streams that drained the eastern slopes of the Rocky Mountains during late Tertiary time. As was the case with the Central Valley aquifer system, details of the depositional processes were complicated leading to variable thicknesses. Uplift of the ancestral Rocky Mountains promoted the deposition of alluvial sediments by eastward flowing streams (Smith et al., 2017). In other words, accommodation-space required for deposition of sediments was created in general by uplift of mountains. However, there are places in southwestern Kansas where local salt dissolution (i.e., salt karst) and faulting led to subsidence, the creation of additional accommodation space, and to thickening of Ogallala deposits (Smith et al., 2017).

At first, the axes of deposition of alluvium were along preexisting valleys that had been eroded in pre-Ogallala rocks. Eventually, fan aprons provided distributed drainage that created blanket deposition widely across the mountain front.

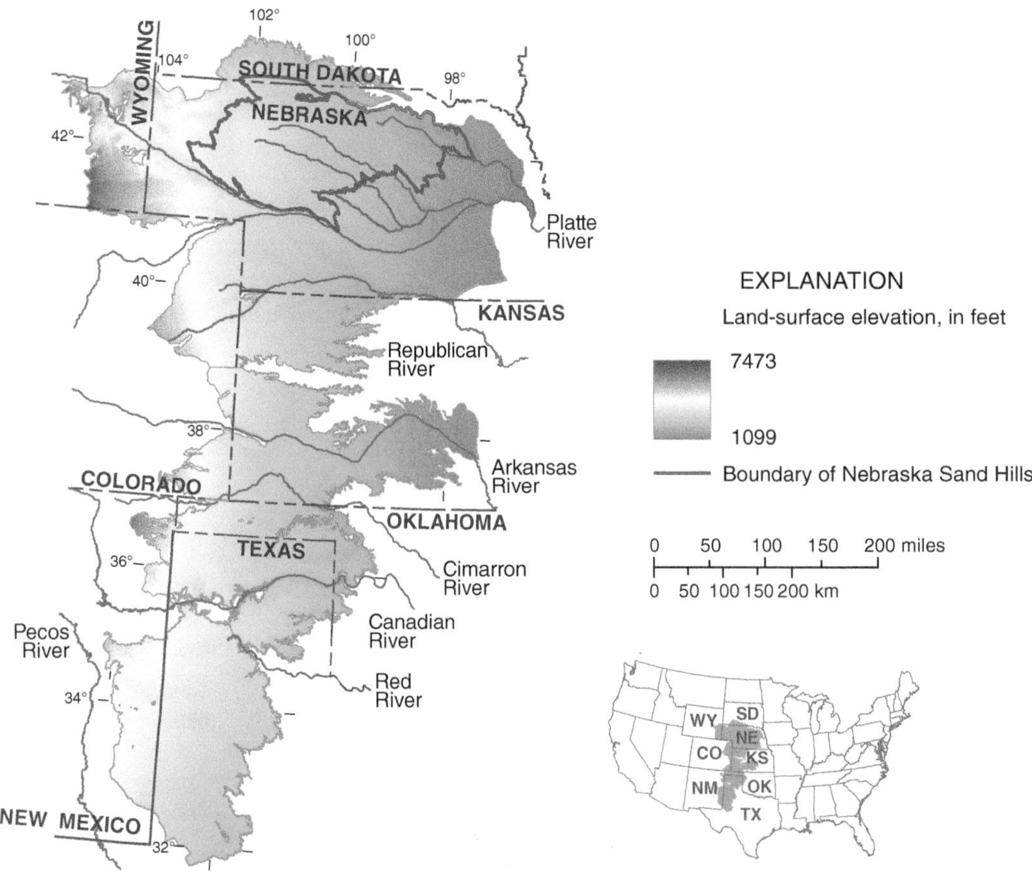

Figure 4.10 The insert map shows the location of the High Plains aquifer system in the Midwest of the United States. The land surface slopes from west to east (U.S. Geological Survey, adapted from Houston et al., 2013/Public domain).

Thus, the aquifer thickness is greatest in places where sediment-filled buried valleys occur together with the overlying blanket of alluvial deposits.

Now, the High Plains aquifer is largely disconnected from the present Rocky Mountains. The modern drainage has rivers flowing eastward across the aquifer to the Mississippi River drainage (Figure 4.10). These rivers are connected to the aquifer hydraulically to the extent that pumping of groundwater for irrigation has been shown to reduce river discharge in the Platte and Canadian Rivers.

4.5.3 Indo-Gangetic Basin Alluvial Aquifer System

The Indo-Gangetic Basin (IGB) aquifer system (Figure 4.11) is perhaps the world's most important aquifer. Irrigation with groundwater is essential for crop production in Pakistan, India, and Bangladesh. Chapter 16 discusses issues concerned with long-term sustainability of groundwater in this aquifer system.

The IGB alluvial aquifer shares similarities with the other examples of alluvial aquifers. Sedimentation into the subsiding Ganges Foreland basin was associated with surface water drainage from the Himalaya. The hydraulic character of this aquifer is similarly complicated by heterogeneity of deposition in both space and time (Bonsor et al., 2017). Near-surface deposits in upper reaches of the Indus and Ganges basins are mostly Early to Late Pleistocene alluvium (Figure 4.12). These deposits are medium-to-coarse-grained oxidized sand (K = 30–50 m/day), but locally varying with coarser channel deposits and fine-grained inter-channel deposits (Bonsor et al., 2017). Braided fluvial sands of Holocene age were redistributed downstream in the Indus and Ganges Rivers.

A unique feature of this aquifer system is so-called megafans, large alluvial fans that formed in middle Late Pleistocene (Shukla et al., 2001). These formed under different climatic and tectonic conditions that promoted sediment transport.

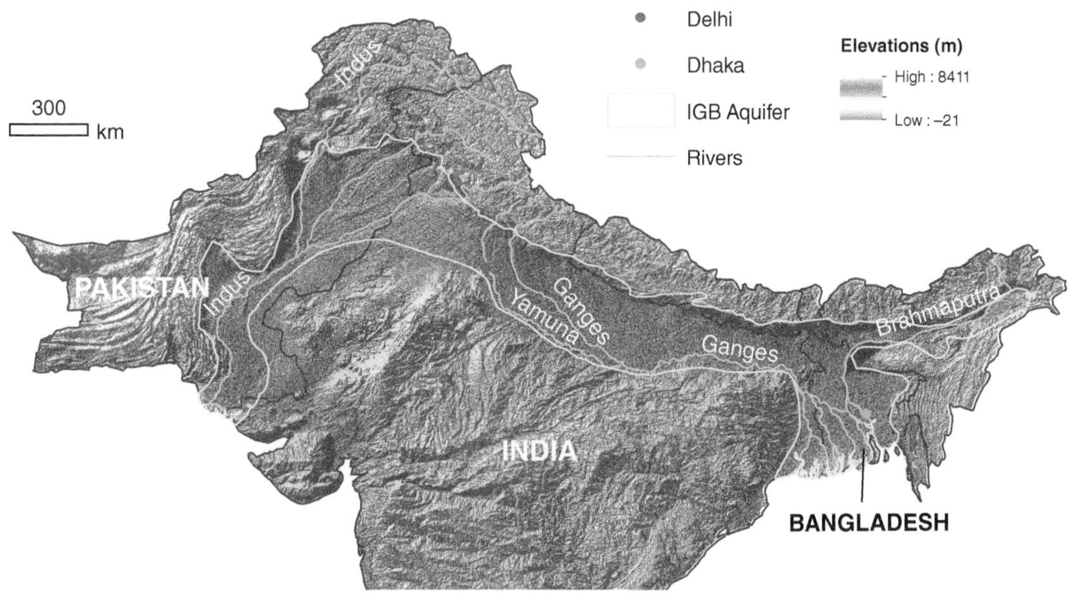

Figure 4.11 The IGB alluvial aquifer has formed from sediments coming off the Himalaya mountains. It occurs as a vast flat plain drained by Indus and Ganges Rivers and tributaries (Courtesy Ganming Liu).

These formed in association with most of the larger ancestral rivers. The megafan that formed from the ancestral Yamuna–Ganges Rivers is approximately 150 km long and 100 km wide (Shukla, 2001). The present-day Ganges River has incised parts of this megafan exposing gravels and cross-bedded sand in the proximal parts of the fan and mud and silts in the distal portions. Present-day alluvial fans are very much smaller in size (Figure 4.12).

4.5.4 Mississippi River Valley Alluvial Aquifer

Another of America's important aquifers is the Mississippi River Valley alluvial (MRVA) aquifer. It covers an area of about 82,875 km^2 across seven states, especially Arkansas, Louisiana, Missouri, and Tennessee (Figure 4.13). The alluvial aquifer varies in width from 125 to 200 km. Historically, water in the aquifer has been used mainly for agriculture, especially for the irrigation of rice. In Arkansas, wells constructed in this aquifer can yield as much as 5000 gpm (1135 m^3/h) with yields of 2000 gpm (450 m^3/h) considered "common" (Kresse et al., 2014).

The aquifer is a typical fluvial unit whose extent is defined by the valley of the Mississippi River. What makes this aquifer unique is the large size of the valley and circumstances that provided a large source of sediments, and water that was able to winnow and transport those materials.

The MRVA aquifer is located within a larger physiographic unit, the Mississippi embayment (Figure 4.13). The unique outline of the Mississippi embayment is created by a syncline that dips southward toward the Gulf of Mexico. The axis of the syncline generally follows the Mississippi River (Figure 4.13).

The fluvial sequence is divided into two units. The aquifer itself is the lowermost unit of the two and consists of coarse sands and gravels. The aquifer developed because the ancestral Mississippi River served as a major drainage way for water and sediments associated with melting of continental glaciers ~14,000 BP. Under conditions of low relief and with a continuous supply of sediments, the river behaved hydraulically as a vast braided stream complex with sediment deposited over large areas. Sediments coarsened toward the bottom of the aquifer and the northern end of the aquifer (Kresse et al., 2014). As deglaciation progressed, meltwater drained through the St Lawrence River system, and the Mississippi River transitioned to a meandering river with much finer deposits and backswamp sediments. This change to a "suspended-load dominated meandering stream" (Kresse et al., 2014) created the overlying fine-grained unit confining the main alluvial aquifer, and the fluvial features that are evident today. Typically, this main unit exhibits a wide range in thickness that varies from 150 ft (46 m) in places to zero where it is absent.

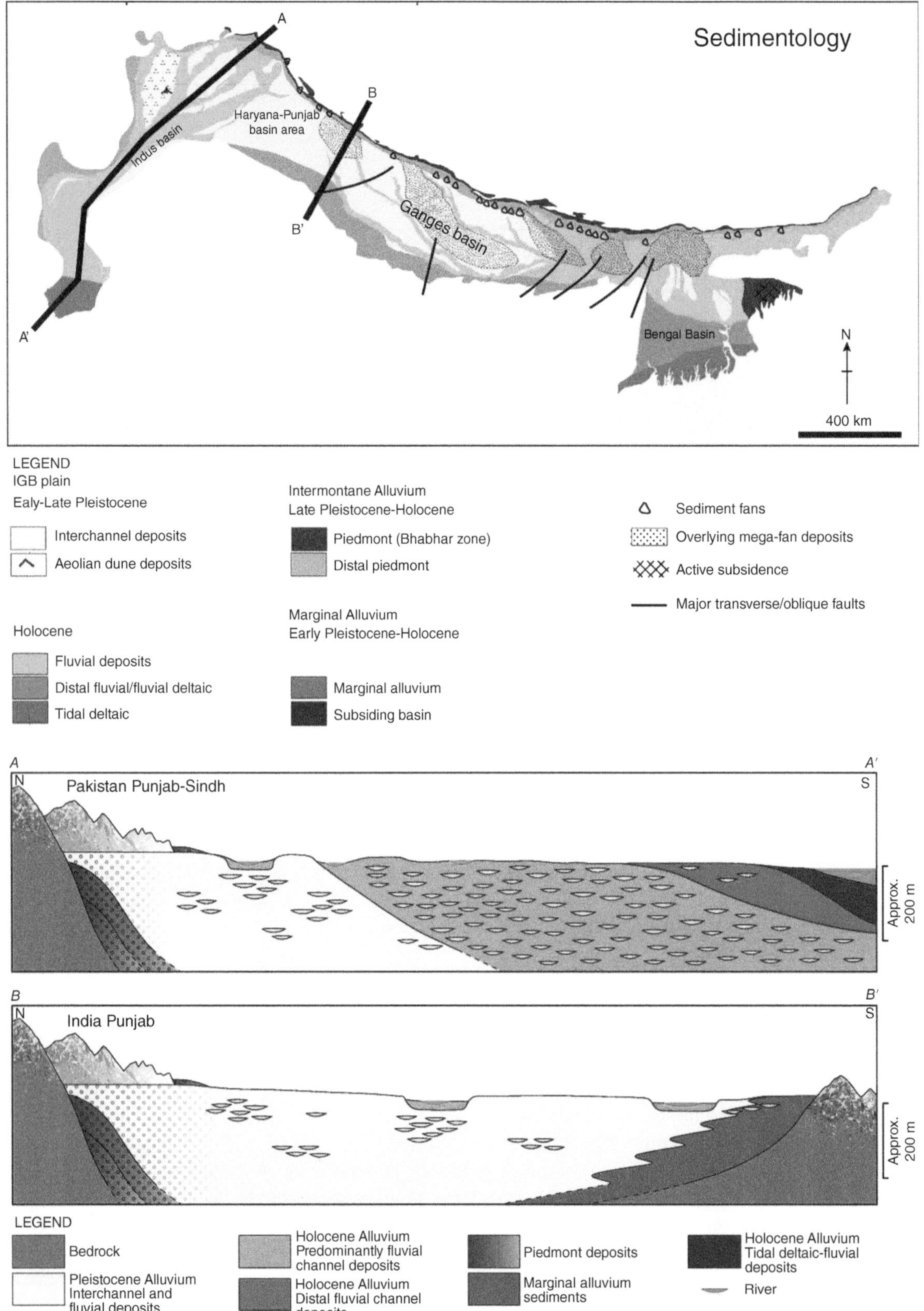

Figure 4.12 Typologies of alluvium sedimentology. Sediment in upper reaches of the Indus and Ganges River basins (yellow) is older (i.e., Early to Late Pleistocene). More recent downcutting has led to deposition of permeable deposits (Holocene) upstream along rivers. The location of cross sections is shown on the map (Bonsor et al., 2017. Contains British Geological Survey material © UKRI 2017 (reproduced under the CC BY 4.0 licence). Coastline outline provided by ESRI).

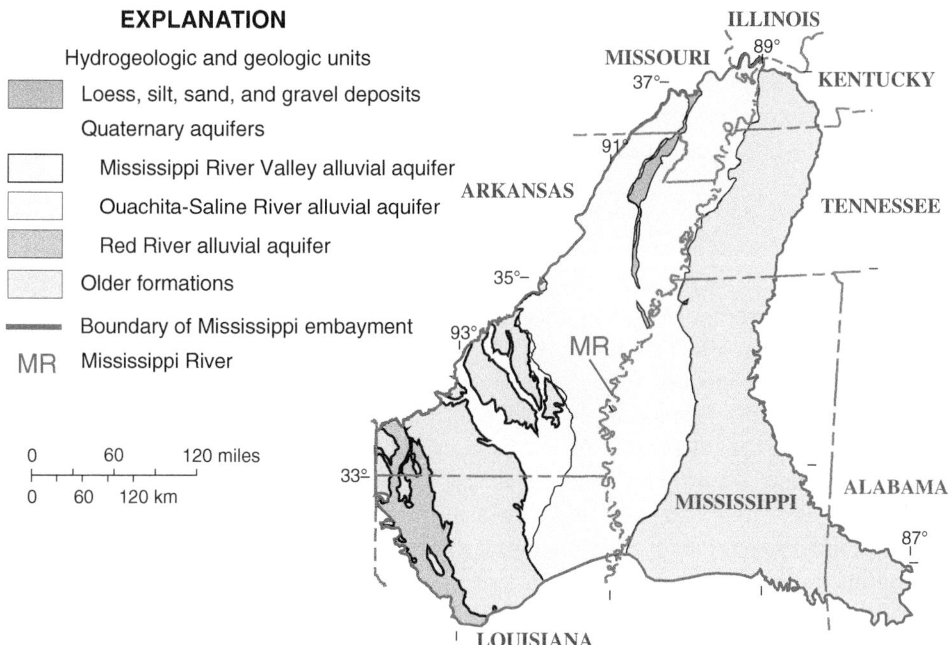

Figure 4.13 This map shows the location of the MRVA aquifer within the larger Mississippi Embayment. There are minor alluvial aquifers associated with tributaries to the Mississippi River (U.S. Geological Survey, adapted from Kresse et al., 2014/Public domain).

Figure 4.13 also includes Arkansas River, the Ouachita-Saline Rivers, and the Red River alluvial aquifers. These somewhat smaller alluvial aquifers are important sources of groundwater in the West Gulf Coastal Plain in southern and southwestern Arkansas (Kresse et al., 2014).

4.5.5 Aquifers Associated with Glacial Meltwater

The glacial environment is another setting that affords opportunities to develop unique fluvial settings. Figure 4.14 illustrates three examples, outwash plains, alluvial deposits in buried bedrock valleys, and alluvial-valley deposits. *Outwash plains* are blanket deposits of sand and gravels (Figure 4.14a) that occur at the front of melting ice and are associated with meltwater. *Bedrock valley* aquifers occur in valleys that are no longer occupied by the streams that cut them (Figure 4.14b). Fluvial stream deposits in valleys ended up being buried by sediments associated with the readvance of nearby glaciers or older deposits buried in a subsequent glacial period. Commonly, deposits are buried by low permeability materials like glacial till, which are intimately associated with the ice (Figure 4.14). Although till is commonly clay-rich, it also contains an assortment of sand, gravel, and rocks, which are the building blocks for these kinds of aquifers.

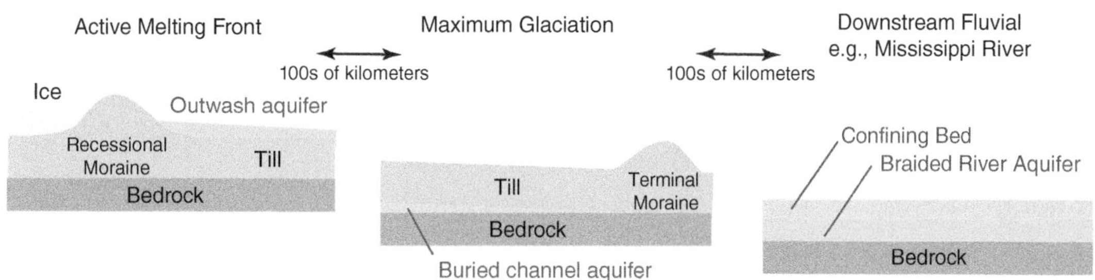

Figure 4.14 Aquifers associated with glacial meltwaters include outwash at the active melting front, buried valleys in glaciated areas, and braided river deposits farther downstream (FWS).

The complexity near an ice margin is illustrated in Figure 4.14. The recessional moraine indicates a pause in the glacial retreat indicated by the creation of a moraine and meltwater deposits such as outwash or lacustrine deposits. Farther away, a terminal moraine marks the furthest ice advance. Buried valley aquifers sometimes can occur due to local ice advances burying alluvial sands and gravels beneath a confining layer of glacial till (Figure 4.14). Downstream, away from the terminal moraine, thick alluvial deposits can develop, as exemplified by the Mississippi River Valley alluvial aquifer.

The Mahomet bedrock valley aquifer (Figure 4.15) provides an interesting example of a fluvial aquifer system in an incised bedrock channel that was completely buried by subsequent glacial advances. The aquifer underlies an area of ~5100 km^2 and is an important water resource for east-central Illinois. This aquifer is associated with the Teays River drainage system that extended from Illinois eastward to West Virginia. The Teays system was an ancestral version of the present-day Mississippi system that developed in late Tertiary time. This aquifer was buried because of subsequent Pleistocene glaciation. Notice in Figure 4.15 how modern river networks have developed independently of the buried bedrock valley because the present-day topography bears no relationship to that in the past.

The Mahomet sand is mainly outwash that was deposited in the bedrock valley system. In this part of Illinois, Mahomet sands range in thickness from 0 to 60 m. The mean thickness is about 30 m. Transmissivities of the Mahomet Valley Aquifer range from 7×10^{-4} to 8×10^{-2} m^2/sec with a mean hydraulic conductivity of 1.4×10^{-3} m/sec. Locally, younger glacial sand units were interbedded within glacial till units. Overall, the pattern of occurrence of the Mahomet Valley Aquifer and overlying aquifers is spatially complex—a point that is made repeatedly in this section.

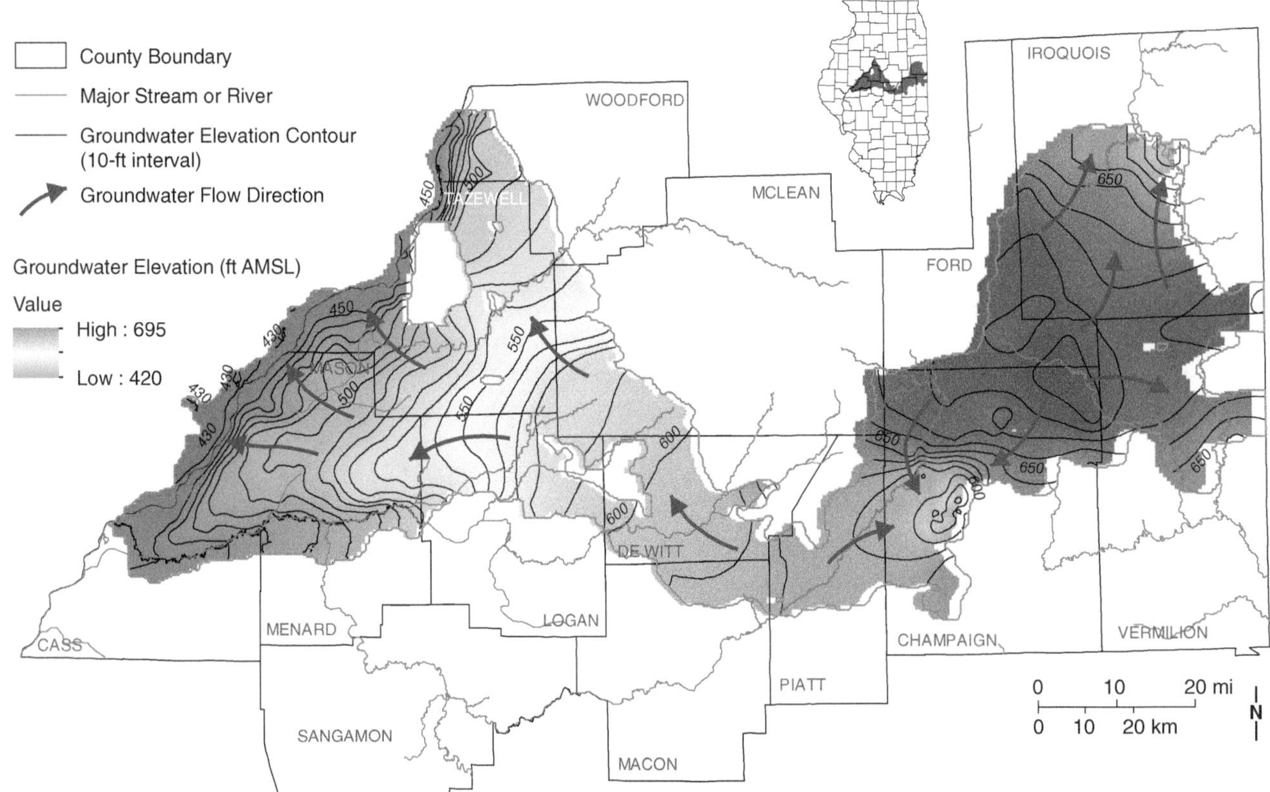

Figure 4.15 The insert map shows location of the Mahomet Valley Aquifer in Central Illinois. It is part of the ancestral Teays River system, valleys of the ancestral Mississippi River system. This map of the potentiometric surface shows a complex pattern of groundwater flow. In Champaign County flow in the aquifer is captured by pumping wells associated with cities of Champaign and Urbana (From Panno, S.V., and W.R. Kelly, 2020, Water quality in the Mahomet aquifer: Chemical indicators of brine migration and mixing: *Illinois State Geological Survey, Special Report 7*. ©2020 University of Illinois Board of Trustees. Used by permission of the Illinois State Geological Survey. Adapted from Roadcap et al., 2011 also used with permission Illinois State Water Survey).

4.6 Aquifers in Semiconsolidated Sediments

This next family of aquifers is semiconsolidated sand interbedded with silt, clay, and minor carbonate rocks. These aquifers were formed in coastal settings, which included fluvial, deltaic, and shallow marine environments. They are associated with the coastal plains along the Atlantic Ocean and Gulf of Mexico. Figure 4.16 shows aquifer systems along the Gulf Coast, which USGS has identified as belonging to this family of sediments, namely the Coastal lowlands aquifer system (1), the Mississippi embayment aquifer system (2), and the Texas coastal uplands aquifer system (3). Farther east and north is the Southeastern Coastal Plain aquifer system, and finally the Northern Atlantic Coastal Plain aquifer system.

These aquifers typically crop out in topographically higher areas where they are unconfined. Moving seaward, they become confined with upward leakage to shallower aquifers or discharge to saltwater bodies. Because flow is sluggish near the ends of long regional flow systems, the aquifers commonly contain unflushed saline water in their deeply buried, downdip parts. When shallow aquifers are pumped near coasts, saltwater intrusion can lead to salinization of the groundwater. The Mississippi River Valley alluvial aquifer is shown overlying these older rocks in the Mississippi embayment.

Figure 4.17 is a cross section (*A–A′* on Figure 4.16) showing the unconsolidated units that comprise the Coastal lowlands aquifer system in Texas. Units are Miocene in age and younger and occur stratigraphically above the Vicksburg–Jackson confining unit. Below the Vicksburg Jackson are units comprising the Texas coastal uplands aquifer system (Figure 4.17). The sand, silt, and clay sediments reflect three depositional sediments—continental (alluvial plain), transitional (delta, lagoon, and beach), and marine (continental shelf). The aquifer system consists of five units, Permeable Units A through E, and two confining beds, the Zone D and Zone E confining units. (Ryder, 1996). As shown by the cross section, units typical thicken toward the Gulf of Mexico. Subsidence in the Gulf Basin has been accompanied by uplift on land to create the wedge-shaped geometry of various units (Figure 4.17). Where the aquifers are confined, storage coefficients are between 10^{-4} to 10^{-3}. In places where units are unconfined, specific yield values vary from 10% to 30%. A model analysis yielded hydraulic conductivity values of 17–21 ft/day (or 5–7 m/day) (Ryder, 1996). The transmissivity of the aquifer system ranges from 5000 to 35000 ft^2/day (or 500–4000 m^2/day).

Fifty percent of total withdrawals from the aquifer system in 1985 was concentrated in and around Houston with production from Permeable units A, B, and C. Large withdrawals of groundwater by 1982 produced a 300-ft (91-m) decline in Permeable Zone B, and >400 ft (121 m) in Zone C. The large declines in water levels contributed to land subsidence that in the Houston area was more than 9 ft (>2.75 m) (Ryder, 1996).

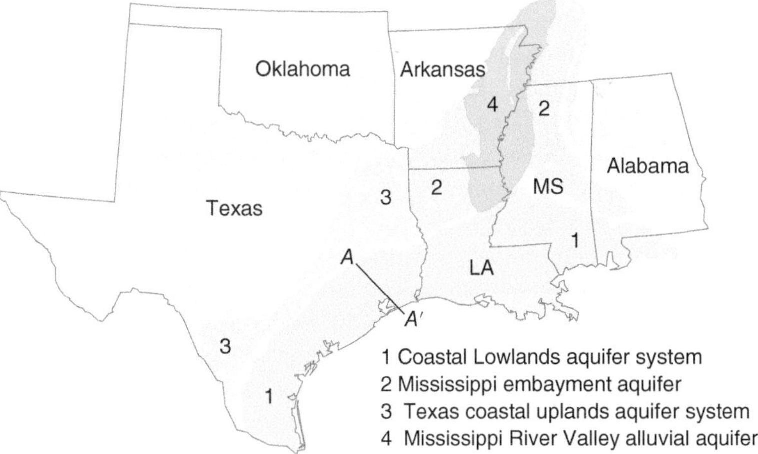

Figure 4.16 Major aquifer systems of the Gulf Coastal plains are shown in shades of yellow. The MRVA aquifer as shown is underlain by aquifers of the Mississippi embayment system (U.S. Geological Survey DeSimone et al., 2014/Public domain).

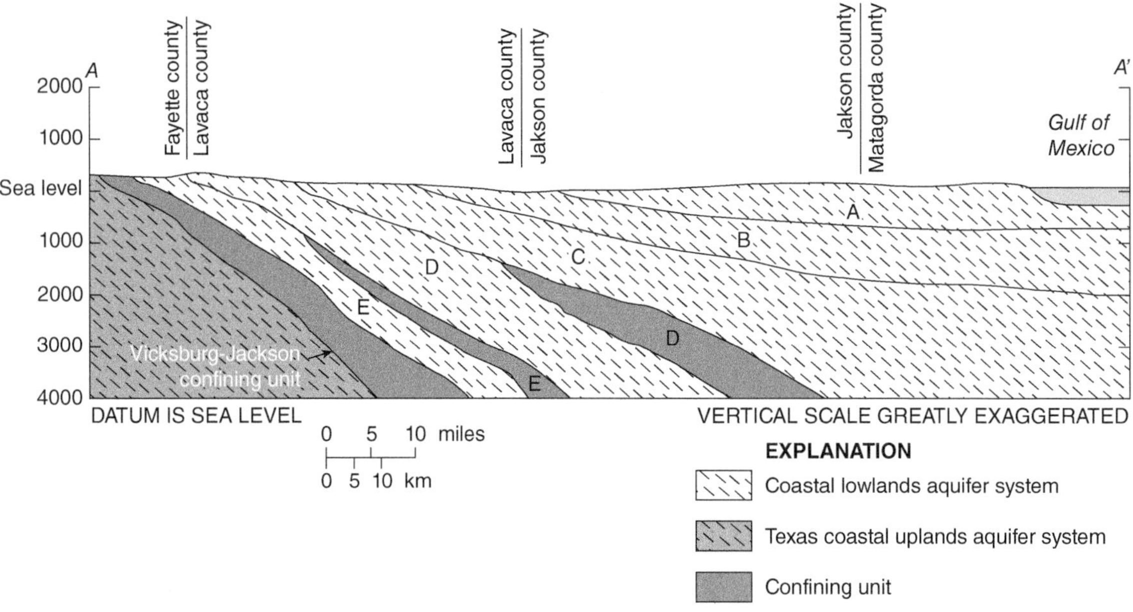

Figure 4.17 Cross section (*A*–*A*′) across the Coastal lowlands aquifer system in Texas showing permeable units, confining units, and the stratigraphic relationship with the Texas coastal uplands aquifer system (U.S. Geological Survey Ryder, 1996/Public domain).

4.7 Sandstone Aquifers

Sandstone is a rock formed from the cementation of sand. It typically retains only a small fraction of the primary porosity of sand. Compaction and cementation together cause a substantial reduction in porosity. Secondary openings, such as joints and fractures, along with bedding-plane partings, facilitate the flow of groundwater in sandstone. Typically, the hydraulic conductivity of sandstone aquifers is low to moderate, but because they extend over large areas, these aquifers can provide large quantities of water. Sandstone aquifers are common across the United States and include, for example, aquifers of the Colorado Plateau, Denver Basin aquifer system, Lower Cretaceous aquifers (Northern Great Plains aquifer system), and others too numerous to mention.

4.7.1 Dakota Sandstone

Discussion in this section will review aspects of Dakota sandstone aquifer, part of the Northern Great aquifer system in the United States, and the Nubian sandstone of North Africa. The Dakota Sandstone is notable for the many studies through the years that led to the modern understanding of classical artesian aquifers and the important role that confining beds play (Bredehoeft et al., 1983). Early studies of what is known as "classical" artesian aquifers included Chamberlin (1885) and Darton (1909). Figure 4.18 illustrates the characteristics of a classic artesian aquifer. The aquifer itself is confined by rocks of much lower permeability, here Cretaceous shale. However, structural deformation of the rocks and subsequent erosion led to exposure to the aquifer in uplands where direct recharge occurs. Because the aquifer is saturated with water at higher elevations, the potentiometric surface might occur above the ground surface, where the aquifer is topographically lower (Figure 4.18). Thus, wells emplaced in the aquifer can flow at the ground surface at least initially, before intensive pumping impacts the aquifer. This simple model assumes that flow is mostly through the aquifer with no flow in beds below and only minor leakage upward. However, both Chamberlain and Darton appreciated that flow was possible through confining beds (Bredehoeft et al., 1983). Studies on this aquifer continued through the years and were the source of fundamental understanding of the sources of water being pumped, the importance of cross-formational flow, and recharge through shale (Bredehoeft et al., 1983).

The aquifer system stretches from the Black Hills of western South Dakota eastward under the entire state (Figure 4.18). There are three major aquifers present in the basin—Mississippian carbonate rocks collectively referred to as the Madison Group, the Inyan Kara Group Sandstones, and the Newcastle Sandstone. These aquifers crop out on

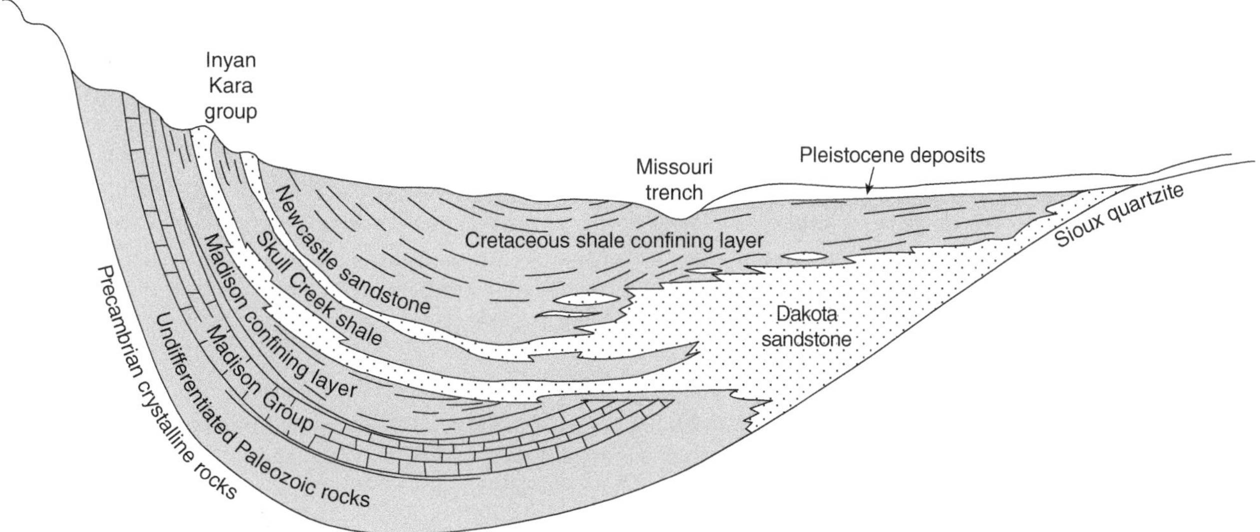

Figure 4.18 West–east cross section showing the most important aquifers and confining beds comprising the Dakota aquifer system (U.S. Geological Survey Bredehoeft et al., 1983/Public domain).

the eastern flank of the Black Hills (Figure 4.18). The confining beds are largely shale and include the Madison Group confining layer, the Skull Creek shale, and the Cretaceous Shale confining layer. The Skull Creek Shale thins eastward and eventually pinches out, permitting the Inyan Kara and New Castle Sandstones to merge and form the Dakota Sandstone of eastern South Dakota. The Madison confining layer also pinches out to the east permitting waters from the Madison Group to enter the basal sands of the Dakota Sandstone. The low permeability confining bed overlying the Dakota-Newcastle Sandstone is the Cretaceous shale confining layer. This unit is mostly shale but contains minor aquifers (Bredehoeft et al., 1983).

Flow modeling with the conceptual model shown in Figure 4.18 indicated a K_z value of 2×10^{-11} ft/sec for the combined Cretaceous shale confining layer, K_{xy} values of 9.6×10^{-5}, 6.4×10^{-5}, and 1×10^{-5} ft/sec for the Inyan Kara Group Sandstones, Dakota-Newcastle Sandstone, and Madison Limestone, respectively. Wells in the aquifer typically yielded 50–500 gpm.

Bredehoeft et al. (1983) used simulation results to calculate steady-state flows. Results showed the aquifer systems behaved much differently than the classic conceptual model of a completely confined aquifer. Total inflows and outflow through the aquifer through the entire system in South Dakota was approximately 80 ft^3/sec. In terms of recharge, only ~25% was associated with upland outcrops of the sandstone units (Newcastle Sandstone and sandstones of the Inyan Kara Group). Recharge to the Madison Group ~29% ended up as flux in the sandstone aquifer because of cross-formational flow at depth. Surprisingly, nearly half the total flow in the aquifer (~46%) came as recharge through the Cretaceous shale confining unit. Although the vertical hydraulic conductivity was low, this unit covers a large area of South Dakota. Most of the outflow (~80%) left the Dakota Sandstone through upper confining bed, rather than the eastern end of the aquifer.

4.8 Carbonate-Rock Aquifers

Most of the developed carbonate aquifers supplying groundwater in the United States are found in the east, particularly in Florida. Examples there include the Floridian aquifer system and Biscayne aquifer. Most carbonate-rock aquifers are limestone, but locally, dolomite and marble yield water. The water-yielding properties of carbonate rocks vary widely; some units yield almost no water and act as confining units, whereas others are among the most productive aquifers known. Outside of Florida, carbonate rock aquifers are common. Here, we will discuss interesting aspects of the Basin and Range carbonates and Edwards-Trinity aquifer, a mixed carbonate and sandstone aquifer.

4.8.1 Enhancement of Permeability and Porosity by Dissolution

Most carbonate rocks originate as sedimentary deposits in marine environments. Compaction, cementation, and dolomitization processes act on the deposits as they lithify. These processes greatly change porosity and hydraulic conductivity. However, in a groundwater context, the most important post-depositional change with carbonate rocks comes from rock dissolution by circulating, slightly acidic groundwater. Such dissolution can enlarge pore spaces, joints, fractures, and bedding-plane partings, leading to increased hydraulic conductivity and porosity. In some cases, solution conduits or caves can develop that connect recharge areas to discharge areas at the downstream end of the flow system. The term *solution conduit* refers to long openings enlarged by groundwater. Such openings are generally considered to be smaller than caves (Palmer, 1990). Where they are saturated, carbonate rocks with well-connected networks of solution openings yield large amounts of water to wells that penetrate the openings. Although the intact rock between the secondary openings is poorly permeable.

Major solutional enhancement of permeability in carbonate aquifers is dependent on the circulation of groundwater. Naturally, chemically aggressive groundwater must be available to recharge the system. Next fractures need to be present to transmit water. Finally, groundwater must be able to drain out of the system (Stringfield and LeGrande, 1966). Commonly, these three conditions develop in unconfined carbonate aquifers.

Creation of the conduits begins as water begins to enlarge the discontinuities close to the recharge area of the flow system (Figure 4.19). Eventually, a few pathways develop, and aggressive groundwater flows rapidly through the rock to the discharge point. At the time that this flow regime develops, the conduits are only about 1 cm wide (Palmer, 1990). When the flow paths become kilometers in length, it is possible for selected conduits to grow to the size of caves, conduits accessible to humans. There is an extensive literature related to the human exploration of caves.

Apart from this network of conduits and caves, other secondary porosity can be present but not enlarged. Thus, significant regions of a carbonate aquifer might have hydraulic conductivities and porosities comparable to fractured rocks.

Not all carbonate rocks develop solution-enhanced porosity and permeability. This happens when the deposits overlying carbonate rocks are of low permeability and are thick, or when the carbonate rock was never elevated into an active groundwater circulation system (Stringfield and LeGrande, 1966). Features of the carbonate rock also can inhibit enhancements in permeability. For example, the required circulation will not develop in a carbonate rock that is essentially unfractured. Similarly, a carbonate rock like chalk can be so porous that flow is diffuse and not concentrated along preferential paths (Brahana et al., 1988). In these cases, pores grow equally.

Brahana et al. (1988) compared properties of 27 carbonate rock units from a variety of different terranes in North America, including the Caribbean. They pointed out the following important facts:

1) "Carbonate rocks serve as significant aquifers throughout North America. They are not limited by location or the age of the formation.
2) Carbonate rocks show a total range of hydraulic conductivity of 10 orders of magnitude, from the tightest confining beds to the most prolific aquifers.

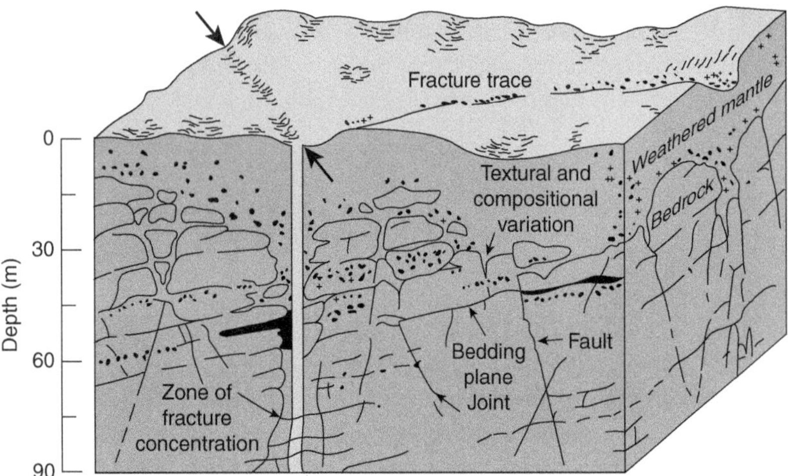

Figure 4.19 Solution enhancement of original porosity of carbonate rock. Highest well yields occur in zones where multiple fractures intersect (Reproduced from Lattman and Parizek, 1964. Copyright 1964 with permission from Elsevier Science).

3) The hydrogeologic response of carbonates at any time is related to the rock permeability, which is affected most by the dynamic process of dynamic freshwater circulation and solution of the rock.
4) The significant control of the dynamic freshwater circulation is the integrity of the hydraulic circuit, which requires that recharge, flow-through, and discharge must be maintained. Without each of these, the system is essentially static and does not act as a conduit. Its evolution rate approaches that of noncarbonate rocks. Secondary controls on dynamic circulation include lithologic, structural, geomorphic, hydrologic, and chronologic aspects.
5) The significant controls on the process of solution are (a) rock solubility, and (b) the chemical character of the groundwater……"

4.8.2 Karst Landscapes

Extensive modifications to carbonate or easily dissolvable rocks have a significant influence on the landscape. The term karst describes landscapes shaped by dissolution processes to provide unique landforms and underground drainage. Sinkholes are probably the most well-known landform associated with karst. Other landforms or features include sinking streams, dry valleys, caves, and large springs (White, 1990; Runkel et al., 2003).

Sinkholes are bowl-shaped depressions on the land surface. Processes at work leading to their formation include dissolution and "suffosion." The term *suffusion* refers to the situation where unconsolidated surface deposits overlying the bedrock are transported into large bedrock openings. Unlike the catastrophic collapse that forms deep sinkholes, suffusion is a relatively slow process. The character of the overburden materials, their thickness, and types of materials play an important role in sinkhole formation (Galloway et al., 1999).

In the setting of shallow carbonate rocks with dissolution features, stream sinks can form (Figure 4.20). *Stream sinks* are places where a small stream simply disappears into the subsurface to follow a connected cave system before rejoining the surface water as a spring.

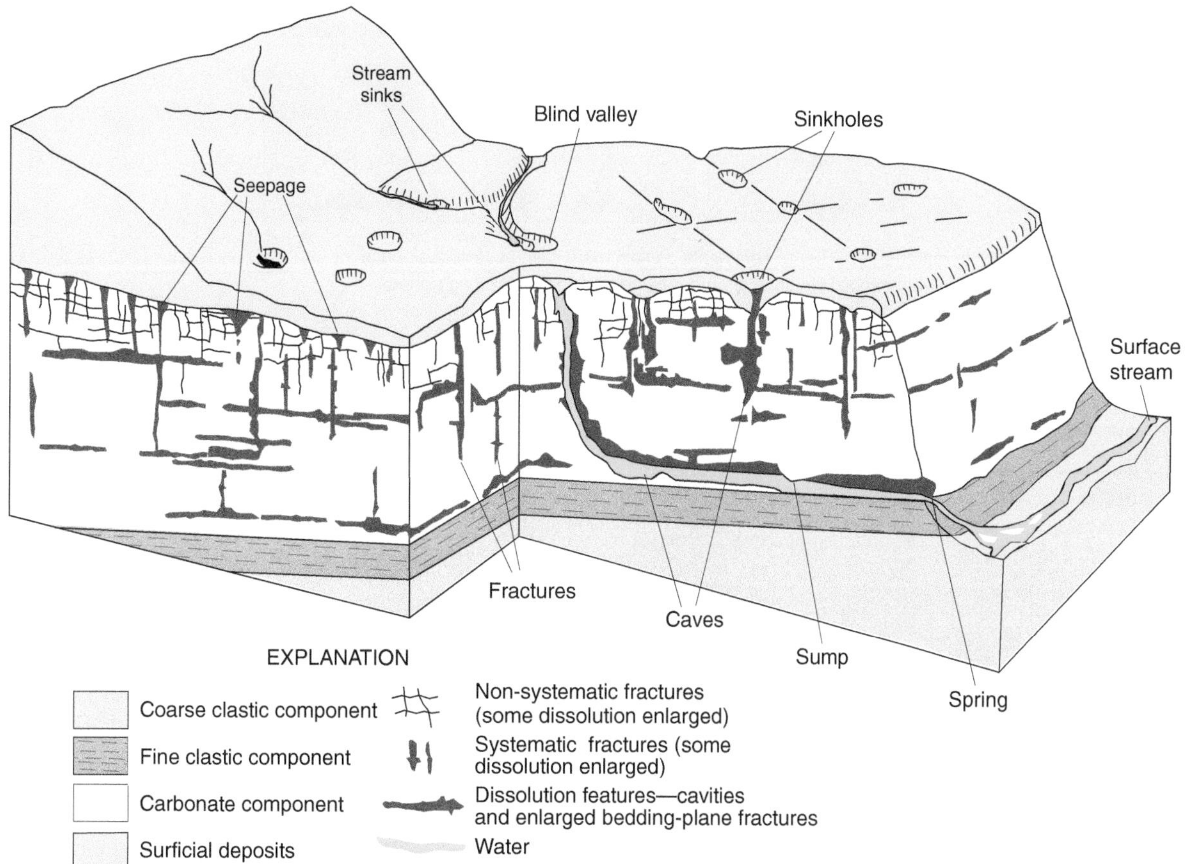

Figure 4.20 Conceptual model showing landforms and other features associated with karst in Minnesota. Carbonate bedrock is overlain by surficial deposits. Groundwater there is present in various types of dissolution features (Adaped from Alexander and Lively, 1995. Used with permission of the Minnesota Geological Survey).

The ability of fracture and cave systems to capture and transmit surface water in various ways sometimes leads to an absence of surface water (i.e., dry valleys) with associated large springs or springs systems.

There are various types of sinkholes described with several classification systems. Galloway et al. (1999) described the types found in Florida (i) dissolution, (ii) cover subsidence, and (iii) cover-collapse sinkholes. Figure 4.21 from Galloway et al. (1999) describes in detail how these different types of sinkholes form. The upcoming discussion of

TYPES OF SINKHOLES

Dissolution of the limestone or dolomite is most intensive where the water first contacts the rock surface. Aggressive dissolution also occurs where flow is focussed in preexisting openings in the rock, such as along joints, fractures, and bedding planes, and in the zone of water-table fluctuation where groundwater is in contact with the atmosphere.

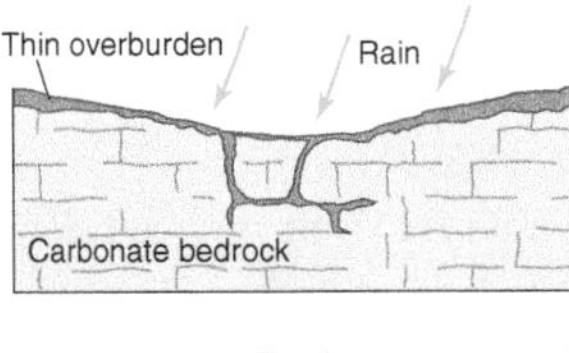

Rainfall and surface water percolate through joints in the limestone. Dissolved carbonate rock is carried away from the surface and a small depression gradually forms.

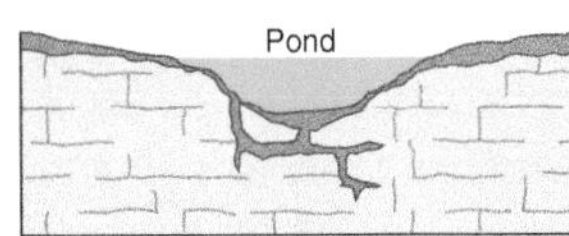

On exposed carbonate surfaces, a depression may focus surface drainage, accelerating the dissolution process. Debris carried into the developing sinkhole may plug the outflow, ponding water and creating wetlands.

Gently rolling hills and shallow depressions caused by solution sinkholes are common topographic features throughout much of Florida.

Cover-subsidence sinkholes tend to develop gradually where the covering sediments are permeable and contain sand.

Granular sediments spall into secondary openings in underlying carbonate rocks.

Column of overlying sediments settles into the vacated spaces (a process termed piping).

Dissolution and infilling tinue with a noticable depression in the land surface.

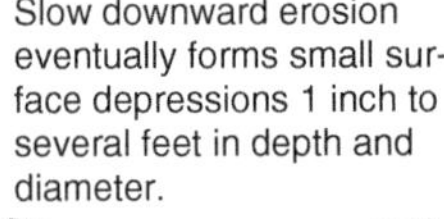

Slow downward erosion eventually forms small surface depressions 1 inch to several feet in depth and diameter.

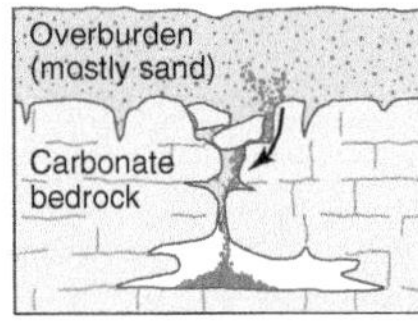

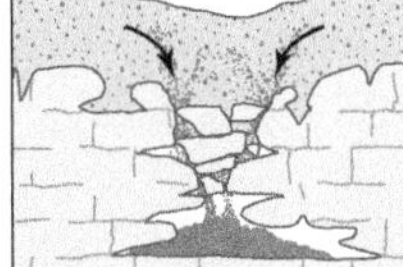

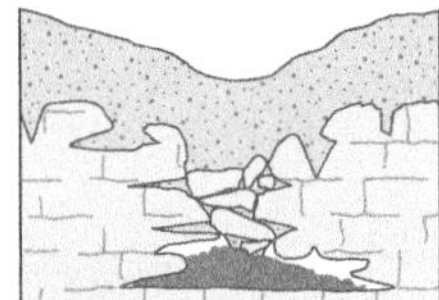

In areas where cover material is thicker or sediments contain more clay, cover-subsidence sinkholes are relatively uncommon, are smaller, and may go undetected for long periods.

Cover-collapse sinkholes may develop abruptly (over a period of hours) and cause catastrophic damages. They occur where the covering sediments contain a significant amount of clay.

Sediments spall into cavity.

As spalling continues, cohesive covering sediments form a structural arch.

The cavity migrates upward by progressive roof collapse.

The cavity eventually breaches the ground surface, creating sudden and dramatic sinkholes.

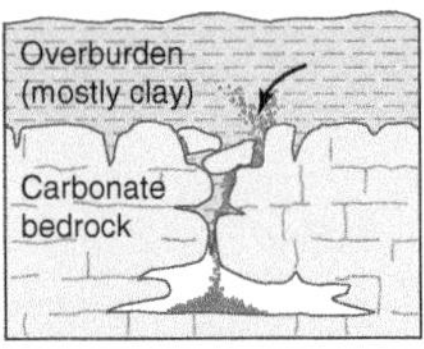

Over time, surface drainage, erosion, and deposition of sediment transform the steep-walled sinkhole into a shallower bowl-shaped depression.

Figure 4.21 There are three different kinds of sinkholes that occur in Florida, aggressive dissolution of carbonate rocks, cover-subsidence sinkholes, and cover-collapse types (U.S. Geological Survey Galloway et al., 1999/Public domain).

the Floridan aquifer system will examine the karst settings and human factors that contribute to sinkhole formation, and the damage that they can cause.

4.8.3 Floridan Aquifer System

The Floridan aquifer system is a major carbonate aquifer that covers much of Florida as well as parts of Alabama, Georgia, and other nearby states (Figure 4.22). Groundwater is extensively developed from this aquifer making it one of the important aquifers in the United States. The Floridan aquifer system consists of two major hydrogeologic units—the Upper Floridan aquifer and the Lower Floridan aquifer. Most groundwater production comes from the Upper Floridian aquifer (Bellino et al., 2018). Hydraulically, these units mostly function as a single aquifer. However, lower permeability units can create hydrologic separation between the Upper Floridan aquifer and Lower Floridan aquifer sub-regionally.

The Floridan aquifer is typically overlain by two units, the surficial aquifer system, and the underlying upper Confining unit. The surficial aquifer system is mainly alluvial sand but includes some carbonate rocks (Williams and Kuniansky, 2016). For example, the Biscayne aquifer located along the southeastern tip of Florida is an important aquifer there. Underlying the surficial aquifer is the upper confining unit. As the name implies, where present, this low-permeability clay-rich unit confines the Floridian aquifer system.

The thickness of the upper confining unit above the Floridian aquifer is important in determining the extent to which active karst formation is underway. For example, Figure 4.22 depicts an area of west central Florida area (Tampa Bay northward) where the aquifer is either unconfined or thinly confined, and active karst formation is occurring. In other areas, the aquifer is confined by >30.5 m of the upper confining unit (Figure 4.22). In these areas, the reduced circulation of groundwater means that karstification is much less active now (Williams and Kuniansky, 2016). However, in areas where the Floridian aquifer is presently confined by thick confining units, karst features developed in the past when sea levels were much lower (Bellino et al., 2018). For example, during the last continental glaciation 18,000 years ago, continental glaciation stored water on the continents producing sea levels some 85–100 m lower around Florida (Galloway et al., 1999).

Much of the active karst development is in west-central Florida (Galloway et al., 1999). As mentioned, the geological setting, particularly the thickness and lithology of the confining unit influenced the type of sinkhole and where it formed.

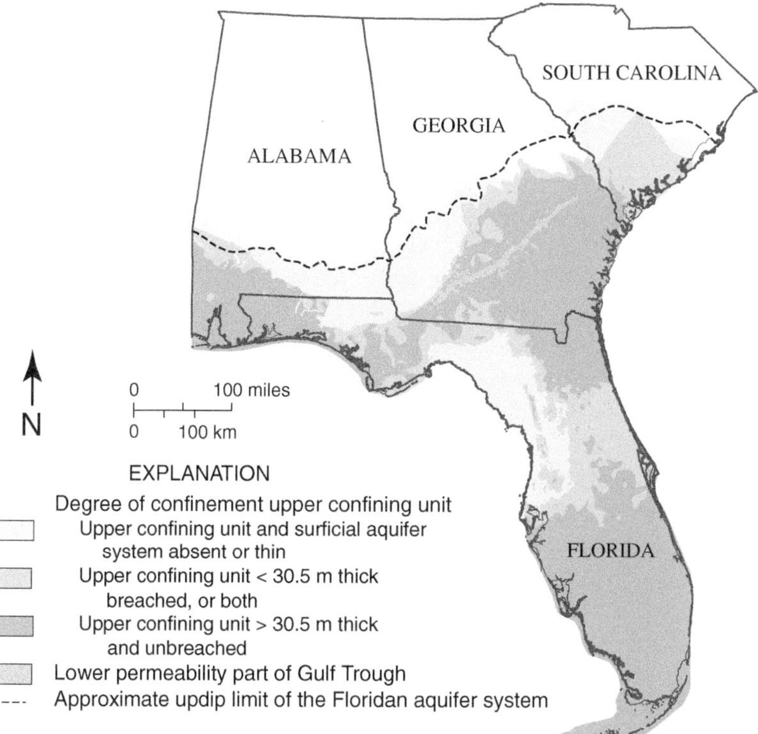

Figure 4.22 Map showing the areal distribution of the Floridan aquifer system and the degrees of confinement by upper confining unit and surficial aquifer (U.S. Geological Survey, adapted from Bellino et al., 2018/Public domain).

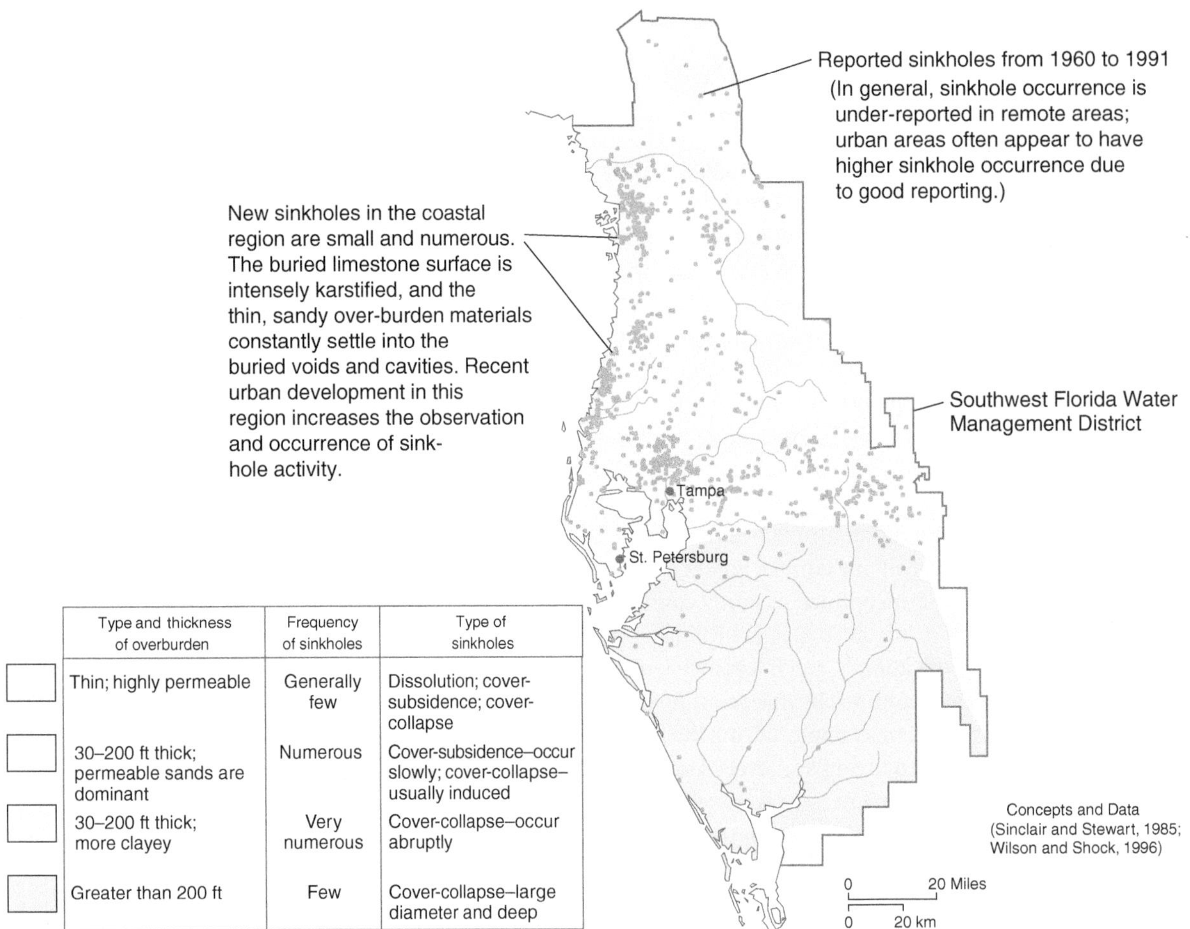

Type and thickness of overburden	Frequency of sinkholes	Type of sinkholes
Thin; highly permeable	Generally few	Dissolution; cover-subsidence; cover-collapse
30–200 ft thick; permeable sands are dominant	Numerous	Cover-subsidence–occur slowly; cover-collapse–usually induced
30–200 ft thick; more clayey	Very numerous	Cover-collapse–occur abruptly
Greater than 200 ft	Few	Cover-collapse–large diameter and deep

Figure 4.23 In west central Florida, the frequency and types of sinkholes that formed are controlled by the type and thickness of overburden (U.S. Geological Survey Galloway et al., 1999 and Sinclair and Stewart, 1985. Data from Wilson and Shock, 1996).

The map and accompanying table (Figure 4.23) from Galloway et al. (1999) showed cover-collapse and cover-subsidence sinkholes to be numerous and very numerous in places where the confining unit ranged from 30 to 200 ft in thickness. The specific lithology of the confining bed—"permeable sands" or "more clayey" created a preference toward cover-subsidence or cover-collapse types, respectively (Galloway et al., 1999). Areas with a thin, highly permeable confining unit or a unit >200 ft had few sink holes (Figure 4.23).

Circular 1182 (Galloway et al., 1999) discusses studies of how human activities have contributed to the creation of new sinkholes (Newton, 1986). The list of activities is long and includes impacts due to (i) groundwater pumping (Sinclair, 1982), (ii) changes to patterns of surface water and (iii) creating impoundments of water, and (iv) various construction activities. For example, the growth of St. Petersburg (Figure 4.23) from the 1930s to 1978 led to the construction of four well fields by 1978 about 20 km north and west of Tampa. Sinkholes formed in association with each of these wellfields. The "Section 21" field, developed in 1963, produced "64 new sinkholes" nearby, most associated with the greatest producing well (Sinclair, 1982; Galloway et al., 1999)

4.8.4 Edwards-Trinity Aquifer System

In a few places in the United States, carbonate rocks are interbedded with an equal thickness of permeable sandstone. Where these two rock types are interbedded, usually carbonate rocks yield much more water than the sandstone. Most carbonate rocks originate as sedimentary deposits in marine environments. Compaction, cementation, and dolomitization processes act on the deposits as they lithify and greatly change their porosity and permeability. However, the principal post-depositional change in carbonate rocks is the enhancement of secondary porosity by circulating groundwater.

Examples of sandstone and carbonate-rock aquifers include Edwards-Trinity aquifer system (Texas and Oklahoma), Valley and Ridge aquifers (eastern U.S.), Mississippian aquifers (central and eastern U.S.), and Paleozoic aquifers (northern Great Plains). Let us examine the Edwards-Trinity aquifer system in more detail.

The Edwards-Trinity aquifer system is an especially important aquifer in Texas. Carbonate rocks, sandstones, and sands are the predominant water-yielding units. The Edward-Trinity aquifer and the Trinity aquifer are stratigraphically equivalent in part and are hydraulically connected in some places.

Near San Antonio, Texas, most groundwater is produced from the saturated, karstic Edwards aquifer within and downgradient of the Balcones Fault zone (Figure 4.24). It is one of several unique features at work to create this productive aquifer (Figure 4.24). The Contributing Zone is a large area on the elevated Edwards Plateau, which is underlain by karstic Edwards limestones and less permeable rocks below. As conceptual model suggests, rain falling across the plateau infiltrates and discharges to the headwaters of streams (Schindel and Gary, 2018). In effect, the Contributing Zone collects water over approximately ~14,200 km^2 and sends it downstream as streamflow. Primary recharge to the Edwards aquifer occurs in the Recharge Zone, a 3160 km^2 area of Balcones fault zone where karstic Edwards limestone is exposed (Schindel and Gary, 2018). Although the Trinity aquifer (Figure 4.24) is less permeable than the Edwards, there are some groundwater contributions to the Recharge and the Confined Zones. Within the Recharge Zone, the Edwards aquifer is unconfined with water levels approximately 70 m below ground surface (Schindel and Gary, 2018). Flow continues along the aquifer to the Confined Zone where faulting caused the aquifer to become confined (Figure 4.24). Water is discharged from springs and production wells. The end of the aquifer is defined by the "bad water line" (Schindel and Gary, 2018) at which point the water is chemically unsuitable for most uses.

Karst processes have created the unique recharge mechanism with the Contributing and Recharge Zones. Karst features evident where the Edwards limestones are exposed include "sinkholes, sinking (losing) streams, caves, springs" (Schindel

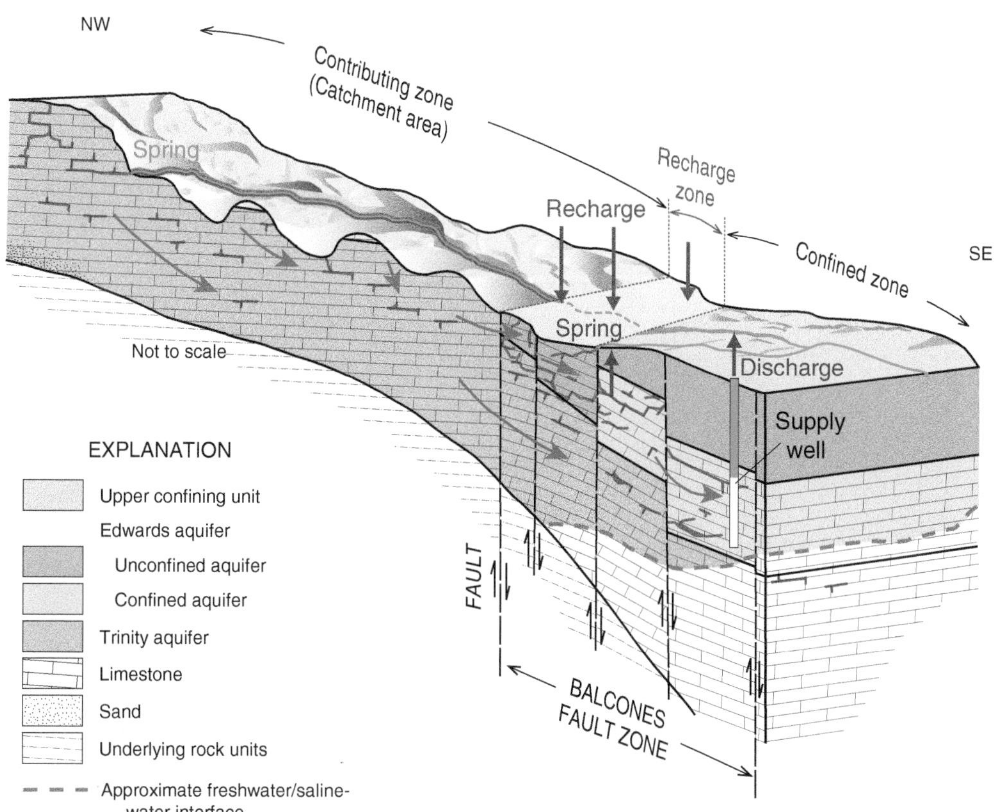

Figure 4.24 Conceptual model of the Edwards-Trinity aquifer along a northwest-to-southeast section near San Antonio, Texas. Key elements of the hydrogeologic framework are shown, such as key zones, generalized flow directions (arrows), and karstic zones (U.S. Geological Survey, adapted from Barker and Ardis, 1996, and Brakefield et al., 2015/Public domain).

and Gary, 2018). Within the Confined (Artesian) Zone, the cavernous porosity creates prolific well yields, as high as 20,000 L/sec and springs capable of discharge rates as high as 9 m^3/sec (Schindel and Gary, 2018).

Karst formation is commonly an epigenetic (near surface) process where rock is dissolved through the increasingly active circulation of water from the land surface. With the Edwards limestone, epigenetic karst processes have been responsible for enhanced porosity within the Recharge Zone and the shallow Confined Zone that is adjacent (Schindel and Gary, 2018). However, the highly permeable, more deeply buried portions of the Confined Zone are puzzling because groundwater flow is upward. A variety of observations point to the likelihood of hypogene karst processes formed due to ascending warm water from below (Schindel and Gary, 2018).

4.8.5 Basin and Range Carbonate Aquifer

An earlier section provided a description of the basin-fill aquifers associated with the Basin and Range region. About two-thirds of this area is also underlain by a sequence of carbonate rocks, the Basin and Range Carbonate aquifer system (Figure 4.25 insert). Groundwater production from this aquifer system is small (Figure 4.4). The most important units are the upper carbonate aquifer unit (UCAU) and the lower carbonate aquifer unit (LCAU), which are separated by a confining bed. As will be apparent, the regional geology is stratigraphically and structurally complicated. Complexity is also compounded by absence of data because associated basins are deep and poorly explored. The UCAU and LCAU have maximum thicknesses of approximately 24,000 ft (7300 m) and 16,500 ft (5000 m). The intervening confining bed can be up to 5000 ft (1525 m) thick. In addition, Cenozoic volcanic rocks are present, which are as much as 3300 m (1000 m) thick. In calderas, these rocks are much thicker (~13,000 ft or 4000 m). Figure 4.25 contains an east–west cross section that illustrates the variability in the geologic settings and the thickness of aquifer units. For example, notice the depths to which the faulted basins extend and the thickness of accumulated alluvial sediments.

Besides being an important aquifer system, the carbonate units in places are connected hydraulically with basin-fill aquifer systems. The individual-filled basins are not continuous within the Basin and Range province. They are encircled by topographic drainage divides, which were classified as one of four types based on similar recharge-discharge relations. The simplest type is the *undrained, closed basin*, a single valley in which the underlying and surrounding bedrock is practically impermeable and does not allow interbasin flow. All recharge is discharged at a sink, a playa near the center of the basin. Basins underlain by permeable bedrock commonly are hydraulically connected as multiple valley systems. The *partly drained, closed basin* is underlain or surrounded by bedrock that is moderately permeable and allows for some groundwater underflow. In this type of basin, some water is evaporated or transpired at the upgradient side of a playa, but most of the water continues to flow past the downgradient side of the playa and leaves the basin. The *drained, closed basin* has a deep

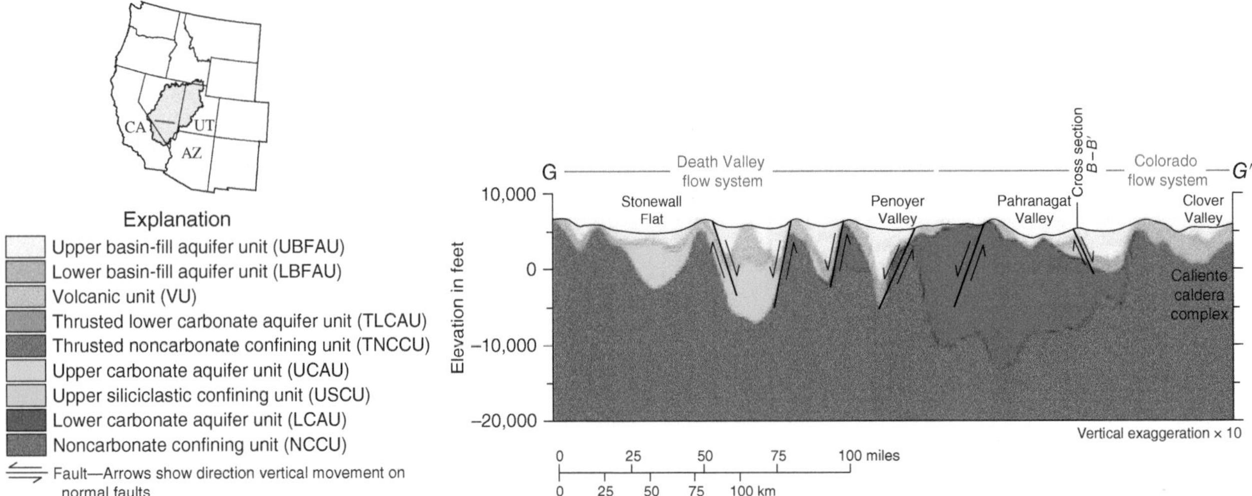

Figure 4.25 The insert map shows the location of that portion of Basin and Range underlain by the Carbonate aquifer system and location of the cross section *G–G′*. The section illustrates the complexity in the thickness and distribution of geological units and faulting that has contributed to the formation of some basin-fill aquifers (U.S. Geological Survey with index map modified from Heilweil and Brooks, 2011 and cross section from Sweetkind et al., 2011/Public domain).

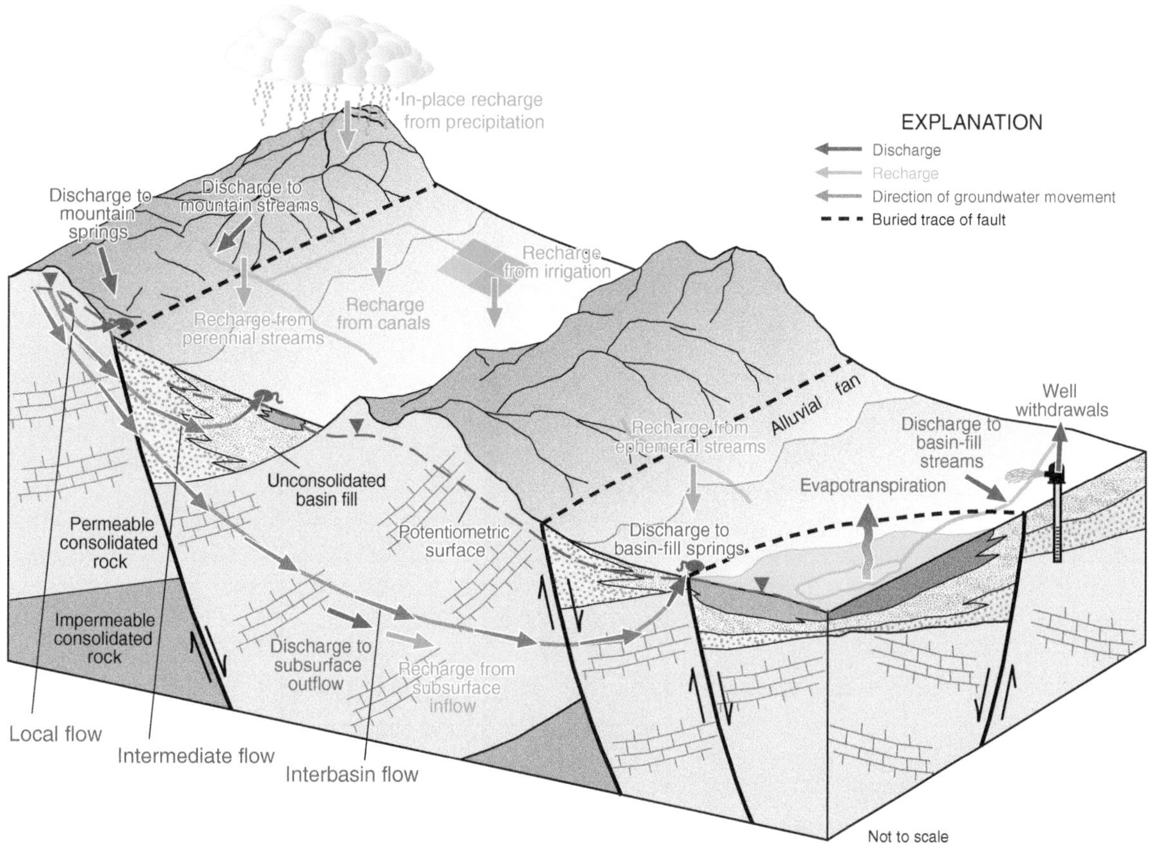

Figure 4.26 Conceptual model that illustrates the inter-relationships between the basin-fill aquifer system and deeper carbonate rock aquifers (U.S. Geological Survey, adapted from Sweetkind et al., 2011/Public domain).

water table that prevents evapotranspiration. The bedrock is sufficiently permeable to allow all recharge to flow through it and out of the basin. The *terminal sink basin* is underlain or surrounded by bedrock that is sufficiently permeable to conduct flow into the basin, and the playa in the basin is the discharge point for recharge from several connected basins.

Figure 4.26 provides an example of a drained, close basin type of basin-fill aquifer upgradient and a terminal basin downstream. Interbasin flow in this case is facilitated by mountain ranges and basin bottoms that are sufficiently permeable for large-scale flow systems to develop (Sweetkind et al., 2011). In the terminal basin, the surface water and groundwater are lost by evapotranspiration from a playa lake.

4.9 Basaltic and Other Volcanic-Rock Aquifers

Aquifers formed from basaltic lava flows are the most productive kinds of aquifers associated with volcanic rocks. Principal aquifers in the continental United States include the Snake River Plain aquifer system (Idaho), and the Columbia Plateau Regional aquifer system. This latter aquifer system underlies ~115,000 km^2, including parts of Oregon, Washington, and Idaho (Figure 4.27*a*). Much of the groundwater produced there supports irrigated agriculture. However, declines in groundwater level have occurred over about one-quarter of the area.

Although Columbia Plateau includes permeable overburden where present, this section will focus on the aquifers developed within the underlying Columbia River Basalt Group. These rocks are Miocene in age and consist of a stack of ~350 basaltic lava flows between 6 and 17 million years in age (Ely et al., 2014). The thickness of individual flows can range from 3 m to more than 100 m (Ely et al., 2014). Most of volcanism on the Columbia plateau took place ~15 million years ago. Subsequent eruptions (14–6.0 million years) were less frequent (Vaccaro, 1986).

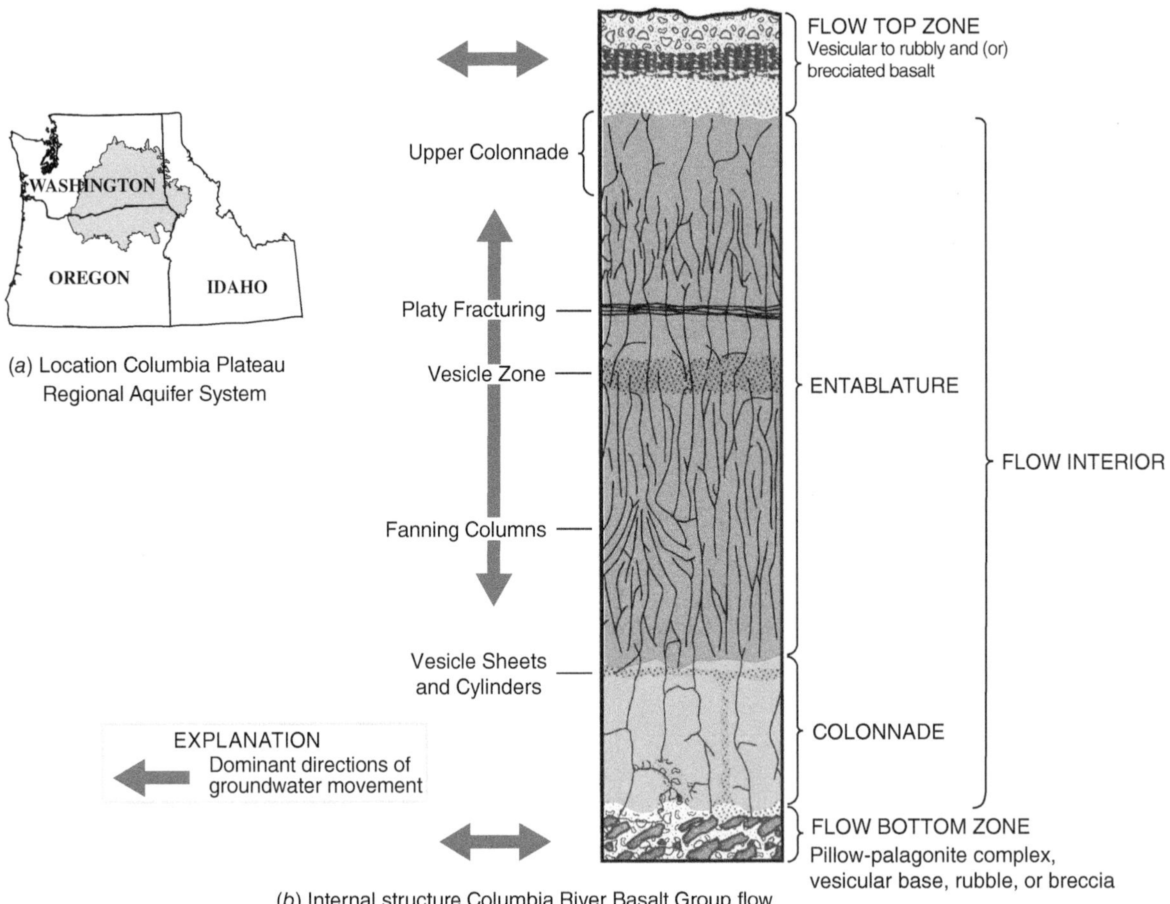

Figure 4.27 Panel (*a*) shows the location of Columbia Plateau Regional aquifer system in the American Northwest. Panel (*b*) illustrates the structure of a basaltic flow unit with permeable flow tops and bottoms forming facilitating flow to wells (U.S. Geological Survey adapted from Ely et al., 2014/Public domain).

These kinds of aquifers are complicated by intricacies and variability in secondary porosity that can develop within a single flow unit. Moreover, sedimentary deposits sometimes developed and remain preserved between flows, i.e., *sedimentary interbeds*, during a period of quiescence in eruption. Hydraulic characteristics like hydraulic conductivity and porosity also vary greatly within and between the individual basalt flows. Figure 4.27*b* is a conceptual model of a single basaltic flow unit. The most permeable zones in a single flow unit occur at the flow top and the flow bottom (Figure 4.27*b*). Permeability developed in the flow tops because rapid cooling of lava flows produced brecciated and (or) vesicular structures (Ely et al., 2014). The flow bottom can also be brecciated and with rubbly textures, which also lead to large values of hydraulic conductivity in the horizontal directions. With a stack of flow units, the permeable top of one unit is juxtaposed with the bottom of the overlying flow to create an *interflow zone*. The interflow zone might also contain sedimentary interbeds.

As the figure suggests with arrows, flow to wells occurs along these thin and permeable interflow zones, which represent about 10% of the total thickness of volcanic units (Ely et al., 2014). Most of the remaining rocks are part of the *flow interior* (Figure 4.27*b*) that consists of the colonnade and entablature, along with other minor textural features. In the flow interiors, what permeability exists is vertical, because of poorly connected vertical fractures (Figure 4.27*b*).

Sedimentary interbeds can be major or minor features within a basaltic terrain. For example, major interbeds separate major basalt aquifer units and are given names. Minor interbeds remain unnamed when they occur within interflow zones. Thus, the Columbia Basalt Group has five hydrogeologic units, Saddle Mountains, Wanapum, and Grande Ronde flow members and interbeds (minor) and two named interbed members that behave as confining beds.

4.10 Hydraulic Properties of Granular and Crystalline Media

Historically, most textbooks try to develop relationships between geology and hydraulic properties, in terms of deposit or aquifer types. For example, knowing the hydraulic conductivity of shale at one site is developed as a basis for understanding similar deposits at other places. In this respect, we have introduced ranges for hydraulic properties in the previous sections. However, the major problem with this approach is the difficulty in ultimately unraveling the influence of geological processes that influence the porosity and hydraulic conductivity. Simply knowing the type of unit or the geological setting does not usually provide enough information to estimate hydraulic parameters. Various geologic processes (chemical weathering and cementation) act to create tremendous variability with each type of deposit or aquifer, which makes generalizations difficult.

Part of the difficulty in creating a broad understanding of Earth materials is that few retain their primary porosity and permeability. As an example, let us consider sediments like clays, carbonate muds, or fluvial sands. When first deposited, they have porosities greater than 30%. In most depositional environments, sediments often become buried, as more and more sediments are added on top. Loading causes the porous medium to consolidate, leading to a loss of porosity and permeability. The greatest loss in porosity as a function of depth occurs in materials containing abundant clay minerals. As loading occurs, the pore structure will change irreversibly as the "house -of-cards" structure of the clay platelets is collapsed. With quartz sands, loading compresses the grains rather than disrupting the entire pore structure. Thus, the effects of loading are less marked in sandstones than in shales. The Frio Sandstone of Texas loses about 1.2% of its porosity for every 1000 ft of burial (Loucks et al., 1984). In the case of shale, porosity decreases from 0.3 to 0.5 near the ground surface, to 0.05–0.15 at depths of 5000 m or more (Davis, 1988).

It is often difficult to separate porosity loss due to compaction alone from the combination of compaction and mineralogical alterations. For example, as temperatures increase, the reactive components in sandstones (e.g., feldspars) can change to clay minerals, which contribute to a more marked loss of porosity with depth. Chemically more mature sandstones (composed of quartz) are less susceptible to porosity losses with depth.

Porosity and permeability are also reduced in sandstone and carbonate rocks through cementation, wherein new minerals form in the pore spaces. Typical cements found in sedimentary rocks include quartz, carbonates, clay minerals, various oxides, sulfides, and sulfates (Schwartz and Longstaffe, 1988). For example, lime sands quickly lose their porosity and permeability due to compaction and cementation.

Our few comments here have just scratched the surface of how processes, collectively referred to as diagenesis, can affect the porosity and permeability of sediments and sedimentary rocks. The topic is of considerable importance in the field of Petroleum Geology, where the formation, migration, and trapping of oil and gas depend upon the porosity and permeability. The key point we would make here is that information on rock types can not replace actual porosity and permeability measurements.

4.10.1 Pore Structure and Permeability Development

One way to understand permeability development is to look at the pore structure rather than the type of rock. The major resistances to flow in a porous network are related to the size distributions of the entrances and exits to a pore and the length of the pore wall. The exact quantitative relationship has remained elusive for more than 60 years. Equations like the well-known Kozeny-family of relationship, however, provide a basis for presenting key relationships in a simplified manner. One form of this equation is given as:

$$k = \frac{c}{M_s^2} \frac{n^3}{(1-n)^2} \tag{4.16}$$

where k is the permeability, c is the Kozeny constant relating shape factor and tortuosity factor to flow path, and M_s is the specific area of solids, defined as the total surface area of solids per unit volume of solid (e.g., m^2/g), and n is porosity. The specific surface is essentially a reflection of the particle surface area found in a given volume of material. Coarse sand, for example, has a specific surface of 0.1 m^2/g. As particle sizes get smaller in a medium, the surface areas become larger. Smectite clay has a specific surface as high as 800 m^2/g.

We can use Eq. (4.16) in a qualitative manner to help explain the hydraulic behavior of some geologic materials. The values of c and $(1-n)^2$ span a relatively limited range so their value will not influence k very much. Values of n^3 and

M_s, however, can range over orders of magnitude and are the key variables controlling permeability. The equation shows those small porosity values and/or large surface areas promote low permeabilities.

For the purposes of this discussion, a few, common, geologic materials are categorized into one of three types. Media-type I could include rocks like salt and granite that has virtually no pore space present in the rock. There might be a few micron-sized spaces present in the rock, between the crystals. These ultra-low porosities coupled with modestly large specific surfaces of these media give these rocks permeabilities of less than 10^{-20} m^2. Media-type II would include fine-grained sediments or sedimentary rocks, comprised of abundant silt and clay-size particles. Typical examples would be shale, glacial till, or silty sediments. These media are porous, but the tiny grains give them a huge specific surface and, thus, a relatively low permeability, such as 10^{-18} to 10^{-20} m^2 for clay-containing media. Media-type III includes media with both relatively high porosities and relatively small specific surfaces. These characteristics promote the high permeabilities that we associate with sands and gravels. For these media, the empirical equations presented in Chapter 3 provide a useful way to estimate k given estimates of particle diameters.

This classification scheme runs into problems because of the tremendous variety of geologic media. The main point was to illustrate how features of the pore system can help to understand permeability development in common media.

4.11 Hydraulic Properties of Fractured Media

In previous sections, we saw how many geologic media were characterized by relatively low permeabilities. To some extent, these observations contradict real-world experience that rocks and sediments are generally permeable, much more so than is the case with sediments and rocks containing fine-grained materials (i.e., silts and clays). In many parts of the world, glacial-till aquifers are developed with dug wells to supply modest domestic needs. In the eastern part of the United States and elsewhere, productive wells are developed in metamorphic rocks, and recrystallized limestone. Pervasive fracturing of rocks and sediments has the potential to increase their hydraulic conductivities significantly with a small increase in porosity.

In the near-surface, rocks contain fractures and joints. A *fracture* is a planar discontinuity or break in a rock or cohesive sediment in response to stress. Commonly, a fracture provides a pathway for flow through a rock. A single fracture can be described in terms of its attitude (orientation) in space, its size (i.e., dimensions), and its aperture. A collection of fractures having roughly the same attitude in space is referred to as a fracture *set*. The *aperture* (b) is a measure of the width of the fracture opening (Figure 4.28). As the figure shows, the apertures of real fractures are quite variable because of the roughness of the fracture walls. Often, fractures contain secondary minerals (e.g., calcite, clay, and minerals) that formed as water continued to flow through the medium. These fracture-filling materials can be important because they can reduce the capability of fractures to transmit water, or even plug the fracture up. Thus, the presence of a fracture does not necessarily mean a permeable pathway for flow exists. A joint is a type of fracture that forms near the ground surface. Joints are described by Trainer (1988) as macrofractures (sometimes you can stick your hand into a joint) along which dilation has occurred without movement parallel to the fracture surface.

Because fractures have a finite size, flow through a rock requires that a connected network of fractures exist. The distance between the two adjacent fractures is referred to as *fracture spacing* (s). The relative abundance of fractures can also be measured as *fracture density*—the number of fractures per volume of material, or as *fracture frequency*—the number of

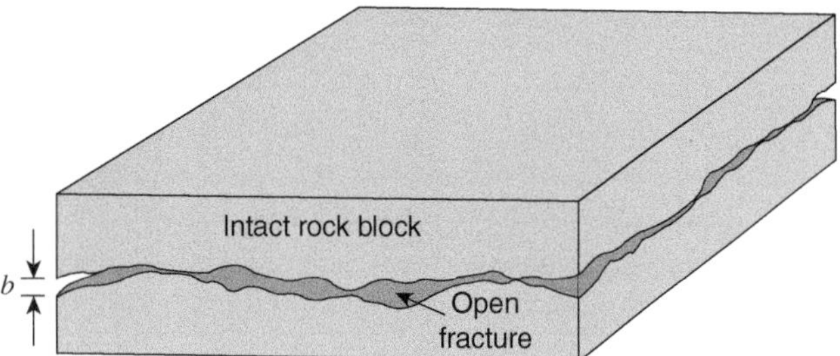

Figure 4.28 Sketch showing a rough-walled fracture with an aperture in rock (Schwartz and Zhang, 2003. © 2003/John Wiley & Sons, Inc.).

fractures intersecting a unit length of borehole. As a rule, the greater the density of open fractures the higher the permeability of the rock.

We reserve the detailed quantitative treatment of flow through fractured media until Chapter 6. However, with the help of the following simple equations, we will illustrate how fracture aperture (b) and fracture spacing (s) controls the permeability and porosity of a fracture network. The ratio of the fracture aperture to the fracture spacing is defined as *fracture porosity*. Consider the fracture network shown in Figure 4.29 with three sets of uniform fractures oriented at right angles to each other. This simple model is also called the "sugar-cube" model—like the network that would be constructed by piling up sugar cubes into a large block.

According to Snow (1968), the permeability (k) and porosity (ϕ) for the fracture set (Figure 4.29) is given as:

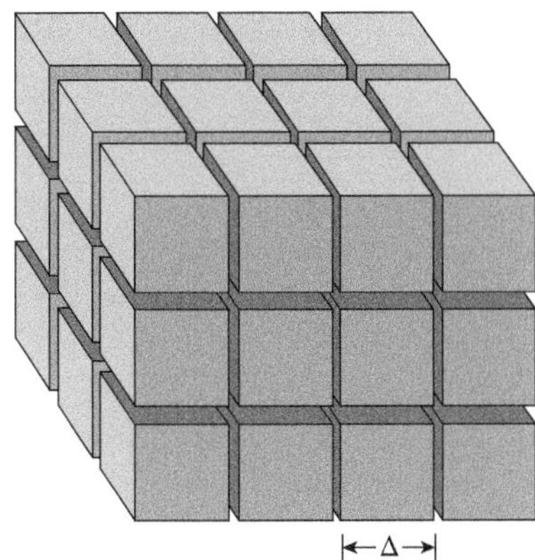

Figure 4.29 "Sugar-cube" model for fractured rocks. Three sets of fractures are assumed to intersect at right angles forming cubic rock blocks (Schwartz and Zhang, 2003/John Wiley & Sons, Inc.).

$$k = \frac{b^3}{12s} \tag{4.17}$$

$$\phi = \frac{b}{s} \tag{4.18}$$

where b is the fracture aperture and s is the distance between fractures. Equation (4.17) shows clearly how permeability is proportional to the cube of the aperture and inversely proportional to the spacing between fractures. Porosity is proportional to aperture and inversely proportional to fracture spacing Eq. (4.18).

Example 4.4 Assume that unfractured granite has a permeability of 1×10^{-17} m^2. Calculate what the permeability and porosity would be for two cubic networks of fractures having the following characteristics: (a) $b = 100$ μm, $s = 0.1$ m; (b) $b = 25$ μm, $s = 1$ m.

Solution

(a) $b = 100$ μm, $s = 0.1$ m.

$$k = \frac{(100 \times 10^{-6}\ \text{m})^3}{12(0.1\ \text{m})} = 8.3 \times 10^{-13}\ \text{m}^2$$

$$\phi = \frac{(100 \times 10^{-6}\ \text{m})}{(0.1\ \text{m})} = 10^{-3}$$

(b) $b = 25$ μm, $s = 1$ m.

$$k = \frac{(25 \times 10^{-6}\ \text{m})^3}{12(1\ \text{m})} = 1.3 \times 10^{-15}\ \text{m}^2$$

$$\phi = \frac{(25 \times 10^{-6}\ \text{m})}{(1\ \text{m})} = 2.5 \times 10^{-5}$$

In both cases, fracturing has substantially increased the permeability of the granite.

4.11.1 Factors Controlling Fracture Development

This section examines some of the key factors influencing the distribution of fractures. We begin here with depth and rock lithology. In many environments, there is a noticeable tendency for the permeability of fractured rocks to decrease with depth. Most explanations point to a reduction in the frequency of fracturing coupled with a decrease in the aperture (b) caused by loading.

Sanford (2017) presented data for fractured rocks, located in upper Potomac River basin near Washington DC. The study set out to determine how permeability is enhanced by declined as a function of depth and to assess how this model-based

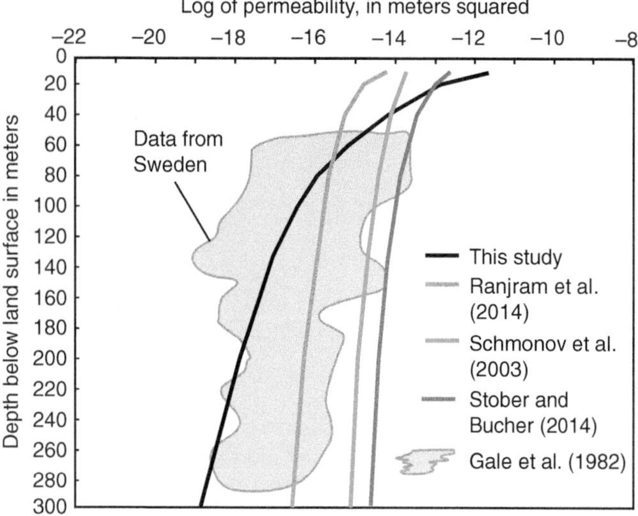

Figure 4.30 The permeability-depth function in fractured rock as compared with other studies (colored lines), and with data from the Stripa site (blue shading) in Sweden (Sanford, 2017. U.S. Geological Survey. © Springer-Verlag Berlin Heidelberg (outside the USA) 2016).

analysis. The basin contained siliciclastic rocks (e.g., sandstones and shales), various carbonates, as well as meta-igneous and meta-sedimentary rocks.

This study found that hydraulic conductivity in the various rock types declined about 5 orders of magnitude or more in the upper 300 m. The cause was a decline in the intensity of fracturing that ranged from "heavily fractured and or regolith" near the ground surface to "very sparsely fractured" at 100 m. Sanford's study suggested that the decline in hydraulic conductivity in hard rocks could be larger than indicated in other studies (Figure 4.30). For rocks at depths between 100 m and 1 km, studies found permeability values of 10^{-15} or 10^{-16} m^2 to be representative. However, these values could be "substantial" overestimates (Sanford, 2017).

It is well known that major fracture zones or fault zones extend to deep depths and can act as conductive pathways for deep flow. Local accidents in the setting of hydraulic fracturing in oil and gas developments are good examples. Sanford (2017) argued that at a much larger scale, these kinds of features are neither sufficiently wide nor sufficiently transmissive. One obvious exception was with the broad development of karst (Sanford, 2017).

Fractures are often highly deformable. When subject to stresses, the apertures decrease, making the fractures and the network less permeable. In granitic rocks, burial to 1000 ft may reduce apertures to cause a 2-order of magnitude reduction in the overall hydraulic conductivity. Thus, both the frequency of fractures and the fracture apertures decline with increasing depth.

Exercises

4.1 The transmissivity calculated from the analysis of aquifer-test data is 0.0134 m^2/sec for a confined aquifer of 13 m thick. Calculate hydraulic conductivity in gpd/ft^2 and intrinsic permeability in Darcy.

4.2 In a groundwater basin of 12 mi^2, there are two aquifers: an upper unconfined aquifer of 500 ft in thickness and a lower confined aquifer with an available hydraulic head drop of 150 ft. Hydraulic tests determined that the specific yield of the upper unit is 0.12 and the storativity of the lower unit is 4×10^{-4}. What is the amount of recoverable groundwater in the basin?

4.3 Calculate the permeability and porosity for a cubic network of fractures having a fracture aperture of 50 μm and a block size of 0.25 m.

References

Alexander, E. C. Jr., and R. S. Lively. 1995. Karst—Aquifers, caves and sinkholes, In R. S. Lively, and N. H. Balaban (eds.), Text Supplement to the Geologic Atlas of Fillmore County, Minnesota. Minnesota Geological Survey County Atlas C-8, pt. C, p. 10–18.

Barker, R. A., and A. F. Ardis. 1996. Hydrogeologic framework of the Edwards-Trinity aquifer system, west-central Texas. U.S. Geological Survey Professional Paper 1421-B, 61 p.

Bellino, J. C., E. L. Kuniansky, A. M. O'Reilly, et al. 2018. Hydrogeologic setting, conceptual groundwater flow system, and hydrologic conditions 1995–2010 in Florida and parts of Georgia, Alabama, and South Carolina. U.S. Geological Survey Scientific Investigations Report 2018–5030, 103 p.

Bertoldi, G. L. 1989. Groundwater resources of the Central Valley of California. U.S. Geological Survey Open-File Report 89-251.

Boggs, S. 1987. Principles of Sedimentology and Stratigraphy. Merrill Pub. Co., Columbus, Ohio, 784 p.

Bonsor, H. C., A. M. MacDonald, K. M. Ahmed et al. 2017. Hydrogeological typologies of the Indo-Gangetic basin alluvial aquifer, South Asia. Hydrogeology Journal, v. 25, no. 5, p. 1377.

Brahana, J. V., J. Thrailkill, T. Freeman et al. 1988. Carbonate rocks, In W. Back, J. S. Rosenshein, and P. R. Seaber (eds.), Hydrogeology, The Geology of North America. Geological Society of America, Boulder, Colorado, O-2, p. 333–352.

Brakefield, L. K., J. T. White, N. A. Houston, et al. 2015. Updated numerical model with uncertainty assessment of 1950–56 drought conditions on brackish-water movement within the Edwards aquifer, San Antonio, Texas. U.S. Geological Survey Scientific Investigations Report 2015–5081, 54 p

Bredehoeft, J. D., C. E. Neuzil, and P. C. D. Milly. 1983. Regional flow in the Dakota Aquifer, a study of the role of confining layers. U.S. Geological Survey Water-Supply Paper 2237, 45 p.

Cant, D. J. 1982. Fluvial facies models and their application, In P. A. Scholle, and D. Spearing (eds.), Sandstone Depositional Environments. American Association of Petroleum Geologists Memoir 37, Tulsa, p. 115–138.

Chamberlin, T. C. 1885. The requisite and qualifying conditions of artesian wells, In Powell (ed.), Fifth Annual report of the United States Geological Survey to the Secretary of the Interior, 1883–1884. United States Geological Survey, Washington, D.C., p. 125–173.

Darton, N. H. 1909. Geology and underground waters of South Dakota. U.S. Geological Survey Water-Supply Paper 227 p.

Davis, G. H., B. E. Lofgren, and S. Mack. 1964. Use of groundwater reservoirs for storage of surface water in the San Joaquin Valley, California. *U.S. Geological Survey Water-Supply Paper* 1618, 125 p.

Davis, S. N. 1988. Sandstones and shales, In W. Back, J. S. Rosenshein, and P. R. Seaber (eds.), Hydrogeology, The Geology of North America. Geological Society of America, Boulder, Colorado, O-2, p. 323–332.

Dennehy, K. F. 2000. High Plains regional groundwater study. *U.S. Geological Survey Fact Sheet* FS-091-00, 6 p.

DeSimone, L. A., P. B. McMahon, and M. R. Rosen. 2014. The quality of our Nation's waters—Water quality in Principal Aquifers of the United States, 1991–2010. *U.S. Geological Survey Circular* 1360, 151 p.

Domenico, P. A. 1972. *Concepts and Models in Groundwater Hydrology*. McGraw-Hill, New York, 405 p.

Domenico, P. A., and F. W. Schwartz. 1998. *Physical and Chemical Hydrogeology*. John Wiley & Sons, New York, 506 p.

Ely, D. M., E. R. Burns, D. S. Morgan, et al. 2014. Numerical simulation of groundwater flow in the Columbia Plateau Regional Aquifer System, Idaho, Oregon, and Washington (ver. 1.1, January 2015). *U.S. Geological Survey Scientific Investigations Report* 2014–5127, 90 p.

Faunt, C. C., ed. 2009. Groundwater Availability of the Central Valley Aquifer, California. *U.S. Geological Survey Professional Paper* 1766, 225 p.

Ferris, J. G., D. B. Knowles, R. H. Brown, et al. 1962. Theory of aquifer tests. *U.S. Geological Survey Professional Paper* 708, 70 p.

Gale, J. E., Wilson, C. R., Witherspoon, P. A., Wilson, C. R. 1982. Swedish American cooperative program on radioactive waste storage inmined caverns in crystalline rock. SKB technical report no. 49. Lawrence Berkeley Laboratory, Berkeley, CA.

Galloway, D., D. R. Jones, and S. E. Ingebritsen. 1999. Land subsidence in the United States. *U.S. Geological Survey Circular* 1182, 177 p.

Heath, R. C. 1989. Basic groundwater hydrology. *U. S. Geological Survey Water-Supply Paper* 2220, 84 p.

Heilweil, V. M., and L. E. Brooks, eds. 2011. Conceptual model of the Great Basin carbonate and alluvial aquifer system. *U.S. Geological Survey Scientific Investigations Report* 2010-5193, 191 p

Houston, N. A., S. L. Gonzales-Bradford, A. T. Flynn, et al. 2013. Geodatabase compilation of hydrogeologic, remote sensing, and water-budget-component data for the High Plains aquifer, 2011. *U.S. Geological Survey Data Series* 777, 12 p

Kresse, T. M., P. D. Hays, K. R. Merriman, et al. 2014. Aquifers of Arkansas—Protection, management, and hydrologic and geochemical characteristics of groundwater resources in Arkansas. *U.S. Geological Survey Scientific Investigations Report* 2014–5149, 334 p.

Lattman, L. A., and R. R. Parizek. 1964. Relationship between fracture traces and the occurrence of groundwater in carbonate rocks. Journal of Hydrology, v. 2, no. 2, p. 73–91.

Loucks, R. G., M. M. Dodge, and W. E. Galloway. 1984. Regional controls on diagenesis and reservoir quality in Lower Tertiary sandstones along the Texas Gulf Coast, In D. A. McDonald, and R. C. Surdam (eds.), Clastic Diagenesis. American Association of Petroleum Geologists Memoir 37, Tulsa, p. 15–45.

Maupin, M. A., and T. L. Arnold. 2010. Estimates for self-supplied domestic withdrawals and population served for selected Principal Aquifers, calendar year 2005. *U.S. Geological Survey Open-File Report* 2010–1223, 10 p.

Maupin, M. A., and N. L. Barber. 2005. Estimated withdrawals from Principal Aquifers in the United States, 2000. *U.S. Geological Survey Circular* 1279, 46 p.

Meinzer, O. E. 1923. Outlines of groundwater in hydrology with definitions. *U.S. Geological Survey* Water-Supply Paper 494, 71 p.

Musgrove, M., and K. Belitz. 2016. Groundwater quality in the Basin and Range basin-fill aquifers, Southwestern United States. *U.S. Geological Survey Fact Sheet* 2016-3080, 4 p.

Newton, J. G. 1986. Development of sinkholes resulting from man's activities in the eastern United States. *U.S. Geological Survey Circular* 968, 54 p.

O'Donnell, J. 2021. Stream Communities and Ecosystems Resource Brief for the Arctic Network. National Parks Service, https://www.nps.gov/articles/000/stream-communities-ecosystems-resource-brief-for-the-arctic-network.htm (accessed on 16 April, 2023).

Palmer, A. N. 1990. Groundwater processes in karst terranes, In C. G. Higgins, and D. R. Coates (eds.), Groundwater Geomorphology, The Role of Subsurface Water in Earth-Surface Processes and Landforms. Geological Society of America, Boulder. Special Paper, 252, p. 177–209.

Panno, S. V., and W. R. Kelly. 2020. Water Quality in the Mahomet Aquifer: Chemical Indicators of Brine Migration and Mixing. Special Report no. 07, Illinois State Geological Survey, 7 p.

Ranjram, M., Gleeson, T., Luijendijk, E. 2014. Is the permeability of crystalline rock in the shallow crust related to depth, lithology or tectonic setting? Geofluids 15 (1-2): 106–119.

Runkel, A. C., R. G. Tipping, E. C. Alexander Jr. et al. 2003. Hydrogeology of the Paleozoic bedrock in southeastern Minnesota. *Minnesota Geological Survey Report of Investigations* 61, 105p., 2 pls.

Runkel, A. C., R. G. Tipping, J. R. Steenberg et al. 2018. A multidisciplinary-based conceptual model of a fractured sedimentary bedrock aquitard: improved prediction of aquitard integrity. Hydrogeology Journal, v. 26, no. 7, p. 2133–2159.

Roadcap, G. S., H. V. Knapp, H. A. Wehrmann, et al. 2011. Meeting east-central Illinois water needs to 2050: Potential impacts on the Mahomet aquifer and surface reservoirs. Illinois State Water Survey Contract Report 2011-08, 179 p.

Ryder, P. D. 1996. Groundwater atlas of the United States, Segment 4, Oklahoma, Texas. *U.S. Geological Survey Hydrologic Investigations Atlas 730-E.*

Sanford, W. E. 2017. Estimating regional-scale permeability-depth relations in a fractured-rock terrain using groundwater-flow model calibration. Hydrogeology Journal, v. 25, no. 2, p. 405–419.

Schindel, G. M., and M. O. Gary. 2018. The Balcones fault zone segment of the Edwards Aquifer of south-central Texas, In K. W. Stafford, and G. Veni (eds.), Hypogene Karst of Texas. Texas Speleological Survey Monograph 3, Austin, p. 78–85.

Schwartz, F. W., and F. J. Longstaffe. 1988. Groundwater and clastic diagenesis, In W. Back, J. S. Rosenshein, and P. R. Seaber (eds.), Hydrogeology, The Geology of North America. Geological Society of America, Boulder, Colorado, O-2, p. 413–434.

Schwartz, F. W., and H. Zhang. 2003. Fundamentals of Groundwater. John Wiley & Sons, 583 p.

Schmonov, V. M., Vitiovtova, V. M., Zharikov, A. V., Grafchikov, A. A. 2003. Permeability of the continental crust: implications of experimental data. Journal Geochemical Exploration 78–79: May 697–699.

Sinclair, W. C. 1982. Sinkhole development resulting from groundwater development in the Tampa area, Florida. *U.S. Geological Survey Water-Resources Investigations Report* 81-50, 19 p.

Sinclair, W. C., and J. W. Stewart. 1985. Sinkhole type, development, and distribution in Florida: *U.S. Geological Survey Map,* Series 110, 1 plate.

Shukla, U. K., I. B. Singh, M. Sharma et al. 2001. A model of alluvial megafan sedimentation: Ganga Megafan. Sedimentary Geology, v. 144, no. 3–4, p. 243–262.

Smith, J. J., G. A. Ludvigson, A. Layzell et al. 2017. Discovery of Paleogene deposits of the central High Plains aquifer in the western Great Plains, USA. Journal of Sedimentary Research, v. 87, no. 8, p. 880–896.

Snow, D. T. 1968. Rock fracture spacings, openings, and porosity. Journal of the Soil Mechanics and Foundations Division Proceedings of the American Society of Civil Engineers, v. 94, no. 1, p. 73–91.

Stober, I., Bucher, K. 2014. Significance of hydraulic conductivity as precondition to fluid flow in crystalline basement and its impact on fluid-rock interaction processes. Proceedings World Geothermal Congress 2015, Melbourne, Australia, April 2015, 5 pp.

Stringfield, V. T., and H. E. LeGrande. 1966. Hydrology of limestone terranes. *Geological Society of America Special Paper 3.*

Sweetkind, D. S., J. R. Cederberg, M. D. Masbruch et al. 2011. Hydrogeologic framework. Chapter B of, In V. M. Heilweil, and L. E. Brooks (eds.), Conceptual Model of the Great Basin Carbonate and Alluvial Aquifer System. U.S. Geological Survey Scientific Investigations Report 2010–5193. Geological Society of America, Boulder, p. 15–50.

Trainer, F. W. 1988. Plutonic and metamorphic rocks, In W. Back, J. S. Rosenshein, and P. R. Seaber (eds.), Hydrogeology, The Geology of North America. Geological Society of America, O-2, p. 367–380.

Thiros, S. A., L. M. Bexfield, D. W. Anning, et al., eds. 2010. Conceptual understanding and groundwater quality of selected basin-fill aquifers in the Southwestern United States. *U.S. Geological Professional Paper* 1781, 288 p

U.S. Bureau of Reclamation. 1995. *Groundwater Manual.* U.S. Government Printing Office, Washington, D.C., 661 p.

Vaccaro, J. J. 1986. Plan of study for the regional aquifer-system analysis, Columbia Plateau, Washington, northern Oregon, and northwestern Idaho. *U.S. Geological Survey Water-Resources Investigations Report* 85-4151, 25 p.

White, W. B. 1990. Surface and near-surface karst landforms, In C. G. Higgins, and D. R. Coates (eds.), Groundwater Geomorphology, The Role Of Subsurface Water in Earth-Surface Processes and Landforms. Geological Society of America Special Paper 252 p. Geological Society of America, Boulder, p. 157–175.

Williams, L.J., and E. L. Kuniansky. 2016. Revised hydrogeologic framework of the Floridan aquifer system in Florida and parts of Georgia, Alabama, and South Carolina (ver 1.1, March 2016). *U.S. Geological Survey Professional Paper 1807*, 140 p., 23 pls.

Wilson, W. L., and E. J. Shock. 1996. New sinkhole data spreadsheet manual (v1.1). Winter Springs, Fla., Subsurface Evaluations, Inc., 31 p., 3 app., 1 disk.

5

Theory of Groundwater Flow

CHAPTER MENU

In this chapter, we describe the theory of flow in saturated, groundwater systems and develop basic equations of groundwater flow. These equations are fundamental to the quantitative treatment of flow and provide the basis for calculating hydraulic heads, given an idealization of some hydrologic system, boundary, and initial conditions. This chapter also provides simple approaches for solving such equations. For example, flow-net theory provides a straightforward graphical way to determine a hydraulic-head distribution and the resulting pattern of flow, especially for problems having a simple geometry. This chapter also shows how analytical solutions are developed and used for solving problems of steady-state flow.

5.1 Differential Equations of Groundwater Flow in Saturated Zones

What sets hydrogeology apart from many of the other geosciences is an emphasis on treating problems mathematically. For example, one might be interested in calculating how much water levels will fall in the vicinity of a well after 10 years of pumping, or how contaminant concentrations change after 5 years of aquifer remediation. These mathematical approaches also help us interpret measurements made in the field (e.g., aquifer tests and slug tests).

Basically, the mathematical approach involves representing a groundwater process by an equation and solving that equation. Let us illustrate this idea with the simple groundwater flow problem shown in Figure 5.1*a*. For this two-dimensional section, assume we know the pattern of layering, the hydraulic conductivity of the various units, and the configuration of the water table. Given this information, can one calculate what the pattern of flow would look like? Developing this problem from a mathematical viewpoint requires (1) finding and using the appropriate equation to describe the flow of groundwater, (2) establishing a domain or region where the equation is to be solved, and (3) defining flow conditions along the boundaries (the so-called boundary conditions) (Figure 5.1*b*). With this information, we can calculate the hydraulic head at a large number of specified locations (x and z) within the domain. In principle, this step is like taking readings from a large number of hypothetical piezometers. Contouring the hydraulic head distribution provides the equipotential distribution, from which we can deduce the patterns of flow (Figure 5.1*c*). This simple example helps to highlight some of the new knowledge that is required for the quantitative treatment of groundwater flow.

Fundamentals of Groundwater, Second Edition. Franklin W. Schwartz and Hubao Zhang.

Companion website: www.wiley.com/go/schwartz/fundamentalsofgroundwater2

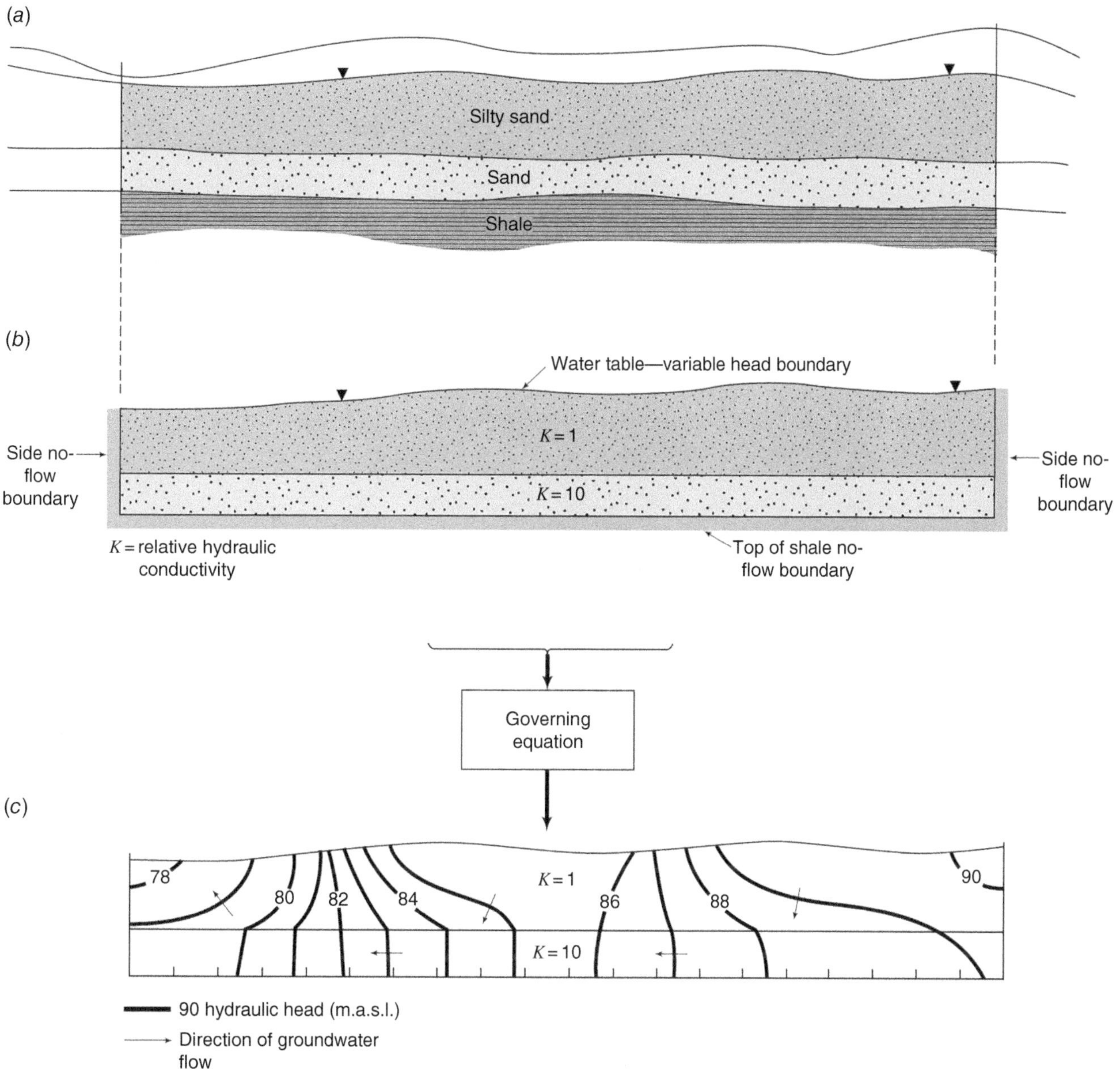

Figure 5.1 A geological problem (*a*) is conceptualized as a formal mathematical problem (*b*), which provides the basis for calculating the hydraulic head distribution (*c*) (Domenico and Schwartz, 1998. Physical and Chemical Hydrogeology. Copyright © 1990, 1998 John Wiley & Sons Inc. All Rights Reserved. Reproduced with permission).

5.1.1 Useful Knowledge About Differential Equations

Many students of groundwater hydrology find dealing with the quantitative aspects of this subject. At first glance, a differential equation describing groundwater flow immediately implies a need to understand advanced mathematical concepts

$$\frac{\partial}{\partial x}\left(T_x\frac{\partial h}{\partial x}\right)+\frac{\partial}{\partial y}\left(T_y\frac{\partial h}{\partial y}\right)+\frac{\partial}{\partial z}\left(T_z\frac{\partial h}{\partial z}\right)=S\frac{\partial h}{\partial t} \tag{5.1}$$

Fortunately, at this introductory level, one does not need to do much with this equation except to recognize it. The basics of hydrogeology are set up to provide simplified approaches for dealing with these equations.

Let us consider the idea that these equations contain recognizable features. Even as toddlers, we all could recognize things. Shown a picture of a rhinoceros we could say—yes, that is a rhino because of the four stubby legs and two horns on an ugly-looking head. In looking at this picture, one would not become consumed with trying to figure out why the legs were stubby and why there were two horns. Let us consider Eq. (5.1). What are its distinguishing features? Well, first, like all equations, it has some unknown that we are trying to evaluate. The unknown is hidden in the derivative terms, for example

$$\frac{\partial h}{\partial x} \text{ or } \frac{\partial h}{\partial y} \text{ or } \frac{\partial h}{\partial z} \tag{5.2}$$

where h is the unknown. In other words, a solution to Eq. (5.1) will be of the form

$$h = \text{...stuff} \tag{5.3}$$

where the "stuff" on the right-hand side are terms that are simple functions of the time and space variables (t, x, y, and z), and various parameters like T and S. For a solution to exist, all the terms on the right-hand side need to be known. Fortunately, for most of our applications, the "stuff" is algebraic in form and easy to evaluate.

Here are some simple steps in examining a differential equation. First look at the equation and determine the unknown. This step is straightforward—an equation containing

h: (hydraulic head) makes the equation a groundwater flow equation,
C: (concentration) makes the equation a mass transport equation, and
T: (temperature) makes the equation an energy-transport equation.

Thus, Eq. (5.1) with h as the unknown is a groundwater flow equation. We know that it is used to apply to aquifers because it also contains the expected hydraulic parameters (T and S). Similarly, a mass transport equation will contain parameters (e.g., D, a dispersion coefficient) related to processes involved with mass transport.

Sometimes, these descriptive parameters (K, T, and S) are not there. What does this mean? A flow equation with no K, T, or S values, implies that K, T, and S everywhere were the same, and for steady-state problems, their values have no bearing on the solution. In other words, the calculated hydraulic head distribution does not depend on the parameter values. In transient problems, the parameter values remain even when values are constant within the domain. In this case, the hydraulic-head values depend on the parameter values.

Equation (5.1) has other distinguishing features. For example, you can look at the space variables to determine the dimensionality of the problem. The dimensionality of a problem describes in how many directions the unknown (hydraulic head) is changing. For example, in Eq. (5.1), there are three space variables, x, y, and z. Three space variables make this equation a three-dimensional equation. Later in this chapter, you will encounter one-dimensional flow equations that imply that values of hydraulic head change in one direction but not the other two.

Next, you need to decide whether hydraulic head changes with time. To figure this out, look and see whether there is a time variable (t) in the equation. In Eq. (5.1), t is there. With hydraulic head changing with time, the equation of flow is transient. If there is no t term, the equation describes a steady-state problem, where hydraulic head does not change with time. A partial differential equation is a concise way to represent hydrogeological processes. By looking at the unknown, parameters, dimensionality, and transient nature, you will understand something about the problem that the equation is trying to portray. Such equations are still difficult to solve, but by looking at a partial differential equation, you will discover plenty of useful information. The following example illustrates how to use these ideas.

Example 5.1 Shown below are two different differential equations that can be applied to groundwater problems. Look at each equation and determine what kind of problem it applies to, what dimensionality is involved, and whether the equation is a transient or steady-state form.

$$\frac{\partial^2 h}{\partial x^2} = 0$$

The unknown is h, therefore it is a groundwater-flow equation. Only one space dimension (x) is included, therefore it is one-dimensional. There is no time term, therefore the equation is a steady-state form. There are no parameters (like K), therefore you can conclude the hydraulic-head distribution, in this case, does not depend on the parameter values.

$$D\frac{\partial^2 C}{\partial x^2} - v_x\frac{\partial C}{\partial x} = \frac{\partial C}{\partial t}$$

The unknown in this equation is C, therefore it is a mass transport equation. With x as the only space dimension and t present it is a one-dimensional transient equation.

5.1.2 More About Dimensionality

This chapter has made the point that the solution to groundwater flow equations describes how hydraulic head varies within some domain or region of interest. Yet, it is not exactly clear how the dimensionality of the equation matches the dimensionality of the flow region. For example, is it necessary to use a three-dimensional form of a flow equation to calculate hydraulic head values in a three-dimensional domain.

In general, there is no requirement that the dimensionality of the equation is the same as the dimensionality of the region. A one-dimensional equation could be applied to a three-dimensional domain. The number of directions in which the groundwater actually moves determines the dimensionality of the equation. For any problem, it is the combination of the region shape along with the boundary conditions and/or heterogeneity, which determines how the groundwater is likely to move and the dimensionality of the equation.

Let us consider some simple examples. Figure 5.2*a* sets up a problem that we will be coming back to later in this chapter. Two parallel rivers fully penetrate a confined aquifer that receives no recharge. The river stages are assumed to be constant but different than each other. This difference in stage sets up a flow through the aquifer from one river to the other. As the arrows in Figure 5.2*b* imply, flow through the groundwater system is only one direction. Thus, this problem can be represented mathematically using a one-dimensional flow equation, even though the aquifer itself has three dimensions.

With a few changes, it is not hard to make the flow be more complicated and require that the dimensionality of governing equation be increased. For example, if the river did not fully penetrate the aquifer, flows out of and into the rivers would have components in both the x and z directions requiring a two-dimensional flow equation. Adding a high permeability lens in the middle of the aquifer would produce flow in all three coordinate directions, requiring a three-dimensional flow equation.

5.1.3 Deriving Groundwater Flow Equations

The differential equations for groundwater flow are developed from principles of mass conservation in a representative elementary volume (REV) (Figure 5.3). In words, such a conservation statement can be written as

$$\text{mass inflow rate} - \text{mass outflow rate} = \text{change of mass storage with time} \tag{5.4}$$

(*a*)

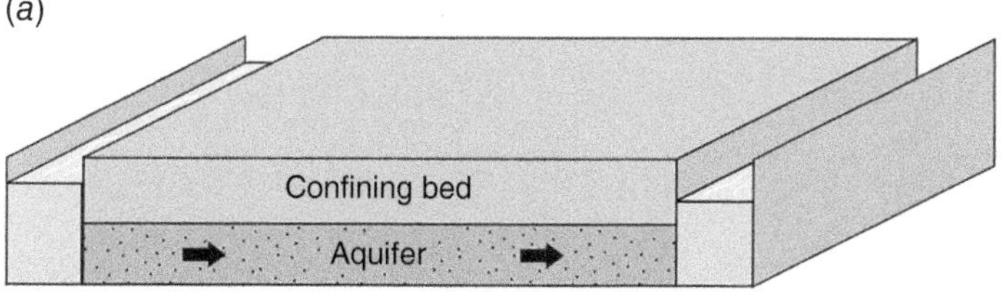

(*b*)

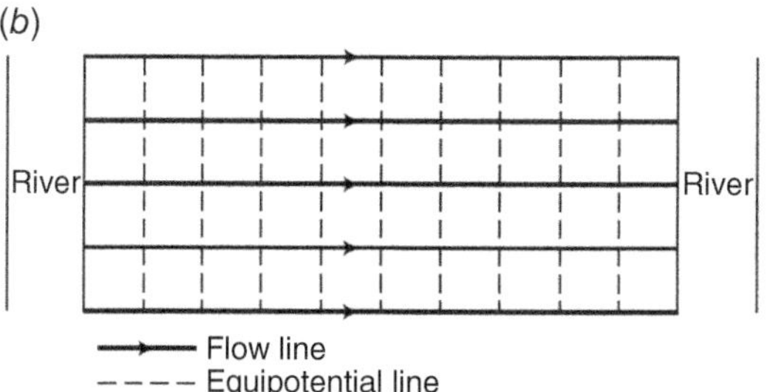

Figure 5.2 Panel (*a*) depicts groundwater flow in an aquifer to inflow and outflow from rivers. Panel (*b*) suggests that although the system is three-dimensional, hydraulic head varies in only one direction. Thus, flow is described by one-dimensional flow equation (Schwartz and Zhang, 2003 / John Wiley & Sons).

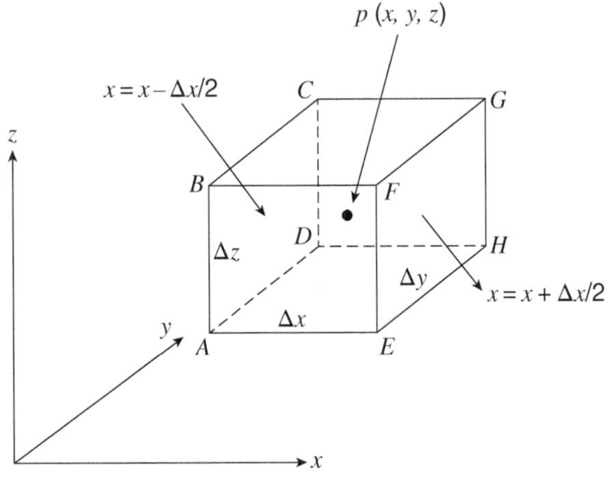

Figure 5.3 Conservation principles applied in relation to this representative elementary volume provide the basis for the development of groundwater flow equations (Schwartz and Zhang, 2003 / John Wiley & Sons).

A REV is defined as a volume exhibiting the average properties of the porous media around a point $P(x, y, \text{and } z)$, which is the center of the volume. For reference purposes, let us label the six faces of the REV as x_1, x_2, y_1, y_2, z_1, and z_2. Now, assume that the water mass fluxes through the six faces are $M_{x1}, M_{x2}, M_{y1}, M_{y2}, M_{z1}$, and M_{z2}, respectively. Furthermore, ρ_w is the fluid density $[ML^{-3}]$ and n is the porosity in the REV. Equation (5.4) may be rewritten in mathematical terms as

$$M_{x1} - M_{x2} + M_{y1} - M_{y2} + M_{z1} - M_{z2} = \frac{\partial}{\partial t}(n\rho_w \Delta x \Delta y \Delta z) \tag{5.5}$$

The mass flux of water in the direction i is expressed as

$$M_i = \rho_w q_i \Delta S_i$$

where q_i is the i (x, y, and z) component of the Darcy velocity vector, and ΔS_i is the area perpendicular to the flow direction i. To obtain the mass inflow and outflow rates in the x direction for the REV, we derive an expression for $(\rho_w q_x)$ across the faces x_1 and x_2. A Taylor's expansion series of $(\rho_w q_x)$ at the outflow face x_2 is given by (Thomas, 1972) as

$$(\rho_w q_x)|x = x + \Delta x/2 = (\rho_w q)\,|x = x + \frac{\partial(\rho_w q_x)}{\partial x}|x = x\left(\frac{\Delta x}{2}\right) + \frac{1}{2}\frac{\partial^2(\rho_w q_x)}{\partial x^2}|x = x\left(\frac{\Delta x}{2}\right)^2 + \cdots \tag{5.6}$$

Similarly, the Taylor's expansion series at the inflow face x_1 is given by

$$(\rho_w q_x)\,|x = x - \Delta x/2 = (\rho_w q)\,|x = x - \frac{\partial(\rho_w q_x)}{\partial x}|x = x\left(\frac{\Delta x}{2}\right) + \frac{1}{2}\frac{\partial^2(\rho_w q_x)}{\partial x^2}|x = x\left(\frac{\Delta x}{2}\right)^2 + \cdots \tag{5.7}$$

By neglecting the terms of higher orders, the mass flow rate out of the REV in x direction is expressed as

$$M_{x2} = \left[\rho_w q_x + \frac{1}{2}\frac{\partial(\rho_w q_x)\Delta x}{\partial x}\right]\Delta y \Delta z \tag{5.8}$$

and the mass inflow rate is rewritten as

$$M_{x1} = \left[\rho_w q_x - \frac{1}{2}\frac{\partial(\rho_w q_x)\Delta x}{\partial x}\right]\Delta y \Delta z \tag{5.9}$$

The net inflow rate into the REV in x direction is the difference between the inflow and the outflow rates.

$$M_{x1} - M_{x2} = -\frac{\partial(\rho_w q_x)\Delta x \Delta y \Delta z}{\partial x} \tag{5.10}$$

Similarly, the net flow rates into the REV in y and z directions are

$$M_{y1} - M_{y2} = -\frac{\partial\left(\rho_w q_y\right)\Delta x \Delta y \Delta z}{\partial y} \tag{5.11}$$

$$M_{z1} - M_{z2} = -\frac{\partial(\rho_w q_z)\Delta x \Delta y \Delta z}{\partial z} \tag{5.12}$$

The sum of water inflow rate minus the sum of water outflow rate for the REV is

$$M_{x1} - M_{x2} + M_{y1} - M_{y2} + M_{z1} - M_{z2} = -\left[\frac{\partial(\rho_w q_x)}{\partial x} + \frac{\partial\left(\rho_w q_y\right)}{\partial y} + \frac{\partial(\rho_w q_z)}{\partial z}\right]\Delta x\Delta y\Delta z \tag{5.13}$$

The change in groundwater storage within the REV is

$$\text{change of water storage per unit time} = \frac{\partial(\rho_w n)}{\partial t}\Delta x\Delta y\Delta z \tag{5.14}$$

According to Eq. (5.4), the net rate of water inflow is equal to the change in storage. Collecting Eqs. (5.13) and (5.14) and dividing both sides by $\Delta x\Delta y\Delta z$ gives

$$-\left[\frac{\partial(\rho_w q_x)}{\partial x} + \frac{\partial\left(\rho_w q_y\right)}{\partial y} + \frac{\partial(\rho_w q_z)}{\partial z}\right] = \frac{\partial(\rho_w n)}{\partial t} \tag{5.15}$$

By making a further assumption that the density of the fluid does not vary spatially, the density term on the left-hand side can be taken out as a constant so that Eq. (5.15) becomes

$$-\left[\frac{\partial q_x}{\partial x} + \frac{\partial q_y}{\partial y} + \frac{\partial q_z}{\partial z}\right] = \frac{1}{\rho_w}\frac{\partial(\rho_w n)}{\partial t} \tag{5.16}$$

The right side of the Eq. (5.16) is related to the specific storage of an aquifer by

$$\frac{1}{\rho_w}\frac{\partial(\rho_w n)}{\partial t} = S_s\frac{\partial h}{\partial t} \tag{5.17}$$

where S_s is the specific storage, and h is the hydraulic head.

Let us recall Eqs. (3.33) through (3.35), which relate the Darcy velocity to the hydraulic head. Making use of these relationships, Eq. (5.16) becomes

$$\frac{\partial}{\partial x}\left(K_x\frac{\partial h}{\partial x}\right) + \frac{\partial}{\partial y}\left(K_y\frac{\partial h}{\partial y}\right) + \frac{\partial}{\partial z}\left(K_z\frac{\partial h}{\partial z}\right) = S_s\frac{\partial h}{\partial t} \tag{5.18}$$

Equation (5.18) is the main equation of groundwater flow in saturated media. It can be written in many forms that apply to a variety of different conditions. Here are some of these alternative equations and the conditions under which they apply.

1) Under steady-state flow conditions ($\frac{\partial h}{\partial t} = 0$), Eq. (5.18) simplifies to

$$\frac{\partial}{\partial x}\left(K_x\frac{\partial h}{\partial x}\right) + \frac{\partial}{\partial y}\left(K_y\frac{\partial h}{\partial y}\right) + \frac{\partial}{\partial z}\left(K_z\frac{\partial h}{\partial z}\right) = 0 \tag{5.19}$$

If the porous medium is isotropic (K_x, K_y, and K_z) and homogeneous ($K_{x,y,z}$ = constant), Eq. (5.19) simplifies the well-known Laplace equation

$$\frac{\partial^2 h}{\partial x^2} + \frac{\partial h}{\partial y^2} + \frac{\partial^2 h}{\partial z^2} = 0 \tag{5.20}$$

2) With the same assumptions about isotropicity and homogeneity, Eq. (5.18) can be rewritten as

$$\frac{\partial^2 h}{\partial x^2} + \frac{\partial^2 h}{\partial y^2} + \frac{\partial^2 h}{\partial z^2} = \frac{S}{K_s}\frac{\partial h}{\partial t} \tag{5.21}$$

3) By dividing both sides of Eq. (5.18) by S_s, the equation is transformed into

$$\frac{\partial}{\partial x}\left(\frac{K_x}{S_s}\frac{\partial h}{\partial x}\right)+\frac{\partial}{\partial y}\left(\frac{K_y}{S_s}\frac{\partial h}{\partial y}\right)+\frac{\partial}{\partial z}\left(\frac{K_z}{S_s}\frac{\partial h}{\partial z}\right)=\frac{\partial h}{\partial t} \tag{5.22}$$

where K_x/S_s, K_y/S_s, and K_z/S_s are called hydraulic diffusivities in x, y, and z directions, respectively. A constant specific storage is assumed in Eq. (5.22). Writing the equation in this form shows that the groundwater flow equation is a form of diffusion equation in which the hydraulic diffusivities and hydraulic gradients in x, y, and z directions are the determining factors.

4) Multiplying both sides of the Eq. (5.14) by the aquifer thickness (b) gives

$$\frac{\partial}{\partial x}\left(T_x\frac{\partial h}{\partial x}\right)+\frac{\partial}{\partial y}\left(T_y\frac{\partial h}{\partial y}\right)+\frac{\partial}{\partial z}\left(T_z\frac{\partial h}{\partial z}\right)=S\frac{\partial h}{\partial t} \tag{5.23}$$

where T is transmissivity, and S is storativity.

$$T_x = K_x b, T_y = K_y b, T_z = K_z b, S = S_s b \tag{5.24}$$

This form of the groundwater flow equation is solved in numerical models like MODFLOW to predict hydraulic head changes due to pumping in complex systems of aquifers and confining beds.

5) If there is no change of hydraulic head in the vertical direction, Eq. (5.23) is simplified to a two-dimensional groundwater flow equation of the following form

$$\frac{\partial}{\partial x}\left(T_x\frac{\partial h}{\partial x}\right)+\frac{\partial}{\partial y}\left(T_y\frac{\partial h}{\partial y}\right)=S\frac{\partial h}{\partial t} \tag{5.25}$$

6) The next equation relies on simplifications stemming from Dupuit's assumption. In words, the direction of groundwater flow is assumed horizontal because the vertical hydraulic gradient is small and negligible (Figure 5.4), the differential equation for groundwater flow in an unconfined aquifer is

$$\frac{\partial}{\partial x}\left(K(x,y)h\frac{\partial h}{\partial x}\right)+\frac{\partial}{\partial y}\left(K(x,y)h\frac{\partial h}{\partial y}\right)=S_y\frac{\partial h}{\partial t} \tag{5.26}$$

where h is the hydraulic head, $K(x,y)$ is the average hydraulic conductivity, and S_y is the specific yield of the water table aquifer. Equation (5.26) is also known as Boussinesq's equation. The equation only applies to groundwater regions where the vertical hydraulic gradient is very small. For general problems of flow in an unconfined aquifer, Eq. (5.14) should be used.

7) If a groundwater source exists, Eq. (5.18) is written as

$$\frac{\partial}{\partial x}\left(K_x\frac{\partial h}{\partial x}\right)+\frac{\partial}{\partial y}\left(K_y\frac{\partial h}{\partial y}\right)+\frac{\partial}{\partial z}\left(K_z\frac{\partial h}{\partial z}\right)+Q(x,y,z,t)=S_s\frac{\partial h}{\partial t} \tag{5.27}$$

where Q is volumetric source rate per unit volume $[L^3T^{-1}\ L^{-3}]$. Equation (5.27) may be expressed in words as

$$\text{Inflow rate} - \text{outflow rate} + \text{source rate} = \text{change of storage} \tag{5.28}$$

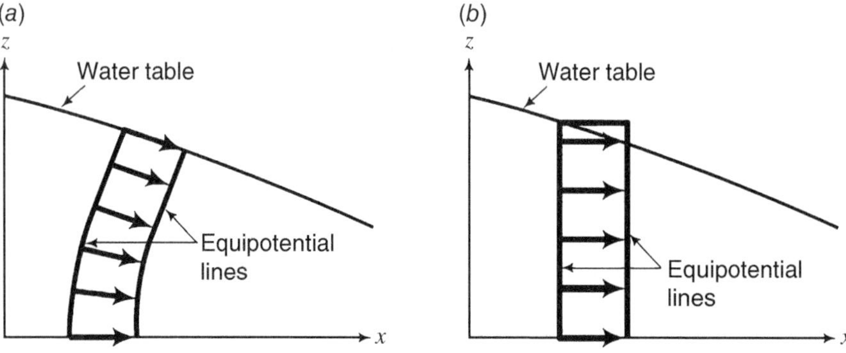

Figure 5.4 This figure illustrates how Dupuit's assumption simplifies flow in an unconfined aquifer. Panel (*a*) depicts the actual flow direction. Panel (*b*) depicts the flow with Dupuit's assumption (Schwartz and Zhang, 2003 / John Wiley & Sons).

5.2 Boundary Conditions

In order to solve a groundwater flow equation, it is necessary to specify boundary conditions. When you define a simulation domain, (e.g., Figure 5.1*b*) you are selecting a small piece of some larger hydrologic system for detailed analysis. Unfortunately, the conditions outside of the domain can influence what is going on inside the domain. The job of the *boundary conditions* is to carry information as to how the simulation domain is impacted by flow conditions outside of the simulation domain. In other words, boundary conditions are the price you pay for attempting to analyze just a piece of a large, continuous system.

Boundary conditions for groundwater-flow problems are one of three types:

1) First-type boundary condition (Dirichlet boundary condition):
   The Dirichlet boundary condition provides a value of hydraulic head at a boundary. Mathematically, this boundary condition can be expressed as

$$h(x,y,z)|_{\Gamma} = h_1(x,y,z,t), (x,y,z) \in \Gamma \tag{5.29}$$

   where $h_1(x,y,z)$ is the specified value of hydraulic head at the boundary Γ. For example, a constant head of 120 m is specified at the face EFGH (Figure 5.5*a*). Physically, a river, a lake, or other locations where hydraulic head in the system is known can provide a constant head boundary.

2) Second-type boundary condition (Neumann boundary condition):
   The Neumann boundary condition gives the water flux at a boundary. This boundary condition is written as

$$q_n|_{\Gamma} = q_1(x,y,z,t), (x,y,z) \in \Gamma \tag{5.30}$$

   or

$$Q_n|_{\Gamma} = Q_1(x,y,z,t), (x,y,z) \in \Gamma \tag{5.31}$$

   where n is an outward direction normal to the boundary, q_n is the outflow Darcy velocity [L/T], Q_n is the volumetric outflow rate [L^3/T], q_1 (x, y, z) is the specified outflow Darcy velocity [L/T], and Q_1 is the specified outflow volumetric flux rate [L^3/T]. An injection well or withdrawal well can also be considered as an inner boundary condition of second type. In this case, the boundary is the wall of the well. An example of a second-type boundary is found at the top of an aquifer where there is recharge or discharge. A no-flow boundary ($q_n = 0$) is a special case of a second-type boundary. Such a condition could occur at the boundary between an aquifer and a low permeability unit, at groundwater divides and boundaries along flow lines where there is no flow perpendicular to the flow line.

3) Third-type boundary condition (Cauchy boundary condition):
   The Cauchy boundary condition relates hydraulic head to the water flux and is expressed as

$$q_n = \frac{K_m}{M}[h(x,y,z,t) - h_m(x,y,z,t)], (x,y,z) \in \Gamma \tag{5.32}$$

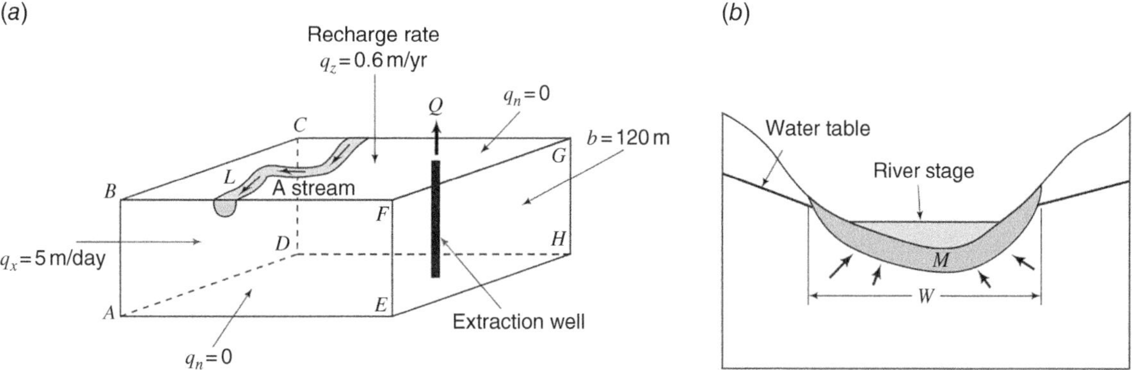

Figure 5.5 This figure illustrates how boundary conditions are applied at the sides and internal to the flow region. Panel (*a*) illustrates Dirichlet and Neumann conditions. Panel (*b*) illustrates a Cauchy boundary condition at the interface between groundwater and surface-water systems (Schwartz and Zhang, 2003 / John Wiley & Sons).

or

$$Q_n = C_m[h(x,y,z,t) - h_m(x,y,z,t)], (x,y,z) \in \Gamma \tag{5.33}$$

where q_n is the outflow Darcy velocity [L/T], Q_n is the volumetric outflow rate [L^3/T], K_m is the hydraulic conductivity of the boundary, and M is the thickness of the boundary, C_m is the conductance of the boundary, h_m is the hydraulic head outside the boundary, and h is the hydraulic head inside the boundary.

One example where a Cauchy boundary condition applies is for stream-aquifer interaction (Figure 5.5*b*). The Darcy velocity from an aquifer to a stream is

$$q_n = \frac{K_m}{M}(h - h_m) \tag{5.34}$$

The volumetric discharge rate from the aquifer to the stream is

$$Q_n = \frac{LW}{M}K_m(h - h_m) \tag{5.35}$$

where K_m is the hydraulic conductivity of the stream bed, L is the length of the stream bed, W is the width of the stream bed, M is the thickness of the stream bed, h is the hydraulic head in the aquifer, and h_m is the river stage.

Example 5.2 Determine the type of the boundary conditions in Figure 5.5*a* and 5.5*b*. Assuming that $L = 500$ m, $W = 20$ m, $M = 0.5$ m, $K_m = 0.005$ m/day, $h = 20$ m, and $h_m = 18$ m, what are the Darcy velocity and the volumetric discharge rate from the aquifer to the stream?

Solution

Dirichlet boundary conditions:

The constant head on the right side of the box: $h = 120$ m

Neumann boundary conditions:

No flow boundaries on front and back sides: $q_n = 0$
Inflow velocity on the left side: $q_x = 5$ m/day
Recharge rate on the top: $q_z = 0.6$ m/yr
Volumetric flow rate in the extraction well in the box: $Q = 20$ gpm

The Darcy velocity and volumetric discharge from the aquifer to the stream are

$$q_n = \frac{K_m}{M}(h - h_m) = \frac{(0.005\ \text{m/day})}{(0.5\ \text{m})}[(20\ \text{m}) - (18\ \text{m})] = 0.02\ \text{m/day}$$

and

$$Q_n = \tfrac{LW}{M}K_m(h - h_m) = \frac{(500\ \text{m})(20\ \text{m})}{(0.5\ \text{m})}(0.005\ \text{m/day})[(20\ \text{m}) - (18\ \text{m})] = 200\ \text{m}^3/\text{day}$$

5.3 Initial Conditions for Groundwater Problems

For solving steady-state equations of groundwater flow, only boundary conditions need to be specified. For transient equations, initial conditions also have to be specified. The *initial condition* provides the hydraulic head everywhere within the domain of interest before the simulation begins (i.e., $t = 0$). The initial condition is written as

$$h(x,y,z,0) = h_0(x,y,z) \tag{5.36}$$

where h_0 is initial hydraulic head in the domain considered. Equation (5.36) states that the hydraulic head at any point (x, y, and z) at time zero should be set equal to h_0. In modeling the response of an aquifer due to pumping, the initial condition would be the original head distribution in the aquifer before pumping began.

5.4 Flow-net Analysis

Having specified a flow equation and boundary/initial conditions, the next logical step is to solve the equation. One of the simplest procedures is a graphical approach that lets sketch the unique set of streamlines and equipotential lines that describe flow within a domain. The *streamlines* (or flow lines) indicate the path followed by a particle of water as it moves through the aquifer in the direction of decreasing head. The *equipotential lines* representing the contours of equal head in the aquifer intersect the streamlines.

5.4.1 Flow Nets in Isotropic and Homogeneous Media

A flow net for an isotropic and homogeneous system provides a graphical solution to the Laplace equation. Recall that the Laplace equation describes steady-state flow in an isotropic and homogeneous media and is written for two dimensions as

$$\frac{\partial^2 h}{\partial x^2} + \frac{\partial^2 h}{\partial y^2} = 0 \tag{5.37}$$

In order to construct a flow net, you need to understand the general features of flow in a two-dimensional domain. These principles form the basis for the set of "rules," which provide guidance in sketching the streamlines and equipotential lines. The distribution of equipotential lines describe the hydraulic head in the domain, which is the unknown in Eq. (5.31).

1) Streamlines are perpendicular to the equipotential lines. If the hydraulic-head drops between the equipotential lines are the same, the streamlines and equipotential lines form curvilinear squares. A *curvilinear square* has curved sides that are tangent to an inscribed circle (Figure 5.6).
2) The same quantity of groundwater flows between adjacent pairs of flow lines provided no flow enters or leaves the region in the internal part of the net. It follows then that the number of flow channels (also known as stream tubes) must remain constant throughout the net.
3) The hydraulic head drop between two adjacent equipotential lines is the same.

Figure 5.6 illustrates these principles with flow in an x–z plane through a pervious rock unit beneath a dam. Flow occurs because hydraulic head of water in the pool above the dam is higher than in the pool below the dam. The bottom surface of the reservoir can be taken as an equipotential line, that is, a constant head boundary across which the flow is directed downward. The bottom of the flow net is the boundary between the impermeable and permeable rock unit. This boundary is an example of a no-flow boundary. Because flow right next to a no-flow boundary must be parallel to the boundary, a no-flow boundary is a flow line. Similarly, the base of the dam is also a flow line. The pool below the dam receives discharge and provides another constant-head boundary or another equipotential line. Because the system is isotropic and homogeneous, flow lines and equipotential lines intersect to form curvilinear squares. Given all this information, the flow net is a theoretical representation of flow beneath the dam.

Bennett (1962) provides strategies for learning to sketch a flow net, which were developed from a paper by Casagrande (1937).

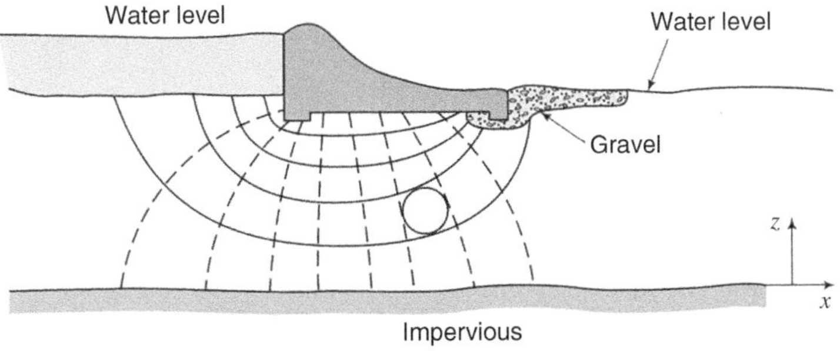

Figure 5.6 Here is an example of a flow net depicting seepage under a dam. It illustrates how flow lines and equipotential lines meet at right angles and form curvilinear squares (Domenico and Schwartz, 1998. Physical and Chemical Hydrogeology. Copyright © 1990, 1998 John Wiley & Sons Inc. All Rights Reserved. Reproduced with permission).

1) Study the appearance of well-constructed flow nets and try to duplicate them by independently reanalyzing the problems that they represent.
2) In a first attempt at sketching, use only four or five flow channels.
3) Observe the appearance of the entire flow net; do not try to adjust details until the entire net is approximately correct.
4) Frequently parts of a flow net consist of straight and parallel lines, which result in uniformly sized true squares. By starting the sketching in such areas, the solution can be obtained more readily.
5) In flow system having symmetry, only a section of the net needs to be constructed, as the other part are images of that section.
6) During the sketching of the net, keep in mind that the size of the rectangle changes gradually; all transitions are smooth and where the paths are curved, are of elliptical or parabolic shape.

It is very useful to keep several simple flow cases in mind before you begin studying complex flow systems. Here are few rules that should help you in preparing a flow net.

1) A no-flow boundary is a streamline.
2) The water table is a streamline, when there is no flow across the water table, that is no recharge or evapotranspiration. When there is recharge, the water table is neither a flow line nor an equipotential line.
3) Streamlines end at extraction wells, drains, and gaining streams, and start from injection wells and losing streams.
4) Lines dividing a flow system into two symmetric parts are streamlines.
5) In natural groundwater systems, streamlines often begin and end at the water table in areas of groundwater recharge and discharge, respectively.

Flow nets usually are drawn only for a two-dimensional flow field. Applying Darcy's equation to one flow channel in the two-dimensional space (Figure 5.7), we get

$$\Delta Q = T\Delta h \frac{\Delta W}{\Delta L} \tag{5.38}$$

where ΔQ is the flow rate between the equipotential lines in the flow channel, T is the transmissivity of the aquifer, Δh is the hydraulic head drop between two equipotential lines, ΔW is the width of the flow channel, and the ΔL is the distance between two equipotential lines. If the equipotential lines and streamlines result in "squares" in a flow net, $\Delta W/\Delta L$ is equal to one. If there are a total number of n_f flow channels and n_d hydraulic head drop, Darcy's equation is written as

$$Q = \frac{n_f}{n_d} T\Delta H \tag{5.39}$$

where Q is the total flow rate and is equal to $n_f\Delta Q$, and ΔH is the total hydraulic head drop and is equal to $n_d\Delta h$.

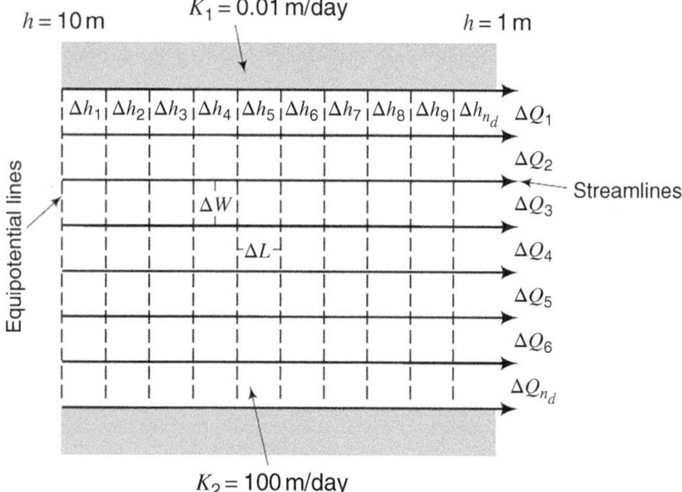

Figure 5.7 Streamlines and equipotential lines under homogeneous and isotropic conditions. The flow is due to the constant heads applied at the left and right sides of the domain (Schwartz and Zhang, 2003 / John Wiley & Sons).

Example 5.3 Bennett and Meyer (1952) found that the average discharge due to groundwater withdrawals from the Patuxent Formation in the Sparrows Point district in 1945 was 1 million ft^3/day (Figure 5.8). Calculate the transmissivity of the formation.

Solution

The number of flow tubes surrounding this area of pumping is 15. The number of hydraulic head drops is 3 between the contour lines 30 and 60 ft for a contour interval of 10 ft. From Eq. (5.39), we get

$$T = \frac{n_d Q}{n_f \Delta H} = \frac{(3)\left(10^6\ \text{ft}^3/\text{day}\right)}{(15)(30\ \text{ft})} = 6670\ \text{ft}^2/\text{day}$$

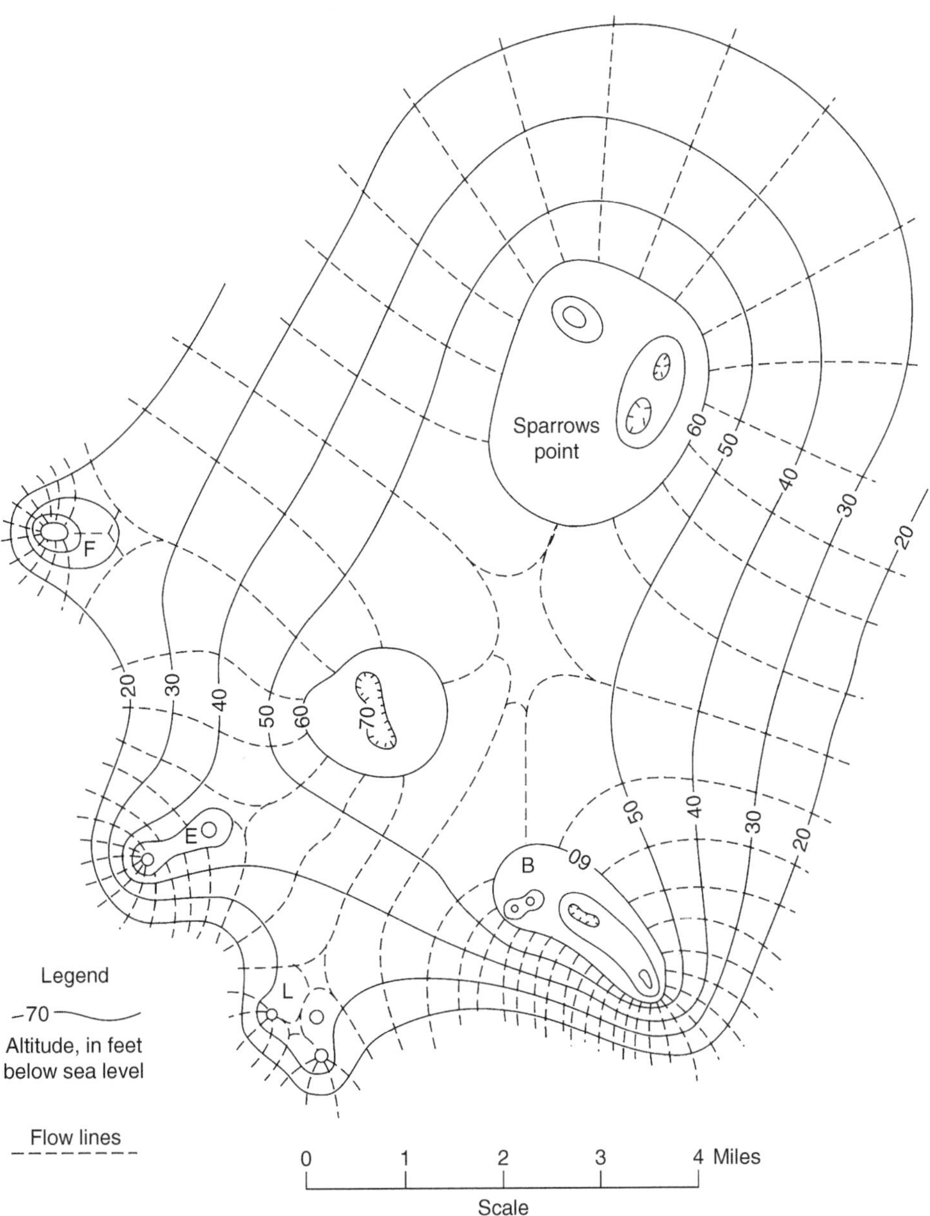

Figure 5.8 Flow net for the Patuxent Formation (Bennett and Meyer, 1952 / Maryland Geological Survey / Public Domain).

5.4.2 Flow Nets in Heterogeneous Media

Equipotential lines and flow lines do not necessarily intersect to form "squares" in heterogeneous media. When flow traverses two adjacent media with differing hydraulic conductivities (Figure 5.9), a condition of constant discharge in a stream tube provides

$$\Delta Q = T_1 \Delta h_1 \frac{\Delta W_1}{\Delta L_1} = T_2 \Delta h_2 \frac{\Delta W_2}{\Delta L_2} \tag{5.40}$$

For the same hydraulic head drop ($\Delta h_1 = \Delta h_2$), Eq. (5.40) is written as

$$\frac{T_1}{T_2} = \frac{\Delta L_1 \Delta W_2}{\Delta L_2 \Delta W_1} \tag{5.41}$$

If the width of the stream tube remains constant, the segment length of the stream tube in the higher transmissivity zone is longer than that in the lower transmissivity zone.

Example 5.4 With reference to Figure 5.9, calculate the flow rate and the length ΔL_2. Assume that $\Delta W_1 = \Delta W_2 = \Delta L_1 = 10$ m and that the thickness of the aquifer is 50 m.

Solution

Flow rate is

$$\Delta Q = T \Delta h \frac{\Delta W}{\Delta L} = (100\ \text{m/day})(50\ \text{m})(1\ \text{m}) \frac{(10\ \text{m})}{(10\ \text{m})} = 5000\ \text{m}^3/\text{day}$$

The segment length (ΔL_2) is

$$\Delta L_2 = \frac{T_2}{T_1} \Delta L_1 = \frac{(200\ \text{m/day})(50\ \text{m})}{(100\ \text{m})(50\ \text{m})}(10\ \text{m}) = 20\ \text{m}$$

When flow crosses interface between media with differing hydraulic conductivities, it is refracted according to the tangent law (see Eq. 3.40) or

$$\frac{K_1}{K_2} = \frac{\tan(\alpha_1)}{\tan(\alpha_2)} \tag{5.42}$$

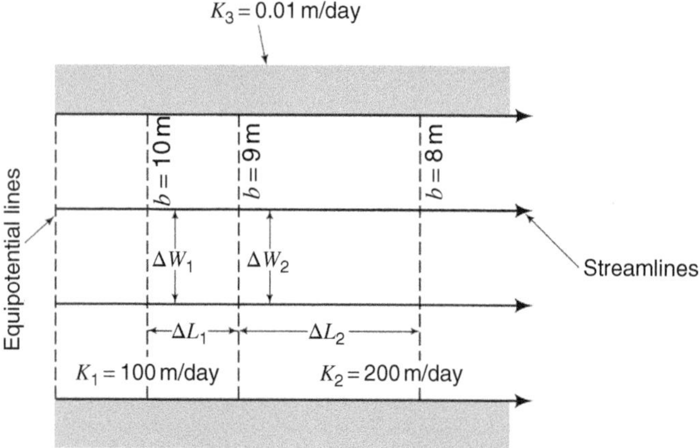

Figure 5.9 An example of flow net in heterogeneous media with a change in hydraulic conductivity in the direction of flow. Note how the distance between the equipotential lines changes across this boundary (Schwartz and Zhang, 2003 / John Wiley & Sons).

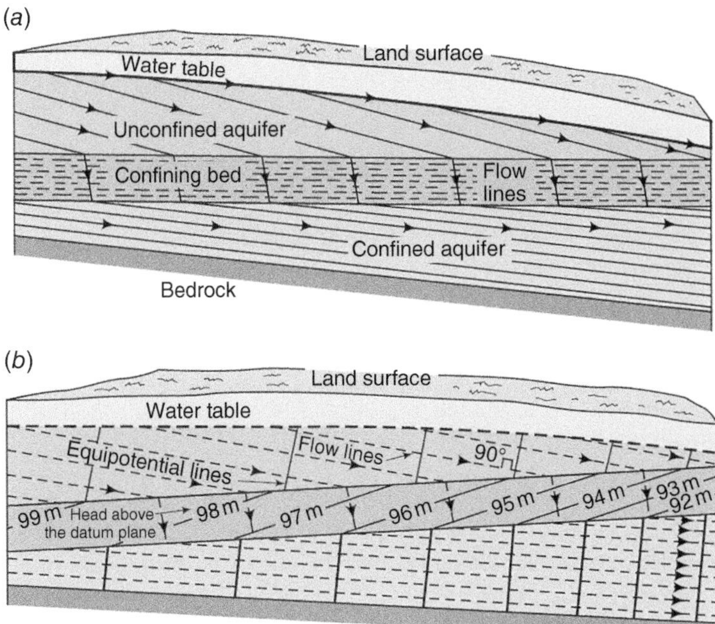

Figure 5.10 Flow net illustrating how flow lines refract in moving across hydraulic conductivity boundaries in systems of aquifers and confining beds U.S. Geological Survey (Heath, 1983 / Public domain).

An example of flow-line refraction is provided by Heath (1983) for flow across aquifers and confining bed, which have quite different hydraulic conductivity values (Figure 5.10). Notice how the flow lines are nearly vertical in the less permeable layer.

For a geological medium consisting of a number of sub-areas, each of which is homogeneous and isotropic, the flow net consists of squares in one sub-area and rectangles in other sub-areas. The streamlines change directions according to the refraction Eq. (5.42).

5.4.3 Flow Nets in Anisotropic Media

In anisotropic media, streamlines do not intercept equipotential lines at right angles except when flow is aligned with one of the principal directions of hydraulic conductivity or transmissivity. Assuming that the principal directions of transmissivity are aligned in the x and y directions, the flow equation in homogeneous, anisotropic media is written as

$$\left(\frac{K_x}{K_y}\right)\frac{\partial^2 h}{\partial x^2} + \frac{\partial^2 h}{\partial y^2} = 0 \tag{5.43}$$

By transforming the horizontal coordinate using $X = (K_y/K_x)^{1/2}\,x$, Eq. (5.43) can be rewritten as

$$\frac{\partial^2 h}{\partial X^2} + \frac{\partial^2 h}{\partial y^2} = 0 \tag{5.44}$$

The result of this mathematical manipulation implies that a flow net in an anisotropic medium may be constructed by transforming the flow field. Details of this procedure are summarized as follows.

1) Determine directions of the maximum and minimum hydraulic conductivity and designate the direction of maximum transmissivity as the x direction and the direction of minimum transmissivity as the y direction.
2) Multiply the dimension in x direction by a factor of $(K_y/K_x)^{1/2}$. Sketch the flow net in the transformed flow domain.
3) Project the flow net back to the original dimension by dividing the x coordinates of the flow net by a factor of $(K_y/K_x)^{1/2}$.

Figure 5.11*a* and 5.11*b* are the flow nets in original and transformed coordinate systems for flow parallel to the principal direction of hydraulic conductivity, respectively. In the transformed system, the flow net consists of squares (Figure 5.11*b*).

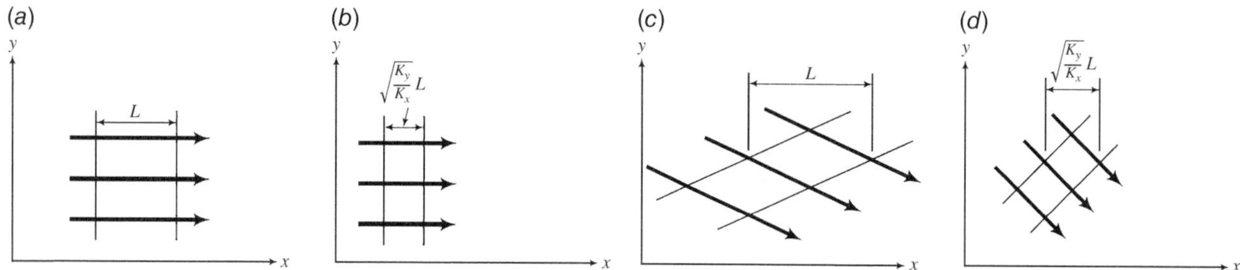

Figure 5.11 Flow net in anisotropic media. In Panel (*a*), the direction of flow is parallel to the principal direction (*x*) of hydraulic conductivity: the horizontal dimension of the porous medium cross section is changed by a ratio of K_y/K_x. In Panel (*b*), a transformed cross section with square nets is formed when the same ratio is applied to horizontal dimension. In Panel (*c*), the direction of flow is at some angle to the principal direction (*x*) of hydraulic conductivity: the angles formed by the streamlines and the equipotential lines are no longer 90°. In Panel (*d*), a transformed cross section with square nets can be formed by applying the ratio of K_y/K_x to the horizontal dimension of the flow net. *Note*: arrows point to the direction of flow (Schwartz and Zhang, 2003 / John Wiley & Sons).

In the original system, the flow net consists of rectangles (Figure 5.11*a*). If the flow direction is not parallel to one of the principal directions, the streamlines are no longer perpendicular to the equipotential lines (Figure 5.11*c*), although the flow net consists of squares in the transformed coordinate system (Figure 5.11*d*).

5.5 Mathematical Analysis of Some Simple Flow Problems

The first half of this book emphasizes the use of flow equations to address problems of flow and the interpretation data from various field tests. Forms of Eq. (5.18) can be solved using both analytical and numerical approaches. Analytical methods are based on classical methods, which have been around for more than 100 years for solving differential equations. They have been used in groundwater applications since the 1930s. For a problem to be amenable to an analytical solution, it needs to be simple. Thus, analytical approaches are usually applied to problems with a regular geometry, homogeneous aquifer properties, and simple initial and boundary conditions. These days, computers are used to help evaluate analytical solutions. However, they still can be treated with a calculator and various tables of functions.

Numerical approaches developed in concert with modern digital computers and require the computational capabilities of these machines. These techniques are tremendously powerful and can be applied to evaluate the most complicated, real-world problems. They can handle variability in hydraulic properties, large numbers of wells, and complicated boundary conditions, which might include variable recharge/evaporation, and groundwater/surface-water interactions.

In this section, we will introduce some simple applications of analytic approaches for solving steady-state flow problems, which are applied to flow in aquifers and flow to wells.

5.5.1 Groundwater Flow in a Confined Aquifer

We illustrate the mathematical analysis with a simple problem of steady-state groundwater flow in a confined aquifer. The flow is one-dimensional, produced by imposing different constant heads on the opposite sides of a rectangular flow domain (Figure 5.12*a*). Assuming that the aquifer is homogeneous and that the system is at steady-state, flow is described by

$$\frac{\partial^2 h}{\partial x^2} = 0 \tag{5.45}$$

with these boundary conditions:

$$h|_{x=0} = h_0 \tag{5.46}$$

and

$$h|_{x=l} = h_L \tag{5.47}$$

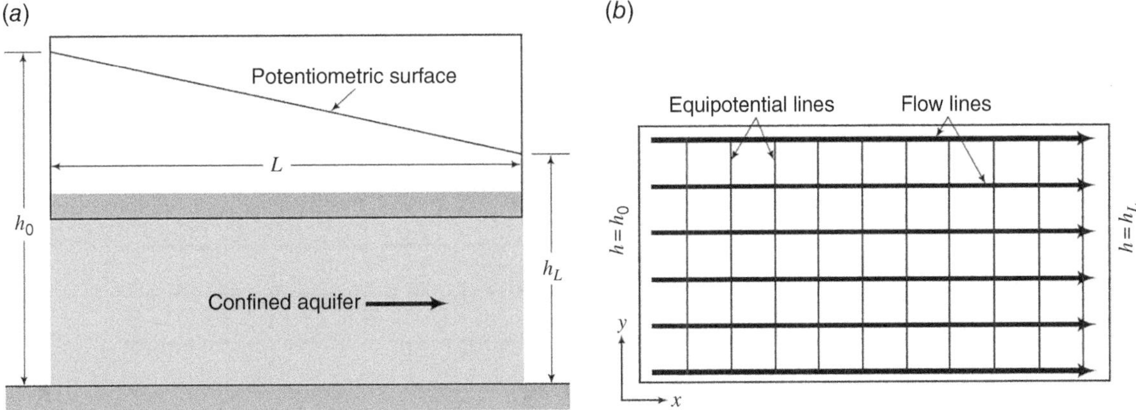

Figure 5.12 Unidirectional flow in a confined aquifer. Panel (*a*) is a cross section showing the region of interest, the aquifer, and the potentiometric surface. Panel (*b*) is a plan view showing the flow net (Schwartz and Zhang, 2003 / John Wiley & Sons).

The resulting analytical solution provides hydraulic head as a function of x-position in the aquifer

$$h = h_0 + (h_L - h_0)\frac{x}{L} \tag{5.48}$$

Because the flow is one-dimensional and the medium is homogeneous, calculation of the Darcy flow velocity is straightforward

$$q_x = K\frac{h_0 - h_L}{L} \tag{5.49}$$

Given the solution in Eq. (5.41), one can construct equipotential lines and sketch flow lines, as shown in Figure 5.12*b*.

Example 5.5 Two rivers 1000 m apart penetrate a confined aquifer 20 m thick. The hydraulic conductivity is 20 m/day. The stages of the two rivers are 500 and 495 m above sea level, respectively. What is the Darcy flow velocity for groundwater in the aquifer? If the reaches of the rivers are 600 m long and there are no pumped wells between them, what is the volume of groundwater outflow/inflow per year from the rivers? If a well were installed at a point located exactly between the rivers, what would be the hydraulic head before any pumping?

Solution

The calculation of inflow/outflow from the rivers starts with the Darcy velocity

$$q = (20\ \text{m/day})\frac{(500\ \text{m}) - (495\ \text{m})}{1000\ \text{m}} = 0.1\ \text{m/day}$$

Given q, the inflow/outflow from the rivers is determined by multiplying q by the area of inflow/outflow areas or

$$Q = (600\ \text{m})(20\ \text{m})(0.1\ \text{m/day})\left(365\frac{\text{days}}{\text{yr}}\right) = 4.38 \times 10^5\text{m}^3/\text{yr}$$

The hydraulic head in the aquifer midway between the two rivers is given from the analytic solution (Eq. 5.48)

$$h = 500\ \text{m} - \frac{(500\ \text{m}) - (495\ \text{m})}{(1000\ \text{m})}(500\ \text{m}) = 497.5\ \text{m}$$

5.5.2 Groundwater Flow in an Unconfined Aquifer

This example is similar to the previous one except now the aquifer is unconfined. The equation for steady-state, groundwater flow in a homogeneous, unconfined aquifer (Figure 5.13*a*) is

$$\frac{\partial}{\partial x}\left(h\frac{\partial h}{\partial x}\right) = 0 \tag{5.50}$$

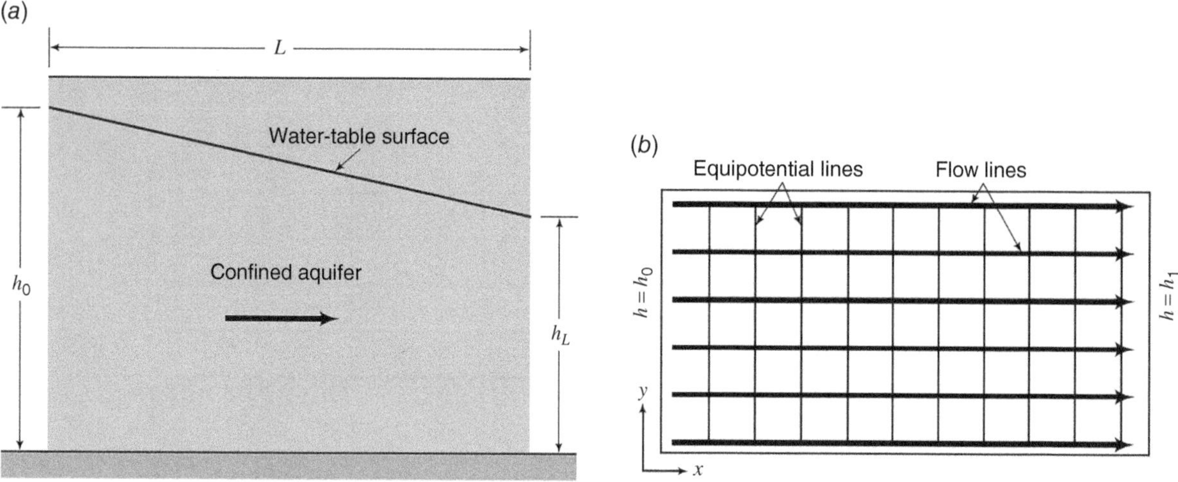

Figure 5.13 Unidirectional flow in an unconfined aquifer. Panel (*a*) is a cross section showing the region of interest and the water-table surface. Panel (*b*) is a plan view showing the flow net (Schwartz and Zhang, 2003 / John Wiley & Sons).

The boundary conditions are the same as before with:

$$h|_{x=0} = h_0 \tag{5.51}$$

and

$$h|_{x=l} = h_L \tag{5.52}$$

The solution with Dupuit's assumption and these boundary conditions is:

$$h = \sqrt{h_0^2 + (h_L^2 - h_0^2)\frac{x}{L}} \tag{5.53}$$

The Darcy flow velocity is

$$q = K\frac{h_0^2 - h_L^2}{2Lh} \tag{5.54}$$

The Darcy velocity will be different for different x locations because in a confined aquifer the thickness of the aquifer changes as a function of the hydraulic head. However, the flow rate through a unit width of the aquifer will be the same.

$$Q_{\text{unit width}} = K\frac{h_0^2 - h_L^2}{2L} \tag{5.55}$$

The solution provides the flow lines and equipotential lines shown in Figure 5.13*b*.

Example 5.6 Return to Example 5.5, but now assume that the aquifer is unconfined. Calculate the hydraulic head in the aquifer at the midpoint, the Darcy flow velocity at $x = 500$ m and the inflow/outflow from the rivers.

Solution

The hydraulic head at the midpoint ($x = 500$ m) is calculated by appropriate substitution into the analytic solution to give

$$h = \sqrt{(500\text{ m})^2 + \left[(495\text{ m})^2 - (500\text{ m})^2\right]\frac{(500\text{ m})}{(1000\text{ m})}} = 497.5\text{ m}$$

The Darcy flow velocity at $x = 500$ m is given as

$$q = (20\text{ m/day})\frac{(500\text{ m})^2 - (495\text{ m})^2}{2(1000\text{ m})(497.5\text{ m})} = 0.1\text{ m/day}$$

Discharge to/from the river is given as

$$Q = (600\ \text{m})(20\ \text{m})(0.1\ \text{m/day})\left(365\frac{\text{days}}{\text{yr}}\right) = 4.38 \times 10^5 \text{m}^3/\text{yr}$$

Comparing the Examples 5.4 and 5.5, the hydraulic characteristics are essentially the same.

5.5.3 Groundwater Flow in an Unconfined Aquifer with Recharge

Here, we examine a problem of flow in an unconfined aquifer with recharge (Figure 5.14). The differential equation for this example is

$$\frac{\partial}{\partial x}\left(K_x h \frac{\partial h}{\partial x}\right) + q_R = 0 \tag{5.56}$$

with boundary conditions:

$$h|_{x=0} = h_0 \tag{5.57}$$

and

$$h|_{x=l} = h_L \tag{5.58}$$

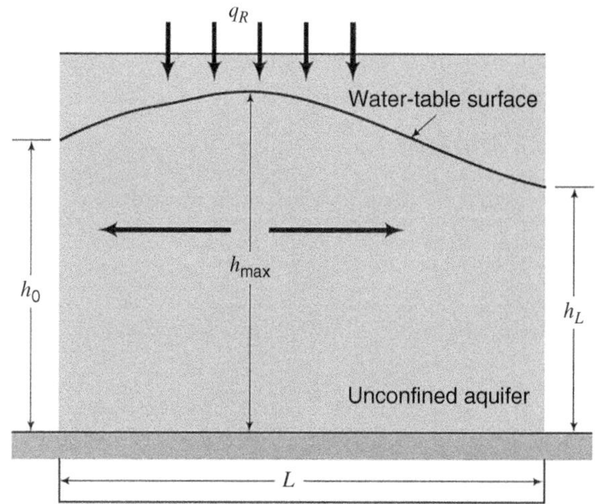

Figure 5.14 This cross section shows groundwater flow in an unconfined aquifer with recharge. Recharge leads to the formation of a drainage divide causing groundwater to flow to each of the rivers (Schwartz and Zhang, 2003 / John Wiley & Sons).

The solution with Dupuit's assumption is:

$$h = \sqrt{h_0^2 + (h_L^2 - h_0^2)\frac{x}{L} + \frac{q_R}{K_x}(L-x)x} \tag{5.59}$$

The solution now has become more complicated with hydraulic head varying in a nonlinear manner in the x-direction. The Darcy velocity is

$$q_x = K\frac{h_0^2 - h_L^2}{2Lh} - \frac{1}{2}q_R\frac{L-2x}{h} \tag{5.60}$$

The Darcy velocity in this case is not only dependent on hydraulic head but also the x-position. The flux through a unit width of the aquifer and its full thickness is

$$Q_{\text{unit width}} = K\frac{h_0^2 - h_L^2}{2L} - \frac{1}{2}q_R(L-2x) \tag{5.61}$$

Comparing Eqs. (5.60) and (5.61), the water flux in a unit width of the aquifer is not constant but changes with x-position. For groundwater flow in the unconfined aquifer with recharge, a groundwater divide exists in the recharge area. The location of the groundwater divide (x_d) is determined by

$$x_d = \frac{L}{2} - \frac{K}{q_R}\frac{h_0^2 - h_L^2}{2L} \tag{5.62}$$

Example 5.7 To avoid the salinization of soil in a farming area, parallel drainage canals are to be installed to maintain the water table at a depth of at least 4 m below the ground surface. It is thought that no evapotranspiration would occur below that depth. This water-table aquifer has hydraulic conductivity of 1 m/day. The canals fully penetrate the aquifer and have a stage of 3 m. The design depth from the ground surface to the bottom of the canals is 10 m. The daily recharge from precipitation and irrigation to the aquifer is 0.015 m. What is the optimal distance between the two drainage canals to achieve the desired water-table control? What is the discharge rate per unit width of aquifer to the canals?

Solution

Assume that the bottom of the canal is the datum and has an elevation of zero feet. The desired hydraulic head at the groundwater divide is

$$h = 10 - 4 = 6\,\text{m}$$

The hydraulic head is equal in the two drainage canals. The location of hydraulic divide is

$$x_d = \frac{L}{2}$$

Considering the hydraulic heads ($h = 6$ m and $h_L = h_0 = 3$ m), the Eq. (5.59) at $x = x_d$ may written as

$$6\,\text{m} = \sqrt{(3\,\text{m})^2 + \frac{0.015\text{m/day}}{1\text{m/day}}\left(L - \frac{L}{2}\right)\frac{L}{2}}$$

The distance between the canals is

$$L = \frac{2}{0.015}\sqrt{(6\,\text{m})^2 - (3\,\text{m})^2} = 693\,\text{m}$$

The discharge rate in the unit width of a canal ($x = 0$) is

$$Q_{\text{unit width}} = -\frac{1}{2}q_R L = \frac{(0.015\,\text{m/day})(693\,\text{m})}{2} = 5.2\,\text{m}^3/\text{day/m}$$

The aquifer discharges 10.4 m^3/day of water per unit length along the two canals.

Exercises

5.1 For the situations indicated below, determine whether a Dirichlet, Neumann, or Cauchy boundary condition is evident

A Surfaces marked by the label "water level" above and below the dam in Figure 5.6.
B The boundary between two media with different hydraulic conductivities through which groundwater is flowing.
C The water-table surface is in Figure 5.14.
D The groundwater divide in Figure 5.14.

5.2 Figure 5.15 shows a dike beside a flow channel. The line IJ is at the bottom of the channel. The dike ABDEF and the sheet BC is very low permeable material. The line EK is the water level in the aquifer and the line GH is the water level in the channel. Draw a flow net that depicts the pattern of flow under the dike.

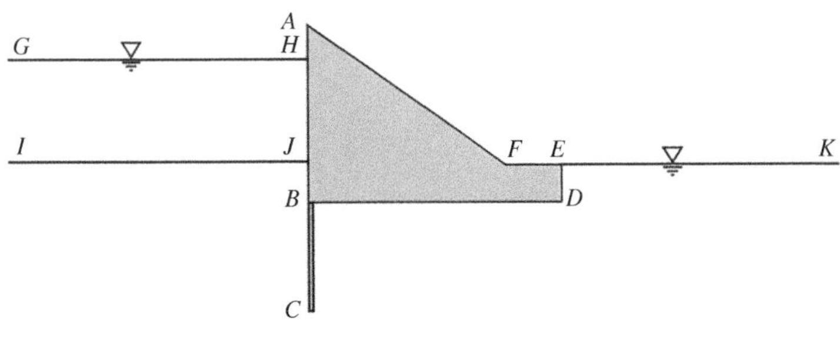

Figure 5.15 Dike along a flow channel (Schwartz and Zhang, 2003 / John Wiley & Sons).

5.3 Figure 5.16 is a map view of an irregularly shaped aquifer bounded by two streams and two no-flow boundaries. The principal directions of hydraulic conductivity are x and y.

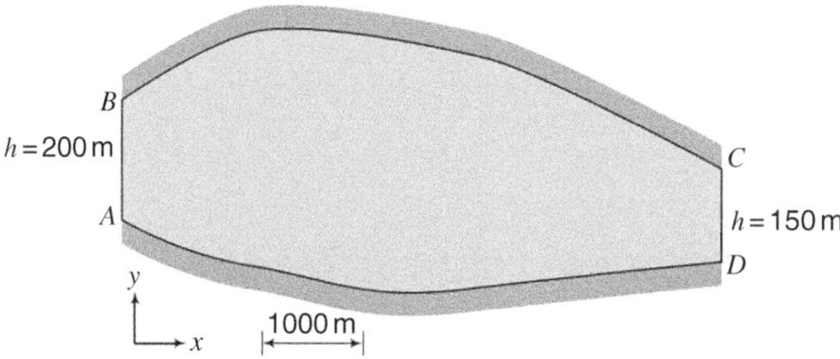

Figure 5.16 Map view of a buried valley aquifer (Schwartz and Zhang, 2003 / John Wiley & Sons).

A Assume that $K_y/K_x = 1$ and transmissivity of the aquifer is 3000 m^2/day, draw a flow net for the aquifer and calculate the total rate of flow from the stream AB to stream CD;
B Assume that $K_y/K_x = 1/9$, draw a flow net for the aquifer.

5.4 Please judge whether the following statements are correct or not.
A In a heterogeneous hydraulic conductivity field, equipotential lines are closer together in areas where the hydraulic conductivity is lower.
B When a streamline crosses into a very low permeability layer bounded by very high permeability formations above and below, the flow in the lower permeability unit tends to be perpendicular to the boundary.
C A no-flow boundary is not a streamline.
D A stream is not a constant head boundary.

5.5 The hydraulic head at a point in a confined aquifer bounded by horizontal formations is 80 m, and the horizontal hydraulic gradient in the aquifer is 0.03. What is the hydraulic head at 300 m down gradient from the point? If the hydraulic conductivity is 20 m/day, what is the Darcy velocity?

5.6 In a water-table aquifer bounded on the ends by rivers, the hydraulic conductivity is 50 m/day. The hydraulic heads at $x = 0$ and $x = 1000$ m are 100 and 90 m, respectively. Calculate the Darcy velocity at $x = 0$, 100, 500, and 1000 m, and the flow rate in a unit width of the aquifer.

5.7 Repeat the calculation in Problem 5.6 with a recharge rate of 0.02 m/day. Given that the streams are located at $x = 0$ and 1000 m determine where the groundwater divide is located.

References

Bennett, R. R. 1962. Flow net analysis, In J. G. Ferris, D. B. Knowles, R. H. Brown, and R. W. Stallman (eds.), Theory of aquifer tests. U.S. Geol. Surv. Prof. Paper 708, 70 p.

Bennett, R. R., and R. R. Meyer. 1952. Geology and Groundwater Resources of the Baltimore Area [Maryland], v. 4. Maryland Dept. Geology, Mines and Water Resour. Bull, 573 p.

Casagrande, A. 1937. Seepage through dams. Harvard Graduate School Eng Pub. 209.

Domenico, P. A., and F. W. Schwartz. 1998. Physical and Chemical Hydrogeology. John Wiley & Sons, New York, 506 p.

Heath, R. C. 1983. Basic groundwater hydrology. U.S. Geological Survey Water-Supply Paper 2220, 84 p.

Schwartz, F. W., and H. Zhang. 2003. Fundamentals of Groundwater. John Wiley & Sons, Hoboken, New Jersey, 583 p.

Thomas, G. B. Jr. 1972. Calculus and Analytic Geometry, 3rd edition. Addison-Wesley Publishing Company, Reading, Massachusetts, p. 635–641.

6

Theory of Groundwater Flow in Unsaturated Zones and Fractured Media

CHAPTER MENU

This chapter continues the exploration of basic principles of groundwater flow. We begin by looking at the vadose or unsaturated zone, which occurs between the water table and the ground surface. The vadose zone is important for the growth of land plants and is the interface between the groundwater and surface hydrologic systems. As will become evident, water flow in the vadose zone is complicated because pores contain both soil gas and water. Parameters, like hydraulic conductivity and storativity, which are hydraulic "constants" in saturated systems become hydraulic functions. In essence, the transmissive and storage properties of an unsaturated medium change dramatically depending upon the relative proportions of soil gas and water. Not surprisingly, then, the governing equations for flow become more complicated to solve because of parameters turn into nonlinear functions.

This chapter also explores issues of fracture flow that we introduced in Chapter 4. It is important to know something about fractured media because they figure so prominently in studies of groundwater resources and contaminant hydrology. This topic is vast and unfortunately, within the scope of this introductory book, we can only begin to scratch the surface.

6.1 Basic Concepts of Flow in Unsaturated Zones

With the exception of parts of the capillary fringe, pores in the unsaturated zone contain both water and soil gases. The quantity of water in a partially saturated medium can be represented in terms of the *volumetric water content* (θ), which is defined as

$$\theta = \frac{V_w}{V_T} \tag{6.1}$$

where V_T is some unit volume of soil or rock, and V_w is the volume of water.

The volumetric water content is a key property of unsaturated media. Wierenga et al. (1993) describe a measurement technique in the laboratory as follows: (1) Place the soil into a can, capping it tightly to prevent evaporation, and weight the soil; (2) Take the lid off, and dry the soil in a forced-draft oven for 10 hours or in a convection oven for 24 hours;

Fundamentals of Groundwater, Second Edition. Franklin W. Schwartz and Hubao Zhang.

Companion website: www.wiley.com/go/schwartz/fundamentalsofgroundwater2

(3) Remove the soil from the oven, replace the lid, and put the soil into desiccating jar, with desiccant, until the soil cools; (4) Weigh the sample again with the lid. The water content θ_w, expressed as a weight fraction is

$$\theta_w = \frac{W_{\text{wet-soil}}}{W_{\text{dry-soil}}} - 1 \tag{6.2}$$

where $W_{\text{wet-soil}}$ and $W_{\text{dry-soil}}$ are the weights of wet and dry soils, respectively. The volumetric water content is related to θ_w as

$$\theta = \theta_w \frac{\rho_b}{\rho_w} \tag{6.3}$$

where ρ_b is the bulk density of the soil, and ρ_w is the density of water. This method for determining the volumetric water content is a direct method. The volumetric water content can be determined indirectly by measuring soil properties related to water content. The possible methods include electrical conductivity, time domain reflectometry, and gamma-ray attenuation (Gardner, 1986).

If the volume of the void space in the sample volume V_T is V_{void}, the *water saturation* (s) is defined as

$$s = \frac{V_w}{V_{\text{void}}} \tag{6.4}$$

In the unsaturated zone, the volume of water present in the sample is usually less than the volume of void space. Thus, the volumetric water content is usually less than the porosity ($0 < \theta < n$), and the water saturation is less than ($0 < s < 1$).

Water in the unsaturated zone can be characterized in terms a hydraulic head that determines the direction of fluid flow. Like its saturated counterpart, hydraulic head also involves two components.

$$h = z + \psi \tag{6.5}$$

where z is the elevation head, and ψ is the pressure head. The pressure head is negative in the unsaturated zone ($\psi < 0$), while the pressure head in the saturated zone is positive ($\psi > 0$) (Figure 6.1). Right at the water table, the pressure head is zero ($\psi = 0$). Water pressures in the unsaturated zone are less than the atmospheric pressure. For this reason, the pressure head in the unsaturated zone is also called tension head or suction head acknowledging the capillary forces that bind water to solids. It is this "negative" pressure head in the unsaturated zone that explains why water present in partially saturated soils cannot flow into a borehole.

A *tensiometer* is a device used to measure pressure head in the unsaturated zone (ASTM, 1992; Stannard, 1986). It consists of a porous ceramic cup connected by a water column to a manometer, or a vacuum gage, or a pressure transducer (Figure 6.2). The very fine pores of the ceramic cup fill with water, which provides a hydraulic connection between the soil water and the water column. As the pressure head changes in the soil, water flows into or out of the tensiometer to maintain hydraulic equilibrium.

Figure 6.3 illustrates what the pressure head and hydraulic head would look like with a constant and continuous rainfall on the soil surface (Figure 6.3*a*). The pressure head distribution is shown in Figure 6.3*b*. Near the bottom of this field is the *water table*, which is defined by the $\psi = 0$ contour. The pressure heads above the water table are negative, and those below the water table are positive. The total head distribution is shown in Figure 6.3*c*. As expected, downward flow through the unsaturated zone reflects total head values that decrease with depth (Figure 6.3*c*).

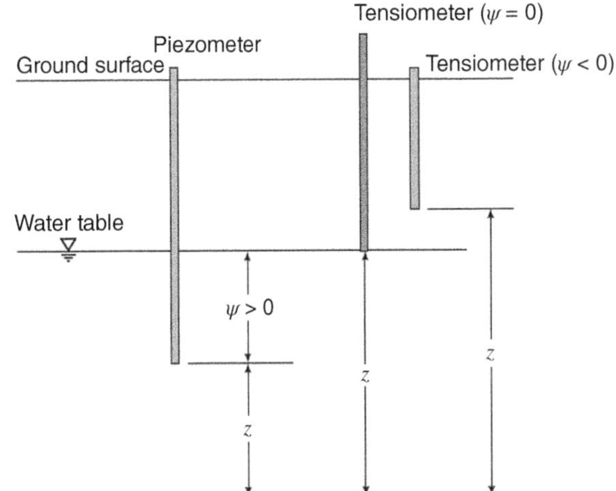

Figure 6.1 Pressure heads are negative in the unsaturated zone, zero along the water table, and positive in the saturated zone. As shown, the total hydraulic head is the algebraic sum of the elevation head and the pressure head (Domenico and Schwartz, 1998. Physical and Chemical Hydrogeology. Copyright © 1990, 1998 John Wiley & Sons, Inc. All Rights Reserved. Reproduced with permission).

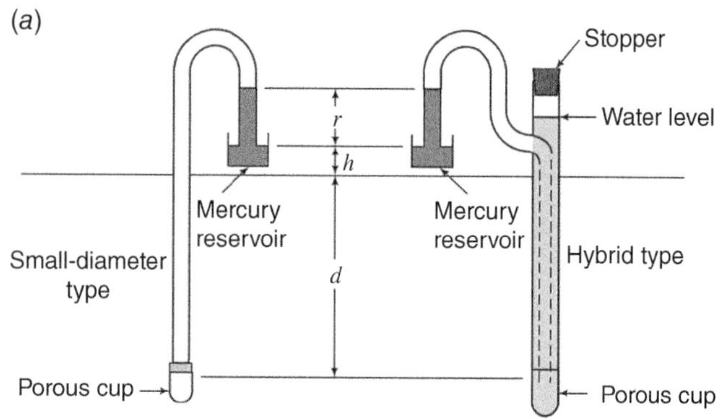

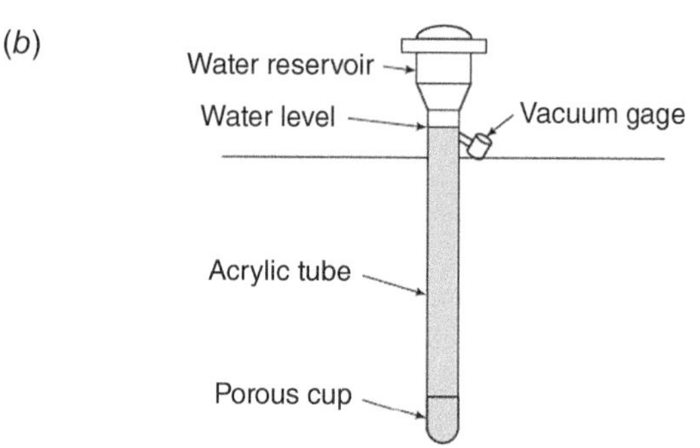

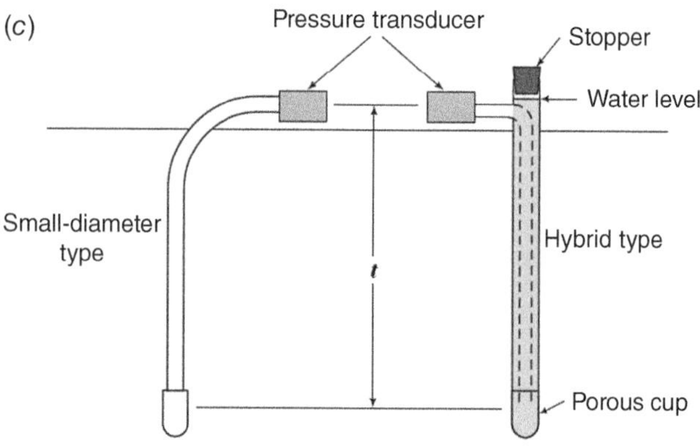

Figure 6.2 Three common types of tensiometers: (*a*) manometer; (*b*) vacuum gage; (*c*) pressure transducer (ASTM, 1992 / Copyright ASTM. Reprinted with permission).

Example 6.1 A soil sample was taken from an unsaturated zone. The wet and dry weights of the sample are 105 and 100 g, respectively. The bulk density of the sample is 1.65 g/cm^3 and the density of water is 1 g/cm^3. What is the volumetric water content?

Solution

The water content expressed as a weight percentage is

$$\theta_w = \frac{(105\,\text{g})}{(100\,\text{g})} - 1 = 0.05 = 5\%$$

The volumetric water content is

$$\theta = (0.05)\frac{(1.65\,\text{g/cm}^3)}{(1.0\,\text{g/cm}^3)} = 0.083$$

6.1.1 Changes in Moisture Content During Infiltration

The water content of an unsaturated medium changes as a function of space and time. We illustrate this idea using an example showing how infiltration from rainfall is redistributed. Water entering the vadose zone increases the water content at early time (Figure 6.4*a*). Capillary forces dominate this initial wetting. With time the infiltration stops, and that water moves downward (Figure 6.4*b*). Both capillary and gravitational forces are in action at this stage. Notice that there is a distinct wetting front evident as the pulse of infiltrated water moves downward. The *wetting front* is defined as the narrow zone that marks the beginning of elevated water contents due to the infiltration event. Once infiltration ceases, evapotranspiration can take place near the ground surface. The final stage of water movement in the unsaturated zone occurs as soil water enters saturated zones (Figure 6.4*c*).

6.2 Characteristic Curves

Flow in the unsaturated zone is complicated by the fact that there are generally two fluid phases (air and water) present together. Both the volumetric moisture content (θ) and the unsaturated hydraulic conductivity (K) depend upon the pressure head or the capillary pressure. The pressure head/moisture content relationship describes how a sample behaves as water is added or removed. With hydraulic conductivity, a relationship between hydraulic conductivity and pressure head would not be surprising. As soil dries out, an increasingly large negative pressure head means that the mostly air-filled system has a large resistance to flow (or small K).

6.2.1 Water Retention or $\theta(\psi)$ Curves

Generally, as the water content of soil decreases, the pressure head becomes more negative, or alternatively, the capillary pressure increases. This response is due to the tendency for the water to find itself located in smaller and smaller voids. The relationship between negative pressure head and volumetric water content for a sample is called a *water retention curve*. An example of one of these curves is shown in Figure 6.5. In practice, a volumetric water content is plotted on an arithmetic scale, while the negative pressure head is plotted on either an arithmetic scale or a logarithmic scale, as shown. The curves are typically nonlinear irrespective of how they are plotted. At both large and small water contents, small changes in water content are accompanied by extremely large changes in pressure head. The behavior at low water contents reflects the fact that soils never lose all of their water. This lower limit in water content is termed the residual volumetric water content (θ_r).

The shape of the water retention curve changes depending upon whether the soil is drying or wetting. The term *hysteretic* is used to describe this effect. *Drying* involves air entering the soil to replace water that is draining. *Wetting* involves the entry of water and the displacement of the air.

The actual shape of the water-retention curve depends upon several factors with pore-size distribution being the most important. Figure 6.6 illustrates curves for sand, fine sand, and silt loam. The sand has the most uniform distribution of large pores, while the silt loam with a broad, grain-size distribution contains small pores, which will lead to small negative pressure heads as the soil dries.

In modeling applications, it is common to represent $\theta(\psi)$ curves using various types of mathematical relationships. Here, we illustrate two of the commonly used relationships, the Brooks–Corey (Brooks and Corey, 1966), and van Genuchten (1980) equations. These equations are written in terms of a dimensionless measure of moisture

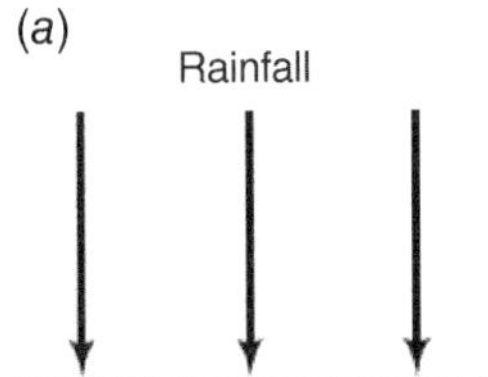

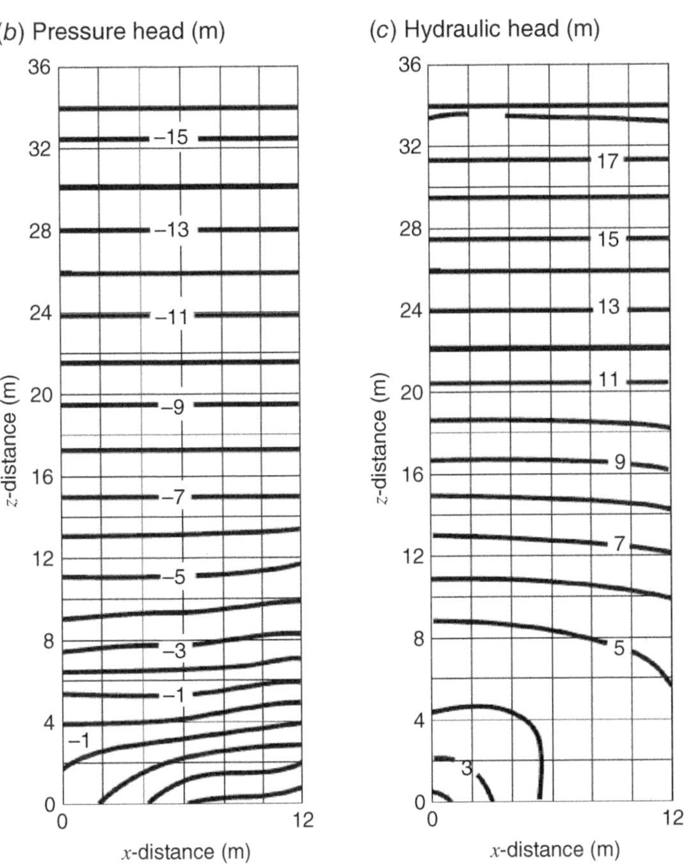

Figure 6.3 Head distribution in the unsaturated zone: (*a*) rainfall; (*b*) pressure head; (*c*) hydraulic head (Schwartz and Zhang, 2003 / John Wiley & Sons).

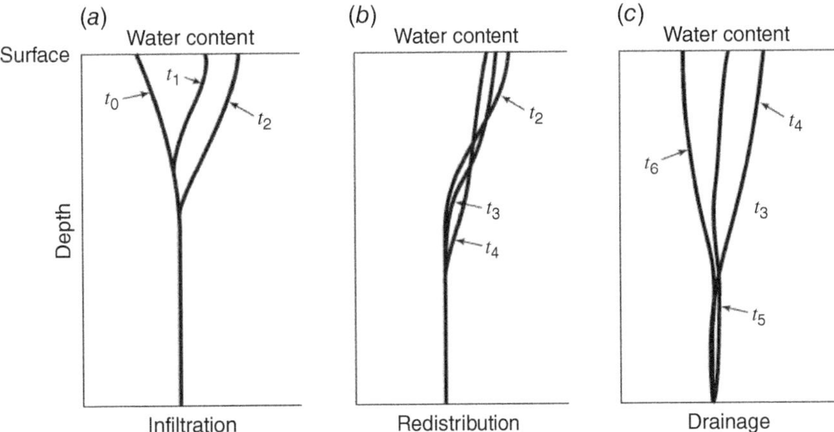

Figure 6.4 Processes of infiltration, redistribution, and recharge (Ravi and Williams, 1998 / United States Environmental Protection Agency).

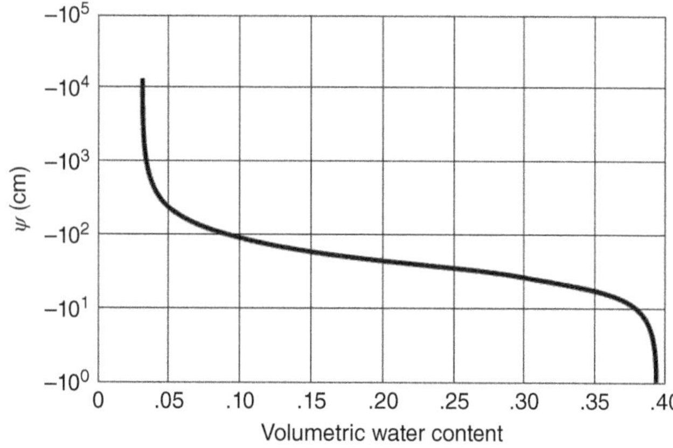

Figure 6.5 Water-retention curve for the Berino fine sandy loam (U.S. Nuclear Regulatory Commission, adapted from Wierenga et al., 1986).

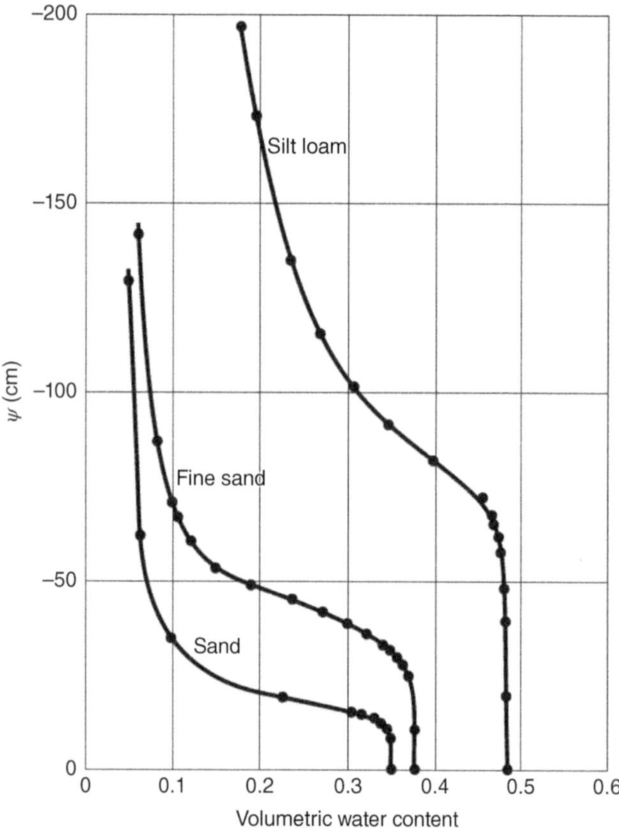

Figure 6.6 Water-retention curve for sand, fine sand, and silt loam (Brooks and Corey, 1966. Copyright by ASCE and reproduced by permission of the publisher, ASCE).

content termed the *effective saturation* (s_e). From the mathematical definition that follows, note how the effective saturation varies between zero and one for every sample

$$s_e = \frac{\theta - \theta_r}{n - \theta_r} \tag{6.6}$$

where s_e is the effective saturation, θ_r is the residual volumetric water content, and n is the porosity.

The Brooks–Corey equation is given by

$$s_e = \begin{cases} \left(\frac{\psi_b}{\psi}\right)^{\lambda}, & \psi < \psi_b \\ 1 \quad , & \psi \geq \psi_b \end{cases} \tag{6.7}$$

where ψ_b is the bubbling or air-entry pressure head [L] and equal to the pressure head to desaturate the largest pores in the medium, and λ is a pore size distribution index.

The van Genuchten equation is expressed as

$$s_e = \frac{1}{\left[1 + (\alpha \mid \psi \mid)^{\beta}\right]^{\gamma}} \tag{6.8}$$

where α is coefficient [1/L], β is the exponent, and $\gamma = 1 - 1/\beta$.

The coefficients ψ_b and λ in the Brooks–Corey equation, and α and β in the van Genuchten equation can be determined by fitting the measured water retention curve with the calculated water retention curves using Eqs. (6.13) through (6.15).

Example 6.2 Water retention curves (Figure 6.7*a* and 6.7*b*) for sand and Yolo light clay were measured. Determine the parameters for Brooks–Corey and van Genuchten models. This example is adapted from Lappala et al. (1987).

Solution

The calculated and measured water retention curves of sand and Yolo light clay for the van Genuchten model are shown in Figure 6.7*a*. The fitted parameters include: K_s = 8.2 m/day, n = 0.435, θ_r = 0.069, α = 3.07 (1/m), and β = 3.9.

The calculated and measured water retention curves of sand and Yolo light clay for the Brooks–Corey model are shown in Figure 6.7*b*. The fitted parameters include: $\theta_r = 0$, $\psi_b = -0.196$ m, and $\lambda = 0.84$.

6.2.2 $K(\psi)$ Curves

For unsaturated media, hydraulic conductivity is not a constant but is strongly dependent upon the degree of saturation. When a medium is near saturation with a pressure head close to zero, the hydraulic conductivity takes on its maximum value. As the water content declines and pores become filled with air, the pressure head becomes more and more negative and the hydraulic conductivity decreases. As the volumetric water constant approaches residual, the water phase may not even be continuous through the sample, providing a hydraulic conductivity that is near zero.

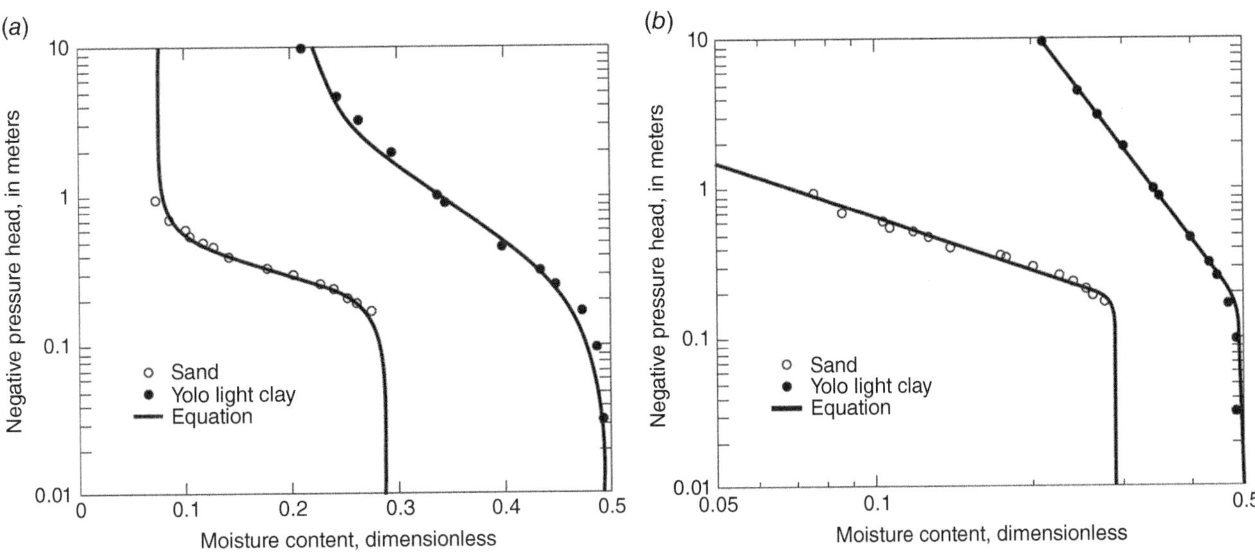

Figure 6.7 Determination of hydraulic properties from water-retention curves (*a*) Water-retention curves of sand and Yolo light clay fitted by van Genuchten equation; (*b*) water-retention curves of sand and Yolo light clay fitted by Brooks–Corey equation. (U.S. Geological Survey, Lappala et al., 1987 / Public domain).

In problems with more than one flowing phase (e.g., air and water), the concept of *relative permeability* or *relative hydraulic conductivity* has proven useful in capturing the relationship existing between hydraulic conductivity (K) and negative pressure head ($-\psi$). The relative hydraulic conductivity of an unsaturated medium is defined as

$$K_r(\psi) = \frac{K(\psi)}{K_s} \tag{6.9}$$

where K_r is the relative hydraulic conductivity, which varies between 0 and 1, K_s is the hydraulic conductivity when the medium is saturated, and K is the unsaturated hydraulic conductivity. The product K_rK_s points out how the unsaturated hydraulic conductivity is really some fraction of the saturated hydraulic conductivity. The specific fraction is determined by $-\psi$.

Not unexpectedly, relative hydraulic conductivity can be written as a function of the pressure head. For the Brooks–Corey model, it is represented by

$$K_r(\psi) = \begin{cases} \left(\dfrac{\psi_b}{\psi}\right)^{2+3\lambda}, & \psi < \psi_b \\ 1, & \psi \geq \psi_b \end{cases} \tag{6.10}$$

For the van Genuchten model, relative hydraulic conductivity is written as

$$K_r(\psi) = \frac{\left\{1-(\alpha\,|\,\psi\,|)^{\beta-1}\left[1+(\alpha\,|\,\psi\,|)^{\beta}\right]^{-\gamma}\right\}^2}{\left[1+(\alpha\,|\,\psi\,|)^{\beta}\right]^{\gamma/2}} \tag{6.11}$$

The relative hydraulic conductivity may also be expressed as a function of saturation. For example, the relative hydraulic conductivity for the van Genuchten model is

$$K_r(S_e) = S_e^{l}\left[1-\left(1-S^{1/\beta}\right)^{\beta}\right]^2 \tag{6.12}$$

where l is pore connectivity and equal to about 0.5 for many soils (Mualem, 1976).

6.2.3 Moisture Capacity or $C(\psi)$ Curves

The storage properties of an unsaturated soil are represented by a parameter called the specific moisture capacity, c_m, which is defined as the change in moisture content divided by the change in pressure head or

$$c_m = \frac{d\theta}{d\psi} \tag{6.13}$$

Mathematically, c_m is the slope of the $\theta(\psi)$ characteristic discussed previously. Qualitatively, an increase in pressure head, for example from −100 cm to −50 cm with the fine sand (Figure 6.6) is accompanied by an increase in the volumetric water content from 0.08 to 0.18. Notice how the increase in storage changes as a function of the material type. In the case of sand, a change in pressure head of this magnitude is accompanied by a minimal change in water storage (Figure 6.6).

As was the case with the other characteristics, specific moisture capacity can be defined as a function of various parameters. For the Brooks–Corey model, specific moisture capacity is given by

$$c_m(\psi) = \begin{cases} -(n-\theta_r)\dfrac{\lambda}{\psi_b}\left(\dfrac{\psi}{\psi_b}\right)^{-(\lambda+1)}, & \psi \le \psi_b \\ 0, & \psi > \psi_b \end{cases} \tag{6.14}$$

For the van Genuchten model, it is expressed as

$$c_m(\psi) = \begin{cases} \dfrac{\alpha\gamma\beta(n-\theta_r)(\alpha\,|\,\psi\,|)^{\beta-1}}{\left[1+(\alpha\,|\,\psi\,|)^{\beta}\right]^{\gamma+1}}, & \psi < 0 \\ 0, & \psi > 0 \end{cases} \tag{6.15}$$

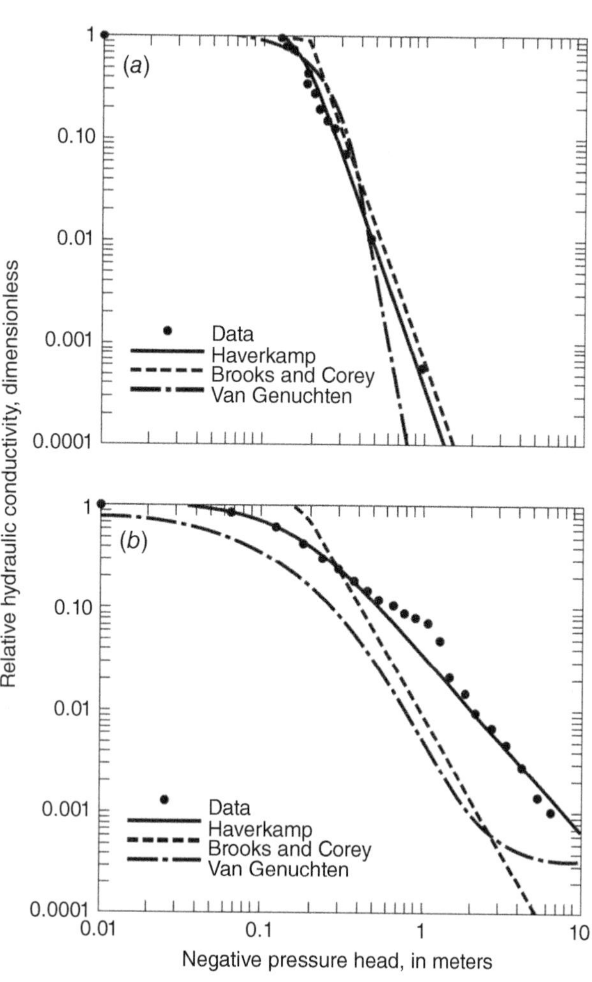

Figure 6.8 Determination of hydraulic properties from relative hydraulic-conductivity curves (*a*) Relative hydraulic-conductivity curve of sand fitted by Havekamp, Brooks–Corey, and van Genuchten equations; (*b*) relative hydraulic conductivity curve of the Yolo light clay fitted by Havekamp, Brooks–Corey, and van Genuchten equations. (U.S. Geological Survey, Lappala et al., 1987 / Public domain).

Example 6.3 The relative hydraulic conductivity curves in Figure 6.8*a* and 6.8*b* were measured for sand and Yolo light clay. Determine the parameters for Brooks–Corey and van Genuchten models and calculate the specific moisture capacity curves. This example is adapted from Lappala et al. (1987).

Solution

The calculated and measured relative hydraulic conductivity curves for the Brooks–Corey model are shown in Figure 6.7*a* and 6.7*b*. The fitted parameters include: K_s = 8.2 m/day, $n = 0.435$, $\theta_r = 0$, $\psi_b = -0.196$ m, and $\lambda = 0.84$. The calculated specific moisture capacity curves are shown in Figure 6.9*a*.

The calculated and measured Relative hydraulic conductivity curves for the van Genuchten model are shown in Figure 6.8*a* and 6.8*b*. The fitted parameters include: $\theta_r = 0.069$, $\alpha = 3.07$ (1/m), and $\beta = 3.9$. The calculated specific moisture capacity curves are shown in Figure 6.9*b*. Also shown in Figure 6.8 are the fitting curves of the Haverkamp model (Haverkamp et al., 1977), which is a simplified form of the van Genuchten model.

Table 6.1 shows the hydraulic properties of eleven soils for Brooks–Corey and van Genuchten models compiled by Lappala et al. (1987).

6.3 Flow Equation in the Unsaturated Zone

The derivation of the partial differential equation for flow in the unsaturated zone is similar to that in the saturated zone. In the unsaturated zone, the Darcy's equation is written as

$$q = -K(\psi)\nabla h \tag{6.16}$$

where q is the Darcy velocity vector [L/T], $K(\psi)$ is hydraulic conductivity tensor [L/T], and h is hydraulic head [L]. The hydraulic conductivity tensor in the unsaturated zone is expressed as

$$K(\psi) = K_r(\psi)K_s \tag{6.17}$$

where $K_r(\psi)$ is the relative hydraulic conductivity [dimensionless] $(0 < K_r < 1)$, and K_s is the saturated hydraulic conductivity tensor as given in (3.29).

$$K = \begin{bmatrix} K_{xx} & K_{xy} & K_{xz} \\ K_{yx} & K_{yy} & K_{yz} \\ K_{zx} & K_{zy} & K_{zz} \end{bmatrix} \tag{3.29}$$

By replacing the porosity n in Eq. (4.12) with soil moisture θ, we obtain the water conservation equation in the unsaturated zone.

$$-\left[\frac{\partial q_x}{\partial x} + \frac{\partial q_y}{\partial y} + \frac{\partial q_z}{\partial z}\right] = \frac{1}{\rho_w}\frac{\partial(\rho_w n)}{\partial t} \tag{6.18}$$

Substitution of Eqs. (6.4) and (6.5) into Eq. (6.6) provides the equation for flow in the unsaturated zone.

$$\frac{\partial}{\partial x_i}\left[K_r(\psi)\left(K_{ij}\frac{\partial h}{\partial x} + K_{iz}\right)\right] = \frac{\partial \theta}{\partial t} \tag{6.19}$$

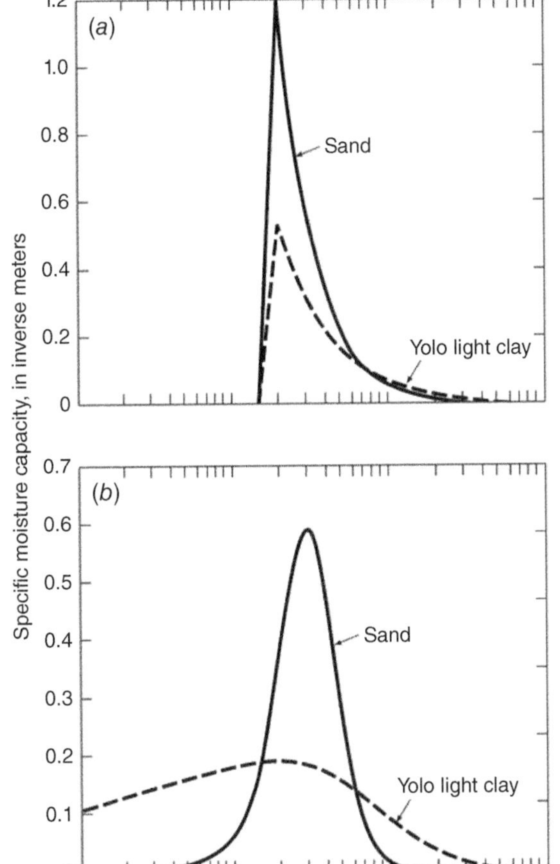

Figure 6.9 Calculated specific moisture capacity as a function of pressure head in sand and yolo light clay (*a*) using Brooks–Corey equation; (*b*) using van Genuchten equation. (U.S. Geological Survey, Lappala et al., 1987 / Public domain).

Table 6.1 Values of hydraulic properties of eleven soils that best fit Brooks–Corey and van Genuchten models.

			Brooks and Corey (1966)			van Genuchten (1980)		
Soil or rock	**K_s (m/day)**	n	θ_r	$-h_b$(m)	Λ	θ_r	α(1/m)	β
Del Monte Sand (20 mesh)	7000	0.36	0.011	0.112	2.5	0.036	7.04	6.3
Fresno medium sand	400	0.375	0	0.149	0.84	0.02	4.31	3.1
Unconsolidated sand	8.5	0.424	0.09	0.114	4.4	0.051	7.46	9
Sand	8.2	0.435	0	0.196	0.84	0.069	3.07	3.9
Fine sand	2.1	0.377	0.063	0.82	3.7	0.072	1.04	6.9
Columbia sandy loam	0.7	0.496	0.11	0.85	1.6	0.15	0.85	4.8
Touchet silt loam	0.22	0.43	0.095	1.45	1.7	0.17	0.51	7
Hygiene sand stone	0.15	0.25	0.13	1.06	2.9	0.15	0.79	10.6
Adelanto loam	0.039	0.42	0.13	1.41	0.51	0.16	0.36	2.06
Limon silt (imbibition data)	0.013	0.449	0	0.338	0.22	0.001	1.54	1.3
Yolo light clay	0.011	0.495	0.055	0.181	0.25	0.175	2.49	1.6

Source: Lappala et al. (1987) / USGS.

where x_1, x_2, and x_3 are x, y, and z coordinates, K_{ij} $(i, j = x, y, \text{and } z)$ are components of hydraulic conductivity tensor, and K_{iz} $(i = x, y, \text{and } z)$ are hydraulic components in z direction. In words, above equation may be expressed as

Inflow rate − Outflow rate = Moisture storage change rate

Equation (6.19) is a modified form of the Richards' equation. In the Eq. (6.19), if the principal directions of hydraulic conductivity are x, y, and z directions the off-diagonal terms in Eq. (6.19) are zero. Compared with the flow equation in the saturated zone, the hydraulic conductivity in the unsaturated zone is a function of pressure head, and the storage change is the change of volumetric water content. The right side of the equation can be rewritten as (Lappala et al., 1987)

$$\frac{\partial \theta}{\partial t} = [c_m + sS_s]\frac{\partial h}{\partial t} \tag{6.20}$$

where c_m is the specific moisture capacity, s is the saturation, and S_s is specific storage given by (4.9).

Combining Eqs. (6.7) and (6.8), we get

$$\frac{\partial}{\partial x}\left(K_r(\psi)K_x\frac{\partial h}{\partial x}\right) + \frac{\partial}{\partial y}\left(K_r(\psi)K_y\frac{\partial h}{\partial y}\right) + \frac{\partial}{\partial z}\left(K_r(\psi)K_z\frac{\partial h}{\partial z}\right) = [c_m + sS_s]\frac{\partial h}{\partial t} \tag{6.10}$$

To solve Eq. (6.10), volumetric water content, hydraulic conductivity, and specific capacity as functions of pressure head need to be known. These functional relations will be discussed in the following sections.

For one-dimensional vertical flow, Eq. (6.7) can be written as

$$\frac{\partial \theta}{\partial t} = \frac{\partial}{\partial z}\left(K_r(\psi)K_z\left(\frac{\partial \psi}{\partial z} + 1\right)\right) \tag{6.11}$$

where z is positive downward.

The linear groundwater velocity in the unsaturated zone is related to the Darcy velocity by

$$v = \frac{q}{\theta} = \frac{q}{sn} \tag{6.12}$$

where v is the linear groundwater velocity or groundwater pore velocity vector, s is the saturation, and n is the porosity.

Example 6.4 The Darcy velocity in the unsaturated zone is 3 cm/h. The porosity of the medium is 0.36. Under steady-state infiltration, the average saturation of soil is 0.8. For an unsaturated zone of 20 m, calculate the time required for a drop of water at the ground surface to travel to the water table.

Solution

The groundwater velocity is the Darcy velocity divided by the volumetric water content.

$$v = \frac{(3\ \text{cm/h})}{(0.35)(0.8)} = 10.7\ \text{cm/h}$$

The time required is

$$t = \frac{(20\ \text{m})}{(10.7\ \text{cm/h})} = 18.7\ \text{h}$$

6.4 Infiltration and Evapotranspiration

Infiltration and evapotranspiration are important processes providing flow in and out of unsaturated zones through the ground surface. To describe these processes, numerous mathematical equations have been developed. The Green–Ampt model (Green and Ampt, 1911) was the first physically-based equation to describe the infiltration of water into a soil. The infiltration rate in the Green–Ampt model can be expressed explicitly as (Salvussi and Entekhabi, 1994)

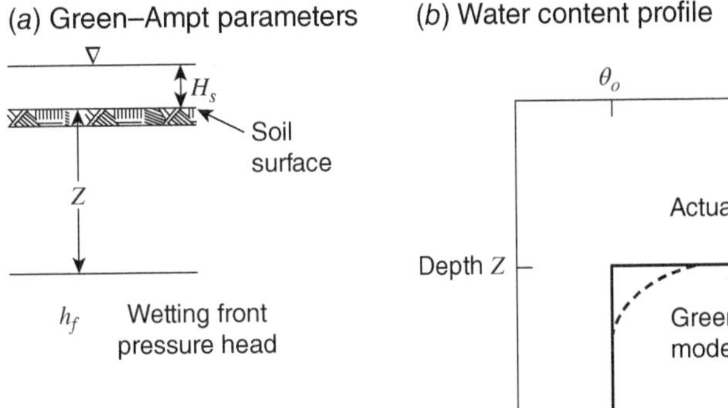

Figure 6.10 Green–Ampt model. (*a*) Green–Ampt model parameters; (*b*) Conceptualized volumetric water content profile in the Green–Ampt model (Vogel et al., 1996 / USDA / Public Domain).

$$q = \left(\frac{\sqrt{2}}{2}\tau^{-1/2} + \frac{2}{3} - \frac{\sqrt{2}}{6}\tau^{1/2} + \frac{1-\sqrt{2}}{3}\tau\right)K_s \tag{6.19}$$

where q is infiltration rate (cm/h), K_s is saturated hydraulic conductivity (cm/h), t is time (h), h_s is ponding depth, or capillary pressure head at the ground surface (cm), h_f is capillary pressure head at the wetting front (cm), θ_s is saturated volumetric water content, and θ_0 is initial volumetric water content. τ is related to time as

$$\tau = \frac{t}{t+\chi} \tag{6.20}$$

with

$$\chi = \frac{(h_s - h_f)(\theta_s - \theta_0)}{K_s} \tag{6.21}$$

The capillary pressure head (h_f) at the wetting front can be estimated from the air entry head or the bubbling pressure head.

$$h_f = \frac{2 + 3\lambda}{1 + 3\lambda}\frac{\psi_b}{2} \tag{6.22}$$

where λ is the exponent in the Brooks–Corey water retention model, and ψ_b is the bubbling-pressure head. The Green–Ampt model parameters and corresponding water content profile are illustrated in Figure 6.10*a* and 6.10*b* (Ravi and Williams, 1998). A sharp wetting front is assumed in the model. As time t becomes large, the τ in Eq. (6.20) is 1 and the infiltration rate is equal to the saturated hydraulic conductivity. This relationship is expressed as

$$q = K_s, \quad \text{for } t \to \infty \tag{6.23}$$

There are several other approaches for modeling infiltration or evapotranspiration rates. For example, the U.S. Soil Conservation Service (SCS) model relates the infiltration rate to precipitation rate, based on site-specific field data (USDA, 1972). Infiltration/exfiltration (evapotranspiration) models (Eagleson, 1978; Philip, 1957) can be used to estimate the water infiltration during wetting season and evapotranspiration during drying season.

Example 6.5 Calculate infiltration rates for the duration of infiltration of 24 hours (adapted from Williams et al., 1997). Assume that $\psi_b = -13.8$ cm, $\lambda = 1.68$, $\theta_s = 0.43$, $\theta_0 = 0.05$, $K_s = 21$ cm/h, and $h_s = 1$ cm.

Solution

The capillary pressure head at the wetting front is

$$h_f = \frac{2 + 3(1.68)}{1 + 3(1.68)}\frac{(-13.8\text{ cm})}{2} = -8.04\text{ cm}$$

Table 6.2 Calculation of infiltration rates using explicit Green–Ampt model.

Time (h)	τ	q (cm/h)	Time (h)	τ	q (cm/h)
1	0.8594	22.937	13	0.9876	21.160
2	0.9244	22.005	14	0.9885	21.149
3	0.9483	21.679	15	0.9892	21.139
4	0.9607	21.513	16	0.9899	21.130
5	0.9683	21.412	17	0.9905	21.123
6	0.9735	21.344	18	0.9910	21.116

Source: Schwartz and Zhang (2003) / John Wiley & Sons.

The χ is calculated as

$$\chi = \frac{[(1\text{ cm}) - (-8.04\text{ cm})](0.43 - 0.05)}{(21\text{cm/h})} = 0.1636\text{ h}$$

The infiltration rates are in Table 6.2 for time = 1, ..., 24 hours.

6.5 Examples of Unsaturated Flow

In practice, solutions to Eqs. (6.7) and (6.10) require the development of powerful finite-element, finite-difference, or other numerical approaches. A variety of codes are available for such applications, including HYDRUS (a one-dimensional finite-element code, Vogel et al., 1996), VS2D (a two-dimensional finite-difference code, Lappala et al., 1987 and Healy, 1990), SWMS_3D (A three-dimensional variable-saturated, finite element code, Simunek et al., 1995), TOUGH2 (Pruess, 1987), and FEHM (Zyvoloski et al., 1997). This section will illustrate examples of several of these codes are used to describe unsaturated flow.

6.5.1 Infiltration and Drainage in a Large Caisson

An experiment was conducted in a caisson 6 m in depth and 3 m in diameter at Los Alamos National Laboratory to study the hydraulic properties of the Bandelier Tuff (Abeele 1984). The experimental data were modeled using HYDRUS (Vogel et al., 1996). Observed $\psi - \theta$ and $K - \theta$ relationships (Figure 6.11) were matched using the simulation model to determine the hydraulic properties in the van Genuchten model. The resultant soil hydraulic parameters are: $K_s = 25.0$ cm/day, $n = 0.3308$, $\theta_r = 0.0$, $\alpha = 0.01433$ (1/cm), and $\beta = 1.506$. Figure 6.12 shows the downward progression of a wetting front

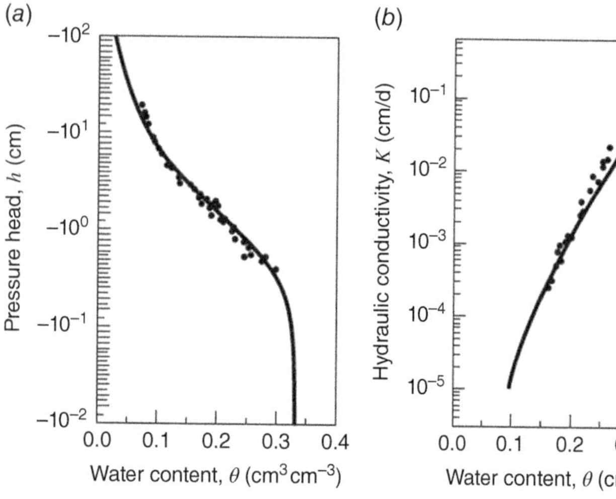

Figure 6.11 Determination of hydraulic properties from the hydraulic testing results for Crushed Bandelier Tuff (Vogel et al., 1996 / USDA / Public Domain). (*a*) Soil water retention curve; (*b*) hydraulic-conductivity curve.

during transient infiltration in the Bandelier Tuff. Note how the pressure head is close to zero behind the wetting front and strongly negative ahead of the front, indicating dry soil. As drainage occurs, the water content declines everywhere over the next 100 days (Figure 6.13).

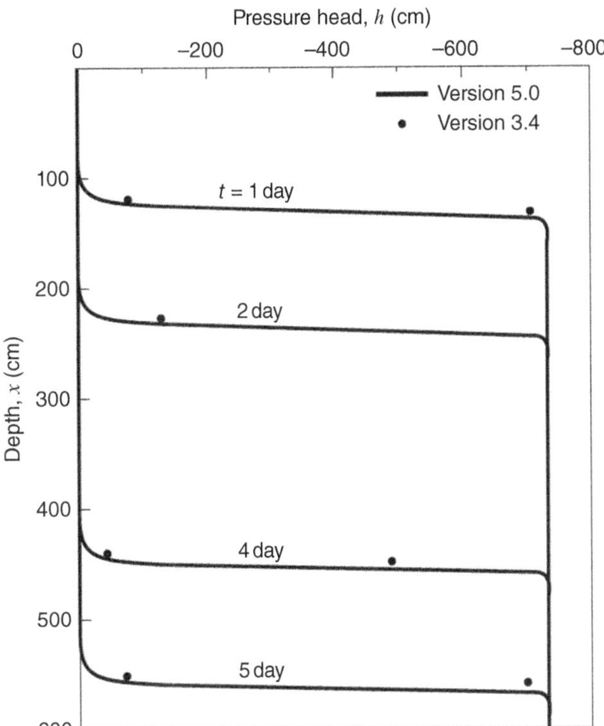

Figure 6.12 Predicted water-content profiles during transient infiltration in Bandelier Tuff (Vogel et al., 1996 / USDA / Public Domain).

6.5.2 Unsaturated Leakage from a Ditch

This example looks at the leakage of water from a ditch 4 ft wide and 10 ft deep. Water infiltrated into the unsaturated zone at a rate of 5 ft/yr. The background soil moisture content is 0.118 produced by a steady annual infiltration rate of 0.011 ft/yr from the ground surface. The space-time variation in pressure head and volumetric water content was simulated using the code VS2DT (Figure 6.14). The pressure head distributions at times of 25, 225, and 525 years are shown in Figure 6.14*a* through 6.14*c*, respectively. Figure 6.14*d* through 6.14*f* depict the calculated distributions in volumetric water content. The hydraulic properties are characteristic of sand: K_s = 16 ft/day, n = 0.33, $\theta_r = 0.072$, $\alpha = 72$ (1/ft), and $\beta = 1.7$ for the van Genuchten model. The ratio of vertical hydraulic conductivity to the horizontal hydraulic conductivity is 0.1. The results illustrate: (1) how high water contents correspond with low negative pressure heads; (2) that flow vertically is larger than flow horizontally; and (3) as time increases, water flows down to the water table.

6.6 Groundwater Flow in Fractured Media

In fractured rocks, the interconnected network of fractures provides the main pathway for fluid flow. The solid rock blocks in most cases are much less permeable than the network.

6.6.1 Cubic Law

A single fracture is commonly represented using the idealized parallel plate model. The plates represent the rock matrix with the fracture defined by the open space between the two plates (Figure 6.15). The thickness of the fracture is described in terms of an aperture *b*. The volumetric flow in a fracture is a function of the aperture cubed—known as the *cubic law* (Romm, 1966):

$$Q = -\frac{\rho_w g b^2}{12\mu}(bw)\frac{\partial h}{\partial L} \tag{6.24}$$

where Q is volumetric flow rate, ρ_w is density of water, g is gravitational acceleration, μ is viscosity, b is aperture opening, w is fracture width perpendicular to the flow direction,

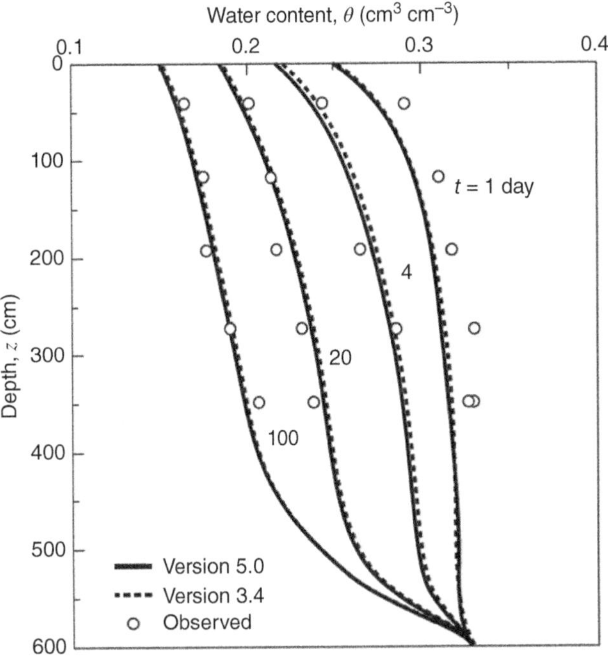

Figure 6.13 Predicted and observed water-content profiles during drainage of Bandelier Tuff (Vogel et al., 1996 / USDA / Public Domain).

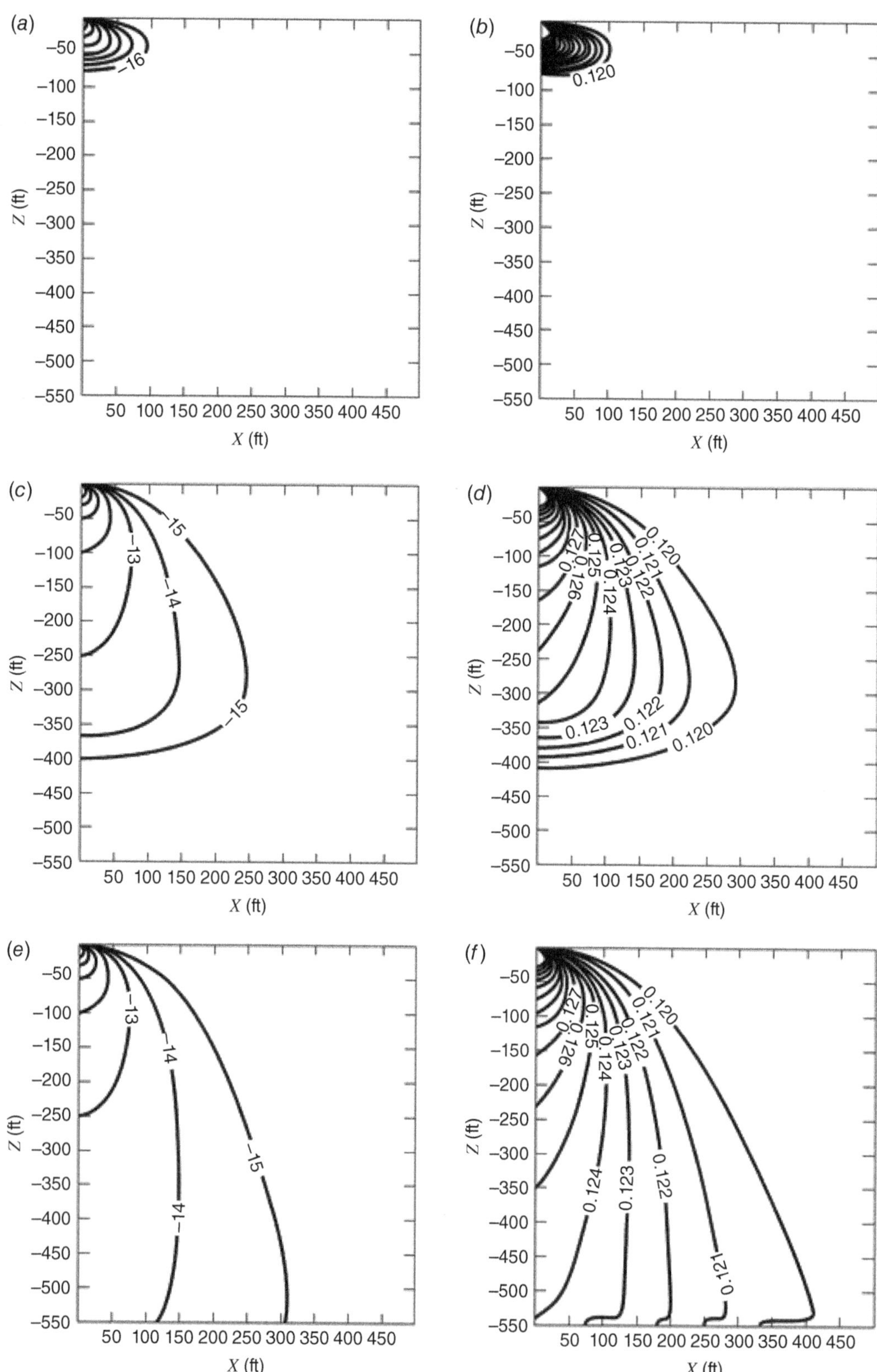

Figure 6.14 Water leakage into the unsaturated zone through a ditch (simulated using VS2DT). (*a*) Pressure head at 25 years; (*b*) pressure head at 225 years; (*c*) pressure head at 525 years; (*d*) volumetric water content at 25 years; (*e*) volumetric water content at 225 years; (*f*) volumetric water content at 525 years (U.S. Geological Survey, Healy, 1990 / Public domain).

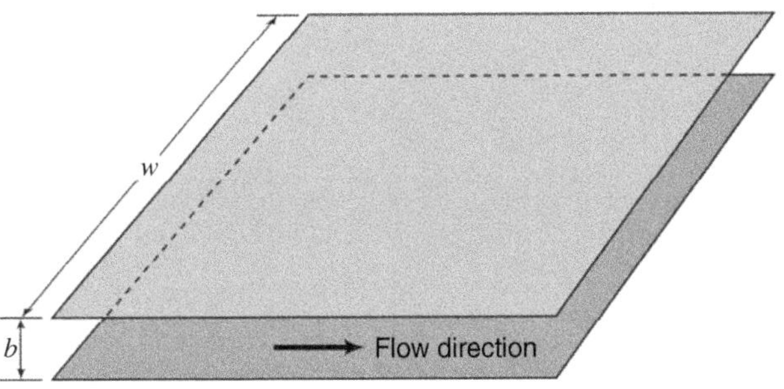

Figure 6.15 A parallel-plate model representing a single fracture (Schwartz and Zhang, 2003 / John Wiley & Sons).

and $\partial h/\partial L$ is head gradient along the fracture. This equation is of the form $Q = KiA$, where i is the gradient ($\partial h/\partial L$) and the area A is (bw). The hydraulic conductivity for this single fracture is

$$K = \frac{\rho_w g b^2}{12\mu} \tag{6.25}$$

Equation (6.25) is confirmed by experiments for smooth optical glass. Real fractures, however, have a variable aperture due to the roughness of the walls of the fracture. In some places, the two walls of the fracture actually touch. In other places, in the same fracture, the aperture can be large. Flow in a fracture network is also commonly reduced by the presence of secondary minerals that form in the fracture and plug it. The hydraulic conductivity for a rough aperture may be written as

$$K = \frac{\rho_w g b^2}{12\mu[1 + C(x)^n]} \tag{6.26}$$

where C is some constant larger than one, x is a group of variables that describe the roughness, and n is some power greater than one. Hence, roughness causes a decrease in hydraulic conductivity.

Example 6.6 The aperture of a fracture is 0.001 m. The fracture itself is 0.1 wide and 10 m long. If the hydraulic gradient in the fracture is 0.001, what is the volumetric flow rate in the fracture?

Solution

The hydraulic conductivity in the fracture is

$$K = \frac{\rho_w g b^2}{12\mu} = \frac{(998.2\text{kg/m}^3)(9.8\ \text{m/s}^2)(0.001\ \text{m})^2}{12(1.002 \times 10^{-3}\text{kg/m/s})} = 0.81\ \text{m/s}$$

The volumetric flow rate in the fracture is

$$Q = -K(bw)\frac{\partial h}{\partial L} = (0.81\ \text{m/s})\left(0.1 \times 10\ \text{m}^2\right)(0.001) = 0.00081\ \text{m}^3/\text{s}$$

6.6.2 Flow in a Set of Parallel Fractures

Calculating the quantity of flow in a complex network of fractures is difficult and usually requires a sophisticated computer model. For scoping calculations, however, there are some simple fracture systems that are amenable for analysis. Examples include networks of equally-spaced parallel fractures with no groundwater flow in the matrix with one (Figure 6.16*a*) or three (Figure 6.16*b*) fracture sets. (Reeves et al., 1986). The density of fractures is defined as the *fracture frequency* (N), which is the number of fractures per unit length. A term called *fracture porosity* is the ratio of the fracture volume over the matrix volume and is written as

$$\varphi_f = \frac{b}{s} \tag{6.27}$$

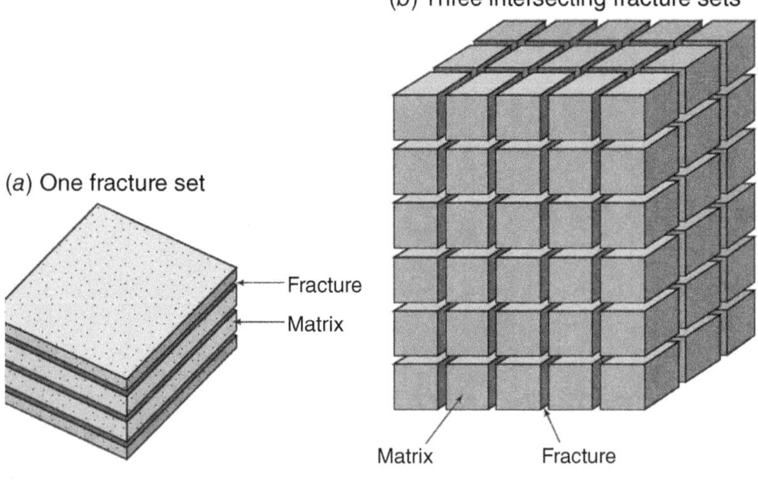

Figure 6.16 A double-porosity model with one (*a*) and three sets of parallel fractures (Reeves et al., 1986 / USDA / U.S. Department of Energy).

where φ_f is fracture porosity, b is fracture aperture, and s is termed fracture spacing which is the distance between fractures. Snow (1968) derived an expression relating the permeability of a fracture network to fracture porosity and fracture aperture.

$$k_f = \frac{b^3}{12}N, \text{or } K_f = \frac{\rho_w g b^3}{12\mu}N \tag{6.28}$$

The fracture spacing and fracture frequency can be converted by

$$N = \frac{\cos(\theta_f)}{s} \tag{6.29}$$

where θ_f is the fracture orientation which is the angle between the fractures and flow directions, or principal directions of hydraulic conductivity. Assuming θ_f is zero, the permeability can be expressed in terms of fracture porosity as

$$k_f = \frac{b^2}{12}\varphi_f, \quad \text{or} \quad K_f = \frac{\rho_w g b^2}{12\mu}\varphi_f \tag{6.30}$$

Equation (6.30) indicates that the permeability of a network of equally-spaced fractures is the product of permeability in a single fracture and fracture porosity.

Example 6.7 Consider a network of parallel fractures with a fracture spacing of 10 m and individual fracture apertures of 0.001 m. The fracture network is oriented parallel to the direction of flow and the geological medium is homogeneous. Calculate the permeability, hydraulic conductivity, and fracture porosity of the medium.

Solution

Assuming that the fractures are parallel to the flow direction, the fracture frequency of the medium is

$$N = \frac{1}{s} = \frac{1}{(10\,\text{m})} = 0.1\,(1/\text{m})$$

The permeability of the medium is

$$k_f = \frac{b^3}{12}N = \frac{(0.001\,\text{m})^3}{12}(0.1\,\text{m}^{-1}) = 8.3 \times 10^{-12}\text{m}^2$$

The hydraulic conductivity is

$$K_f = \frac{\rho_w g k_f}{\mu} = \frac{(1000\text{kg/m}^3)(9.8\text{m/s}^2)(8.3 \times 10^{-12}\,\text{m}^2)}{(1.002 \times 10^{-3}\text{kg/m/s})} = 8.1 \times 10^{-5}\text{m/s}$$

The fracture porosity is

$$\varphi_f = \frac{b}{s} = \frac{(0.001\ \text{m})}{(10\ \text{m})} = 1.0 \times 10^{-4}$$

6.6.3 Equivalent-Continuum Approach

So far, we have considered fracture flow in terms of single fractures or a network of parallel fractures. For most problems, it is not practical to worry about flow in terms of the characteristics of individual fractures or simple sets. More commonly, a fractured medium is represented as an equivalent porous medium. In other words, if the medium is sufficiently fractured and one "stands back" far enough, the medium looks like a porous medium with large grains. The equivalent permeability of the continuum in each principal direction can be obtained from the permeability of fracture and matrix network.

$$k = \frac{k_m + \frac{A_f}{A_m} k_f}{1 + \frac{A_f}{A_m}} \tag{6.31}$$

where k_m is the permeability of the matrix, k_f is the permeability of fracture network, and A_m and A_f are the cross-sectional contact areas of matrix and fractures in the principal directions, respectively. For a fractured rock with fracture components in three coordinate directions (Figure 6.16*b*), Eq. (6.31) applies to each direction, respectively.

Example 6.8 In a network of fractures, the permeability of the fractures is 1.0×10^{-8} m^2, and permeability of the matrix is 1.0×10^{-14} m^2. The ratio of the cross-sectional contact areas between a fracture and matrix block is 1.0×10^{-6}. What is the equivalent permeability of the medium? If the ratio is 10^{-4}, what is the equivalent permeability?

Solution

The equivalent permeability for $A_f/A_m = 1.0 \times 10^{-6}$:

$$k = \frac{k_m + \frac{A_f}{A_m} k_f}{1 + \frac{A_f}{A_m}} = \frac{(1.0 \times 10^{-14}\ \text{m}^2) + (1.0 \times 10^{-6})(1.0 \times 10^{-8}\text{m}^2)}{1 + 1.0 \times 10^{-6}} = 2.0 \times 10^{-14}\ \text{m}^2$$

The equivalent permeability of the effective continuum for $A_f/A_m = 1.0 \times 10^{-4}$:

$$k = \frac{k_m + \frac{A_f}{A_m} k_f}{1 + \frac{A_f}{A_m}} = \frac{(1.0 \times 10^{-14}\ \text{m}^2) + (10^{-4})(1.0 \times 10^{-8}\ \text{m}^2)}{1 + 10^{-4}} = 10^{-12}\text{m}^2$$

Exercises

6.1 The density of water is 1.0 g/cm^3, the bulk density of soil is 1.6 g/cm^3, and the grain density is 2.67 g/cm^3. Calculate the porosity of the soil. If the weight of dry soil is 200 g and the weight of the wet soil is 230 g, calculate the volumetric water content and water saturation of the sample.

6.2 The porosity, volumetric water content, and residual water content of a sample are 0.45, 0.21, and 0.05, respectively. Calculate the water saturation and the effective water saturation.

6.3 Infiltration is very important for water flow in the unsaturated zone. For a two-hour precipitation event, the precipitation rate is 5 in./h. The infiltration rate attains steady after 10 minutes. The saturated hydraulic conductivity is 5 ft/day. Draw a plot of infiltration rate versus time in the unsaturated zone. Assume the infiltration rate is zero before the storm.

6.4 Compare water flow Eq. (4.14) in the saturated zone with water flow Eq. (6.10) in the unsaturated zone. Write down and explain the similarity and the difference between the two equations.

6.5 Calculate and plot the water retention curves, relative hydraulic conductivity curves, and specific moisture capacity curves for Del Monte sand, Columbia sandy loam, and Limon silt in Table 6.1 using Brooks–Corey and van Genuchten models. Discuss the characteristics of the curves for different geological materials and different models.

6.6 Numerical models are available for the simulation of water flow in the unsaturated zone. An infiltration experiment was carried out in the farming field to study the potential pesticide contamination of groundwater. The infiltration rate is 10 in./h in a control area. The thickness of the unsaturated zone is 20 ft. Assuming that one-dimensional model will be used for this project, what boundary and initial conditions would you like to specify? What hydraulic parameters do you need in the model? What do saturation and pressure head profiles look like in the unsaturated zone?

6.7 If the model is two-dimensional vertical section, answer the same questions in Problem 6.6.

6.8 For an unconsolidated sand (Table 6.1), calculate the infiltration rate using the Green–Ampt model for a ponding depth of 2 cm. Assume that a precipitation event lasts two hours. If the precipitation rate is 40 cm/h, plot precipitation rate, infiltration rate, and run-off rate on the same graphic paper.

6.9 Briefly describe the discrete fracture, the double-porosity, and the effective continuum approaches for dealing with water flow in fractured geological media.

6.10 If the aperture of a fracture is 0.001 m, what is the permeability and hydraulic conductivity in the fracture? For a network of equally-spaced fractures with fracture porosity of 1.0×10^{-4}, calculate the permeability and the hydraulic conductivity of the fractured media.

6.11 If the permeabilities of fracture and matrix blocks are 10^{-4} and 10^{-8} m^2, respectively, what is the effective permeability if an effective continuum model is used to describe the fractured medium? Assume that the ratio of cross-sectional contact areas between the fracture and the matrix blocks is 10^{-3}.

References

Abeele, W. V. 1984. Hydraulic testing of rushed Bandelier Tuff. Report No. LA-10037-MS, Los Almos National Laboratory, Los Almos, New Mexico.

ASTM. 1992. ASTM Standards on groundwater and vadose zone investigations. ASTM Publication Code Number (PCN): 03-418192-38, Philadelphia, Pennsylvania, 166 p.

Brooks, R. H., and A. T. Corey. 1966. Properties of porous media affecting fluid flow. Journal of the Irrigation and Drainage Division, v. 72, no. IR2, p. 61–88.

Domenico, P. A., and F. W. Schwartz. 1998. Physical and Chemical Hydrogeology. John Wiley & Sons, New York, 506 p.

Eagleson, P. S. 1978. Climate, soil, and vegetation. 3. A simplified model of soil moisture movement in the liquid phase. Water Resources Research, v. 14, no. 5, p. 722–730.

Gardner, W. H. 1986, In A. Klute (ed.), Water Content in Methods of Soil Analysis. American Society of Agronomy, Madison, Wisconsin, p. 493–544.

Green, W. H., and C. A. Ampt. 1911. Studies on soil physics. I. The flow of air and water through soils. Journal of Agricultural Sciences, v. IV, no. Part I 1911, p. 1–24.

Haverkamp, R., M. Vauclin, J. Tovina, P. J. Wierenga, and G. Vachaud. 1977. A comparison of numerical simulation models for one-dimensional infiltration. Soil Science Society of America Proceedings, v. 41, p. 285–294.

Healy, R.W. 1990. Simulation of solute transport in variably saturated porous media with supplemental information on modification to the U.S. Geological Survey's computer program VS2D. U.S. Geological Survey Water-Resources Investigations Report 90-4025, 125 p.

Lappala, E. G., R. W. Healy, and E. P. Weeks. 1987. Documentation of computer program VS2D to solve the equations of fluid flow in variably saturated porous media, U.S. Geological Survey Water-Resources Investigations Report 83-4099, 184 p.

Mualem, Y. 1976. A new model for predicting the hydraulic conductivity of unsaturated porous media. Water Resources Research, v. 12, no. 3, p. 513–522.

Philip, J. R. 1957. The theory of infiltration. 4. Sorptivity and algebraic infiltration equations. Soil Sciences, v. 84, p. 257–264.

Pruess, K. 1987. TOUGH user's guide. LBL-20700, NUREG/CR-4645, Lawrence Berkeley Laboratory, Berkeley, California.

Ravi, V., and J. R. Williams. 1998. Estimation of infiltration rate in the vadose zone. V. I: Compilation of simple mathematical models. U.S. Environmental Agency, EPA/600/R-97/128a, 26 p.

Reed, J. E. 1980. Type curves for selected problems of flow to wells in confined aquifers. U.S. Geological Survey Water-Resources Investigations, Chapter B3, 106 p.

Reeves, M., D. S. Ward, N. D. Johns, and R. M. Cranwell. 1986. Theory and implementation for SWIFT II, The Sandia Waste-Isolation Flow and Transport Model (SWIIFT) Release 4.81. NUREG/CR-2324 and SAND81-2516, Sandia National Laboratories, Albuquerque, New Mexico.

Romm, E. S. 1966. Flow characteristics of fractured rocks (in Russian), Nedra, Moscow.

Salvussi, G. D., and D. Entekhabi. 1994. Explicit expressions for Green–Ampt (delta function diffusivity) infiltration rate and cumulative storage. Water Resources Research, v. 30, no. 9, p. 2661–2663.

Schwartz, F. W., and H. Zhang. 2003. Fundamentals of Groundwater. John Wiley & Sons, Hoboken, New Jersey, 583 p.

Simunek, J., K. Huang, and M. Th. Van Genuchten. 1995. The SWMS_3D code for simulating water flow and solute transport in three-dimensional variable-saturated media. Version 1.0, Research Report No. 139, U.S. Salinity Laboratory, Riverside California.

Snow, D. T. 1968. Rock fracture spacings, openings, and porosity. Journal of the Soil Mechanics and Foundations Division, v. 94, p. 73–91.

Stannard, D. I. 1986. Theory, construction and operation of simple tensiometers. Groundwater Monitoring and Remediation, v. 6, p. 70–78.

USDA-SCS. 1972. National Engineering Handbook, Hydrology Section 4. USDA, Washington, DC.

Van Genuchten, M. T. 1980. A closed-form equation for predicting the hydraulic conductivity of unsaturated soils. Soil Science of American Proceedings, v. 44, no. 5, p. 892–898.

Vogel, T., K. Huang, R. Zhang, and M. Th Van Genuchten. 1996. The HYDRUS code for simulating one-dimensional water flow, solute transport, and heat movement in variable-saturated media. Version 5.0, Research Report No. 140, U.S. Salinity Laboratory, Riverside, California.

Wierenga, P. J., M. H. Young, G. W. Gee, R. G. Hills, C. T. Kincaid, T. J. Nicholson, and R. E. Cady. 1993. Soil characterization methods for unsaturated low-level waste sites. U.S. Nuclear Regulatory Commission, NUREG/CR-5988 (PNL-8480), Washington, DC.

Wierenga, P. J., L. W. Gelhar, C. S. Simmons, G. W. Gee, and T. J. Nicholson. 1986. Validation of stochastic flow and transport models for unsaturated soil: A comprehensive field study. U.S. Nuclear Regulatory Commission, NUREG/CR-4622.

Williams, J. R., Y. Ouyang, J. Chen, and V. Ravi. 1997. Estimation of infiltration rate in the vadose zone. V. II: Application of selected mathematical models. U.S. Environmental Agency, EPA/600/R-97/128b, 44 p.

Zyvoloski, G. A., B. A. Robinson, Z. V. Dash, and L. L. Trease. 1997. Summary of the models and methods for FEHM Application—a finite-element heat- and mass-transfer code. NTIS, U.S. Department of Commerce, Virginia.

7

Geologic and Hydrogeologic Investigations

CHAPTER MENU

Groundwater studies depend on an ability to investigate the subsurface and make key measurements. This chapter examines basic techniques for hydrogeologic investigations. It begins by looking at methods for describing the hydrogeologic setting. Experience shows that it is important to get this part of any study right. The geologic framework provides the foundation for subsequent investigations. The chapter continues with a discussion of methods for making hydraulic head measurements in the field. The last part of this chapter provides an overview of several geophysical approaches and a discussion of the overall site-investigation process.

The discussion of field measurements of hydraulic parameters will involve a series of chapters beginning with Chapter 9.

7.1 Key Drilling and Push Technologies

Field investigations depend on drilling holes in the ground or pushing sampling/measurement probes into the ground without creating cuttings. Various types of drilling rigs and newly emerging "push" technologies accomplish these tasks. Here, we focus on auger drilling and mud/air rotary drilling, which are used most. Readers interested in other methods can refer to a detailed summary in Barcelona et al. (1985) or Campbell and Lehr (1973).

7.1.1 Auger Drilling

Augers can be used to drill holes in unconsolidated sediments or in other words, materials other than rock. The rig rotates an auger with a drill bit attached into the ground (Figure 7.1). The augers carry material cut by the drill bit up the hole to the ground surface. The hole is deepened by adding augers to the string of augers already in the ground. There are two main types of augers. *Solid-stem augers* look like large versions of the drill bits for drilling large holes in wood. They work best in cohesive deposits (like clays) that will stay open once the augers are removed. The open hole provides access for side-wall sampling, geophysical logging, or for installing a piezometer. Certain saturated deposits, like sand, are non-cohesive. When solid-stem augers are withdrawn from such materials, the hole caves in, blocking further access.

Hollow-stem augers provide a nifty solution to the problem of caving. The drill bit is constructed in two parts, with the central part held in place by a set of small diameter drilling rods that are inserted through the hollow core of the augers.

Fundamentals of Groundwater, Second Edition. Franklin W. Schwartz and Hubao Zhang.

Companion website: www.wiley.com/go/schwartz/fundamentalsofgroundwater2

When it comes time to sample or to install a piezometer, the core of the bit is pulled out at the end of the drill rods to provide an opening from the surface to the porous medium below the augers. The augers act as a temporary casing that holds the hole open. Drilling begins again once the bit is reassembled.

This feature is ideal for collecting periodic soil samples as drilling continues. A coring device, like a split spoon sampler, can be added to the end of a drill rod, that is run down the hole, and hammered into the porous medium just below the tip of the auger to provide the sample. As the photo shows (Figure 7.2), the sample becomes accessible by taking the sample tube apart.

This procedure simply involves unscrewing the steel drive shoe (bottom) from one end of the sampler and the drill rod coupler from the other end (top). These two fittings are what hold the two pieces of the tube together during sampling. With sand or other difficult-to-sample sediments, a sample retainer can be added internally.

A Shelby tube sampler can also collect a core from unlithified sediments. This sampler is basically a piece of steel pipe with one end sharpened around the circumference to create a cutting shoe. It is fastened to the drill rod and is pushed or hammered ahead of an auger to collect the sample. At the surface, the tube is clamped, and the sample is extruded with a hydraulic ram.

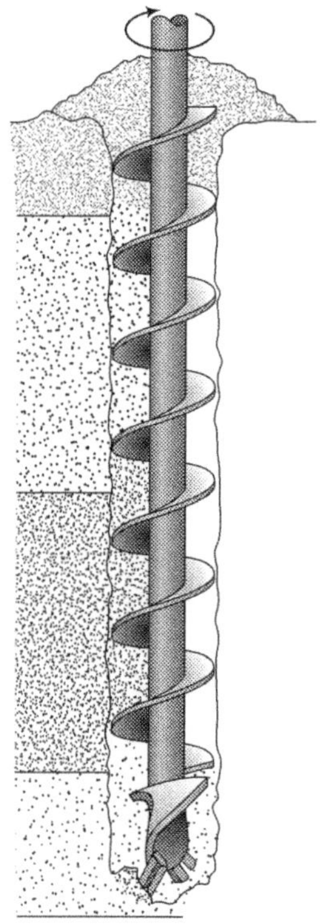

Figure 7.1 A solid-stem auger in the process of drilling a hole. The bit dislodges sediment at the bottom of the hole, which is carried up the borehole by the auger (U.S. Environmental Protection Agency Scalf et al., 1981/Public Domain).

7.1.2 Mud/Air Rotary Drilling

Mud/air rotary drilling involves turning a casing (pipe) string with a bit attached (Figure 7.3). There are several different styles of bits, selected depending upon the medium being drilled. The tricone bit is a familiar style with three "gear-like" rollers attached. Rotating the bit produces cuttings that are up and out of the borehole. A *drilling fluid*, i.e., air or water, is circulated under pressure down the drill casing and through the drill bit. The drill cuttings are brought back up the hole with the return flow of drilling water or air. Drilling with air has the advantage of not needing to haul water for drilling, which can be problematic in arid settings. In addition, cleanup is often easier because the cuttings are not mixed with large volumes of water.

With mud rotary rigs, circulation of the drilling fluid is occasionally "lost" when water or air goes into a permeable zone rather than back up the hole. In this case, it is necessary to plug this permeable zone using drilling fluid thickened by additives (e.g., bentonite). This *drilling mud* must be removed eventually if the zone is subsequently developed to provide groundwater.

Figure 7.2 Upon recovery, a split spoon sampler is broken in half along its length to obtain the sample (U.S. Geological Survey Kappel and Teece, 2007/Public domain).

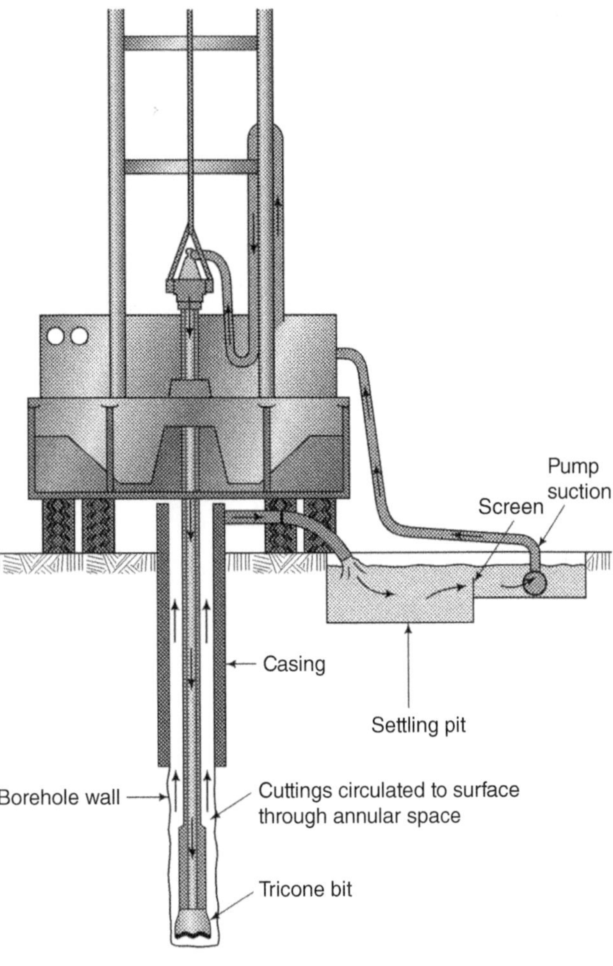

Figure 7.3 Example of a rotary drill rig. The arrows illustrate how drilling mud circulate from the mud pit down the casing, and out the bit. The fluid returning up the borehole carries cuttings up and out of the hole to the mud pit (NWWA, 1984 used with permission Australian Drilling Industry Association).

Rotary drill rigs are capable of drilling through both unconsolidated sediment and bedrock. This capability makes it the rig of choice for installing wells deep into bedrock. With a mud rotary rig, the drilling mud usually keeps the hole open in unconsolidated units. Drilling with air works best units are consolidated or semi-consolidated. When air rotary is used for "soft" formations, a surface casing might be needed to keep upper part of the hole open.

Drill cuttings returned with the mud can be examined to provide a geologic log. It is also possible to collect split spoon samples with a rotary rig. Commonly, geophysical logs are run in these boreholes to provide a more complete picture of the stratigraphy. One drawback with air rotary rigs is the possibility for small quantities of oil from the compressor to be introduced to the subsurface. Filters can be attached to avoid this problem.

7.1.3 Direct-Push Rigs

Direct-push technologies provide useful alternatives for hydrogeological investigations in unlithified materials particularly. Dedicated cone *penetrometer rigs* provided some of the first push-type measurements (e.g., Figure 7.4). The *cone penetrometer test* (CPT) system uses a large capacity hydraulic ram to physically push the cone penetrometer into the ground using small diameter steel rods. For the hydraulic ram to work, it needs to be anchored in a way to provide a deadweight downforce, such as the ~200 kilonewtons [kN] or ~22 tons, provided by the heavy truck shown in Figure 7.4.

Otherwise, the hydraulic ram would simply lift the vehicle. These days, there are smaller, lighter, and less expensive rigs capable of making CPT measurements. Earth-anchoring systems make it possible to use lighter equipment, able to push above their mass. The newer equipment not only maintains the functionality for CPT measurements but also capabilities for conventional drilling.

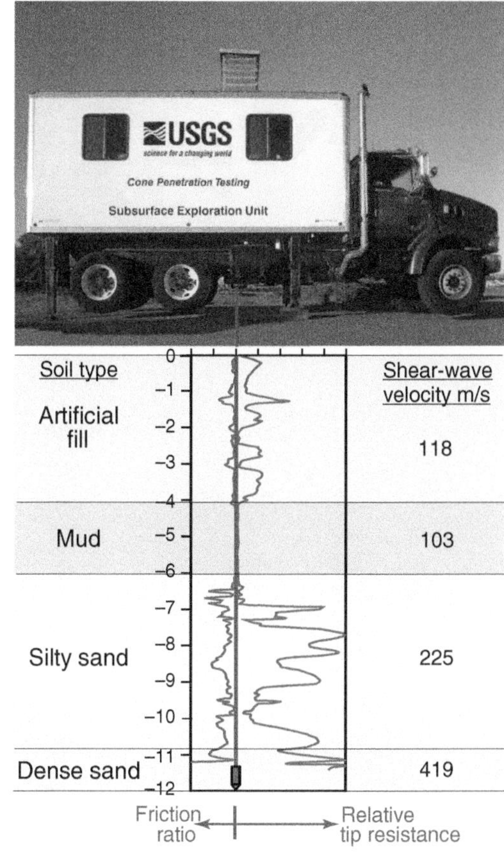

Figure 7.4 Example of a heavy truck equipped with a CPT system. Measurements from the probe can provide assessments of site stratigraphy and in this case seismic velocities (U.S. Geological Survey Noce and Holzer, 2003/Public domain).

Two electronic sensors on the cone penetrometer measure various shear resistances (Chiang et al., 1989). One sensor measures the force on the cone tip (also called cone-end bearing resistance), and the second measures the shear resistance along the side of the penetrometer (Strutynsky and Sainey, 1990). The cone tip resistance has SI units of kiloPascals, where 1 kPa = 1 kN/m^2, or megaPascals where 1 kPa = 1000 kN/m^2 (Mayne, 2007). Shear resistance along the sleeve of the probe is measured as kPa. Occasionally, resistances are in English units, tons per square foot where 1 tsf ≈ 1 bar = 100 kPa = 0.1 MPa (Mayne, 2007).

As the string of steel rods with the cone penetrometer on the end is pushed into the ground, the two shear resistances (tip and side) are measured, along with the penetration depth. The ratio of the shear or frictional resistance to the cone-end bearing resistance is the *friction ratio*. The friction ratio is related to the fines content of the medium—typically lower in sands (e.g., 0.02% or 2%) and high in clays (e.g., 0.05% or 5%). Typically, values range between 1% and 10%. The cone end-bearing resistance increases exponentially with grain size. Typical values of cone-end bearing resistances are 0.7–1.5 MPa (7–15 tsf) for stiff clay and 15–30 MPa (150–300 tsf) for dense sand (Strutynsky and Sainey, 1990). The cone end-bearing resistance can respond to layers just a few centimeters thick. The frictional resistance has a resolution of about 15 cm (Strutynsky and Sainey, 1990).

The kinds of CPT measurements just described commonly assist in geotechnical applications, such as the design of foundations and pilings. Additionally, the downhole resistance measurements can be used with graphical templates to provide continuous stratigraphic data at the site of investigation. For example, the four geologic layers in Figure 7.4 were identified based on the friction ratio combined with the tip resistance to develop information on the stratigraphy. There are many published templates for translating penetrometry measurements to types of geologic materials. The simplest use is tip resistance plotted versus friction ratio to provide an indication of the types of material (Figure 7.5). More sophisticated approaches use additional measured parameters such as normalized pore-water pressures (Robertson et al., 1986). For major investigations, the choice of an optimum interpretive template is made by tests on nearby sites with well-characterized core measurements.

Figure 7.6 illustrates the typical results available from a cone penetrometer carrying the two resistance sensors. The measurements were part of a USGS study on the east side of San Francisco Bay, California south of Oakland (Bennett et al., 2009).

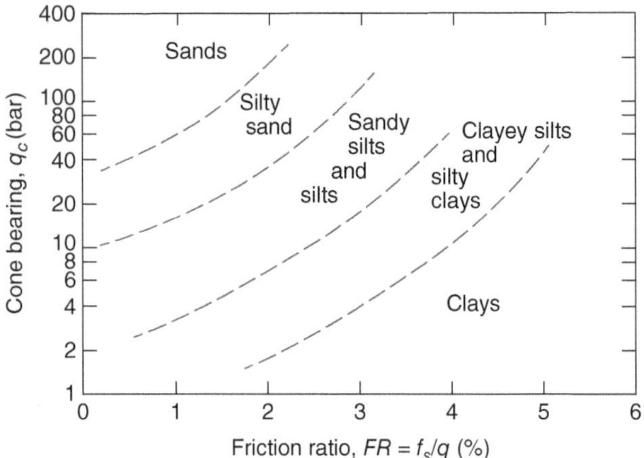

Figure 7.5 Information from a cone penetrometer test can be interpreted to provide a geostratigraphic profile using one of many classification schemes. Shown here is one of the simplest schemes (Robertson and Campanella, 1983. With permission Canadian Science Publishing. All Rights Reserved.).

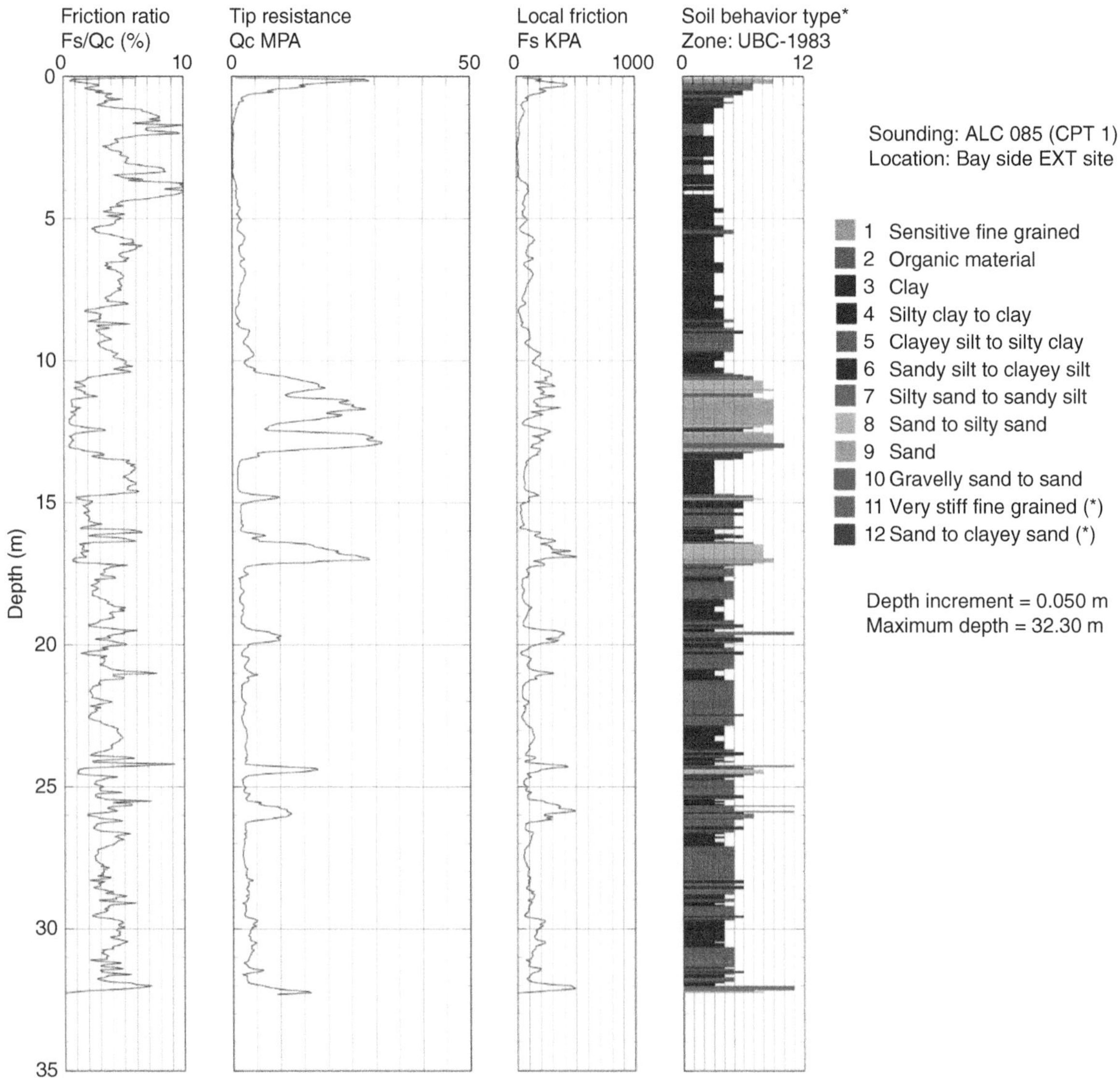

Friction ratio Fs/Qc (%) 0 10

Tip resistance Qc MPA 0 50

Local friction Fs KPA 0 1000

Soil behavior type* Zone: UBC-1983 0 12

Depth (m) 0 5 10 15 20 25 30 35

Sounding: ALC 085 (CPT 1)
Location: Bay side EXT site

1 Sensitive fine grained
2 Organic material
3 Clay
4 Silty clay to clay
5 Clayey silt to silty clay
6 Sandy silt to clayey silt
7 Silty sand to sandy silt
8 Sand to silty sand
9 Sand
10 Gravelly sand to sand
11 Very stiff fine grained (*)
12 Sand to clayey sand (*)

Depth increment = 0.050 m
Maximum depth = 32.30 m

Figure 7.6 Example of a CPT log that shows friction ratio, tip resistance, local (side friction) friction resistance, and an interpreted geostratigraphic column (U.S. Geological Survey Bennett et al., 2009/Public domain).

The CPT logs available were friction ratio (%), tip resistance (MPa), and local friction or (side friction resistance) (kPa). The 12-zone geostratographic profile (Figure 7.6) utilized the classification scheme from Robertson (1990). The study also provided a regional context for the stratigraphic interpretations. For example, the dense sand unit and associated gravel between 10 and 13 m correlated regionally with the Pleistocene Merritt Sand (Bennett et al., 2009).

In other applications, individual geostratigraphic profiles collected along a line facilitated construction of hydrogeologic cross sections. Benefit of this application of CPT include the consistency provided by geophysical parameters collected in the same way and the high vertical resolution coming from continuous measurements.

Refinements of push technologies have extended to in situ, water-pressure measurements (yielding hydraulic heads) as a function of depth, and hydraulic conductivities at specific depths, using the rate of pore-pressure dissipation (Chiang et al., 1989) or estimates from correlations with the types of materials. The available uses are numerous and include soil sampling, measurements of shear-wave velocities, continuous profiling of temperature, electrical resistivity, laser-induced fluorescence, abundance of volatile organic compounds, and more (Noce and Holzer, 2003). Interested readers can find instructive web resources on the use and application of these methodologies. Here is a link to a Geoprobe Systems® webinar on how their hydraulic profiling tool (HPT) yields information on pore pressures and hydraulic conductivities (https://www.youtube.com/watch?v=UMVQdeXflP8).

A second approach to direct-push assessment uses the static weight of the vehicle together with percussive hammering to advance a string of rods into the ground. This approach has much in common with CPT methods. The ease of use, flexibility, and cost-effectiveness make it ideal for the investigation of contaminated sites. Typical applications include soil sampling and collecting samples of soil gas and water. Unlike auger drilling, this type of push technology does not produce cuttings, which might require disposal when they are contaminated.

These rigs (e.g., Figure 7.7) can push or hammer a casing string to about 25 m under good conditions. They also provide access to congested sites that might include the insides of buildings. They are the tool of choice for rapid sampling of solids at contaminated sites with capabilities of installing permanent monitoring wells for gas and water sampling.

Figure 7.7 Soil-probing machine mounted in a pickup truck can easily be transported and set up at a site. At this site on the floodplain of Fourche Renault Creek, it was being used for sediment sampling (U.S Geological Survey / https://www.usgs.gov/media/images/geoprobe-used-flood-plain-sampling / last accessed under 5 May 2023).

7.2 Piezometers and Water-Table Observation Wells

One of the main reasons for drilling at a site is to install piezometers or water-table observation wells. As we saw in Chapter 3, a *piezometer* in its simplest form is a standpipe that is installed to some depth below the water table (Figure 7.8). Water entering the piezometer, gradually fills the casing up to some stable elevation. The elevation of the water in the casing is the hydraulic head at the point of measurement—the midpoint of the screen at the bottom of the standpipe (Figure 7.8).

A *water-table observation well* has a slightly different design and provides somewhat different information. The standpipe is screened across the water table (Figure 7.8). The water-level measurement in this well provides the elevation of the water table and the hydraulic head at the top of the groundwater system. The point of measurement is the water table, which is the top of the saturated groundwater system. Thus, as the water table rises or falls, the point of measurement also moves.

7.2.1 Basic Designs for Piezometers and Water-Table Observation Wells

Real piezometers do not look much like the idealization in Figure 7.8. A practical design (Figure 7.9) provides:

i) a screen to create a relatively large surface area for water to enter the standpipe
ii) a sand pack around the screen to increase the effective size of the screen and to support material placed above
iii) a seal above the sand pack to prevent water from leaking along the casing
iv) screen and casing materials that will not react with the groundwater or contaminants carried in the groundwater and
v) a casing protector to finish the top of the piezometer and to prevent unauthorized access.

Let us describe a piezometer typically installed in the field. It is constructed with a screen at the end of the casing. A screen is a piece of the casing with holes or slots cut to let water flow into the casing. Manufactured screens are designed both to be strong and to maximize the open area. The screen is surrounded by a *sand pack* that supports the screen structurally and provides a foundation for all the other materials that are added to the borehole above.

For water levels to change in a piezometer, some volume of water must flow into or out of the standpipe. The larger the surface area of the intake, the faster these inflows or outflows occur and the more rapidly water levels in well adjust to changing flow conditions. There is a big difference in surface area provided by a 5.1 cm (2 in.) diameter open end in a piece of pipe as compared to a 1.5 m long screen packed with sand in a 15.2 cm (6 in.) borehole (Figure 7.10). The surface area with a sand-packed screen is about 3500 times larger (Figure 7.10). Adding a sand pack makes the effective diameter of the screen the same as the diameter of the borehole.

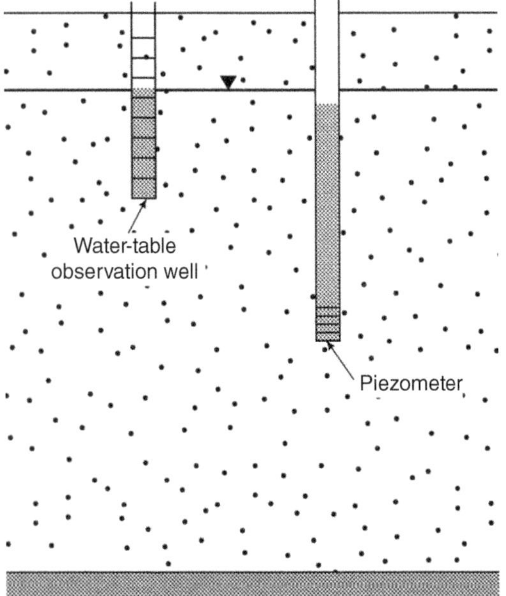

Figure 7.8 Schematic of a piezometer and a water-table observation well (Schwartz and Zhang, 2003/© John Wiley & Sons).

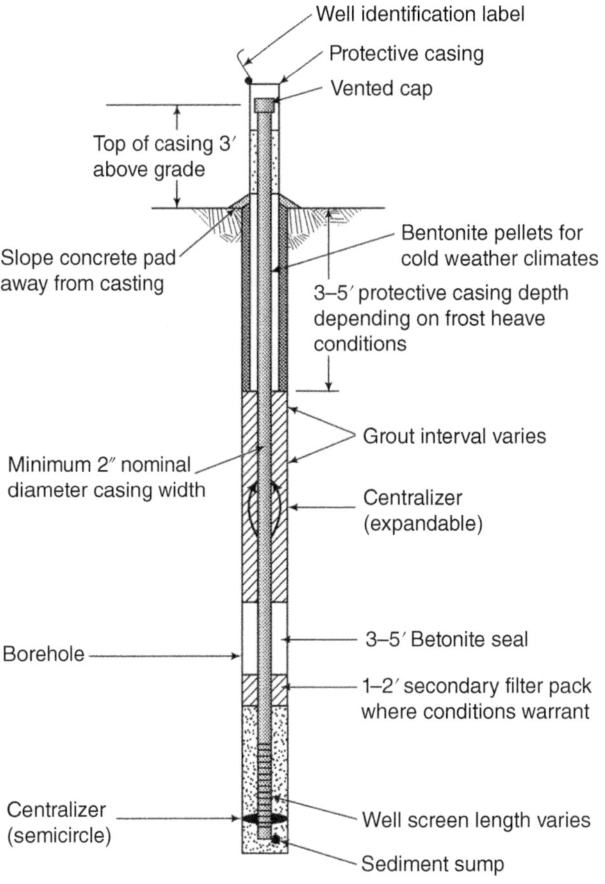

Figure 7.9 This figure shows the construction details for a basic standpipe piezometer (Nielsen, 1996. Reproduced by permission of the National Groundwater Association. Copyright © 1996. All rights reserved.).

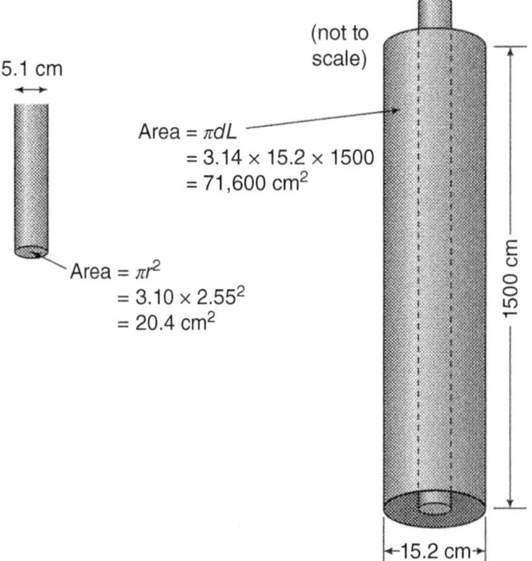

Figure 7.10 The surface area afforded by the open end of a 5.1 cm (2 in.) pipe as compared to the open area of a 1.5 m long, 15.2 cm (6 in.) diameter sand-packed borehole (Schwartz and Zhang, 2003/ © John Wiley & Sons).

A screen that is one meter long is adequate for most applications. In practice, however, screens are often constructed 3–5 m long, with piezometers resembling small wells. These longer screens are not inherently bad except that a hydraulic head measurement is supposed to be "point" measurement rather than a formation average. With wells used primarily for water-quality monitoring, the long screens give one an ability to "detect" contaminants, but not to estimate "point" concentration values.

A piezometer only works properly if the water moving into or out of the standpipe comes from the zone adjacent to the intake. If water leaks down the borehole from zones above the intake, then hydraulic head and water-quality measurements can be erroneous. Piezometers usually incorporate a seal and backfill above the sand pack, which prevent leakage down the borehole. The *seal* is usually constructed from *granular bentonite*, a clay mineral that expands and has a very low hydraulic conductivity once it absorbs water. Above the seal, the hole can be backfilled with bentonite grout or appropriate natural materials that support the casing and plug the borehole.

The upper part of the borehole is completed using concrete and a metal casing protector. The concrete seal prevents surface drainage from moving down the old borehole and anchors the casing protector. The casing protector can be locked to prevent unauthorized access to the piezometer and sometimes can prevent damage from collisions with vehicles. In high-traffic areas, piezometers are completed below grade level to prevent damage.

Piezometers are usually constructed with a nominal casing diameter of 5.1 cm (2 in.). This diameter is large enough to facilitate development following installation, measurement of hydraulic head with an electric tape, and sampling with small pumps or bailers. Common casing materials include polyvinyl chloride (PVC), or stainless steel. PVC pipes with threaded joints sealed by O-rings are used commonly. Applications involved with water-quality monitoring might necessitate other materials. On occasion, piezometers will be constructed with 15.2 cm (6 in.) casings. However, as we will discuss, these piezometers can be insensitive in low-permeability units.

A water table monitoring well (Figure 7.12) has many of the same characteristics as a piezometer. The main difference, however, is in the length of the screen. The water-table observation well has a long screen that extends above and below the water table. Thus, if the water-table level rises or falls, it will remain within the screened section. Providing a seal with this well is less critical than a piezometer because the zone above the screen is unsaturated. Good practice would provide a low permeability backfill to prevent surface drainage from moving down the old borehole and a casing protector.

7.3 Installing Piezometers and Water-Table Wells

The description of the basic design for wells and piezometers does not really explain how they are installed. Moreover, the point needs to be made that many different variations on these basic designs are implemented to respond to specific issues related to hole depth, geological setting, and requirements of the study. For example, it is difficult to provide good seals in formations that cave (that is, saturated sands and silts), or at depths greater than about 50 m.

7.3.1 Shallow Piezometer in Non-Caving Materials

The designs for piezometers or water-table observation wells (Figures 7.9 and 7.11) are typically used for non-caving materials. In rock or clayey sediments (e.g., lacustrine clay or glacial till), a borehole will remain open once the rotary drill stems or augers are removed. The screen and casing can be lowered down the hole and proper amounts of filter sand, bentonite, and backfill added in order. In a shallow borehole, these materials can be poured down the hole. A tape with a heavy weight attached is used to measure where these materials end up in the borehole. Once holes become deeper than about 30 m, it is difficult to get the materials to the bottom of the hole. For these deeper holes, alternative approaches are necessary.

7.3.2 Shallow Piezometer in Caving Materials

Caving materials, like water-saturated sands or silts, pose problems, especially with auger drilling techniques. Mud rotary holes often will stay open in these materials but the mud present in the hole may contaminate subsequent water samples collected for chemical analysis. Solid-stem augers are also problematical. Once the augers are removed, the borehole will collapse making it virtually impossible to install the casing. One approach with caving materials is to drill a hole to the desired depth with a hollow-stem auger, install the casing down the hollow center of the auger, and back out the auger. The hole will collapse around the casing to provide a natural seal. These types of piezometers are commonly used in non-regulatory applications or research investigations, where cost is an issue. Most regulatory agencies find these wells without seals unsatisfactory because the leakage along the casing cannot be ruled out.

Techniques are available with a hollow-stem auger or soil probing machines to install both a sand pack and a seal. However, it is difficult to send sand down to the screen, along the middle of a hollow stem auger. New technologies offer a potential solution to the installation of piezometers in caving materials. Access to the interval of interest is provided via a hollow stem auger or probe rods driven to depth with a push rig. A PVC screen, prepacked with sand, is lowered down inside the augers or probe rods attached to a standard 5.1 cm (2 in.) PVC riser pipe. This approach assures that filter material is

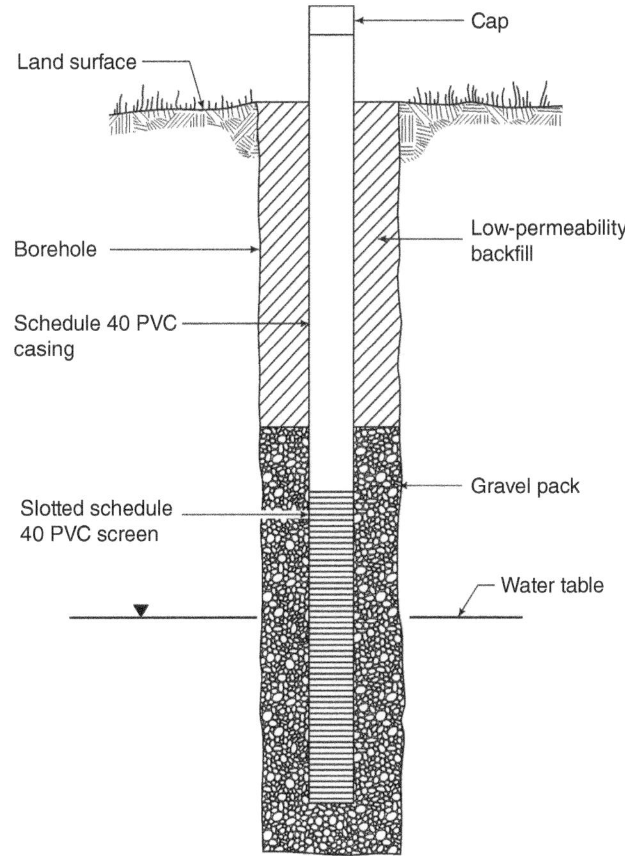

Figure 7.11 Construction details for a basic water-table observation well (Nielsen, 1996. Reproduced by permission of the National Groundwater Association. Copyright © 1996. All rights reserved.).

properly placed together with the screen. With the screen in place, the augers are retracted to expose the formation. A short section of sand is installed to provide a grout barrier and the borehole above the screen is grouted with granular bentonite or bentonite slurry. Piezometers of this type have been installed to depths of >20 m. Figure 7.12 illustrates a variation in this approach. In addition to a prepacked screen, one can also screw on a prepacked foam seal and prepacked bentonite seals, which are in turn installed to the bottom of the PVC casing (Figure 7.12*a*). Once the probe rods reach the appropriate depth and the expendable drive point is removed, the PVC casing with the attached screens and seals run down the inside of the probe rods. The probe rods are withdrawn and as water contacts the foam seal, it expands quickly seal the anulus of the borehole above the screen (Figure 7.12*b*). The bentonite seals expand more slowly, but also work to seal the anulus above the screen. Finally, grout can be pumped down to seal the upper portion of the borehole (Figure 7.12*c*).

In very complex settings, it is often necessary to case off caving intervals and to provide access for deeper holes. For example, at some sites, auger holes are cased to bedrock to keep the hole open for core drilling into bedrock. Temporary casing used for this purpose can be jacked out of the ground once the piezometer is completed. Cable-tool drilling is ideally suited for problems where a temporary casing needs to be installed while drilling. The advantage of a advantageous as compared to a hollow-stem auger is (i) a larger diameter hole to work in, and (ii) the ability to drill through rock. Cable tool rigs are, however, often slow in drilling.

7.3.3 Deep Piezometers

Installing piezometers to depths greater than 30 m in rock is difficult. There is less of a problem with caving because the drilling technique usually can be counted on to keep the hole open. The greater problem is in placing the sand pack and grout seal. One approach uses no sand pack, where the seal and backfill are kept away from the screen by a funnel-shaped metal or plastic basket attached to the casing above the screen. The seal is emplaced as pumped grout from above. These types of piezometers can be installed to about 100 m.

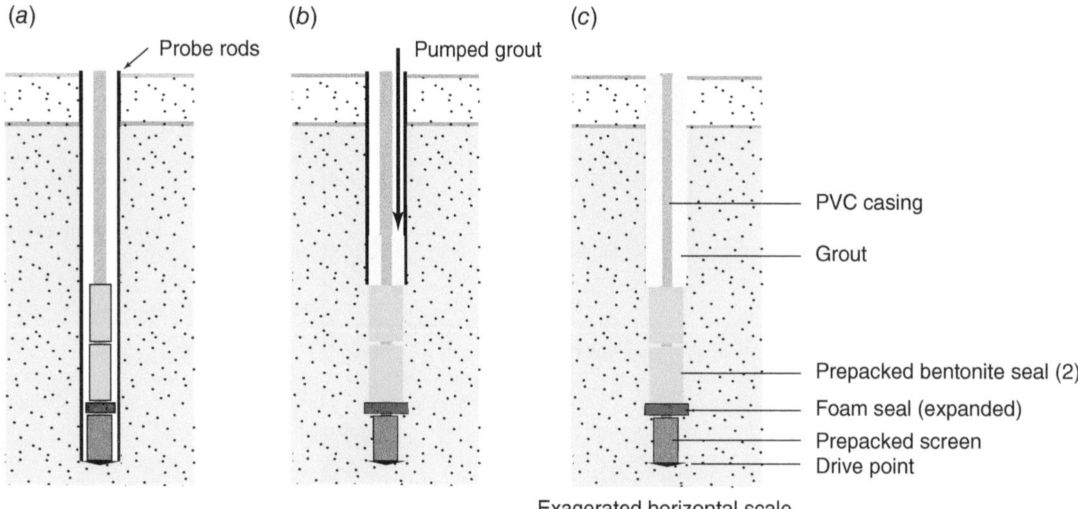

Figure 7.12 Piezometers can be installed using push rigs. (*a*) Probe rods with an expendable drive point are emplaced to the depth of interest. A prepacked well screen, short foam seal, and bentonite seal(s) are threaded onto a PVC pipe and lowered to the bottom. (*b*) As the probe rods are withdrawn the various seals above the screen expand and bentonite grout is pumped. (*c*) The completed piezometer is depicted with the various seals emplaced (FWS).

Some applications, for example nuclear waste storage, CO_2 sequestration and mine development could require piezometers up to 1000 m deep. Westbay® Instruments (https://www.westbay.com/) has equipment suited to these needs.

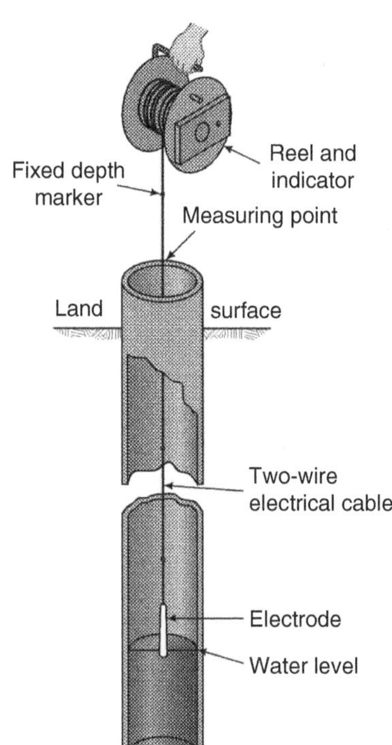

Figure 7.13 Measuring the depth to water from top of casing using an electric tape. This depth is subtracted from the elevation of the measuring point at the top of the well to provide the elevation of the water level in the well (U.S. Geological Survey Heath, 1989/Public domain).

7.4 Making Water-Level Measurements

There are several techniques available to measure the elevations of water levels in observation wells or piezometers. Commonly, an electric tape is used to measure the distance from a fixed measurement point on the well casing at the top of the hole to the water. An electric tape is a plastic tape (feet or meters), which has electrical wires running inside and a weighted electrode at the end. It is lowered down the well. When the electrode touches the water surface, an electrical circuit is completed turning on a buzzer or light (Figure 7.13). The electrode is constructed so that touching the mineralized water completes the electrical circuit. In air, the circuit remains open and the buzzer is silent.

The actual elevation of the water surface is determined by subtracting the measurement on the tape from the elevation of the fixed measurement point at the top of the casing. Usually, a surveying crew accurately measures the well-top elevation in relation to nearby benchmarks.

This technique works well in most applications. Care must be taken to completely clean the tape between measurements to avoid cross-contaminating piezometers. One disadvantage is the time required to lower the tape and make an actual measurement. During the initial stages of an aquifer test with several wells, it may be difficult for a single observer to measure all the wells frequently. One way to deal with this problem is to provide essentially continuous water-level measurements in all the wells at the same time. Modern systems use a pressure transducer placed in the well to measure the pressure of a column of water above it (Figure 7.14). Recall that knowledge of the pressure (converted to pressure head) and the elevation head yield the hydraulic head (see example calculations in Figure 7.14).

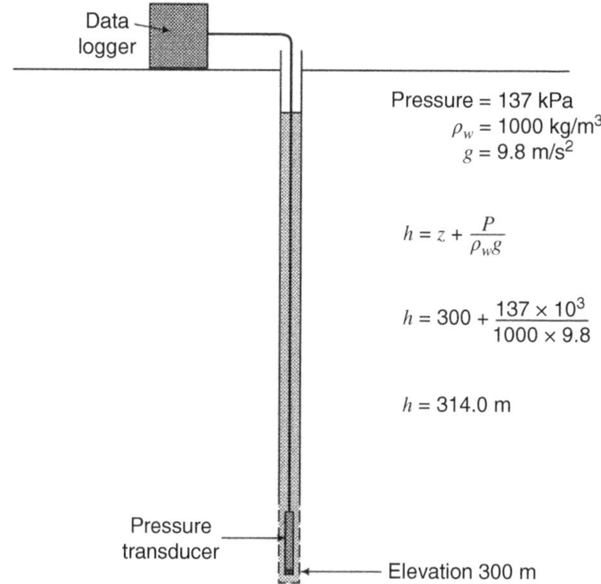

Figure 7.14 Hydraulic heads can also be determined in a well or piezometer using a pressure transducer and data logger. The equation illustrates how the elevation of the transducer, and the measured pressure are used to calculate hydraulic head (Schwartz and Zhang, 2003/© John Wiley & Sons.).

Output from the transducer is monitored at specified time intervals, as small as a few seconds. This information is stored in specialized data-acquisition systems or transmitted directly to the office. For example, In-Situ Inc. (https://in-situ.com/us/) markets Troll® brand data loggers for measuring and logging water-level and pressure information in wells. Their Hermit® brand pump-test kit facilitates data collection with an aquifer test involving many wells. Data collected in digital form is amenable for plotting water-level hydrographs or other types of analyses.

7.5 Geophysics Applied to Site Investigations

A variety of geophysical techniques are used in site investigations. Geophysical methods are based upon physical properties of materials below Earth surface. Surface geophysical methods are commonly used to map features of the geological setting and the location of abandoned hazardous-waste disposal sites. Borehole geophysical methods provide useful ways to augment stratigraphic and hydrogeologic data.

Electrical resistivity, electromagnetics, and gravity techniques are commonly applied surface geophysical methods that are used in groundwater investigations. In the following sections, we will describe applications of the various methods and explain how they are used in groundwater studies.

7.5.1 Electric Resistivity Method

Electrical methods are useful in describing the characteristics of aquifers and patterns of groundwater contamination. The various electrical measurements usually indirectly measure *electrical conductivity*, the ability of a material to conduct electricity, through measurements of *resistivity*, the reciprocal of electrical conductivity. Rocks and sediments conduct electricity because of (1) ions in solution in the groundwater, (2) the presence of clay minerals, and (3) rarely, metallic minerals that can conduct electricity. For typical groundwater applications, increasing the total dissolved solids content of the groundwater in water reduces the resistivity (i.e., increasing electrical conductivity). Increasing the clay content of sediment with the same pore-water chemistry also causes a reduction in resistivity.

We will begin by describing various surface measurements. The simplest approach is resistivity soundings using four steel rods (i.e., electrodes) in a line with specified distances between them (Figure 7.15). Users have the option of choosing from several named geometries. For example, electrode arrangement in Figure 7.15*a* is a Schlumberger array. The resistivity instrument introduces known electrical current (I) to the subsurface using current electrodes (I_1 and I_2). With current flowing

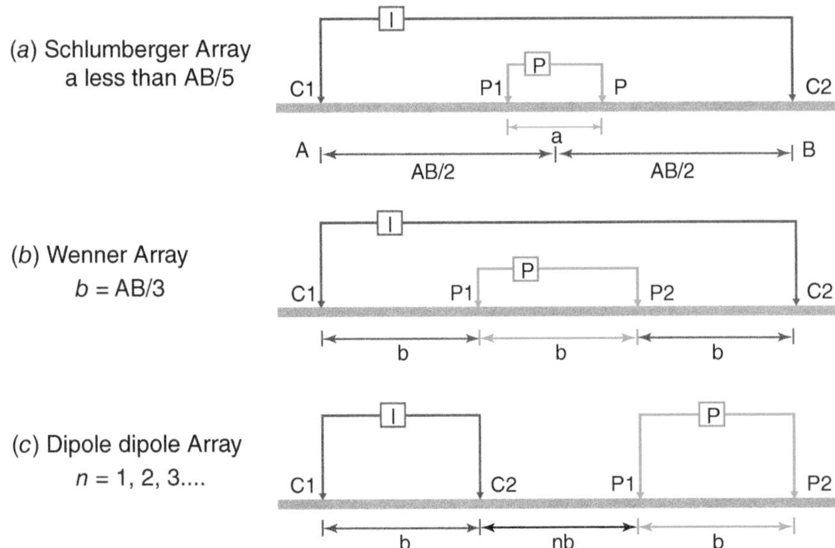

Figure 7.15 Show here are three arrangements of electrodes, which include (*a*) Schlumberger, (*b*) Wenner, and (*c*) Dipole-Dipole arrays. C1 and C2 are current electrodes and P1 and P2 are potential electrodes (adapted from U.S. Environmental Protection Agency, https://archive.epa.gov/esd/archive-geophysics/web/html/resistivity_methods.html (accessed 2 May 2023)).

through the subsurface, it is possible to measure the electric potential voltage (V) at the potential electrodes (P_1 and P_2). Each pair of current electrodes and potential electrodes is a dipole.

The Wenner and dipole-dipole arrays are other common examples of four-electrode arrays. They differ in terms of electrode spacings and how the dipoles are arranged (Figure 7.15*b* and 7.15*c*). Nevertheless, the various four-electrode array of electrodes have similar applications with sounding and profiling. The idea with soundings is to explore deeper formations at one site by increasing the electrode spacings. Profiling uses a constant configuration of electrodes along a line or transects to examine the spatial variability in resistivity at a fixed depth. As will be evident, modern equipment usually provides for both sounding and profiling.

At each measurement station, the equipment provides the measurements of the applied current and measured voltage. The horizontal location of the measurement point normally coincides with the mid-point of the array. The vertical position coincides with the mean of the depth of investigation. The depth of investigation for the three arrays (Figure 7.15) is approximately 0.19 of the size of array spacing (AB) for the Schlumberger array, 0.17, for the Wenner array, and 0.20 for the dipole-dipole. With the help of Ohm's Law, apparent resistivity can be calculated as

$$\rho_a = K\frac{\Delta V}{I} \tag{7.1}$$

where ρ_a is the apparent resistivity, ΔV is the measured voltage difference, I is the electric current induced into the ground, and K is a geometric factor that depends on the geometry of the electrodes and patterns of electrode spacing, and other factors. These calculations of apparent resistivity provide the first step in processing. Eventually, data inversion finds the appropriate electrical model for the subsurface in terms of layering and apparent resistivity.

Advances in instrumentation with 10s to 100s of electrodes facilitate both sounding and profiling together. These advances have depended upon the development of smart electrodes, which provide for automatic selection without having to physically connect and disconnect the electrodes. Figure 7.16 shows a closeup of one of these electrodes with another in the background (Straub, 2017). Once a hundred or more electrodes are installed along a line, the system automatically selects four electrodes at time, following a process that accounts for the type of array and adjustments to spacings along the line necessary to provide for deeper penetration.

This automated process provides for the creation of an array of apparent resistivity values along a two-dimensional profile or pseudosection. Figure 7.17 is an example of potential measurement points associated with an inverse Schlumberger array with 25 electrodes (Lucius et al., 2008). The figure shows the electrode configuration, C1, C2 and P1, P2, that yielded the measurement point at the circled plus sign on the figure. Notice that horizontally this measurement point is midway

Figure 7.16 The foreground shows an example of a smart or configurable electrode. Many electrodes installed along a line facilitate rapid resistivity measurements with the help of modern instruments (Straub, 2017. Reprinted by permission of the National Groundwater Association. Copyright (2017). All Rights reserved.).

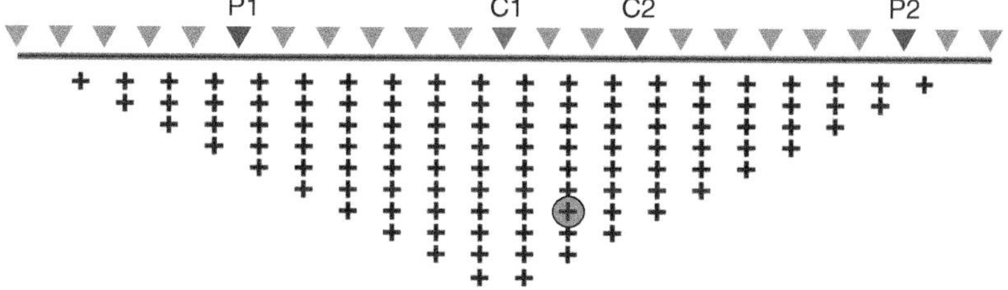

Figure 7.17 Twenty-five electrodes measured using an inverse Schlumberger array will provide a pseudosection with the following measurement points. The large separation from P1 to P2 facilitates penetration as shown by the red circle (U.S. Geological Survey, adapted from Lucius et al., 2008/Public domain).

between C1 and C2. Achieving the vertical depth of penetration at this measurement point requires a large separation of the potential electrodes P1 and P2 (Figure 7.17). This arrangement of four electrodes can only be shifted to the right two more times before P2 is at the end of the line. Thus, at this depth, the last apparent resistivity measurement will be available for the second measurement point to the right of the red circle. This is the reason why this array of measurement points takes on

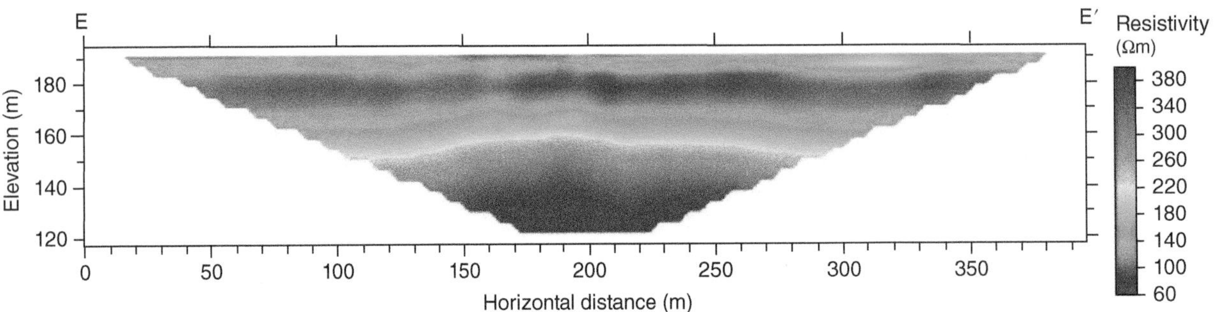

Figure 7.18 Example of an electrical resistivity profile obtained at the site at Columbus, Indiana (U.S. Geological Survey adapted from Ellefsen et al., 2007/Public domain).

this triangular shape. In other words, deep measurements along the line of electrodes are only possible around the midpoint of the line of electrodes.

Pseudosections provide a distorted representation of resistivity along the section. Thus, one last step, an inversion of the 2D data is necessary to provide an accurate picture of subsurface conditions (Loke, 2014).

Figure 7.18 shows interpreted results of the resistivity survey at London, Indiana (Ellefsen et al., 2007) with 100 electrodes along their cross section E–E′. The line of section was ~398 m long with a maximum depth of penetration of about 65 m. The inversion of the data provided an interpretation of the hydrostratigraphic setting. At this site, there are two major units. The upper unit at elevations between ~193 and 160 masl has a relatively low resistivity, indicated by purple and blue colors (Figure 7.18). Ellefsen et al. (2007) interpreted this unit as an unlithified alluvial deposits, including mixtures of sands and gravels with some clays. With smaller electrode spacings, sand, and gravel lenses were evident (Ellefsen et al., 2007). The second major unit has a higher resistivity as indicated by the orange and red colors and was interpreted as Devonian limestone bedrock.

7.5.2 Capacitively Coupled Resistivity Profiling

Another approach to measuring resistivity is similar in concept to systems using in-ground electrodes. However, the "electrodes" consists of insulated wire cables (line antennas) or steel plates laid on the ground surface. A transmitter dipole sends an alternating-current signal that is measured by one or more receiver dipoles. With fast automated switching and data acquisition, and without need to hammer electrodes into the ground, measurements along a relatively long section are collected rapidly. An all-terrain vehicle (ATV) or car can drag the transmitters and receivers along the ground at 3–5 km/h (Burton et al., 2014).

For this application, a dipole-dipole array is utilized with one transmitter dipole and several receiver dipoles arranged in a line. Figure 7.19 shows an example of this setup with a transmitter and five receiver diodes that are linked by a nonconductive rope. A global positioning system on the ATV provides spatial information along the line of section.

This geometry with five receivers provides simultaneous apparent resistivity measurements at five depths at each measurement point along the line of section being tracked by the ATV. The depth of penetration depends upon the geometry of

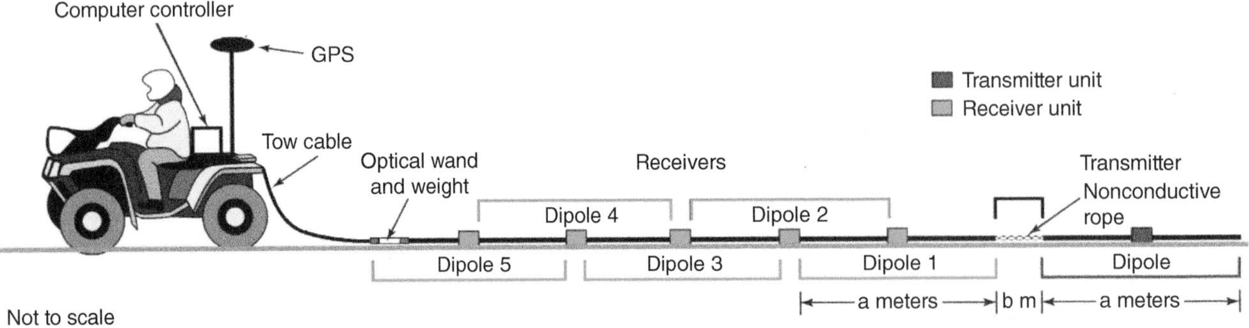

Figure 7.19 Example of a capacitively-coupled resistivity system with a transmitter and five receivers towed by an ATV (U.S. Geological Survey, adapted from Ball et al., 2006; Burton et al., 2014/Public domain).

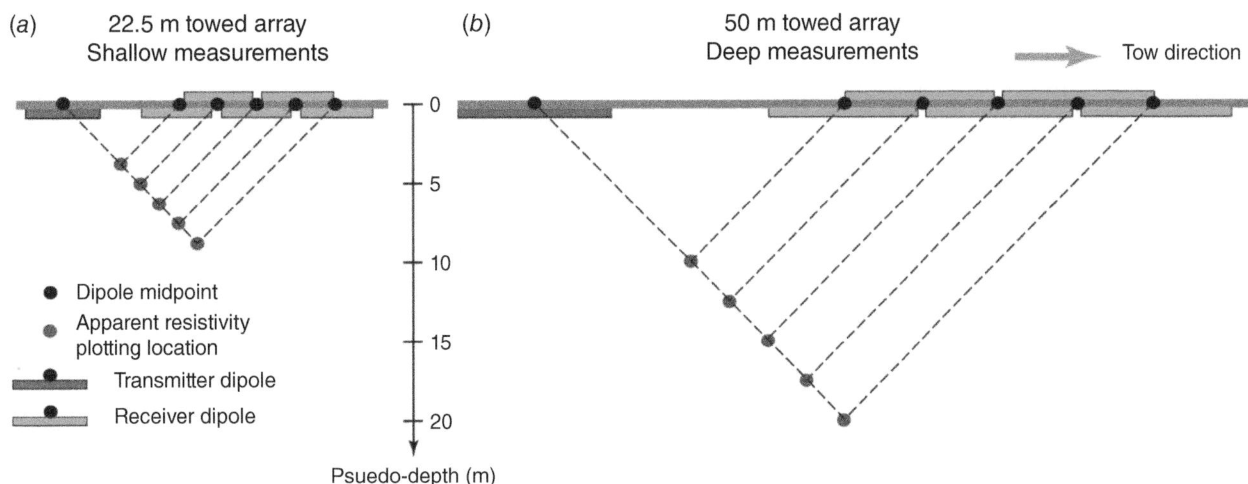

Figure 7.20 Examples of dipole-dipole arrays that illustrate theoretical plotting positions for apparent resistivity values measured at (*a*) a shallow depth and (*b*) a deep depth (FWS).

the dipoles, the dipole length ("a" meters), and the length of nonconductive rope. Deeper probing along the line of section typically requires a second pass along the line with a larger "a"-spacing.

A study of floodplain deposits along the American River in California (Burton et al., 2014) used system in Figure 7.19 with a Geometrics OhmMapper TR5 system (Geometrics, Inc., San Jose, California, USA). They acquired 8.3 km of resistivity measurements along bike and footpaths. Each line was sampled twice using two different array geometries to provide deep measurements. The shallow measurements utilized an "a"-spacing of 5 m with a 2.5 m rope length ("b") (Burton et al., 2014). For the deeper measurements, the dipole length was 10 m with a rope length of 10 m. Thus, the towed array was 22.5 m long for the shallow measurements and 50 m long with for the deep measurements.

Figure 7.20 illustrates how changing the "a" and "b" spacings of the towed array leads to a deeper penetration depth. The apparent plotting location of the apparent resistivity value along the pseudosection for each of the transmitter-receiver pairs (brown dots, Figure 7.20) is midway between the dipole midpoints (black dots) at a depth equal to one-half the distance between each pair of dipole midpoints. The apparent plotting positions in Figure 7.20*a* and 7.20*b* are an idealization of those created by a towed arrays of 22.5 and 50 m used by Burton et al. (2014) in California.

7.5.3 Electromagnetic Methods

Electromagnetic instruments have been developed for ground and aerial resistivity surveys. Electromagnetic methods induce a current in the ground with an alternating current transmitting coil without the necessity of ground contact. The magnetic field around the coil induces an electrical field in the Earth to depths that are largely controlled by the background properties of the medium, the moisture content, and the relative difference in the conducting properties of the medium and the target. An example of a frequency domain instrument for shallow site surveys is the Geonics EM31-MK2 that has two coils mounted 3.7 m apart on a rigid boom. In addition to providing apparent conductivity measurements as millisiemens per meter (mS/m), this instrument can measure magnetic susceptibility in-phase, the ratio of the secondary to primary magnetic field in parts per thousand (ppt). Magnetic susceptibility is the extent to which some target can be magnetized in an external magnetic field.

The equipment is portable, and an operator can collect many closely spaced electrical conductivity measurements in a short time. The depth of signal penetration is about 0.75 to 1.5 times the coil separations (Zalasiewicz et al., 1985). Other instruments from Geonics and other manufacturers provide larger coil separations for greater signal penetration.

Terrain conductivity and magnetic susceptibility methods have been frequently applied to subsurface mapping, as well as the investigation of irrigation problems or contaminated sites. Stewart and Gay (1986) have evaluated electromagnetic soundings for deep detection of conducting fluids, while Greenhouse and Slaine (1983) have applied the technique to mapping contaminant distributions in the subsurface.

Electromagnetic surveying with an EM31 device can detect buried objects or other features at abandoned facilities, waste disposal sites, or infrastructure at active facilities. Examples of buried objects have included waste trenches or lagoons, buried steel drums, and abandoned underground piping or storage tanks. Field experiments set out to determine the efficacy of various geophysical instruments for the identification of common subsurface objects, i.e., pipes comprised of various materials (steel, copper, and plastic), and steel drums (Bjella et al., 2010). These objects were planted in vacant test plots at known locations.

Ground penetrating radar was able to detect all the various objects, given a mode of operation that produces wave reflections from dense materials. The electromagnetic instruments (e.g., EM31) did a good job with the iron targets (i.e., steel pipes and drums), especially in terms of magnetic susceptibility. Other objects were not magnetically susceptible and small resistivity targets.

An early application of electromagnetic surveying of a contaminated site with a handheld device illustrated the usefulness of the method in a conductivity (resistivity) mode (Jordan et al., 1991). The conductivity survey collected 4823 EM data points along lines 3.8 m apart in areas suspected of contamination. Stations were about 0.6–0.9 m apart. The resulting conductivity map indicated a broad range of conductivity. Suspected lagoon locations were indicated by conductivity values greater than 50 mS/m (i.e., milliSiemens per meter), while near-zero conductivities are thought to represent areas of buried metallic debris. The locations of the lagoons correspond with locations determined from old aerial photographs.

7.5.4 Large-Scale, Airborne Electromagnetic Surveys

Large-scale, airborne electromagnetic (AEM) surveying has become increasingly prevalent in groundwater investigations in Australia and the United States. Although relatively expensive, the large and "robust" datasets from AEM surveying represent an important technological "next step" for the investigation of shallow alluvial deposits (Minsley et al., 2021). large example in the assessment of shallow sand and gravel resources.

Usually, construction of regional-scale, aquifer models has depended upon driller's logs that vary in quality and tend to be scattered (Minsley et al., 2021). AEM surveying provides consistent data that have proven useful in mapping the spatial variability in geophysical properties as a basis for estimating hydrogeologic information, such as hydraulic conductivity, recharge rates, and the thickness of shallow units. Aerial surveys conducted by the USGS in Nebraska have redefined the geometry of bedrock surfaces that influence estimates of aquifer thickness and the location of paleochannels (Abraham et al., 2012).

There are several obvious advantages of airborne surveys. First, they can cover large areas encompassing the largest alluvial aquifers in the United States. One of the largest AEM surveys in the United States involved the Mississippi Alluvial Plain (MAP), which included the prolific Mississippi River Valley alluvial aquifer system (Minsley et al., 2021). This investigation involved >48,000 flight-line kilometers over an area of ~140,000 km^2. The resulting spatial coverage (including depth) far exceeded the scope of any conventional ground-based surveying methods or scientific drilling programs in terms the systematic coverage and data densities. The use of helicopter or fixed-wing aircraft also facilitates surveying of rugged and inaccessible areas. Surveys of the MAP also involved approximately 3000 flight-line kilometers of data along streams and rivers (Minsley et al., 2021).

There are a variety of different instruments available for this type of surveying. Figure 7.21 illustrates the RESOLVE AEM system that features a long cylindrical sensor slung below a small helicopter. Surveys are conducted with a 30 m terrain clearance at a speed of 130 km/h at depths ranging from 60 to 100 m with the highest resolution in the upper 2–5 below the ground surface (Minsley et al., 2021). Precise geolocations are provided by altimeters and a GPS system. The MAP study also used the Tempest AEM system. This system is flown with an airplane and provides data to depths of 200–300 m with moderate resolution at shallow depths of 5–10 m (Minsley et al., 2021).

Surveys involved flying a series of parallel lines back and forth across the area of interest. The combination of the two survey tools across the MAP yielded flight-line spacings of ~3 km over most of the region with smaller spacings over selected areas (Minsley et al., 2021). Readers interested in additional technical details and inversion methods can refer to Minsley et al. (2021). There is also a "Geophysical Survey Story Map" that provides an excellent visual display of the sampling approaches, flight lines, and various data products (Hoogenboom, 2021) https://www2.usgs.gov/water/lowermississippi-gulf/map/regional_SM.html. Figure 7.22 is the cover photograph for the story map that illustrates three of the data products over different parts of the MAP (Hoogenboom, 2021).

The surface connectivity map (Figure 7.22) makes use of information on the "thickness and facies classification of the shallowest layer" to provide an indication of the connection across the shallow subsurface with respect to recharge. The red colors indicate better hydraulic connectivity, and the blue colors indicate worse (Hoogenboom, 2021). The pastels colors are areas between the two extremes. The middle part of the visualization (Figure 7.22) provides an example of the vertically

Figure 7.21 AEM surveying with a CGG RESOLVE helicopter system in Greenwood, Mississippi (U.S. Geological Survey / https://www.usgs.gov/news/state-news-release/media-alert-flights-above-mississippi-alluvial-plain-continue-aquifer / last accessed under 5 May 2023).

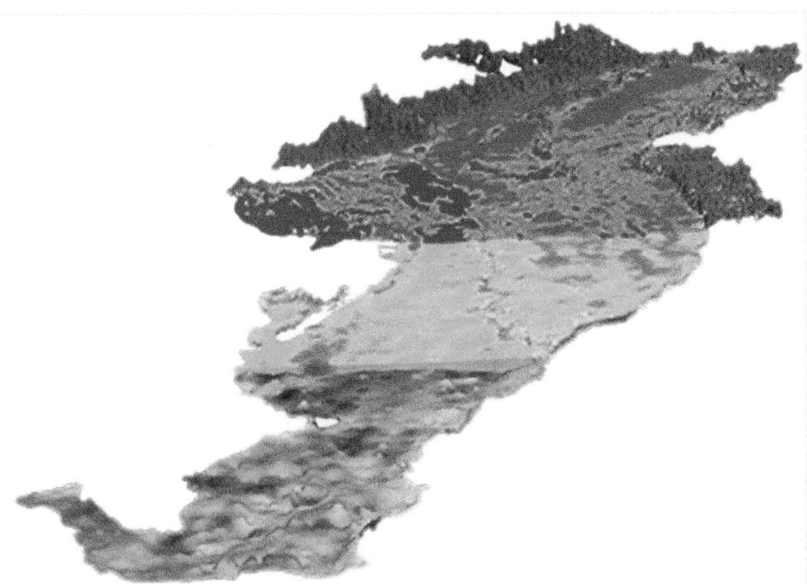

Figure 7.22 Three-dimensional ARM-based visualization of the Mississippi Alluvial Plain aquifer that illustrates examples of three data products together: a surface connectivity map (top), the average aquifer resistivity map (middle), and base conductivity with underlying units (U.S. Geological Survey / https://www2.usgs.gov/water/lowermississippigulf/map/regional_SM.html / last accessed under 5 May 2023).

averaged aquifer resistivity across the saturated aquifer thickness (Hoogenboom, 2021). Yellow–red colors represent relatively higher resistivity values (more permeable) and green–blue colors represent relatively low resistivity values (less permeable). Definition of the aquifer bottom involved conventional hydrogeologic data along with a supervised machine learning approach using the resistivity data (Minsley et al., 2021). The bottom third of the visualization illustrates how well the Mississippi River Valley alluvial aquifer connects to deeper aquifers. The base of aquifer connectivity represents integrated resistivity values 10 m above and below the aquifer bottom, as conductance (i.e., resistivity inverse). The resulting connectivity map is draped on the 3D map of the aquifer-bottom topography (Hoogenboom, 2021). Yellow colors indicate high hydraulic connection and blue/purple colors indicate low hydraulic connection (Figure 7.22).

In closing, here we mention some of the useful other products and study results (Minsley et al., 2021). For example, AEM surveying formed the basis for streambed-conductance maps, which will contribute to a better understanding of aquifer recharge from surface water. Resistivity facies mapping should prove helpful in next-generation aquifer modeling. This study also demonstrated the potential usefulness of integrated land-based and aerial surveys, and the possibilities for identifying subsurface faults (Minsley et al., 2021).

7.5.5 Borehole Geophysical and Flow Meter Logging

Borehole geophysical techniques, associated with exploration and production problems in the petroleum industry, are available for groundwater applications. Unfortunately, the economics of the groundwater industry are such that the most sophisticated tools are unaffordable in most studies. Similarly, the equipment for groundwater applications is commonly much less sophisticated than that available in the petroleum industry.

There are many types of logs that could be used to provide useful information about different aspects of the subsurface. Keys (1990) discusses these techniques in detail. In our brief survey, we will discuss four logging techniques, caliper logs, resistivity/SP logs, and the natural gamma log.

The *caliper log* provides a continuous record of the variation in the diameter of the casing and borehole with depth. According to Dyck et al. (1972) caliper logs are used

- to aid in the interpretation of other logging methods that are affected by hole size,
- to determine details of well construction,
- to provide information on fracture distributions and lithologies, and
- to estimate quantities of cement or gravel required to complete a water well.

A typical probe has three arms (120° apart) that maintain contact with the sides of the borehole, as the probe is raised the hole. Figure 7.23*a* illustrates a caliper logging tool with spring-loaded arms that open and close as the caliper is pulled up the borehole (USGS, 2007). The associated log (Figure 7.23*b*) illustrates the typical response to the presence of an open, water-bearing log, as well as the casing length and diameter (USGS, 2007).

Resistivity/SP logs are usually run together. There are a variety of different types of resistivity tools, the simplest of which is the so-called single-point resistance device. This tool measures the change in resistance between a lead electrode in the borehole and a fixed electrode at the surface. Long and short-normal resistivity tools use two current electrodes and two potential electrodes to measure resistivity. The 64-in. long-normal sensor makes measurements further into the rock

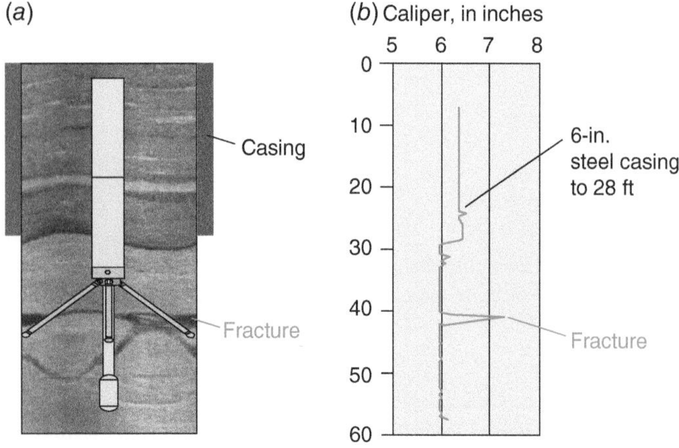

Figure 7.23 A caliper tool being raised up a borehole in fractured (*a*) detects a fracture as a thin zone where the diameter (*b*) of the borehole is slight larger. The casing is reflected on the caliber log as a zone where the borehole diameter is constant (U.S. Geological Survey USGS, 2007/Public domain).

as compared to a 16-in. short-normal sensor (Grauch et al., 2015). Measurements close to the borehole are possible using a lateral resistivity sensor.

The resistivity logs provide information useful in distinguishing different types of lithologies (such as sand versus lacustrine clay or types of rock) and help in correlating units between boreholes. Typically, sand and gravel units have relatively high resistivities, as compared to clay-rich units, and result in log deflection to the right. Fine-grained deposits containing clay minerals (e.g., glacial till or shale) are much less resistive. An upcoming example should make these behaviors clear.

Figure 7.24 shows a caliper log together with a lateral resistivity log for a borehole drilled in slightly fractured bedrock in crystalline bedrock (Grauch et al., 2015). Flow into the borehole was due to the presence of open fractures that intersected the borehole. The caliper log (Figure 7.24*a*) indicated the presence of open fractures at a depth range between ~250 and 300 ft (Grauch et al., 2015). The lateral resistivity log in Figure 7.24*b* indicated this water-bearing zone with a section of lower resistivity. Resistivity values for unfractured crystalline rock in this borehole were much higher, ~300 Ωm (Grauch et al., 2015).

The *spontaneous potential or SP log* measures natural electrical potentials or voltages that develop in boreholes at the contact between clay (also shale) beds and sands (also sandstones) (Keys, 1990). These currents develop because of differences in lithologies and differences in chemistry between the drilling water and the formation water. With drilling water that is less saline than the formation water, the presence of a sand unit causes a deflection in the SP log to the left. As mentioned, we will illustrate the unique sand signal provided by resistivity/SP log pair.

The SP log is helpful in identifying geologic units and in correlating units between boreholes. It also forms the basis for estimating the resistivity of formation water around the borehole, when the resistivity and temperature of the drilling mud are known (Vonhof, 1966). Thus, it has been used in water-quality investigations, particularly in basins where oil-exploration wells have been drilled and logged.

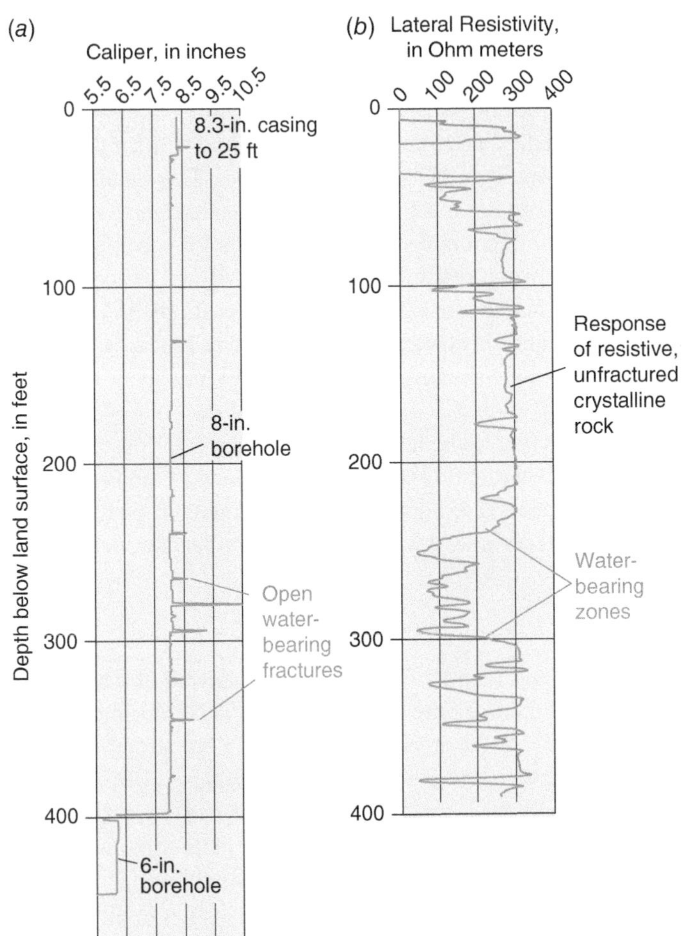

Figure 7.24 Examples of a caliper log (*a*) and a lateral resistivity log (*b*) run in a borehole drilled in a fractured crystalline unit (U.S. Geological Survey, adapted from USGS, 2007/Public domain).

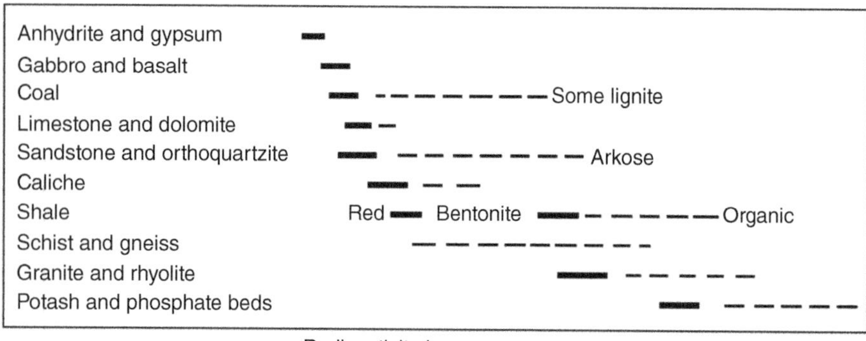

Figure 7.25 Relative radioactivity of some common rocks (U.S. Geological Survey Keys, 1990/Public domain).

The *natural gamma log* measures the total intensity of natural gamma radiation present in rocks or sediments (Dyck et al., 1972). The most important radionuclides found in rocks are potassium-40 and daughter products of the thorium and uranium decay series (Keys, 1990). The relative abundance of these radionuclides in various rock types is summarized in Figure 7.25.

For these logs, the American Petroleum Institute (API) gamma-ray unit is preferred, although there are many "older" units used like counts/second. Unlike the other logs discussed here, except for an induction resistivity tool, gamma logs can be run in wells cased with plastic.

The USGS completed a 326 ft (~100 m) test hole as part of a water-resource investigation of an area that includes the Great Sand Dunes National Park, located in south-central Colorado within San Luis Valley (Grauch et al., 2015). Here, we focus on a few of the basic logs discussed earlier. However, the study also included a detailed characterization of the lithology of shallow, unconsolidated sediments from drilling samples, and core along with the other geophysical logs, i.e., sonic velocity and compensated density logs. Logging also described fossils, other biogenic material, and biological indicators.

The lithologic log (Figure 7.26) indicated the presence of various sand deposits from zero to ~77 ft bgs (23.5 m), followed by a fining downward and interbedded sequence or sand, silt, and clay, from 77 to 232 ft (23.5–70.7 m) bgs. This middle unit is underlain in turn by several thick clay units, separated by fine sands from 232 to 326 ft (70.7–99.4 m) (Grauch et al., 2015).

Figure 7.26 contains two resistivity logs, a 16-in. normal log and an induction log. The induction tool is capable of measurements through the plastic casing at the top of the borehole. The normal resistivity curve was corrected to account for borehole effects, e.g., borehole diameter and mud resistivity, and smoothed to remove spikes (Grauch et al., 2015). Using the resistivity data together with the SP log, it is possible to recognize sand and clay layers. We have added red shading with red arrows to illustrate signal changes caused by the presence of sand and blue shading with blue arrows for clay or clay-rich units (Figure 7.26).

The natural gamma log often behaves similarly to the SP log—deflecting to the left with quartz sand (i.e., lower gamma values) and to the right (i.e., higher gamma values) with clay units that commonly contain higher concentrations of radionuclides. For the illustrative log suite from Colorado, the natural gamma log did not behave in this manner. Grauch et al. (2015) suggested volcanic sand units might exhibit higher gamma levels associated with the presence of potassium feldspar.

7.5.6 Flowmeter Logging

Various techniques, such as caliper logging or borehole camera logs, can identify fractures in the sidewall of a borehole. However, it can be difficult to determine which fractures are open and contribute water to a well. This information can be important because water moving through shallow fracture networks could be contaminated. One straightforward approach to understand the behavior of fracture systems in an open borehole is with flowmeter logging (USGS, 2007). In concept, this approach is similar systematic streamflow gaging to establish losing and gaining of stream reaches. The borehole flowmeter measures vertical flow rates in the borehole, as well as the direction. Thus, it is helpful in identifying fractures that are active in promoting inflow to or outflow from the well bore. Flow in a borehole might occur naturally because of vertical hydraulic gradients between deep and shallow sections of the borehole. In some cases, it might be necessary to pump the well to create a measurable flow in the well.

Figure 7.27 illustrates how appropriate positioning of the flowmeter permits allocation of measured inflows to various fractures (USGS, 2007). The figure depicts the location of the flowmeter at two discrete measurement locations (B and E).

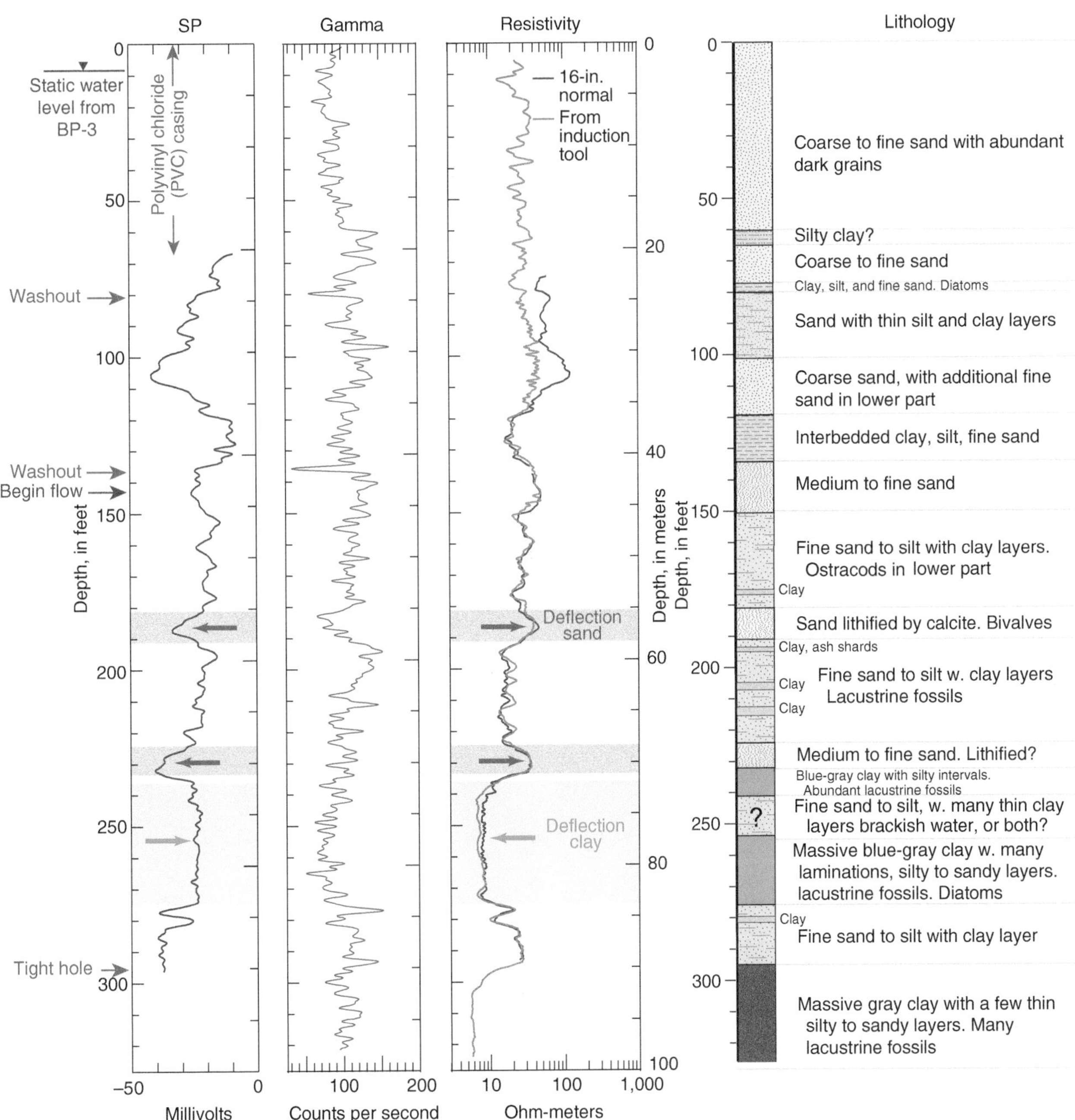

Figure 7.26 Examples of SP, natural gamma, and resistivity logs for a borehole completed in sand, and other fine-grained deposits including massive clays. For the shaded zones on the logs, the arrows indicate expected deflections for sand and clay units (U.S. Geological Survey Grauch et al., 2015/Public domain).

The hypothetical measurements from these points and four others comprised the flowmeter log (USGS, 2007). Measurements A and B above and below fracture (1) (Figure 7.27) were the same indicating no flows due to that fracture. With fracture (2), there is an obvious increase from ~0 gpm at C to ~2.5 gpm at D, which points to inflow from that fracture. There is a similar inflow due to fracture (3), between D and E, which provided an added ~1.5 gpm to the well-bore flow of the well. The last fracture (4) contributed no additional flow. In wells with many fractures, the flowmeter can be raised slow to provide a continuous log (Paillet, 2000).

The most used logging tools are electromagnetic and heat-pulse flow meters. Readers interested in a more technical description should refer to Paillet (2000).

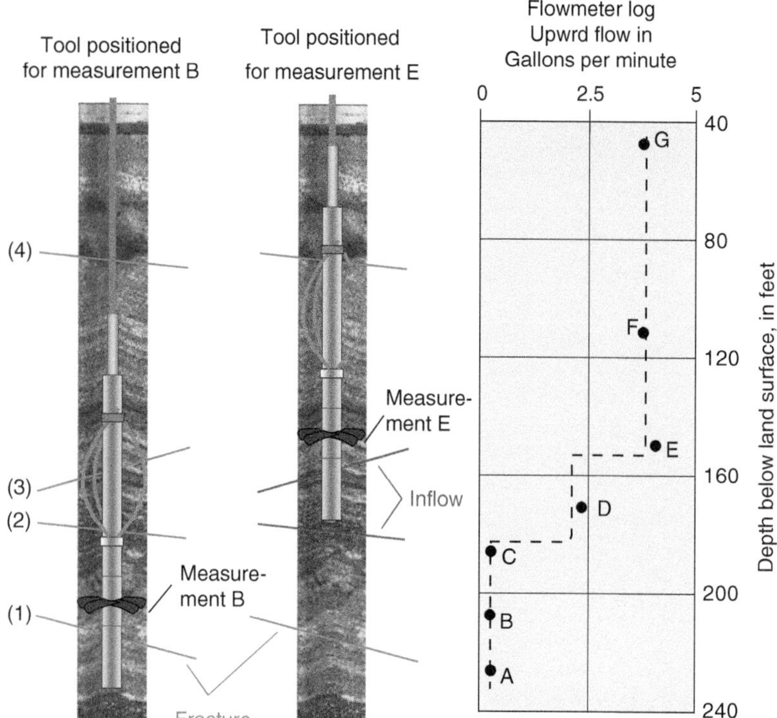

Figure 7.27 Example of a flowmeter log illustrating how a discrete set of flow meter measurements can quantify inflows for 4 hypothetical fractures (U.S. Geological Survey USGS, 2007/Public domain).

7.6 Groundwater Investigations

Groundwater investigations can be carried out at a variety of scales. A report by the U.S. EPA (1990) distinguishes the following sorts of investigations:

regional investigation—study may encompass hundreds or thousands of square miles and typically provides an overall evaluation of groundwater conditions,

local investigation—the area of interest typically involves a few tens or hundreds of square miles, and the study provides a more detailed look at the geology, hydrogeology, and water quality, and

site investigation—a detailed look at a specific site of concern such as a well field, a leaking refinery, or an abandoned industrial site. Usually, site investigations are conducted in conjunction with other investigations, such as risk assessments, air-quality monitoring, and establishing patterns of usage of hazardous chemicals.

The type of investigative tools used in a study change as a function of scale. The regional scale is studied using reconnaissance-type approaches that usually emphasize the synthesis of existing information concerning the climate, geology, and hydrogeology, with selected field operations to fill in data gaps or to obtain more specialized information (for example, age dates from water samples). The site scale is studied using through field studies, like those presented in this chapter.

Operationally, all investigations should involve the same set of six steps (U.S. EPA, 1990); (1) establishing objectives, (2) preparing a workplan, (3) data collection, (4) data interpretation, (5) develop conclusions, and (6) present results. The first two steps in the process are designed to make clear exactly what the goals and objectives of the study are, and how exactly these will be achieved. A workplan is commonly employed to make all those involved aware of how the study will be conducted and what approaches will be followed. Workplans lay out in detail the standard operating procedures (SOPs) to be followed for example, in installing wells, measuring water levels, collecting water samples, and maintaining chain-of-custody of samples. Issues of worker health and safety and emergency procedures are also addressed in these plans. Step 3 encompasses the process of data collection. A later section in this chapter will summarize the different approaches that one can follow in assembling relevant data for an investigation. The last three steps involve assembling the data and presenting a logical set of conclusions.

Table 7.1 Sources of information in groundwater investigations.

Existing information	Geologic test data	Hydrogeologic test data
Topographic maps	Outcrop mapping	Hydraulic head
Soil maps	Borehole logs	Hydraulic conductivity – slug tests
Geologic maps	Core samples	
Aerial photographs	Geophysical logs	– aquifer tests
Satellite images	Textural analyses	– permeameter
Well/water-quality data		Tracer tests
Climate data		
Stream-discharge data		
Reports previous studies		
Reports government agencies		
Personal interviews		

Source: Based on U.S. Environmental Protection Agency EPA (1990).

This chapter should have shed some light on how one makes and interprets basic field information relating to geology and hydrogeology. It should also be evident that much will remain to be considered as the book moves on to review details related to topics of hydraulic testing, water-quality monitoring, and mathematical modeling.

7.6.1 Investigative Methods

This book is concerned throughout with techniques for investigating groundwater problems at a variety of scales. Some of these approaches have been discussed in this chapter. Others come from previous course experience. Many investigations at a regional scale depended on information compiled in previous geologic or hydrogeologic surveys, and data existing in the form of well drillers' reports and water-quality reports for individual domestic or industrial wells. Table 7.1 provides examples of information sources that might be useful in hydrogeologic investigations.

Not included in Table 7.1 are the chemical and isotope tools that provide very useful information about groundwater systems. These will be discussed later in the book.

Exercises

7.1 Given the following data on resistances from a cone penetrometer test, calculate the USCS soil type.

Cone resistance (TSF)	Friction resistance (TSF)
200	1.25
15	0.75

7.2 An electric tape was utilized to measure water level for determining hydraulic head in a piezometer. The elevation of the measurement point is 735 m a.s.l. and the casing length to the mid-point of the screen is 20 m. Given the depth of water in the well is 12.2 m, calculate

A the hydraulic head for the piezometer, and

B the pressure at the mid-point of the screen.

References

Abraham, J. D., J. C. Cannia, P. A. Bedrosian, et al. 2012. Airborne electromagnetic mapping of the base of aquifer in areas of western Nebraska. U.S. Geological Survey Scientific Investigations Report 2011–5219, 38 p.

Ball, L. B., W. H. Kress, G. V. Steele, J. C. Cannia, and M. J. Andersen. 2006. Determination of canal leakage potential using continuous resistivity profiling techniques, Interstate and Tri-State Canals, western Nebraska and eastern Wyoming, 2004. U.S. Geological Survey Scientific Investigations Report 2006–5032, 53 p.

Barcelona, M. J., J. P. Gibb, J. A. Helfrich, and E. E. Garske. 1985. Practical guide for groundwater sampling. Illinois State Water Survey, ISWS Contract Report 374, 94 p.

Bennett, M. J., M. Sneed, T.E., Noce, and J. Tinsley. 2009. Cone penetration test and soil boring at the Bayside Groundwater Project in San Lorenzo, Alameda County, California. U.S. Geological Survey Open-File Report 2009-1050, 25 p.

Bjella, K. L., B. N. Astley, and R. E. North. 2010. Geophysics for Military Construction Projects. U.S. Army Engineer Research and Development Center ERDC TR-10-9, 67 p.

Burton, B. L., M. H. Powers, and L. B. Ball. 2014. Characterization of subsurface stratigraphy along the lower American River floodplain using electrical resistivity, Sacramento, California, 2011. U.S. Geological Survey Open-File Report 2014–1242, 62 p.

Campbell, M. D., and J. H. Lehr. 1973. Water well technology. McGraw Hill Book Company, New York, 681 p.

Chiang, C. Y., K. R. Loos, R. A. Klopp, M. C. Beltz. 1989. A real-time determination of geological/chemical properties of an aquifer by penetration testing. Proceedings NWWA/API Conference Petroleum Hydrocarbons and Organic Chemicals in Groundwater – Prevention, Detection, and Restoration, National Water Well Association, p. 175–189.

Dyck, J. H., W. S. Keys, and W. A. Meneley. 1972. Application of geophysical logging to groundwater studies in southern Saskatchewan. Canadian Journal of Earth Science, v. 9, no. 1, p. 78–94.

Ellefsen, K. J., B. L. Burton, J. E. Lucius, et al. 2007. Field demonstrations of five geophysical methods that could be used to characterize deposits of alluvial aggregate. U.S. Geological Survey Scientific Investigations Report 2007-5226, 20 p.

Grauch, V. J. S., G. L. Skipp, J. V. Thomas, J. K. Davis, and M. E. Benson. 2015. Sample descriptions and geophysical logs for cored well BP-3-USGS, Great Sand Dunes National Park and Preserve, Alamosa County, Colorado. U.S. Geological Survey Data Series 918, 53 p.

Greenhouse, J. P., and D. D. Slaine. 1983. The use of reconnaissance electrical methods to map contaminant migration. Groundwater Monitoring Review, v. 3, no. 5, p. 47–49.

Heath, R. C. 1989. Basic groundwater hydrology. U.S. Geological Survey Water-Supply Paper 2220, 84 p.

Hoogenboom, B. 2021. Mississippi alluvial plain: regional geophysical survey. U.S. Geological Survey, https://www2.usgs.gov/water/lowermississippigulf/map/regional_SM.html (accessed 6 April 2023).

Jordan, T.E., Leask, D.G., Slain, D. et al. 1991. The use of high resolution electromagnetic methods for reconnaissance mapping of buried waste. Proceedings of the Fifth National Outdoor Action Conference on Aquifer Restoration: Groundwater Monitoring and Geophysical Methods. National Groundwater Association, p. 849–862.

Kappel, W. M., and M. A. Teece. 2007. Paleoenvironmental assessment and deglacial chronology of the onondaga trough, Onondaga County, New York. U.S. Geological Survey Open-File Report 2007-1060, 12 p.

Keys, W. S. 1990. Borehole geophysics applied to groundwater investigations. U.S. Geological Survey Techniques of Water-Resources Investigations, Chapter E-2, 150 p.

Loke, M. H. 2014. Tutorial: 2-D and 3-D electrical imaging surveys: Penang, Malaysia, Geotomo Software, 173 p. http://www.geotomosoft.com/downloads.php (accessed 15 May 2023).

Lucius, J. E., J. D. Abraham, and B. L. Burton. 2008. Resistivity profiling for mapping gravel layers that may control contaminant migration at the Amargosa Desert Research Site, Nevada. U.S. Geological Survey Scientific Investigations Report 2008-5091, 30 p.

Mayne, P. W. 2007. Cone penetration testing. NCHRP Synthesis of Highway Practice 368, Transportation Research Board, 125 p.

Minsley, B. J., J. R. Rigby, S. R. James et al. 2021. Airborne geophysical surveys of the lower Mississippi Valley demonstrate system-scale mapping of subsurface architecture. Communications Earth & Environment, v. 2, no. 1, p. 1–14.

Nielsen, D. M. 1996. Design and construction of contaminant monitoring wells to improve performance and cut monitoring costs. Workshop Notebook, Tenth National Outdoor Action Conference and Exposition, National Groundwater Association, p. 34–42.

Noce, T. E., and T.L. Holzer. 2003. Subsurface exploration with the cone penetration testing truck. U.S. Geological Survey Fact Sheet 028-03, 2 p.

NWWA (National Water Well Association of Australia). 1984. Drillers Training and Reference Manual. National Water Well Association of Australia, St. Ives, South Wales, p. 267.

Paillet, F. L. 2000. Flow logging in difficult boreholes—Making the best of a bad deal. Proceedings of the 7th International Symposium on Borehole Geophysics for Minerals, Geotechnical, and Groundwater Applications, Denver, Colorado, p. 125–135.

Robertson, P. K. 1990. Soil classification using the CPT. Canadian Geotechnical Journal, v. 27, no. 1, p. 151–158.

Robertson, P. K., and R. G. Campanella. 1983. Interpretation of cone penetration tests. Part I: Sand. Canadian Geotechnical Journal, v. 20, no. 4, p. 718–733.

Robertson, P. K., R. G. Campanella, D. Gillespie et al. 1986. Use of piezometer cone data, In S. P. Clemence (ed.), Use of In-Situ Tests in Geotechnical Engineering, GSP 6. American Society of Civil Engineers, p. 1263–1280.

Scalf, M.R., J.F. McNabb, W.J. Dunlap, and R. L. Cosby. 1981. Manual of groundwater sampling procedures. U.S. Environmental Protection Agency, Reports EPA 660/2-81-160, 93 p.

Schwartz, F. W., and H. Zhang. 2003. Fundamentals of Groundwater. John Wiley & Sons, 583 p.

Stewart, M. T., and M. C. Gay. 1986. Evaluation of transient electromagnetic soundings for deep detection of conductive fluids. Groundwater, v. 24, no. 3, p. 351–356.

Straub, R. L. 2017. Field notes – Electrical resistivity imaging. Water Well Journal, v. 71, no. 2, p. 24–27.

Strutynsky, A. I., and T. J. Sainey. 1990. Use of cone penetrometer testing and penetrometer groundwater sampling for volatile organic contaminant plume detection. Proceedings NWWA/API Conference Petroleum Hydrocarbons and Organic Chemicals in Groundwater—Prevention, Detection, and Restoration, National Water Well Association, p. 71–84.

USEPA. 1990. Handbook, Groundwater, v. 1. Groundwater and Contamination. USEPA Office of Research and Development EPA 625/6-90/016a, 144 p.

USGS. 2007. Borehole geophysical logging of water-supply wells in the Piedmont, Blue Ridge, and Valley and Ridge, Georgia. U.S. Geological Survey Fact Sheet 2007-3048, 4 p.

Vonhof, J. A. 1966. Water quality determination from spontaneous-potential electric log curves. Journal of Hydrology, v. 4, p. 341–347.

Zalasiewicz, J. A., S. J. Mathers, and J. D. Cornwell. 1985. The application of ground conductivity measurements to geological mapping. Quarterly Journal of Engineering Geology and Hydrogeology, v. 18, no. 2, p. 139–148.

8

Regional Groundwater Flow

CHAPTER MENU

Groundwater hydrologists have generally studied flow at two different scales. Historically, a major emphasis has been on small-scale problems, concerned with the transient response of an aquifer due to pumping from one or several closely spaced wells. There has been a less intensive but nevertheless important effort to examine the manifestations of natural groundwater flow over large regions. This topic is referred to as regional groundwater flow. Here, we begin with Hubbert's (1940) classical study and touch on the important theoretically based studies of Tóth (1962, 1963) and Freeze and Witherspoon (1966, 1967). This body of science forms the foundation for the modern study of natural groundwater flow at a regional scale.

This chapter also will take up issues of how groundwater systems interact with other components of the hydrologic cycle. We explain the processes of recharge and discharge, which operate to add and remove water from active flow systems and also examine features of how groundwater and surface water systems interact. Finally, we describe the special case where seaward moving groundwater meets an ocean and the unique flow physics that are involved.

8.1 Groundwater Basins

In the study of regional groundwater flow, the groundwater basin provides a convenient unit for analysis. The *groundwater basin* was defined by Freeze (1969a) as a three-dimensional closed system, which contains the entire flow paths followed by all the water recharging the basin. Let us examine this concept in more detail by looking at M. King Hubbert's famous figure (Figure 8.1) showing the regional steady-state flow of groundwater from an upland to nearby streams. This figure also describes one of the two complete groundwater basins that are present. This basin is *closed* in the sense that no flow lines pass through the left- and right-side boundaries or the bottom (Figure 8.1). Water enters the basin at the water table, as recharge in the upland area, and leaves as discharge in topographically low areas. Although this figure is only two-dimensional, this same idea applies to a three-dimensional system.

An interesting feature in Figure 8.1 is that the water table is a subdued replica of the ground surface. In other words, the water table is higher under topographically high areas and lower under topographically low areas. However, the overall relief on the water table is somewhat less marked than the relief on the ground surface. This tendency for the water table to follow the topography of the ground surface but with smaller ups and downs is what is meant by "subdued replica." Because of the similarity between the water table and the ground surface, investigators often use the ground surface as a proxy to the water-table configuration. This approximation works best in areas with plenty of recharge. In arid areas, the configuration of the water table can be different than the ground surface.

Fundamentals of Groundwater, Second Edition. Franklin W. Schwartz and Hubao Zhang.

Companion website: www.wiley.com/go/schwartz/fundamentalsofgroundwater2

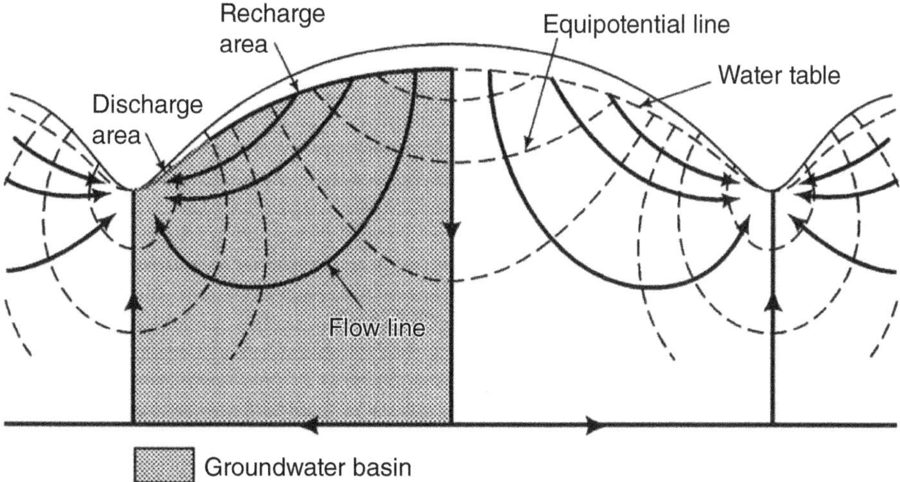

Figure 8.1 Topographically controlled flow pattern. The shaded area defines a groundwater basin (reproduced from The Journal of Geology, by permission of the University of Chicago Press. Copyright © 1940).

The land-surface topography is helpful in defining groundwater basins when the water table is a replica of the ground surface. As an initial assumption, a major topographic high is taken as a groundwater divide, as are major topographic lows. In effect, the surface-water divides and groundwater divides are the same. For example, in Figure 8.1, if information is only available on the topography of the ground surface, you could have done a good job in defining the groundwater basins. We make use of topographic data in many studies to imply what the groundwater flow conditions are likely to look like when groundwater data are sparse.

The term *recharge* refers to water that percolates down through the unsaturated zone and enters the dynamic groundwater flow system (Freeze, 1969a). In a *recharge area,* the flow of groundwater is downward away from the water table. The recharge area in Figure 8.1 encompasses most of the upland area of the system. The term *discharge* refers to groundwater that is lost from a dynamic groundwater system by means of stream baseflow, springs, seepage areas, or evapotranspiration. A *discharge area* is an area where the flow of water is directed upward with respect to the water table (see Figure 8.1).

8.2 Mathematical Analysis of Regional Flow

Mathematical models have proved to be invaluable in developing an understanding of regional groundwater flow and a capability for analyzing real problems quantitatively. A quantitative approach applied to regional groundwater flow requires a groundwater flow equation, a simulation domain, and boundary conditions. The groundwater system is usually assumed to be at steady state so that no initial conditions are required. The simulation domain is a groundwater basin. Conveniently, the vertical side boundaries are no-flow boundaries because a groundwater basin has no flow through the sides. The side boundaries are flow lines, which in many model studies are adopted as no-flow boundaries. The bottom boundary is also a no-flow boundary. This boundary is more easily understood in the sense that if you go deep enough into the Earth, you will eventually find a unit with such a low permeability that essentially no flow is moving into it. The top boundary of the system is a specified head boundary. It is assumed that hydraulic heads are fixed values all along this boundary. Figure 8.2 shows how a simulation domain with boundary conditions can be extracted from Hubbert's figure (Figure 8.1).

8.2.1 Water-Table Controls on Regional Groundwater Flow

One of the important contributions, illustrating the power of mathematical tools was Tóth's (1962, 1963) work on the influence of water-table configuration on flow in groundwater basins. His analytical approach let him test different water-table configurations and determine what types of groundwater flow patterns emerged.

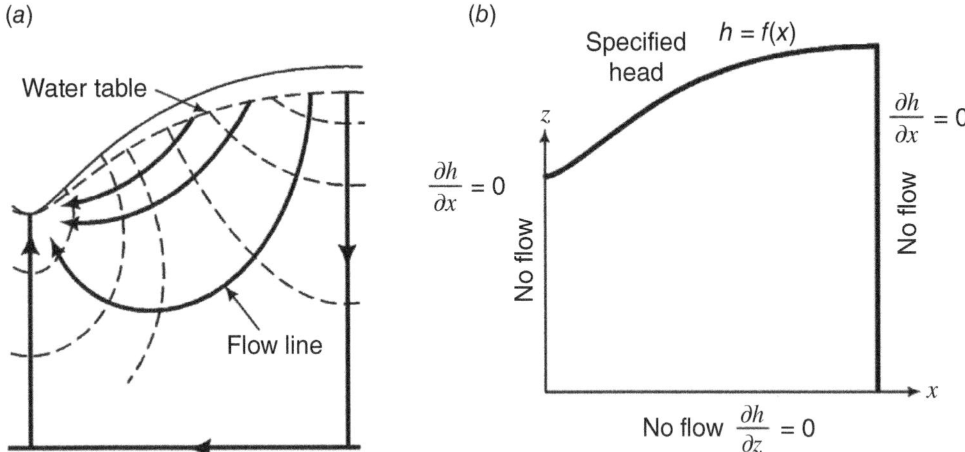

Figure 8.2 Mathematical representation of a groundwater basin. The groundwater basin provides the simulation domain (Panel *a*). The boundary conditions for the analysis are shown in Panel *b* (Schwartz and Zhang, 2003/© John Wiley & Sons).

The first case that Tóth considered was flow in a basin with a linear, sloping water table. This kind of water-table configuration would develop in a groundwater basin with a gently sloping ground surface. Tóth (1962) formulated this problem for a two-dimensional cross-section oriented parallel to the likely direction of regional flow in a groundwater basin (Figure 8.3). To simplify the mathematics, he assumed that the simulation domain was rectangular. However, as is evident in Figure 8.3, hydraulic head varies along the top boundary. Further, he assumed that the porous medium was homogeneous and isotropic.

Hydraulic head is calculated for this problem as a solution of a two-dimensional form of Laplace's equation:

$$\frac{\partial^2 h}{\partial x^2} + \frac{\partial^2 h}{\partial z^2} = 0 \tag{8.1}$$

The boundary conditions have no flow boundary conditions at left, right, and bottom boundaries:

$$\left.\frac{\partial h}{\partial x}\right|_{x=0} = 0, \quad \left.\frac{\partial h}{\partial x}\right|_{x=s} = 0, \quad \left.\frac{\partial h}{\partial z}\right|_{z=0} = 0 \tag{8.2}$$

The sloping linear water table along the top boundary is represented by a specified head boundary with

$$h = z_0 + cx \tag{8.3}$$

where z_0 is the hydraulic head at $x = 0$, and c is the slope of the water table.

An analytical solution to this problem was given by Tóth (1962) and provides the steady-state distribution of hydraulic head in the two-dimensional domain.

$$h = z_0 + \frac{cs}{2} - \frac{4cs}{\pi^2}\sum_{m=0}^{\infty}\frac{\cos[(2m+1)\pi x/s]\cosh[(2m+1)\pi z/s]}{(2m+1)^2\cosh[(2m+1)\pi z_0/s]} \tag{8.4}$$

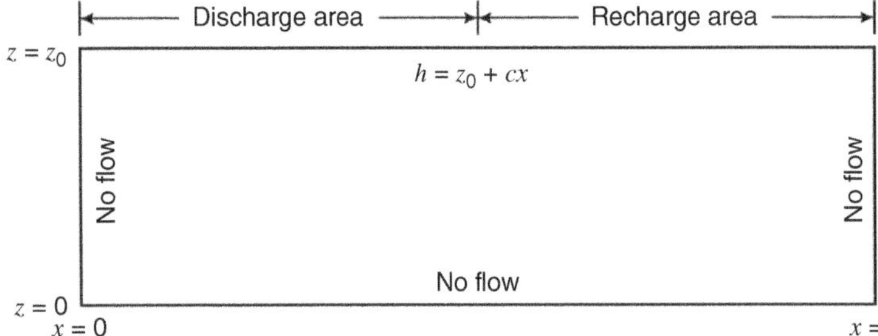

Figure 8.3 Two-dimensional region with boundary conditions for regional flow (Schwartz and Zhang, 2003/© John Wiley & Sons).

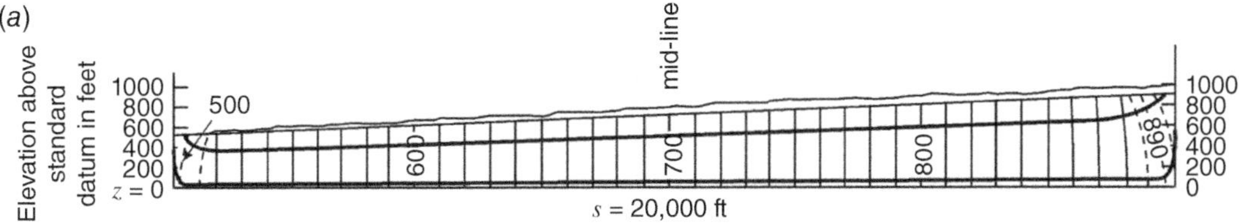

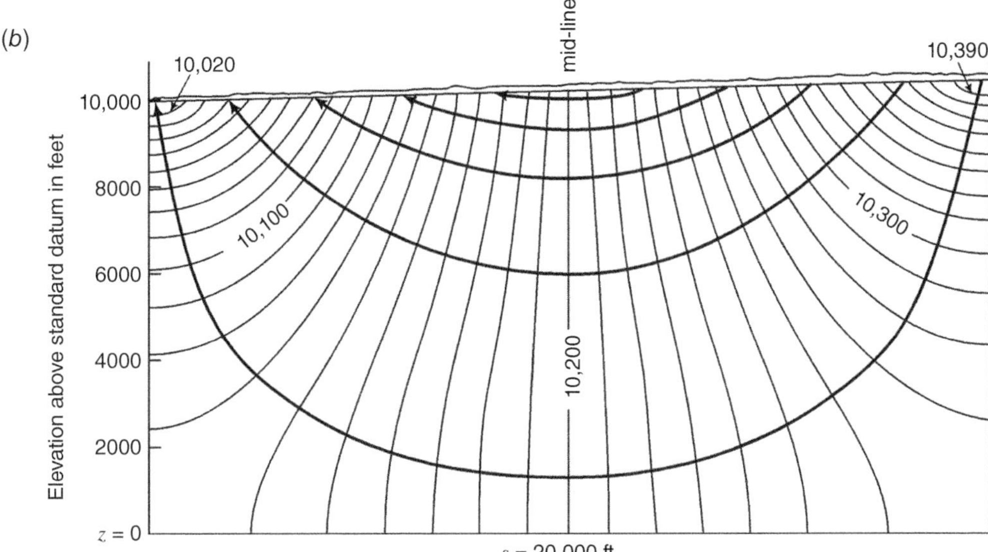

Figure 8.4 Examples of topographically flow patterns derived as a solution to Eq. (8.4). The two panels illustrate how the regional flow changes as a function of region shape; (*a*) a shallow basin with extensive lateral and (*b*) a deep basin with significant vertical flow components (Tóth, 1962/Journal of Geophysical Research 67(11): 4375–4388. Copyright 1962 by American Geophysical Union).

Figure 8.4*a* and 8.4*b* are examples of the contoured hydraulic head field that comes from evaluating the solution to Eq. (8.4). With this water-table configuration, only one large flow system develops in the domain. For a positive value of c, the right side of the flow domain is the recharge area and the left side is the discharge area. The *hinge line*, a line separating the recharge and discharge areas, is located at $x = s/2$. The hydraulic head is between z_0 and $z_0 + cs$.

Example 8.1 A regional flow system is characterized by a sloping linear water table. The hydraulic heads at left and right ends of the water table are 20 and 40 m, respectively. The total length of the domain is 2000 m. Find all the parameters in Eq. (8.4) so that the hydraulic head in the domain can be calculated.

Solution

It is known that $s = 2000$ m and $z_0 = 20$. The parameter c may be calculated by solving the following equation.

$$40\text{ m} = 20\text{ m} + c \times 2000\text{ m}$$

The result is $c = 0.01$.

In a subsequent study, Tóth (1963) examined how topography on the water table influenced patterns of flow. Superimposing a sinusoidal fluctuation on a regional slope (Figure 8.5) provided this idealized water table. Hydraulic head along the water table is given as

$$h(x, z_0) = z_0 + x \tan\alpha + a\frac{\sin(bx/\cos\alpha)}{\cos\alpha} \tag{8.5}$$

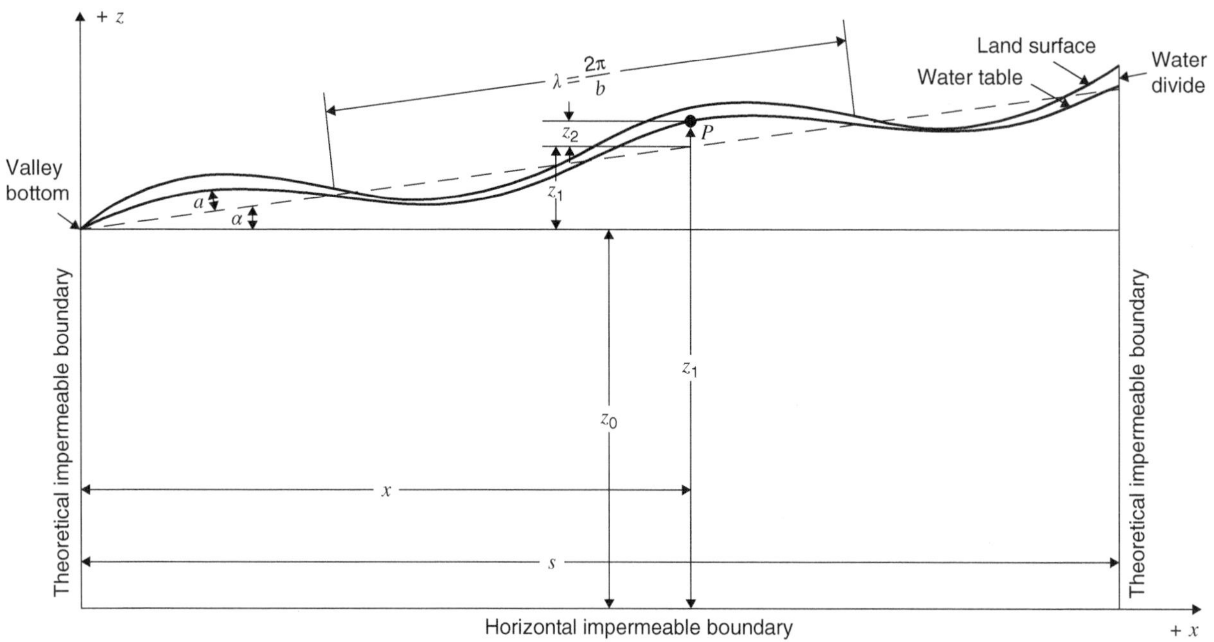

Figure 8.5 Two-dimensional simulation domain for analyzing how a sinusoidal water table with a regional slope affected groundwater flow (Tóth, 1963/Journal of Geophysical Research 68(16): 4795–4812. Copyright 1963 by American Geophysical Union).

where z_0 is the depth of the flow domain, x is the horizontal coordinate, α is the average slope of the water table, a is the amplitude of the sine wave, $b = 2\lambda/\pi$ is the frequency, and λ is the period of the sine wave. The second term on the right-hand side corresponds to the regional slope and the third term is the local relief superposed on the regional slope. This third term, s/λ, defines the number of sine waves for a flow domain of length s. A simplified form of Eq. (8.5) is written as

$$h(x, z_0) = z_0 + c'x + a' \sin(b'x) \tag{8.6}$$

where $c' = \tan\alpha$, $a' = a/\cos\alpha$, and $b' = b/\cos\alpha$. With combinations of z_0, c', a', and b', a variety of different flow fields could be generated.

Tóth (1963) summarized results for various water-table configurations. He found that a hierarchical pattern of flow systems developed, which he termed local, intermediate, and regional (Figure 8.6). A *local flow system* has its recharge area at a topographic high and its discharge area in the adjacent topographical low. An *intermediate flow system* has one or more topographic lows intervening between recharge and discharge areas. A *regional flow system* has its recharge area at the highest part of the basin and its discharge area in the lowest part of the basin.

Tóth (1963) recognized the presence of stagnation points at the juncture of flow systems (Figure 8.6). A *stagnation point* is a point where groundwater flows into two flow systems with opposite flow directions and equal flow magnitudes. Therefore, the groundwater velocity is zero at the stagnation point. Based on his set of simulation experiments, Tóth provided a general set of conclusions with respect to regional flow.

1) For an extended flat area, groundwater flow is very slow. The discharge of water would occur mainly by means of evapotranspiration. Water in these areas will have high concentration of dissolved mineral matter.
2) When local relief is negligible ($a = 0$), the general slope will create a regional flow system by itself. Tóth (1963) cited the case of groundwater flow in southeastern Nassau County, Long Island, New York, as an example (Figure 8.7).
3) Local flow systems develop when the local relief is well defined. The higher the relief, the deeper the local flow system (Figure 8.8).
4) The flow velocity in a local flow system is often much higher than in a regional system. In addition, major streams in a basin receive most of their groundwater inflows from local flow systems, as compared to regional flow systems. Groundwater at shallow depths will likely be influenced by seasonal recharge and discharge, whereas water at depth is relatively stagnant.

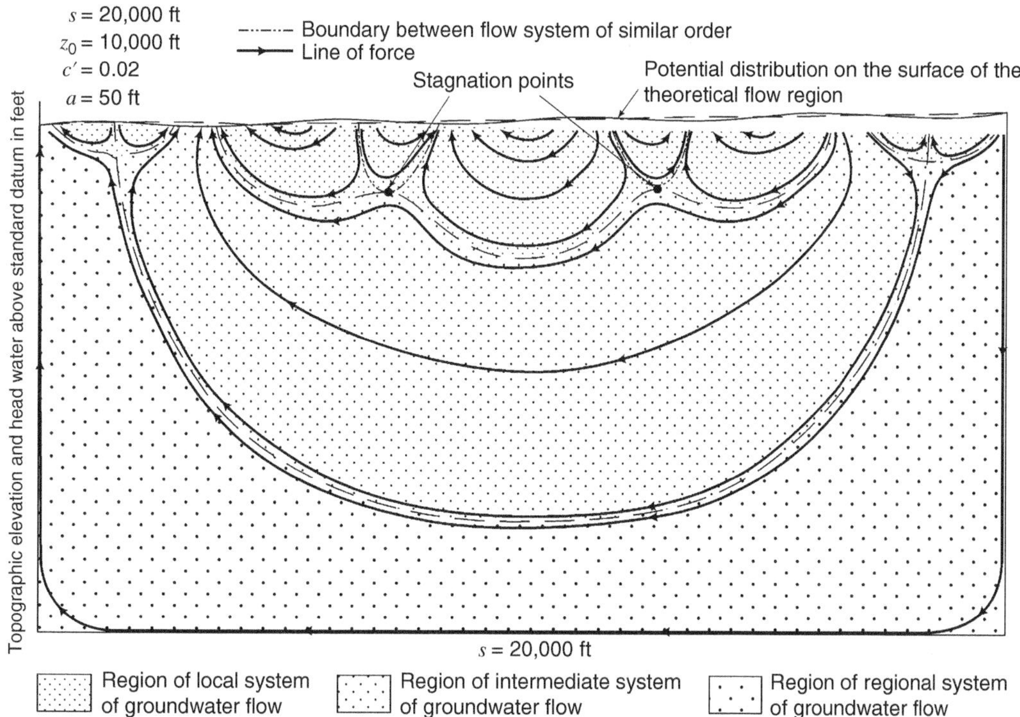

Figure 8.6 Two-dimensional isotropic flow model showing the distribution of local, intermediate, and regional groundwater flow systems (Tóth, 1963/Journal of Geophysical Research 68(16): 4795–4812. Copyright 1963 by American Geophysical Union).

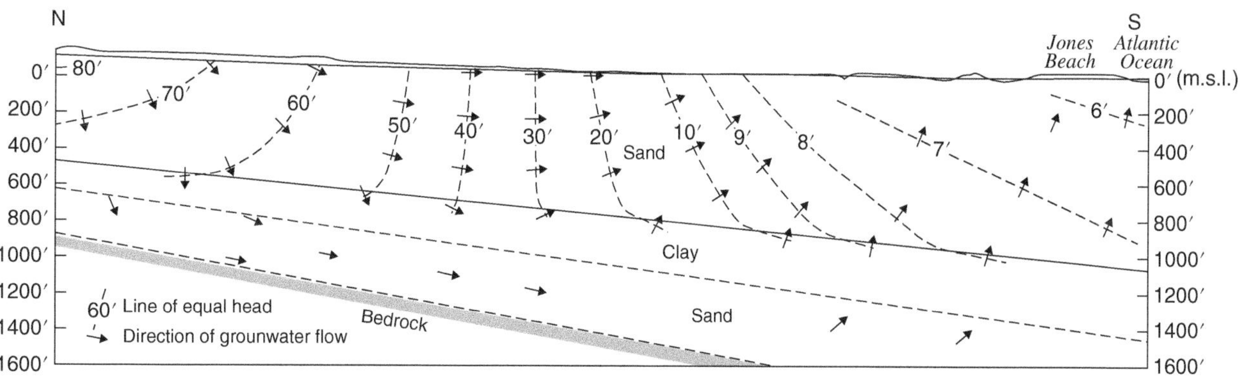

Figure 8.7 Groundwater flow in southeastern Nassau County, Long Island New York (Tóth, 1963/Journal of Geophysical Research 68(16): 4795–4812. Copyright 1963 by American Geophysical Union).

8.2.2 Effects of Basin Geology on Groundwater Flow

In the mid-1960s, Freeze and Witherspoon (1966, 1967) extended Tóth's model-based studies of regional flow. They examined how complexities in the hydraulic conductivity distribution together with water-table configuration influenced regional flow. These complexities were produced by changing patterns of hydraulic-conductivity layering and the anisotropic character of some units (i.e., $K_h \neq K_v$).

Their first set of simulation trials looked at how differing water-table configurations in layered systems impacted patterns of flow. Figure 8.9*a* shows the pattern of flow that developed with a simple segmented water table. The absence of local topography on the water table promoted the development of a regional flow system. Near the discharge area, the steeper water table focused discharge at the valley bottom. Figure 8.9*b* shows the complexity that developed in the head field by adding local topography on the water table, and by creating a flattened discharge area. Now, local, intermediate, and

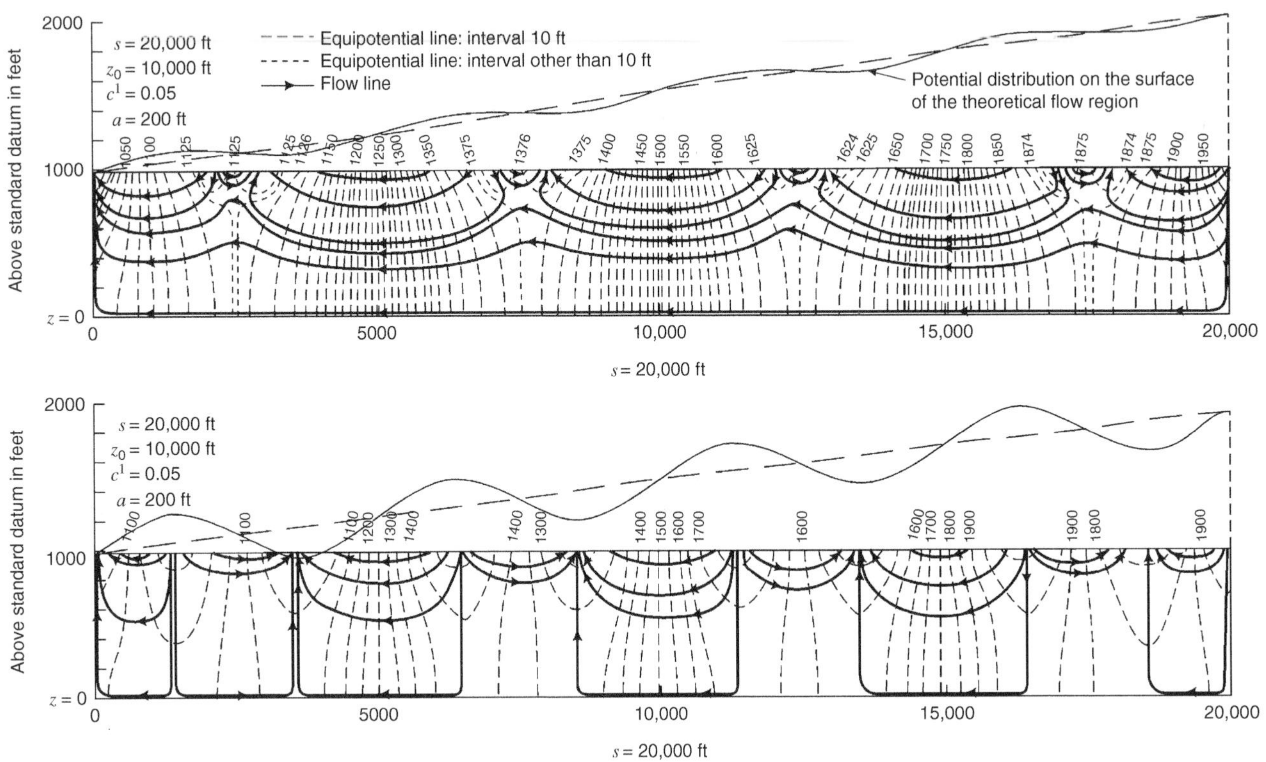

Figure 8.8 The amplitude of fluctuations in the water table controls the vertical size of flow cells that develop (Tóth, 1963/Journal of Geophysical Research 68(16): 4795–4812. Copyright 1963 by American Geophysical Union).

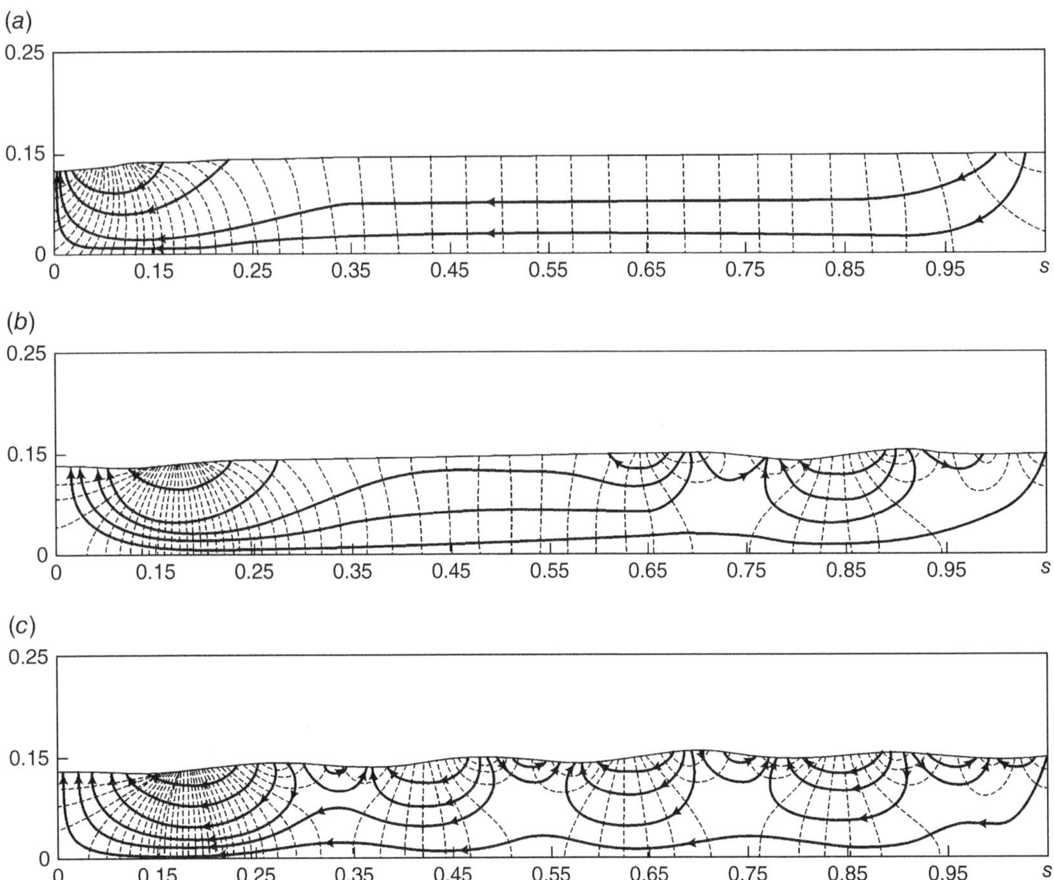

Figure 8.9 Effects of water-table configuration on regional groundwater flow through homogeneous isotropic media. With (*a*), the generally smooth water table created regional flow, With (*b*), areas with a hummocky water table were associated with local flow systems. With (*c*), the hummocky water table resulted in a proliferation of local flow systems (Freeze and Witherspoon, 1967/Water Resources Research3(2): 623–634. Copyright 1967 by American Geophysical Union).

regional systems develop. The final panel (Figure 8.9c) shows how local topography caused the development of a set of flow cells, like those that Tóth found.

The next set of simulations examined the effects of hydraulic conductivity contrasts among layers on the pattern of flow (Figure 8.10). In the first group of trials (Figure 8.10*a*, 8.10*b*, and 8.10*c*), a low *K* layer was assumed to be overly a high *K* layer. The contrast in hydraulic conductivity between the layers was adjusted by increasing the hydraulic conductivity of the

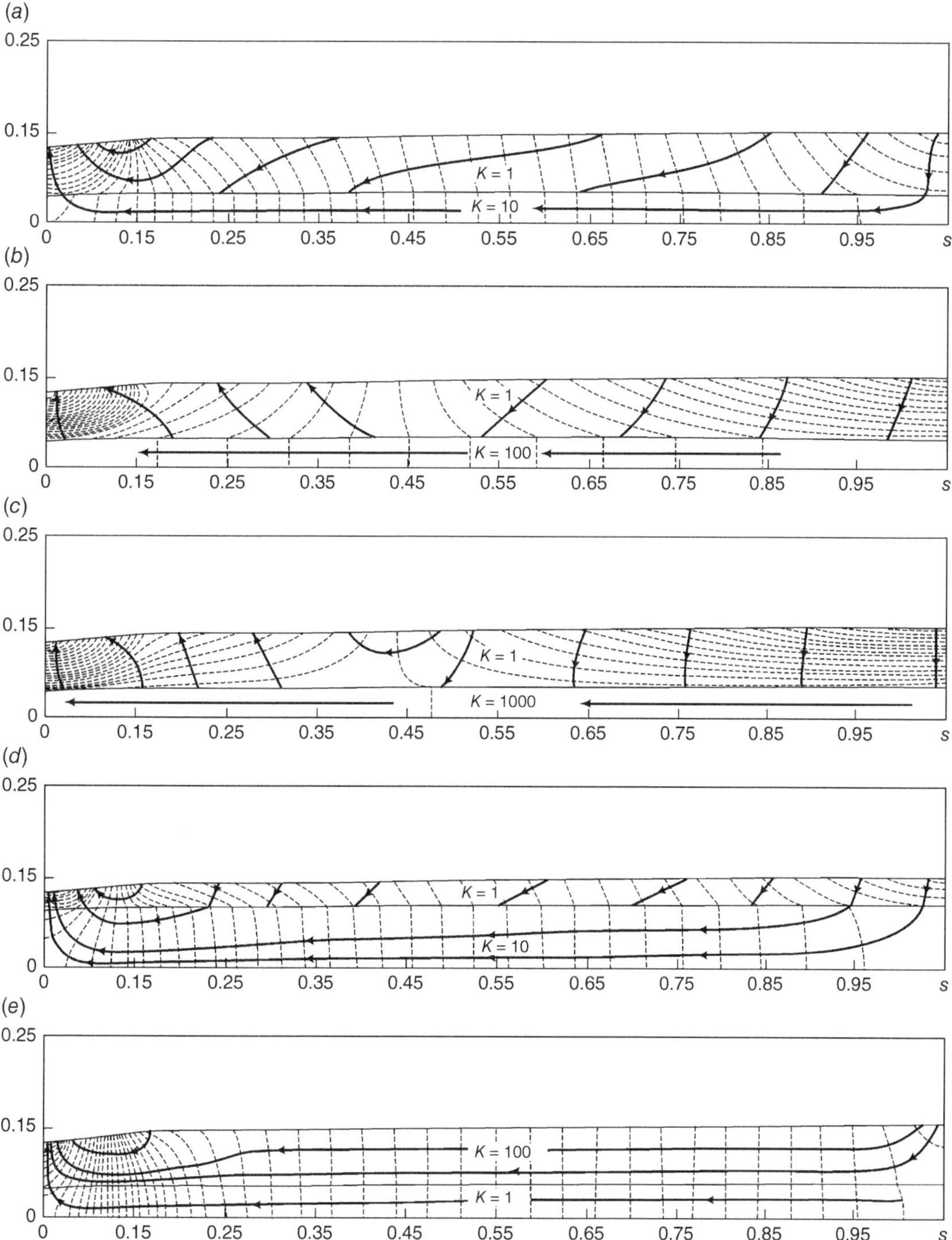

Figure 8.10 Effect of permeability contrast on regional groundwater flow in horizontally layered media. As the contrast in hydraulic conductivity between increased from 1:10 (*a*) to 1:100 (*b*) to 1:1000 (*c*), flow in the uppermost layer became increasingly vertical. With the thinning of the upper layer (*d*) flow in the lower layer exhibited a vertical flow component. With a more permeable upper layer (100:1) (*e*), most of the flow occurred in that layer (Freeze and Witherspoon, 1967/Water Resources Research 3(2): 623–634. Copyright 1967 by American Geophysical Union).

lower, more permeable unit. Increasing the contrast in hydraulic conductivity with this pattern of layering resulted in an increase in the vertical hydraulic head gradient in the upper layer and a decrease in the horizontal hydraulic head gradient in the more permeable layer (Figure 8.10*a* through 8.10*c*). Increasing the thickness of the lower layer did not change the pattern of flow appreciably (compare Figure 8.10*a* and 8.10*d*). There was mainly downward flow in the low *K* units and lateral flow in the high *K* units. When the pattern of layering has a high *K* unit overlying a low *K* unit, the flow patterns were not sensitive to layering. Mainly lateral flow develops in both layers, as illustrated in Figure 8.10*e*.

Their third set of simulations showed the effect of lenticular bodies of high permeability on flow (Figure 8.11). In general, irrespective of whether the lens was upstream (Figure 8.11*a*), downstream (Figure 8.11*b*), or in the middle (Figures 8.11*c* and 8.11*d*), it tended to gain flow on their up-gradient segments and lose flow on their down gradient segments. Often discharge areas form in upland settings when lenticular bodies terminated there (e.g., Figure 8.11*a* and 8.11*c*).

By looking at the simulation results of Freeze and Witherspoon (1966, 1967), one can gradually develop a conceptual understanding of how the pattern of layering influences flow. Flow will proceed from recharge areas to discharge areas by following the most permeable pathways. Thus, flow is mainly lateral in high hydraulic conductivity layers and locally vertical (either up or down) as flow moves across low hydraulic conductivity layers.

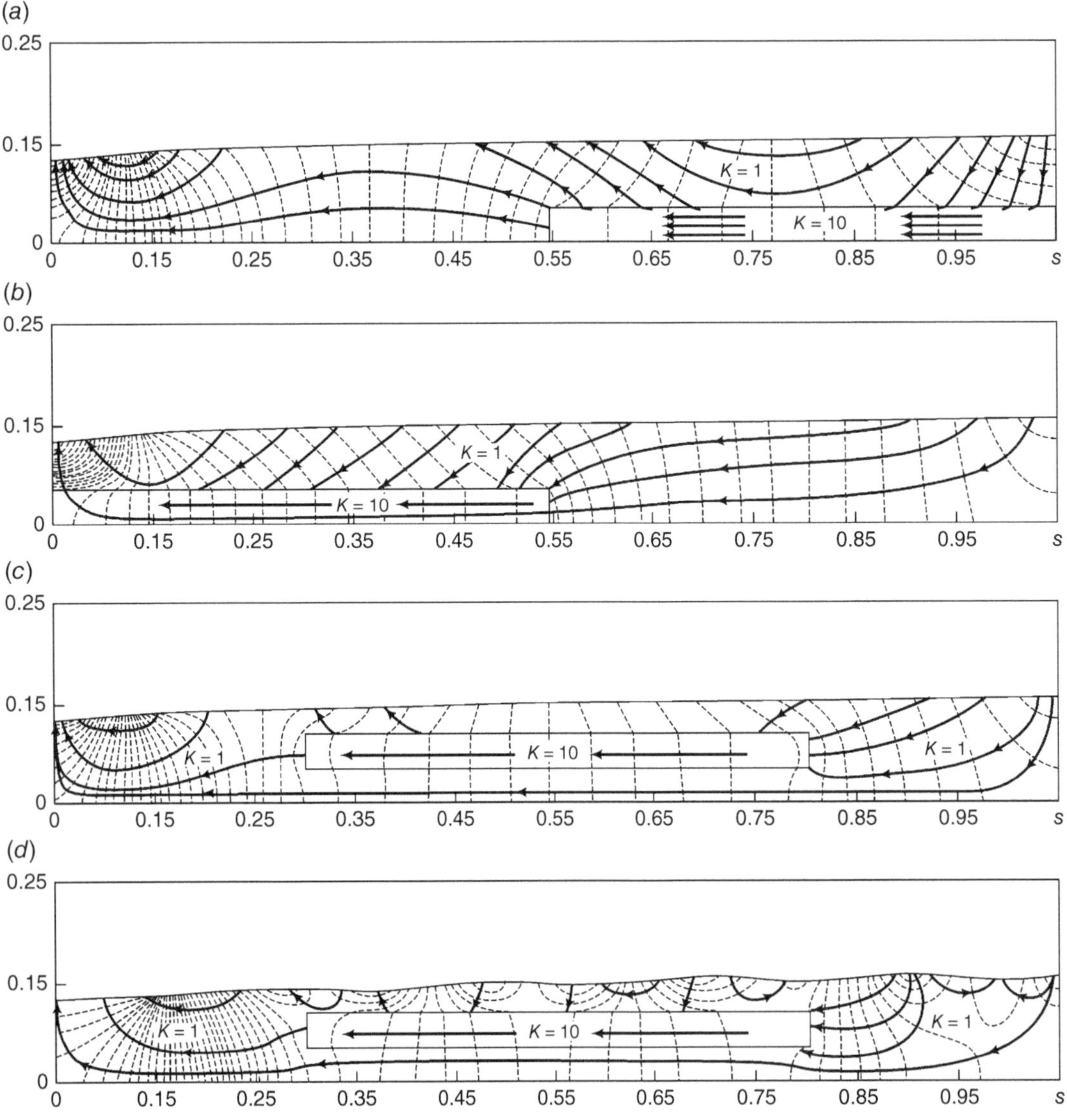

Figure 8.11 Effects of high-permeability bodies on regional groundwater flow. The presence and location of permeable units (*a*, *b*, *c*) often can determine the position of groundwater discharge areas. Local flow systems created due to a hummocky water table (*d*) only extend downward to the top of the higher conductivity units (Freeze and Witherspoon, 1967/ Water Resources Research 3(2): 623–634. Copyright 1967 by American Geophysical Union).

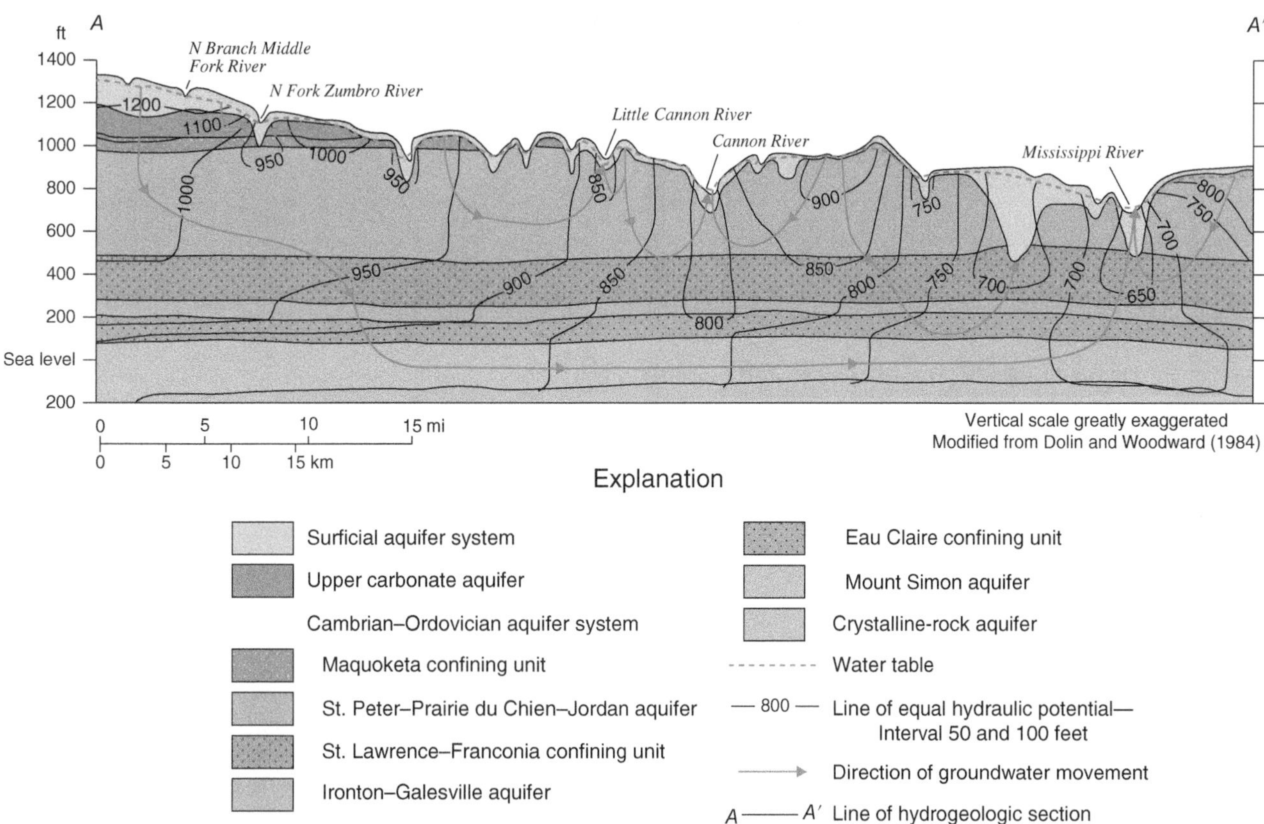

Figure 8.12 Hydraulic head distribution and patterns of regional groundwater flow in the Hollandale embayment, southeastern Minnesota (U.S. Geological Survey Delin and Woodward, 1984; Olcott, 1992/Public domain).

Studies in Minnesota, USA highlighted many of the features of regional flow that we have examined in this section. Figure 8.12 shows the equipotential distributions and patterns of groundwater flow in the Cambro-Ordovician system along an approximately N–S cross section in the southeastern tip of the state. The sequence of aquifers and confining beds shown in the cross section create patterns of layering somewhat similar but obviously more complicated than that treated by Freeze and Witherspoon. The hummocky topography/water table promotes the development of local and intermediate flow systems in the upper part of the stratigraphic section. There is a regional system developed from the topographic high in the basin to the valley of the Mississippi River. Flow is mainly vertical through confining units like the St. Lawrence-Franconia or the Eau Claire. Lateral flow is evident in the most permeable units, like the St. Peter-Prairie du Chen-Jordan aquifer and the Mt. Simon aquifer (Figure 8.12).

8.3 Recharge

Recharge is the process by which water moves downward from the ground surface through the vadose zone, eventually joining an active groundwater flow system. Recharge is promoted by water standing on the ground surface, a relatively shallow water table, and relatively permeable materials in the vadose zone. The following examples will illustrate how these factors work together to control recharge. We will begin with a discussion of recharge in relatively arid settings and progress to moist settings.

8.3.1 Desert Environments

In desert settings, rates of potential evaporation often greatly exceed precipitation. What is surprising is that any recharge at all reaches the groundwater. Usually, two factors work together in promoting recharge—significant but infrequent storms that bring water to very dry areas and mechanisms that locally accumulate this water.

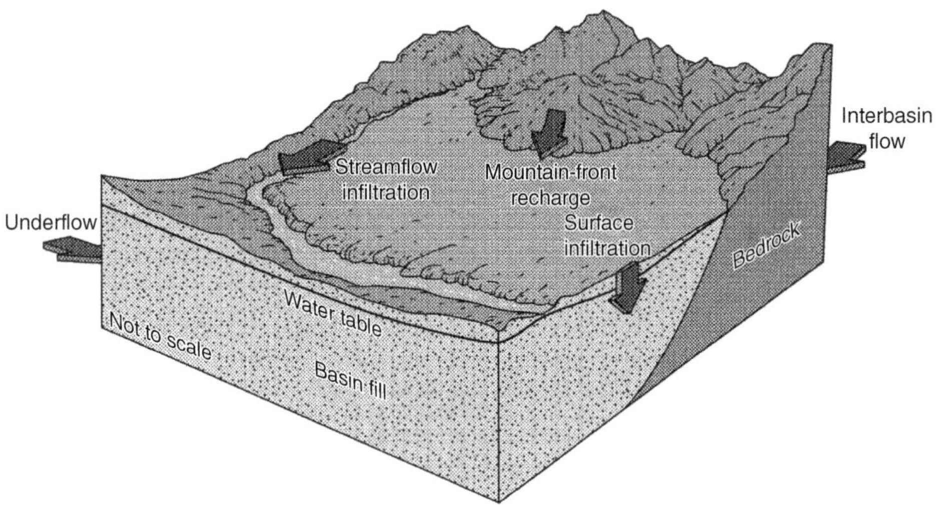

Figure 8.13 In desert settings, recharge is often localized along ephemeral streams and the upstream ends of alluvial fans at the base of mountains. Surface infiltration can occur infrequently (U.S. Geological Survey Robson and Banta, 1995/Public domain).

The Yucca Mountain area in Nevada is a case in point. The mean annual precipitation is approximately 160 mm (about 6 in.) per year. With so little precipitation and high rates of potential evapotranspiration, there is no recharge in most years. However, on occasion, every few years, a single storm could yield several inches of precipitation. With this precipitation, normally dry riverbeds carry water for a week or so, with some of that water becoming available for recharge due to leakage out of the stream. Figure 8.13 develops the conceptual model of recharge from ephemeral streams in such settings in arid southwest of the United States.

Clearly, not every dry area is this arid. Precipitation can come and go seasonally, with stream flow persisting over months. With water present on the landscape for longer times, there is potential for much more significant quantities of recharge.

Recharge in dry settings can also be promoted by the presence of mountains. Typically, precipitation is greater at higher elevations, and the relief promotes rapid and efficient surface runoff. Water running off mountains effectively infiltrates alluvial fans bordering mountain ranges (Figure 8.13). Alluvial-fan sediments are highly permeable and infiltrating water can move downward rapidly, away from the reach of surface evaporation.

8.3.2 Semi-Arid Climate and Hummocky Terrain

Another type of recharge occurs across the semi-arid Great Plains of the United States and Canada. Although the quantities of precipitation are relatively low and for most of the year less than potential evapotranspiration, there is often a window of opportunity for recharge in late spring. The melting of accumulated snow during a time of low temperatures and no plant growth can in some years provide the excess moisture necessary for recharge. Recharge is also promoted by poorly drained glacial and dune landscapes, formed from closely spaced hills and depressions. Because of the poorly organized surface drainage, snow-melt runoff accumulates and is retained in small pothole lakes and small wetlands. When these lakes and wetlands are in topographically high settings (Figure 8.14), they are a source of recharge through hot summer months,

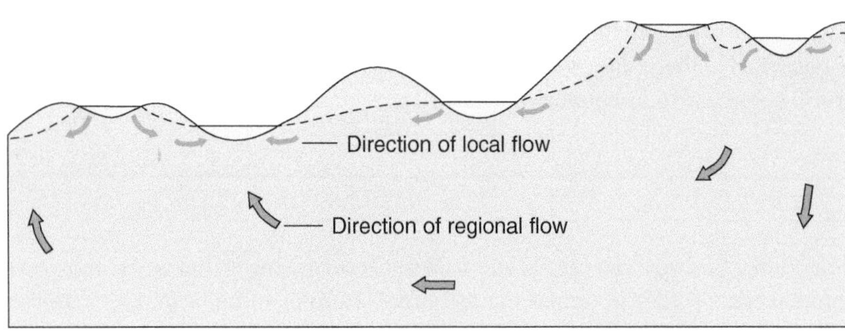

Figure 8.14 Lakes and wetlands in topographically high positions are often a significant source of recharge to groundwater. In North America, these conditions are associated with glacial and dune terrains (U.S. Geological Survey Winter et al., 1998/Public domain).

until they eventually dry up. Such water bodies are important in sustaining migratory waterfowl in the central region of the North American continent (Liu et al., 2016).

Water bodies at lower elevations last longer as groundwater discharge contributes to inflow. In dune settings, the relatively high hydraulic conductivity of these deposits promotes relatively active flow or groundwater between depressions (Figure 8.14). Lakes and wetlands are extremely sensitive to both long and short-term climatic fluctuations because the water available to these water bodies can be markedly different in wet versus dry years. Therefore, stages fluctuate markedly.

8.3.3 Recharge in Structurally Controlled Settings

With crystalline rocks, the structural setting exerts a major influence on groundwater flow at both local and regional scales (Whitehead, 1996). Fault zones can conduct and store groundwater. Where permeable fault zones extend to great depths, groundwater can circulate downward and may become heated. Groundwater in carbonate rocks of the Valley and Ridge aquifers flows along fractures and bedding planes (Figure 8.15) (Trapp and Horn, 1997).

Precipitation recharges carbonate aquifers through infiltration through alluvium and regolith. Commonly, groundwater dissolves carbonate rocks and creates a network of large and interconnected openings in carbonate aquifers. Precipitation that falls on the valley recharges the carbonate rocks as well as runoff from adjacent ridges.

When underlain by karst, recharge can be tremendously effective. Fractures and sinkholes create a direct connection between the land surface and deeper aquifers. Streams are also an important source of recharges for carbonate aquifers.

8.3.4 Distributed Recharge in Moist Climates

In many parts of the world, there is abundant precipitation. With the opportunity for water often to be present across the landscape, recharge occurs in a distributed fashion. Recharge, then, is much more aerially extensive without a need for special situations to concentrate surface water from larger areas. In many more northerly settings, spring snowmelt and accompanying rainstorms provide more water than most systems can accommodate. Thus, in many eastern U.S. states and Canadian provinces, there is a condition where potential recharge is rejected because the water table is so close to the ground surface in spring months. Through the warmer summer months, there is much less recharge due to the increasing utilization of water by growing plants.

8.3.5 Approaches for Estimating Recharge

In water resources investigations, there is often a need to estimate the quantity of recharge added to flow systems. A variety of empirical and measurement-based approaches are available to help answer this question. One simple empirical approach

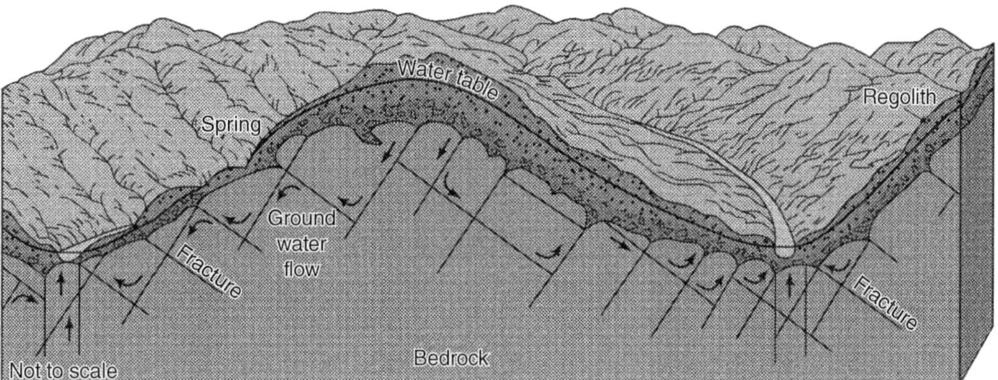

Figure 8.15 Recharge and discharge in crystalline rock and undifferentiated sedimentary rock aquifers (U.S. Geological Survey Trapp and Horn, 1997/Public domain).

is to estimate recharge based on precipitation. The rule-of-thumb is that about 5% of the total precipitation could be available as recharge. This kind of estimate is essentially an educated guess.

Other more rigorous approaches are available. For example, water balance calculations represent yet another way to estimate recharge rates. Simple water balance calculations for soils, based on simple climatic information, let one determine rates of precipitation, and potential/actual evapotranspiration. With knowledge of the soil moisture storage capacity of the soil, calculated soil moisture surpluses represent water that is available for recharge (Freeze, 1969b).

Freeze (1969a) described how quantitative flow nets for regional groundwater systems can be used to estimate recharge. As we discussed previously, given a flow net with curvilinear squares, one can calculate the discharge in a collection stream tube using the Darcy equation

$$Q = \frac{n_f}{n_d} K \cdot \Delta H \tag{8.7}$$

where Q is the discharge from a collection of flow tubes, K is hydraulic conductivity, and ΔH is the total head drop where ΔH is the product of the head drop between equipotential lines, Δh, times n_d, the number of head drops. In the third dimension, which is into and out of the cross section, the section is assumed to have unit thickness. For example, measuring in meters, the flow system would have a thickness of one meter. We illustrate this approach using an example based on the approach of Freeze (1969a).

Example 8.2 Figure 8.16 shows a flow net with flow lines and equipotential lines at a contour interval of 5 m above sea level. The six flow tubes are numbered. Assume that the hydraulic conductivity of the unit is 0.25 m/day. Calculate the total quantity of water recharging this flow system.

Solution

This system has six flow tubes, each with the same constant quantity of flow. This calculation uses one curvilinear square between two equipotential lines with Δh equal to 5 m asl and an n_d value of 1 to estimate the flow in that flow tube.

$$Q = \frac{n_f}{n_d} K \cdot \Delta H = \frac{n_f}{n_d} K \cdot \Delta h \cdot n_d = 6 \times 0.25 \times 5.0 = 7.5\text{m}^3/\text{day}$$

Remember, this recharge calculation applies to a slice 1 m thick into and out of the section.

Chemical approaches have also proven to be useful for estimating recharge. Readers can refer to Allison and Hughes (1978), Sharma and Hughes (1985), and Phillips et al. (1988) for a discussion of chloride mass balance approaches. Allison et al. (1983) discuss the use of environmental isotopes, deuterium, and oxygen-18. More recently, bomb-pulse tritium and chlorine-36 methods have provided ways to estimate recharge (Zimmerman et al., 1966).

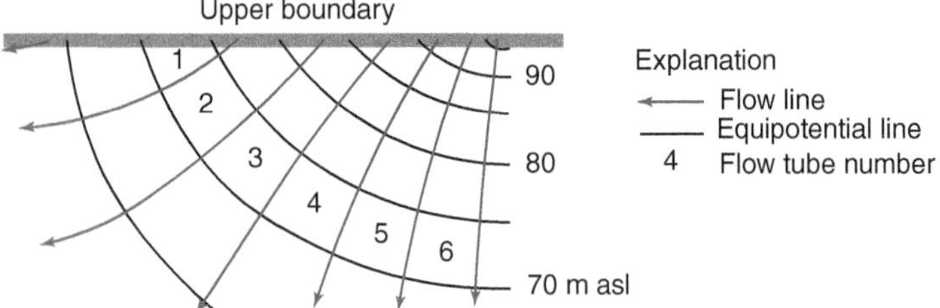

Figure 8.16 In this recharge area, there are six flow tubes moving water downward from the top boundary (FWS).

8.4 Discharge

Discharge is a process by which water leaves a groundwater flow system and returns to the surface hydrologic cycle. The common modes of natural discharge include: (1) outflow to rivers, lakes, and wetlands, (2) flow as springs and seeps to the ground surface, and (3) evapotranspiration in settings with a shallow water table.

8.4.1 Inflow to Wetlands, Lakes, and Rivers

In the case of wetlands or lakes, groundwater inflow occurs as subsurface seepage or springs (Figure 8.17). Seepage measurements in lakes show that typically most of the seepage occurs near the shore. In the case of rivers, the discharge situation can be complicated by the fact that the water in a permeable streambed is moving downstream along with the surface water.

Groundwater entering this so-called hyporheic zone mixes with the streambed water before discharging to the stream. The *hyporheic zone* is the subsurface zone where stream water flows through short segments of its adjacent bed and banks (Figure 8.18) (Winter et al., 1998). Because this water is a mixture of stream water and discharging groundwater, it can have a unique chemical and biological character. Given the importance of the interactions between groundwater and surface waters, we will discuss this topic in more detail in Section 8.6.

8.4.2 Springs and Seeps

A *spring* is a concentrated flow of groundwater issuing from the subsurface into a body of surface water or onto the land surface at a rate sufficient to form a current. Springs can be subdivided into two main groups, gravity springs, and artesian springs. *Gravity springs* normally occur in simple unconfined systems and can be one of three types; *filtration* or *depressional springs*, *contact springs*, and *fracture springs*. Figure 8.19 shows examples of the first two types (Medler and Eldridge, 2021). *Depressional springs* form where a change in the land surface causes the water table to intersect the land surface. *Contact springs* form when downward flow intersects a low permeability unit that creates lateral flow and eventual discharge. In the North Dakota example, contact springs formed above and below the low permeability unit (Figure 8.19). Typically, these two types of springs yield relatively small quantities of water because of relatively small hydraulic gradients and relatively low hydraulic conductivities. Springs associated with fractured rocks, particularly carbonate rocks in karst settings, may be capable of extremely large storm flows.

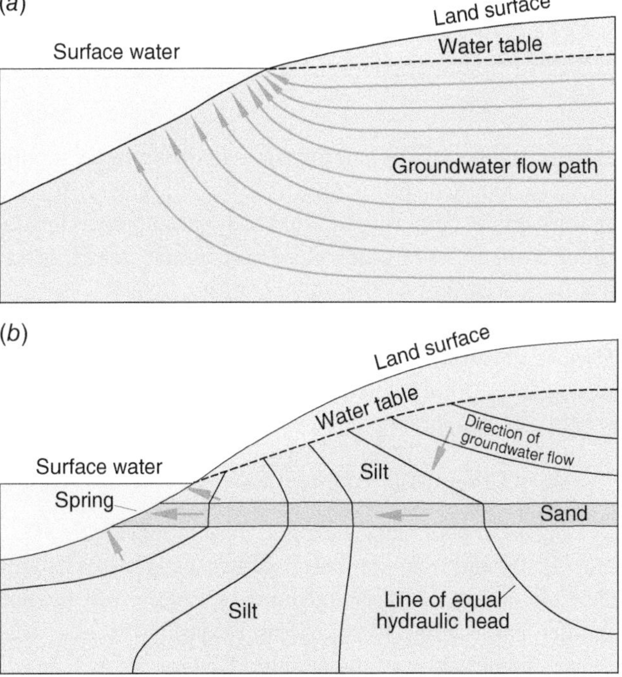

Figure 8.17 Panel (*a*) illustrates the pattern of groundwater seepage into surface water. Most of the inflow is concentrated near the shore. Panel (*b*) illustrates how subaqueous springs can lead to discharge into lakes. In this case, the pattern of geologic layering promoted the concentration of discharge (U.S. Geological Survey Winter et al., 1998/Public domain).

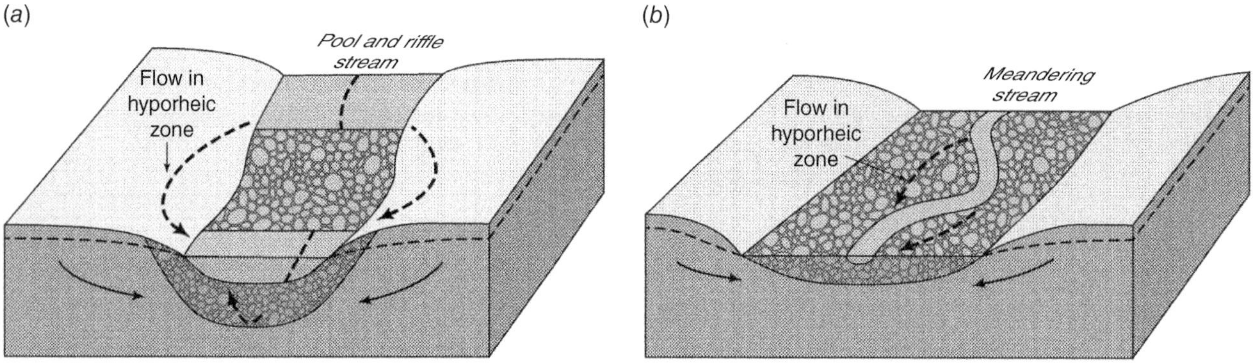

Figure 8.18 The exchange of groundwater in surface water in the hyporheic zone is associated with abrupt changes in streambed slope (*a*) and streambed meanders (*b*) (U.S. Geological Survey Winter et al., 1998/Public domain).

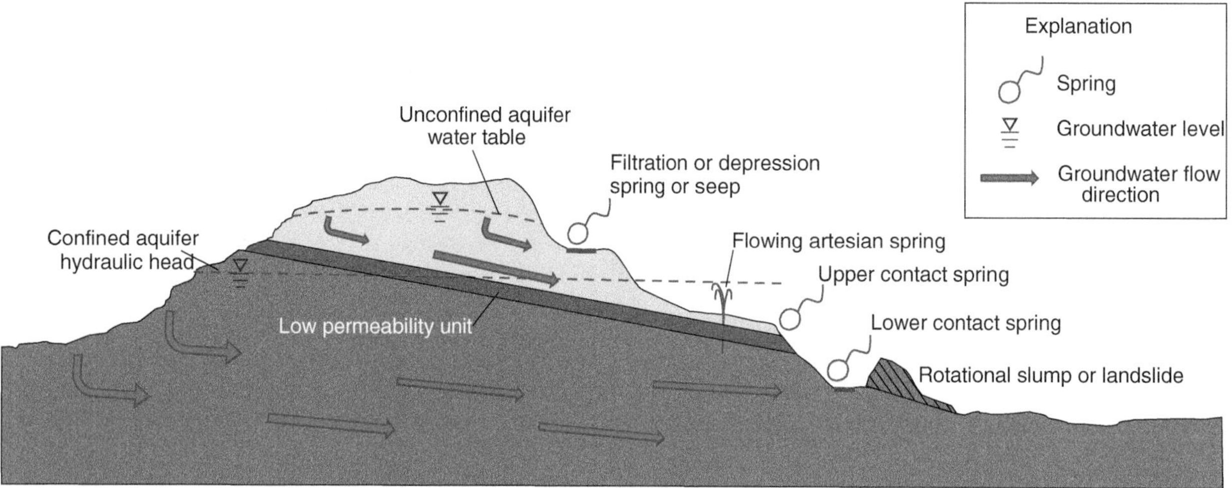

Figure 8.19 Conceptual model of springs in Theodore Roosevelt National Park, North Dakota (U.S. Geological Survey, adapted from Medler and Eldridge, 2021)/Public domain).

Artesian springs form when the water in the spring is supplied by a confined aquifer in which water can be significantly pressurized. Figure 8.19 illustrates a flowing artesian spring (Medler and Eldridge, 2021). Its location coincides with fracture zone through the confining bed that facilitates upward flow to the ground surface. In some places, these springs can yield significant quantities of water when the confined aquifer is transmissive. Springs are also classified as cold springs or thermal springs according to their temperatures.

Discharge of a spring depends on many factors including, the aperture of the fractures within the rock, the hydraulic head, the size of the groundwater drainage area which supplies water to the spring, and the quantity of rainfall. Groundwater discharges at a spring when the hydraulic head in an aquifer (h_{aq}) is higher than the elevation of the spring (z_{sp}). The spring flow rate can be approximated by

$$Q_{sp} = C(h_{aq} - z_{sp}) + D(h_{aq} - z_{sp})^2 \tag{8.8}$$

where Q_{sp} is the volumetric flow rate from the spring [L^3/T], C is the hydraulic conductance between the aquifer and the spring [L^2/T], and D is a proportionality constant. D is usually assumed to be zero for laminar flow.

The time variation in discharge from a spring can provide useful information about changing hydraulic conditions within an aquifer. For example, discharge from a spring will cease when the hydraulic head in the aquifer is lowered by pumping to less than the elevation of the spring. Likewise, a dry year with minimal recharge can lower hydraulic heads, causing a reduction in spring flow. A wet year can lead to an increase in spring flow.

Around the world, many springs have ceased to flow because of lowered potentiometric surfaces. For example, Jinan City of Shangdong province, China, has been known as the “City of Springs” for thousands of years. However, many of the springs have stopped flowing in last few decades because of the overdevelopment of groundwater. In the United States, many springs in western Kansas have stopped flowing due to the overproduction of groundwater from the High Plains aquifer system.

Example 8.3 The estimated conductance between an aquifer and a spring is between 10 and 100 m^2/day. The spring has an elevation of 500 m above sea level. The hydraulic head in the aquifer is 505 m above sea level. Calculate the range of the spring flow rates.

Solution

The range of spring flow rate is

$$Q_{sp} = (10{\sim}100)(\mathrm{m}^2/\mathrm{day})\,(505\ \mathrm{m} - 500\ \mathrm{m}) = (50{\sim}500)\mathrm{m}^3/\mathrm{day}$$

8.4.3 Evapotranspiration

In many settings, groundwater discharges through transpiration and evaporation of soil water. Transpiration involves the utilization of groundwater by plants. One group of plants called phreatophytes is particularly important in this respect. Meinzer (1927) defined *phreatophytes* as plants that habitually obtain their water supply from the zone of saturation, either directly or through the capillary fringe. However, they also can grow under conditions of abundant soil moisture. Phreatophytes are particularly noticeable in dry regions, where locally a shallow water table often provides the stable water supply necessary to sustain the growth of these plants. When a community of phreatophytes is established, it can use tremendous quantities of water, which represents an important mechanism for natural discharge.

According to Meyboom (1967), evidence that phreatophytes use water is provided by a record of daily water-level fluctuations. In general, the water table declines during the day when transpiration by the plants is at a maximum. Water levels bottom out in the early evening and begin to rise as surrounding groundwater moves back into the zone of depressed hydraulic head caused by the plants. Phreatophytes depend upon this nightly recovery of water levels. If it were not for this cycle of recovery, the water table would fall below the root system and the vegetation would perish (Meyboom, 1967). Phreatophytes are common in groundwater discharge areas because in these areas the water table is often close to the ground surface and able to recover rapidly. Phreatophytes growing along major rivers not only use groundwater but also surface water through infiltration of river water.

On the Canadian Prairies, the types phreatophytes include Manitoba maple, salt grass, Baltic rush, alfalfa, poplar, and buffalo berry. In the arid southwestern U.S., typical phreatophytes include salt grass, greasewood, and mesquite. Trees that depend upon groundwater include the willow, cottonwood, and sycamore. Apart from alfalfa, which is used for animal forage, phreatophytes have typically been regarded as nuisance plants that waste water. However, communities of phreatophytes found along perennial streams form unique and important ecological environments. These unique environments found along streams in arid regions are called *riparian zones*.

Any time that mineralized groundwater gets within a few meters of the ground surface in an arid climate, the potential exists for water to discharge by evaporation. Evaporative rates can be sufficiently high that evaporation is the principal mode of discharge. Evaporation concentrates dissolved constituents in the remaining water and eventually salts precipitate to form saline soils. The following example documents these interesting discharge conditions.

Saline soils that sometimes form in arid areas can be indicators of groundwater discharge. An example of this process leading to soil salinization was documented in studies in the Red River Basin of the North in western North Dakota (USA) (Stoner et al., 1998; Strobel and Haffield, 1995). Shown on the map in Figure 8.20 is a relatively large area of saline seepage related to the discharge of saline groundwater flowing eastward out of the sedimentary basin to the west. Associated with areas of seepage are saline soils (Figure 8.20), which formed from the evaporative discharge of shallow groundwater at the ground surface (Stoner et al., 1998; Strobel and Haffield, 1995). The upward flow gradient from deeper bedrock through lower permeability glacial sediments is indicated by flowing wells. In effect, the hydraulic heads at depth are higher than the water table and the ground surface. The more localized discharge of saline groundwater has also led to the formation of small lakes and wetlands (Strobel and Haffield, 1995), and the salinization of surficial aquifers and surface waters.

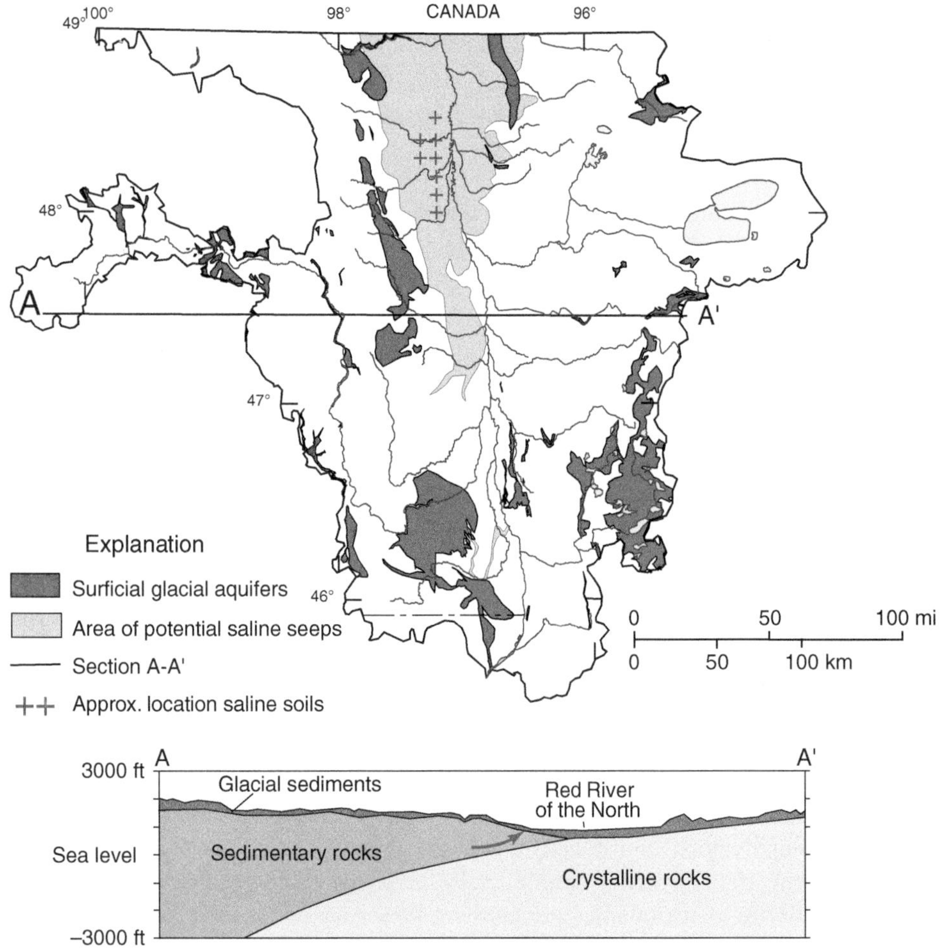

Figure 8.20 Lowlands along the northward-flowing Red River of North in North Dakota are areas of evaporative discharge for groundwater flowing out of the sedimentary basin to the west as shown by the red arrow. The evaporative discharge has led to areas of soil salinity west of the Red River as indicated on the map (U.S. Geological Survey, adapted from Stoner et al., 1998/Public domain).

Human activities associated with farming can result in soil salinization. For example, replacing natural grasslands by farm fields can increase the recharge leading to a rise in water table in lowland areas. Over time, soil salinization prevents the growth of agricultural plants. Similarly, long-term irrigation of crops in arid areas with water containing dissolved constituents can also lead to soil salinization.

8.5 Groundwater Surface-Water Interactions

One important feature of streams and rivers is their ability to gain or lose water through the groundwater system. A gaining stream receives groundwater discharge; a losing stream provides groundwater recharge. Once hydraulic head data are available in the vicinity of a stream, one can determine whether the stream is gaining (Figure 8.21*a*) or losing (Figure 8.21*b*). The stage of a gaining stream is lower than that of hydraulic heads in the aquifer immediately adjacent. The stage in a losing stream is higher than adjacent hydraulic heads. The resulting flow patterns show that groundwater flows toward a gaining stream, and away from a losing stream (compare Figure 8.21*a* and 8.21*b*). Numerous studies have shown that the hydraulic conductivity of a streambed, sometimes called a clogging layer, is lower than that in the aquifer immediately below the streambed. Thus, a situation arises where the stream is not all that well connected to the groundwater system and flow has difficulties moving in or out.

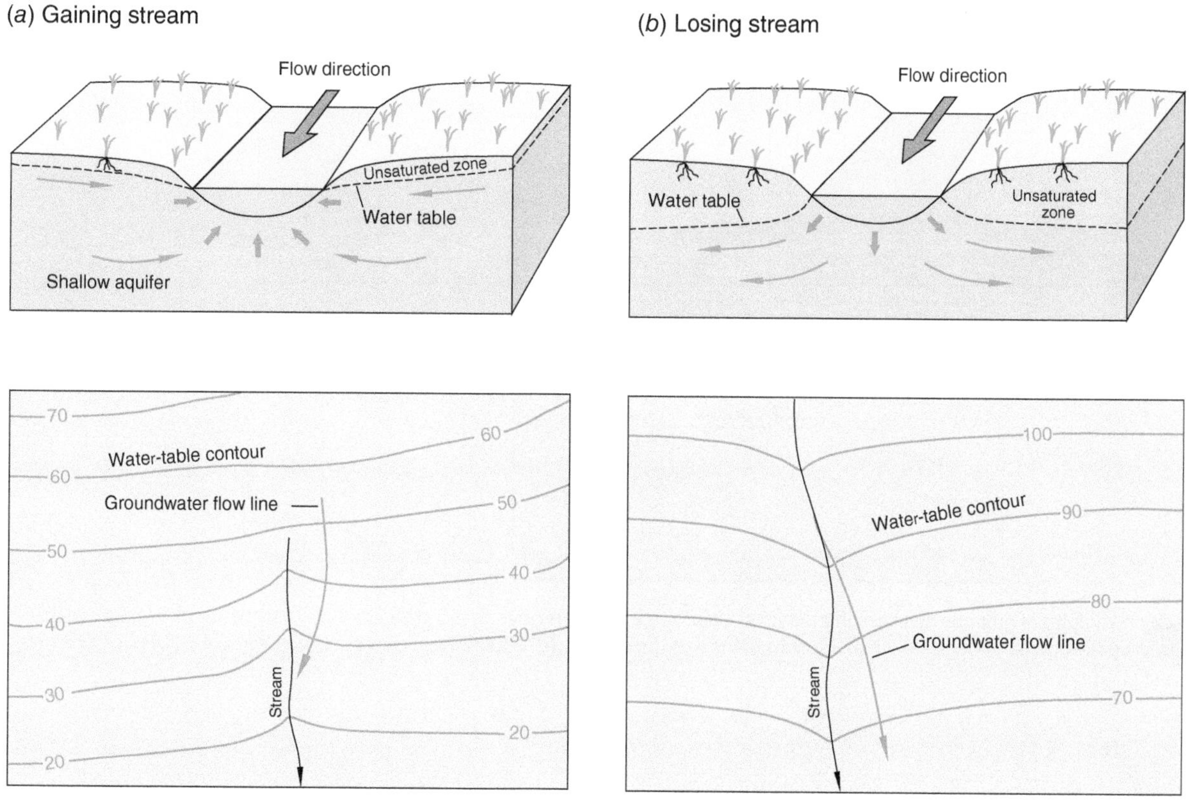

Figure 8.21 With a gaining stream (*a*), groundwater discharges to the stream as shown by the schematic and the configuration of the water table in the vicinity of the stream. In (*b*), the stream is a source of recharge to the groundwater (U.S. Geological Survey Winter et al., 1998/Public domain).

Peterson and Wilson (1988) classified streams as connected gaining streams, connected losing streams, disconnected losing streams with a shallow water table, and disconnected losing streams with a deep water table. Their studies showed that even if a stream and an aquifer are disconnected, the unsaturated flow beneath the stream depends on the depth of water table for a disconnected losing stream with shallow water table. Overall, features of the unsaturated flow affect the discharge rate from stream to aquifer.

Lakes, like rivers, commonly interact with groundwaters are classified as discharge lakes, recharge lakes, or flow-through lakes depending upon their mode of interaction with the groundwater. A discharge lake receives groundwater discharge throughout their entire bed (Figure 8.22*a*) (Winter et al., 1998). A recharge lake loses water as seepage to a groundwater flow system (Figure 8.22*b*). Most lakes are through-flow types, receiving inflow over part of their bed and losing water elsewhere (Figure 8.22*c*).

Much has been learned about the interaction between aquifers and lakes using computer models (Winter 1976; Winter and Pfannkuch 1984). Figure 8.23 illustrates how a discharge lake receives inflow from local flow systems, controlled by the position of the water table in the vicinity of the lake. A stagnation point exists for a discharge lake when a local flow system recharges the lake. At the stagnation point, the hydraulic head is higher than the water level in the discharge lake.

Wetlands are common in countries all around the world. They form when groundwater discharges to land surface or when conditions are such that surface water is slow to drain away. The development of wetlands is often intimately related to regional flow conditions. For example, the upland wetland in Figure 8.24*a* forms in a groundwater discharge area, which itself is a manifestation of the complexities of the regional groundwater flow (Winter et al., 1998). The wetland in Figure 8.24*b* formed at a break in slope, which fostered groundwater discharge as seepages or springs. Another type of wetland forms along a stream (Figure 8.24*c*). A riverine wetland depends primarily on the stream and incidentally on groundwater discharge for water (Winter et al., 1998). In some settings, wetlands receive precipitation and surface runoff and function as groundwater recharge areas (Figure 8.24*d*). Often in these circumstances, water is slow to leak from the wetlands because the organic sediments in the wetlands do not conduct water well (Winter et al., 1998).

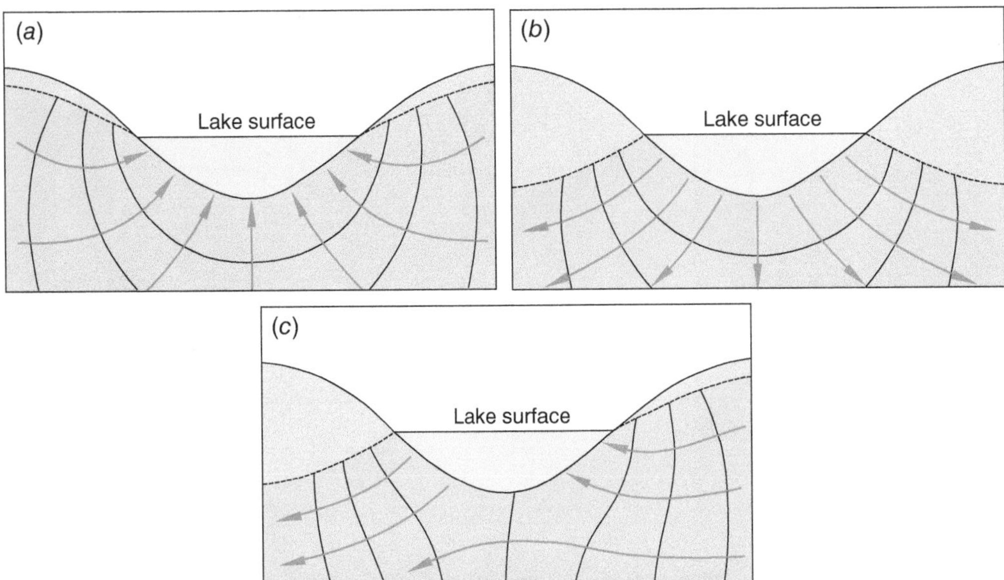

Figure 8.22 Examples of groundwater/lake interactions. A discharge lake (*a*) receives groundwater inflow. A recharge lake (*b*) loses water as seepage to groundwater. A through-flow lake (*c*) both gains and loses water (U.S. Geological Survey, Winter et al., 1998/Public domain).

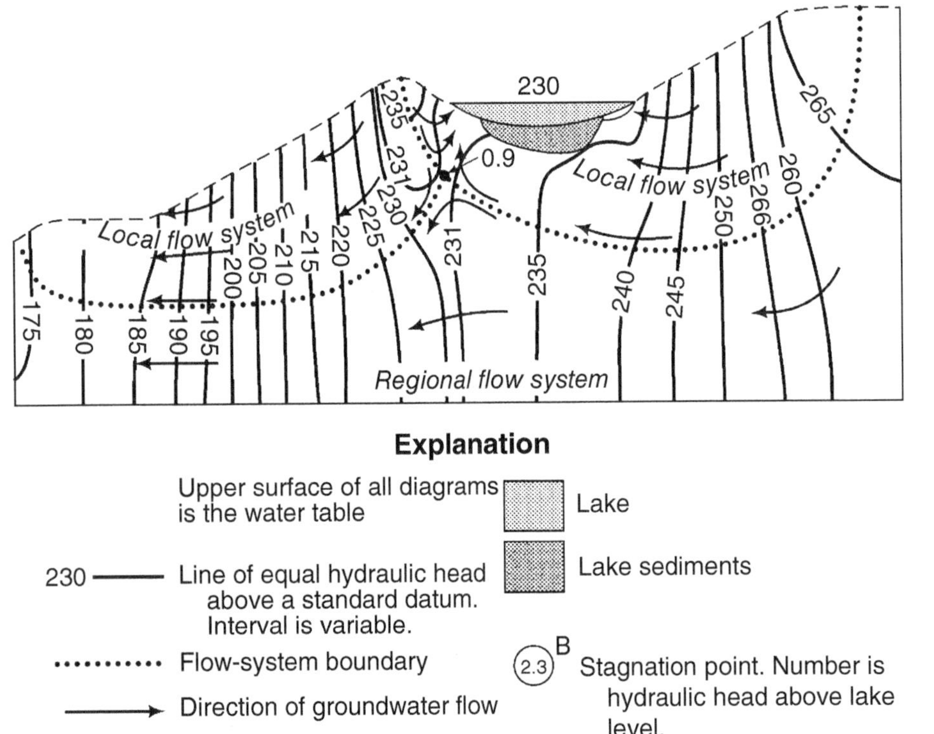

Figure 8.23 Simulation result that illustrates the pattern of flow in the vicinity of a lake and the relationship to local and intermediate flow systems. A stagnation point is evident in the vicinity of the lake (Winter and Pfannkuch, 1984. Reproduced from Journal of Hydrology 75(1-4): 239-253. Copyright 1984. With permission from Elsevier Science).

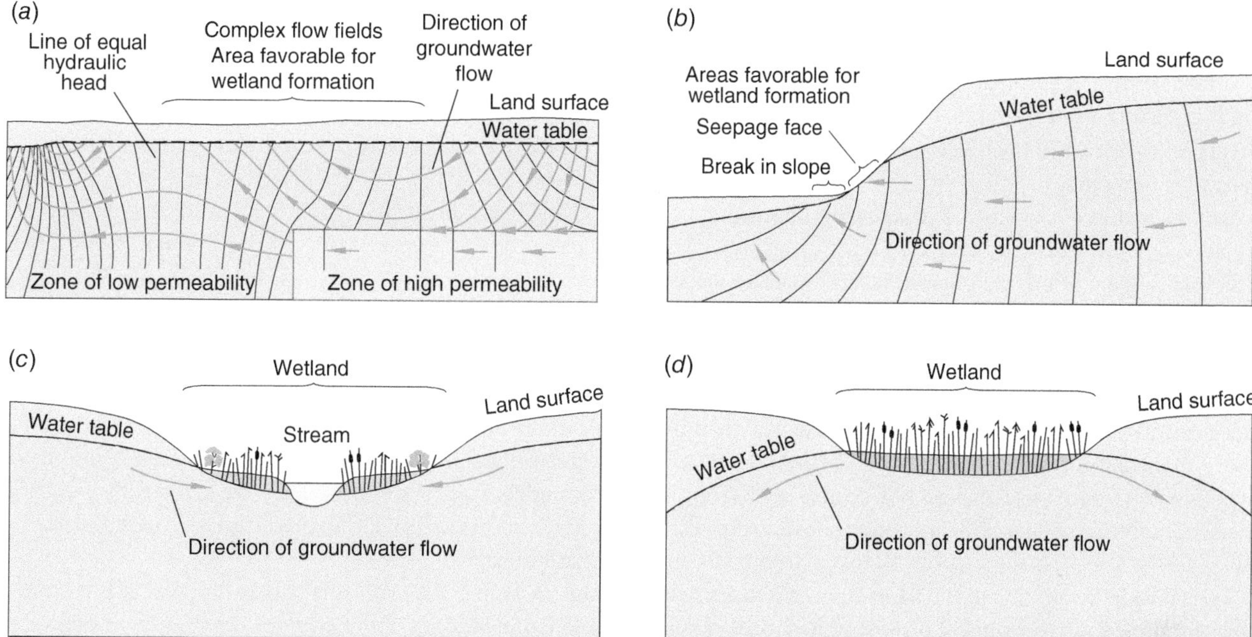

Figure 8.24 Types of wetlands. Panel (*a*) shows groundwater discharge to the ground surface from a complex groundwater flow system. Panel (*b*) illustrates groundwater discharge through a seepage face at a break-in slope. In Panel (*c*) riverine wetlands form along streams. The wetlands in Panel (*d*) are formed by the accumulation of precipitation in a depression (U.S. Geological Survey Winter et al., 1998/Public domain).

8.6 Freshwater/Saltwater Interactions

Freshwater and salt water interact with each in a variety of hydrogeological settings. In coastal areas, less dense, freshwater tends to override denser, salt water (Figure 8.25). Between the two water masses there is a zone of mixing, sometimes called the zone of diffusion, where the fluid has an "in-between" chemical composition. Thus, the actual interface between freshwater and saltwater can be more than 1000 ft wide (Figure 8.25), although most mathematical analyses are based on an assumption of a sharp interface. The 1000 mg/L isochlor line commonly delineates the saltwater front (Figure 8.25).

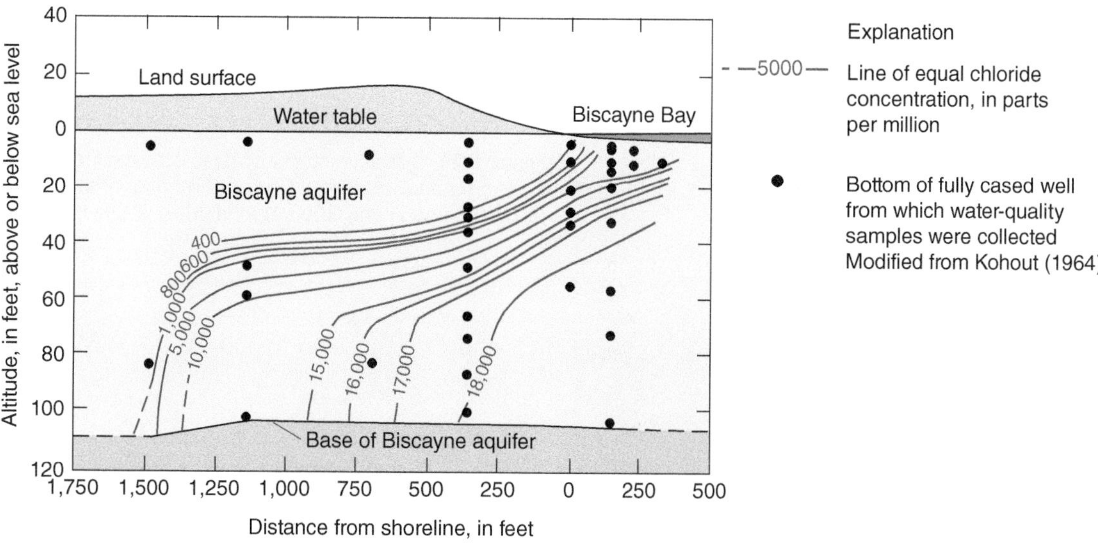

Figure 8.25 Cross section of seawater intrusion into the Biscayne aquifer near Miami, Florida. The zone of mixing is reflected by the variation in chloride concentrations as determined by monitoring wells (U.S. Geological Survey, Barlow, 2003 modified from Kohout, 1964/Public domain).

An island in an ocean provides a slight variation in the distribution of water. In this case, a lens of freshwater sits on top of the salt water, which extends beneath the island. The position of the mixing zone varies depending upon the quantity of freshwater moving through the system. With greater recharge raising the water table, the interface moves deeper. With smaller quantities of recharge, the interface rises up beneath the island. It follows then that in coastal and island settings, diversion of seaward-moving groundwater by pumping can cause the interface to retreat and produce a condition known as saltwater intrusion.

On the continents, deep geological basins often contain salt water in permeable units. When these units crop out and are invaded by freshwater, the condition is set up again where freshwater and saltwater are juxtaposed. Although the setting is different, a zone of mixing often develops between the two fluids.

8.6.1 Locating the Interface

Because of the potential for saltwater intrusion to impair the quality of coastal aquifers, studies have continued for more than half a century to understand the behavior of freshwater/saltwater systems and their interfaces. These days, groundwater hydrologists have a powerful array of computational models at their disposal. They are sufficiently powerful to account for the details of the most complicated settings. Our objective here is to provide an understanding of the processes involved, by studying some of the simpler, historically important, approaches.

Let us begin by looking at the freshwater/saltwater system shown in Figure 8.26. For this simple analysis, it is assumed that the system is hydrostatic. This assumption means that the weight of a column of freshwater extending from the water table to a point on the interface is the same as a column of salt water extending from sea level to the same depth (Figure 8.26). This condition can be expressed mathematically as

$$\rho_s g z = \rho_f g(h_f + z) \tag{8.9}$$

where ρ_f is the density of freshwater, ρ_s is the density of salt water, z is the height of the saltwater column, or the depth below sea level to a point on the interface, h_f is the hydraulic head above sea level, and $h_f + z$ is the height of the freshwater column. Equation (8.9) can be rearranged as

$$z = \frac{\rho_f}{\rho_s - \rho_f} h_f \tag{8.10}$$

If the density of freshwater is taken as 1.0 g/cm^3 and seawater as 1.025 g/cm^3, then

$$z = 40h_f \tag{8.11}$$

In words, the depth of the interface is approximately 40 times the height of the water table above sea level. Equation (8.11) is referred to as the Ghyben–Herzberg relation because it was first determined by the two scientists independently (Ghyben 1899; Herzberg 1901).

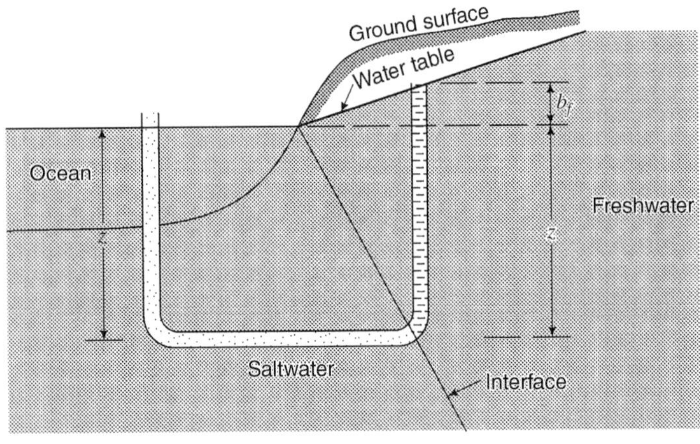

Figure 8.26 Idealization of a coastal saltwater/freshwater system for analysis by the Ghyben–Herzberg relation (Schwartz and Zhang, 2003/© John Wiley & Sons).

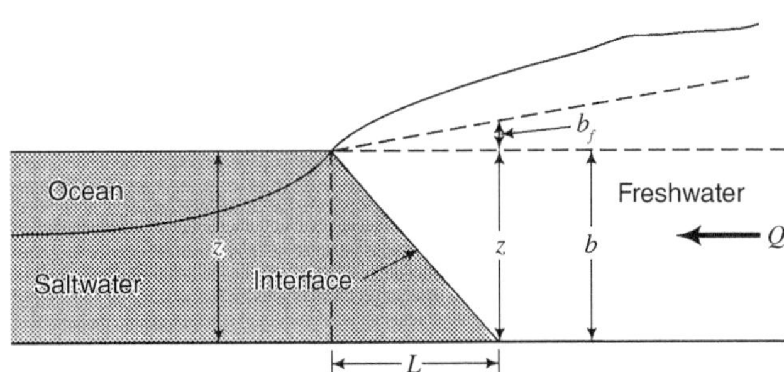

Figure 8.27 Geometry of a coastal aquifer for calculations of the position of the interface and length of the saltwater wedge for flowing conditions (Domenico and Schwartz, 1998, Physical and Chemical Hydrogeology. Copyright © 1990, 1998 John Wiley & Sons, Inc. All Rights Reserved. Reproduced with permission.).

This simple theory can be applied to examine some of the features of saltwater/freshwater interactions involving a confined aquifer of thickness b (Figure 8.27). Let us assume we know a little more about the system, like the discharge rate per unit length of coastline (Q'). An improved estimate of the position of the freshwater/saltwater interface is given as

$$x = \frac{1}{2}\frac{\left(\rho_s - \rho_f\right)Kz^2}{\rho_f Q'} \tag{8.12}$$

The form of Eq. (8.12) indicates that the freshwater-saltwater interface is now no longer a straight line. For an aquifer with thickness of b (Figure 8.27), the length of the saltwater protrusion wedge (L) is expressed as

$$L = \frac{1}{2}\frac{\left(\rho_s - \rho_f\right)Kb^2}{\rho_f Q'} \tag{8.13}$$

where K is the hydraulic conductivity of the aquifer, and b is the thickness of the aquifer. Thus, the length of protrusion of salt water under natural conditions is directly proportional to the hydraulic conductivity and thickness squared and inversely proportional to the flow of fresh water to the sea.

Example 8.4 The densities of fresh and saltwater are 1.0 and 1.025 g/cm^3, respectively. The water levels in two monitoring wells far from the shoreline are 0.5 and 1.0 m above sea level. The distance between the two wells is 1000 m. The hydraulic conductivity of the coast aquifer is 10 m/day. If the aquifer thickness is 50 m, calculate the length of saltwater wedge and the interface between the freshwater and the saltwater using Eqs. (8.12) and (8.13).

Solution

The discharge from the aquifer to the sea per unit length of shoreline ($z = b$) is

$$Q' = Kb\frac{dh}{dx} = (10\ \text{m/day})(50\ \text{m})\frac{(1.0\ \text{m}) - (0.5\ \text{m})}{(1000\ \text{m})} = 0.25\left(\text{m}^3/\text{day/m}\right)$$

The equation describing the interface is

$$x = \frac{1}{2}\frac{\left(\rho_s - \rho_f\right)Kz^2}{\rho_f Q'} = \frac{1}{2}\frac{(1.025\ \text{g/cm}^3) - (1.0\ \text{g/cm}^3)}{(1.0\ \text{g/cm}^3)}\frac{(10.0\ \text{m/day})}{(0.25\ \text{m}^2/\text{day})}z^2 = 0.5z^2$$

The length of protrusion of salt water into the aquifer is

$$L = \frac{1}{2}(0.025)\frac{(10\text{m/day})(50)^2}{(0.25\text{m}^2/\text{day})} = 1250\ \text{m}$$

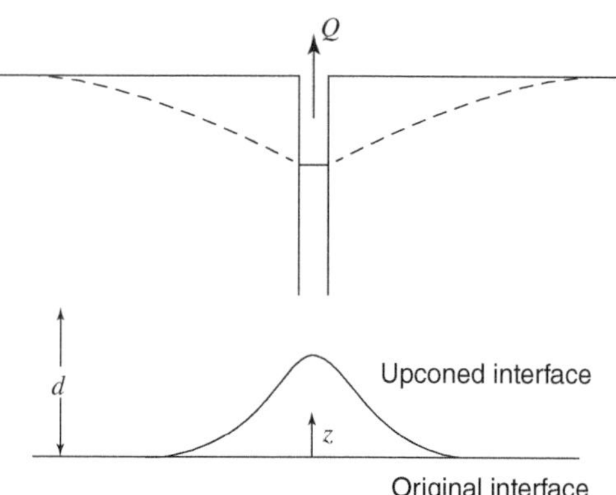

Figure 8.28 Upconing of the saltwater/freshwater interface in response to pumping (Domenico and Schwartz, 1998, Physical and Chemical Hydrogeology. Copyright © 1990, 1998 John Wiley & Sons, Inc. All Rights Reserved. Reproduced with permission.).

8.6.2 Upconing of the Interface Caused by Pumping Wells

The Ghyben–Herzberg relationship shows how lowering of water levels causes the interface to rise. In this vicinity of pumping wells where water levels are most affected, the interface can rise, a phenomenon referred to as *upconing* (Figure 8.28). The extent of upconing can be calculated with the following equation from Schmorak and Mercado (1969)

$$z = \frac{Q\rho_f}{2\pi dK\left(\rho_s - \rho_f\right)} \tag{8.14}$$

where z is the vertical change of the freshwater-saltwater interface, Q is the pumping rate, and d is the distance from the base of the well to the original (prepumping) interface (Figure 8.28).

Both laboratory and field observations suggest that this relationship holds only for very small rises in the interface and there exists a critical elevation at which the interface is no longer stable and saltwater flows to the well (Schmorak and Mercado, 1969).

Dagan and Bear (1968) suggest that the interface will be stable for upconed heights that do not exceed one-third of d given in Figure 8.28. Thus, if z is taken as $0.3d$, the maximum permitted pumping rate should not exceed

$$Q_{\max} \leq 0.6\pi d^2 K \frac{\left(\rho_s - \rho_f\right)}{\rho_f} \tag{8.15}$$

Example 8.5 The distance from the base of a pumping well to the freshwater-saltwater interface is 100 m, the pumping rate is 3000 m^3/day, and the hydraulic conductivity is 10 m/day. What will be in the position of the freshwater-saltwater interface? What is the maximum permitted pumping rate?

Solution

The rise in the freshwater-saltwater interface is

$$z = \frac{(3000\ \text{m}^3/\text{day})(1.0\ \text{g/cm}^3)}{(2 \times 3.14.16)(100\ \text{m})(10\ \text{m/day})[(1.025\ \text{g/cm}^3) - (1.0\ \text{g/cm}^3)]} = 13.26\ \text{m}$$

The maximum permitted pumping rate is

$$Q_{\max} \leq 0.6(3.1416)(100\ \text{m})^2(10\ \text{m/day})(0.025) = 4.7 \times 10^3 \text{m}^3/\text{day}$$

Exercises

8.1 The flux rate per unit coastline from an aquifer to the sea is 10 m^3/day/m. The hydraulic conductivity and thickness of the aquifer are 10 m/day and 100 m, respectively. Calculate the position of freshwater/saltwater interface.

8.2 A saltwater interface is stable at a position of about 40 m beneath the base of a well in a formation with a hydraulic conductivity of 1×10^{-1} cm/s. Calculate the maximum pumping rate so as not to cause unstable upconing of the interface.

References

Allison, G. B., and M. W. Hughes. 1978. The use of environmental chloride and tritium to estimate total recharge to an unconfined aquifer. Australian Journal of Soil Research, v. 16, no. 2, p. 181–195.

Allison, G. B., C. J. Barnes, M. W. Hughes, and F. W. J. Leaney. 1983. The effect of climate and vegetation on oxygen-18 and deuterium profiles in soils. Proceedings International Symposium on Isotope Hydrology and Water Resource Development, No. IAEA-SM-270/20, IAEA, Vienna.

Barlow, P. M. 2003. Groundwater in fresh water-salt water environments of the Atlantic Coast. U.S. Geological Survey Circular 1262, 113 p.

Dagan, G., and J. Bear. 1968. Solving the problem of local interface upconing in a coastal aquifer by the method of small perturbations. Journal of Hydraulic Research, v. 6, no. 1, p. 15–44.

Delin, G. N., and D. G. Woodward. 1984. Hydrogeologic setting and the potentiometric surfaces of regional aquifers in the Hollandale embayment, southeastern Minnesota, 1970-80. U.S. Geological Survey Water-Supply Paper 2219, 56 p.

Domenico, P. A., and F. W. Schwartz. 1998. Physical and Chemical Hydrogeology. John Wiley & Sons, New York, 506 p.

Freeze, R. A. 1969a. Theoretical analysis of regional groundwater flow. Department of Energy Mines and Resources, Inland Waters Branch, Scientific Series No. 3, 147 p.

Freeze, R. A. 1969b. Regional groundwater flow—Old Wives Lake Drainage basin, Saskatchewan. Department of Energy Mines and Resources, Inland Waters Branch, Scientific Series No. 5, 243 p.

Freeze, R. A., and P. A. Witherspoon. 1966. Theoretical analysis of regional groundwater flow. I: Analytical and numerical solutions to the mathematical model. Water Resources Research, v. 2, no. 4, p. 641–656.

Freeze, R. A., and P. A. Witherspoon. 1967. Theoretical analysis of regional groundwater flow. II: Effects of water table configuration and subsurface permeability variations. Water Resources Research, v. 3, no. 2, p. 623–634.

Ghyben, W.B. 1899. Nota in verband met de voorgenomen putboring nabij Amsterdam [Notes on the behaviour of wells near Amsterdam]. *Tijdschrift het koninklijk Institut voor Ingenieurs, KIVI, The Hague*, p. 8-22.

Herzberg, B. 1901. Die Wasserversovgung einiger Nordseebaser. Journal of Gasbeleucht and Wasserversov, v. 44, p. 815–819.

Hubbert, M. K. 1940. The theory of groundwater motion. The Journal of Geology, v. 48, no. 8, p. 785–944.

Kohout, F. A. 1964. The flow of fresh water and salt water in the Biscayne aquifer of the Miami area, Florida. U.S. Geological Survey Water-Supply Paper 1613-C, p. 12–32.

Liu, G., F. W. Schwartz, C. K. Wright, and N. E. McIntyre. 2016. Characterizing the climate-driven collapses and expansions of wetland habitats with a fully integrated surface–subsurface hydrologic model. Wetlands, v. 36, no. Suppl 2, p. 287–297.

Medler, C. J., and W. G. Eldridge. 2021. Spring types and contributing aquifers from water-chemistry and multivariate statistical analyses for seeps and springs in Theodore Roosevelt National Park, North Dakota, 2018. U.S. Geological Survey Scientific Investigations Report 2020-5121, 48 p.

Meinzer, O. E. 1927. Plants as indicators of groundwater. U.S. Geological Survey Water-Supply Papers 577.

Meyboom, P. 1967. Mass transfer studies to determine the groundwater regime of permanent lakes in hummocky moraine of western Canada. Journal of Hydrology, v. 5, p. 117–142.

Olcott, P. 1992. Groundwater atlas of the United States, Segment 9, Iowa, Michigan, Minnesota, and Wisconsin. U.S. Geological Survey Hydrologic Investigations Atlas 730-J, 31 p.

Peterson, D. M., and J. L. Wilson. 1988. Field study of ephemeral stream infiltration and recharge. New Mexico Water Resources Research Institute, Technical Completion Report No. 228, New Mexico State University, New Mexico.

Phillips, F. M., J. L. Mattick, T. A. Duval, D. Elmore, and P. W. Kubik. 1988. Chlorine-36 and tritium from nuclear weapons fallouts as tracer for long-term liquid and vapor movement in desert soils. Water Resources Research, v. 24, no. 11, p. 1877–1891.

Robson, S. G., and E. R. Banta. 1995. Groundwater atlas of the United States, Segment 2: Arizona, Colorado, New Mexico, Utah. U.S. Geological Survey Hydrologic Investigations Atlas 730-C, 32 p.

Schmorak, S., and A. Mercado. 1969. Upconing of freshwater-seawater interface below pumping wells. Water Resources Research, v. 5, no. 6, p. 1290–1311.

Schwartz, F. W., and H. Zhang. 2003. Fundamentals of Groundwater. John Wiley & Sons, New York, 583 p.

Sharma, M. L., and M. W. Hughes. 1985. Groundwater recharge estimation using chloride, deuterium and Oxygen-18 profiles in the deep coastal sands of western Australia. Journal of Hydrology, v. 8, no. 1–2, p. 93–109.

Stoner, J. D., D. L. Lorenz, R. M. Goldstein, M. E. Brigham, and T. K. Cowdery. 1998. Water Quality in the Red River of the North Basin, Minnesota, North Dakota, and South Dakota, 1992-95. *U.S. Geological Survey Circular* 1169, 33 p.

Strobel, M. L., and N. D. Haffield. 1995. Salinity in surface water in the Red River of the North basin, northeastern North Dakota. U.S. Geological Survey Water-Resources Investigations Report 95-4082, 14 p.

Tóth, J. 1962. A theory of groundwater motion in small drainage basins in central Alberta, Canada. Journal of Geophysical Research, v. 67, no. 11, p. 4375–4388.

Tóth, J. 1963. A theoretical analysis of groundwater flow in small drainage basins. Journal of Geophysical Research, v. 68, no. 16, p. 4795–4812.

Trapp, H., Jr., and M. A. Horn. 1997. Groundwater atlas of the United States, Segment 11: Delaware, Maryland, New Jersey, North Carolina, Pennsylvania, Virginia, West Virginia. U.S. Geological Survey Hydrologic Investigations Atlas 730-L, 24 p.

Whitehead, R. L. 1996. Groundwater atlas of the United States, Segment 8: Montana, North Dakota, South Dakota, Wyoming. U.S. Geological Survey, Hydrologic Investigations Atlas 730-I, 24 p.

Winter, T. C. 1976. Numerical simulation analysis of the interaction of lakes and groundwaters. U.S. Geological Survey Professional Paper 1001, 45 p.

Winter, T. C., and H. O. Pfannkuch. 1984. Effect of anisotropy and groundwater system geometry on seepage through lakebeds, 2. Numerical simulation analysis. Journal of Hydrology, v. 75, no. 1–4, p. 239–253.

Winter, T. C., J.W. Harvey, O.L. Franke, and W. M. Alley. 1998. Groundwater and surface water, A single Resource. U.S. Geological Survey Circular 1139, 79 p.

Zimmerman, U., D. Enhalt, and K. O. Munnich. 1966. Soil-water movement and evapotranspiration: Changes in the isotopic composition of the water. Isotopes in Hydrology, Proceedings of the IAEA Symposium 1966, Vienna, p. 567–584.

9

Response of Confined Aquifers to Pumping

CHAPTER MENU

One of the important jobs for groundwater hydrologists is to find and develop water supplies. High capacity wells installed in productive aquifers are capable of providing thousands of gallons of water per minute. Research in the 1940s and 1950s produced quantitative analytical tools to predict how pumping would impact hydraulic head in the aquifers and to interpret the results of hydraulic tests. From the middle 1960s through early 1980s, powerful numerical codes like MODFLOW became available to help analyze much more complicated systems.

There are fundamentally two types of problems related to an aquifer's responses to pumping. The so called forward problem is concerned with predicting what the hydraulic head distribution will be in an aquifer at times in the future, given boundary conditions, initial conditions, and information about transmissivity, storativity, and pumping rate. As we saw in Chapter 3, one can calculate this hydraulic this head distribution by solving a groundwater flow equation. Forward modeling is essential to designing well systems, analyzing whether drawdowns caused by a well are impacting other wells, and in designing dewatering systems.

The inverse problem, as applied to well problems, involves using measurements of hydraulic head in an aquifer as a function of time to calculate values of transmissivity, storativity, specific yield etc. In other words, the mathematical theory provides the basis for interpreting the results of an aquifer test. Here and in Chapters 10–13, you will learn how to make drawdown predictions for various types of aquifers and how to interpret the results of aquifer tests.

9.1 Aquifers and Aquifer Tests

The study of well hydraulics is complicated. Every situation where the type of aquifer, the pattern of layering of aquifers, or the length of the screen in relation to the aquifer thickness changes, we need a different analytical solution. If our goal was to treat every pumping situation that one might conceivably encounter in the field, we could end up showing you 40 of 50 different solutions. Fortunately, we have set a less ambitious goal of looking at only a few of the most common situations. This chapter is focused on a confined aquifer that is homogeneous, isotropic, and infinite in extent.

We have used the term aquifer test without discussing it in detail. An *aquifer test* involves pumping a well for the purpose of determining aquifer parameters like T and S. A test typically involves a pumping well, and one or more observation wells. A *pumping well* has a relative large diameter casing, and is screened across all or part of the aquifer (Figure 9.1). A large diameter casing is necessary because a pump and piping system needs to be installed down in the well. *Observations wells* are

Fundamentals of Groundwater, Second Edition. Franklin W. Schwartz and Hubao Zhang.

Companion website: www.wiley.com/go/schwartz/fundamentalsofgroundwater2

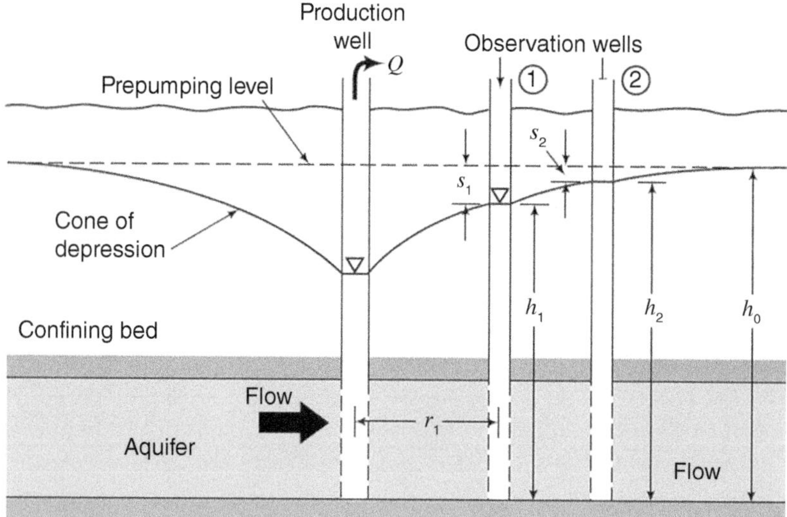

Figure 9.1 A confined aquifer from which groundwater is being withdrawn at a constant rate Q. The cone of depression spreads away from the well and produces drawdowns s_1 and s_2 (Schwartz and Zhang, 2003 / John Wiley & Sons).

located at varying distances from the pumping well. They commonly are smaller in diameter and again are screened across all or part of the aquifer. Before an aquifer test is begun, water levels in all of the wells are measured to provide the *pre-pumping or static water levels* (h_0). In other words, these measurements provide hydraulic heads in the wells at time zero. A test begins by beginning to pump water from the well. The *pumping rate* (Q) is the volume of water pumped from a well per unit time [L^3/T].

As the aquifer is pumped, water levels are measured periodically in the pumping and observation wells. Water levels are measured frequently at first, because at early times they change rapidly. The term *pumping water level* (h) is the term used to describe the water level in a well during a test.

Convention is to work with the change in water levels through the test rather than water levels. The term *drawdown* ($s = h - h_0$) is the difference between the static water level and the pumping water level (Figure 9.1). Once the impact of pumping becomes evident at a well, drawdowns usually increase with time. The zone around the well in which there is a measurable water-level change is called the cone of depression. The *cone of depression* is a water-level low in water table or potentiometric surface, which has the shape of an inverted cone, centered on the pumped well Away from the cone of depression, drawdown caused by pumping is undetectable. The *radius of influence* (R) is the distance from a pumped well to the edge of cone of depression. Under steady-state conditions, the water discharged by a well is assumed coming from sources beyond the radius of influence. Under transient-flow conditions, the water discharged by a well is assumed coming from the aquifer storage within the radius of influence and sources beyond the radius of influence.

9.1.1 Units

The analytical solutions presented in this and following chapters are all developed in terms of consistent units. Thus, it does not matter what units of length (for example feet or meters) or time (seconds, day) you use, as long as all the units are consistent. For example, if meters and days are selected as the consistent units, discharge would have units of m^3/day, distances would be in meters, transmissivities in m^2/day and so on. Readers will often find it necessary to convert units before using any of the equations in the following chapters. Conversion equations for transmissivity and hydraulic conductivity are introduced in Chapters 3 and 4. The following equations will help provide pumping rates as a consistent unit from the English unit.

$$1\,\text{gpm} = 192.5\text{ft}^3/\text{day} = 5.45\text{m}^3/\text{day} = 6.3 \times 10^{-5}\,\text{m}^3/\text{s} \tag{9.1}$$

$$1\text{ft}^3/\text{day} = 5.19 \times 10^{-3}\text{gpm} = 2.832 \times 10^{-2}\text{m}^3/\text{day} = 3.28 \times 10^{-7}\text{m}^3/\text{s} \tag{9.2}$$

$$1\text{m}^3/\text{day} = 35.31\text{ft}^3/\text{day} = 0.1835\text{ gpm} = 1.1574 \times 10^{-5}\text{ m}^3/\text{s} \tag{9.3}$$

$$1\text{ m}^3/\text{s} = 3.051 \times 10^6\text{ft}^3/\text{day} = 1.58 \times 10^4\text{ gpm} = 8.64 \times 10^4\text{m}^3/\text{day} \tag{9.4}$$

9.2 Thiem's Method for Steady-State Flow in a Confined Aquifer

Historically, one of the first quantitative approaches for looking at flow in a confined aquifer is that of Theim (1906). This theory applies to a homogeneous and isotropic aquifer that is infinite in extent. The analysis also assumes that pumping has been sufficiently that the groundwater system achieves steady state. In other words, water levels in the wells do not change with time (Figure 9.2*a*). The map view in Figure 9.2*b* shows that the flow in this case is radial, toward the well with hydraulic heads increasing away from the pumping well.

The hydraulic head in the aquifer can be determined as a solution to a groundwater flow equation, like those we presented in Chapter 3. However, in this case, it is beneficial to use radial coordinates, where distances (r) are measured from the well to some point of interest. A solution to the flow equation with appropriate boundary conditions is

$$h = h_0 + \frac{Q}{2\pi T} \ln \frac{r}{R} \tag{9.5}$$

where h is the hydraulic head at a distance r from a pumped well, h_0 is the prepumping hydraulic head, Q is the pumping rate of the well (positive for withdrawal, and negative for injection), T is the transmissivity of the aquifer, and R is the radius of influence of the pumped well. You can determine the hydraulic head in the actual pumping well by assuming that the radius to the observation point is equal to the radius of the pumped well, as

$$h_W = h_0 + \frac{Q}{2\pi T} \ln \frac{r_w}{R} \tag{9.6}$$

where r_w is the radius of the pumped well. In a situation with two observation wells, the hydraulic heads in the wells are related by

$$h_2 = h_1 + \frac{Q}{2\pi T} \ln \frac{r_2}{r_1} \tag{9.7}$$

Remember Eq. (9.7) because it will be used as the basis for a field test to estimate transmissivity. The radius of influence can be estimated from the hydraulic head measurements.

$$R = \frac{h_0 - h_1}{h_2 - h_1} \ln r_2 - \frac{h_0 - h_2}{h_2 - h_1} \ln r_1 \tag{9.8}$$

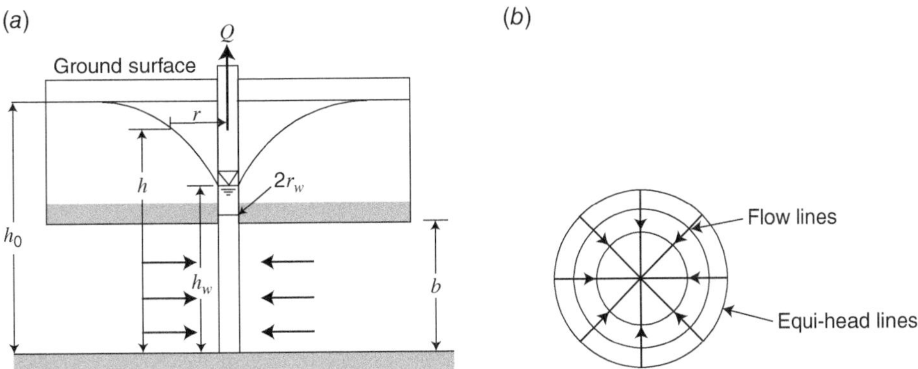

Figure 9.2 Steady-state cone of depression in a confined aquifer, (*a*) sectional view, (*b*) map view (Schwartz and Zhang, 2003 / John Wiley & Sons).

Example 9.1 A well is extracting water from a confined aquifer 50 m thick. At steady state, the hydraulic heads at distances of 100 and 200 m away from the well are 81.7 and 87.2 m, respectively. The hydraulic head before pumping began was 100 m. Calculate the radius of influence.

Solution

The radius of influence is calculated as

$$R = \frac{100-81.7}{87.2-81.7}\ln 200 - \frac{100-87.2}{87.2-81.7}\ln 100 = 1000\text{ m}$$

9.2.1 Interpreting Aquifer Test Data

The Theim equation is rarely used for forward calculations because often the requirement that the system be at steady state is not met. However, the theory is used on occasion to calculate aquifer transmissivity from aquifer test data. Because the Theim equations are developed for steady-state conditions, it is not possible to calculate storativity values. Knowing the pumping rate (Q), and steady-state water levels, h_1 and h_2, for two observation wells located at distances r_1 and r_2, respectively (Figure 9.1), transmissivity for a confined aquifer is calculated as

$$T = \frac{Q}{2\pi(h_2-h_1)}\ln\frac{r_2}{r_1} = \frac{2.3Q}{2\pi(h_2-h_1)}\log\frac{r_2}{r_1} \tag{9.9}$$

where Eq. (9.9) comes by rearranging Eq. (9.7). The drawdowns are related to hydraulic heads by

$$s_1 = h_0 - h_1 \text{ and } s_2 = h_0 - h_2 \tag{9.10}$$

where h_0 is the static water level, and s_1 and s_2 are the drawdowns at distances r_1 and r_2, respectively. Thus, Eq. (9.9) can be written in terms of drawdowns as

$$T = \frac{2.3Q}{2\pi(s_1-s_2)}\log\frac{r_2}{r_1} \tag{9.11}$$

Remember that in applying Eq. (9.11) several assumptions need to be met. First, the aquifer should be homogeneous, isotropic, uniform in thickness and infinite in extent. You will find with experience that there is wiggle room in meeting these assumptions because otherwise you would never be able to use Eq. (9.11). Second, the pumping well must be screened across the entire aquifer (that is fully penetrating) and pumped at a constant discharge rate. Third, sufficient time has elapsed that water levels are no longer changing and the system is at steady state. The following example illustrates how aquifer test data are used to estimate transmissivity.

Example 9.2 Table 9.1 lists drawdowns measured in five observation wells at steady state in a confined aquifer. The well is being pumped at a constant rate of 100 m^3/day. Use the test data to calculate the transmissivity of the aquifer.

Table 9.1 Drawdowns in a confined aquifer under steady-state flow conditions.

r (m)	*s* (m)
30	2.79
50	2.38
100	1.83
200	1.28
400	0.73

Source: Schwartz and Zhang (2003) / John Wiley & Sons.

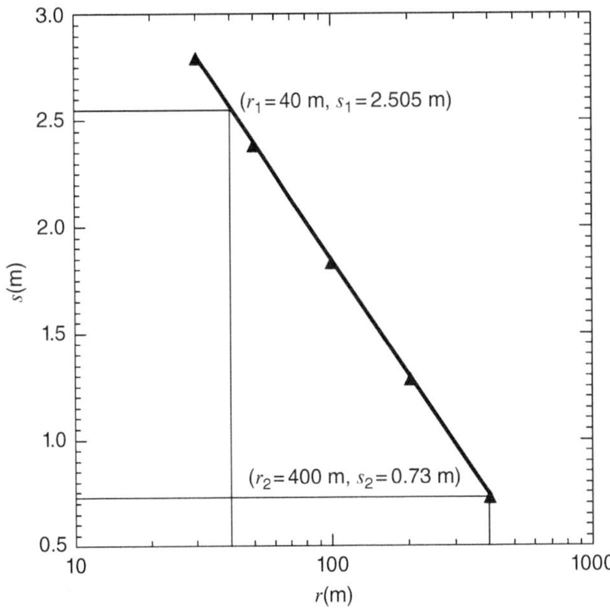

Figure 9.3 Determination of transmissivity using drawdowns measured at multiple observation wells (Schwartz and Zhang, 2003 / John Wiley & Sons).

Solution

A plot of drawdown versus the logarithm of distance is prepared (Figure 9.3). A best-fit line is drawn through the data points. Two points are chosen to calculate the transmissivity.

$$T = \frac{2.3\left(\frac{100\ \mathrm{m}^3}{\mathrm{day}}\right)}{2(2.505 - 0.73)\mathrm{m}} \log(400/40) = 65\ \mathrm{m}^2/\mathrm{day}$$

9.3 Theis Solution for Transient Flow in a Fully Penetrating, Confined Aquifer

Theis theory lets us evaluate the behavior of a well pumping in a confined aquifers, which is under transient (that is nonsteady-state conditions). The flow equation describing hydraulic head in a confined aquifer (Figure 9.4) can be written in polar coordinates as

$$\frac{\partial^2 h}{\partial r^2} + \frac{1}{r}\frac{\partial h}{\partial r} = \frac{S}{T}\frac{\partial h}{\partial t} \tag{9.12}$$

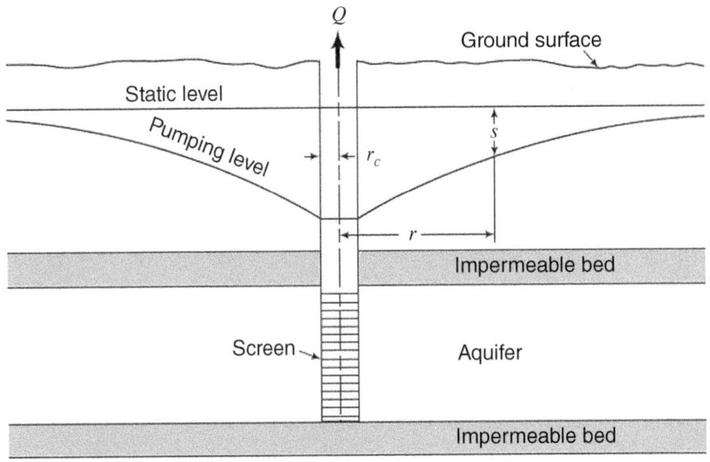

Figure 9.4 Illustration of a nonleaky, confined aquifer being pumped by a fully penetrating well (U.S. Geological Survey Reed, 1980 / Public domain).

where h is the hydraulic head, r is the radial distance from a pumped well to an observation well, t is time since the pumping started, S is the storativity of the aquifer, and T is the transmissivity of the aquifer. The following initial and two boundary conditions apply to this problem

$$h(r,0) = h_0$$

$$h(\infty, t) = h_0$$

$$\lim_{r \to 0} \left(r \frac{\partial h}{\partial r} \right) = \frac{Q}{2\pi T}$$

In words, the first condition means that at time zero and any distance r from the well, the head is equal to zero. The second condition means that at an infinite radius (one boundary) for all time the hydraulic head is fixed at h_0. The third condition provides a constant withdrawal rate at the pumping well (another boundary).

The solution of the Eq. (9.12) was first derived by Theis (1935) and is expressed as

$$h_0 - h = s = \frac{Q}{4\pi T} W(u) \tag{9.13}$$

where Q is the pumping rate, and T is the transmissivity of the aquifer. The well function $W(u)$ and the dimensionless variable u are expressed as:

$$W(u) = \int_u^\infty \frac{e^{-y}}{y} dy = -0.577216 - \ln(u) + u - \frac{u^2}{2!2} + \frac{u^3}{3!3} - \frac{u^4}{4!4} + \ldots \tag{9.14}$$

and

$$u = \frac{r^2 S}{4Tt} \tag{9.15}$$

Even though $W(u)$ is a complicated function, it can be evaluated using well function tables (Table 9.2) or approximated by computer programs. The Theis solution is based on the following assumptions:

1) The pumping well is fully penetrating with a constant discharge rate, infinitesimal diameter, and negligible storage.

Table 9.2 Values of well function $W(u)$.

u	1.0	2.0	3.0	4.0	5.0	6.0	7.0	8.0	9.0
×1	0.219	0.049	0.013	0.0038	0.0011	0.00036	0.00012	0.000038	0.000012
$\times 10^{-1}$	1.82	1.22	0.91	0.70	0.56	0.45	0.37	0.31	0.26
$\times 10^{-2}$	4.04	3.35	2.96	2.68	2.47	2.30	2.15	2.03	1.92
$\times 10^{-3}$	6.33	5.64	5.23	4.95	4.73	4.54	4.39	4.26	4.14
$\times 10^{-4}$	8.63	7.94	7.53	7.25	7.02	6.84	6.69	6.55	6.44
$\times 10^{-5}$	10.94	10.24	9.84	9.55	9.33	9.14	8.99	8.86	8.74
$\times 10^{-6}$	13.24	12.55	12.14	11.85	11.63	11.45	11.29	11.16	11.04
$\times 10^{-7}$	15.54	14.85	14.44	14.15	13.93	13.75	13.60	13.46	13.34
$\times 10^{-8}$	17.84	17.15	16.74	16.46	16.23	16.05	15.90	15.76	15.65
$\times 10^{-9}$	20.15	19.45	19.05	18.76	18.54	18.35	18.20	18.07	17.95
$\times 10^{-10}$	22.45	21.76	21.35	21.06	20.84	20.66	20.50	20.37	20.25
$\times 10^{-11}$	24.75	24.06	23.65	23.36	23.14	22.96	22.81	22.67	22.55
$\times 10^{-12}$	27.05	26.36	25.96	25.67	25.44	25.26	25.11	24.97	24.86
$\times 10^{-13}$	29.36	28.66	28.26	27.97	27.75	27.56	27.41	27.28	27.16
$\times 10^{-14}$	31.66	30.97	30.56	30.27	30.05	29.87	29.71	29.58	29.46
$\times 10^{-15}$	33.96	33.27	32.86	32.58	32.35	32.17	32.02	31.88	31.76

Source: U.S Geological Survey, adapted from Wenzel and Fishel 1942 / Public domain.

2) The aquifer is a confined, infinite in extent, homogeneous, and isotropic.
3) All water pumped by the well is entirely from the storage and discharged instantaneously with the decline in head.

9.4 Prediction of Drawdown and Pumping Rate Using the Theis Solution

The drawdown in an observation well at some future time can be calculated directly using Eq. (9.14) for known hydraulic parameters. For other problems, we might need to know what pumping rate provides a specified drawdown at a fixed place and time in the future. This calculation requires transforming the Theis equation into the following form.

$$Q = \frac{4\pi T\, s}{W(u)} \tag{9.16}$$

Values of well function $W(u)$ is tabulated in Table 9.2.

Example 9.3 The transmissivity and storativity of a confined aquifer are 1000 m^2/day and 0.0001, respectively. An observation well is located 500 m away from a pumping well. For a pumping period of 220 minutes, calculate (a) the drawdown at the observation well if the discharge rate is 1000 m^3/day; (b) the pumping rate required to provide a drawdown of 1 m at that well after 220 minutes.

Solution

Drawdown can be calculated using the Theis equation. All of the parameters on the RHS of Eq. (9.13) are known except for $W(u)$. To evaluate $W(u)$, we first calculate u as

$$u = \frac{r^2 S}{4\,T\,t} = \frac{500 \times 500\ \text{m}^2 \times 0.0001}{4 \times 1000\dfrac{\text{m}^2}{\text{day}} \times \dfrac{1}{1440}\dfrac{\text{day}}{\text{min}} \times\ 220\ \text{min}} = 0.041$$

The well function $W(u)$ at $u = 0.041$ is

$$W(0.041) = 2.66$$

For a pumping rate of 1000 m^3/day, the drawdown is calculated as

$$s = \frac{Q}{4\pi T} W(u) = \frac{1000\text{m}^3/\text{day}}{4\ \times\ 3.14 \times 1000\text{m}^2/\text{day}}(2.66) = 0.21\ \text{m}$$

For a drawdown of 1 m, the pumping rate is calculated as

$$Q = \frac{4\pi T\, s}{W(u)} = \frac{4 \times 3.14 \times 1000\ \text{m}^2/\text{day}\ \ \times 1\ \text{m}}{(2.66)} = 4.72 \times 10^3 \text{m}^3/\text{day}$$

9.5 Theis Type-Curve Method

Another important use of the Theis solution is in the determination of transmissivity and storativity from data collected from an aquifer test. There is a variety of aquifer testing approaches. We begin here with the so-called type-curve matching technique, which is widely used in practice. The test data are a series of drawdown values in an observation well, each matched with a time since pumping began. The approach involves plotting the field data on one graph, which is overlain on a type curve plotted at the same scale. Here are the details:

1) Create the type curve by plotting the well function $W(u)$ versus $1/u$ on log-log graph paper (Figure 9.5). Usually, you can buy a copy of this curve.

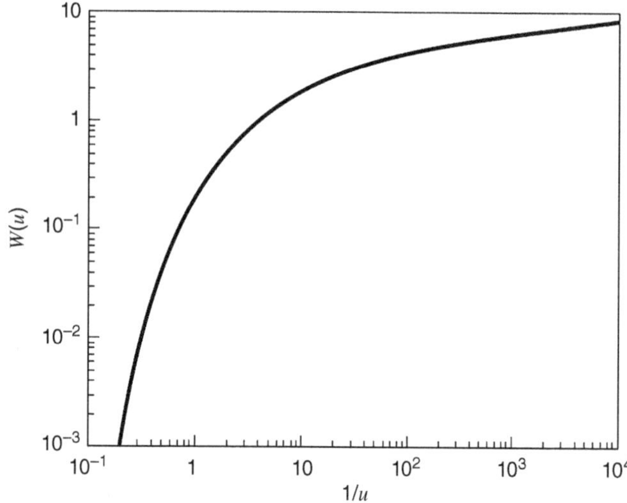

Figure 9.5 A plot of the well function $W(u)$ versus $1/u$ on log–log paper (Schwartz and Zhang, 2003 / John Wiley & Sons).

2) Define a match point on the type curve. This match point only serves as a reference and can be located anywhere on the graph. However, the math works out best if you choose a match point with "simple" coordinates like $W(u) = 1$ and $1/u = 10^3$, 10^2, or 10.
3) Prepare a transparent overlay with drawdown (s) plotted versus time (t) on log–log graph paper. This graph paper must be the same as the type curve. This step is where you use the set of field data from the observation well.
4) Superimpose the transparent graph of the field data onto the type curve. Adjust the field curve until the collection of field points appear to fall along the type curve underneath. You must keep the axes of the two graphs parallel to each other.
5) Mark the point on the field curve that exactly corresponds with the match point on the type curve underneath. Now you will have points marked on both graphs with coordinates $W(u)$, $1/u$, and $s(t)$, t. These pairs of values will be substituted in step (6).
6) Calculate T and S using the following equations.

$$T = \frac{Q}{4\pi s} W(u) \tag{9.17}$$

and

$$S = \frac{4\,T\,t\,u}{r^2} \tag{9.18}$$

In the case of Eq. (9.17), Q is known from the pump-test data, $W(u)$ is the coordinate value of the match point on the type curve, and s is coordinate value for the match point on the curve of the field data. In the case of Eq. (9.18), T is known from the previous calculation, r is known from the setup of the aquifer test, and $1/u$ and t are the match point coordinates obtained via curve matching. Here is an example that illustrates these steps.

Example 9.4 In a test of a confined aquifer, the pumping rate was 500 m^3/day. Drawdown/time data were collected at an observation well 300 m away (Table 9.3). Use the type-curve method to determine hydraulic conductivity and storativity of the aquifer.

Solution

Figure 9.6 is the plot of drawdown versus time. Superimposing the field curve on the type curve as shown in Figure 9.7, gives the match point coordinates $1/u = 10$, $W(u) = 1.0$, $t = 22$ min, and $s = 0.78$ m. Thus,

$$T = \frac{Q}{4\pi s} W(u) = \frac{(500\ \text{m}^3/\text{day})(1)}{(4\pi)(0.78\ \text{m})} = 51\ \text{m}^2/\text{day}$$

Table 9.3 Drawdowns measured at an observation well 300 m away.

Time (min)	*S* (m)
1.00	0.03
1.27	0.05
1.61	0.09
2.04	0.15
2.59	0.22
3.29	0.31
4.18	0.41
5.30	0.53
6.72	0.66
8.53	0.80
10.83	0.95
13.74	1.11
17.43	1.27
22.12	1.44
28.07	1.61
35.62	1.79
45.20	1.97
57.36	2.15
72.79	2.33
92.37	2.52
117.21	2.70
148.74	2.89
188.74	3.07
239.50	3.26
303.92	3.45
385.66	3.64
489.39	3.83
621.02	4.02
788.05	4.21
1000.0	4.39

Source: Schwartz and Zhang (2003) / John Wiley & Sons.

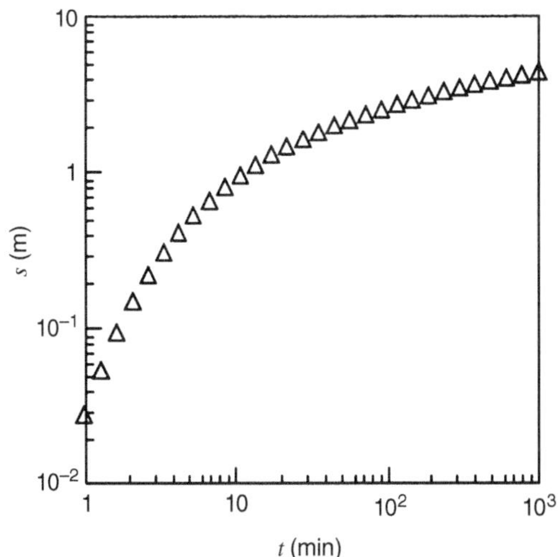

Figure 9.6 A plot of measured drawdown versus time on log–log paper (Schwartz and Zhang, 2003 / John Wiley & Sons).

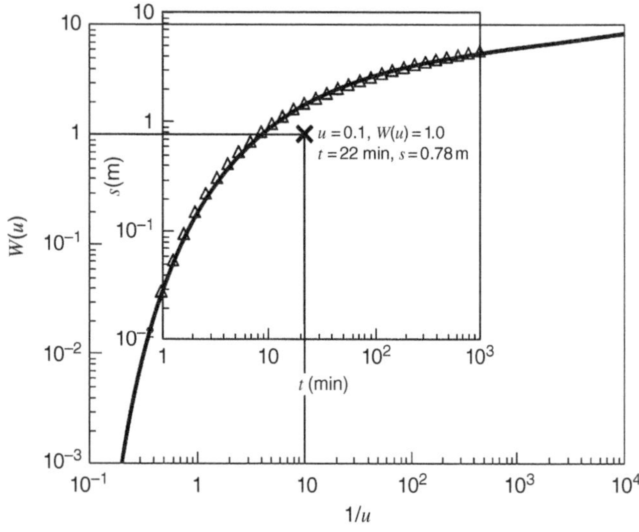

Figure 9.7 Illustration of the method of superposition used to determine transmissivity and storativity (Schwartz and Zhang, 2003 / John Wiley & Sons).

and

$$S = \frac{4Ttu}{r^2} = \frac{4(51\ \text{m}^2/\text{day})\left(22\ \text{min}\ \frac{1\ \text{day}}{1440\ \text{min}}\right)(0.1)}{(300\ \text{m})^2} = 3.46 \times 10^{-4}$$

9.6 Cooper–Jacob Straight-Line Method

Using the Theis solution (Eq. 9.13) is complicated by the fact that it contains the function $W(u)$, which is an exponential integral. The value of $W(u)$ is the sum of an infinite series (see Eq. 9.14), which we normally evaluate using the table of well functions. However, when values of u are small, less than 0.01, the higher-order terms of the infinite series are negligible and can be ignored (Cooper and Jacob, 1946; Jacob, 1940). With the Cooper Jacob's assumption, drawdown is calculated as

$$s(t) = \frac{Q}{4\pi T}(-0.577216 - \ln(u)) = \frac{2.3Q}{4\pi T}\log\frac{2.25\,T\,t}{r^2 S} \tag{9.19}$$

The result of this simplification is that integral expression in the Theis solution is replaced by a much simpler function. It can be evaluated using a calculator without the need for the table of well functions.

Another benefit of the Cooper and Jacob (1946) modification of the Theis equation is that it leads to a simple, graphical approach to evaluating aquifer test data. If drawdown/time data are plotted as drawdown versus the logarithm of time, Cooper–Jacob theory predicts the data will fall along a straight line. By extracting two numbers from the graph, we can solve equations to determine transmissivity and storativity. Here is a summary of the steps involved, followed by an example.

1) Plot drawdown versus time on semi-log graph paper, with time on the x-axis as a logarithmic scale and drawdown on the y-axis as an arithmetic scale. By convention, zero drawdown is at the top of the y-axis.
2) Fit a straight line through the data points. If there is difficulty, use the later-time points.
3) Select two points (t_1, s_1 and t_2, s_2) on the line. The equation that is needed can be derived from Eq. (9.19) by writing one equation in terms of s_2 and one equation in terms of s_1 and subtracting them from each other. After some manipulation, the resulting equation is

$$\Delta s = s_2 - s_1 = \frac{2.3Q}{4\pi T}\log\frac{t_2}{t_1} \tag{9.20}$$

4) Choose t_1 and t_2 one log cycle apart, for example, $t_1 = 10$ minutes and $t_2 = 100$ minutes, to give Δs or drawdown per log cycle. This choice simplifies the math. For example, with $t_1 = 10$ minutes and $t_2 = 100$ minutes, $\log(t_2/t_1) = \log(100/10) = 1$. The log term in Eq. (9.20) becomes one and the equation simplifies to

$$T = \frac{2.3\,Q}{4\pi\Delta s} \tag{9.21}$$

where Δs is the drawdown per log cycle. All the terms on the RHS of Eq. (9.21) are known, so it is a simple matter to calculate transmissivity.

5) Find the value of t_0 on the graph by extending the straight line to intersect the line of zero drawdown ($s = 0$). The corresponding time (t_0) is in effect the time that it takes for the cone of depression to reach the observation well. With this value of t_0, all values on the RHS of Eq. (9.21) are known, and storativity can be calculated as

$$S = \frac{2.25\,Tt_0}{r^2} \tag{9.22}$$

You can derive Eq. (9.22) by setting the drawdown in Eq. (9.19) to zero and rearranging terms.

6) The final step is a final check to make sure that the Cooper–Jacob simplification applies to this problem or in other words whether $u = rS/(4Tt) < 0.01$. Hopefully, the check will be successful and the problem solved.

Example 9.5 Determine the transmissivity and storativity of the aquifer in Example 9.3 using the Cooper–Jacob straight-line method.

Solution

Figure 9.8 shows a plot of drawdown versus log(t) for the data set. A line is fitted to a late-time section of the curve. On the figure, we find drawdowns corresponding to times a factor of 10 different. For example, with $t_1 = 100$ min, $s_1 = 2.58$ m, and with $t_2 = 1000$ min, $s_2 = 4.39$ m. Thus, the drawdown per log cycle 4.39 – 2.58 or 1.81 m

$$T = \frac{2.3\,Q}{4\pi\Delta s} = \frac{(2.3)(500\text{m}^3/\text{day})}{4\pi(1.81\text{ m})} = 51\text{ m}^2/\text{day}$$

The next step is to substitute the value of t_0, 3.4 minutes (determined from the graph) in Eq. (9.22), along with the other known parameters and to calculate S

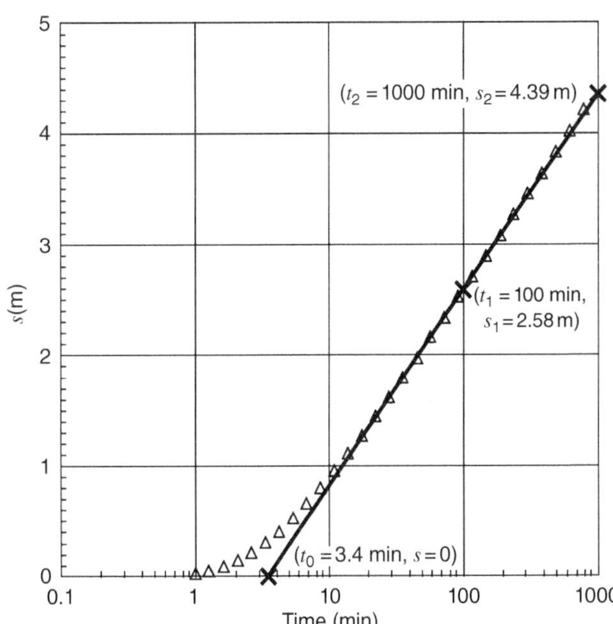

Figure 9.8 Illustration of how the Cooper–Jacob straight-line method is used with observation well data (Schwartz and Zhang, 2003 / John Wiley & Sons).

$$S = \frac{2.25\ Tt_0}{r^2} = \frac{(2.25)(51\ \mathrm{m^2/day})\left(3.4\ \mathrm{min}\ \dfrac{1\ \mathrm{day}}{1440\ \mathrm{min}}\right)}{(300\ \mathrm{m})^2} = 3.0 \times 10^{-6}$$

The last step is to use the calculated T and S values with other parameters to see whether u is appropriate for the Cooper–Jacob assumption

$$u = \frac{r^2 S}{4Tt} = \frac{(300)^2 \mathrm{m^2}(3.0 \times 10^{-6})}{4(51\mathrm{m^2/day})\left(100\ \mathrm{min}\ \dfrac{1\ \mathrm{day}}{1440\ \mathrm{min}}\right)} = 0.002$$

In this example, the maximum u is less than 0.01 and it is acceptable to calculate T and S values using the Cooper–Jacob method.

9.7 Distance-Drawdown Method

The Cooper–Jacob simplification is also useful in determining transmissivity and storativity values in aquifer tests if water levels are measured in two or more observation wells at the same time. For example, assume that two observation wells are located at distances r_1 and r_2 from the pumping well. Knowing the drawdowns at these two wells are s_1 and s_2, respectively, at some time t, we can write this equation based on Eq. (9.19)

$$s_1 = \frac{2.3Q}{4\pi T} \log \frac{2.25\ T\ t}{r_1^2 S} \tag{9.23}$$

and

$$s_2 = \frac{2.3Q}{4\pi T} \log \frac{2.25\ T\ t}{r_2^2 S} \tag{9.24}$$

The combination of Eqs. (9.23) and (9.24) yields

$$s_1 - s_2 = \frac{2.3Q}{2\pi T} \log \frac{r_2}{r_1} \tag{9.25}$$

The procedures for determining the transmissivity and storativity for an aquifer using the distance-drawdown method are as follows:

1) For a selected time, plot the drawdown and distance information for the observation wells on semi-log graph paper. Distance is plotted as a logarithmic scale on the x-axis and drawdown is plotted on a linear scale on the y-axis.
2) Fit a straight line through the data points.
3) Select two points (r_1, r_2) on the line, one log cycle apart and determine the drawdown Δs. The transmissivity is calculated by

$$T = \frac{2.3Q}{2\pi(\Delta s)} \tag{9.26}$$

4) Extend the straight line to $s = 0$, and determine the distance r_0. The storativity is calculated by

$$S = \frac{2.25\ T\ t}{r_0^2} \tag{9.27}$$

Example 9.6 A confined aquifer is pumped at 220 gpm. At time = 220 minutes, drawdowns were recorded in nine observation wells (Table 9.4). Calculate the transmissivity and storativity of the aquifer.

Solution

Drawdown versus distance is plotted in Figure 9.9. The pumping rate in gpm is first converted to ft³/day using Eq. (9.1).

$$Q = (220\text{ gpm})\left(192.5\frac{\text{ft}^3/\text{day}}{\text{gpm}}\right) = 42350\text{ ft}^3/\text{day}$$

Selecting r_1 as 10 ft and $r_2 = 100$ ft, the drawdown per log cycle is $s_1 - s_2$ or $35.2 - 19.5 = 15.7$ ft. The value of $r_0 = 1900$ ft. Thus,

$$T = \frac{2.3Q}{2\pi(\Delta s)} = \frac{2.3(42350\text{ ft}^3/\text{day})}{2\pi(15.7)\text{ft}} = 987\text{ft}^2/\text{day}$$

and

$$S = \frac{2.25(987\text{ ft}^2/\text{day})(220\text{ min})\left(\frac{1}{1440}\frac{\text{day}}{\text{min}}\right)}{1900^2} = 9.4\times10^{-5}$$

Table 9.4 Values of drawdown versus distance measured at time = 220 minutes.

r(ft)	*s*(ft)
10	35.20
50	24.35
100	19.68
150	16.96
200	15.03
250	13.54
300	12.32
400	10.42
500	8.97

Source: Schwartz and Zhang (2003) / John Wiley & Sons.

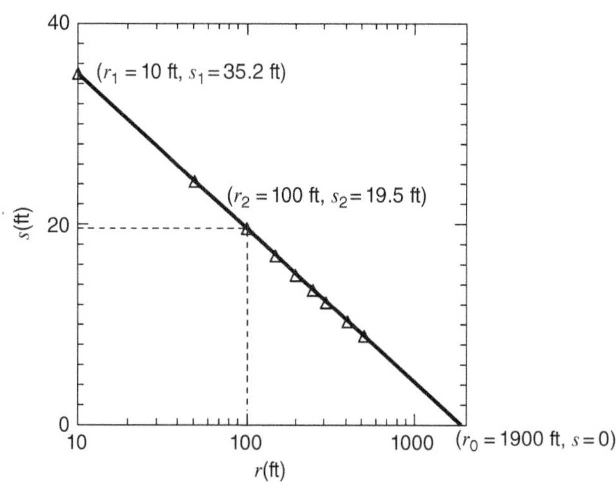

Figure 9.9 Drawdown data for nine observation wells at different radial distances from a pumping well are fit with a straight line to provide s and r_0 (Schwartz and Zhang, 2003 / John Wiley & Sons).

9.8 Estimating *T* and *S* Using Recovery Data (Theis, 1935)

If pumping of a well is halted, theory predicts that the water level in the aquifer will return to its prepumping level, h_0. Water-level data obtained during this recovery phase of a test provides a basis for determining the transmissivity of the aquifer. The residual drawdown during the recovery period for a confined aquifer is expressed as

$$s' = \frac{Q}{4\pi T}\left[\int_u^\infty \frac{e^{-u}}{u}du - \int_{u'}^\infty \frac{e^{-u'}}{u'}d'u\right] \tag{9.28}$$

where

$$u = \frac{r^2 S}{4\,T\,t} \tag{9.29}$$

and

$$u' = \frac{r^2 S}{4\,T\,\{t'} \tag{9.30}$$

where t is the time since pumping starts, t' is time since pumping stops, and s' is the residual drawdown. Because recovery measurements are made in the pumped well or in a nearby observation well the radius to the measurement point, r is typically small. A small r usually leads to a small value u', which enables us to take advantage of the Cooper–Jacob simplification. Under the Cooper–Jacob assumption, Eq. (9.28) reduces to

$$s' = \frac{2.3\,Q}{4\pi T}\log\left(\frac{t}{t'}\right) \tag{9.31}$$

and transmissivity can be determined as

$$T = \frac{2.3\,Q}{4\pi s'}\log\left(\frac{t}{t'}\right) \tag{9.32}$$

The procedure for determining transmissivity using recovery data is as follows:

1) Plot the residual drawdown (s') on an arithmetic scale versus the time ratio (t/t') on a logarithmic scale.
2) Choose two points on the graph. Again it helps to select the two points one log cycle apart. The transmissivity is obtained by

$$T = \frac{2.3Q}{4\pi\Delta s'} \tag{9.33}$$

where $\Delta s'$ is the change in residual drawdown over one log cycle (t/t').

Storativity can be calculated if the recovery data are collected in an observation well rather than the pumping well. The drawdown (s_P) when the pump is turned off at time (t_P) is expressed as

$$s_P = \frac{2.3Q}{4\pi T}\log\frac{2.25\,Tt_P}{r^2 S} \tag{9.34}$$

Once T is known, the storativity is obtained by

$$S = \frac{2.25Tt_P}{r^2}10^{-\frac{4\pi T s_P}{2.3Q}} \tag{9.35}$$

Example 9.7 In an aquifer test reported by USBR in 1995, drawdowns are recorded in the pumped well (Table 9.5) and an observation well (Table 9.6). In both tables, the first column is time since the pumping started, the second column is the drawdown during the pumping period, the third column is the time since pumping stopped, the fourth column is the time ratio, and the last column is the residual drawdown. A constant pumping rate of 162.9 ft^3/min was maintained during the pumping part of the test. The observation well is 100 ft away from the pumped well. Calculate the hydraulic parameters using the recovery data.

Solution

Drawdown data for the pumped well and the observation well are plotted in Figures 9.10 and 9.11, respectively. A straight line approximates the drawdown versus time curve in each figure. We have derived $\Delta s' = 0.86$ ft and 0.84 ft for the pumped and observation wells, respectively. Therefore, the transmissivity determined from the recovery data in the pumped well is

$$T = \frac{2.3Q}{4\pi\Delta s'} = \frac{2.3(162.9\text{ft}^3/\text{min})}{4\pi(0.86\text{ ft})} = 34.7\text{ft}^2/\text{min} = 5.0 \times 10^4 \text{ ft}^2/\text{day}$$

and transmissivity determined from the recovery data in the observation well is

$$T = \frac{2.3Q}{4\pi\Delta s'} = \frac{2.3(162.9\text{ft}^3/\text{min})}{4\pi(0.84\text{ ft})} = 35.5\text{ft}^2/\text{min} = 5.1 \times 10^4 \text{ ft}^2/\text{day}$$

Using $t_P = 800$ min, $s_P = 1.86$ ft, $Q = 162.9$ ft^3/min, $r = 100$ ft, and $T = 34.7$ ft^2/min, storativity of the aquifer can be determined from the residual drawdown in the observation well.

$$S = \frac{2.25\,Tt_P}{r^2}10^{-\frac{4\pi Ts_P}{2.3Q}} = \frac{2.25(34.7\text{ ft}^2\text{ min})(800\text{ min})}{(100\text{ ft})^2}10^{-\frac{4\pi(34.7\text{ ft}^2/\text{min})(1.86\text{ ft})}{2.3(162.9\text{ ft}^3/\text{min})}} = 0.043$$

It is not possible to determine the storativity of the aquifer using drawdown data from the pumped well.

Table 9.5 Aquifer test information from a pumped well.

t(min)	s(ft)	t′(min)	t/t′	s′(ft)
3	10.2	3	267.67	−20
8	10.6	8	101.00	−5
13	10.8	13	62.54	−0.5
20	11.3	20	41.00	1.5
80	11.6	80	11.00	1
140	11.8	140	6.71	0.8
195	11.8	195	5.10	0.69
255	11.8	255	4.14	0.59
315	12	315	3.54	0.51
375	12.2	375	3.13	0.49
435	12.2	435	2.84	0.46
495	12.2	495	2.62	0.38
560	12.2	560	2.43	0.34
616	12.3	616	2.30	0.33
668	12.4	668	2.12	0.33
737	12.5	727	2.10	0.22
800	12.5	800	2.00	0.22

Source: Modified from USBR (1995).

Table 9.6 Aquifer test information at an observation well.

t(min)	*s*(ft)	*t'*(min)	*t/t'*	*s'*(ft)
5	0.08	5	161.00	1.78
10	0.22	10	81.00	1.64
15	0.33	15	54.33	1.53
20	0.41	20	41.00	1.45
25	0.5	25	33.00	1.37
30	0.55	30	27.67	1.32
40	0.66	40	21.00	1.22
50	0.73	50	17.00	1.15
60	0.8	60	14.33	1.09
70	0.86	70	12.43	1.03
80	0.92	80	11.00	0.97
90	0.96	90	9.89	0.94
100	1	100	9.00	0.9
110	1.04	110	8.27	0.87
120	1.07	120	7.67	0.85
180	1.24	180	5.44	0.7
240	1.35	240	4.33	0.61
300	1.45	300	3.67	0.54
360	1.52	360	3.22	0.49
420	1.59	420	2.90	0.46
480	1.65	480	2.67	0.4
540	1.71	540	2.48	0.36
600	1.73	600	2.33	0.36
660	1.77	660	2.21	0.34
720	1.81	720	2.11	0.31
800	1.86	800	2.00	0.29

Source: Modified from USBR (1995).

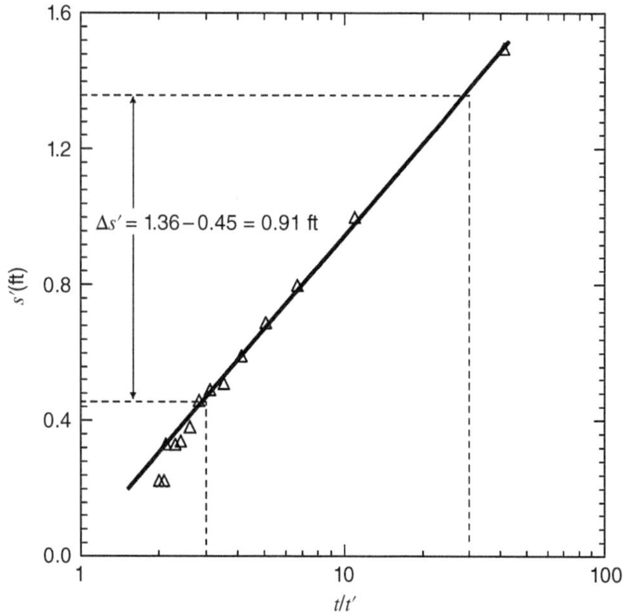

Figure 9.10 This plot illustrates the straight-line methods for determining transmissivity using residual drawdown data from the pumped well (Schwartz and Zhang, 2003 / John Wiley & Sons).

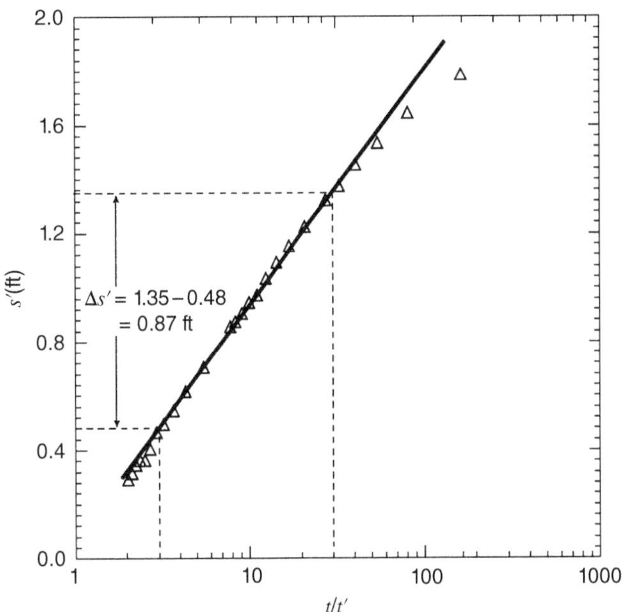

Figure 9.11 This plot illustrates the straight-line methods for determining transmissivity using residual drawdown data from the observation well (Schwartz and Zhang, 2003 / John Wiley & Sons).

By mathematically manipulating Eq. (9.28), analytical methods developed for a pumping scheme can also be used to determine the hydraulic parameters from recovery measurements. Equation (9.28) can be rewritten as

$$s^* \frac{Q}{4\pi T} W(u') = \frac{Q}{4\pi T} W(u) - s' \tag{9.36}$$

where s^* is the recovery drawdown, the first term on the right side of Eq. (9.36) is the drawdown during the pumping period projected to time t', and s' is the residual drawdown. The same techniques for interpreting the drawdown-time curves during a pumping period can be used to interpret the recovery drawdown (s^*) versus recovery time (t') curves. Figure 9.12 shows the relationship between drawdown, residual drawdown, and recovery drawdown.

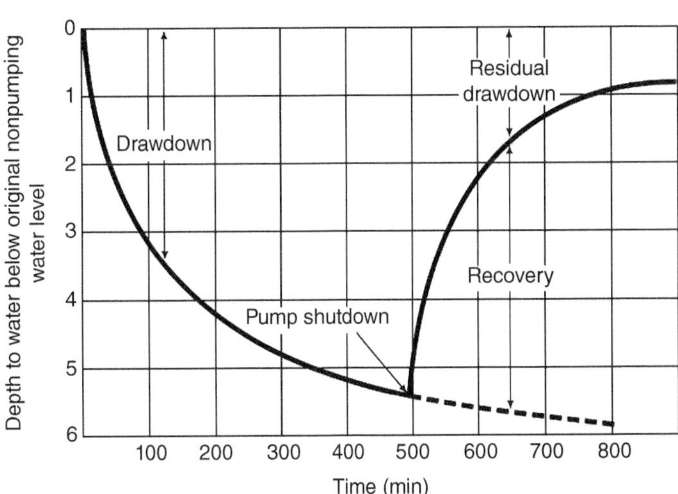

Figure 9.12 Arithmetic plot showing the shape of the drawdown/recovery curve versus time (Domenico and Schwartz, 1998. Physical and Chemical Hydrogeology. Copyright ©1998 by John Wiley & Sons, Inc. All Rights Reserved. Reproduced with permission).

Exercises

9.1 For transmissivity (T) = 2500 m^2/day, storativity (S) = 1.0×10^{-3}, and a pumping rate (Q) = 500 m^3/day, calculate drawdowns in a confined aquifer for r = 10, 50, and 100 m at t = 150 minutes.

9.2 Through your calculations, answer the following questions: (a) How will be the cone of depression changed if the transmissivity is increased while other parameters are kept constant? (b) How will be the cone of depression changed if the storativity is increased while other parameters are kept constant?

9.3 Values of drawdown and time in a pumping test are listed in Table 9.7. Determine the transmissivity and storativity of a confined aquifer using the Theis type-curve technique. Both the pumped and observation wells are fully penetrating. A pumping rate of 10000 m^3/day is used in the test. The observation well is located 150 m away from the pumped well. You will find two template files for this exercise in the directory Chapter 9 on the book website. The template file "temp-9a.doc" contains the Theis type curve. The other template file "temp-9.doc" is a graphic template in the same log–log scale as the type curves. Print the template files, plot the values of drawdown versus time on the graphic template, and superimpose the two graphic papers to determine hydraulic parameters.

Table 9.7 Values of drawdown versus time in an aquifer test with a confined aquifer.

Time (min)	Drawdown (m)	Time (min)	Drawdown (m)
14.4	0.62	254	1.08
18	0.66	316.7	1.11
22.4	0.69	394.9	1.15
27.9	0.73	492.5	1.18
34.8	0.76	614.1	1.22
43.4	0.8	765.8	1.25
54.2	0.83	955	1.29
67.5	0.87	1190.9	1.33
84.2	0.9	1485.1	1.36
105	0.94	1852	1.4
131	0.97	2309.5	1.43
163.3	1.01	2880	1.47
203.6	1.04		

Source: Schwartz and Zhang (2003) / John Wiley & Sons.

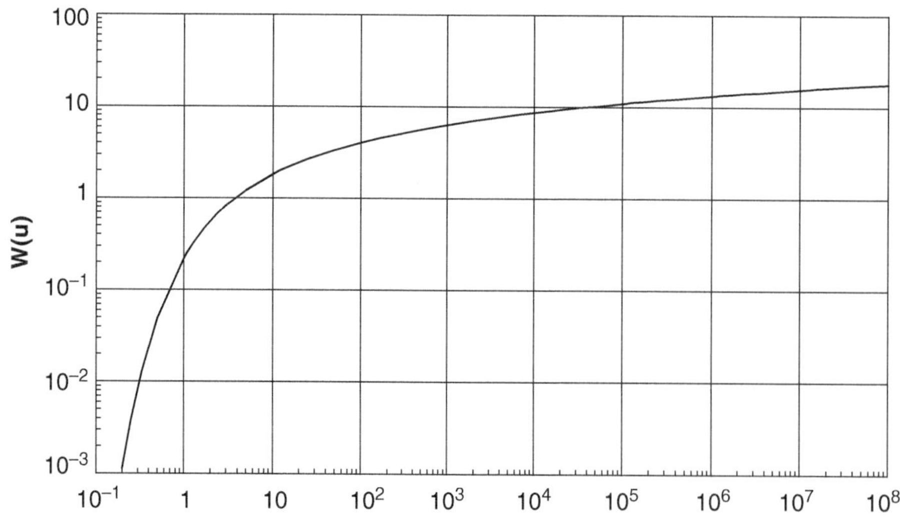

9.4 Use the values of time and drawdown in Table 9.7 and determine the aquifer parameters by the time straight-line technique.

9.5 Values of drawdown and distance in a pumping test in a confined aquifer are listed in Table 9.8 Determine hydraulic parameters of the aquifer using the distance straight-line technique.

Table 9.8 Values of drawdown versus time in an aquifer test with a confined aquifer.

Distance (m)	Drawdown (m)
25	1.51
50	1.29
100	1.07
150	0.94

Source: Schwartz and Zhang (2003) / John Wiley & Sons.

9.6 Table 9.9 lists drawdowns during a pumping period and residual drawdowns during a recovery period in a confined aquifer. The pumping rate is 1.0×10^4 m^3/day. The distance from the observation well to the pumped well is 100 m. Calculate the hydraulic parameters using residual drawdown data.

Table 9.9 Values of drawdown versus time in an observation well.

t(min)	s(ft)	t′(min)	s′(ft)	t(min)	s(ft)	t′(min)	s′(ft)
1.44	0.089	1.44	1.052	53.08	0.595	53.08	0.508
1.64	0.102	1.64	1.032	60.38	0.615	60.38	0.489
1.86	0.115	1.86	1.012	68.68	0.635	68.68	0.470
2.12	0.130	2.12	0.993	78.13	0.656	78.13	0.452
2.41	0.145	2.41	0.973	88.87	0.676	88.87	0.433
2.74	0.160	2.74	0.954	101.09	0.696	101.09	0.415
3.12	0.176	3.12	0.934	114.99	0.717	114.99	0.396
3.55	0.193	3.55	0.915	130.80	0.737	130.80	0.378
4.04	0.210	4.04	0.895	148.78	0.757	148.78	0.360
4.59	0.227	4.59	0.875	169.24	0.778	169.24	0.343
5.22	0.245	5.22	0.856	192.51	0.798	192.51	0.325
5.94	0.263	5.94	0.836	218.98	0.819	218.98	0.308
6.76	0.281	6.76	0.817	249.09	0.839	249.09	0.291
7.69	0.300	7.69	0.797	283.33	0.860	283.33	0.275
8.74	0.318	8.74	0.778	322.29	0.880	322.29	0.259
9.95	0.337	9.95	0.758	366.60	0.900	366.60	0.243
11.31	0.356	11.31	0.739	417.01	0.921	417.01	0.227
12.87	0.376	12.87	0.719	474.35	0.941	474.35	0.212
14.64	0.395	14.64	0.700	539.57	0.962	539.57	0.198
16.65	0.415	16.65	0.681	613.76	0.982	613.76	0.184
18.94	0.435	18.94	0.661	698.15	1.003	698.15	0.170
21.54	0.454	21.54	0.642	794.14	1.023	794.14	0.157
24.51	0.474	24.51	0.623	903.33	1.044	903.33	0.145
27.87	0.494	27.87	0.603	1027.53	1.064	1027.53	0.133
31.71	0.514	31.71	0.584	1168.81	1.085	1168.81	0.122
36.07	0.534	36.07	0.565	1329.52	1.105	1329.52	0.112
41.03	0.554	41.03	0.546	1440.00	1.106	1440.00	0.105
46.67	0.574	46.67	0.527				

Source: Schwartz and Zhang (2003) / John Wiley & Sons.

9.7 Use Excel spreadsheet to interpret hydraulic parameters. In the folder, "Curving-Fitting Using Excel Spreadsheet," of the book website, there is a user guide on how to use Excel spreadsheet to interpret pumping test data for confined aquifers. Teachers of this book will be able to use the Excel file, Confined.xlsm, to show students a modern tool to solve aquifer testing problem.

References

Bureau of Reclamation, U.S. Department of Interior. 1995. Groundwater Manual. A Water Resources Technical Publication, 2nd edition. Am. Geophys. Union, 661 p.

Cooper, H. H., and C. E. Jacob. 1946. A generalized graphical method for evaluating formation constants and summarizing well field history. Transactions of the American Geophysical Union, v. 27, p. 526–534.

Domenico, P. A., and F. W. Schwartz. 1998. Physical and Chemical Hydrogeology. John Wiley & Sons, New York, 506 p.

Jacob, C. E. 1940. On the flow of water in an elastic artesian aquifer. Eos, Transactions American Geophysical Union, v. 22, no. 2, p. 574–586.

Reed, J. E. 1980. Type curves for selected problems of flow to wells in confined aquifers. U.S. Geological Survey Techniques of Water-Resources Investigations, Chapter B3, 106 p.

Schwartz, F. W., and H. Zhang. 2003. Fundamentals of Groundwater. John Wiley & Sons, Hoboken, New Jersey, 583 p.

Theis, C. V. 1935. The relation between the lowering of the piezometric surface and rate and duration of discharge of a well using groundwater storage. Eos, Transactions American Geophysical Union, v. 16, no. 2, p. 519–524.

Thiem, G., 1906. Hydrologische Methoden [Hydrological methods]. PhD Thesis, University of Stuttgart, Stuttgart, Germany.

United States Bureau of Reclamation. 1995. Groundwater manual. U.S. Department of the Interior, Bureau of Reclamation, Groundwater—480 p, https://www.usbr.gov/tsc/techreferences/mands/mands-pdfs/GndWater.pdf.

Wenzel L. K., and V. C. Fishel. 1942. Methods for determining permeability of water-bearing materials, with special reference to discharging-well methods, with a section on direct laboratory methods and bibliography on permeability and laminar flow. Water Supply Paper 887.

10

Leaky Confined Aquifers and Partially-Penetrating Wells

CHAPTER MENU

When a confined aquifer is pumped, there sometimes can be leakage into the aquifer from adjacent units. This leakage is caused by flow across the confining unit from an adjacent aquifer (Figure 10.1). Besides flow across the confining bed, another component of leakage is the release of water from storage in the confining bed. The net effect of this leakage is to reduce the drawdown in the confined aquifer that is expected from a Theis-type response.

This chapter looks at the response of a leaky confined aquifer to pumping. The first case we examine is flow across a confining bed without storage. An idealization of this setting is presented in Figure 10.1. Later, we consider situations where storage in the confining beds is also a source of leakage.

During the initial stage of an aquifer test, the change in water level in an observation well exhibits a Theis-type response. At early times, water flowing to the pumped well mainly comes from storage in the confined aquifer. At late times, water being pumped from well comes from leakage, causing a deviation from the Theis response.

10.1 Transient Solution for Flow Without Storage in the Confining Bed

Hantush and Jacob (1955) derived an analytical solution for flow in a leaky confined aquifer system (Figure 10.1). The governing equation, describing drawdown in the pumped aquifer, is

$$\frac{\partial^2 s}{\partial r^2} + \frac{1}{r}\frac{\partial s}{\partial r} - \frac{K'}{T'b'}s = \frac{S}{T}\frac{\partial s}{\partial t} \tag{10.1}$$

where s is the drawdown, t is time, r is the distance from the pumping well to an observation well, T is the transmissivity of the aquifer, S is the storativity of the aquifer, K' is the hydraulic conductivity of the confining bed, and b' is the thickness of the confining bed. The extra term in this equation represents leakage into the aquifer through the confining bed from a neighboring aquifer. The solution to Eq. (10.1) provides the drawdown in a leaky confined aquifer due to pumping off a fully penetrating well as (Hantush and Jacob, 1955)

$$s = \frac{Q}{4\pi T} W\left(u, \frac{r}{B}\right) \tag{10.2}$$

where $B = (Tb'/K')^{1/2}$ and $W(u,r/B)$ is expressed as

$$W\left(u, \frac{r}{B}\right) = \int_u^\infty \frac{e^{-z-\frac{r^2}{4B^2 z}}}{z} dz \tag{10.3}$$

Fundamentals of Groundwater, Second Edition. Franklin W. Schwartz and Hubao Zhang.

Companion website: www.wiley.com/go/schwartz/fundamentalsofgroundwater2

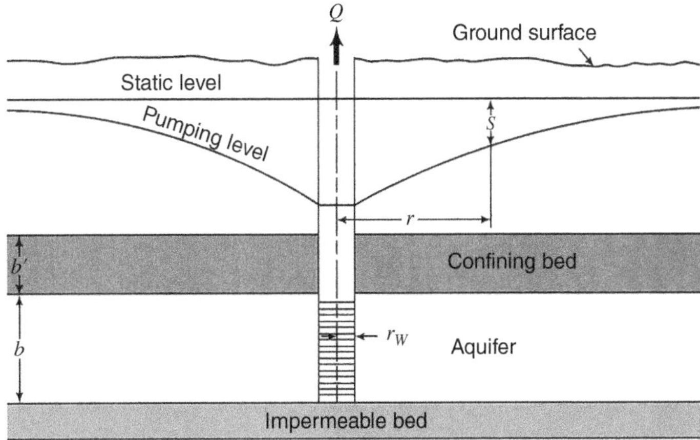

Figure 10.1 Schematic illustration of a leaky, confined aquifer, which is being pumped by a well at a constant rate of discharge (U.S. Geological Survey adapted from Reed, 1980 / Public domain).

Values of $W(u, r/B)$ are tabulated in Table 10.1 (Hantush, 1956). The function $W(u, r/B)$ is plotted in Figure 10.2. Unlike the last type of curve, there is now a family of curves, each labeled with a separate r/B value. The different r/B curves represent differences in the amounts of leakage across the confining bed. For example, the curve for an r/B value of zero applies to a situation where the confining bed is impermeable. This type of curve is same as the $W(u)$ curve presented previously in Chapter 9 and explains the label Theis-type curve in Figure 10.2. Curves with small r/B values (e.g., 0.05) exemplify minor leakage, hence minor deviation from the Theis curve. Curves with a large r/B value (e.g., 1) exemplify significant leakage with a much greater deviation from the Theis type curve.

Before proceeding further, let us formally set down the assumption implicit in the solution, Eq. (10.2).

1) The pumping well has a constant rate of discharge and fully penetrates the aquifer.
2) The pumping and observation wells have an infinitesimally small diameter.
3) The confining bed that overlies or underlies the aquifer has a uniform hydraulic conductivity (K') and thickness (b').
4) Storage in the confining bed is negligible.
5) Leakage across the confining bed comes from an aquifer whose head is assumed to remain constant during the test.
6) Flow is vertical in the confining bed and radial in the confined aquifer.

Equation (10.2) can be applied to the prediction of drawdown with time and the interpretation of aquifer test data in which leaky behavior is evident. The following example illustrates the forward calculation.

Example 10.1 A leaky confined aquifer has a transmissivity of 2.5×10^{-2} m^2/s and a storativity of 2.7×10^{-4}. The vertical hydraulic conductivity (K') of the 3-m thick confining bed is 1.8×10^{-8} m/s. The well is pumped at a rate of 0.063 m^3/s. What is the drawdown in an observation well 150 m away after 1000 minutes?

Solution

Make the units consistent by changing time from minutes to seconds—1000 minutes is 6×10^4 s. Calculate u and r/B.

$$u = \frac{r^2 S}{4Tt} = \frac{(150\text{ m})^2 \times 2.7 \times 10^{-4}}{4 \times 2.5 \times 10^{-2}\text{m}^2/\text{s} \times 6 \times 10^4\text{s}} = 0.001$$

$$\frac{r}{B} = r\left(\frac{K'/m'}{T}\right)^{1/2} = 150\text{ m}\left(\frac{1.8 \times 10^{-8}\text{m/s}/3\text{m}}{2.5 \times 10^{-2}\text{m}^2/\text{s}}\right)^{1/2} = 0.075$$

Determine $W(u, r/B) = W(0.001, 0.075) = 5.30$ from Table 10.1. Calculate the drawdown as

$$s = \frac{Q}{4\pi T} W(u, r/B) = \frac{0.063\text{ m}^3/\text{s}}{4\pi \times 2.5 \times 10^{-2}\text{m}^2/\text{s}} \times 5.30 = 1.06\text{ m}$$

Table 10.1 Values of $W(u, r/B)$.

	r/B									
u	0.01	0.015	0.03	0.05	0.075	0.10	0.15	0.2	0.3	0.4
0.000001										
0.000005	9.4413									
0.00001	9.4176	8.6313								
0.00005	8.8827	8.4533	7.2450							
0.0001	8.3983	8.1414	7.2122	6.2282	5.4228					
0.0005	6.9750	6.9152	6.6219	6.0821	5.4062	4.8530				
0.001	6.3069	6.2765	6.1202	5,7965	5,3078	4.8292	4.0595	3.5054		
0.005	4.7212	4.7152	4.6829	4.6084	4.4713	4.2960	3.8821	3.4567	2.7428	2.2290
0.01	4.0356	4.0326	4.0167	3.9795	3.9091	3.8150	3.5725	3.2875	2.7104	2.2253
0.05	2.4675	2.4670	2.4642	2.4576	2.4448	2.4271	2.3776	2.3110	1.9283	1.7075
0.1	1.8227	1.8225	1.8213	1.8184	1.8128	1.8050	1.7829	1.7527	1.6704	1.5644
0.5	0.5598	0.5597	0.5596	0.5594	0.5588	0.5581	0.5561	0.5532	0.5453	0.5344
1.0	0.2194	0.2194	0.2193	0.2193	0.2191	0.2190	0.2186	0.2179	0.2161	0.2135
5,0	0.0011	0.0011	0.0011	0.0011	0.0011	0.0011	0.0011	0.0011	0.0011	0.0011

	r/B								
a	0.5	0.6	0.7	0,8	0.9	1.0	1.5	2.0	2.5
0.000001									
0.000005									
0.00001									
0.00005									
0.0001									
0.0005									
0.001									
0.005									
0.01	18486	1.5550	1.3210	0.1307					
0.05	1.4927	1.3955	1.2955	1.1210	0.9700	0.8409			
0.1	1.4422	1.3115	1.1791	1.0505	0.9297	0.8190	0.4271	0.2278	
0.5	0.5206	0.5044	0,4860	0,4658	0.4440	0.4210	0.3007	0.1944	0.1174
1.0	0.2103	0.2065	0.2020	0.1970	0.1914	0.1855	0.1509	0,1139	0,0803
5.0	0.0011	0.0011	0,011	0,0011	0.0011	00011	0.0010	00010	00009

Source: Hantush (1956) / John Wiley & Sons.

10.1.1 Interpreting Aquifer-Test Data

Chapter 9 introduced a curve-matching procedure for estimating transmissivity and storativity in a confined aquifer from an aquifer test. The same procedure can be applied with modifications to leaky-confined aquifer systems. One extra benefit is that a test also provides estimates of the vertical hydraulic conductivity of the confining bed K'. The steps are as follows:

1) Create or find type curves for the well function $W(u, r/B)$ versus $1/u$ on log–log graph paper (Figure 10.2). Mark a match point on this graph with $W(u, r/B) = 1$ and $1/u$ as 10, 100, or 1000.
2) Plot the observed drawdown versus time on a transparent overlay using log–log paper with the same scale.
3) Move the plotted drawdown versus time curve on the type curves and find which of the family of type curves matches best. Record the r/B value for that curve.

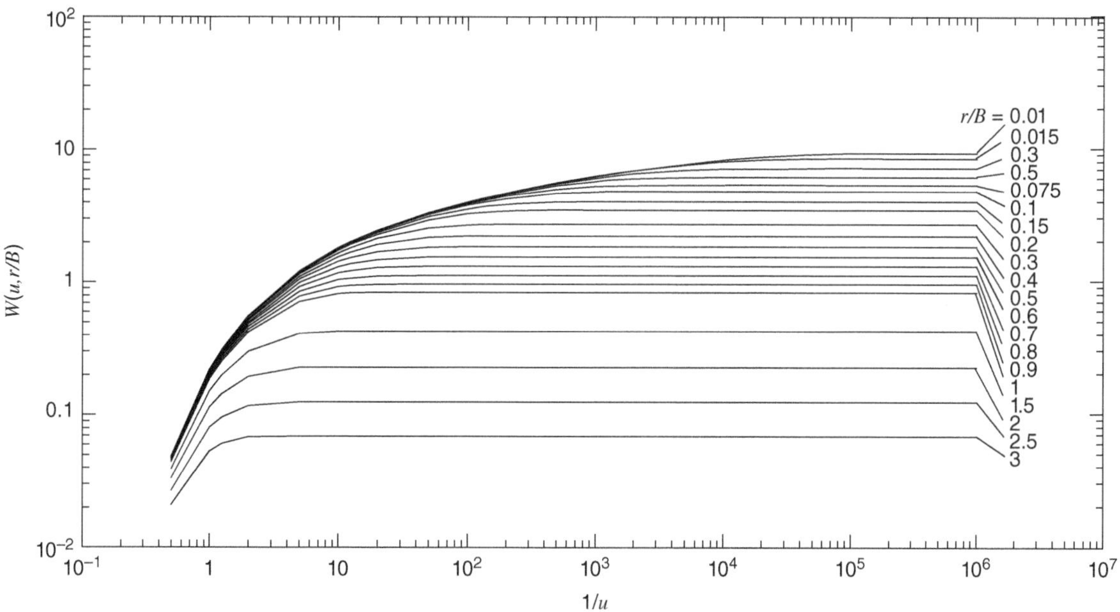

Figure 10.2 Type curve, $W(u, r/B)$, for a leaky, confined aquifer without storage in the confining bed (Schwartz and Zhang, 2003 / John Wiley & Sons).

4) Plot the match point on the field curve and record the coordinates s and t. T and S are calculated by substituting the appropriate values of $W(u, r/B)$, $1/u$, s, and t as needed in

$$T = \frac{Q}{4\pi s} W(u, r/B) \tag{10.4}$$

and

$$S = \frac{4uTt}{r^2} \tag{10.5}$$

5) Record r/B. The hydraulic conductivity of the confining bed is determined by

$$K' = \frac{Tb'(r/B)^2}{r^2} \tag{10.6}$$

Example 10.2 A test of a leaky confined aquifer was conducted at a pumping rate of 7.55 m^3/min. Drawdown versus time measurements were collected at an observation well 154 m away (Table 10.2). Calculate the transmissivity and storativity of the aquifer. Assuming the thickness of the semi-confining bed is 20 m, what is the vertical hydraulic conductivity of the confining bed?

Solution

Plot values of drawdown versus time on log-log graph paper at the same scale as the type curves (Figure 10.3). Overlay the plot of the field data on the family of type curves. Find the best match between the field data and one of the type curves (Figure 10.4). The match point coordinates are $1/u = 100$, $W(u, r/B) = 10$, $t = 6.2$ min, and $s = 6.3$ m. The data best fit the type curve with $r/B = 0.15$. Substitute known values into the appropriate equations

$$T = \frac{Q}{4\pi s} W(u, r/B) = \frac{7.55\ \text{m}^3/\ \text{min}}{4\pi(6.3\ \text{m})}(10) = 0.95\ \text{m}^2/\ \text{min}\ = 1373\ \text{m}^2/\text{day}$$

Table 10.2 Drawdown versus time in a leaky confined aquifer test.

t(min)	*s*(m)	*t*(min)	*s*(m)
0.10	0.27	11.72	2.54
0.14	0.39	16.10	2.59
0.19	0.53	22.12	2.63
0.26	0.68	30.39	2.64
0.36	0.85	41.75	2.65
0.49	1.03	57.36	2.65
0.67	1.21	78.80	2.65
0.92	1.39	108.26	2.65
1.27	1.58	148.74	2.65
1.74	1.76	204.34	2.65
2.40	1.93	280.72	2.65
3.29	2.09	385.66	2.65
4.52	2.23	529.83	2.65
6.21	2.36	727.90	2.65
8.53	2.46	1000.00	2.65

Source: Schwartz and Zhang (2003) / John Wiley & Sons.

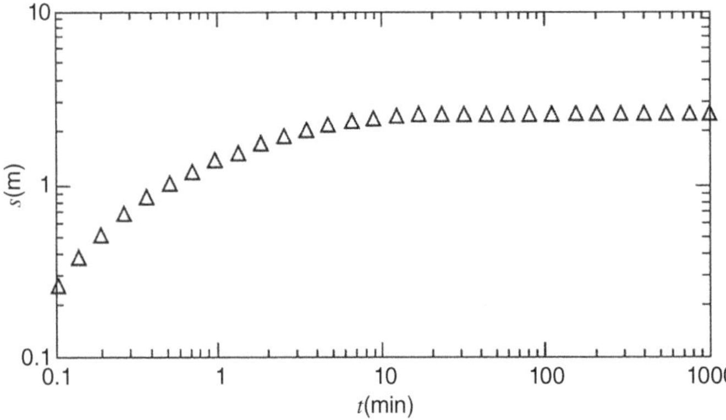

Figure 10.3 Plot of drawdown versus time on a log–log scale from an aquifer test of a leaky, confined aquifer (Schwartz and Zhang, 2003 / John Wiley & Sons).

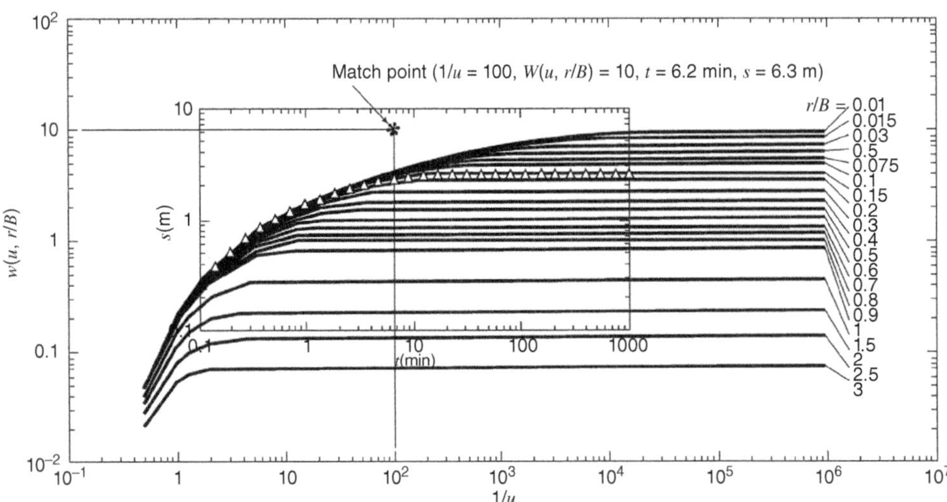

Figure 10.4 Graphical procedure for determining the hydraulic parameters of a leaky, confined aquifer using the type-curve matching technique (Schwartz and Zhang, 2003 / John Wiley & Sons).

and

$$S = \frac{4uTt}{r^2} = \frac{(4)(0.01)(0.95\text{m}^2/\min)(6.2\ \min)}{(154\ \text{m})^2} = 1.0 \times 10^{-5}$$

and

$$K' = \frac{Tb'(r/B)^2}{r^2} = \frac{(0.95\ \text{m}^2/\min)(20\ \text{m})(0.15)^2}{(154\ \text{m})^2} = 1.8 \times 10^{-5}\ \text{m}/\min = 0.026\ \text{m/day}$$

10.2 Steady-State Solution

At steady state, the drawdown in a leaky, confined aquifer is expressed as (Jacob, 1940)

$$s = \frac{Q}{2\pi T} K_0\left(\frac{r}{B}\right) \tag{10.7}$$

where $K_0(x)$ is the zero-order modified Bessel function of the second kind. This equation can be used to predict drawdown in a leaky confined aquifer at steady state. Tabulated values of $K_0(x)$ are available in the literature for this application (Table 10.3).

As before, this solution also provides the basis for evaluating the results of an aquifer tests. Steady-state values of drawdown at a set of observation wells can be used to derive a transmissivity value for the aquifer and a vertical hydraulic conductivity value for the leaky confining bed. The steps are as follows:

1) Create a type curve by plotting $K_0(x)$ versus x on log–log graph paper (Figure 10.5). Provide a match point on the type curve with simple coordinates (e.g., $x = 0.1$; $K_0(x) = 1.0$).
2) Plot measured values of drawdown versus distance on a transparent overlay the same, log–log scale.
3) Superimpose the field curve on the type curve. Match the two curves keeping the axes parallel.
4) Plot the match point on the field curve and determine the coordinates r and s. The transmissivity is calculated as

$$T = \frac{Q}{2\pi s} K_0(x) \tag{10.8}$$

and the vertical hydraulic conductivity is

$$K' = \frac{x\, b' T}{r^2} \tag{10.9}$$

Table 10.3 Values of $K_0(x)$ for values of x.

x	$K_0(x)$	x	$K_0(x)$	x	$K_0(x)$
0.01	4.7212	0.09	2.531	0.8	0.5653
0.015	4.3159	0.1	2.4271	0.9	0.4867
0.02	4.0285	0.15	2.03	1	0.421
0.03	3.6235	0.2	1.7527	1.5	0.2138
0.04	3.3365	0.3	1.3725	2	0.1139
0.05	3.1142	0.4	1.1145	3	0.0347
0.06	2.9329	0.5	0.9244	4	0.0112
0.07	2.7798	0.6	0.7775	5	0.0037
0.08	2.6475	0.7	0.6605		

Source: Hantush (1956) / John Wiley & Sons.

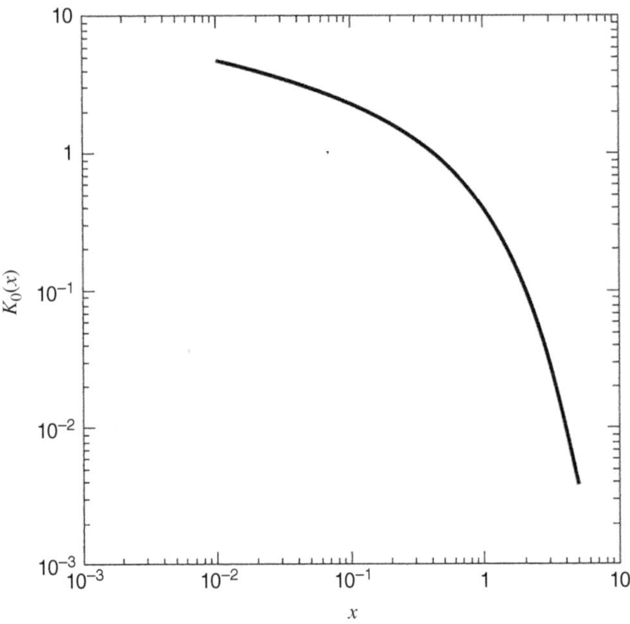

Figure 10.5 Type curve of $K_0(x)$ for steady-state flow in a leaky, confined aquifer (Schwartz and Zhang, 2003 / John Wiley & Sons).

Example 10.3 An aquifer test is run with a leaky, confined aquifer at a pumping rate of 220 gpm. At a late time, drawdowns were recorded for a number of wells (Table 10.4). The thickness of the confining bed is 10 ft. Calculate the hydraulic conductivity for the aquifer and K' for the confining bed.

Solution

Drawdown versus distance data for the observation wells are plotted on a log–log scale in Figure 10.6. Figure 10.7 shows the field curve superimposed on the type curve. The coordinates of the match points are $x = 0.1$, $K_0(x) = 1.0$, $r = 140$ ft, and $s = 7$ ft. Convert the pumping rate from gpm to ft^3/day.

$$Q = (220\ \text{gpm})\left(192.5\frac{\text{ft}^3/\text{day}}{\text{gpm}}\right) = 42,350\ \text{ft}^3/\text{day}$$

Table 10.4 Values of drawdown versus distance.

r(ft)	*S*(ft)
10	35.2
50	24.352
100	19.683
150	16.957
200	15.027
250	13.535
300	12.321
400	10.42
500	8.965

Source: Schwartz and Zhang (2003) / John Wiley & Sons.

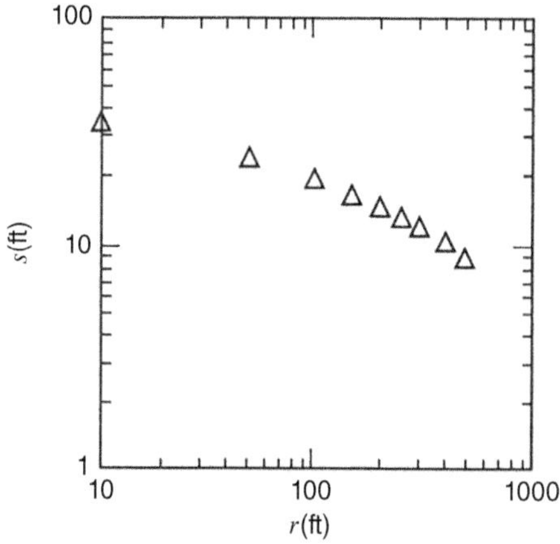

Figure 10.6 Plot of drawdown versus distance for steady-state flow in a leaky, confined aquifer (Schwartz and Zhang, 2003 / John Wiley & Sons).

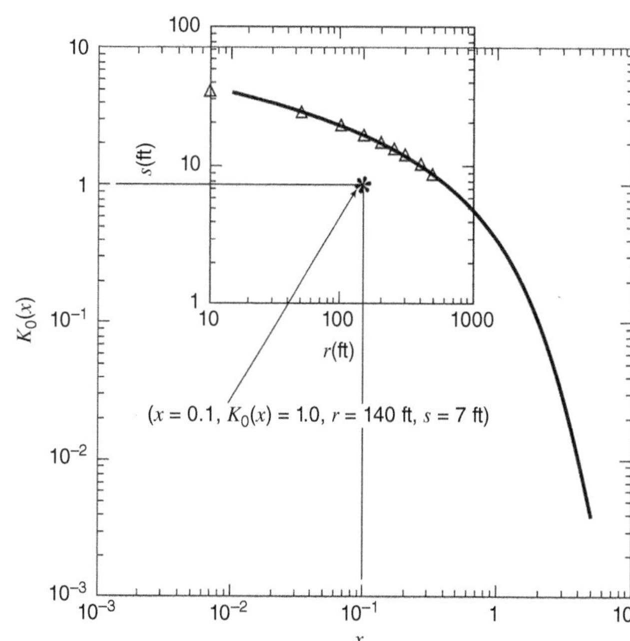

Figure 10.7 Graphical procedure for determining the hydraulic parameters of a leaky, confined aquifer using the type-curve matching technique (Schwartz and Zhang, 2003 / John Wiley & Sons).

From Eqs. (10.8) and (10.9),

$$T = \frac{(42,350\ \text{ft}^3/\text{day})}{2\pi(7\ \text{ft})} 1.0 = 963\ \text{ft}^2/\text{day}$$

and

$$K' = \frac{(0.1)(10\ \text{ft})(963\ \text{ft}^2/\text{day})}{140^2} = 0.05\ \text{ft/day}$$

10.3 Transient Solutions for Flow with Storage in Confining Beds

Hantush (1960) considered three situations with storage in confining beds overlying or underlying the aquifer (Figure 10.8). Here are the cases: (1) both of the confining beds are bounded by source units having constant heads; (2) both of the confining beds are bounded by impermeable layers; (3) the upper confining bed is bounded by a source unit with a constant

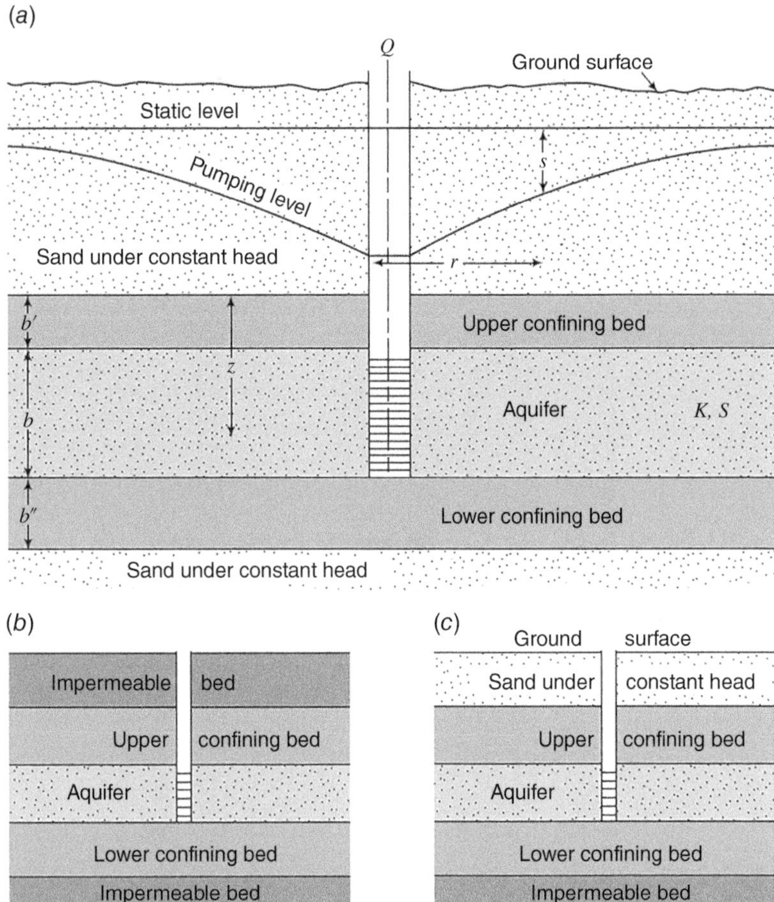

Figure 10.8 Three different geometries of a leaky, confined aquifer with storage in the confining beds. (*a*) Case 1 has constant-head plane sources above and below; (*b*) Case 2 has impermeable beds above and below; (*c*) Case 3 has a constant-head plane above and impermeable bed below (Schwartz and Zhang, 2003 / John Wiley & Sons).

head and the lower confining bed is bounded by an impermeable layer. With assumptions comparable to those for leaky confined aquifers, Hantush (1960) derived the following analytical solution for early time drawdown.

$$s = \frac{Q}{4\pi T} H(u, \beta), \text{ for } t < \frac{b'S'}{10\,K'}, t < \frac{b''S''}{10\,K''} \tag{10.10}$$

where

$$H(u, \beta) = \int_u^\infty \frac{e^{-y}}{y} erfc\left(\frac{\beta\sqrt{u}}{\sqrt{y(y-u)}}\right) dy \tag{10.11}$$

$$u = \frac{r^2 S}{4Tt} \tag{10.12}$$

$$\beta = \frac{r}{4}\left(\sqrt{\frac{K'S'}{b'TS}} + \sqrt{\frac{K''S''}{b''TS}}\right) \tag{10.13}$$

$$erfc(x) = \frac{2}{\sqrt{\pi}} \int_x^\infty e^{-y^2} dy \tag{10.14}$$

where T and S are the transmissivity and the storativity of the confined aquifer, respectively, T' and S' are the transmissivity and the storativity of the upper confining layer, respectively, and T'' and S'' are the transmissivity and the storativity of the lower confining layer, respectively. Tabulated values of well function $H(u, \beta)$ are provided in Table 10.5 (Hantush, 1961b). Figure 10.9 shows the resulting family of type curves, with each curve having its own β value.

Table 10.5 values of $H(u, \beta)$ for selected values of u and β.

	β							
u	**0.03**	**0.1**	**0.3**	**1**	**3**	**10**	**30**	**100**
1×10^{-9}	12.3088	11.1051	10.0066	8.8030	7.7051	6.5033	5.4101	4.2221
2	11.9622	10.7585	9.6602	8.4566	7.3590	6.1579	5.0666	3.8839
3	11.7593	10.5558	9.4575	8.2540	7.1565	5.9561	4.8661	3.6874
5	11.5038	10.3003	9.2021	7.9987	6.9016	5.7020	4.6142	3.4413
7	11.3354	10.1321	9.0339	7.8306	6.7337	5.5348	4.4487	3.2804
1×10^{-8}	11.1569	9.9538	8.8556	7.6525	6.5558	5.3578	4.2737	3.1110
2	10.8100	9.6071	8.5091	7.3063	6.2104	5.0145	3.9352	2.7858
3	10.6070	9.4044	8.3065	7.1039	6.0085	4.8141	3.7383	2.5985
5	10.3511	9.1489	8.0512	6.8490	5.7544	4.5623	3.4919	2.3662
7	10.1825	8.9806	7.8830	6.6811	5.5872	4.3969	3.3307	2.2159
1×10^{-7}	10.0037	8.8021	7.7048	6.5032	5.4101	4.2221	3.1609	2.0591
2	9.6560	8.4554	7.3585	6.1578	5.0666	3.8839	2.8348	1.7633
3	9.4524	8.2525	7.1560	5.9559	4.8661	3.6874	2.6469	1.5966
5	9.1955	7.9968	6.9009	5.7018	4.6141	3.4413	2.4137	1.3944
7	9.0261	7.8283	6.7329	5.5346	4.4486	3.2804	2.2627	1.2666
1×10^{-6}	8.8463	7.6497	6.5549	5.3575	4.2736	3.1110	2.1051	1.1361
2	8.4960	7.3024	6.2091	5.0141	3.9350	2.7857	1.8074	.8995
3	8.2904	7.0991	6.0069	4.8136	3.7382	2.5984	1.6395	.7725
5	8.0304	6.8427	5.7523	4.5617	3.4917	2.3661	1.4354	.6256
7	7.8584	6.6737	5.5847	4.3962	3.3304	2.2158	1.3061	.5375
1×10^{-5}	7.6754	6.4944	5.4071	4.2212	3.1606	2.0590	1.1741	.4519
2	7.3170	6.1453	5.0624	3.8827	2.8344	1.7632	.9339	.3091
3	7.1051	5.9406	4.8610	3.6858	2.6464	1.5965	.8046	.2402
7	6.6553	5.5113	4.4408	3.2781	2.2619	1.2664	.5643	.1300
1×10^{-4}	6.4623	5.3297	4.2643	3.1082	2.1042	1.1359	.4763	963(−4)
2	6.0787	4.9747	3.9220	2.7819	1.8026	.8992	.3287	494(−4)
3	5.8479	4.7655	3.7222	2.5937	1.6380	.7721	.2570	315(−4)
5	5.5488	4.4996	3.4711	2.3601	1.4335	.6252	.1818	166(−4)
7	5.3458	4.3228	3.3062	2.2087	1.3039	.5370	.1412	103(−4)
1×10^{-3}	5.1247	4.1337	3.1317	2.0506	1.1715	.4513	.1055	390(−5)
2	4.6753	3.7598	2.7938	1.7516	.9305	.3084	551(−4)	169(−5)
3	4.3993	3.5363	2.5969	1.5825	.8006	.2394	355(−4)	713(−6)
5	4.0369	3.2483	2.3499	1.3767	.6498	.1677	190(−4)	205(−6)
7	3.7893	3.0542	2.1877	1.2460	.5589	.1292	120(−4)	821(v7)

(Continued)

Table 10.5 (Continued)

	β							
u	**0.03**	**0.1**	**0.3**	**1**	**3**	**10**	**30**	**100**
1×10^{-2}	3.5195	2.8443	2.0164	1.1122	.4702	955(−4)	695(−5)	274(−7)
2	2.9759	2.4227	1.6853	.8677	.3214	487(−4)	205(−5)	226(−8)
3	2.6487	2.1680	1.4932	.7353	.2491	308(−4)	888(−6)	
5	2.2312	1.8401	1.2535	.5812	.1733	160(−4)	261(−6)	
7	1.9558	1.6213	1.0979	.4880	.1325	982(−5)	106(−6)	
1×10^{-1}	1.6667	1.3893	.9358	.3970	966(−4)	552(−5)	365(−7)	
2	1.1278	.9497	.6352	.2452	468(−4)	149(−5)	307(−8)	
3	.8389	.7103	.4740	.1729	281(−4)	592(−6)		
5	.5207	.4436	.2956	.1006	130(−4)	151(−6)		
7	.3485	.2980	.1985	646(−4)	714(−5)	534(−7)		
1×1	.2050	.1758	.1172	365(−4)	337(−5)	151(−7)		
2	458(−4)	395(−4)	264(−4)	760(−5)	487(−6)			
3	122(−4)	106(−4)	707(−5)	196(−5)	102(−6)			
5	108(−5)	934(−6)	624(−6)	167(−6)	672(−8)			
7	109(−6)	941(−7)	629(−7)	165(−7)				
1×10	391(−8)	339(−8)	227(−8)					
2								
3								
5								
7								

Source: Hantush, 1961b / New Mexico Institute of Mining and Technology.

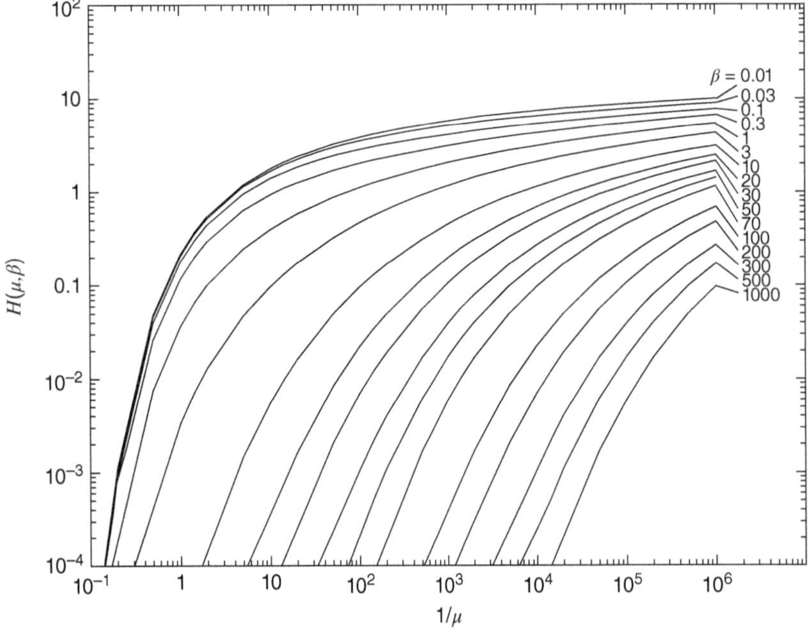

Figure 10.9 Early-time type curves of $H(u, \beta)$ in a leaky, confined aquifer with storage in confining beds (U.S. Geological Survey, adapted from Reed, 1980 / Public domain).

This theory can be applied to interpret the results of an aquifer test following the general scheme for superposition developed previously.

1) Create the family of type curves by plotting $H(u, \beta)$ versus $1/u$ on a log–log graph paper (Figure 10.9). Choose a match point that works.
2) Plot the field curve of drawdown versus time on a log–log paper at the same scale as the family of type curves.
3) Superimpose the two graphs and find the best match between the field curve and one of the family of type curves.
4) Record β and $1/u$, $H(u, \beta)$, t and s for the match point.
5) The transmissivity of the leaky confined aquifer is determined by

$$T = \frac{Q}{4\pi s} H(u, \beta) \tag{10.15}$$

6) The storativity of the leaky confined aquifer is

$$S = \frac{4T\,t\,u}{r^2} \tag{10.16}$$

7) If $T''S'' = 0$, then

$$K'S' = \frac{16\beta^2 b' TS}{r^2} \tag{10.17}$$

8) If $T'S' = T''S''$, then

$$K'S' = \frac{16\beta^2 TS}{r^2} \frac{b'b''}{b' + b'' + 2\sqrt{b'b''}} \tag{10.18}$$

Example 10.4 A test of a leaky confined aquifer was run at a pumping rate of 750 gpm (Lohman, 1972). A confining layer 6 ft thick overlies the aquifer. Above the confining layer is an unconfined aquifer 200 ft thick. At an observation well 1400 ft away, values of drawdown versus time were recorded (Table 10.6). Calculate hydraulic parameters using the Hantush's method.

Solution

The drawdown versus time data are plotted as shown in Figure 10.10. This field curve (Figure 10.10) is overlain on the type curves in Figure 10.9. The best match comes with the field data and the $\beta = 3$ type curve. With the match point transferred to the field curve (Figure 10.11), the match-point coordinates are $1/u = 10$, $H(u, \beta) = 1.0$, $t = 90$ min, and $s = 6.5$ ft. The pumping rate is converted from gpm to ft^3/day.

$$Q = (750\text{ gpm})\left(192.5\frac{\text{ft}^3/\text{day}}{\text{gpm}}\right) = 144,375\text{ ft}^3/\text{day}$$

The transmissivity is

$$T = \frac{Q}{4\pi s} H(u, \beta) = \frac{(144,375\text{ ft}^3/\text{day})}{4\pi(6.5\text{ ft})} 1.0 = 1768\text{ ft}^2/\text{day}$$

The storativity is

$$S = \frac{4T\,t\,u}{r^2} = \frac{4(1768\text{ ft}^2/\text{day})(90\text{ min})\left(\frac{1}{1440}\frac{\text{day}}{\text{min}}\right)(0.1)}{(1400\text{ ft})^2} = 2.3 \times 10^{-5}$$

For $T''S'' = 0$ in this example, the hydraulic properties of the upper confining are calculated with Eq. (10.17).

$$K'S' = \frac{16(3)^2(6\text{ ft})(1768\text{ ft}^2/\text{day})(2.3 \times 10^{-5})}{(1400\text{ ft})^2} = 3 \times 10^{-5}\text{ ft/day}$$

Table 10.6 Values of drawdown versus time in a leaky-confined aquifer.

t(min)	s(ft)	t(min)	s(ft)	t(min)	s(ft)
6.37	0.01	41	0.33	315	1.83
8.58	0.02	44	0.36	335	1.87
10.23	0.03	47	0.38	365	1.99
11.9	0.04	50	0.42	390	2.1
12.95	0.05	54	0.46	410	2.13
14.42	0.06	60	0.52	430	2.2
15.1	0.07	65	0.56	450	2.23
16.88	0.08	70	0.6	470	2.29
17.92	0.1	80	0.65	490	2.32
21.35	0.12	90	0.75	510	2.39
21.7	0.13	100	0.82	560	2.48
22.7	0.14	137	1.04	740	2.92
23.58	0.15	150	1.12	810	3.05
24.65	0.17	160	1.17	890	3.19
29	0.21	173	1.24	1255	3.66
30	0.22	184	1.27	1400	3.81
32	0.24	200	1.35	1440	3.86
34	0.26	210	1.4	1485	3.9
36	0.28	278	1.68		
38	0.3	300	1.76		

Source: U.S. Geological Survey Lohman, 1972 / Public domain.

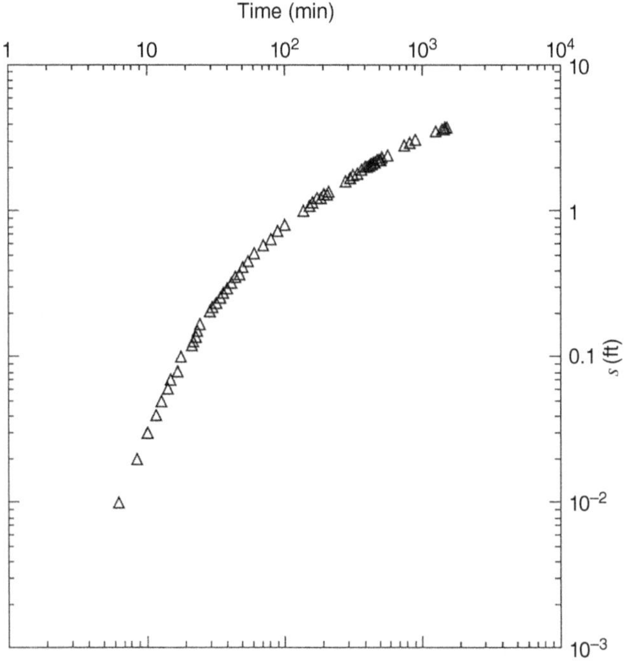

Figure 10.10 Plot of drawdown versus time in a leaky, confined aquifer with storage in the confining beds (Schwartz and Zhang, 2003 / John Wiley & Sons).

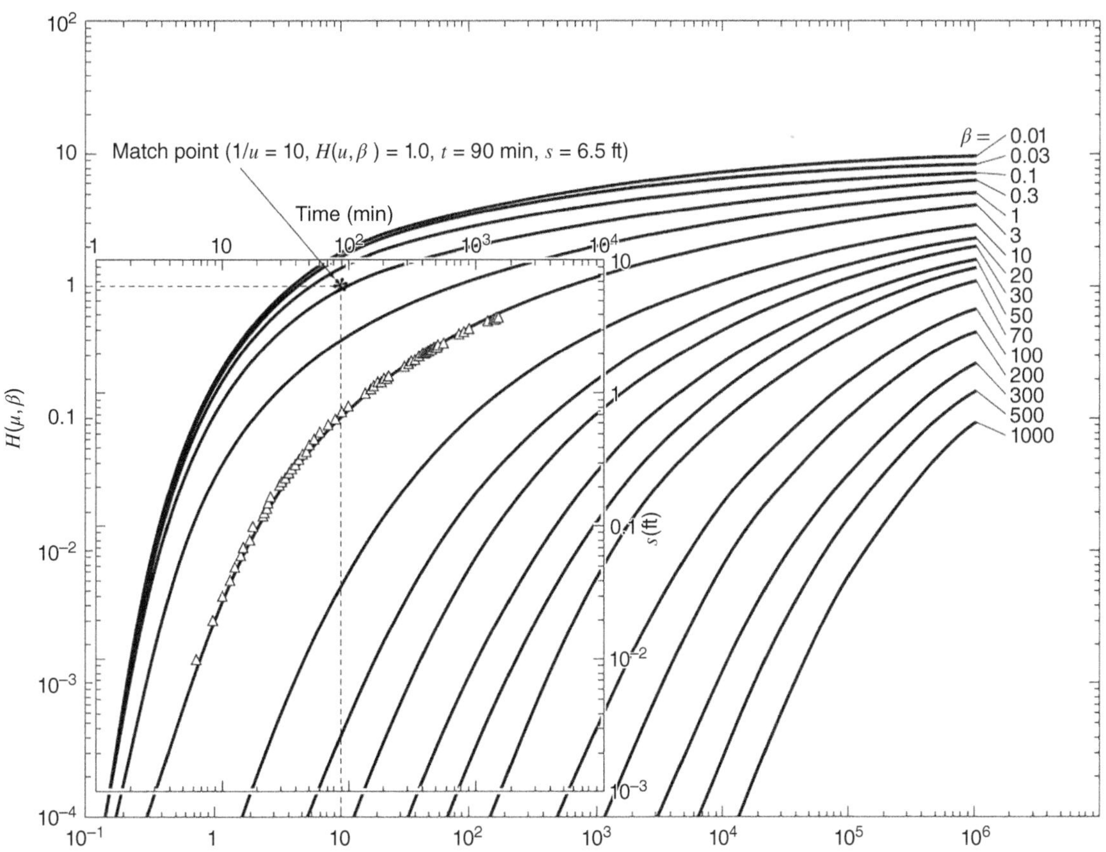

Figure 10.11 Determination of hydraulic parameters in a leaky, confined aquifer with storage in confining beds (Schwartz and Zhang, 2003 / John Wiley & Sons).

10.4 Effects of Partially Penetrating Wells

When the screened or open section of a well casing does not coincide with the full thickness of the aquifer it penetrates, the well is referred to as partially penetrating. With many aquifers, this well design is the rule rather than the exception. Under such conditions, the flow toward the pumping well (or observation point) will be three-dimensional because of vertical flow components (Figure 10.12). In practice, once you get far enough away from a pumping well, the effects of partial penetration become unimportant. The following equation describes when the effects of partial penetration become negligible

$$r > 1.5\,b\left(\frac{K_r}{K_z}\right)^{\frac{1}{2}} \tag{10.19}$$

where b is the thickness of the aquifer, K_z and K_r are the horizontal and vertical hydraulic conductivities of the aquifer, respectively.

The topic of partial penetration has been the subject of numerous papers (Hantush, 1961c, 1964; Muskat, 1937; Neuman, 1972). Hantush (1964) has conducted most of the work in this area and provides some general guidelines for confined aquifers. The Hantush's correction equation of drawdown in a piezometer for a partially penetrating pumping well (Figure 10.13) can be expressed as

$$s_{\text{partially}}(t) = s_{\text{fully}}(t) + \frac{Q}{4\pi T}f\left(u, \frac{ar}{b}, \frac{l}{b}, \frac{d}{b}, \frac{z}{b}\right) \tag{10.20}$$

where t is the time, $u = r^2S/(4Tt)$, $a = sqrt(K_z/K_r)$, r is the radial distance from the pumping well to the observation well, b is the aquifer thickness, l is the depth of the pumped well in the aquifer, z_1 and z_2 are the bottom and top z coordinates of the screen in the observation well, respectively. For a confined aquifer

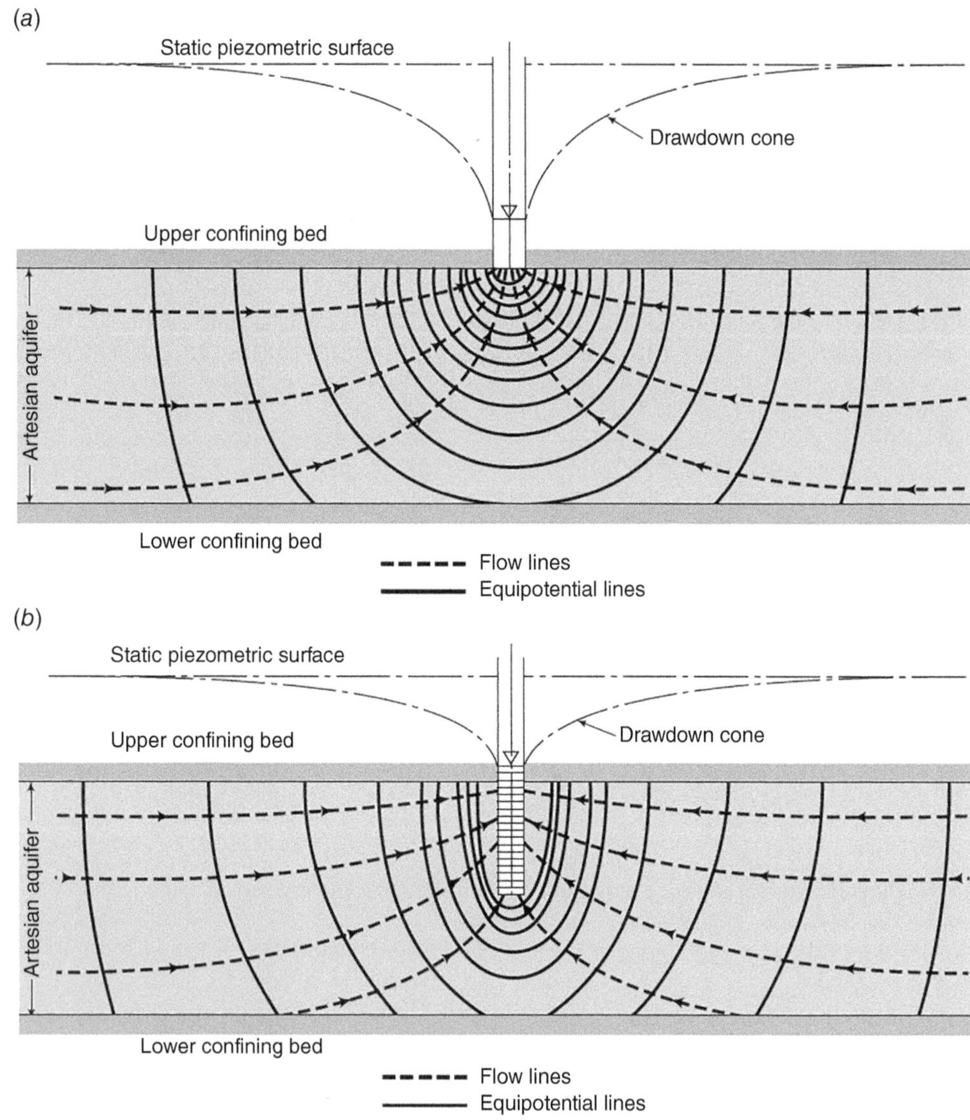

Figure 10.12 Patterns of flow in a confined aquifer to various partially penetrating wells (U.S. Bureau of Reclamation, 1995 / Public Domain). In Panel (*a*), the well penetrates the top of the confined aquifer. In Panel (*b*), the well is screened in the top 50% of the aquifer.

$$s_{\text{full}} = \frac{Q}{4\pi T} W(u) \tag{10.21}$$

and

$$f\left(u, \frac{ar}{b}, \frac{l}{b}, \frac{d}{b}, \frac{z}{b}\right) = \frac{2}{\pi(l/b - d/b)} \sum_{n=1}^{\infty} \frac{1}{n}\left(\sin\frac{n\pi l}{b} - \sin\frac{n\pi d}{b}\right)\cos\frac{n\pi z}{b} W\left(u, \frac{n\pi a\, r}{b}\right) \tag{10.22}$$

where $W(u)$ is the Theis well function defined in Section 6.1 and $W(u, \beta)$ is the Hantush and Jacob well function defined in Section 10.2. For a leaky confined aquifer

$$s_{\text{full}} = \frac{Q}{4\pi T} W(u, r/B) \tag{10.23}$$

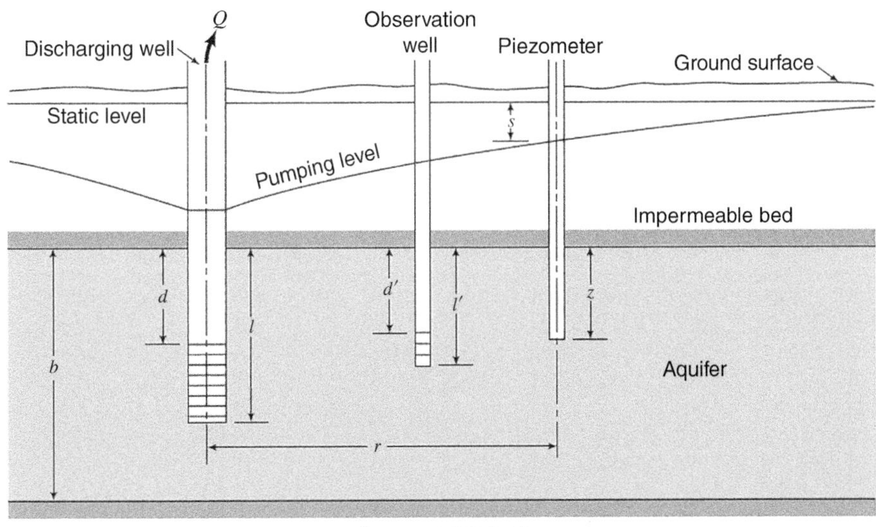

Figure 10.13 Cross section of a confined aquifer with partially penetrating pumping and observation wells (Schwartz and Zhang, 2003 / John Wiley & Sons).

and

$$f\left(u, \frac{ar}{b}, \frac{l}{b}, \frac{d}{b}, \frac{z}{b}\right) = \frac{2}{\pi(l/b - d/b)} \sum_{n=1}^{\infty} \frac{1}{n}\left(\sin\frac{n\pi l}{b} - \sin\frac{n\pi d}{b}\right)\cos\frac{n\pi z}{b} W\left(u, \sqrt{\beta^2 + \left(\frac{n\pi a\, r}{b}\right)^2}\right) \tag{10.24}$$

For an observation well, the Hantush correction term is the average hydraulic head along a well screen.

$$f\left(u, \frac{ar}{b}, \frac{l}{b}, \frac{d}{b}, \frac{z_1}{b}, \frac{z_2}{b}\right) = \frac{1}{(z_2 - z_1)} \int_{z_1}^{z_2} f\left(u, \frac{ar}{b}, \frac{l}{b}, \frac{d}{b}, \frac{z}{b}\right) dz \tag{10.25}$$

Exercises

10.1 An aquifer test was run with a leaky confined aquifer without storage in the confining bed at a pumping rate of 1440 m^3/day. Drawdown-time data from an observation well that is located 40 m away from the pumping well are tabulated in Table 10.7. Calculate the transmissivity and storativity of the aquifer. If the thickness of the confining bed is 5 m, what is the vertical hydraulic conductivity of the confining bed? You will find a template file for this exercise in the directory Chapter 10 on the book resource website. The template file "temp-10a.doc" contains type curves for a leaky confined aquifer. The template file "temp-10.doc" in the directory Chapter 10 provides a log–log graph at the same scale as the type curves. Use these templates and a curve-matching procedure to answer this question.

Table 10.7 Values of drawdown versus time obtained from an aquifer test of a leaky confined aquifer.

t(min)	s(m)	t(min)	s(m)
0.10	0.21	11.25	0.57
0.13	0.23	14.25	0.58
0.16	0.25	18.05	0.60
0.20	0.27	22.85	0.61
0.26	0.29	28.94	0.62
0.33	0.30	36.65	0.63
0.41	0.32	46.42	0.64
0.52	0.34	58.78	0.65
0.66	0.36	74.44	0.65
0.84	0.38	94.27	0.65
1.06	0.40	119.38	0.66
1.34	0.42	151.18	0.66
1.70	0.43	191.45	0.66
2.15	0.45	242.45	0.66
2.73	0.47	307.03	0.66
3.46	0.49	388.82	0.66
4.38	0.50	492.39	0.66
5.54	0.52	623.55	0.66
7.02	0.54	789.65	0.66
8.89	0.55	1000.00	0.66

Source: Schwartz and Zhang (2003) / John Wiley & Sons.

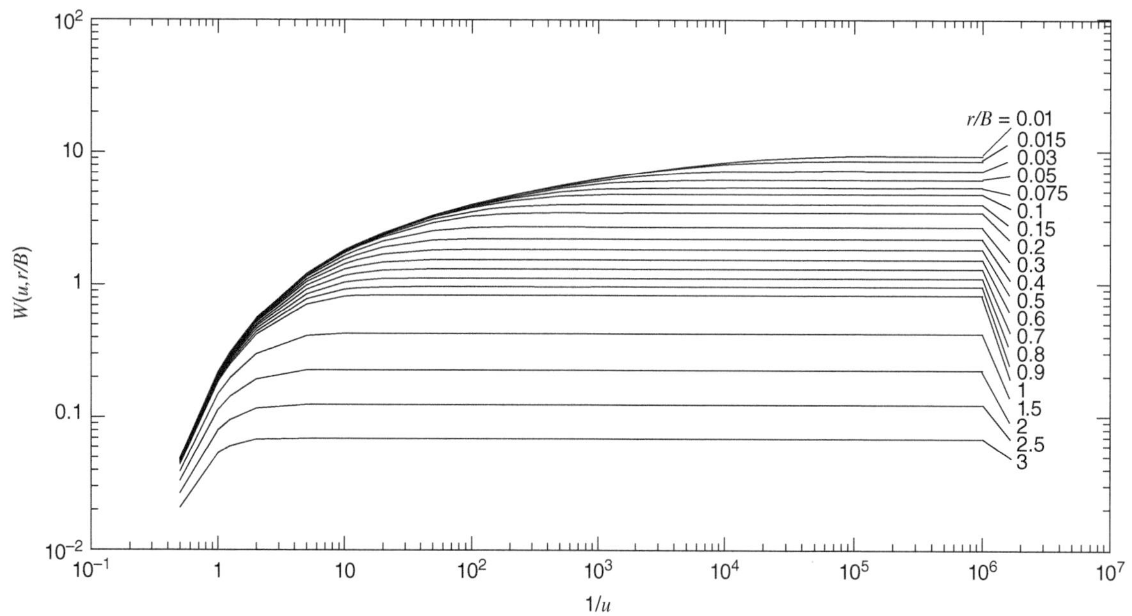

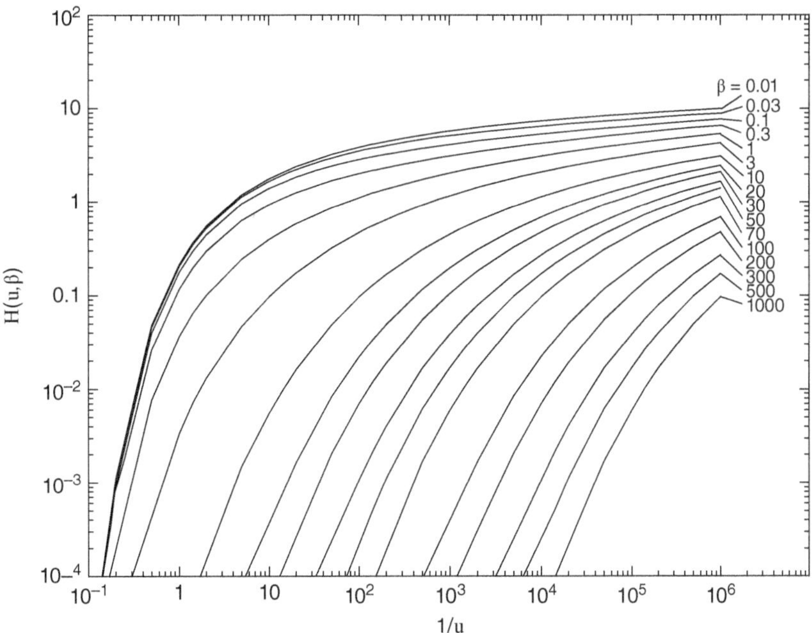

10.2 An aquifer test is conducted with a leaky confined aquifer by pumping at a constant rate of 2 m^3/min. The thickness of the upper confining bed is 4 m. Eventually, steady-state conditions are achieved and drawdowns are measured for a set of observation wells (Table 10.8). Calculate the hydraulic parameters for the aquifer and the confining bed.

Table 10.8 Values of drawdown versus distance from an aquifer test of a leaky confined aquifer at steady state.

r(m)	*s*(m)
20	0.469022
40	0.185689
80	0.038312
100	0.018347
120	0.008944
150	0.003113

Source: Schwartz and Zhang (2003) / John Wiley & Sons.

10.3 This aquifer test involves a leaky confined aquifer with storage in the confining beds. The well is pumped at a constant rate of 2880 m^3/day, and drawdowns are measured in an observation well located 50 m away from the pumping well (Table 10.9). The thickness of the upper and lower confining beds is 10 m. Assuming that $K'S' = K''S''$, calculate the transmissivity, storativity of the aquifer, and the hydraulic parameters of the confining beds. The template file "temp-10b.doc" contains type curves for a leaky confined aquifer with storage in the confining beds.

Table 10.9 Values of drawdown versus time from an aquifer test in a leaky confined aquifer with storage in the confining beds.

t(min)	*s*(m)	*t*(min)	*s*(m)
0.10	0.09	14.15	0.40
0.14	0.11	19.69	0.43
0.19	0.12	27.39	0.45
0.27	0.14	38.11	0.47
0.37	0.16	53.02	0.50
0.52	0.18	73.76	0.52
0.73	0.20	102.62	0.55
1.01	0.22	142.76	0.57
1.40	0.24	198.61	0.60
1.95	0.26	276.31	0.63
2.72	0.28	384.41	0.65
3.78	0.31	534.79	0.68
5.26	0.33	744.01	0.70
7.31	0.35	1035.07	0.73
10.17	0.38	1440.00	0.75

Source: Schwartz and Zhang (2003) / John Wiley & Sons.

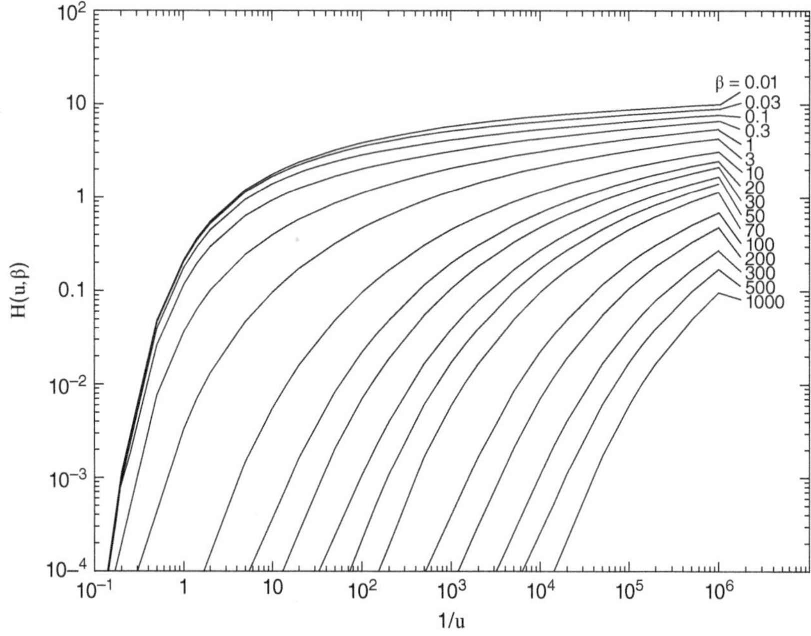

References

Bureau of Reclamation, U.S. Department of Interior. 1995. Groundwater Manual. A Water Resources. Technical Publication, 2nd edition. Bureau of Reclamation, U. S. Department of Interior, p. 661.

Hantush, M. S. 1956. Analysis of data from pumping tests in leaky aquifers. Transactions of the American Geophysical Union, v. 37, no. 2, p. 702–714.

Hantush, M. S. 1960. Modification of the theory of leaky aquifers. Journal of Geophysical Research, v. 65, no. 11, p. 3713–3725.

Hantush, M. S. 1961b. Tables of the function $H(u,\beta) = \int_u^\infty \frac{e^{-y}}{y} erfc\left(\frac{\beta\sqrt{u}}{\sqrt{y(y-u)}}\right) dy$. New Mexico Institute of Mining and Technology Professional Paper 103, 12 p.

Hantush, M. S. 1961c. Aquifer tests on partially penetrating wells. Proceedings of the American Society of Civil Engineers, v. 87, no. 5, p. 171–195.

Hantush, M. S. 1964. Hydraulics of wells, In V. T. Chow (ed.), Advances In Hydroscience. Academic Press, New York, p. 281–432.

Hantush, M. S., and C. E. Jacob. 1955. Nonsteady radial flow in an infinite leaky aquifer. Transactions of the American Geophysical Union, v. 36, no. 1, p. 95–100.

Jacob, C. E. 1940. On the flow of water in an elastic artesian aquifer. Transactions of the American Geophysical Union, v. 21, no. 2, p. 574–586.

Lohman, S. W. 1972. Groundwater hydraulics. U.S. Geological Survey Professional Paper 708, 70 p.

Muskat, M. 1937. The Flow of Homogeneous Fluids Through Porous Media. McGraw-Hill, New York.

Neuman, S. P. 1972. Theory of flow in unconfined aquifers considering delayed response of the water table. Water Resources Research, v. 8, no. 4, p. 1031–1045.

Reed, J. E. 1980. Type curves for selected problems of flow to wells in confined aquifers. U.S. Geological Survey Techniques of Water-Resources Investigations, Chapter B3, 106 p.

Schwartz, F. W., and H. Zhang. 2003. Fundamentals of Groundwater. John Wiley & Sons, Hoboken, New Jersey, 583 p.

11

Response of an Unconfined Aquifer to Pumping

CHAPTER MENU

The response of an unconfined aquifer to pumping is complicated. Once a cone of depression forms, it naturally decreases the aquifer thickness and transmissivity because the upper boundary of the aquifer is the water table. Also, the way in which water comes out of storage in the aquifer changes with time. At early time, when the well is first turned on, water is released from storage due to compression of the matrix and expansion of the water. This response is the same Theis behavior that applies to confined aquifers (Chapter 9). Thus, early time/drawdown data, plotted on log–log graph paper, follow the Theis-type curve (Figure 11.1). Storativity values would be comparable to those for confined aquifers, 10^{-4} or 10^{-5}.

As pumping continues, water comes from the slow gravity drainage of water from pores as the water table falls near the well. The pattern of drawdown depends on the vertical and horizontal hydraulic conductivity and the thickness of the aquifer. Once this delayed drainage begins, drawdown data deviate from the Theis curve. The drawdown is less than expected, resembling the pattern of drawdown in a leaky aquifer (Figure 11.1). Eventually, the contribution of water from delayed drainage ceases. Flow in the aquifer is mainly radial and drawdown/time data again fall on a Theis-type curve (Figure 11.1). The storativity of the aquifer now is the same as the specific yield (S_y). The *specific yield* is the ratio of the volume of water that drains from a rock or sediment by gravity to the volume of the rock or soil. The curve-fitting approach that is developed in this chapter for interpreting aquifer-test data requires being able to recognize these differing aquifer responses.

11.1 Calculation of Drawdowns by Correcting Estimates for a Confined Aquifer

Because of the intricacies in the response of an unconfined aquifer to pumping, a theory developed for confined aquifers is not generally transferable. However, if the point of interest is far away from the pumped well, one can simply apply confined-aquifer theory without too much error. A slightly more sophisticated approach is to calculate drawdown assuming the aquifer to be confined and correcting the drawdown value with the following equation

$$s = b - \left(b^2 - 2s'b\right)^{1/2} \tag{11.1}$$

where s is the drawdown for the unconfined aquifer, s' is the drawdown for the equivalent confined aquifer, and b is the original thickness of the aquifer. The following example illustrates this calculation.

Fundamentals of Groundwater, Second Edition. Franklin W. Schwartz and Hubao Zhang.

Companion website: www.wiley.com/go/schwartz/fundamentalsofgroundwater2

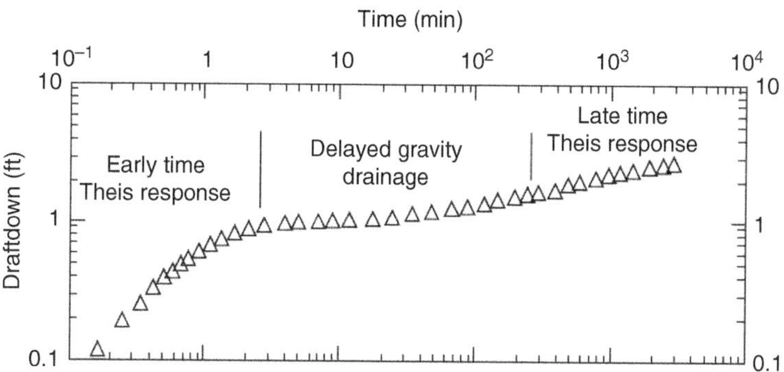

Figure 11.1 An example of the characteristic "S"-shaped drawdown curve obtained from a test with an unconfined aquifer (Schwartz and Zhang, 2003/ John Wiley & Sons).

Example 11.1 A well in an unconfined aquifer is pumped at a constant rate of 99 gpm. The aquifer has a hydraulic conductivity of 388 ft/day, an initial saturated thickness of 20 ft, and a specific yield of 0.01. Calculate what the drawdown will be at an observation well 51 ft away after 3180 minutes.

Solution

The first step in solving this problem is to adjust the units to feet and days to make them consistent. Also, T needs to be calculated.

$$Q = 99\ \text{gpm} = 99 \times 192.5\ \text{ft}^3/\text{day} = 19{,}060\ \text{ft}^3/\text{day}$$

$$t = 3180\ \text{minutes} = 3180/1440 = 2.21\ \text{days}$$

$$T = Kb = 388\ \text{ft/day} \times 20\ \text{ft} = 7760\ \text{ft}^2/\text{day}$$

Calculate the drawdown at the observation point assuming the aquifer to be confined.

$$u = \frac{r^2 S}{4Tt} = \frac{(51\ \text{ft})^2 \times 0.01}{4 \times 7760\ \text{ft}^2/\text{day} \times 2.21\ \text{days}} = 3.79 \times 10^{-4};\ W(u) = 7.31$$

$$s' = \frac{Q}{4\pi T} W(u) = \frac{19,060\ \text{ft}^3/\text{day}}{4 \times 3.1417 \times 7760\ \text{ft}^2/\text{day}} \times 7.31 = 1.43\ \text{ft}$$

Correct the drawdown with Eq. (11.1) because the aquifer is unconfined.

$$s = b - \left(b^2 - 2s'b\right)^{1/2} = 20\ \text{ft} - \left\{(20\ \text{ft})^2 - 2 \times 1.43\ \text{ft} \times 20\ \text{ft}\right\}^{1/2} = 1.49\ \text{ft}$$

In this case, the difference between the confined and unconfined estimates is quite small.

These same ideas also can be applied to the interpretation of aquifer tests. One simply converts the observed drawdowns in an unconfined aquifer to drawdowns in an equivalent confined aquifer with this equation

$$s' = s\ \frac{s^2}{2b} \tag{11.2}$$

where s is the observed drawdown in an unconfined aquifer and s' is the drawdown in an equivalent confined aquifer. These equivalent values of drawdown can be analyzed by using the techniques presented previously for a confined aquifer. However, this simple approach is typically used with late-time drawdown data, once the contribution from delayed yield ceases (Neuman, 1974). Weeks (1969) suggested the following criteria for estimating the time at which delayed yield becomes negligible (Peters, 1987):

$$t = b\frac{S_y}{K_z},\ \text{if}\ \ r < 0.4b\left(\frac{K_r}{K_z}\right)^{1/2} \tag{11.3}$$

or

$$t = b\frac{S_y}{K_z}\left[0.5 + 1.25\frac{r}{b}\left(\frac{K_z}{K_r}\right)^{1/2}\right],\ \text{if}\ \ r \geq 0.4b\left(\frac{K_r}{K_z}\right)^{1/2} \tag{11.4}$$

11.2 Determination of Hydraulic Parameters Using Distance/Drawdown Data

The distance–drawdown method, introduced in Chapter 9, also can be applied to interpreting test data from unconfined aquifers. This section discusses how values of equivalent drawdown as a function of distance can be used for this purpose. Equations (9.1) and (9.2) can be rewritten in terms of corrected drawdowns as

$$T = \frac{2.3Q}{2\pi(s_1{}' - s_2{}')} \log \frac{r_2}{r_1} \tag{11.5}$$

$$S_y = \frac{2.25\,T\,t}{r_0^2} \tag{11.6}$$

where $s_1{}'$ and $s_2{}'$ are equivalent or corrected drawdowns at distances r_1 and r_2, respectively. Remember, Eqs. (11.5) and (11.6) only apply for late-time drawdown data.

The procedures for using the corrected drawdown versus distance data are summarized as follows.

1) Calculate corrected drawdowns using Eq. (11.2).
2) Plot corrected drawdown s' against r on semi-log graph paper.
3) Approximate the curve s' versus log (t) with a straight line.
4) Determine hydraulic parameters using Eqs. (11.5) and (11.6).

Example 11.2 Table 11.1 lists drawdown values measured in wells at various distances from a pumping well in an aquifer test near Grand Island, Nebraska (Wenzel, 1936). Measurements were made after the well had been pumped continuously for 48 hours at 540 gpm. The thickness of the unconfined aquifer is 100 ft. Determine the transmissivity and the specific yield.

Solution

Table 11.1 shows the calculations of equivalent drawdowns. These values are plotted in Figure 11.2 as a function of distance on a semi-log scale. The pumping rate needs to be converted from gallons per min to ft^3/day.

$$Q = 540 \times 192.5\text{ft}^3/\text{day} = 1.04 \times 10^5\text{ft}^3/\text{day}$$

Table 11.1 Drawdown versus distance data for an aquifer test near Grand Island, Nebraska, after 48 hours of continuous pumping at 540 gpm.

r(ft)	s(ft)	$S^2/(2b)$ (ft)	s'(ft)
24.9	4.01	0.08	3.93
59.9	2.79	0.04	2.75
114.4	2.03	0.02	2.01
164.2	1.61	0.01	1.60
229	1.14	0.01	1.13
354	0.65	0.00	0.65
429	0.52	0.00	0.52
479	0.44	0.00	0.44
604	0.26	0.00	0.26
755	0.16	0.00	0.16
904	0.11	0.00	0.11

Source: U.S. Geological Survey, adapted from Wenzel, 1936/Public domain.

Figure 11.2 Estimation of hydraulic parameters using a distance–drawdown approach. The y-axis is the drawdown of a well in an equivalent confined aquifer (Schwartz and Zhang, 2003 / John Wiley & Sons).

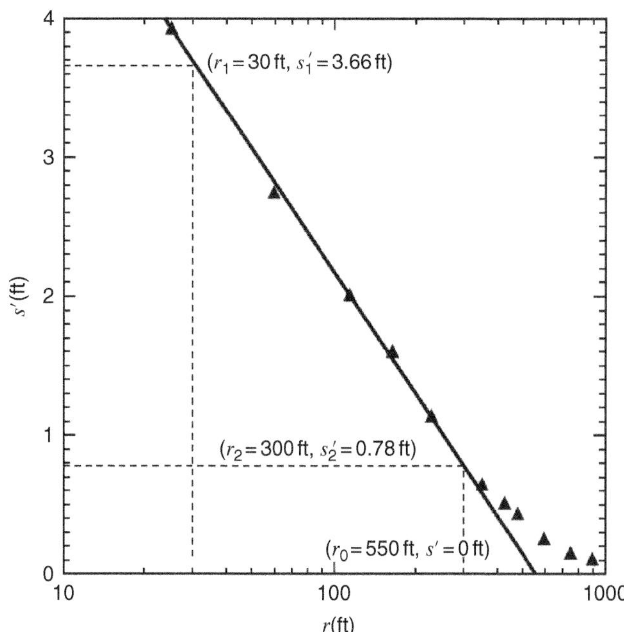

Two points are selected on the straight line: $r_1 = 30$ ft, $s_1{}' = 3.66$, $r_2 = 300$ ft, and $s_2{}' = 0.78$ ft. The transmissivity is

$$T = \frac{2.3\left(1.04 \times 10^5\ \text{ft}^3/\text{day}\right)}{2\pi(3.66 - 0.78)\text{ft}}(1) = 1.32 \times 10^4 \text{ft}^2/\text{day}$$

The specific yield is

$$S_y = \frac{2.25\left(1.32 \times 10^4\ \text{ft}^2/\text{day}\right)(2\text{day})}{(550\ \text{ft})^2} = 0.20$$

11.3 A General Solution for Drawdown

Methods for determining hydraulic parameters in an unconfined aquifer were first introduced by Boulton (1954, 1955, and 1963). They have been improved since then through the effort of numerous investigators (Boulton, 1970; Dagan, 1967; Neuman, 1972, 1973, 1974, and 1975; Prickett, 1965; Streltsova, 1972; Streltsova and Rushton, 1973). In this section, we examine three solutions that differ from each other as a function of where in the aquifer the drawdown is calculated (Figure 11.3). For fully penetrating pumping and observation wells in an unconfined aquifer (Figure 11.3), drawdown is given as

$$s = \frac{Q}{4\pi T} W(u_A, u_B, \beta) \tag{11.7}$$

where $W(u_A, u_B, \beta)$ is the well function for the unconfined aquifer, s is the drawdown, Q is the pumping rate, and T is the transmissivity (Neuman, 1975). The dimensionless parameters (u_A, u_B, and β) are defined as

$$\frac{1}{u_A} = \frac{Tt}{Sr^2} \text{(for early time data)} \tag{11.8}$$

$$\frac{1}{u_B} = \frac{Tt}{S_y r^2} \text{(for later data)} \tag{11.9}$$

$$\beta = \frac{K_z r^2}{K_r b^2} \tag{11.10}$$

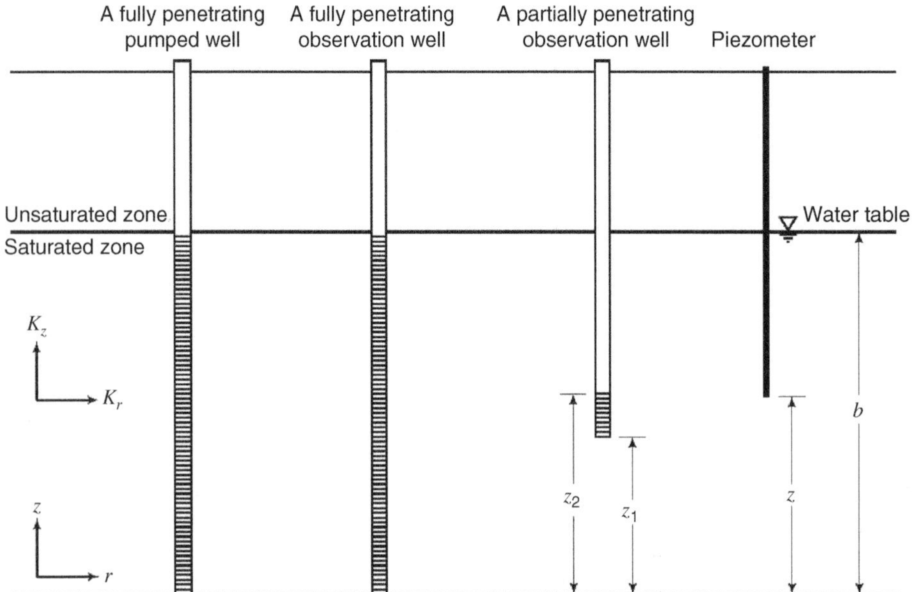

Figure 11.3 This sketch shows a fully penetrating pumping well along with various types of observation wells in an unconfined aquifer (Schwartz and Zhang, 2003 / John Wiley & Sons).

where b is the initial saturated thickness, K_r is the hydraulic conductivity in the radial horizontal direction, and K_z is the vertical hydraulic conductivity. Neuman (1975) provides tables for the function $W(u_A, u_B, \beta)$.

When drawdown is calculated for a piezometer in an unconfined aquifer that is being pumped at a constant rate by a fully penetrating well, the solution depends upon where in the aquifer the piezometer is completed (Figure 11.3). Close to the well, the head changes vertically because there is a downward flow component. This analytical solution is

$$s = \frac{Q}{4\pi T} W(u_A, u_B, \beta, z_D) \tag{11.11}$$

where $z_D = z/b$, z is the coordinate of the piezometer, and b is the thickness of the aquifer. The drawdown in an observation well is the average of the solution expressed by Eq. (11.11) along the screen.

$$W(u_A, u_B, \beta, z_{D_1}, z_{D_2}) = \frac{Q}{4\pi T} \frac{1}{z_{D_1} - z_{D_2}} \int_{z_{D_1}}^{z_{D2}} W(u_A, u_B, \beta, z_D) dz_D \tag{11.12}$$

where $z_{D_1} = z_1/b$, $z_{D_2} = z_2/b$, and z_2 and z_1 are the top and bottom coordinates of the observation well, respectively. Neuman (1972, 1973, 1974, and 1975) tabulates these well functions.

The evaluation of the Neuman's analytical solutions requires a large amount of computational time as noted by Moench (1993). To improve the efficiency and accuracy of the calculation, Moench (1993, 1995, and 1996) computed Neuman's solutions by numerical inversion of the Laplace transform solution (Moench and Ogata, 1984; Stehfest, 1970).

The solutions expressed by Eqs. (11.7), (11.11), and (11.12) are based on the following assumptions (Neuman, 1974; Moench, 1993): (1) the aquifer is homogeneous, infinite in extent, with the principal directions of the hydraulic conductivity tensor oriented parallel to the coordinate directions; (2) the aquifer is bounded by a confining layer at the bottom and a free surface at the top; (3) the diameters of the pumping well, observation wells or piezometers are infinitesimal; (4) the well is pumped at a constant rate; (5) water is released instantaneously in a vertical direction from a zone above the water table in response to decline in the elevation of the water table; (6) the change of the saturated thickness is small compared to the initial saturated thickness.

11.4 Type-Curve Method

There are two common approaches for using aquifer-test data to determine the hydraulic parameters of an unconfined aquifer. Both apply to the case of fully penetrating well and observation wells. The first is a curve-fitting procedure, which we discuss in this section. The simpler straight-line procedure is explained in Section 11.5.

The curve-fitting approach is quite similar to that used for confined aquifers. Added complexity stems from the fact that the analysis uses the early time part of the response (including the delayed drainage) and the later time part separately. The steps in the approach are summarized by Boulton (1963), Prickett (1965), and Neuman (1975).

1) Begin by finding the appropriate type curve. The Neuman (1975) curves $W(u_A, u_B, \text{and } \beta)$ versus $1/u_A$, and $1/u_B$ work in this case (Figure 11.4). The early/intermediate and late sequences of curves are referred to as the Type-A and the Type-B curves, respectively. The Type-A curves lie to the left of the labels for the β values (Figure 11.4). The Type-B curves lie to the right. The Type-A and Type-B curves are connected by horizontal asymptotes. The distance between the Type-A and Type-B curves depends on the ratio S/S_y. When this ratio is zero, the two sets of curves are located at the two ends of infinite horizontal asymptotes. For this reason, the two curves are usually plotted on different scales. The top horizontal coordinate axis, $1/u_A$, is the horizontal axis for Type-A curves and the bottom horizontal coordinate axis, $1/u_B$, is the horizontal axis for Type-B curves.
2) Plot the field drawdown (s) versus time (t) on (Neuman, 1974)semi-transparent log–log graph paper.
3) Select a convenient match point on the type curves and superimpose the s–t curve on the type-B curves $W(u_B, \beta)$. The match provides values of $W(u_B, \beta)$, $1/u_B$, s, and t.
4) The transmissivity of the aquifer is determined from

$$T = \frac{Q}{4\pi} \frac{W(u_B, \beta)}{s} \tag{11.13}$$

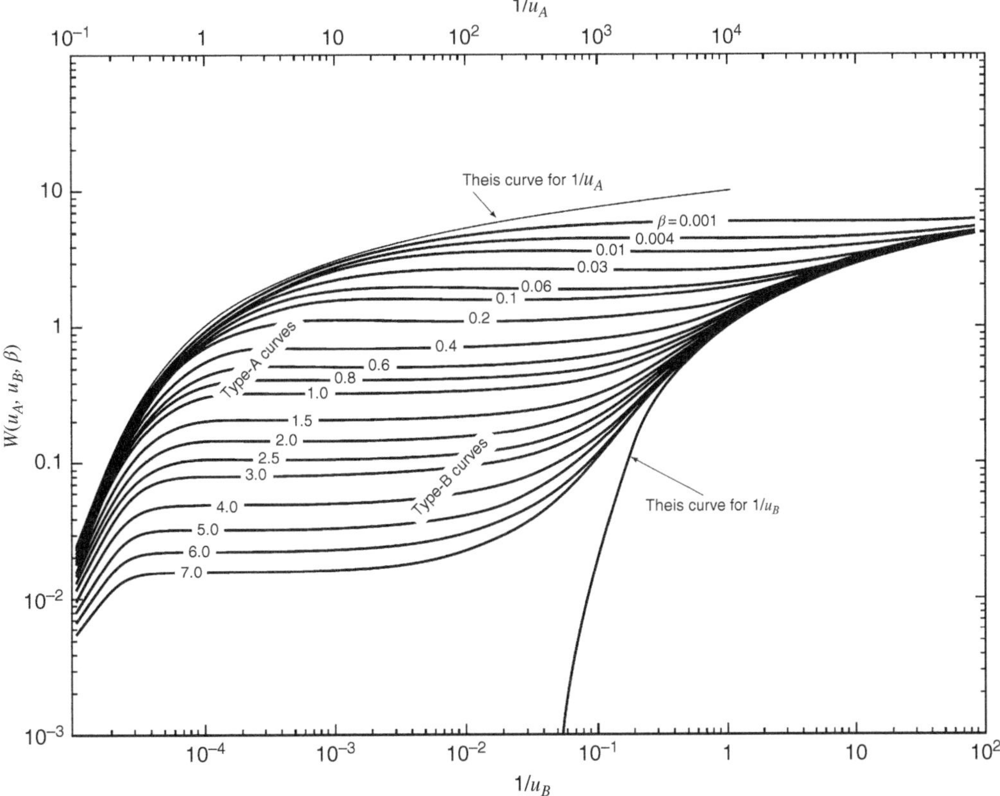

Figure 11.4 Type curves for drawdown in an unconfined aquifer with fully penetrating wells (Neuman, 1975. Water Resources Research 11(2): 329–342. Copyright 1975 by the American Geophysical Union).

5) From Eq. (11.9), the specific yield is

$$S_y = \frac{Ttu_B}{r^2} \tag{11.14}$$

6) Superimpose a field curve of early drawdown versus time curve on the type-A curves.
7) The match provides values $W(u_A, \beta)$, $1/u_A$, s, and t values as well as a β value from one of the family of curves. The transmissivity value can be calculated from Eq. (11.13) and should be very close to that calculated from step (4). The storativity is

$$S = \frac{Ttu_A}{r^2} \tag{11.15}$$

8) The horizontal hydraulic conductivity is the ratio of transmissivity over aquifer thickness ($K_r = T/b$). The vertical hydraulic conductivity is given by

$$K_z = \frac{\beta K_r b^2}{r^2} \tag{11.16}$$

Example 11.3 A test was conducted with an unconfined aquifer near Fairborn Ohio (Lohman, 1972). The well was pumped at a constant rate of 1080 gpm. The drawdowns, measured in an observation well 73 ft away, are listed in Table 11.2. The aquifer thickness is 78 ft. Assuming fully penetrating wells, calculate the hydraulic parameters for the aquifer using the type-curve method.

Table 11.2 Drawdowns data were obtained from an observation well 73 ft away from the pumping well.

Time(min)	s(ft)	Time(min)	s(ft)	Time(min)	s(ft)
0.165	0.12	2.5	0.91	60	1.22
0.25	0.195	2.65	0.92	70	1.25
0.34	0.255	2.8	0.93	80	1.28
0.42	0.33	3	0.94	90	1.29
0.5	0.39	3.5	0.95	100	1.31
0.58	0.43	4	0.97	120	1.36
0.66	0.49	4.5	0.975	150	1.45
0.75	0.53	5	0.98	200	1.52
0.83	0.57	6	0.99	250	1.59
0.92	0.61	7	1	300	1.65
1	0.64	8	1.01	400	1.7
1.08	0.67	9	1.015	500	1.85
1.16	0.7	10	1.02	600	1.95
1.24	0.72	12	1.03	700	2.01
1.33	0.74	15	1.04	800	2.09
1.42	0.76	18	1.05	900	2.15
1.5	0.78	20	1.06	1000	2.2
1.68	0.82	25	1.08	1200	2.27
1.85	0.84	30	1.13	1500	2.35
2	0.86	35	1.15	2000	2.49
2.15	0.87	40	1.17	2500	2.59
2.35	0.9	50	1.19	3000	2.66

Source: U.S. Geological Survey, adapted from Lohman, 1972/Public domain.

Solution

The drawdown (s) versus time (t) on a log–log scale is shown in Figure 11.5. The curve is first matched with Type B in Figure 11.6. At the match point, $1/u_B = 1$, $W(u_B, \beta) = 1$, $t = 28$ min, $s = 0.47$ ft. The pumping rate is converted from gpm to ft^3/day using Eq. (9.1).

$$Q = 1080\,\text{gpm} = 1080 \times 192.5\text{ft}^3/\text{day} = 2.08 \times 10^5\text{ft}^3/\text{day}$$

The transmissivity (T) is obtained by

$$T = \frac{Q}{4\pi}\frac{W(u_B,\beta)}{s} = \frac{(2.08 \times 10^5\ \text{ft}^3/\text{day})}{4\pi}\frac{1}{0.47\ \text{ft}} = 3.5 \times 10^4\text{ft}^2/\text{day}$$

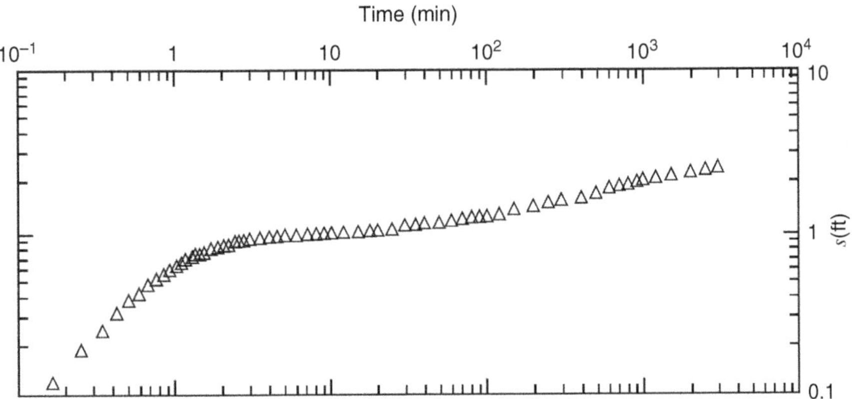

Figure 11.5 Drawdown (s) versus time (t) data plotted on a log-log scale for an aquifer test near Fairborn, Ohio (U.S. Geological Survey, adapted from Lohman, 1972/Public domain).

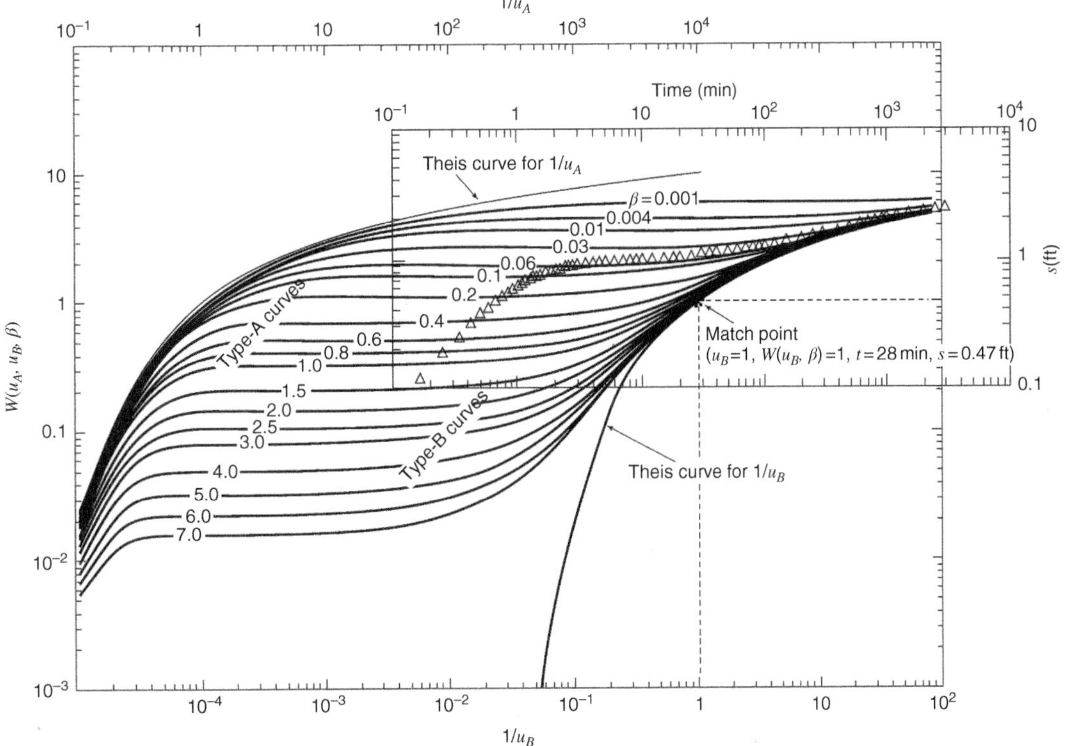

Figure 11.6 The match of the late-time data with the Type-B curve for an aquifer test near Fairborn, Ohio (U.S. Geological Survey, adapted from Lohman, 1972/Public domain).

and the specific yield is

$$S_y = \frac{Tt}{r^2}u_B = \frac{(3.5 \times 10^4\ \text{ft}^2/\text{day})(28\ \text{min})\left(\dfrac{1}{1440}\ \dfrac{\text{day}}{\text{min}}\right)}{(73\ \text{ft})^2}(1) = 0.13$$

Superimpose the early-time drawdown curve on the type A curves and determine the match point coordinates (Figure 11.7). At the match point, record $1/u_A = 1$, $W(u_A, \beta) = 1$, $t = 0.57$ min, and $s = 0.5$ ft. The field data match the $\beta = 0.06$ curve. From Eq. (11.15),

$$S = \frac{Tt}{r^2}u_A = \frac{3.5 \times 10^4\ \text{ft}^3/\text{day}\ (0.57\ \text{min})\left(\dfrac{1}{1440}\ \dfrac{\text{day}}{\text{min}}\right)}{(73\ \text{ft})^2}(1) = 2.6 \times 10^{-3}$$

and the vertical hydraulic conductivity is

$$K_z = \frac{\beta K_r b^2}{r^2} = \frac{(0.06)\left(\dfrac{3.5 \times 10^4\ \text{ft}}{78}\ \text{ft/day}\right)(78\ \text{ft})^2}{(73\ \text{ft})^2} = 31\ \text{ft/day}$$

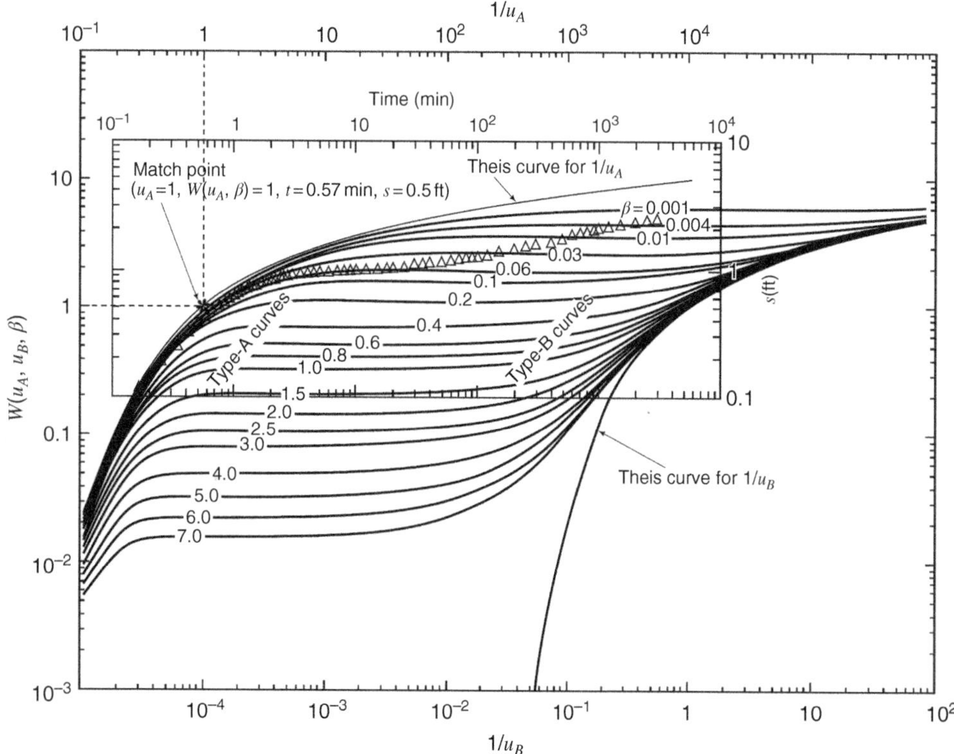

Figure 11.7 The match of the early-time data with the Type-A curves for an aquifer test near Fairborn, Ohio (U.S. Geological Survey, adapted from Lohman, 1972/Public domain).

11.5 Straight-Line Method

Neuman (1975) illustrated a somewhat simpler method for determining the hydraulic parameters of an unconfined aquifer by straight-line fits to early, intermediate, and late-time drawdown data. The scheme is summarized as follows.

1) Plot values of drawdown versus time on semi-log graph paper.
2) From the late-time segment of the s-log (t) curve, the transmissivity and specific yield of an unconfined aquifer are determined as

$$T = \frac{2.3\,Q}{4\pi\Delta s'} \tag{11.17}$$

$$S_y = \frac{2.25\,Tt_{0y}}{r^2} \tag{11.18}$$

3) From the early-time segment of the s-log (t) curve, the transmissivity and specific yield of an unconfined aquifer are determined by

$$T = \frac{2.3\,Q}{4\pi\Delta s'} \tag{11.19}$$

$$S = \frac{2.25\,Tt_{0S}}{r^2} \tag{11.20}$$

4) Fit a straight line through the intermediate part of the s-log (t) curve. The line intercepts the Theis curve at $t = t_\beta$. If the type curves are prepared as a semi-logarithmic plot, the intermediate part of the Type A and B curves form a family of straight lines (Figure 11.8). The interception of the straight line and the Theis curve for $1/u_B$ is at $u_{B\beta}$. Neuman (1975) indicated that the $u_{B\beta}$ is related to β by (Neuman, 1975)

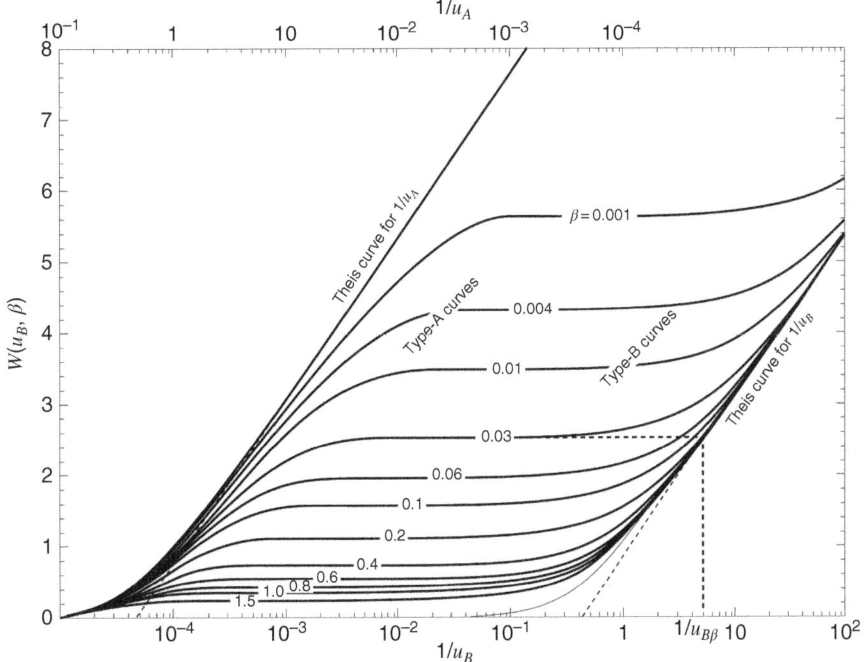

Figure 11.8 Type-A and Type-B curves plotted on a semi-logarithmic scale (Neuman, 1975. © by John Wiley & Sons.).

$$\beta = 0.195 u_{B\beta}^{1.1053} = 0.195 \left[\frac{S_y r^2}{T t_\beta}\right]^{1.1053} \quad \text{for } 4 \le u_{B\beta} \le 100 \tag{11.21}$$

Once β is known, the vertical hydraulic conductivity is determined by

$$K_z = \frac{\beta K_r b^2}{r^2} = 0.195 \left[\frac{S_y r^2}{T t_\beta}\right]^{1.1053} \frac{K_r b^2}{r^2} \tag{11.22}$$

However, Eq. (11.7) was derived from the assumption that the decline of the water table is small in comparison to the saturated thickness of an unconfined aquifer. For a large drawdown, Jacob's correction (11.2) should be used with the late drawdown measurements.

Example 11.4 Let us apply this method to the data from the aquifer test near Fairborn Ohio (Lohman, 1972). Information about the test was presented in Example 11.3. Determine hydraulic parameters for the aquifer using Neuman's straight-line method.

Solution

Values of drawdown versus time are plotted on semi-log paper in Figure 11.9. The plot reflects the three types of drawdown responses that we described in the introduction. At early time, the aquifer behaves as though it is confined with data falling along a straight line. At some intermediate time, it deviates from this straight line due to delayed gravity drainage. At late time, data again follows a straight line as the aquifer exhibits a Theis-type response with S equivalent to S_y.

The transmissivity and specific yield of the aquifer are determined by fitting a straight line through the late-time data. The early-time response provides another estimate of transmissivity and a confined storativity value for the aquifer can be obtained. The pumping rate is converted from gpm to ft³/day using Eq. (9.1).

$$Q = 1080 \text{ gpm} = 1080 \times 192.5 \text{ft}^3/\text{day} = 2.08 \times 10^5 \text{ft}^3/\text{day}$$

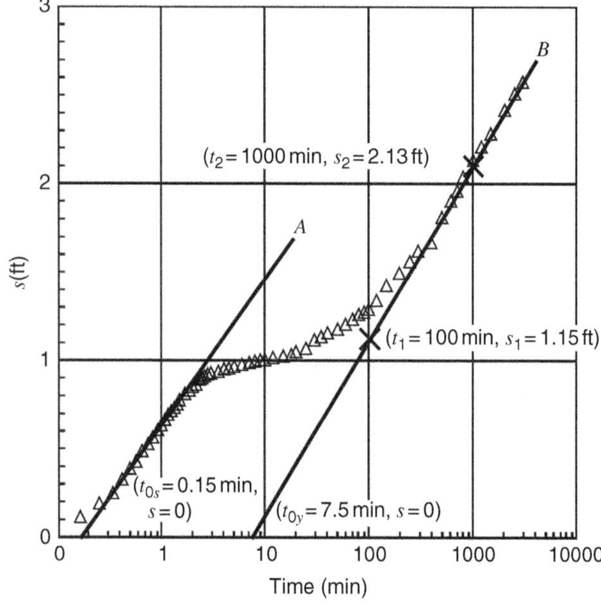

Figure 11.9 Drawdown (s) versus time (t) data plotted on a semi-log scale for an aquifer test near Fairborn, Ohio (U.S. Geological Survey, adapted from Lohman, 1972/Public domain). Straight lines fitted through the data as shown form the basis for the calculation of hydraulic parameters.

From Eq. (9.15), the transmissivity of the aquifer is

$$T = \frac{2.3\,Q}{4\pi\Delta s} = \frac{2.3\left(2.08\times10^{5}\ \text{ft}^{3}/\text{day}\right)}{4\pi(2.13-1.15)\ \text{ft}} = 3.9\times10^{4}\text{ft}^{2}/\text{day}$$

From Eq. (9.16), the specific yield is

$$S_y = \frac{2.25\,Tt_{0y}}{r^2} = \frac{2.25\left(4.1\times10^{4}\ \text{ft}^{2}/\text{day}\right)\left(7.5\ \text{min}\ \frac{1\ \text{day}}{1440\ \text{min}}\right)}{(73\ \text{ft})^{2}} = 0.09$$

From Eq. (9.18), the confined storativity is

$$S = \frac{2.25\,Tt_{0S}}{r^2} = \frac{2.25\left(4.1\times10^{4}\ \text{ft}^{2}/\text{day}\right)\left(0.15\ \text{min}\ \frac{1\ \text{day}}{1440\ \text{min}}\right)}{(73\ \text{ft})^{2}} = 1.8\times10^{-3}$$

β is calculated by

$$\beta = 0.195\left[\frac{(0.09)(73\ \text{ft})^{2}}{\left(4.1\times10^{4}\ \text{ft}^{2}/\text{day}\right)(80\ \text{min})\left(\frac{1}{1440}\ \frac{\text{day}}{\text{min}}\right)}\right]^{1.1053} = 0.035$$

The vertical hydraulic conductivity is

$$K_z = \frac{\beta K_r b^2}{r^2} = \frac{(0.035)\left(4.1\times10^{4}\ \text{ft}^{2}/\text{day}\right)(78\ \text{ft})}{(73\ \text{ft})^{2}} = 21\ \text{ft}/\text{day}$$

11.6 Aquifer Testing with a Partially-Penetrating Well

The well function in Figure 11.4 is valid for fully penetrating discharge and observation wells in an unconfined aquifer. Similar type curves can be calculated for a partially-penetrating discharge well (Figure 11.10). The type-curve and straight-line methods will apply to a hydraulic testing with partially-penetrating wells when type curves A and B are recalculated to account for the effects of partial penetration.

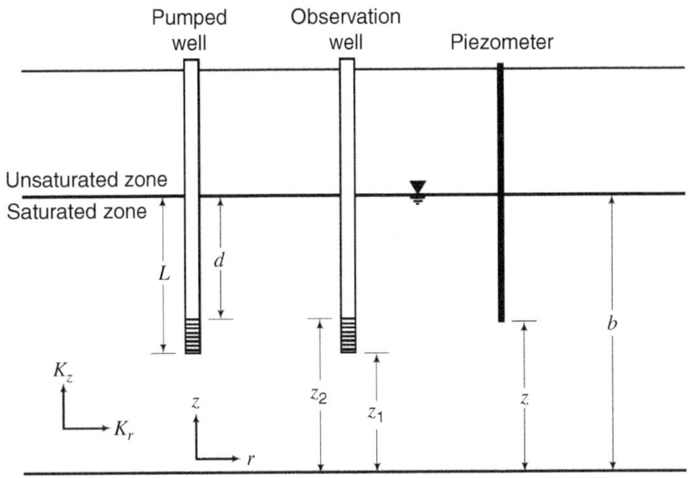

Figure 11.10 Sketch showing partially penetrating pumping and observation wells in an unconfined aquifer (Schwartz and Zhang, 2003 / John Wiley & Sons).

Exercises

11.1 Drawdown versus distance data are provided in Table 11.3 for an aquifer test near Grand Island, Nebraska. The data were collected after 48 hours of continuous pumping at 540 gpm (Wenzel, 1936). Calculate hydraulic parameters using Jacob's correction Eq. (11.2). The saturated thickness of the aquifer is 100 ft.

Table 11.3 Values of drawdown versus distance for an aquifer test near Grand Island, Nebraska after 48 hours of continuous pumping at 540 gpm.

r(ft)	*s*(ft)
40.1	3.15
95.1	2.24
144.7	1.71
214	1.24
324	0.77
423	0.51
448	0.46
573	0.28
73	0.15
872	0.10
1073	0.06
1197	0.05

Source: U.S. Geological Survey, adapted from Wenzel, 1936/Public domain.

11.2 Table 11.4 lists drawdown versus time data collected at an observation well 50 m away from a well being pumped at 2 m^3/min. The pumped well and observation wells are fully penetrating. The original saturated thickness of the aquifer is 50 m. Calculate the hydraulic parameters using the type-curve method. You will find a template file for this exercise in the directory Chapter 11 on the enclosed CD. The template file "temp-11a.doc" contains type curves for fully penetrating wells in an unconfined aquifer. The template file "temp-11.doc" in the directory Chapter 11 provides log–log paper at the same scale as the type curves. With the help of these template files, determine hydraulic parameters for the aquifer.

Table 11.4 Drawdown at an observation of 50 m away from a well pumped at 2 m^3/min.

Time(min)	*s*(m)	Time(min)	*s*(m)	Time(min)	*s*(m)	Time(min)	*s*(m)
0.10	0.00	19.36	1.06	1.39	0.64	269.34	1.12
0.13	0.00	25.19	1.06	1.81	0.75	350.47	1.14
0.17	0.01	32.78	1.06	2.36	0.85	456.03	1.16
0.22	0.02	42.65	1.06	3.07	0.93	593.38	1.19
0.29	0.06	55.49	1.07	3.99	0.99	772.10	1.24
0.37	0.11	72.21	1.07	5.19	1.04	1004.66	1.29
0.49	0.18	93.96	1.08	6.75	1.05	1307.26	1.36
0.63	0.28	122.26	1.08	8.79	1.05	1701.01	1.44
0.82	0.39	159.08	1.09	11.43	1.06	2213.34	1.55
1.07	0.51	206.99	1.10	14.88	1.06	2880.00	1.68

Source: Schwartz and Zhang (2003) / John Wiley & Sons.

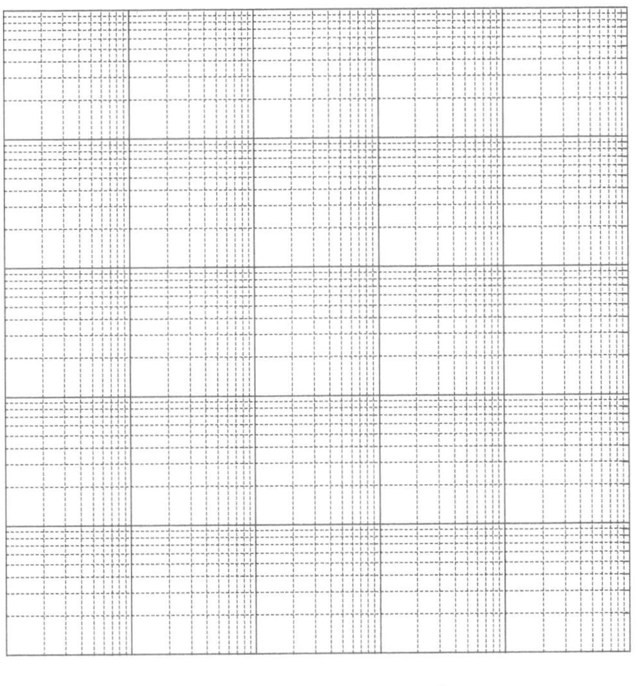

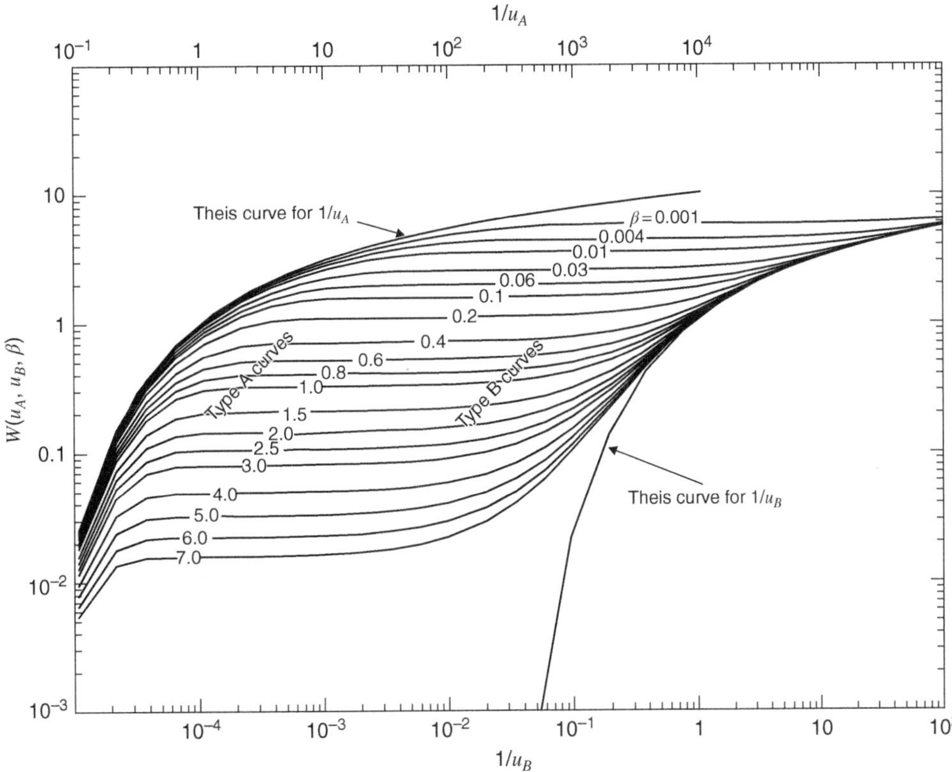

11.3 Recalculate the hydraulic parameters for the test data in Table 11.4 using the straight-line method.

References

Boulton, N. S. 1954. The drawdown of the water table under nonsteady conditions near a pumped well in an unconfined formation. Proceedings of the Institution of Civil Engineers, v. 3, no. 4, p. 564–579.

Boulton, N. S. 1955. Unsteady radial flow to a pumped well allowing for delayed yield from storage. International Association of Hydrological Sciences, v. 37, p. 472–477.

Boulton, N. S. 1963. Analysis of data from nonequilibrium pumping test allowing for delayed yield from storage. Proceedings of the Institution of Civil Engineers, v. 26, no. 3, p. 469–482.

Boulton, N. S. 1970. Analysis of data from pumping test in unconfined anisotropic aquifers. Journal of Hydrology, v. 10, no. 4, p. 369.

Dagan, G. 1967. A method of determining the permeability and effective porosity of unconfined anisotropic aquifers. Water Resourses Research, v. 3, no. 4, p. 1059–1071.

Lohman, S. W. 1972. Groundwater hydraulics. U.S. Geological Survey Professional Paper 708, 70 p.

Moench, A. F. 1993. Computation of type curves for flow to partially penetrating wells in water-table aquifers. Groundwater, v. 31, p. 966–971.

Moench, A. F. 1995. Combining the Neuman and Boulton models for flow to a well in an unconfined aquifer. Groundwater, v. 33, p. 378–384.

Moench, A. F. 1996. Flow to a well in a water-table aquifer: an improved Laplace transform solution. Groundwater, v. 34, p. 593–596.

Moench, A., and A. Ogata. 1984. Analysis of constant discharge wells by numerical inversion of Laplace transform solutions, In J. S. Rosenshein, and G. D. Bennett (eds.), Groundwater Hydraulics: Water Resources Monograph 9. American Geophysical Union, Washington D.C., p. 146–170.

Neuman, S. P. 1972. Theory of flow in unconfined aquifers considering delayed response of the water table. Water Resources Research, v. 8, no. 4, p. 1031–1045.

Neuman, S. P. 1973. Supplementary comments on 'Theory of flow in unconfined aquifers considering delayed response of the water table'. Water Resources Research, v. 9, no. 4, p. 1102–1103.

Neuman, S. P. 1974. Effect of partial penetration on flow in unconfined aquifers considering delayed response of the water table. Water Resources Research, v. 9, no. 2, p. 1102–1103.

Neuman, S. P. 1975. Analysis of pumping test data from anisotropic unconfined aquifers considering delayed gravity response. Water Resources Research, v. 11, no. 2, p. 329–342.

Peters, J. G. 1987. Description and comparison of selected models for hydrologic analysis of ground-water flow, St. Joseph River Basin, Indiana. U.S. Geological Survey Water-Resources Investigations Report 86-4199, p. 125.

Prickett, T. A. 1965. Type curve soslution to aquifer tests under water table conditions. Groundwater, v. 3, no. 3, p. 5–14.

Schwartz, F. W., and H. Zhang. 2003. Fundamentals of Groundwater. John Wiley & Sons, Hoboken, New Jersey, 583 p.

Stehfest, H. 1970. Numerical inversion of Laplace tranforms. Communications of. ACM., v. 13, no. 1, p. 47–49.

Streltsova, T. D. 1972. Unsteady radial flow in an unconfined aquifer. Water Resources Research, v. 8, no. 4, p. 1059–1066.

Streltsova, T. D., and K. Rushton. 1973. Water table drawdown due to a pumped well. Water Resources Research, v. 9, no. 1, p. 236–242.

Weeks, E. P. 1969. Determining the ratio of horizontal to vertical permeability by aquifer test analysis. Water Resources Research, v. 5, no. 1, p. 196–214.

Wenzel, L. K. 1936. The Thiem method for determining permeability of water-bearing materials and its application to the determination of specific yield, results of investigations in the Platte River Valley, Nebraska. U.S. Geological Survey Water-Supply Paper 679-A, 57 p.

12

Slug, Step, and Intermittent Tests

CHAPTER MENU

The "slug test" is a method for determining the hydraulic conductivity of a geological unit using an observation well or piezometer. It involves displacing the water level in a well away from some equilibrium position by adding water or withdrawing a *slug* from the well, which causes an immediate decline in water level (Figure 12.1). A slug is a solid cylinder of metal 5 ft or more long that can be hung in a well on a cable for an appropriate time period before the test. Upon removal, the water level is lowered downward instantaneously from its equilibrium level. Monitoring the return of the water level back to the pretest level provides the data for the analysis. Typically, water levels can be measured using an electric tape. In some cases, where manual measurements cannot be made quickly enough, electrical monitoring systems are used. When water-quality sampling is contemplated, it is advisable not to add water to the well in a slug test.

Hvorslev originally developed this method in 1951. Since then, more rigorous variants of the methods have been derived for a wide range of test conditions (Cooper et al., 1967; Papadopulos and Cooper, 1967; Papadopulos et al., 1973; Bouwer and Rice, 1976; Bouwer, 1989). In practice, these tests are used when quick or inexpensive estimates of hydraulic conductivity are required (Thompson, 1987). They are much simpler than a conventional aquifer test and will work with relatively small diameter wells or piezometers.

Hydraulic conductivity values obtained from this test are considered to be less representative than an aquifer test. Much small volumes of water are displaced in a slug test as compared to a conventional aquifer test. Thus, the slug test reflects the hydraulic conductivity of a small volume of the medium near the well. Also, most slug tests do not provide estimates of storativity.

12.1 Hvorslev Slug Test

In the classical Hvorslev approach, the deviations in water level from the pretest water level are measured in terms of a so-called drawdown ratio (H_t). The *drawdown ratio* is the ratio of the drawdown at any time t to the maximum drawdown when the test is begun or $H_t = s_t/s_0$ (Figure 12.1). In effect, using the drawdown ratios normalizes the drawdown between zero and one. Hvorslev found that the return of the water level to equilibrium is exponential—the change per unit time starts out relatively large and slows down. However, the time required to return to equilibrium depends on the hydraulic conductivity. In high K units, water levels return to equilibrium in seconds or minutes. In low K units, the recovery might take months. The recovery rate also depends on how the piezometer is designed. Wells having a large area for water to enter the casing recover more rapidly than wells with a small open area. Thus, the equation for calculating hydraulic

Fundamentals of Groundwater, Second Edition. Franklin W. Schwartz and Hubao Zhang.

Companion website: www.wiley.com/go/schwartz/fundamentalsofgroundwater2

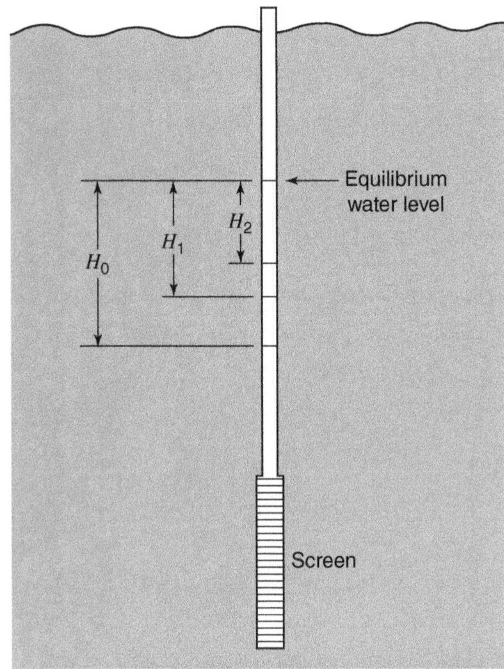

Figure 12.1 In a slug test, the water level in a well or piezometer is displaced away from its equilibrium position. The return of water levels to equilibrium is monitored as a function of time (Adapted from Thompson, 1987. Reprinted by permission of the National Groundwater Assoc. © 1989. All rights reserved).

conductivity must account for the construction details of the well in interpreting the recovery rates. Look carefully at the following form of Hvorslev's equation

$$K = \frac{A}{F}\frac{1}{t_2 - t_1}\ln\frac{H_1}{H_2} \tag{12.1}$$

where K is the hydraulic conductivity of an aquifer, A is the cross-section area of the well, and F is a shape factor related to the list in Table 12.1 (U. S. Department of Navy, 1992), H_1 and H_2 are the draw-down ratios at time t_1 and t_2, respectively. Information about the well construction is contained in the shape factor F and the area A. The shape factor, as will be evident, can take on different forms as the pictures in Table 12.1 show. Thus, Eq. (12.1) takes on different forms depending upon the shape factor. To find the correct equation, you need to go to Table 12.1 and find the well design that best matches your piezometer or well. Table 12.1 does not contain all the well designs one might encounter. Readers interested in seeing the complete suite of shape factors should refer to Cedergren (1967).

The steps in running the test and determining hydraulic conductivity using the Hvorslev method are summarized as follows:

1) Before displacing the water level from equilibrium, record the initial depth of water. The test begins when the water level is changed by adding water or removing a slug from the well. Measure the water level periodically as the water level returns to the pretest level. Keep collecting data until 90% plus of the initial waterlevel displacement is recovered.
2) From the water-level measurements, calculate drawdowns, s_0, s_1, ... s_n. Determine the drawdown ratios H_0, H_1, ... H_n by dividing each of the drawdowns by s_0, the drawdown (maximum) at t_1, for example, $H_1 = s_1/s_0$.
3) Plot the head ratios (H_t) on the log scale and time (t) on the linear scale of semi-log graph paper.
4) Fit the best straight line through the set of data points.
5) Choose two points on the straight line and record t_1, H_1, t_2, and H_2.
6) Calculate K using the appropriate form of the equation. In Table 12.1, D is the depth of the well measured from water table to the bottom of the well, L is the length of well screen or well open, S is the thickness of saturated permeable material above an underlying confining bed, and T is the thickness of the confined aquifer. The C_s can be determined as (Thompson, 1987)

$$C_s = \frac{2\pi(L/R)}{\ln(L/R + 1.36)} \tag{12.2}$$

This example illustrates how slug-test data are interpreted to provide an estimate of K.

Example 12.1 The slug test data in Table 12.2 were reported by Cooper et al. (1967). The well has a diameter of 7.6 cm with an open hole 98 m in length. The design is similar to F-3 in Table 12.1. Calculate hydraulic conductivity using Hvorslev's method.

Solution

The working equation comes from Table 12.1 and is written as

$$K = \frac{R^2\ln(R_0/\mathrm{R})}{2L_3(t_2 - t_1)}\ln\frac{H_1}{H_2} \tag{12.3}$$

Table 12.1 Shape factors, calculational equations, and notes on applicability for various piezometer designs.

	Condition	Diagram	Shape factor, F	Permeability, K by variable head test	Applicability
Observation well or piezometer in saturated isotropic stratum of infinite depth	A) Uncased hole	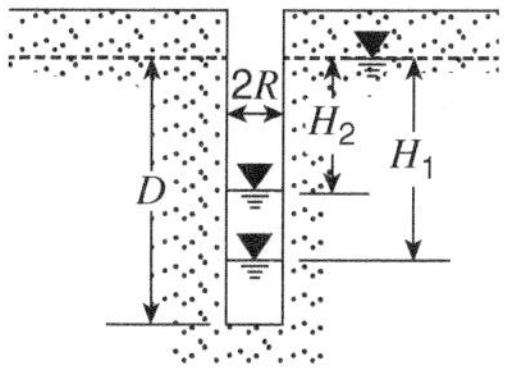 $2R$, H_2, H_1, D	$F = 16\pi DSR$	(For observation well of constant cross section) $K = \frac{R}{16DS} \times \frac{(H_2 - H_1)}{(t_2 - t_1)}$ For $\frac{D}{R} < 50$	Simplest method for permeability determination. Not applicable in stratified soils
	B) Cased hole, soil flush with bottom	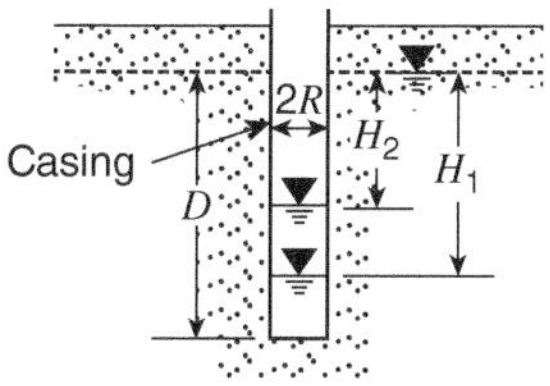 Casing, $2R$, H_2, H_1, D	$F = \frac{11R}{2}$	$K = \frac{2\pi R}{11(t_2 - t_1)} \ln \frac{(H_1)}{(H_2)}$ For $6'' \leq D \leq 60''$	Used for permeability determination at shallow depths below the water table. May yield unreliable results in falling head test with silting of bottom of hole
	C) Cased hole, uncased or performed extension of length "L"	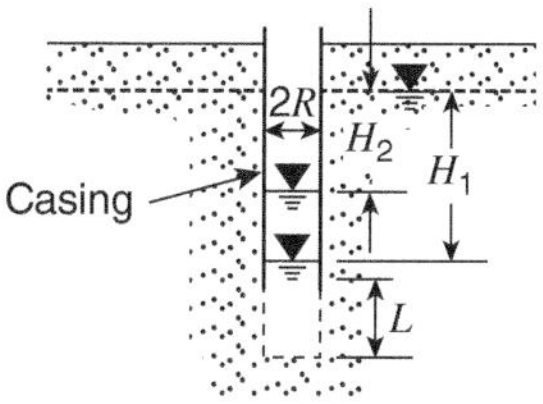 Casing, $2R$, H_2, H_1, L	$F = \frac{2\pi L}{\ln\left(\frac{L}{R}\right)}$	$K = \frac{R^2}{2L(t_2 - t_1)} \ln \frac{(L)}{(R)} \ln \frac{(H_1)}{(H_2)}$ For $\frac{L}{R} > 8$	Used for permeability determination at greater depths below water table
	D) Cased hole, opening flush with upper boundary of aquifer of infinity depth	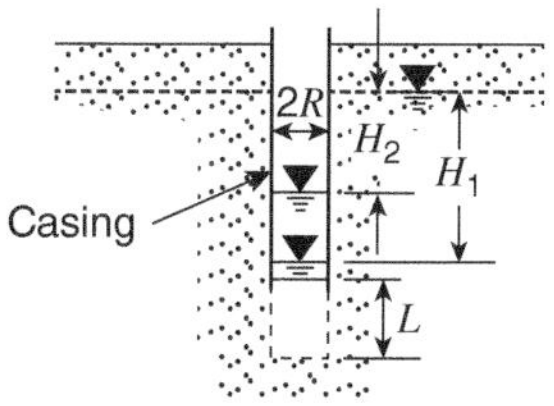 Casing, $2R$, H_2, H_1, L	$F = \frac{11\pi R^2}{2\pi R + 11L}$	$K = \frac{2\pi R + 11L}{11(t_2 - t_1)} \ln \frac{(H_1)}{(H_2)}$	Principal use is for permeability in vertical direction in anisotropic soils
	E) Cased hole, opening flush with upper boundary of aquifer of infinite depth	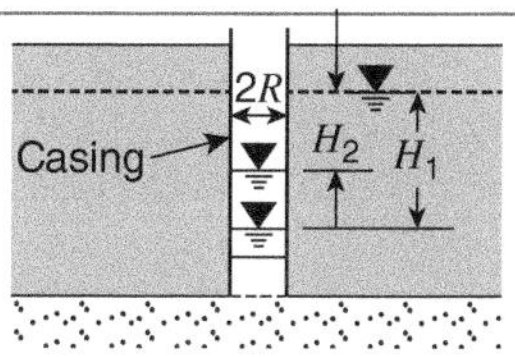 Casing, $2R$, H_2, H_1	$F = 4R$	(For observation well of constant cross section) $K = \frac{\pi R}{4(t_2 - t_1)} \ln \frac{(H_1)}{(H_2)}$	Used for permeability determination when surface impervious layer is relatively thin. May yield unreliable results in falling head test with silting of bottom of hole
Observation well or piezometer in aquifer with impervious upper layer	F) Cased hole, uncased or perforated extension into aquifer of finite thickness: 1) $\frac{L_1}{T} \leq 0.2$ 2) $0.2 < \frac{L_2}{T} < 0.85$ 3) $\frac{L_3}{T} = 1.00$ Note: R_0 equals effective radios to source at constant head	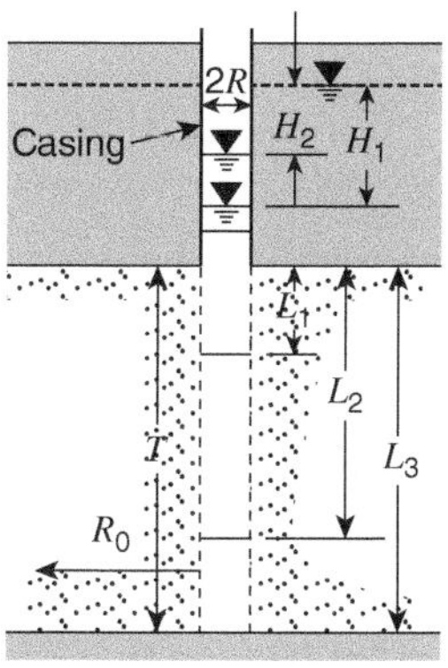 Casing, $2R$, H_2, H_1, L_1, L_2, L_3, T, R_0	(1) $F = C_s R$	$K = \frac{\pi R}{C_3(t_2 - t_1)} \ln \frac{(H_1)}{(H_2)}$	Used for permeability determinations at depths greater than about 5 ft
			(2) $F = \frac{2\pi L_2}{\ln(L_2/R)}$	$K = \frac{R^2 \ln\left(\frac{L_2}{R}\right)}{2L_2(t_2 - t_1)} \ln \frac{(H_1)}{(H_2)}$ For $\frac{L}{R} \geq 8$	Used for permeability determinations at greater depths and for fine-grained soils using porous intake point of piezometer
			(3) $F = \frac{2\pi L_3}{\ln\left(\frac{R_0}{R}\right)}$	$K = \frac{R^2 \ln\left(\frac{R_0}{R}\right)}{2L_3(t_2 - t_1)} \ln\left(\frac{H_1}{H_2}\right)$	Assume value of a $R_0/R = 200$ for estimates unless observation wells are made to determine actual value of R_0

Source: U. S. Department of the Navy, Naval Facilities Engineering Command (1992).

Table 12.2 Slug test result.

Time (sec)	s (m)	s/s_0
0	0.560	1
3	0.457	0.816
6	0.392	0.7
9	0.345	0.616
12	0.308	0.55
15	0.280	0.5
18	0.252	0.45
21	0.224	0.4
24	0.205	0.366
27	0.187	0.334
30	0.168	0.3
33	0.149	0.266
36	0.140	0.25
39	0.131	0.234
42	0.112	0.2
45	0.108	0.193
48	0.093	0.166
51	0.089	0.159
54	0.082	0.146
57	0.075	0.134
60	0.071	0.127
63	0.065	0.116

Source: Adapted from Cooper et al. (1967). Water Resources Research 3(1): 263–269. Copyright 1967 by the American Geophysical Union.

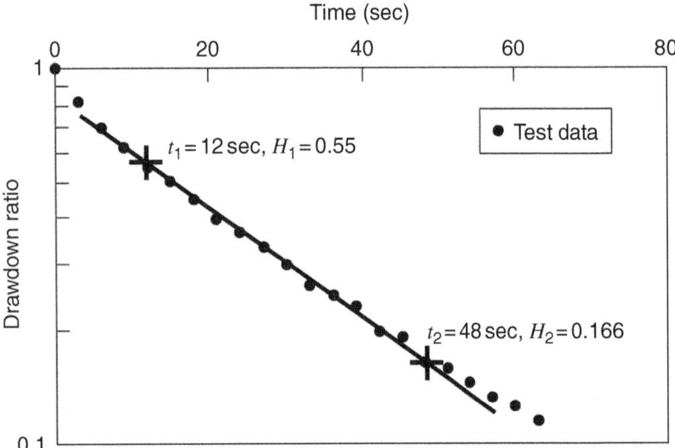

Figure 12.2 A plot of the log of the drawdown ratio versus time provides H_1, t_1 and H_2, t_2 values required to calculate hydraulic conductivity with the Hvorslev theory (Adapted from Cooper et al., 1967. Water Resources Research 3(1): 263–269. Copyright 1967 by the American Geophysical Union).

Values for L, 98 m, and r_w, 7.6 cm, are given. The value for $\ln(R_0/R)$ is given as a default parameter on the table with ln(1/200). Figure 12.2 shows the semi-log plot of the drawdown ratios versus time with a straight line fitted through the points. We selected two points on the line, $t_1 = 12$ sec, $H_1 = 0.55$ m, $t_2 = 48$ sec, and $H_2 = 0.166$ m. Thus,

$$K = \frac{(7.6^2\text{cm}^2)(\ln(200))}{2(9800\text{ cm})(48-12)\text{ sec}} \ln\left(\frac{0.55}{0.166}\right) = 5.2 \times 10^{-4}\text{ m/sec}$$

The transmissivity is

$$T = Kb = \left(5.2 \times 10^{-4}\,\text{cm/sec}\right)(9800\,\text{cm}) = 5.1\text{cm}^2/\text{sec}$$

12.2 Cooper–Bredehoeft–Papadopulos Test

The Cooper–Bredehoeft–Papadopulos method (Cooper et al., 1967; Papadopulos and Cooper, 1967; Papadopulos et al., 1973) provides a more sophisticated approach to single-well testing. The test setup is shown in Figure 12.3. The following analytical solution provides the drawdown ratio in a confined aquifer for a fully penetrating well.

$$\frac{s}{s_0} = F(\beta, \alpha) = \frac{8\alpha}{\pi^2}\int_0^\infty \frac{e^{\beta u^2/\alpha}}{u\left([uJ_0(u) - 2\alpha J_1(u)]^2 + [uY_0(u) - 2\alpha Y_1(u)]^2\right)}du \tag{12.4}$$

where s is the head change at time t, and s_0 is the initial head change after the injection or withdrawal. The parameters β and α are expressed as

$$\beta = \frac{Tt}{r_c^2} \tag{12.5}$$

and

$$\alpha = \frac{r_s^2 S}{r_c^2} \tag{12.6}$$

where r_c is the radius of the casing, T is the transmissivity, S is the storativity, and r_s is the effective radius of the well. Tabulated values of $F(\beta, \alpha)$ are available for generating type curves (Figure 12.4). The general approach to applying the

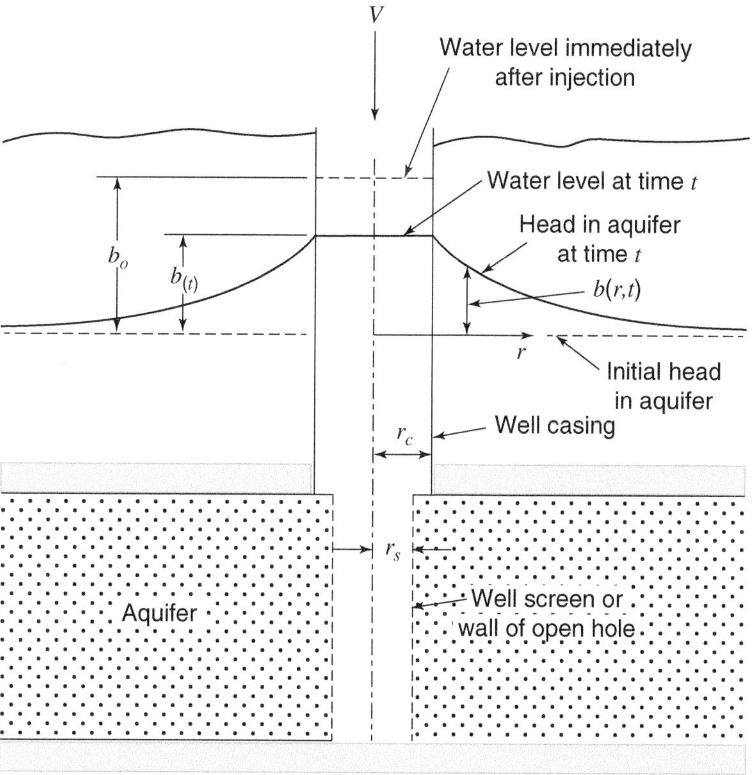

Figure 12.3 Basic geometry for the Cooper–Bredehoeft–Papadopulos slug test (Adapted from Cooper et al., 1967. Water Resources Research 3(1): 263–269. Copyright 1967 by the American Geophysical Union).

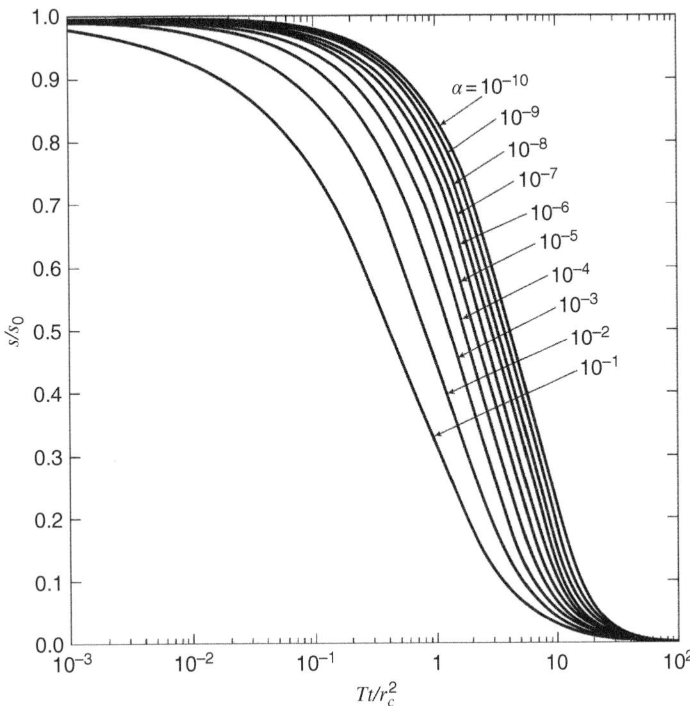

Figure 12.4 Type curves for the Cooper–Bredehoeft–Papadopulos slug test (Adapted from Cooper et al., 1967. Water Resources Research 3(1): 263–269. Copyright 1967 by the American Geophysical Union).

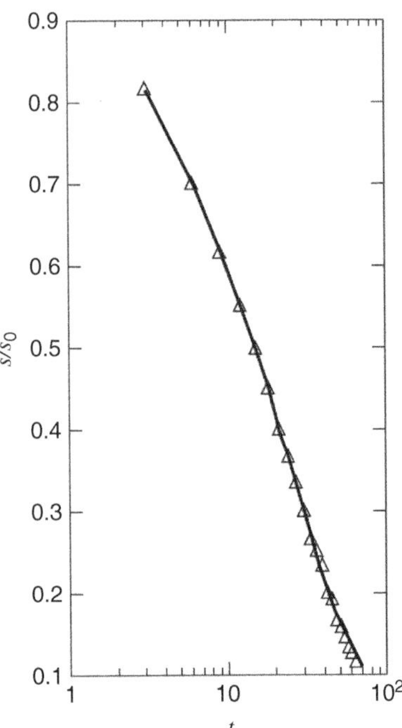

Figure 12.5 A plot of the slug-test result as the log of the drawdown ratio versus the log of time (Adapted from Cooper et al., 1967. Water Resources Research 3(1): 263–269. Copyright 1967 by the American Geophysical Union).

Cooper–Bredehoeft–Papadopulos method is similar to the curve-matching schemes of previous chapters. Here is a summary of the steps:

1) Plot the drawdown ratio on a linear scale versus time on a log scale at the same scale as the type curves.
2) Superimpose the plot of field data on the type curves and find the best fit with one of the type curves.
3) Select β (usually 1) and find the corresponding t. Transmissivity can be calculated by Eq. (12.4).

$$T = \frac{1.0r_c^2}{t} \tag{12.7}$$

4) Record α. The storativity can be calculated by Eq. (12.5).

$$S = \frac{r_c^2\alpha}{r_s^2} \tag{12.8}$$

Example 12.2 Use the data from Example 12.1 and calculate the transmissivity and the storativity by the Cooper–Bredehoeft–Papadopulos method.

Solution

The measured drawdown ratios are plotted versus time in Figure 12.5. The plot is superimposed on the type curve (Figure 12.6). The best match between the field curve and the type curves is with $\alpha = 10^{-3}$. For a β value of 1.0, the corresponding time (t) is 12 seconds. It is also known $r_c = r_s = 7.6$ cm. Thus,

$$T = \frac{1.0r_c^2}{t} = \frac{(1.0)(7.6^2\ \text{cm}^2)}{(12\ \text{sec})} = 4.8\text{cm}^2/\ \text{sec}$$

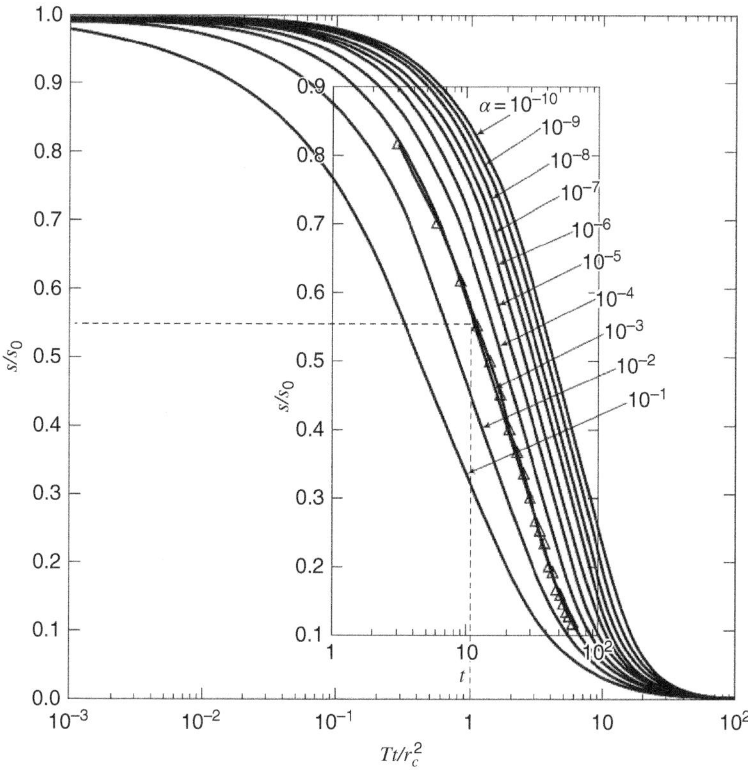

Figure 12.6 Using a type-curve matching procedure to determine α and β values for the Cooper–Bredehoeft–Papadopulos analysis (Adapted from Cooper et al., 1967. Water Resources Research 3(1): 263–269. Copyright 1967 by the American Geophysical Association).

and

$$S = \frac{r_c^2 \alpha}{r_s^2} = \frac{(7.6\ \text{cm})^2 10^{-3}}{(7.6\ \text{cm})^2} = 10^{-3}$$

12.3 Bower and Rice Slug Test

Bouwer and Rice (1976) developed a technique for determining the hydraulic conductivity of an unconfined aquifer with a fully or a partially penetrating well. Bouwer (1989) later extended the technique to a confined aquifer. The approach is similar to Hvorslev's method but involves using a set of curves to determine the radius of influence. The rate of water level change (drawdown or mound) in a slug test (Figure 12.7) is expressed as

$$\frac{ds}{dt} = \frac{Q}{\pi r_c^2} \tag{12.9}$$

where r_c is the casing of the well, Q is the inflow or outflow rate of water into or out of the well after a sudden water level change, and s is the head change. At steady state, the flow rate is

$$Q = 2\pi K L_e \frac{s}{\ln(R_e/r_w)} \tag{12.10}$$

where K is the hydraulic conductivity of the aquifer, L_e is the screen length, R_e is the radius of influence, and r_w is the radius of the well. By inserting

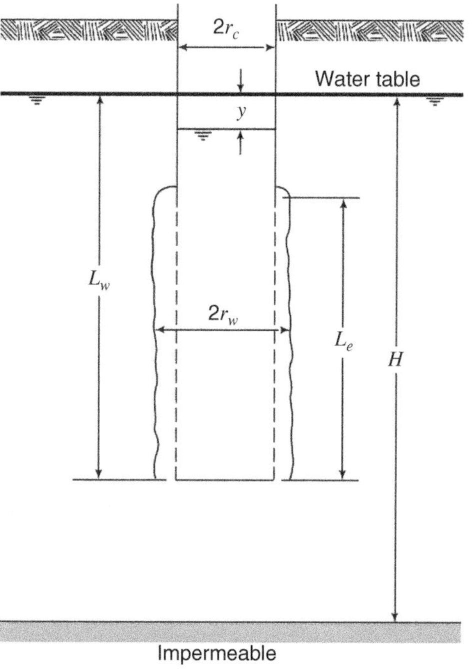

Figure 12.7 Basic geometry for the Bouwer and Rice slug test (Bouwer, 1989. Reproduced by permission of the National Groundwater Assoc. Copyright © 1989. All rights reserved).

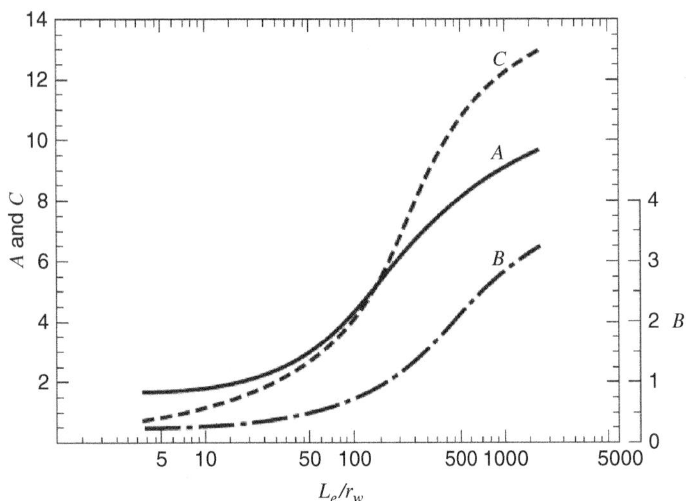

Figure 12.8 Dimensionless parameters *A*, *B*, and *C* as a function of L_e/Rw for the Bouwer and Rice slug test (Reproduced by permission of the National Groundwater Assoc. Copyright © 1989. All rights Reserved).

Eq. (12.10) into (12.9) and integrating, a working equation is obtained for calculating the hydraulic conductivity as

$$K = \frac{r_c^2 \ln(R_e/r_w)}{2L_e} \frac{1}{t} \ln \frac{s_0}{s} \tag{12.11}$$

where s_0 is the initial water level change, and s is the water level change at t. An empirical equation relates $\ln(R_e/r_w)$ to the geometry of the system.

$$\ln \frac{R_e}{r_w} = \left[\frac{1.1}{\ln(L_w/r_w)} + \frac{A + B\ln[(H - L_w)/r_w]}{L_e/r_w} \right]^{-1} \tag{12.12}$$

where L_w is the length of the well in the aquifer, D is the thickness of the aquifer, A and B depend on the ratio L_e/r_w (Figure 12.8), and H is the thickness of the saturated material. When $L_w = H$, a simpler form of the equation is written as

$$\ln \frac{R_e}{r_w} = \left[\frac{1.1}{\ln(L_w/r_w)} + \frac{C}{L_e/r_w} \right]^{-1} \tag{12.13}$$

where C is a function of L_e/r_w (Figure 12.8).

The steps in determining the hydraulic conductivity using the Bouwer and Rice method are as follows:

1) Plot water level change s on a log scale versus time t on a linear scale using semi-log graph paper.
2) Approximate the straight portion of the plotted curve by a straight line and extend the line to $t = 0$.
3) Calculate $\ln[(H - L_w)/r_w]$ for $L_w \neq H$. If $\ln[(H - L_w)/r_w] > 6$, set $\ln[(H - L_w)/r_w] = 6$.
4) Find A and B for $H \neq L_w$ or C for $H = L_w$ in Figure 12.8.
5) Calculate $\ln(R_e/r_w)$ using Eq. (12.12) or (12.13).
6) Record s_0, and s and t for one other point on the line. Calculate K using Eq. (12.11).

Although this technique is developed for an unconfined aquifer, it can also be used for a confined aquifer that receives water from an overlying confining layer (Bouwer, 1989).

Example 12.3 Use the data set from the previous problem to calculate hydraulic conductivity with the Bouwer and Rice method.

Solution

Recall that we are given that $r_w = r_c = 7.6$ cm, $L_e = L_w = H = 98$ m. Determine the ratio L_e/r_w, which is equal to 1290. We use this ratio with Figure 12.8 to determine that $C = 12.4$. Thus,

$$\ln \frac{R_e}{r_w} = \left[\frac{1.1}{\ln(9800/7.6)} + \frac{12.4}{9800/7.6} \right]^{1} = 6.13$$

Plot the drawdown versus time in Figure 12.9 and record $s_0 = 0.49$ m, $t = 33$ sec, and $s = 0.149$ m. These latter coordinates are points on the line. The hydraulic conductivity is

$$K = \frac{7.6^2 \text{cm}^2 (6.13)}{2(9800 \text{ cm})} \frac{1}{(33 \text{ sec})} \ln \frac{0.49}{0.149} = 6.5 \times 10^{-4} \text{ cm/sec}$$

with $H = 98$ m, the transmissivity is

$$T = HK = (9800 \text{ cm})(6.5 \times 10^{-4} \text{ cm/sec}) = 6.4 \text{ m}^2/\text{sec}$$

This value of transmissivity value compares favorably to values of 5.2 and 4.8 cm^2/sec, calculated from the Hvorslev and the Cooper–Bredehoeft–Papadopulos methods.

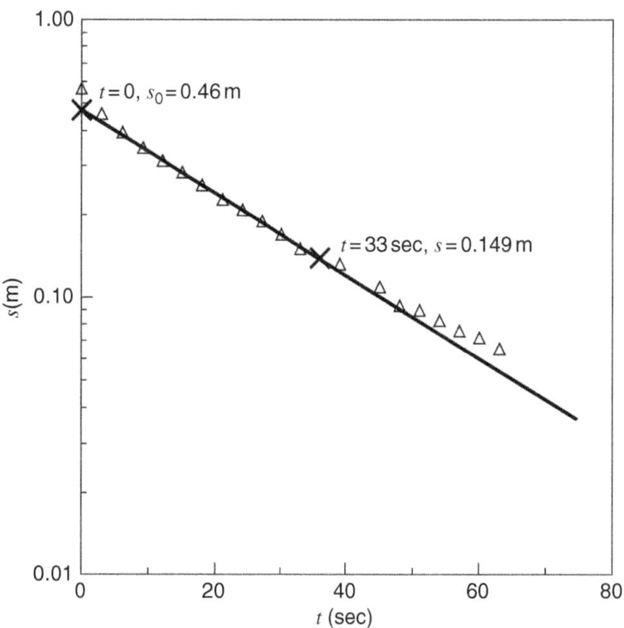

Figure 12.9 Selecting points on the best-fit line for estimating hydraulic conductivity with the Bouwer and Rice slug test (Adapted from Cooper et al., 1967. Water Resources Research 3(1): 263–269. Copyright 1967 by the American Geophysical Union).

12.4 Step and Intermittent Drawdown Tests

Step and intermittent drawdown tests are useful in determining the aquifer parameters and the efficiency of a pumped well (Brown, 1963; Theis, 1963; Harril, 1971; and Birsoy and Summers, 1980). The intermittent drawdown tests consist of a number of pumping and recovery periods with variable pumping rates (Figure 12.10) while step drawdown tests consist of a pumping period with variable pumping rates and a recovery period (Figure 12.11). Birsoy and Summers (1980) derived a convenient equation to express the drawdown for the step and intermittent drawdown tests.

$$\frac{s}{Q_n} = \frac{1}{4\pi T}\log\left[\beta_n(t)\frac{t-\tau_n}{t-\tau'_n}\right] \quad \text{for } t > \tau'_n \text{ during recovery} \tag{12.14}$$

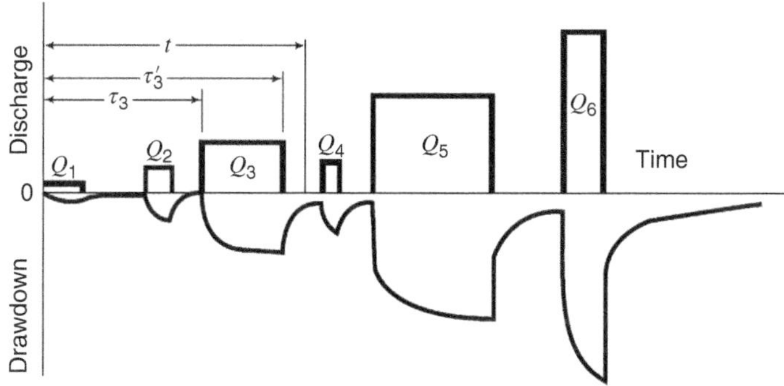

Figure 12.10 Time history of pumping rates for an intermittent drawdown test (from Birsoy and Summers, 1980. Reproduced by permission of the National Groundwater Assoc.

and

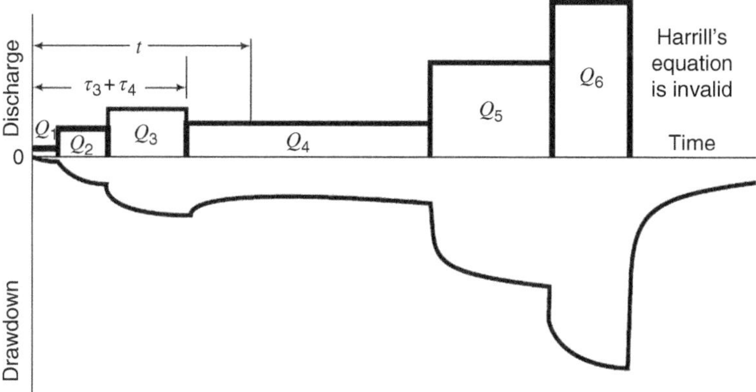

Figure 12.11 Time history of pumping rates for a step drawdown test (Birsoy and Summers, 1980. Reproduced by permission of the National Groundwater Assoc.

$$\frac{s}{Q_n} = \frac{1}{4\pi T} \log\left[\frac{2.25T}{r^2 S}\beta_n(t)(t-\tau_n)\tau_n\right] \quad \text{for } t > \tau_n \text{ during pumping} \tag{12.15}$$

where n is the number of pumping rates, Q_i is the ith rate τ_i and τ_i' are the times when the ith pumping rate starts and ends, respectively. For $n = 1$, $\beta_n(t) = 1$. For $n > 1$, $\beta_n(t)$ is expressed as

$$\beta_n(t) = \prod_{i=1}^{n-1} \left(\frac{t-\tau_i}{t-\tau_i'}\right)^{\frac{Q_i}{Q_n}} = \left(\frac{t-\tau_1}{t-\tau_1'}\right)^{\frac{Q_1}{Q_n}} \cdots \left(\frac{t-\tau_{n-1}}{t-\tau_{n-1}'}\right)^{\frac{Q_{n-1}}{Q_n}} \quad \text{for } n > 1 \tag{12.16}$$

12.4.1 Determination of Transmissivity and Storativity

The procedure for determining the transmissivity and storativity of an aquifer using the step and intermittent drawdown tests is as follows:

1) Calculate $\beta_n(t)$ at each measurement time for the entire period.
2) Plot s/Q_n in linear scale versus adjusted time $\beta_n(t)(t - \tau_n)$ (or adjusted dimensionless time $\beta_n(t)(t - \tau_n)/(t - \tau_i'))$ on the logarithmic scale of semi-log graph paper. The points should fall on a straight line. Selecting two points separated by one log cycle, we calculate the transmissivity as

$$T = \frac{2.3}{4\pi\Delta(s/Q)} \tag{12.17}$$

3) To determine storativity you need to find the intercept of the straight line connecting the plotted points and $s/Q_n = 0$ and substituting it into the following equation.

$$S = \frac{2.25T\beta_n(t_0)(t_0 - \tau_n)}{r^2} \tag{12.18}$$

where $\beta_n(t_0)(t_0 - \tau_n)$ is the intercept.

Example 12.4 In a step drawdown test reported by Birsoy and Summers (1980), the initial pumping rate of 2.3 gpm was maintained for 30 min. Pumping rates, drawdowns at the pumped well, starting and ending times for the other four steps are summarized in Table 12.3. The length of the well screen is 200 ft. Calculate the transmissivity of the aquifer.

Solution

The adjustment time factor $\beta_n(t)$, and adjusted time $\beta_n(t)(t - \tau_n)$ are calculated using Eq. (12.16) and listed in Table 12.3. Because the aquifer is unconfined, Jacob's correction is applied to the measured drawdowns. The corrected drawdown on a linear scale versus adjusted time on a logarithmic scale is shown in Figure 12.12. Theoretically, the data should fall on a straight line. The departure from the straight line may be caused by the assumption that the Jacob modification of Theis equation is only valid for large time, or the well was still developing at early times. Late-time data are used for the determination of transmissivity. In one log cycle, the drawdown difference is 0.28 ft. The transmissivity is calculated by

$$T = \frac{2.3}{4\pi\Delta(s/Q)} = \frac{2.3}{4\pi(0.28 \text{ ft/gpm})} = 126 \text{ ft}^2/\text{day}$$

For the pumping well, Eq. (12.18) cannot be used to obtain storativity because the effective radius of the well is not known.

Table 12.3 Step pumping test data.

Step 2: τ_2 = 30 min, τ'_2 = 60 min, Q_2 = 14.7 gpm						Step 4: τ_4 = 90 min, τ'_4 = 120 min, Q_4 = 38.5 gpm					
T (min)	S (ft)	$\beta(t)$	$\beta(t-\tau_n)$ (min)	s/Q (ft/gpm)	$[s-s^2/(2b)]/Q$ (ft/gpm)	t (min)	s(ft)	$\beta(t)$	$\beta(t-\tau_n)$ (min)	s/Q (ft/gpm)	$[s-s^2/(2b)]/Q$ (ft/gpm)
30	1.04					90	14.41				
30.5	2.05	1.90	0.95	0.139	0.139	90.5	14.99	41.86	20.93	0.389	0.375
31	2.53	1.71	1.71	0.172	0.171	91	15.36	23.65	23.65	0.399	0.384
31.5	2.85	1.61	2.42	0.194	0.192	91.5	15.61	17.00	25.51	0.405	0.390
32	3.09	1.54	3.09	0.210	0.209	92	15.84	13.49	26.99	0.411	0.395
32.5	3.29	1.49	3.73	0.224	0.222	92.5	16.04	11.30	28.26	0.417	0.400
33	3.45	1.46	4.37	0.235	0.233	93	16.17	9.80	29.39	0.420	0.403
33.5	3.58	1.42	4.98	0.244	0.241	93.5	16.31	8.69	30.42	0.424	0.406
34	3.71	1.40	5.59	0.252	0.250	94	16.45	7.85	31.38	0.427	0.410
34.5	3.81	1.38	6.19	0.259	0.257	94.5	16.57	7.17	32.29	0.430	0.413
35	3.91	1.36	6.78	0.266	0.263	95	16.67	6.63	33.15	0.433	0.415
35.5	4.01	1.34	7.36	0.273	0.270	95.5	16.77	6.18	33.98	0.436	0.417
36	4.1	1.32	7.94	0.279	0.276	96	16.86	5.80	34.77	0.438	0.419
36.5	4.18	1.31	8.51	0.284	0.281	96.5	16.95	5.47	35.54	0.440	0.422
37	4.25	1.30	9.08	0.289	0.286	97	17.03	5.18	36.29	0.442	0.424
37.5	4.32	1.29	9.65	0.294	0.291	97.5	17.12	4.94	37.02	0.445	0.426
38	4.39	1.28	10.21	0.299	0.295	98	17.19	4.72	37.74	0.446	0.427
38.5	4.46	1.27	10.77	0.303	0.300	98.5	17.27	4.52	38.44	0.449	0.429
39	4.51	1.26	11.32	0.307	0.303	99	17.41	4.35	39.13	0.452	0.433
39.5	4.56	1.25	11.87	0.310	0.307	99.5	17.44	4.19	39.81	0.453	0.433
40	4.63	1.24	12.42	0.315	0.311	100	17.51	4.05	40.47	0.455	0.435
41	4.72	1.23	13.51	0.321	0.317	101	17.64	3.80	41.78	0.458	0.438
42	4.82	1.22	14.60	0.328	0.324	102	17.78	3.59	43.06	0.462	0.441
43	4.92	1.21	15.68	0.335	0.331	103	17.9	3.41	44.31	0.465	0.444
44	5.01	1.20	16.75	0.341	0.337	104	18.03	3.25	45.54	0.468	0.447
45	5.11	1.19	17.81	0.348	0.343	105	18.13	3.12	46.75	0.471	0.450
46	5.19	1.18	18.87	0.353	0.348	106	18.24	3.00	47.95	0.474	0.452
47	5.27	1.17	19.93	0.359	0.354	107	18.33	2.89	49.13	0.476	0.454
48	5.35	1.17	20.99	0.364	0.359	108	18.41	2.79	50.30	0.478	0.456
49	5.43	1.16	22.04	0.369	0.364	109	18.56	2.71	51.46	0.482	0.460
50	5.51	1.15	23.08	0.375	0.370	110	18.65	2.63	52.61	0.484	0.462
52	5.64	1.14	25.17	0.384	0.378	112	18.84	2.49	54.89	0.489	0.466
54	5.79	1.14	27.25	0.394	0.388	114	19.03	2.38	57.13	0.494	0.471
56	5.91	1.13	29.32	0.402	0.396	116	19.22	2.28	59.35	0.499	0.475
58	6.04	1.12	31.38	0.411	0.405	118	19.39	2.20	61.55	0.504	0.479
60	6.16	1.11	33.44	0.419	0.413	120	19.56	2.12	63.73	0.508	0.483

Table 12.3 (Continued)

Step 3: τ_3 = 60 min, τ'_3 = 90 min, Q_3 = 32.3 gpm						Step 5: τ_5= 150 min, τ'_5 = 150 min, Q_5 = 46.0 gpm					
t (min)	S (ft)	$\beta(t)$	$\beta(t-\tau_n)$ (min)	s/Q (ft/gpm)	$[s-s^2/(2b)]/Q$ (ft/gpm)	t (min)	s (ft)	$\beta(t)$	$\beta(t-\tau_n)$ (min)	s/Q (ft/gpm)	$[s-s^2/(2b)]/Q$ (ft/gpm)
60	6.16					120	19.56				
60.5	7.66	6.82	3.41	0.237	0.233	120.5	20.42	58.24	29.12	0.444	0.421
61	8.4	5.01	5.01	0.260	0.255	121	20.86	32.83	32.83	0.453	0.430
61.5	8.93	4.19	6.29	0.276	0.270	121.5	21.16	23.55	35.32	0.460	0.436
62	9.3	3.70	7.40	0.288	0.281	122	21.39	18.64	37.28	0.465	0.440
62.5	9.62	3.37	8.42	0.298	0.291	122.5	21.6	15.57	38.92	0.470	0.444
63	9.9	3.12	9.36	0.307	0.299	123	21.8	13.46	40.37	0.474	0.448
63.5	10.16	2.93	10.24	0.315	0.307	123.5	21.96	11.91	41.67	0.477	0.451
64	10.35	2.77	11.08	0.320	0.312	124	22.09	10.72	42.87	0.480	0.454
64.5	10.53	2.64	11.89	0.326	0.317	124.5	22.23	9.78	43.99	0.483	0.456
65	10.7	2.53	12.67	0.331	0.322	125	22.35	9.01	45.05	0.486	0.459
65.5	10.86	2.44	13.42	0.336	0.327	125.5	22.47	8.37	46.05	0.488	0.461
66	10.99	2.36	14.16	0.340	0.331	126	22.49	7.83	47.01	0.489	0.461
66.5	11.11	2.29	14.88	0.344	0.334	126.5	22.68	7.37	47.93	0.493	0.465
67	11.23	2.23	15.58	0.348	0.338	127	22.77	6.97	48.82	0.495	0.467
67.5	11.35	2.17	16.27	0.351	0.341	127.5	22.87	6.62	49.68	0.497	0.469
68	11.47	2.12	16.95	0.355	0.345	128	22.96	6.31	50.52	0.499	0.470
68.5	11.57	2.07	17.61	0.358	0.348	128.5	23.03	6.04	51.33	0.501	0.472
69	11.68	2.03	18.27	0.362	0.351	129	23.12	5.79	52.13	0.503	0.474
69.5	11.78	1.99	18.92	0.365	0.354	129.5	23.19	5.57	52.91	0.504	0.475
70	11.88	1.96	19.56	0.368	0.357	130	23.28	5.37	53.67	0.506	0.477
71	12.07	1.89	20.82	0.374	0.362	131	23.42	5.01	55.16	0.509	0.479
72	12.24	1.84	22.05	0.379	0.367	132	23.58	4.72	56.60	0.513	0.482
73	12.41	1.79	23.27	0.384	0.372	133	23.58	4.46	58.01	0.513	0.482
74	12.55	1.75	24.46	0.389	0.376	134	23.68	4.24	59.38	0.515	0.484
75	13.13	1.71	25.65	0.407	0.393	135	23.78	4.05	60.72	0.517	0.486
76	12.86	1.68	26.81	0.398	0.385	136	23.96	3.88	62.04	0.521	0.490
77	13	1.65	27.97	0.402	0.389	137	24.11	3.73	63.34	0.524	0.493
78	13.13	1.62	29.12	0.407	0.393	138	24.23	3.59	64.61	0.527	0.495
79	13.26	1.59	30.25	0.411	0.397	139	24.35	3.47	65.87	0.529	0.497
80	13.39	1.57	31.38	0.415	0.401	140	24.46	3.36	67.12	0.532	0.499
82	13.62	1.53	33.61	0.422	0.407	142	24.69	3.16	69.57	0.537	0.504
84	13.84	1.49	35.82	0.428	0.414	144	24.91	3.00	71.97	0.542	0.508
86	14.02	1.46	38.01	0.434	0.419	146	25.11	2.86	74.34	0.546	0.512
88	14.22	1.43	40.18	0.440	0.425	148	25.32	2.74	76.67	0.550	0.516
90	14.41	1.41	42.33	0.446	0.430	150	25.51	2.63	78.97	0.555	0.519

Source: Birsoy and Summers (1980). Reproduced with permission of the National Groundwater Assoc.

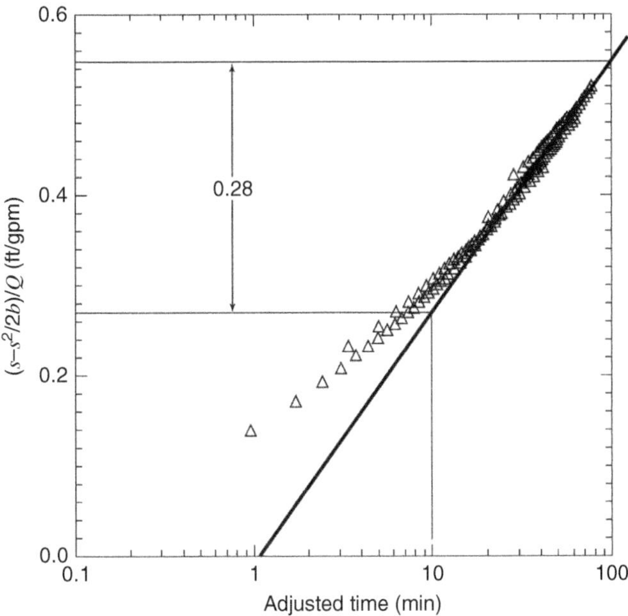

Figure 12.12 Plotting corrected drawdown versus the log of adjusted time to analyze data from a step drawdown test (Birsoy and Summers, 1980. Reproduced by permission of the National Groundwater Assoc. Copyright © 1980. All rights reserved).

12.4.2 Estimating Well Efficiency

As a well is pumped, water moves from the aquifer, through the gravel pack and screen, and finally into the casing. Usually, the large volume of water moving toward the well has difficulties in passing through the gravel pack and screen, creating a substantial drop in hydraulic head between the aquifer and the inside of the well. Just how well this water passes through gravel pack and screen is measured by the *well efficiency*. Well efficiency (E) is defined as the ratio of the drawdown (s_a) in the aquifer at the radius just outside of the pumped well to the drawdown (s_t) inside the well (Heath, 1983)

$$E = \frac{s_a}{s_t} \times 100 \tag{12.19}$$

The theoretical drawdown (s_a) can be calculated from known aquifer parameters obtained from other sources. Once s_a is known, the well efficiency can be calculated from s_a and drawdown in the pumping well.

Good practice in well construction is to provide maximum efficiency. A well that is efficient can be pumped at a greater rate than the same, inefficient well. In effect, the productivity of the well depends on minimizing head losses through the gravel pack and the screen. Using "best" well construction and development techniques, a well efficiency of 80% is the maximum possible and well efficiency of 60% may be achievable under unfavorable conditions for most screened wells (Heath, 1983). The use of a proper well screen with a large open area for flow and appropriate slot sizes, and complete development of the gravel pack (removal of fines) are key.

Step drawdown tests are used in the field to estimate well efficiency. The theoretical drawdown (s_a) is linearly proportionally to the pumping rate. The total drawdown in a pumped well is the summation of head loss in the pumped well and head loss in the aquifer.

$$s_t = BQ + f(Q) \tag{12.20}$$

where B is a function of time for a given aquifer and pumping conditions. The first term BQ represents the drawdown s_a. Different forms of $f(Q)$ were proposed to account for the head loss in the pumped well (Jacob, 1947; Rorabaugh, 1953; Lennox, 1966). According to Jacob (1947), the total head loss (s_t)

$$s_t = BQ + CQ^2 \tag{12.21}$$

where C is a constant. The step drawdown test is commonly applied to derive constants B and C. Once B and C are obtained, well efficiency may be calculated from such an estimate. The estimated well efficiency is called apparent well efficiency and is defined as

$$E' = \frac{BQ}{BQ + CQ^2} \tag{12.22}$$

For convenience, Eq. (13.9) may be rewritten as

$$\frac{s_t}{Q} = B + CQ \tag{12.23}$$

The procedures for determining B and C are summarized as follows:

1) Plot s_t/Q versus Q at the same corrected time $\beta_n(t)(t - \tau_n)$ in linear scales on graph paper.
2) Approximate the plot by a straight line. B is the intercept at where $Q = 0$ and C is the slope of the plot.

Example 12.5 Calculate B, C, and apparent well efficiency using the data in Example 13.1.

Solution

Table 12.4 Pumping rates and drawdowns at corrected time $\beta_n(t)(t - \tau_n)$ = 31.38 min.

Q (gpm)	$[s - s^2/(2b)]/Q$ (ft/gpm)
14.700	0.405
32.300	0.401
38.500	0.410
46.000	0.426

Source: Adapted from Birsoy and Summers (1980). Reproduced by permission of the National Groundwater Assoc.

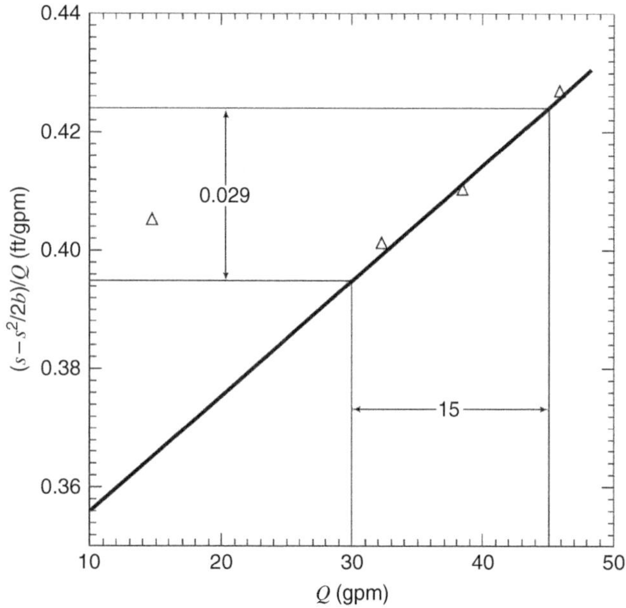

Figure 12.13 Plotting measurement data needed to calculate the apparent well efficiency using step drawdown tests (Adapted from Birsoy and Summers, 1980. Reproduced by permission of the National Groundwater Association. Copyright 1980. All rights reserved).

The pumping rates and drawdowns at corrected time $\beta_n(t)(t - \tau_n) = 31.38$ min is listed in Table 12.4 and plotted in Figure 12.13. The drawdown in test step 5 is calculated through linear interpolation. The data in step 3 deviates from the straight line. Only data in steps 3, 4, and 5 are utilized. From the graph, we get

$$B = 0.356 \text{ ft/gpm}$$

and

$$C = \frac{0.29 \text{ ft/gpm}}{15 \text{ gpm}} = 0.002 \text{ ft/(gpm)}^2$$

and

$$E' = \frac{BQ}{BQ + CQ^2}$$

$$= \frac{0.356 \text{ ft/gpm}}{0.356 \text{ ft/gpm} + \left(0.002 \frac{\text{ft}}{\text{gpm}^2}\right)(46 \text{ gpm})}$$

$$= 79\%$$

Exercises

12.1 A slug test was carried in a 6-in. well by providing an instantaneous injection of 5.2 ft^3 slug of water. The residual drawdown with time is listed in Table 12.5. Assume that the aquifer is an unconfined aquifer. Calculate hydraulic conductivity using the Hvorslev technique.

Table 12.5 Data for a slug injection test at Speedway City, Indiana.

Time (min)	Residual head (ft)	Time (min)	Residual head (ft)
1.25	0.26	3.87	0.08
1.33	0.25	4.10	0.08
1.50	0.20	4.33	0.08
1.92	0.17	4.52	0.08
2.17	0.16	4.58	0.07
2.30	0.15	4.72	0.07
2.37	0.14	5.17	0.07
2.42	0.14	5.28	0.06
2.67	0.12	5.45	0.06
2.72	0.12	6.10	0.06
2.77	0.12	6.40	0.05
2.92	0.11	6.83	0.05
3.00	0.11	7.17	0.05
3.22	0.10	7.75	0.04
3.28	0.10	8.58	0.04
3.33	0.10	9.37	0.04
3.40	0.09	10.12	0.03
3.47	0.09	11.00	0.03
3.55	0.09	12.5	0.03
3.67	0.09	13.0	0.03
3.77	0.09		

Source: U.S. Geological Survey Ferris and Knowles (1963)/Public domain.

12.2 Table 12.6 provides slug test results for a confined aquifer. The radius of the borehole is 7.6 cm. Calculate the transmissivity of the aquifer using the Cooper–Bredehoeft–Papadopulos slug test technique. The following first figure shows the type curve for the Cooper–Bredehoeft–Papadopulos slug test technique. The following second figure provides graph paper at the same scale for plotting values of head ratio versus time.

Table 12.6 Slug test results in a confined aquifer.

Time (min)	H/H_0	Time (min)	H/H_0
0.010	0.985	0.356	0.803
0.013	0.983	0.452	0.765
0.016	0.980	0.574	0.721
0.020	0.977	0.728	0.671
0.026	0.973	0.924	0.614
0.033	0.967	1.172	0.552
0.042	0.962	1.487	0.485
0.053	0.954	1.887	0.415

(Continued)

Table 12.6 (Continued)

Time (min)	H/H_0	Time (min)	H/H_0
0.067	0.945	2.395	0.344
0.085	0.934	3.039	0.275
0.108	0.920	3.857	0.211
0.137	0.905	4.894	0.156
0.174	0.885	6.210	0.111
0.221	0.862	7.880	0.076
0.281	0.835	10.000	0.051

Source: Schwartz and Zhang (2003) / John Wiley & Sons.

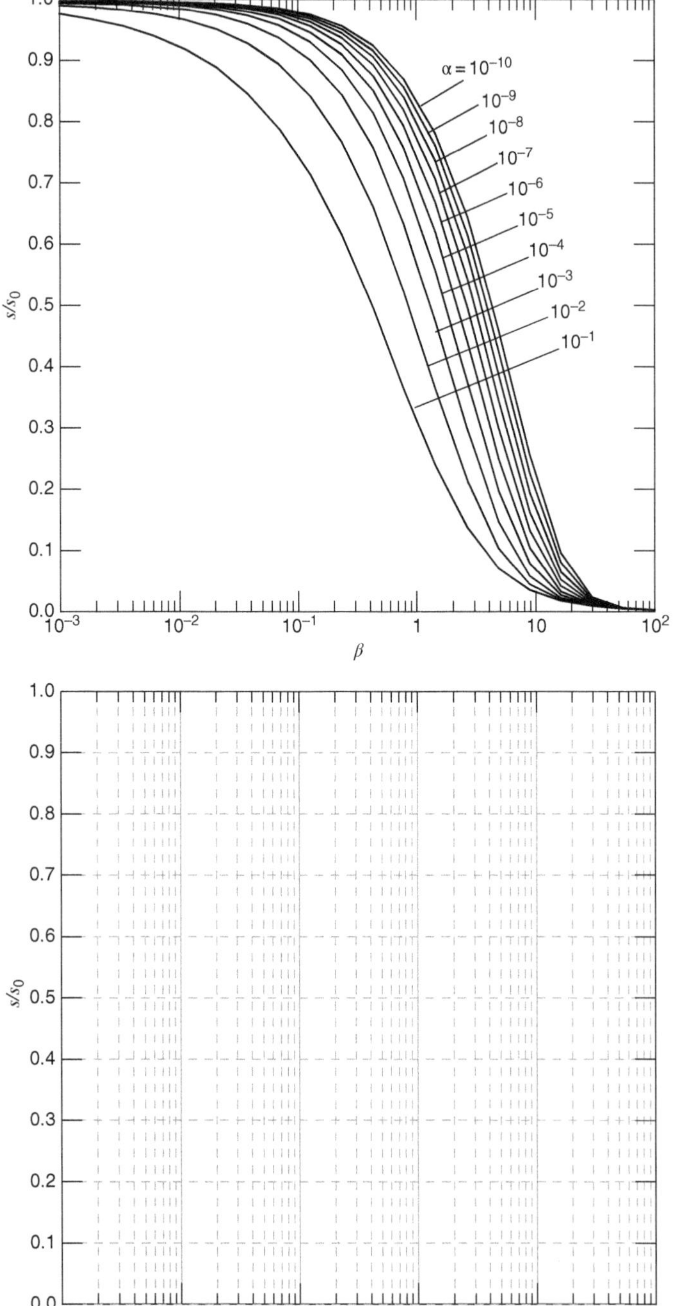

12.3 Calculate hydraulic parameters for slug tests in Tables 12.3 and 12.4 using the Bower and Rice technique.

12.4 Table 12.7 gives values of time and drawdown in a pumped well for a step-drawdown test reported by Clark (1977). Calculate the transmissivity and storativity of the aquifer. It is assumed that the effective well radius is 0.3 m for the test.

12.5 Calculate the apparent well efficiency using the data in Table 12.7.

Table 12.7 Values of time and drawdown from a step-drawdown test.

Step 1: Q = 1306 m^3/day		Step 2: Q = 1693 m^3/day		Step 3: Q = 2423 m^3/day	
Time (min)	**Drawdown (m)**	**Time (min)**	**Drawdown (m)**	**Time (min)**	**Drawdown (m)**
10	3.521	190	5.74	370	8.672
12	3.592	192	5.81	372	8.663
14	3.627	194	5.81	374	8.698
16	3.733	196	5.824	376	8.733
18	3.768	198	5.845	378	8.839
20	3.836	200	5.81	380	8.874
25	3.873	205	5.824	385	8.874
30	4.013	210	5.824	390	8.979
35	3.803	215	5.881	395	8.979
40	4.043	230	6.092	400	8.994
60	4.12	235	6.092	405	9.05
70	4.12	240	6.176	410	9.05
80	4.226	250	6.162	415	9.12
90	4.226	260	6.176	420	9.12
100	4.226	270	6.169	430	9.155
120	4.402	280	6.169	440	9.191
150	4.402	330	6.374	450	9.191
180	4.683	360	6.514	460	9.226
				480	9.261
				510	9.367
				540	9.587

Step 4: Q = 3261 m^3/day		Step 5: Q = 4094 m^3/day		Step 6: Q = 5016 m^3/day	
Time (min)	**Drawdown (m)**	**Time (min)**	**Drawdown (m)**	**Time (min)**	**Drawdown (m)**
550	12.325	730	16.093	910	20.917
552	12.36	732	16.198	912	20.952
554	12.395	734	16.268	914	21.022
556	12.43	736	16.304	916	21.128
558	12.43	738	16.374	918	21.163
560	12.501	740	16.409	920	21.198
565	12.508	745	16.586	925	21.304
570	12.606	750	16.621	930	21.375

(Continued)

Table 12.7 (Continued)

Step 4: Q = 3261 m³/day		Step 5: Q = 4094 m³/day		Step 6: Q = 5016 m³/day	
Time (min)	Drawdown (m)	Time (min)	Drawdown (m)	Time (min)	Drawdown (m)
575	12.712	755	16.691	935	21.48
580	12.747	760	16.726	940	21.551
585	12.783	765	16.776	945	21.619
590	12.818	770	16.797	950	21.656
595	12.853	775	16.938	960	21.663
600	12.853	780	16.973	970	21.691
610	12.888	790	17.079	980	21.762
620	12.923	800	17.079	990	21.832
630	12.994	810	17.114	1000	21.903
640	12.994	820	17.219	1020	22.008
660	13.099	840	17.325	1050	22.184
690	13.205	870	17.395	1080	22.325
720	13.24	900			

Source: Adapted from Clark (1977).

References

Birsoy, Y. K., and W. K. Summers. 1980. Determination of aquifer parameters from step tests and intermittent pumping data. Groundwater, v. 18, no. 2, p. 137–146.

Bouwer, H., and R. C. Rice. 1976. A slug test for determining hydraulic conductivity of unconfined aquifers with completely or partially penetrating wells. Water Resources Research, v. 12, no. 3, p. 423–428.

Bouwer, H. 1989. The Bouwer and Rice slug test – an update. Groundwater, v. 27, no. 3, p. 304–309.

Brown, R. H. 1963. Drawdown resulting from cyclic intervals of discharge. U. S. Geol. Surv. Water-Supply Paper 1536-I, p. 324-330.

Cedergren, H. R. 1967. Seepage, Drainage, and Flow Nets. John Wiley and Sons, New York, p. 163.

Clark, L. 1977. The analysis and planning of step-drawdown tests. Quarterly Journal of Engineering Geology and Hydrogeology, v. 10, no. 2, p. 125–143.

Cooper, H. H., J. D. Bredehoeft, and I. S. Papadopulos. 1967. Response of a finite diameter well to an instantaneous charge of water. Water Resources Research, v. 3, no. 1, p. 263–269.

Ferris, J. G., and D. B. Knowles. 1963. The slug-injection test for estimating the coefficient of transmissibility of an aquifer. U. S. Geological Survey Water-Supply Paper 1563-I.

Harril, J. R. 1971. Determining transmissivity from water-level recovery of a step-drawdown test. U. S. Geol. Surv. Prof. Paper 700-C, p. C212–C213.

Heath, R. C. 1983. Basic groundwater hydrology. U. S. Geological Survey Water-Supply Paper 2220, 84 p.

Jacob, C. E. 1947. Drawdown test to determine effective radius of artesian well. Transactions, ASCE, v. 112, paper 2321, p. 1047–1070.

Lennox, D. H. 1966. Analysis and application of step-drawdown test. ASCE Proceedings, v. 92, no. 6.

Papadopulos, I. S., and H. H. Cooper. 1967. Drawdown in a well of large diameter. Water Resources Research, v. 3, no. 1, p. 241–244.

Papadopulos, I. S., J. D. Bredehoeft, and H. H. Cooper. 1973. On the analysis of slug test data. Water Resources Research, v. 9, no. 4, p. 1087–1089.

Rorabaugh, M. I. 1953. Graphical and theoretical analysis of step-drawdown test of artesian well. Proceedings Separate No. 362, ASCE, v. 79, no. 362, p. 1–23.

Schwartz, F. W., and H. Zhang. 2003. Fundamentals of Groundwater. John Wiley & Sons, Hoboken, New Jersey, 583 p.

Theis, C. V. 1963. Drawdowns resulting from cyclic rates of discharge. In R. Bentall, compiler, Methods of determining permeability, transmissivity, and drawdown. U.S. Geol. Surv. Water-Supply Paper 1536-I, p. 319-329.

Thompson, D. B. 1987. A microcomputer program for interpreting time-lag permeability test. Groundwater, v. 25, no. 2, p. 212–218.

U. S. Department of the Navy, Naval Facilities Engineering Command. 1992. Soil Mechanics, NAVFAC Design Manual 7.1.

13

Calculations and Interpretation of Hydraulic Head in Complex Settings

CHAPTER MENU

Most well-hydraulics problems are not nearly as simple as the examples presented so far. Complexities arise because more than one well can be pumping at the same time or because aquifers are not infinite in extent. This chapter provides the theoretical basis for accommodating these complexities within the framework of traditional analytical solutions for well-hydraulics problems and introduces commonly used numerical modeling software.

13.1 Multiple Wells and Superposition

As a well is pumped, a cone of depression forms and grows with time. When pumping wells are spaced so closely that their cones of depression overlap, they will interfere with each other. *Interference* means that there is more drawdown than expected in each of the pumping wells because the water level decline in a pumping well is due not only to its actual pumping but also to drawdown caused by nearby wells. Heath (1983) illustrates how the cones of depression due to two pumping wells (Figure 13.1*a*) combine to produce a zone of enhanced drawdown (Figure 13.1*b*).

It is simple to analyze this multi-well problem, at least for confined aquifers, using the *principle of superposition*. The total drawdown at any location of interest is the sum of the drawdowns due to each well by itself. In practice, you first calculate the drawdown (s_1) at the point of interest (e.g., an observation well) due to the first well a distance r_1 away. Next, calculate s_2 for the second well at a distance r_2. The total drawdown at the observation well is the sum $s_1 + s_2$. The principle of superposition can be applied to confined aquifers because the parameters T and S remain constant. In effect, the differential equations are linear for this case, making the solutions additive. In the case of unconfined aquifers, where T is a function of drawdown, the equations are nonlinear and this approach does not hold.

For some problems, both pumping and injection wells may be operating together in the same confined aquifer. The principle of superposition also holds in this case. The total drawdown is the algebraic sum of the drawdowns due to the pumping wells ($+s$) and the buildup due to the injection wells ($-s$). Here is an example that illustrates the principle of superposition for a problem involving a pumping well and an injection well.

Fundamentals of Groundwater, Second Edition. Franklin W. Schwartz and Hubao Zhang.

Companion website: www.wiley.com/go/schwartz/fundamentalsofgroundwater2

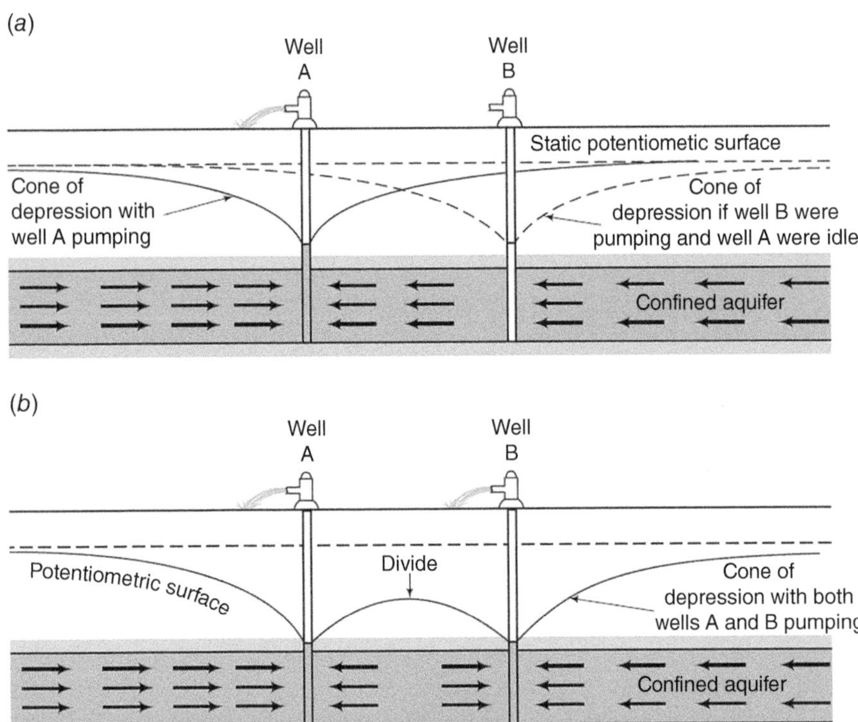

Figure 13.1 Panel (*a*) shows the cones of depression at wells *A* and *B* assuming that one of the wells is idle. Panel (*b*) shows the resulting cone of depression with both wells pumping together. The interference causes significant lowering and the formation of a divide between the two wells (U.S. Geological Survey Heath, 1983/Public domain).

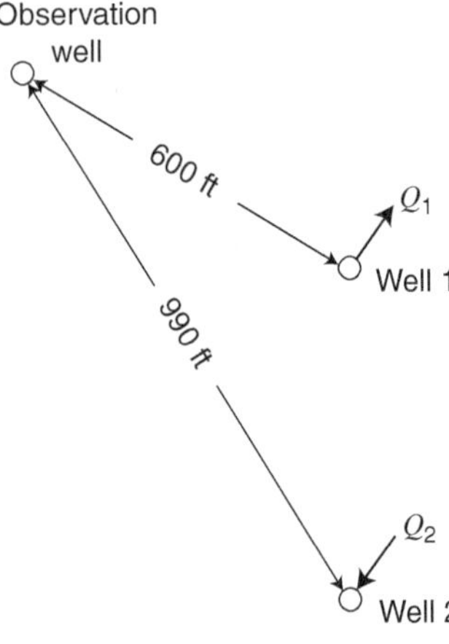

Figure 13.2 Map showing the location of wells in Example 13.1 (Schwartz and Zhang, 2003 / John Wiley & Sons).

Example 13.1 Figure 13.2 shows the location of a withdrawal well and an injection well in a confined aquifer having a transmissivity of 130 ft^2/day and a storativity of 5×10^{-4}. Water is pumped from the withdrawal well (#1) at 19,000 ft^3/day. Water is injected at 8000 ft^3/day at the injection well (#2). Calculate the drawdown at the observation well shown in Figure 13.2 after one year of pumping and injection.

Solution

Calculate the drawdown at the observation well due to pumping at Well #1 and injection at Well #2 separately using the Cooper–Jacob approximation.

$$s_1 = \frac{2.3Q_1}{4\pi T} \log \frac{2.25Tt}{r_1^2 S} = \frac{2.3 \times 19,000\ \text{ft}^3/\text{days}}{4 \times 3.14 \times 1500\ \text{ft}^2/\text{days}} \log \frac{2.25 \times 1500\ \text{ft}^2/\text{days} \times 365\ \text{days}}{(600\text{ft})^2 \times 0.0005}$$

$$= 8.9\ \text{ft}$$

$$s_2 = \frac{2.3 - Q_2}{4\pi T} \log \frac{2.25Tt}{r_2^2 S} = \frac{2.3 \times (-8000)\ \text{ft}^3/\text{days}}{4 \times 3.14 \times 1500\ \text{ft}^2/\text{days}} \log \frac{2.25 \times 1500\ \text{ft}^2/\text{days} \times 365\text{days}}{(990\text{ft})^2 \times 0.0005}$$

$$= -3.3\ \text{ft}$$

Using the principle of superposition, the drawdown at the observation well is the sum of the drawdown due to Well #1 and the buildup due to Well #2.

$$s_{\text{tot}} = s_1 + s_2 = 8.9\ \text{ft} + (-3.3\ \text{ft}) = 5.6\ \text{ft}$$

13.2 Drawdown Superimposed on a Uniform Flow Field

In all of the well-hydraulics problems that we have considered so far, the initial water level in the aquifer is assumed to be a constant. However, for most aquifers, there is a gradient in the hydraulic head because the groundwater is flowing. The principle of superposition can be used to determine what the resultant head distribution looks like with two flow elements—the uniform hydraulic head distribution and the steady-state cone of depression due to a pumping well. Let us illustrate this process conceptually. We start with a uniform flow field with a regular distribution of equipotential lines (Figure 13.3*a*). A well, shown on the figure, is pumped for a long time providing a cone of depression (Figure 13.3*b*). The resulting flow field (Figure 13.3*c*) is determined by subtracting the drawdown from the original head field.

Now, while straightforward in principle, this procedure is tedious to do by hand because the drawdown due to the well needs to be calculated at a large number of points. Here is a steady-state mathematical approach for calculating the effects of

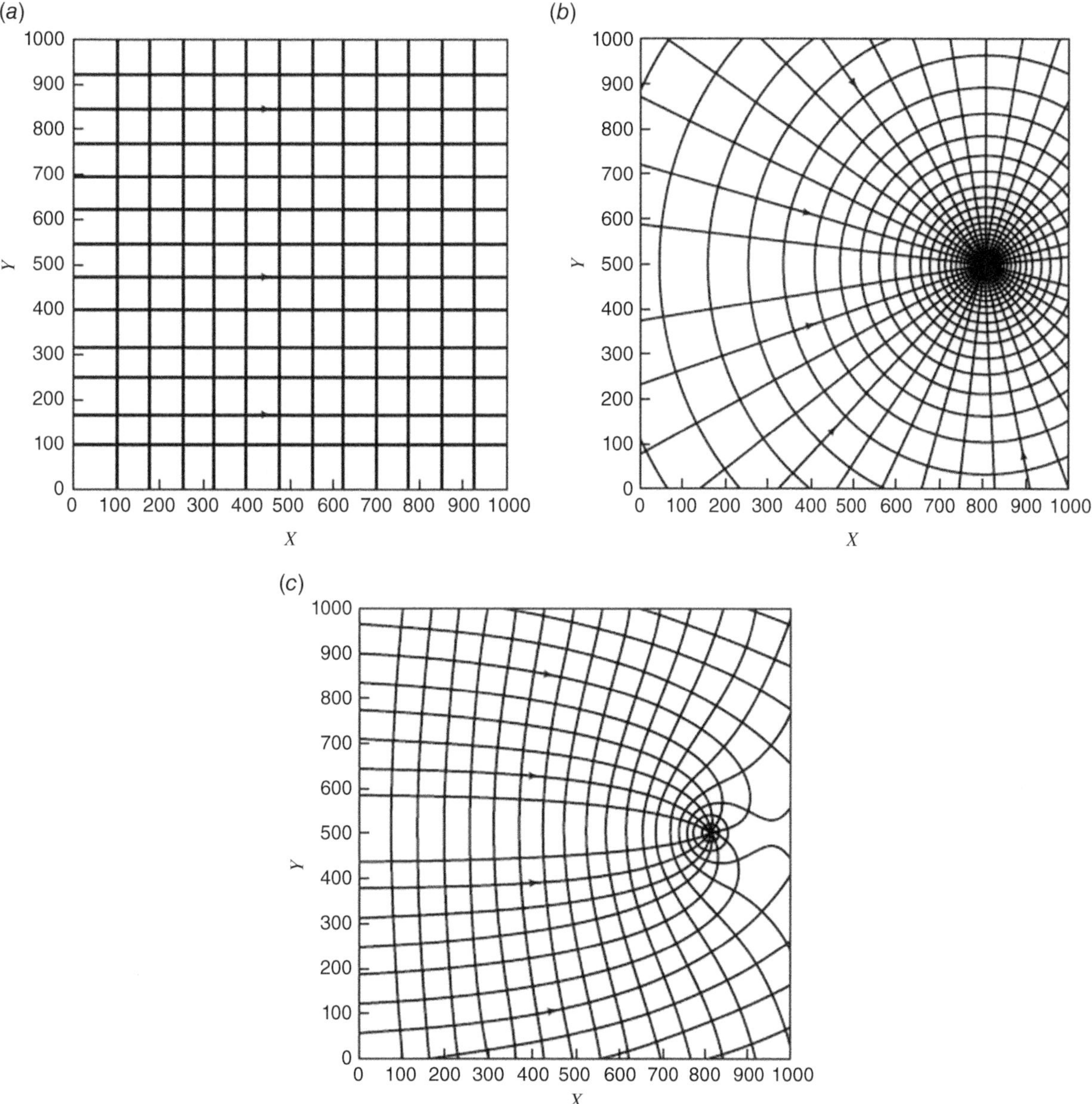

Figure 13.3 This figure illustrates the principle of superposition that applies to a uniform flow field (Panel *a*) and the steady-state cone of depression due to a pumping well (Panel *b*). The resulting flow field (Panel *c*) shows stream tubes being captured by the well and others bypassing the well (Schwartz and Zhang, 2003 / John Wiley & Sons).

pumping superimposed on a uniform flow field. Assume that the regional Darcy velocity is q_0 in the x-direction, hydraulic conductivity of the aquifer is K, and the pumping rate of the well is Q. If the pumped well with a zero initial hydraulic head is located at $x = 0$, the total hydraulic head is (Bear, 1979; Javandel et al., 1984)

$$h = -\frac{q_0}{K}x + \frac{Q}{2\pi Kb}\ln\frac{r}{r_w} \tag{13.1}$$

where b is the thickness of the aquifer, q_0 is the regional Darcy velocity, and r_w is the radius of the pumped well. The distance from the pumped well to any location in the domain is expressed as

$$r = \sqrt{x^2 + y^2} \tag{13.2}$$

where x and y are coordinates. Looking at Eq. (13.1), we see how the superposition is accomplished. The first term on the RHS defines the hydraulic head in the flow system as a function of hydraulic head. The second term describes the steady-state drawdown due to pumping.

13.3 Replacing a Geologic Boundary with an Image Well

Hydraulic theory introduced so far has assumed that the aquifer of interest is infinite in a lateral extent. However, this condition is the exception rather than the rule. Aquifers are not infinite because they can be cut by tight faults or end abruptly due to changes in geology. These impermeable boundaries effectively halt the spread of the cone of depression and influence the pattern of drawdown related to the well. Similarly, a surface water body (e.g., a fully-penetrating stream or lake), or an adjacent segment of aquifer, having a significantly higher transmissivity or storativity, can halt the spread of a cone of depression by providing a source of recharge to the aquifer. Such recharge boundaries also can influence the pattern of drawdown in the vicinity of the pumping wells.

In general, well-hydraulics theory cannot cope with the presence of one of these aquifer boundaries. It is possible, however, to get rid of the boundary by adding an imaginary well-known as an *image well* (Ferris et al., 1962; Stallman, 1952). The following sections illustrate how image-well theory helps in both calculating drawdowns in aquifers containing flow boundaries and using the results of aquifer tests to estimate the location of boundaries.

13.3.1 Impermeable Boundary

Let us begin by looking at an aquifer with a geologic boundary that forms an impermeable boundary (Figure 13.4*a*). Figure 13.4*b* and 13.4*c* illustrate how to replace this boundary by adding an image well. This well is placed across the boundary at the same distance from the boundary as the original well (Figure 13.4*c*). Adding the pumping well makes the impermeable boundary disappear. Drawdowns in the aquifer can then be calculated using the principle of superposition with two pumping wells. Mathematically, the drawdown due to a single pumping well in a confined aquifer with an impermeable boundary is

$$s = s_r + s_i = \frac{Q}{4\pi T}[W(u_r) + W(u_i)] \tag{13.3}$$

where

$$u_r = \frac{r_r^2 S}{4Tt} \tag{13.4}$$

and

$$u_i = \frac{r_i^2 S}{4Tt} \tag{13.5}$$

where r_r and r_i are the distances from the pumped and image wells to an observation well, respectively. For smaller u_r and u_i, the equation can be approximated by

$$s = \frac{2.30Q}{4\pi T}\left[\log\left(\frac{2.25Tt}{r_r^2 S}\right) + \log\left(\frac{2.25Tt}{r_i^2 S}\right)\right] \tag{13.6}$$

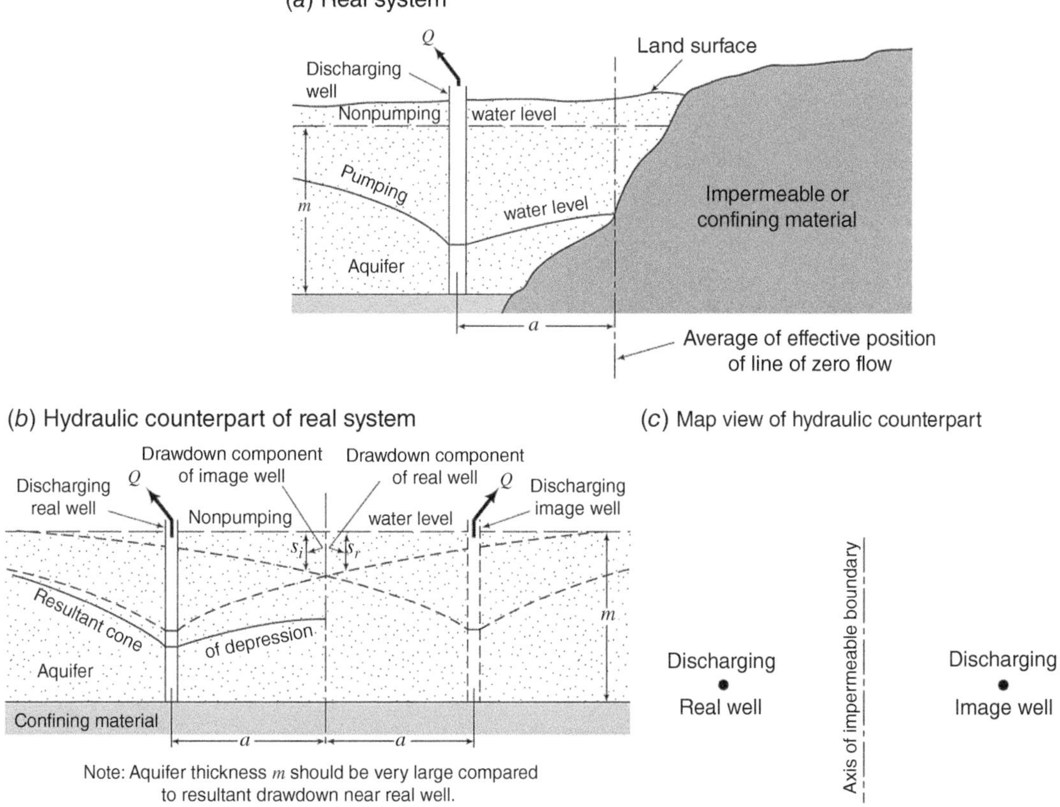

Figure 13.4 A cross section illustrating an aquifer with an impermeable boundary. Panel (*a*) shows the real system; (*b*) the hydraulic counterpart of the real system (U.S. Geological Survey, adapted from Ferris et al., 1962/Public domain); and (*c*) a map view of the hydraulic counterpart.

In an aquifer with an impermeable boundary, aquifer test data can be interpreted not only to provide hydraulic parameters but also the location of the boundary. Let us consider the simple case of a single straight-line impermeable boundary with a single pumping well. Assume that the observation well where the data have been collected is much closer to the real pumping well than the image well. When drawdown at the observation well is plotted versus the log of time, there should be two distinct slopes. The slope at early time corresponds to the drawdown change caused by the pumped well only. The slope at later times is different because drawdown is due to both the real well and the image well. Drawdown change affected by the combination of the pumped and image wells in one log cycle of time will be twice that of the pumped well alone (Figure 13.5). Following the Cooper–Jacob method, a straight line fitted to the early-time data will yield the aquifer parameters.

Next is an approach to determine the location of the boundary. From Eqs. (13.4) and (13.5), we know that u_r and u_i must be equal in order to generate the same drawdown in an observation well for the pumped well and the image well, respectively. Thus,

$$\frac{r_i^2}{t_i} = \frac{r_r^2}{t_r} \tag{13.7}$$

where t_r and t_i are the times for the pumped and the image to generate the same amount drawdowns in an observation well, respectively. Equation (13.7) may be rewritten as

$$r_i = r_r\sqrt{\frac{t_i}{t_r}} \tag{13.8}$$

Knowing the radial distance from the observation well to the image well begins to define where the boundary is located—except that several observation wells are required. Here are the steps.

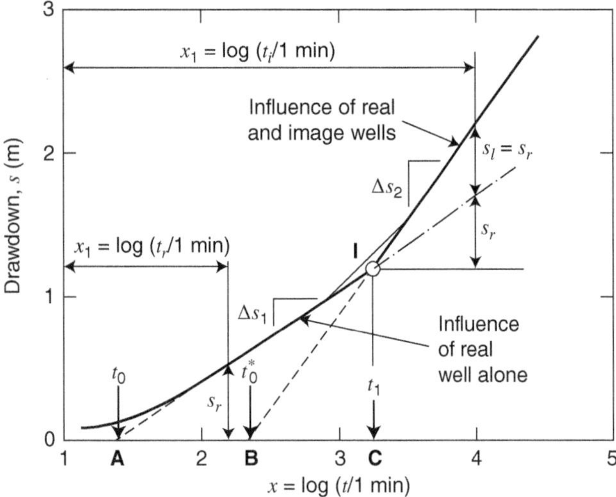

Figure 13.5 Plot of drawdown versus time on a semi-log scale for hydraulic testing near an impermeable boundary in a confined aquifer (Chapuis, 1994a. Reproduced by permission of the National Groundwater Assoc. Copyright ©1994. All rights reserved.).

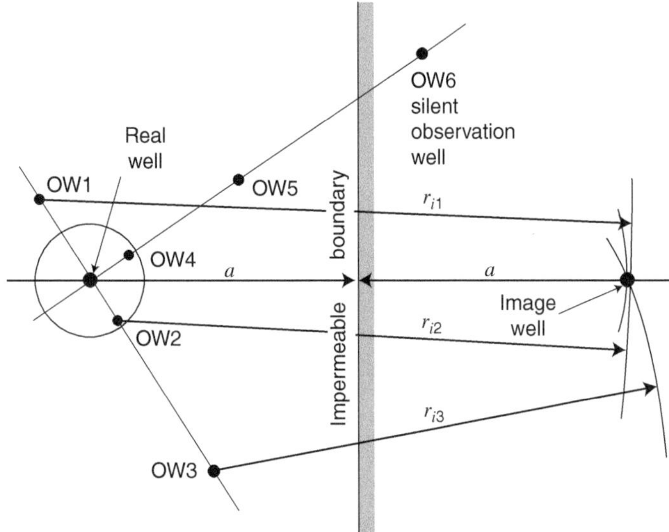

Figure 13.6 Determining the location of an impermeable boundary with at least three observation wells (Chapuis, 1994a. Reproduced by permission of National Groundwater Assoc. Copyright ©1994. All rights reserved).

1) Install at least four observation wells near the pumped well for the aquifer test (Figure 13.6).
2) Plot the observed drawdown from each of the four wells on a linear scale versus the log of time (Figure 13.7).
3) Determine the aquifer parameters using the Cooper–Jacob time straight-line method. Use the early-time drawdown data when the boundary effect is negligible.
4) Calculate and plot the theoretical Theis drawdown for each observation well using the derived aquifer parameters, or simply extending the early-time straight line out to the last measurement time (Figure 13.7).
5) Select measured times (t_{i1}, t_{i2}, t_{i3}, and t_{i4}) when the boundary effect is significant at each observation well (Figure 13.7).
6) Calculate the drawdown contribution from the image well only (s_{i1}, s_{i2}, s_{i3}, and s_{i4}) by subtracting the calculated Theis drawdown from the measured drawdown at the selected times (t_{i1}, t_{i2}, t_{i3}, and t_{i4}) (Figure 13.7).
7) Record the times (t_{r1}, t_{r2}, t_{r3}, and t_{r4}) when the Theis drawdown is the same as the drawdown contribution by the image well at each observation well (Figure 13.7).
8) Calculate the distances (r_{i1}, r_{i2}, r_{i3}, and r_{i4}) from the image well to all of the observation wells using Eq. (13.8).
9) Draw arcs centered at observation wells with radii of r_{i1}, r_{i2}, r_{i3}, and r_{i4}. The interception of the arcs is the location of the image well. The impervious boundary is located halfway between the image well and the real well (Figure 13.6).

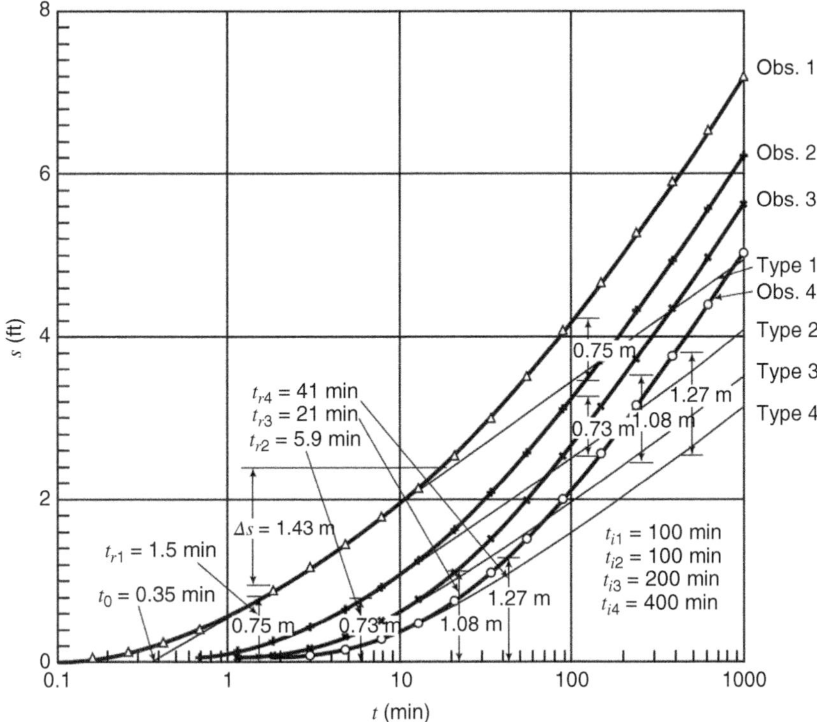

Figure 13.7 An example of drawdown data and Theis fits for observation wells placed near an impervious boundary in a confined aquifer (Chapuis, 1994a. Reproduced by permission of the National Groundwater Assoc. Copyright © 1994. All rights reserved).

Example 13.2 The arrangement of wells is shown in Figure 13.8 for a hypothetical aquifer test affected by an impermeable boundary. The pumping rate in the well is 7.52 m^3/min. The drawdowns measured at four observation wells are listed in Table 13.1. Calculate the aquifer properties and locate the impermeable boundary.

Solution

Plots of drawdown versus time plots are shown in Figure 13.7 for the four observation wells. Use the early part of drawdown plot for observation well 1 to determine the aquifer parameters. We recorded $\Delta s = 1.43$ m and $t_0 = 0.35$ min. The transmissivity is obtained by

$$T = \frac{2.3Q}{4\pi\Delta s} = \frac{2.3(7.52\ \text{m}^3/\ \text{min})}{4\pi(1.43\ \text{m})} = 0.975\text{m}^2/\ \text{min} = 1.4 \times 10^3\ \text{m}^2/\text{day}$$

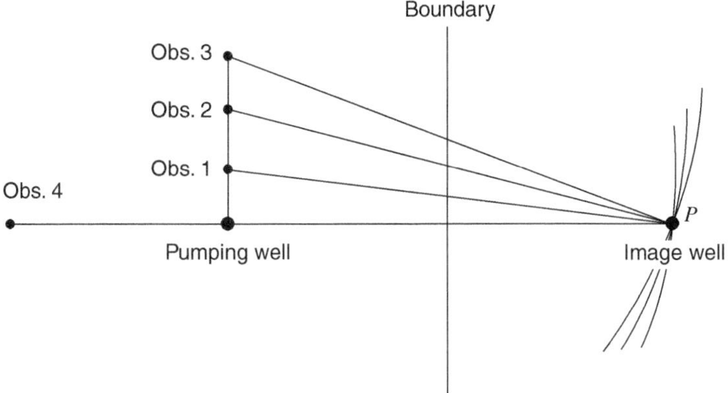

Figure 13.8 Determination of the distances from each of the observation wells to the pumped well provides a basis for locating the image well and the position of the impermeable boundary (Schwartz and Zhang, 2003 / John Wiley & Sons).

Table 13.1 Drawdown measurements in four hypothetical observation wells.

Time(min)	s_1(m)	s_2(m)	s_3(m)	s_4(m)
0.10	0.00	0.00	0.00	0.00
0.16	0.03	0.00	0.00	0.00
0.26	0.09	0.00	0.00	0.00
0.43	0.20	0.01	0.00	0.00
0.70	0.37	0.03	0.00	0.00
1.13	0.59	0.10	0.01	0.00
1.83	0.85	0.22	0.05	0.01
2.98	1.14	0.40	0.13	0.04
4.83	1.44	0.63	0.28	0.11
7.85	1.76	0.90	0.48	0.25
12.74	2.11	1.22	0.74	0.44
20.69	2.52	1.60	1.08	0.71
33.60	2.98	2.04	1.50	1.07
54.56	3.50	2.55	1.99	1.50
88.59	4.07	3.11	2.53	2.00
143.84	4.67	3.70	3.11	2.56
233.57	5.29	4.32	3.73	3.15
379.27	5.93	4.95	4.36	3.77
615.85	6.57	5.60	5.00	4.41
1000.00	7.23	6.25	5.65	5.05

Source: Schwartz and Zhang (2003) / John Wiley & Sons.

The storativity is

$$S = \frac{2.25(0.975m^2/\min)(0.35\min)}{(50m)^2} = 3.1 \times 10^{-4}$$

Theis drawdowns are calculated for each observation well using derived T and S (Figure 13.7). The calculated drawdown is the contribution from the pumped well, assuming the aquifer to be infinite. The drawdown contributions from the image well ($s_{i1} = 0.75$ m, $s_{i2} = 0.73$ m, $s_{i3} = 1.08$ m, and $s_{i4} = 1.28$ m) at selected times ($t_{i1} = 100$ min, $t_{i2} = 100$ min, $t_{i3} = 200$ min, and $t_{i4} = 400$ min) are calculated and shown in Figure 13.7. Times are marked on the semi-log drawdown plots for the same contribution from the real and image wells. The calculation of distances from the image well to the various observation wells is summarized in Table 13.2. Four arcs are drawn, centered at the observation wells. The image well is located at point P in Figure 13.7.

Table 13.2 Distances from the pumped and image wells to observation wells and times to achieve the same drawdowns.

Observation well	No. 1	No. 2	No. 3	No. 4
r_r (m)	50.0	100.0	150.0	200.0
t_i (min)	100.0	100.0	200.0	400.0
t_r (min)	1.5	5.9	21.0	41.0
r_i (m)	408.2	411.7	462.9	624.7

Source: Schwartz and Zhang (2003) / John Wiley & Sons.

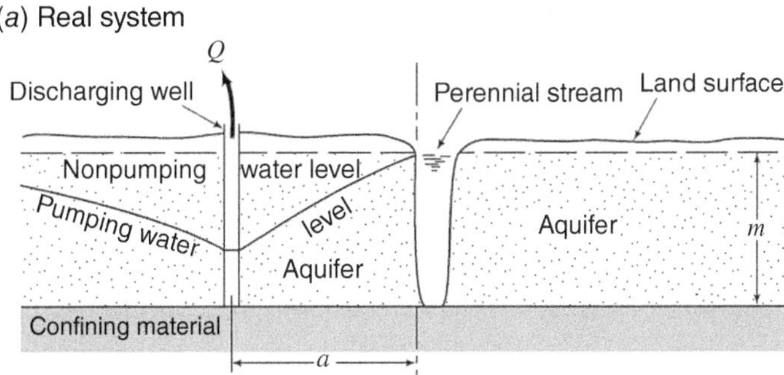

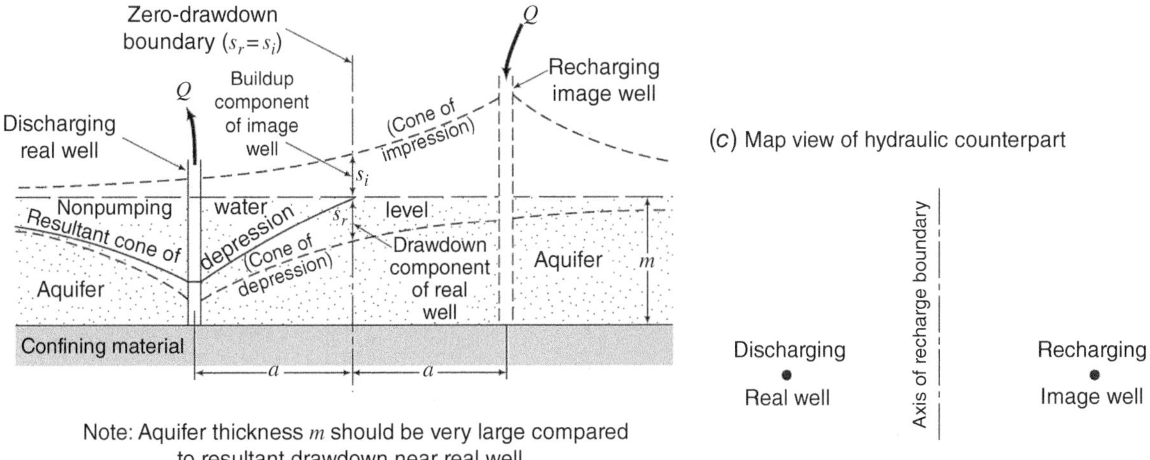

Figure 13.9 A cross section illustrating an aquifer with a recharge boundary. Panel (*a*) shows the real system; (*b*) the hydraulic counterpart of the real system; and (*c*) a map view of the hydraulic counterpart (U.S. Geological Survey Ferris et al., 1962/Public domain).

13.3.2 Recharge Boundary

The procedure for handling a recharge boundary (e.g., a fully penetrating stream; Figure 13.9*a*) involves replacing the boundary with one or more wells. The image well is located across the boundary, the same distance away as the pumped well (Figure 13.9*b* and 13.9*c*). At early time, drawdown in an observation well close to the real well is due only to the real well pumping. Eventually, the total drawdown is the combination of drawdown caused by the pumped well and buildup produced by the image well (Figure 13.10). Along the boundary, the drawdown is zero because the drawdown and the buildup cancel each other. The drawdown in an observation well in this case is expressed as

$$s = s_r - s_i = \frac{Q}{4\pi T}[W(u_r) - W(u_i)] \tag{13.9}$$

As before, time variation in the pattern of drawdown at an observation well can be interpreted to provide hydraulic parameters for the aquifer and the location of the recharge boundary. In calculating aquifer parameters, again use the initial (early time) slope of the drawdown versus the log of time plot for observation wells. Under the influence of a recharge boundary, the buildup of the head attributable to the image well is the difference between the measured drawdown and that due to the calculated Theis drawdown caused by the real well (s_i in Figure 13.10). Equation (13.7) also applies to drawdown produced by the real well and buildup due to the image well. In practice, at least four observation wells are necessary to determine the location of the recharge boundary (Figure 13.11).

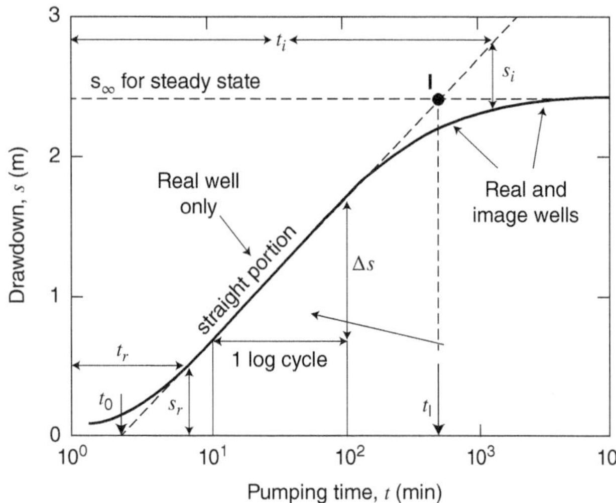

Figure 13.10 A plot of drawdown versus time and the early-time Theis fit for an observation close to a recharge boundary in a confined aquifer (adapted from Chapuis, 1994a. Reproduced by permission of the National Groundwater Assoc. Copyright © 1994. All rights reserved).

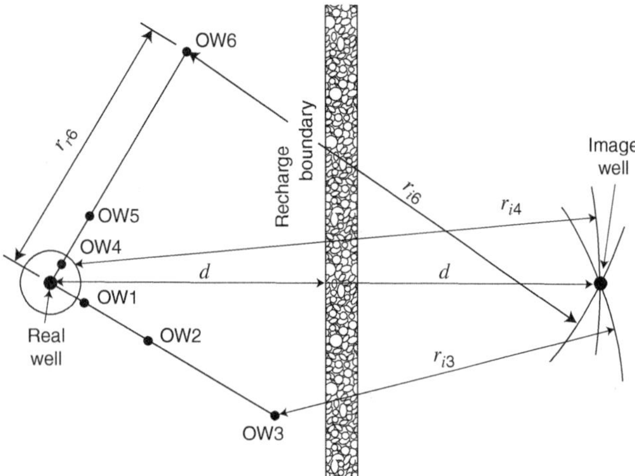

Figure 13.11 Determining the location of a recharge boundary with at least three observation wells (adapted from Chapuis, 1994b. Reproduced by permission of the National Groundwater Assoc. Copyright © 1994. All rights reserved).

13.4 Multiple Boundaries

Image theory is also applicable to the analysis of flow in an aquifer with multiple boundaries. With more than one boundary present, multiple image wells are usually needed to replace the boundaries. In locating the image well, each boundary should be considered to determine the type of image well and where it is located. For example, consider a case where a recharge stream and an impervious boundary intersect at right angles (Figure 13.12*a*). First, we consider the recharge boundary by adding an injection well (image well 1) and the impermeable boundary by adding a discharge well (image well 2). Although the image well 1 will create zero drawdown along the recharge boundary, it will not provide for no-flow conditions at the impermeable boundary. Another injection well (image well 3) is needed to create no flow at the impermeable boundary. The resultant drawdown will be

$$s = \frac{Q}{4\pi T}[W(u_r) - W(u_{i1}) + W(u_{i2}) - W(u_{i3})] \tag{13.10}$$

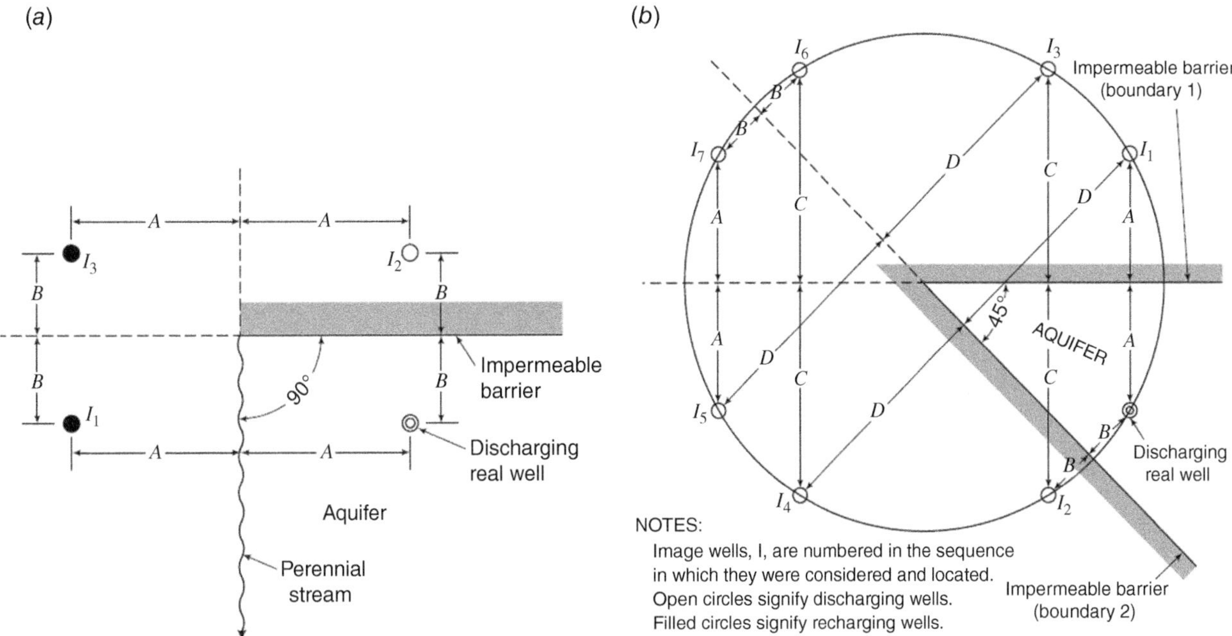

Figure 13.12 The arrangement of image wells for various arrangements of aquifer boundaries. In Panel (*a*) a fully penetrating stream and an impermeable boundary intersect at right angles. In Panel (*b*), two impermeable boundaries provide a wedge-shaped aquifer. Dealing with these boundaries for a single withdrawal well requires seven image wells (U.S. Geological Survey Ferris et al., 1962/Public domain).

For a wedge shape boundary bounding an aquifer pumped by a single discharge well (Figure 13.12*b*), the number of image wells is given by (Ferris et al., 1962)

$$n = \frac{360^\circ}{\theta} - 1 \tag{13.11}$$

where θ is the wedge angle.

13.5 Calculation and Interpretation of Hydraulic Problems Using Computers

We have derived an equation that describes the groundwater flow in Chapter 5:

$$\frac{\partial}{\partial x}\left(K_x \frac{\partial h}{\partial x}\right) + \frac{\partial}{\partial y}\left(K_y \frac{\partial h}{\partial y}\right) + \frac{\partial}{\partial z}\left(K_z \frac{\partial h}{\partial z}\right) + Q(x, y, z, t) = S_s \frac{\partial h}{\partial t} \tag{5.27}$$

We have introduced a number of groundwater flow solutions from Chapter 8 to Chapter 13. These simple solutions are called analytical solutions or analytical models. They can be solved by simple computer programs or Excel spreadsheets. In this section, we will describe some numerical models used in groundwater simulations:

13.5.1 Numerical Models for Groundwater Simulations

There are several types of numerical methods that are applied to groundwater simulations. USGS developed a finite difference model, MODFLOW-2005, available on USGS website (MODFLOW-2005: USGS Three-Dimensional Finite-Difference Groundwater Model I U.S. Geological Survey). The model was subsequently mass transported in groundwater,

MODFLOW 6 (MODFLOW 6: USGS Modular Hydrologic Model | U.S. Geological Survey). This computer software can be downloaded from the above link. Basic idea is to discretize the simulation domain into cells and calculate hydraulic head in each cell (Figure 13.13).

Simulation domain normally consists of consisting of layers, rows, and columns. Hydraulic head or groundwater rate is specified on the boundaries of the simulation domain. MODFLOW 6 allows a general unstructured grid (Figure 13.14).

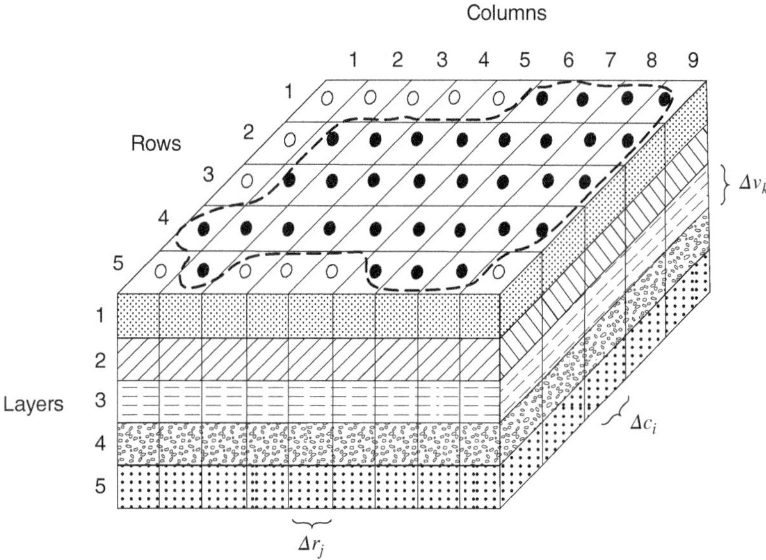

Figure 13.13 Simulation domain of MODFLOW software (MODFLOW-2005: USGS Three-Dimensional Finite-Difference Groundwater Model) (U.S. Geological Survey Harbaugh et al. 2017/Public domain).

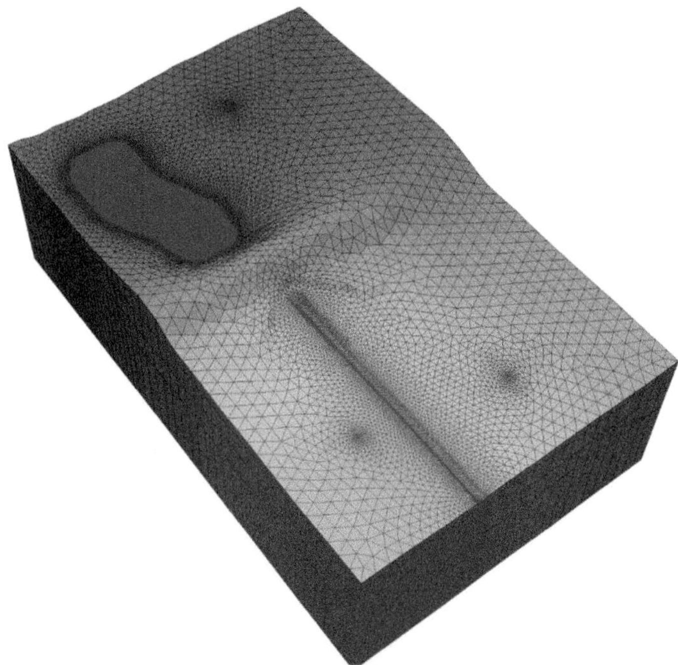

Figure 13.14 A triangular grid in which the size of the triangular cells is reduced in areas with relatively large hydraulic gradients, such as around the shoreline of a lake, near pumping wells, and along a stream. This type of layered grid can be represented using the Discretization by Vertices (DISV) Package in MODFLOW 6 (MODFLOW-2005: USGS Three-Dimensional Finite-Difference Groundwater Model) (U.S. Geological Survey Langevin et al. 2022/Public domain).

Another suite of software is TOUGH (TOUGH2 website). The **TOUGH** ("**T**ransport **O**f **U**nsaturated **G**roundwater and **H**eat") suite of software codes are multi-dimensional numerical model for simulating the coupled transport of water, vapor, noncondensible gas, and heat in porous and fractured media. The computer software is more complex and is widely used for solving a number of groundwater problems. For example, TOUGH 2 or 3 can be used to simulate unsaturated flow while MODFLOW is limited to saturated flow simulation.

13.5.2 Interpreting Aquifer Tests

Personal computers are useful for interpreting aquifer tests. The interpretative schemes basically involve minimizing the difference between the measured drawdown in a well and the calculated drawdown in the same well. Eventually, a unique set of aquifer parameters (e.g., T and S) will emerge from this analysis. This approach is cast mathematically as minimizing the mean squared error (MSE) between measured and calculated drawdowns

$$\varepsilon = \frac{1}{n}\sum_{i=1}^{n} f_i^2 = \frac{1}{n}\sum_{i=1}^{n}\left[s(t_i,\overline{P}) - s_p(t_i,\overline{P})\right]^2 \tag{13.12}$$

where ε is the MSE, n is the number of measurement times, s and s_p are measured and predicted drawdowns at time t_i, respectively, and f_i is the difference between measured and predicted drawdowns at time t_i. $\overline{P}$ is a vector of length m with

$$\overline{P} = (p_1, p_2, \dots, p_m) \tag{13.13}$$

where m is the number of parameters to be determined. The process of determining the parameter vector P is called *inversion*. The MSE error is called the *objective function* in an inversion algorithm. The inversion process finds a set of aquifer parameters, which minimize the objective function.

Exercises

13.1 In a hydraulic test near a recharge boundary in a confined aquifer, the pumping rate is 7.52 m^3/min. The locations of the observations are shown in Figure 13.15 and the values of time versus drawdown in the four observation wells are listed in Table 13.3. Calculate the hydraulic parameters of the aquifer and determine the recharge boundary.

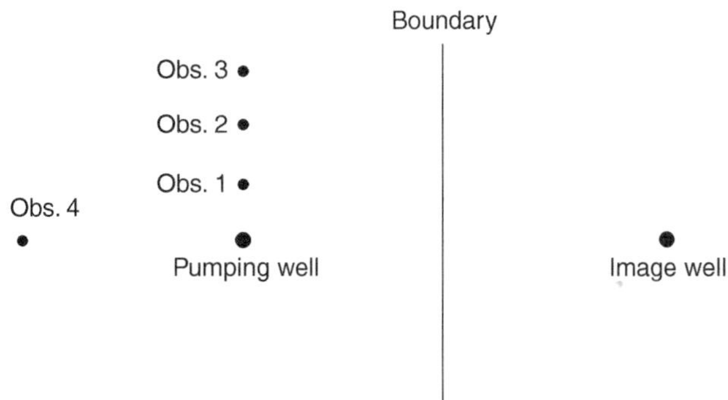

Figure 13.15 A recharge stream and an impermeable boundary intersect at right angles (Schwartz and Zhang, 2003 / John Wiley & Sons).

Table 13.3 Values of drawdown vs. time for a pumping test in a confined aquifer near a recharge boundary.

Time (min)	$r = 5$ m	$r = 100$ m	$r = 150$ m	$r = 200$ m
0.10	0.00	0.00	0.00	0.00
0.16	0.03	0.00	0.00	0.00
0.26	0.09	0.00	0.00	0.00
0.43	0.20	0.01	0.00	0.00
0.70	0.37	0.03	0.00	0.00
1.13	0.59	0.10	0.01	0.00
1.83	0.85	0.22	0.05	0.01
2.98	1.14	0.40	0.13	0.04
4.83	1.44	0.63	0.27	0.11
7.85	1.74	0.88	0.47	0.24
12.74	2.03	1.14	0.68	0.42
20.69	2.27	1.37	0.88	0.61
33.60	2.46	1.55	1.05	0.79
54.56	2.59	1.68	1.18	0.94
88.59	2.69	1.77	1.27	1.05
143.84	2.75	1.83	1.33	1.12
233.57	2.79	1.87	1.37	1.17
379.27	2.81	1.89	1.39	1.20
615.85	2.83	1.91	1.40	1.22
1000.00	2.84	1.92	1.41	1.23

Source: Schwartz and Zhang (2003) / John Wiley & Sons.

References

Bear, J. 1979. Hydraulics of Groundwater. McGraw-Hill, New York, p. 569.

Chapuis, R. P. 1994a. Assessment of methods and conditions to locate boundaries: 1. One or two straight impervious boundaries. Groundwater, v. 32, no. 4, p. 576–582.

Chapuis, R. P. 1994b. Assessment of methods and conditions to locate boundaries: 1. One straight recharge boundary. Groundwater, v. 32, no. 4, p. 583–590.

Ferris, J. G., D. B. Knowles, R. H. Brown, and R. W. Stallman. 1962. Theory of aquifer tests. U.S. Geological Survey Professional Paper 708, 70 p.

Harbaugh, A. W., C. D. Langevin, J. D. Hughes, R. N. Niswonger, and L. F. Konikow. 2017. MODFLOW-2005 version 1.12.00, the U.S. Geological Survey modular groundwater model: U.S. Geological Survey Software Release, 03 February 2017, http://dx.doi.org/10.5066/F7RF5S7G

Heath, R. C. 1983. Basic groundwater hydrology. U.S. Geological Survey Water-Supply Paper 2220, 84 p.

Javandel, I., C. Doughty, and C. Tsang. 1984. Groundwater Transport: Handbook of Mathematical Models. American Geophysical Union, Washington DC, 228 p.

Langevin, C. D., J. D. Hughes, A. M. Provost, et al. 2022. MODFLOW 6 Modular Hydrologic Model version 6.4.1. U.S. Geological Survey Software Release, 9 December 2022, https://doi.org/10.5066/P9FL1JCC.

Schwartz, F. W., and H. Zhang. 2003. Fundamentals of Groundwater. John Wiley & Sons, Hoboken, New Jersey, 583 p.

Stallman, R.W. 1952. Nonequilbrium type curves for two well systems. U.S. Geological Survey Groundwater Notes. U.S. Geological Survey Open file Report 3.

14

Depletion of Groundwater Resources

CHAPTER MENU

This chapter describes and explains key problems that come along with the overpumping of aquifers, large and small. The most serious concern is the decline of the groundwater resource by pumping stored water out of aquifer faster than it is being replenished. Overpumping may also come along with a collection of other serious problems such as land subsidence, diminished streamflow, devastation of riparian ecology, and chemical destruction caused by seawater intrusion.

Several different strategies exist to evaluate impacts related to the overpumping of aquifers. The first is a data-based approach that involves long-term monitoring of groundwater and surface-water systems. Most of the examples presented in this chapter illustrate this approach, as practiced in the United States through the long-term efforts of the U.S. Geological Survey. Groundwater modeling (Section 14.6) is a second strategy to identify issues associated with overpumping and to understand problems likely to develop in the future.

14.1 Water-Level Declines from Overpumping

Just how much water can be pumped from an aquifer without causing problems? Answering this simple question has motivated the work of hydrogeologists for more than a century. As will become evident in this chapter, there are now theories and tools available to answer this question. These advances form the bases of the modern scientific management of aquifers.

However, most groundwater users do not need complicated theories to understand the symptoms of a depleting aquifer system. Wells no long yield as much water to the point where they become dry. Deepening of wells may be helpful to a certain extent but there are practical constraints based on costs, where the bottom of the aquifer is located, or some unacceptable deterioration in water quality.

One fundamental idea from well hydraulics is that in the absence of recharge or nearby sources of water, such as rivers or lakes, the water pumped by a well comes out of aquifer storage. The practical implications may not be so obvious. Analytical theories in well hydraulics are built on the assumption of an aquifer that is infinite in extent with a cone of depression that can reach out farther and farther to access stored water forever. In the real world, an aquifer has a

Fundamentals of Groundwater, Second Edition. Franklin W. Schwartz and Hubao Zhang.

Companion website: www.wiley.com/go/schwartz/fundamentalsofgroundwater2

finite size (length, width, and thickness) that implies some finite quantity of stored water. Thus, in recharge-poor settings, tens to hundreds of thousands of water users with high-capacity wells can remove most of the stored water in large aquifers in a hundred years or less.

Before the development of aquifers with wells, regional flow systems will be evident with zones of recharge, lateral flow, and discharge. As an aquifer is exploited by wells, the predevelopment distribution of hydraulic head transitions to one where hydraulic head distribution is determined by the groundwater withdrawals from wells. In effect, the collection of pumped wells creates new discharge areas. Recharge to the aquifer moves through a perturbed flow system to the pumping wells.

If many wells are distributed across some aquifer, for example, with irrigation for agriculture, there can be a broad lowering of water levels with large cones of depression. With towns and cities, groundwater pumping is often localized around municipal or industrial well fields where pumping is significantly higher than surrounding domestic wells. Figure 14.1 illustrates a large aquifer system where flow conditions have changed due to extensive urban pumping, especially at pumping centers. The schematic (Figure 14.1) is a potentiometric-surface map for part of the Memphis aquifer, at Memphis Tennessee (Taylor and Alley, 2001; modified from Kingsbury, 1996).

What existed before pumping began was a simple flow system with uniform flow, approximately east to west, away from the outcrop area for the aquifer. The map shows a regional cone of depression (>50 km north to south) created by broadly distributed urban pumping. At five well fields, focused municipal and industrial pumping depressed water levels even further. The addition of flow arrows to the schematic provides an indication of just how much the regional flow adjusted in response to pumping. After development, groundwater flowed from all directions toward the middle of the large cone of depression.

The development of this large regional cone of depression was indicative of a situation where pumping was much greater than available recharge. Much of the water being pumped came from aquifer storage that promoted a decline in water levels and the growth in the cone of depression. The relatively large size of this aquifer contributed to the longevity of pumping.

The following sections provide a more detailed look at some of the other problems associated with aquifers. We begin with the problem of subsidence (Section 14.2) where the ground surface sinks because of the decline in pore pressures that

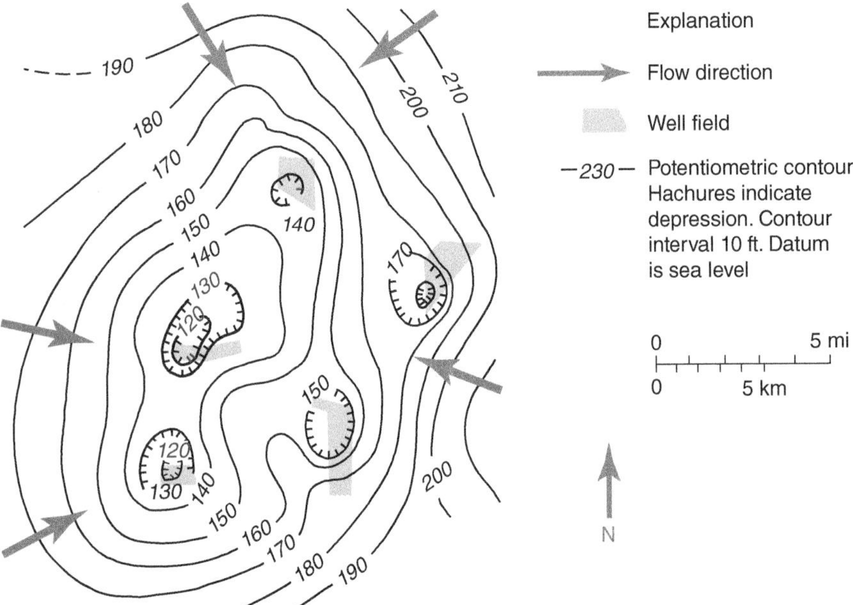

Figure 14.1 What started out as simple flow system with groundwater flowing from east (right) to west was modified by extensive pumping from various well fields. Flow ended up controlled by cones of depression that attracted flow from all directions (U.S. Geological Survey, adapted from Taylor and Alley, 2001; Kingsbury, 1996/Public domain).

accompany overpumping of groundwater. It is a serious and expensive worldwide problem especially when the subsidence occurs in major cities. Pumping for irrigation may sometimes create subsidence, as was the case in the Central Valley of California.

Section 14.3 discusses the implications of connected groundwater and surface water systems. Overpumping of groundwater can sometimes lead directly to declines in the flows of perennial streams or even the complete loss of flow except for storm flows. These kinds of problem often escape notice because studies need to integrate both groundwater and surface water data over decadal time intervals. In many countries, these kinds of long-term data are unavailable. One area that has been studied this respect is the High Plains aquifer system, the largest aquifer in the United States in terms of water withdrawals. One manifestation of the continuing large withdrawals of ground there has been surface-water impacts, which have been studied in the United States in Texas, Oklahoma, and Nebraska. The combination of pumping close to streams and declines in streamflow have the potential to harm vegetation growing along stream (Section 14.4).

The usability of groundwater as a resource many also impaired because of chemical problems, caused by human activities (e.g., groundwater contamination), the presence of natural contaminants (e.g., geogenic arsenic and fluoride), or some combination of the two. The problems of contamination will be the focus of later chapters. However, here in Section 14.5, we discuss the common problem of seawater intrusion and its potential impact on coastal aquifers. This chapter concludes with an overview of groundwater modeling. As the case studies will show, modeling is particularly useful in analyzing problems in overpumping.

14.1.1 Challenges in the Investigation of Water-level Changes

The practical problems discussed in this chapter are among the most difficult faced by hydrogeologists. The reasons are varied but commonly relate to data issues, inherent complexity in hydrogeologic settings and processes, limitations in the availability and capabilities of models, and capacity to conduct needed studies in terms of the technical expertise of the workforce and government commitment to the mission. In each of the sections, we will highlight what strategies are needed to succeed in understanding and solving problems.

One challenge associated with problems of overpumping is an absence of water level and associated hydrogeological data. Often these data gaps are related to spatial and temporal scales over which problems can develop. In America, the important aquifer systems, High Plains, Central Valley, and Mississippi River Valley, have been intensively developed for about 100 years over extremely large areas. The High Plains Aquifer that is discussed in Section 14.3 extends over 450,000 km^2 with evident variability in the aquifer thickness and the distribution of pumping. Historical and continuing measurements of water levels are essential to fully understand the extent of water level declines and patterns in their development. Here in the United States, long-term efforts of four or five generations of researchers with the United States Geological Survey (USGS) and their reports have been essential to addressing groundwater problems.

Taylor and Alley (2001) summarized ten potential use of water-level data, such as monitoring effects of climate, groundwater and surface interactions, and regional effects of groundwater development. Seven of those uses required efforts in data collection that stretched over years and decades. In many countries, these kinds of data simply do not exist or if they do, they are not curated in a coherent manner. To a large extent, illustrative examples in this section are based on works of the USGS over more than a century.

With time, satellite-remote sensing with gravity-based measurements, for example GRACE (Feng et al., 2018; Long et al., 2017; Rodell et al., 2009, 2018), may have a role to play in monitoring water-level changes in large aquifers. However, comprehensive monitoring of wells will be essential for aquifer-related work.

14.2 Land Subsidence

Land subsidence is a serious problem associated with the overpumping of aquifers containing significant thicknesses of clay units, which might occur as lenses and/or confining beds. With time, the land surface falls relative to the sea-level datum. Local variability in subsidence occurs over horizontal distances of less than 1 m with ground cracks, to several hundreds of kilometers with broad regional declines.

Table 14.1 Summary of the different problems associated with land subsidence due to groundwater withdrawals and important impacts in cities and agricultural areas.

Subsidence problems	Urban impacts	Irrigated agriculture
Ground cracking	Damage to roads, buildings, sewers	Damage to highways, aqueducts
Differential settlement	Structural damages to buildings, foundations	Water distribution to fields, ponding
	Disruptions to high-speed rail damage to tracks, bridges	
	Loss of function with storm, sanitary sewers	
Large subsidence basin	Coastal flooding from ocean storms	Reduced flows along aqueducts
	Floods behind dikes from rainstorms	Expensive modifications to raise side walls on aqueducts
	Increased flooding from disruptions to gradients of drains, streams	
Places	Dhaka, Ho Chi Min City, Jakarta, Mexico City	San Joaquin Valley

Source: FWS.

The problems are pervasive worldwide. The deltas of Asia's rivers are being impacted because of the coincidence of clay-rich aquifers and large groundwater withdrawals, which support large populations and irrigated agriculture. Some of the affected areas include Shanghai on the Yangtze River Delta, the Mekong River Delta in Vietnam, the Ganges–Brahmaputra Delta in Bangladesh, and Chao Phraya Delta including Bangkok.

Other high-risk areas include alluvial plains and basins. Mexico City and California's San Joachim Valley are among the best-known examples in the world. In Mexico City, the land surface has declined more than 10 m over the last century with present-day declines of 30 cm/yr. Irrigated lands in the San Joachim Valley have experienced local declines of nearly 10 m in the first half of the 20th century, which are continuing during periods of drought.

Table 14.1 summarizes how various manifestations of subsidence influence impact cities and areas that depend upon groundwater for irrigation. Ground cracking affects both urban and rural infrastructure, including sewers and aqueducts. Problems associated with differential settlement are often obvious in cities. For example, in Mexico, the longevity of historical buildings is of particular concern, such as the Basilica of Our Lady of Guadalupe. Subsidence there has broken roads and disrupted underground services like sewers. Subsidence was implicated in damages to the above-ground metro system (Kornei, 2017). Differential settlement in farm fields (Table 14.1) in the San Joaquin valley requires attention from farmers to regrade fields to prevent flooding and to maintain flows in irrigation canals and collector drains (Galloway et al., 1999).

At larger scales, other subsidence-related problems, e.g., subsidence basins (Table 14.1) have created even more expensive problems for cities. Subsidence of the coastal areas of cities, such as Jakarta, Indonesia, and Santa Clara, California has necessitated the construction of dikes to prevent ocean flooding. Parts of Jakarta below sea level, experience prolonged flooding on a regular basis from extreme rainfall events on land. Flood flows in streams passing through the city accumulate in subsidence basins close to the coast. The channels of streams flowing into San Francisco Bay at Santa Clara, California required rebuilding at a higher elevation to facilitate discharge into the Bay over the dikes (Galloway et al., 1999).

In the San Joaquin valley, surface water is imported via aqueduct/canal from Northern California for irrigation. Subsidence basins along the route of these system have reduced flows of imported water and required costly repairs to keep water flowing.

14.2.1 Conceptual Model

This section examines four key concepts needed to understand land subsidence. Let us begin by considering some volume of saturated porous medium below the water table. The water in the pores of this volume is pressurized. A piezometer installed into that volume has both an elevation head and a pressure head, the latter a function of the pore pressure. The volume also supports the load of material found above it, up to the ground surface.

The grains making up a porous medium obviously play a role in supporting all the other grains above. The system of grains is known as the "granular" skeleton of the porous medium (Galloway et al., 1999). The 1st important concept is the idea that pressure of the pore fluid is supporting the load as well (Galloway et al., 1999). With an overlying load on the volume, which

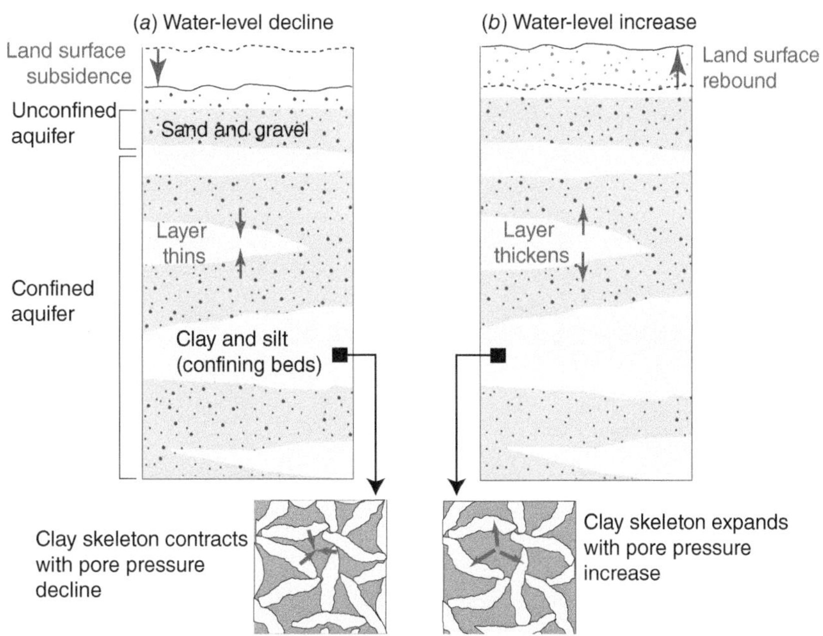

Figure 14.2 (*a*) illustrates how a reduction in water levels produces a reduction in pore pressure and subsidence of the land surface. This decline causes more of the load to be carried by the skeleton of the aquifer which compresses and reduces the porosity slightly, see blowup for (*a*). With a rise in the water table (*b*), the porosity increases, see the blowup, and the land surface returns to its original level. This behavior illustrates elastic deformation (U.S. Geological Survey Galloway et al., 1999/Public domain).

is constant, a decline in pore pressure requires the granular skeleton to carry the extra load, creating a slight reduction in the porosity (Figure 14.2) and subsidence. The opposite effect occurs when the fluid pressure in the element increases, causing the land surface to rise.

Consider the simple analogy of your car tires. The load of a typical tire is carried by the rubber skeleton of the tire and the fluid (i.e., air). Decrease the air pressure in the four tires and the roof of the car is noticeably lowered, because the rubber carcass of a typical tire is compressible. If you have "run-flat tires," the sidewall of the tire is strong (i.e., much less compressible) and can support the car without much air. Thus, the decline in the elevation of the roof is much less noticeable.

The second important concept is that a decline in the ground surface caused by a decrease in pore pressure in media like sand or sand and gravel is completely reversible by increasing the pore pressure. In other words, the deformation of the sand grains can be reversed—so-called elastic deformation (Galloway et al., 1999). Similarly, this outcome is the same in car tires. If pressure declines in a tire, you can simply repressurize it. Aquifers experience this kind of cyclical deformation on a seasonal basis, which produces small (i.e., 2–3 cm) changes in land surface elevations due to fluctuations of the water table (Galloway et al., 1999). A falling water table decreases the pore pressure at depth and leads to a small decline in land-surface elevation (Figure 14.2*a*). A rising water table causes an increase in pore pressure, leading to a slight porosity increase and small increase in the elevation of the ground surface (Figure 14.2*b*).

More severe cases of land subsidence can cause subsidence as much as 10 m or more. These situations develop when two conditions exist together, large drawdowns due pumping of groundwater from permeable zones in a confined aquifer and an abundance of clay units either as confining beds or as clay lenses (Galloway et al., 1999). Declines in hydraulic head of 100s of meters shifts support of overlying materials substantially from the pressurized pore fluid to the skeleton of the granular media. The 3rd important concept is that in a sequence of alluvium with sand and clay units subsidence is due almost entirely to compaction of the clay units. By virtue of its unique, platy structure, clay is compressible to the extent that substantial loading of the clay-rich skeleton causes a large reduction in porosity. Compaction of the clay units physically squeezes the pore water out into adjacent permeable units (Galloway et al., 1999). Estimates are this water provides a one-time increase of 10–30% of the water pumped (Galloway et al., 1999).

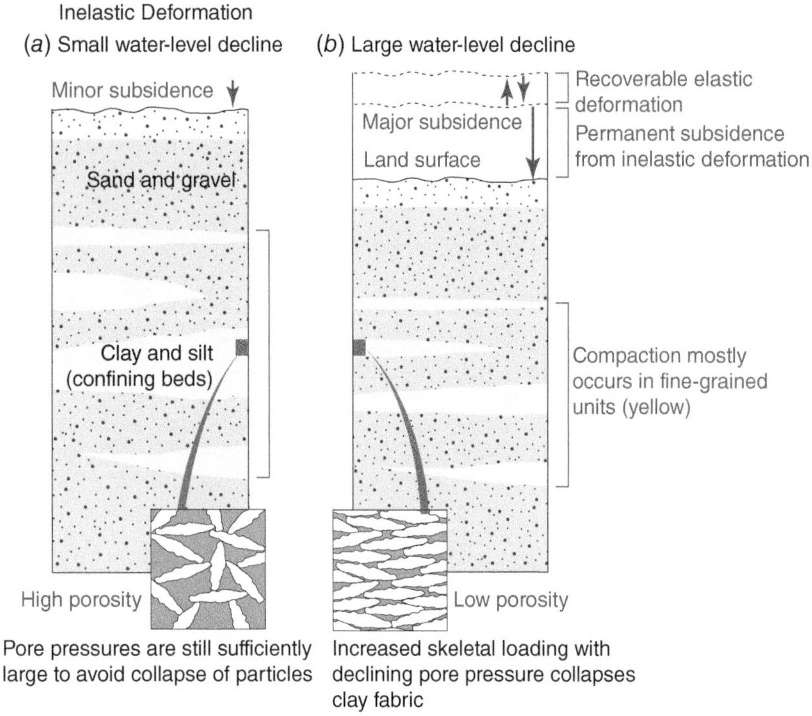

Inelastic Deformation
(*a*) Small water-level decline (*b*) Large water-level decline
Minor subsidence
Sand and gravel
Clay and silt (confining beds)
High porosity
Pore pressures are still sufficiently large to avoid collapse of particles
Recoverable elastic deformation
Major subsidence
Land surface
Permanent subsidence from inelastic deformation
Compaction mostly occurs in fine-grained units (yellow)
Low porosity
Increased skeletal loading with declining pore pressure collapses clay fabric

Figure 14.3 This figure provides illustrates the irreversible compaction of fine-grained lenses or confining beds. (*a*) illustrates a stratigraphic column with fine-grained units having a "house-of cards-structure" as shown in blowup. With significant and continued pumping (*b*), the load shifted to the grain skeleton of fine-grained units is sufficient to destroy the microstructure (see blowup). The porosity reduction and subsidence is substantial with most of that subsidence unrecoverable (U.S. Geological Survey Galloway et al., 1999/ Public domain).

Unlike the situation for sand, substantial compaction of a clay unit is irreversible. The 4th important concept is that even when pore pressures begin to increase from natural or managed aquifer recharge this clay-related subsidence is not recovered. The reason is that the compaction destroys the original, high-porosity clay fabric. Consider the example of a vertical section of alluvium that contains permeable sands and gravels, as well as clay-rich confining beds and lenses (Figure 14.3). Before pumping began (left stratigraphic column), the clays in the confining bed had a "house-of-cards" fabric with a relatively large, saturated porosity (Galloway et al., 1999). Notice that the unit indicated as a "confining bed" was relatively thick. As pumping began, the pore pressure in the clay-rich units declined with minor subsidence (Figure 14.3*a*). Eventually, large reductions in pore pressure led to compaction that destroyed the fabric by crushing the house-of-cards structure (Figure 14.3*b*).

The clay unit ended up with a much smaller porosity and a substantially reduced thickness. This compaction was irreversible because repressurization could not recreate the original clay fabric. However, as shown in Figure 14.3*b*, some elastic compression was recovered.

There are several additional points to understand. The irreversible compaction of clay-rich units reduces the storage capacity of the aquifer. The confining beds and lenses, by virtue of their relatively high compressibility and porosity, comprise a significant part of the water stored in an aquifer (Galloway et al., 1999). Water released as the clay-rich units compact is a one-time release and their ability to store significant quantities of water is lost.

Land subsidence can continue even though corrective action might arrest water-level declines in the aquifer. Because of their low permeability, higher pore pressures in the interior of thick clay-rich will dissipate slowly (Galloway et al., 1999). This time lag in subsidence occurs because slow drainage of low-permeability clay units.

14.2.2 Terzaghi Principle of Effective Stress

The total vertical stress acting on a horizontal plane at some depth due to the weight of overlying water and sediment is exactly balanced by the effective stress, which represents the load supported by grain-to-grain contact within the aquifer

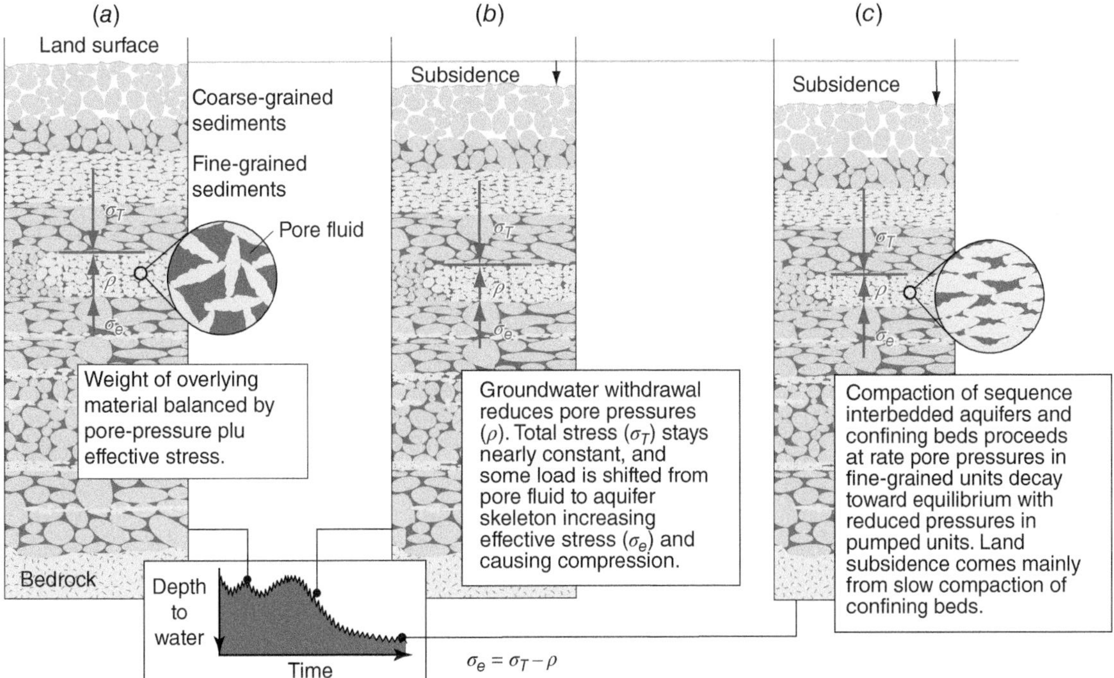

Figure 14.4 This figure illustrates the principle of effective stress. It shows a sequence of aquifer and fine-grained units where the depth to water gradually declines with time. The total effective stress of overlying units is balanced equally by the effective stress and fluid pressure (*a*). That balance shifts in (*b*) and eventually much of the weight of the overburden is borne by the grain skeleton (*c*) (U.S. Geological Survey, adapted from Sneed et al., 2018; Galloway et al., 1999/Public domain).

skeleton, and the pore pressure that is sometimes referred to as the neutral stress. Mathematically, this relationship is expressed as:

$$\sigma_T = \sigma_e + \rho \tag{14.1}$$

where σ_T is the total effective stress, σ_e is the effective stress, and ρ is the pore fluid stress (Galloway et al., 1999).

Figure 14.4 provides three snapshots in time showing the distribution of stresses due to pumping at different stages of depressurization of a confined aquifer containing lenses thick and thin. The small insert of a water-level hydrograph shows that Panel *a* (Figure 14.4*a*) represents the situation before any significant water-level lowering has occurred. Panel *b* shows stress distributions after modest declines in water levels, and Panel *c* after large declines. The plane of reference is shown as a red line on the top of the thick lens in the aquifer (Figure 14.4*a*).

Looking at Figure 14.4*a*, the total stress (σ_T) is balanced by two components of upward stress, ρ and σ_e. The arrows are drawn to scale, so in this case, the aquifer skeleton and fluid pressure contribute equally to balancing total stress due to the weight of overlying material (Galloway et al., 1999). As pumping gets underway (Figure 14.4*b*), the decline in fluid pressure requires that σ_e increases as ρ decreases. The increasing effective stress (σ_e) leads to some compaction of the clay-rich lens and subsidence of the ground surface. With continued a more significant decline in water levels (Figure 14.14*c*), fluid pressures decline markedly. Shifting the load collapses the clay matrix, magnifying the subsidence.

This theory of effective stress has relevance in understanding to variety of other groundwater-related problems. Examples include the origin of earthquakes caused by the injection of wastewater (Hsieh and Bredehoeft, 1981) and the mechanics of overthrust faulting (Hubbert and Rubey, 1959).

14.2.3 Subsidence in the San Joaquin Valley of California

This section focuses on problems land subsidence in the San Joaquin Valley of central California and the long history of associated scientific work (e.g., Bull, 1964; Poland, 1961; Poland and Davis, 1956, etc.). Later chapters discuss water quality issues in the Valley around salinity, nitrate, and geogenic (natural) contaminants. The San Joaquin Valley is part of the

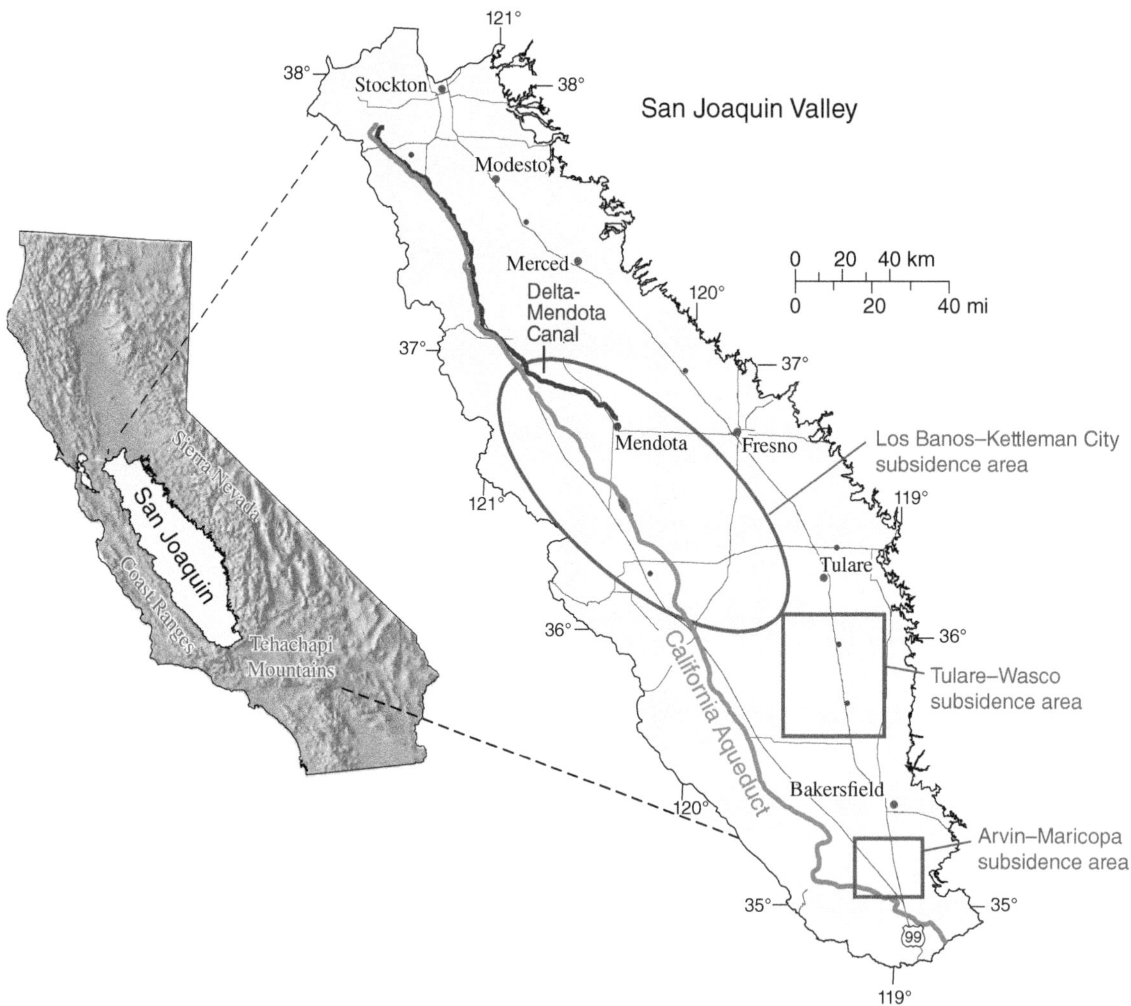

Figure 14.5 The San Joaquin Valley comprises the lower two-thirds of the larger Central Valley of California between the Sierra Nevada Mountains and Coast Ranges. The larger scale map shows important towns and cities. The areas of historical subsidence can be associated with aqueducts supplying imported surface water to the San Joaquin Valley (U.S. Geological Survey Sneed et al., 2018/ Public domain).

larger Central Valley that is located between Coast Ranges to the west and the Sierra Nevada Mountains to the east (Figure 14.5). It extends approximately 400 km from Stockton in the north to the Tehachapi Mountains north of Los Angeles.

The region is important for agriculture in the United States, amounting to approximately 50% of California's production (Hanak et al., 2019). Key commodities include fruits, table vegetables, dairy products, and nuts. Precipitation in the valley is modest with annual precipitation ranging from ~450 in the north to ~150 mm around Bakersfield in the south. Crop production has depended significantly upon irrigation across approximately 22,000 km^2 (Hanak et al., 2019). Before 1960, groundwater was the main source of water for irrigation, which caused significant subsidence across about the valley. The three principal areas of historical subsidence were outlined by Poland and coworkers. Subsidence slowed with the importation of surface water with the Delta-Mendota Canal in 1950, and the California Aqueduct in 1970 (Figure 14.5)

The Central Valley is a structural trough, bounded on the east by the Sierra Nevada Mountains and to the west by the Coast Range (Galloway et al., 1999). The thick (~5000 m) assemblage of sediments ranges in age from Jurassic to Holocene (Faunt, 2009).

Figure 14.6*a* is a conceptual model of the middle portion of the San Joaquin Valley before development of groundwater resources (Belitz and Heimes, 1990; Galloway et al., 1999). The aquifer system consists of sediments accumulated in the structural trough between mountain ranges. Notice the extensive clay units present as lenses and confining beds

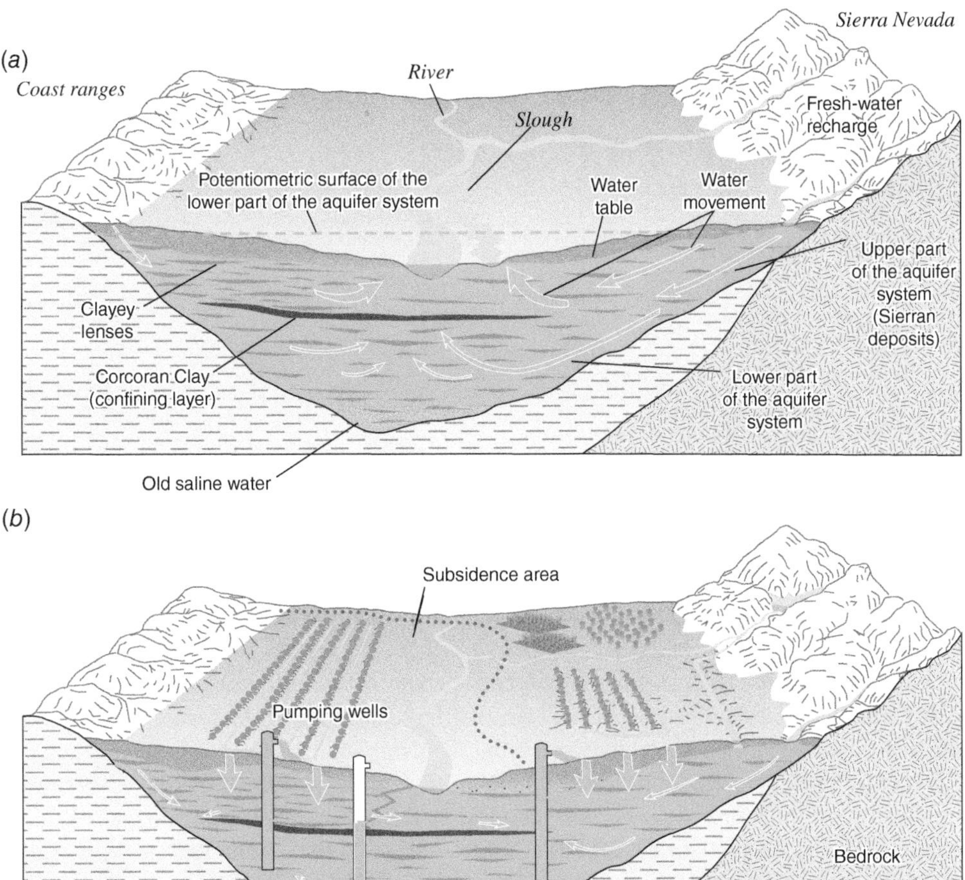

Figure 14.6 (*a*) is a conceptual hydrogeological model showing conditions before significant development of the groundwater in the central spart of the San Joaquin Valley. (*b*) shows the development of groundwater and surface water for irrigation. Much of the water-level declines occurred toward the west (Figure 14.5*c*; left). Drawdowns were greatest in permeable zones of the confined parts of the aquifer below the Corcoran Clay (U.S. Geological Survey Faunt, 2009 as modified from Belitz and Heimes, 1990; Galloway et al., 1999/Public domain).

(Figure 14.6*a*). Freshwater occurred at relatively shallow depths (i.e., upper 400–900 m) in a shallow, unconfined (water table) aquifer or partially confined zones. The unconfined aquifers near the margins of the valley and near the toes of younger alluvial fans facilitated deeper recharge. The Corcoran Clay is a thick (up to 50 m) and laterally extensive lacustrine clay unit, which occurs in the central and western valley (Galloway et al., 1999). The Corcoran Clay confines a deeper aquifer system that contains both fine and coarse-grained units.

Before development, snow-melt runoff from the Sierra Nevada Mountains to the east (Figure 14.6*a*) recharged the aquifers. Groundwater recharged near the mountain front, flowed downgradient and discharged to wetlands and sloughs in the valley bottom or by evaporation. Deeper units had relatively high hydraulic heads to the extent that early wells completed there were artesian flowing wells (Figure 14.6*b*) (Galloway et al., 1999). Now, that streamflow is an integral part of the agriculture water supply used for irrigation, especially early in the growing season.

Most of the historical subsidence occurred on the western side of the valley due to the large withdrawals of water from wells completed in the confined aquifer (Galloway et al., 1999) (Figure 14.6*b*). There tends to be less drawdown on the east side of the valley because irrigators there had access to seasonal surface runoff from the Sierra Nevada Mountains.

Figure 14.7 shows the cumulative depletion of groundwater in the Central Valley from 1900 to 2008 (Konikow, 2013). Cumulatively, ~150 km^3 of groundwater has been pumped from the aquifer. The notes on the figure summarize points of interest along the timeline (from Konikow, 2013).

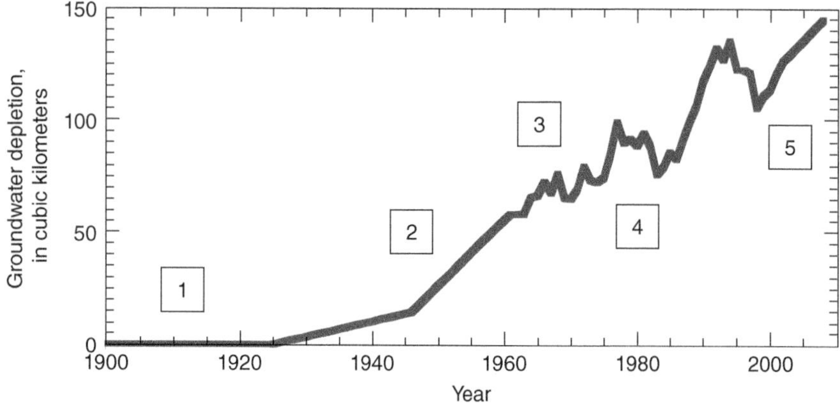

Figure 14.7 Cumulative groundwater depletion in the Central Valley aquifer system versus time from 1900 to 2008 (U.S. Geological Survey, modified from Konikow, 2013 with notes excerpted from Konikow, 2013/Public domain).

The greatest depletion of groundwater occurred from the mid-1940s, into the late 1960s with withdrawal rates of 14.2 km^3/yr in the 1960s. Subsequently, the importation of surface water reduced the reliance on groundwater; but pumping commonly spiked during periods of drought (Figure 14.7). Groundwater remains an important source of water for irrigation.

Land subsidence depends upon two conditions, clay-rich rich units within the aquifer system and large withdrawals of groundwater leading to significant pore pressure declines. Both conditions existed within the San Joaquin Valley. Figure 14.8*a* is a map of drawdown in the deep confined aquifer (Galloway et al., 1999), from 1860 to 1961. Head declines >24 ft (7.3 m) were evident across the southern San Joaquin Valley (Figure 14.8*a*). The largest drawdowns occurred along the western side of the valley (Figure 14.8*a*; dark grey shading), southwest of Fresno where drawdowns exceeded 300 ft (90 m).

Subsidence of >1 ft (0.3 m) was evident over half of the San Joaquin Valley (Figure 14.8*b*). Southwest of Fresno, subsidence was >24 ft (~7.9 m), with a maximum greater than 28 ft (8.5 m) (Galloway et al., 1999).

In the late 1960s, it became apparent that such intensive pumping of water for irrigation was unsustainable. The solution was to replace some of groundwater withdrawals with surface water. To that end, surface water was imported via canals and aqueducts from northern California and water from the San Joaquin River (Faunt et al., 2016). With a reduction in groundwater levels, water levels in the deep aquifer system recovered and rates of subsidence declined. However, periodic droughts have remained a significant vulnerability because of reduced availability of surface-water supplies. As we discuss further in Chapter 16, during droughts farmers returned to groundwater as a reliable source for irrigation water. Another factor leading to increased groundwater utilization during droughts is changes in land use toward "permanent" crops, such as vineyards or orchards (Faunt et al., 2016), which require irrigation water on a continuing basis. For these reasons, groundwater uses for irrigation became variable, ranging from 30% of water requirements in wet years to approximately 70% during years that are extremely dry, such as the California drought from 2012 to 2015 (Faunt et al., 2016).

In the San Joaquin Valley, damages due to subsidence have affected infrastructure, which includes "aqueducts, levees, dams, roads, bridges, pipelines, and well casings" (Sneed et al., 2018). Estimates of historical damages (1955–1972) were US $1.3 billion (Borchers et al., 2014) with expenditures continuing. Canals and aqueducts have required expensive repairs to infrastructure to fix declining flows. This infrastructure is sensitive to subsidence because aqueducts and canals were constructed to move water downhill by gravity (Sneed et al., 2018). Within the San Joaquin Valley, aqueducts were constructed with low flow gradients and have suffered flow reductions because of variability in the magnitude of land surface declines. For example, sags change gradients locally, leading to a reduction in flows and freeboards (the distance between the water surface and the top of the concrete channel liner). Fixes have involved raising aqueduct liners, the associated levees, and other associated infrastructure, i.e., canal service turnouts, drain inlets, bridges, etc. (Borchers et al., 2014; Sneed et al., 2018). Subsidence has also caused cracking of the concrete liner and leakage.

Subsidence is continuing in central San Joaquin Valley through periods of droughts. Over a 17-month period from May 2015 to September 2016, a NASA report (NASA EO, 2017) indicated areas with subsidence as high as 60 cm in the vicinity of

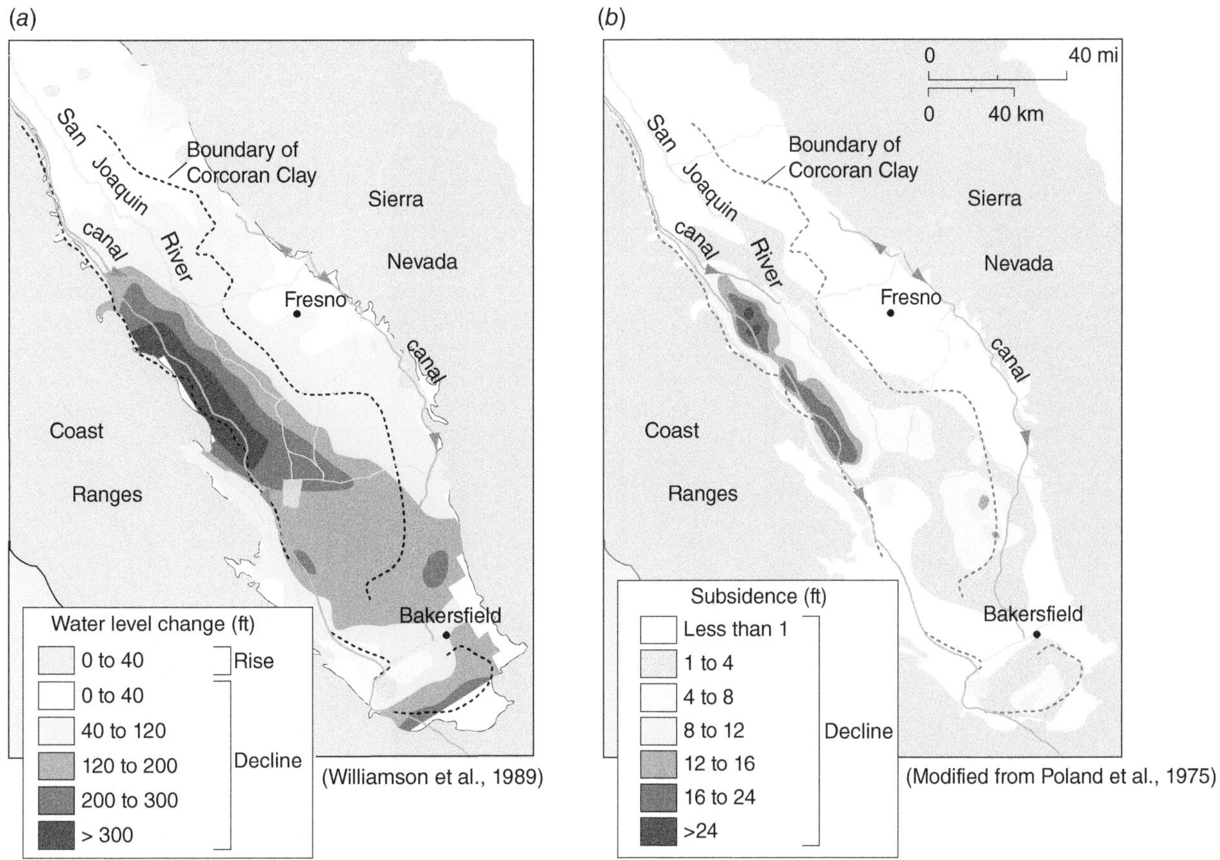

Figure 14.8 Water-level declined from 1861 to 1961 across the San Joaquin Valley in California (*a*). The largest drawdowns exceeded 300 ft (90 m) along the western margin of the basin. Subsidence from 1926 to 1970 (*b*) coincides with the development of the large drawdowns. The areas with the greatest subsidence are generally underlain by the Corcoran Clay (U.S. Geological Survey, (*a*) modified from Galloway et al., 1999; Williamson et al., 1989; (*b*) modified from Galloway et al., 1999; Poland et al., 1975/Public domain).

Corcoran, north of Bakersfield (Figure 14.9), as well comparable areas in the vicinity of Chowchilla, south of Merced, and Tranquility. Unpublished data indicated that from 2015 to 2020, subsidence at Chowchilla ranged from 60 to 90 cm and >90 cm at Corcoran.

14.2.4 Challenges in the Investigation of Subsidence

Understanding the historical development of subsidence due to the overpumping of aquifers requires, water-level data, hydrostratigraphic data, and subsidence measurements through time. As discussed, water-level data often may not exist, especially back in time. High-quality, hydrostratigraphic data are necessary to assess the spatial variability in the distribution of clay lenses and confining beds. Such data are often unavailable. Finally, and most importantly, there is a need for repeated surveys of the decline in the elevation of the land surface. These measurements are relatively uncommon in a hydrogeological context and new technologies are increasingly sophisticated.

Sneed et al. (2018) provided an overview of techniques used for measuring elevations and aquifer-system compaction. The early studies of subsidence in the San Joaquin (1926–1970) were carried out using topographic maps, spirit-level surveying and borehole extensometers. The more recent studies (e.g., Sneed et al., 2018) used Interferometric Synthetic Aperture Radar (InSAR). This method can detect cm-scale changes in surface elevation for many small pixels (e.g., 90 m) over hundreds of square kilometers. It is a radar-based approach that works by recording the travel time from the satellite to the ground surface and back. A second approach involved Continuous Global Positioning (CGPS) using a network of eight stations that for the San Joaquin were originally designed to monitor the behavior of plate boundaries. Developing countries often lack the in-country capacity for this monitoring. However, researchers in other countries have to provide this capability with the broad availability of satellite data.

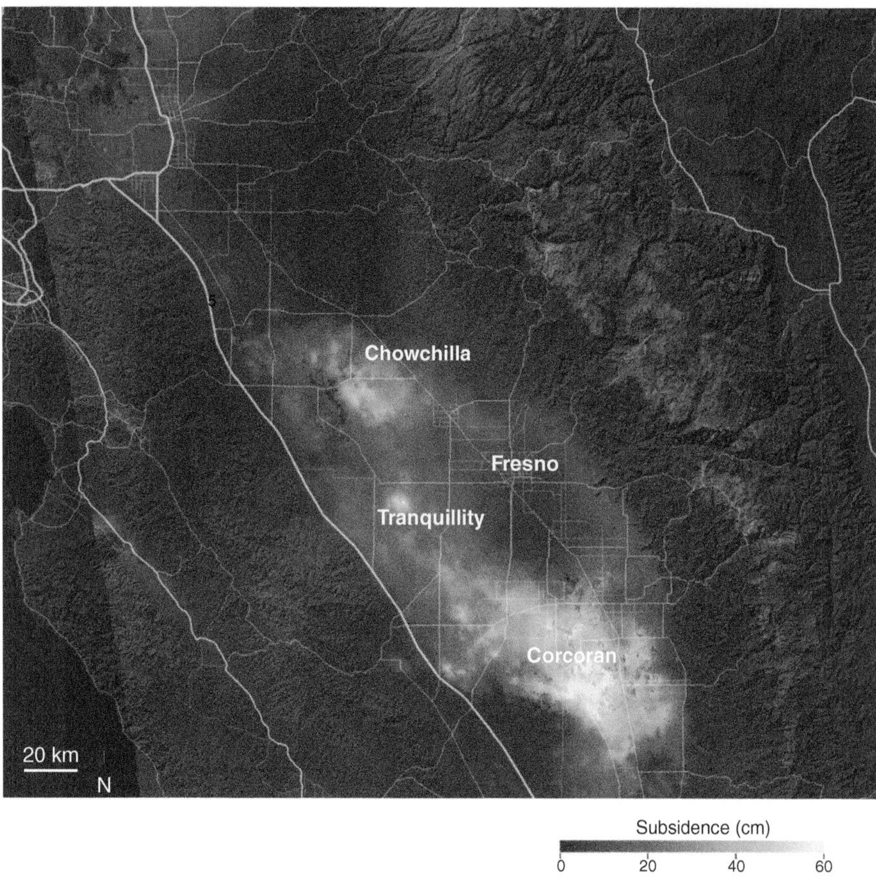

Figure 14.9 Subsidence across the central San Joaquin Valley from May 2015 to September 2017 (Joshua Stevens, 2017/NASA).

14.3 Connected Groundwaters and Surface Waters

Historically, scientists working in water areas saw their areas of expertise as exclusive. Water flowing on the ground surface was the domain of surface-water hydrologists. Soil physicists took care of problems of the unsaturated zone, from the water table up to the ground surface. Groundwater hydrologists took care of problems from the water table downward. This partitioned view of groundwater and surface water was also reflected in laws governing the use of water. For example, in parts of the United States, the use of groundwater was commonly unregulated. On the one hand, landowners, for example, in parts of Texas and Arizona could withdraw all the groundwater they needed. On the other hand, the rights to pump water from a stream flowing were strictly enforced. However, the reality is that groundwater and surface water resources are coupled. They are in effect “a single resource” (Winter et al., 1998). The days of studying groundwater in isolation of surface waters are long gone.

14.3.1 Declines in Streamflow

Groundwater-surface water interactions are difficult to characterize because of the subtleties in the way impacts develop. As the groundwater resources of a basin are developed, problems of overproduction in some areas transition from being a groundwater issue to a surface-water one. Wells pumping in an aquifer in the vicinity of a stream can reduce streamflow in two ways. First, a pumping well can lower water levels in the vicinity of a stream to the extent the water flows out of the stream to the well. This process is termed “induced infiltration of streamflow” (Barlow and Leake, 2012).

Streamflow can also be affected when a well “captures” water that would otherwise discharge into the stream. The cone of depression of a well pumping in an active flow system can reach out and capture some of that flow. Groundwater will still be discharging into the stream but a lesser quantity than expected. Thus, unlike the case of induced infiltration, the stream reach will be "gaining." Given that both mechanisms lead to streamflow reductions, Barlow and Leake (2012) use the general term "*streamflow depletion*" to describe the potential loss in flow.

With both these mechanisms, water wells start out producing water from storage but transition to other sources by capturing induced infiltration and/or recharge (Barlow and Leake, 2012). Early in the pumping history, the cone of depression grows radially until it reaches one of these other two sources of water. Eventually, a steady state is established, and all the water being pumped by the well is from that source. A point of interest in this transition process is the time when more than half of the water being pumped comes from streamflow depletion. Barlow and Leake (2012) termed this time as t_{dss} where "dss" means depletion-dominated supply. The time to steady state with no further drawdown has been called the "*time to full capture.*"

14.3.2 Induced Infiltration of Streamflow

The time needed for full capture of induced streamflow due to a well pumping in the vicinity of stream might be decades, centuries, or more (Barlow and Leake, 2012). Two key parameters control the time to full capture, the distance of the well from the stream (d) and the *hydraulic diffusivity* of the aquifer (D), where hydraulic diffusivity is the aquifer transmissivity divided by the storativity or specific yield or $D = T/S$ with units like m^2/s.

Figure 14.10 illustrates how distance of the pumping well from the stream influences the time response of induced infiltration (Barlow and Leake, 2012). The example assumes that water in the stream is the only potential alternative to aquifer storage. From the pie diagram, note how stream flow depletion is developing much more quickly with well B (closer to stream) than with well A (farther away). After 50 years, approximately two-thirds of the water being pumped by well B is coming from the stream. With well A, most of the water at 50 years is still being provided by aquifer storage. The difference in the time response of the two wells is explained by differences in the growth of the cones of depression. Water from the stream is not available until the appropriately large gradients for flow develop between the well and the stream. The cone of depression for well B being closer to the stream provides for much steeper hydraulic gradients and more efficient transfer of water to the well. With well A, a larger cone of depression needs to develop before water becomes available at the stream. Induced infiltration occurs along a much longer reach of the stream with well A.

The second parameter controlling the time response for induced infiltration of streamflow is hydraulic diffusivity. One can think about hydraulic diffusivity as a parameter describing the expansion of the cone of depression away from a pumping well. A large value of hydraulic diffusivity results in rapid, expansive spreading of the cone of the depression away from the well. Large diffusivity values are associated with confined aquifers, which are characterized by small storativity values. With pumping from an unconfined aquifer, the cone of depression expands more slowly because an unconfined aquifer has much more water in storage as compared to a confined aquifer.

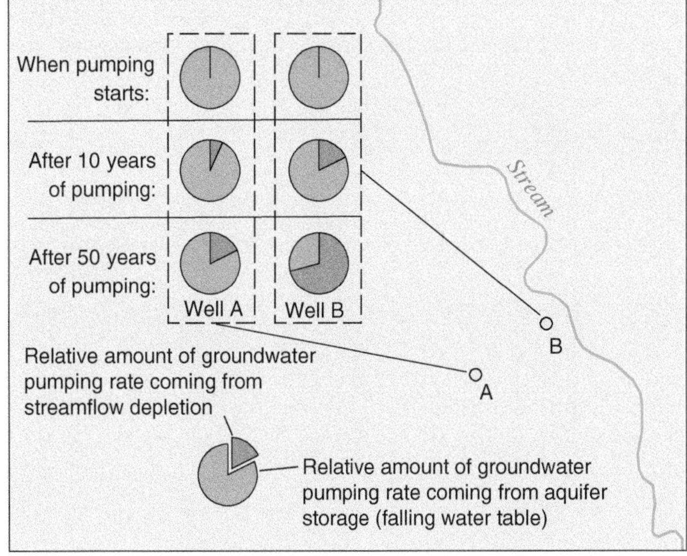

Figure 14.10 In similar aquifers, well B located closer to streams end up transitioning to stream flow depletion much more quickly than well A farther away (U.S. Geological Survey adapted from Leake and Haney, 2010/ Public domain).

Example 14.1 A confined aquifer and an unconfined aquifer have the same transmissivity of 1000 m^2/day. The storativity value for the unconfined aquifer is 1×10^{-4}, and the specific yield for the unconfined aquifer is 1×10^{-1}. Calculate the hydraulic diffusivity for each and qualitatively compare rates of expansion of the cone of depression.

Solution

$$\text{Confined aquifer: } D = \frac{1000}{0.0001} = 1 \times 10^{7}\ \text{m}^2/\text{day} - \text{realtively faster}$$

$$\text{Unconfined aquifer: } D = \frac{1000}{0.1} = 1 \times 10^{4}\text{m}^2/\text{day} - \text{relatively slower}$$

Two parameters are important in determining when water pumped from a well adjacent to a stream will include a significant proportion of stream water. These parameters are the distance of the well from the stream and the hydraulic diffusivity for the aquifer. Equation (14.2) defines the so-called *streamflow depletion factor* (SDF) as a function of these parameters (Jenkins 1968a,b).

$$SDF = \frac{d^2}{D} \text{ or } \frac{d^2 \cdot S}{T} \text{ time} \tag{14.2}$$

where d is distance, D is hydraulic diffusivity, S is storativity or specific yield, and T is transmissivity. SDF has units of time that reflect the lag before significant depletion of streamflow due to that pumping will be evident. The following example, modified from Barlow and Leake (2012), illustrates this calculation and how to make a map showing the spatial variability of SDF. Notice on the map that values of SDF increase nonlinearly away from the stream.

Example 14.2 An unconfined aquifer occurs adjacent to a stream, idealized as a straight line. There are two hypothetical wells A and B that are located 75 and 150 m away from the stream respectively (Figure 14.11). Assume the aquifer has a transmissivity of 95 m^2/day and a specific yield of 0.1. Calculate the stream depletion factor, SDF, for each of the wells.

Solution

Substituting the distance and aquifer parameters into Eq. (14.2), yields the following:

$$\text{Well A } SDF_A = \frac{{d_A}^2 \cdot S}{T} = \frac{75 \times 75 \times 0.1}{95} = 5.9 \text{ days}$$

$$\text{Well B } SDF_B = \frac{{d_B}^2 \cdot S}{T} = \frac{150 \times 150 \times 0.1}{95} = 23.7 \text{ days}$$

As expected, the calculation shows streamflow impacts due to pumping at well A would be expected in several days, as compared to well B.

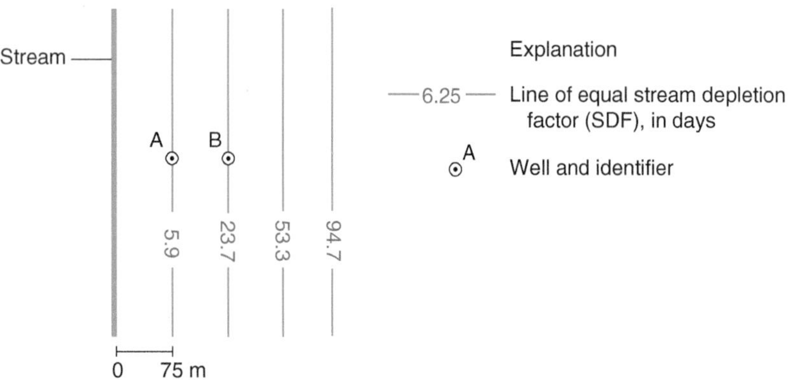

Figure 14.11 For an idealized stream, calculated values of the stream deletion factor (SDF) away from the stream yields contour lines (U.S. Geological Survey, adapted from Barlow and Leake, 2012/Public domain).

As implied with the contouring on Figure 14.11, one can map the stream depletion factor away from a stream. For the simple example, contours of SDF are lines parallel to the stream. There are of SDF maps for actual systems (Hurr and Schneider, 1972) and comparable maps derived by modeling. A key takeaway from this example is that time lags are variable as a function of well locations. In the settings of unconfined aquifers, it might be a long time before effects of pumping would be fully reflected by the depletion in streamflow. Finally, remember that this theory does not account for natural recharge to the aquifer.

Using an analytical solution developed by Glover and Balmer (1954), one can use calculated values of SDF to estimate the stream's contribution (Q_s) to the total production for the well (Q_w). As with all analytical solutions, theirs comes with simplifying assumptions (Barlow and Leake 2012; Jenkins 1968a). Briefly, the aquifer is assumed to be semi-infinite away from the stream, homogeneous, isotropic, and constant in thickness with a fully penetrating well that is pumped at a constant rate. Water is released instantaneously from storage. The infinitely long stream fully penetrates the aquifer and is perfectly connected with it.

Thus, given the pumping rate of the well (Q_w) and parameters (d, T, and S, or D), the solution provides an estimate of streamflow depletion (Q_s) as:

$$Q_s = Q_w \cdot erfc(z) \tag{14.3}$$

where $erfc(z)$ is the complimentary error function, which is a tabulated function for a numerical value of z, where $z = \sqrt{(d^2 S)/(4Tt)}$ or in terms of the streamflow depletion factor (DSF), as $z = \sqrt{DSF/4t}$ (Miller et al. 2007). Equation (14.3) has a variety of uses. For example, at a single well location, one can calculate the response function describing the transition from pumping mostly groundwater to mostly streamflow. Another possibility is to create a series of spatial maps map at t_1, $t_2 ... t_n$ showing the area impacting the stream depletion.

Example 14.3 Using the information from Example 14.2 and given that pumping rate is 4400 m³/day, calculate the surface water contribution to water being withdrawn by a well located first at A and then at location B at 100 days (modified from Barlow and Leake, 2012).

Solution

Well at A $\quad z = \sqrt{(d^2 \cdot S)/(4Tt)} = 0.122$ and $erfc(0.122) = 0.863$

$$Q_s = Q_w \cdot erfc(z) = 4400 \cdot 0.863 = 3800 \text{ m}^3/\text{day}$$

or

$$Q_s/Q_w \approx 0.86$$

Well at B $\quad z = \sqrt{(d^2 \cdot S)/(4Tt)} = 0.243$ and $erfc(0.243) = 0.731$

$$Q_s = Q_w \cdot erfc(z) = 4400 \cdot 0.731 = 3215 \text{ m}^3/\text{day}$$

or

$$Q_s/Q_w \approx 0.73$$

As expected at 100 days, a well located at location B will be pumping 13% less stream water than at location A.

Example 14.3 improves understanding about the stream deletion factor (SDF). In Example 14.3, t was 100 days. What if time (t) in both cases, location A and location B, was chosen to be equal to SDF or 5.9 days and 23.7 days, respectively? In both cases, using the SDF (time) in the Q_s/Q_w calculation determines that slightly less than 50% of the water being pumped would be from the stream or $Q_s/Q_w \approx 0.5$. Using a time value of $0.1 \times$ SDF, Q_s/Q_w is small, with approximately 1.6% of water supplied by the stream. Interestingly, with larger times $10 \times$ SDF or $100 \times$ SDF, reaching $Q_s/Q_w = 1$ can be slow (Miller et al., 2007).

The depletion in streamflows associated with such pumping and significant localized drawdowns are especially problematic in arid areas with sensitive riparian ecosystems. Importantly, these impacts on surface waters scale regionally so that distributed and massive pumping for irrigation in a large aquifer system can deplete the flow of large rivers.

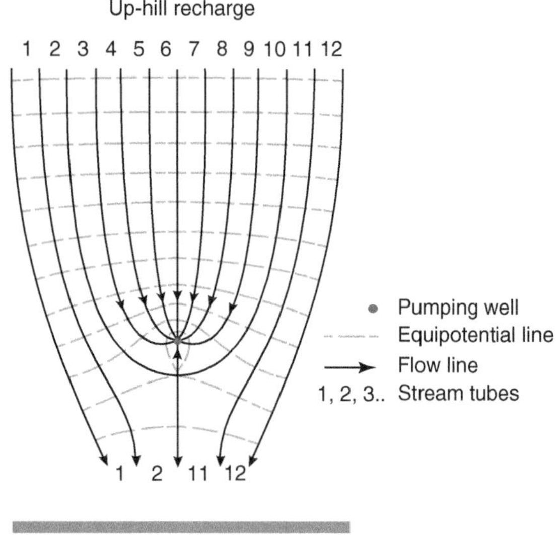

Figure 14.12 A well pumping at a location away from a stream can reduce the stream discharge by capturing some fraction of the groundwater that otherwise would discharge there (FWS).

14.3.3 Capture Zone for a Well

A *capture zone* is the region in the vicinity of a well that supplies water to the well. For a well, pumping in a simple, infinite, artesian aquifer the capture zone is circular corresponding to the depletion of storage within the cone of depression. For the more realistic situation of a well far from a stream, capturing sources of recharge, the capture zone has an oblong shape, extending from the vicinity of the well far up gradient. This more complicated shape develops because the hydraulic head field in the vicinity of the well is a function of the head distribution related to flow of water through the aquifer and drawdown due to pumping.

Figure 14.12 illustrates how an irrigation well located far away from a stream can capture groundwater flow that otherwise would discharge into the stream. Before pumping, the 12 stream tubes would extend from some up-hill recharge area to the stream. After pumping begins, this simple uniform flow system becomes more complicated. Flow tubes 3 through 10 (Figure 14.12) end at the well, effectively contributing to water being pumped. This set of stream tubes define the extent of the capture zone for this well. In the case of flow tubes 1, 2 and 11, 12, although attracted toward the well, they bypass it and continue to the distant stream. Given that the quantity of flow is the same in each of the stream tubes, pumping reduces the groundwater contribution to the stream from this segment of the flow system by 2/3.

The pattern of flow for this simple problem can be calculated using the principle of superposition and a simple analytical solution drawdown due to pumping from a single well. The drawdown created by the pumping well is subtracted from the hydraulic head distribution for the undisturbed flow system before pumping began.

14.3.4 Pumping of the High Plains Aquifer System and Streamflow Reduction

The High Plains aquifer system in the Midwest of the United States continues to be impacted by unsustainable groundwater withdrawals. It is also associated with declining streamflows associated with widespread lowering in water levels. It is also noteworthy as the largest aquifer in the United States. It extends from South Dakota to the Texas Panhandle with an area of approximately 450,000 km^2 (Figure 14.13). The High Plains aquifer system stores ~4000 km^3 of groundwater (McGuire et al., 2003). Present recharge to the southern two-thirds of the High Plains aquifer is small. The stored water there is old, a function of higher recharge when climatic conditions were wetter during to last glacial period. Groundwater pumped from the High Plains aquifer system is used mostly for irrigation of agricultural crops. For example, Kansas is among the top cattle-producing area of the United States, which is made possible by corn produced locally and irrigated by groundwater.

The High Plains is referred to as an "aquifer system" because in places it is comprised of several permeable units of different geological age. The Ogallala Formation, Miocene in age, comprises much of the High Plains aquifer system (McGuire et al., 2003). This formation consists of sands and gravels that were deposited in coalescing alluvial fans by streams flowing from the Rocky Mountains. In the north, the High Plains aquifer system, also includes Pleistocene and younger deposits of dune sand and loess. Because of erosion, the aquifer system is no longer connected to the Rocky Mountains. The thickness of the aquifer decreases from >300 m in Nebraska southward to Kansas and Texas where it is 100 m or commonly less. Locally, the aquifer system is thicker along buried drainage systems.

The original saturated thickness has been declining due to massive pumping for irrigation. Since 1950, the cumulative depletion in storage reached ~341 km^3 (Figure 14.13; Konikow, 2013). This total represents the largest depletion of any aquifer in the United States. From 2001 to 2008 alone, the depletion was 82 km^3. Severe droughts in 2012 and 2013 appear to have increased withdrawals. Depletion is continuing, because here there is no readily available supplies of surface water to supplement groundwater.

There are large and continuing declines in water levels especially south of the Arkansas River in Colorado (Figure 14.14). In the Texas Panhandle and southwestern Kansas, water-levels declines in several places have exceeded 150 ft (46 m), and

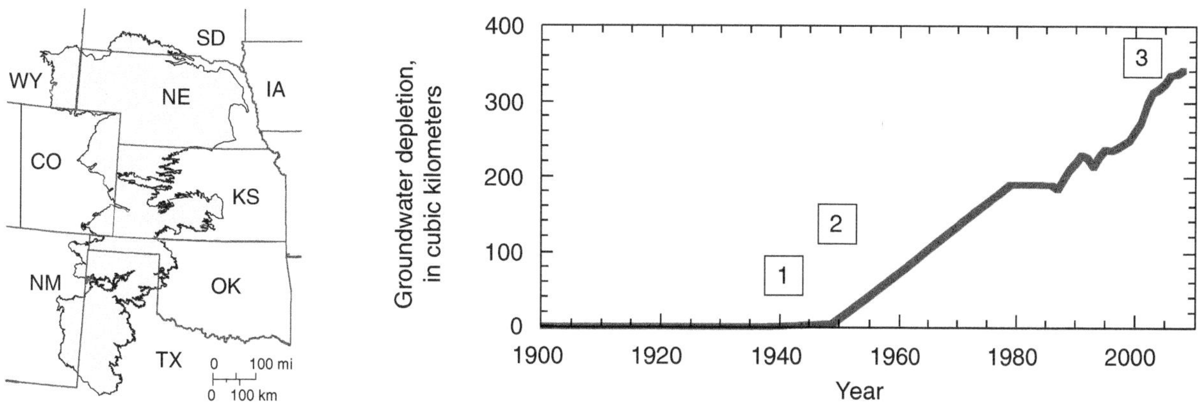

Notes 1. Pumping for irrigation began in 1940s
2. Pumping rates high, especially Texas, 4.9 km^3/yr 1949
3. Depletion yet to stabilize up to present time

Figure 14.13 Groundwater depletion in the High Plains aquifer system versus time from 1900 to 2008 (U.S. Geological Survey, adapted from Konikow, 2013 with notes excerpted from Konikow, 2013/Public domain).

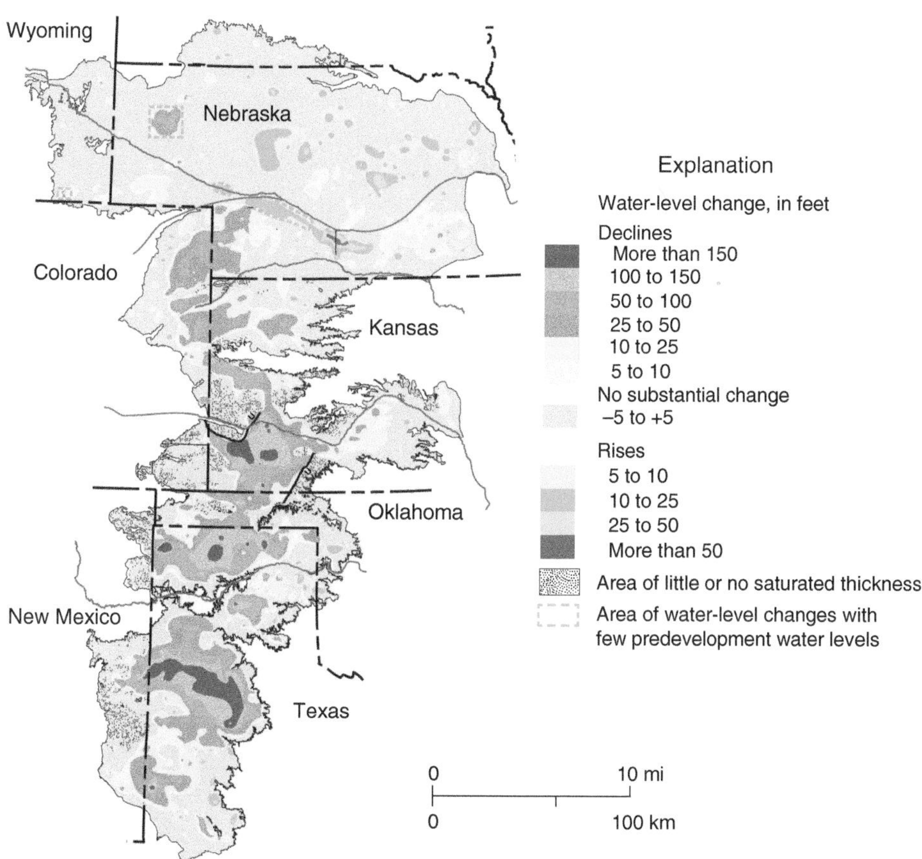

Figure 14.14 Map of water-level changes in the High Plains aquifer system from predevelopment (~1950) to 2014 (U.S. Geological Survey, adapted from McGuire, 2017/Public domain).

>50 ft (15 m) occur over relatively large areas. In Nebraska, recharge is larger so that cumulative water-level declines are smaller and in places even increasing. The localization of the areas of greatest drawdown (Figure 14.14) points to a somewhat disconnected aquifer system with a variable thickness (Alley et al., 1999). The large rates of withdrawal are unsustainable and potentially able to drain the aquifer. In some places, conservation practices have been implemented to reduce the rate of decline, but at this point, hydrogeologists are simply monitoring the decline.

There are estimates of the likely lifespan of the aquifer in supporting high-intensity, irrigated agriculture in western Kansas (Buchanan et al., 2015). In some areas in western Kansas, the aquifer already has been exhausted. These areas coincide with places where the saturated thickness was originally relatively small. There are other areas with a remaining lifetime of 25 years. The best good news story is that the south-central portion of the aquifer in western Kansas has a predicted lifetime of >250 years.

14.3.5 Streamflow Declines in Beaver-North Canadian River Basin

Problems of streamflow declines have been noted across northern half of the High Plains aquifer. For example, Kustu et al. (2010) examined streamflow data across the Great Plains and found a decrease in annual and dry-season (mean July–August) streamflow, together with an increase in the number of low-flow days. Other studies (e.g., Szilagi, 1999) reported similar problems in the Republican River basin in Nebraska. A study by Perkin et al. (2017) in an area of Colorado, Kansas, and Nebraska determined a loss 558 km of streams with important impact on fish populations, and the likelihood of similar trends extending to 2060.

This section describes an example of streamflow depletion in the Beaver-North Canadian River basin of Texas and Oklahoma, which is associated with pumping from the High Plains aquifer. Barlow and Leake (2012) summarized work by Wahl and Tortorelli (1997) on the Beaver-North Canadian River basin (Figure 14.15). As the small insert-map shows, the basin sits atop the High Plains aquifer system. A look back at Figure 14.14 shows that within the Beaver-Canadian basin, particularly in Texas, the aquifer has experienced significant historical drawdowns. The hydrograph for a well located in western Oklahoma (Figure 14.16a) shows a decline in water level of approximately 24 ft (~7.2 m) from 1956 to 1995.

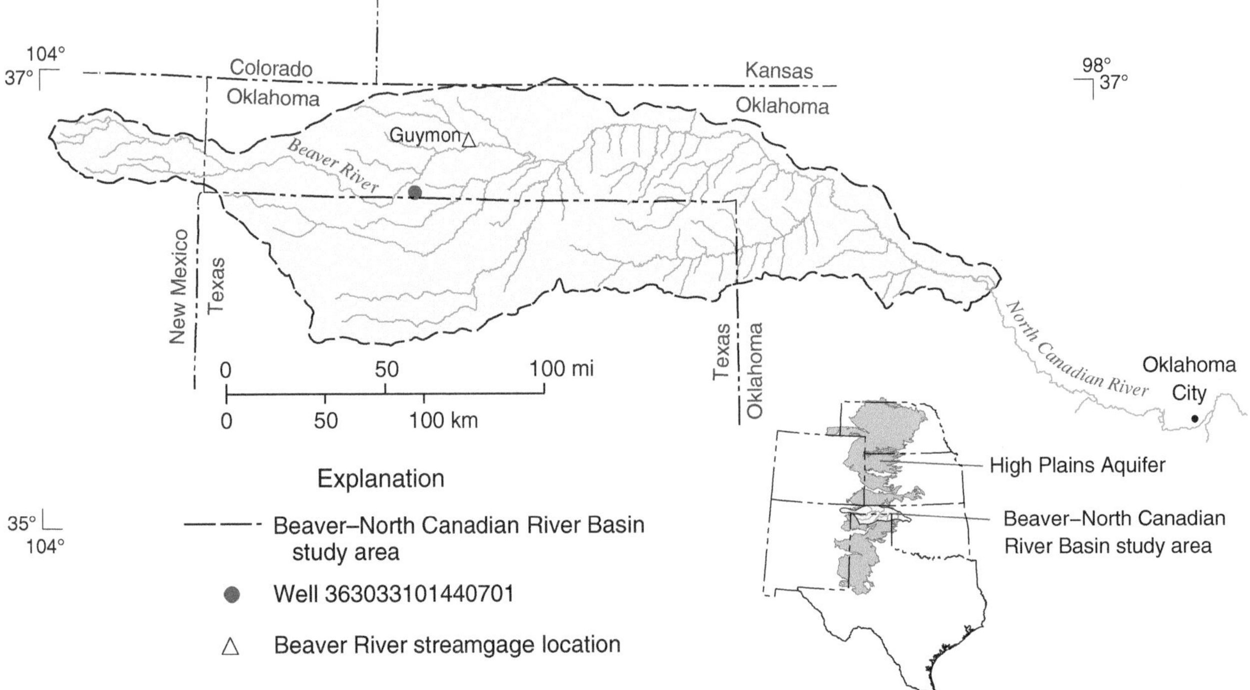

Figure 14.15 The Beaver-North Canadian River basin overlies the High Plains aquifer system. The larger map shows the drainage network, a reference well in western Oklahoma and a gaging station on the Beaver River at Guymon Oklahoma (U.S. Geological Survey, adapted from Barlow and Leake, 2012, as modified from Wahl and Tortorelli, 1997).

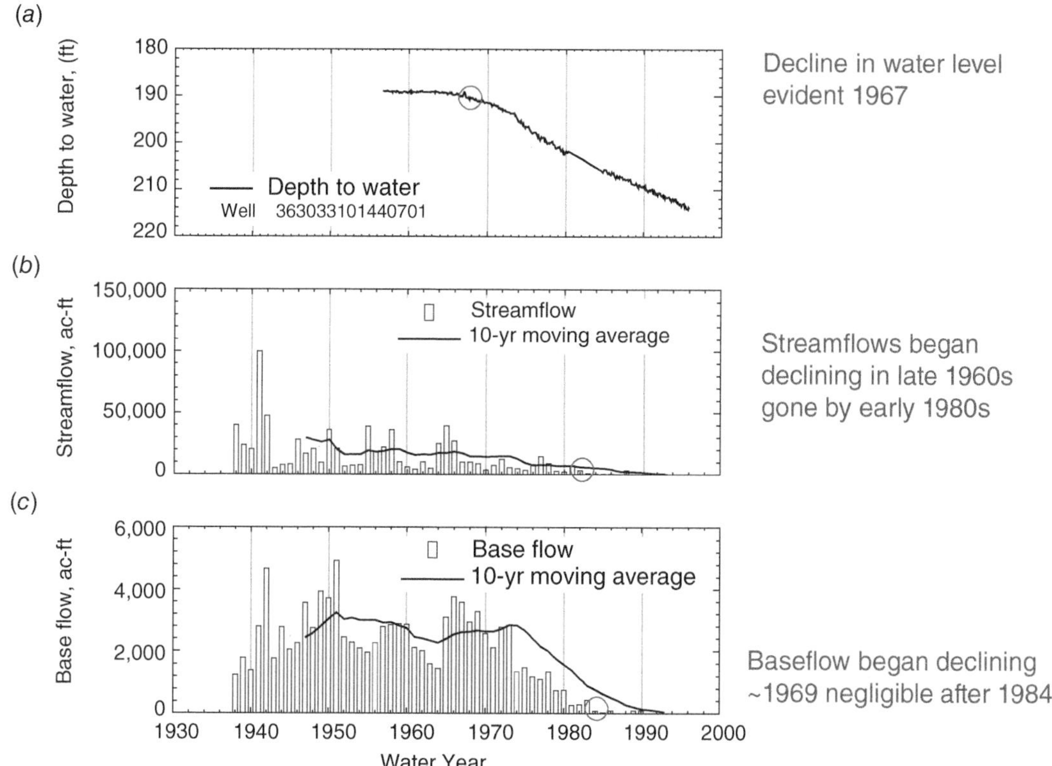

Figure 14.16 The historical water-level decline is shown for the reference well in western Oklahoma (*a*). The declines streamflow (*b*) and baseflow (*c*) are for a gaging station on the Beaver River at Guymon, Oklahoma. The declines in flows track the decline in groundwater levels (U.S. Geological Survey, adapted from Barlow and Leake, 2012; as modified from Wahl and Tortorelli, 1997/Public domain).

As water levels in the aquifer declined, so did annual discharge on the Beaver River (Figure 14.16*b*), near Guymon, Oklahoma (1938–1993). This loss in streamflow was of concern because water from the North Canadian River in the 1990s provided about half of the supply for Oklahoma City (Wahl and Tortorelli, 1997). Identification of the behavior of the baseflow component of the hydrograph for Guymon provides a more explicit picture of the impacts due to groundwater pumping. By the early 1980s, it is apparent that there is essentially no groundwater contribution to flow (Figure 14.16*c*). Statistical analyses point to depletion in groundwater due to pumping for irrigation as the most reasonable explanation for the marked reductions in streamflow in the basin.

14.3.6 Challenges in the Investigation of Streamflow Loss

Studies that we highlighted concerned with declining streamflow depended upon the availability of long-term groundwater-level data for the High Plains aquifer system and streamflow data from the Great Plains. Within the United States, many surface-water gaging sites have long records more than 50 years old. For example, the record for the site on the Beaver River at Guymon began in 1938 and ended in 1993 as flow disappeared. Records for key rivers in Nebraska extend over 90 years. Outside of the United States, Brazil, and Europe, historical and modern records of stream flows are not available, which is a major impediment for the study of problems of streamflow depletion.

14.4 Destruction of Riparian Zones

Another impact of overpumping is the potential to destroy riparian areas, which are environments that occur between aquatic ecosystems (i.e., lakes and streams) and terrestrial ecosystems away from water (National Research Council, 2002). More formally, *riparian* zones are areas adjacent to stream channels between low and high-water levels (Cooper and Merritt, 2012). Riparian zones can occur in both wet and arid settings. Regular rainfall fosters the development of

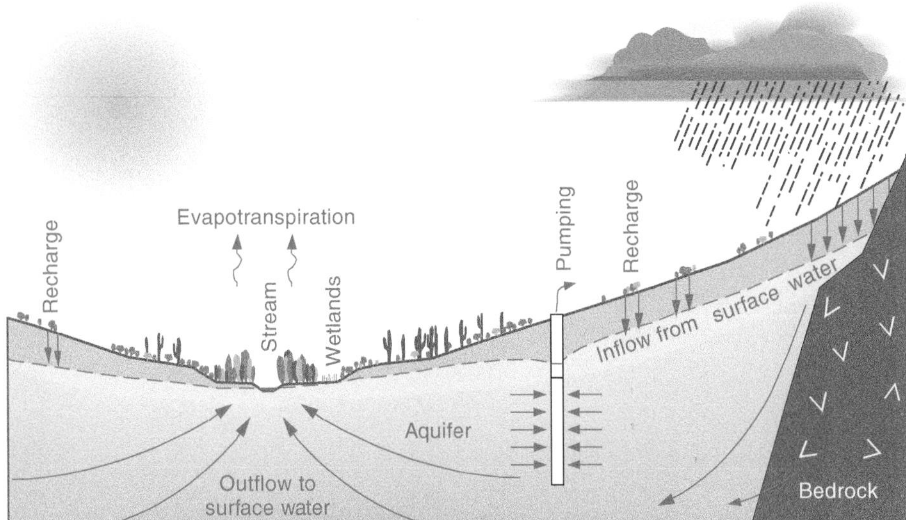

Figure 14.17 Conceptual model illustrating the occurrence of groundwater in relation to riparian zones for desert basins of the American Southwest. Thick alluvial aquifers are developed in basins between mountains and uplands. Recharge occurs at higher elevations with discharge to surface waters at lower elevations. Pumping captures groundwater and impacts the local water cycle of riparian zones (U.S. Geological Survey Leake et al., 2000/Public domain).

perennial streams that exert "a more constant and more dominant control on ecological function" (Cooper and Merritt, 2012) through flooding and a reliable presence of groundwater.

Hydrogeologists are particularly concerned with riparian zones in arid settings, such as the southwest of the United States. Figure 14.17 is a conceptual model of a riparian area supported by an alluvial aquifer system in an arid setting (Leake et al., 2000). Desert valleys typically contain thick sequences of permeable alluvium containing both potable and saline groundwater. Runoff from adjacent uplands or mountain areas of higher precipitation provides much of the aquifer recharge. Natural groundwater discharges as springs, stream baseflow, and evapotranspiration.

In arid settings, surface water supplies for irrigated agriculture or cities are often minimal and less reliable than groundwater. However, intensive groundwater withdrawals are commonly associated with declines to stream and river flows in riparian areas. Also, withdrawals of groundwater adjacent to riparian areas can directly lower water levels along floodplains and reduce the availability of water for riparian plants.

Riparian areas are extremely important ecologically. They support birds, small mammals, and amphibians in large numbers. The framework for riparian areas of the American Southwest is provided by woody plants, such as willow, cottonwood, mesquite, and salt cedar (Figure 14.18). To survive in arid settings, this riparian vegetation requires a reliable source of groundwater. For example, willow and cottonwood trees are obligate phreatophytes, plants that extend deep roots below the capillary fringe for groundwater. They typically will not survive a significant decline in groundwater levels. Mesquite and salt cedar are facultative phreatophytes. These plants use groundwater when it is available, but they can survive without groundwater by switching sources, for example, from groundwater to rainfall or runoff.

Riparian areas in the arid Southwestern USA have been particularly impacted by human activities. Besides groundwater withdrawals, other hydrologic impacts come from (1) dam construction, (2) surface-water diversions for irrigation, and (3) loss of small stream ponds from trapping of beaver in the 1800s (Zaimes, 2007). Biological impacts include (1) loss of natural riparian areas to farming, and (2) invasive species like salt cedar that reduce the abundance of natural riparian species. Stromberg et al. (2004) sum up the loss in riparian areas:

> *"Only after an eye is cast over the past can we fully appreciate how dramatically and rapidly riparian conditions in the Southwest have declined. More than 100 years ago, the lower reaches of desert rivers such as the Colorado, Gila, San Carlos, San Pedro, Rio Salado, Verde, Santa Cruz, Rio Yaqui, Rio Mayo, Agua Fria, and Hassayampa annually carried millions of acre-ft of water to support extensive riverside forests, palustrine wetlands and cienegas....Over the last century, however, streamflow characteristics have been significantly altered and riparian conditions have changed profoundly. Drought- and salt-tolerant shrubs or cultivated and urban lands have replaced the riparian forests, marshlands, and grasslands that once dominated the bottomlands"* (page 100)

Figure 14.18 This small riparian zone is developed within an entrenched bedrock channel of Silver Creek, in northern Arizona. In this arid setting, rainfall is less than 15 cm per year. The vegetation there is supported by groundwater discharge from the Coconino aquifer that is seen there in outcrop (FWS).

14.5 Seawater Intrusion

Near seacoasts, freshwater flowing in an aquifer tends to override more dense saltwater (Figure 14.19). Because freshwater and seawater are miscible, a zone of mixing develops between them, more concentrated toward the ocean and less concentrated toward the land. The thickness of the mixing zone varies depending upon aquifer thicknesses and character of the dynamic changes causing sea-level fluctuations (Barlow and Reichard, 2010).

Out of convenience, the assumption with most conceptual models is that the fluids are immiscible with a sharp boundary between them, whose position is determined by the Guyben-Herzberg principle. For example, if there is 10 m of freshwater above sea level, there will be 400 m of freshwater below sea level. What this means is that the interface between freshwater and saltwater is relatively deep, approaching the coast. Thus, as implied in Figure 14.19*a*, wells installed close to the ocean should find freshwater. Of course, real systems are obviously more complicated.

Overpumping of coastal aquifers disturbs the interface between freshwater and seawater, effectively letting saltwater invade landward. Looking at Figure 14.19*b*, it is easy to understand why. Creation of a cone of depression reduces the height of freshwater above sea level. Using the previous example, a 5 m decline in the water table will cause the interface directly below to rise 200 m. Thus, widespread lowering of water levels near a seacoast will cause an invasion of seawater accompanying the rise of the freshwater-saltwater interface.

A second problem is associated with a high capacity well close to the ocean. Close to the pumping well, water levels are significantly lowered. The interface between freshwater and saltwater locally responds by rising toward the pumping well in a process known as upconing. Various analyses suggested after a small rise the interface becomes unstable and saltwater would break through into the well (Domenico and Schwartz, 1998).

Experience with real systems shows that saltwater can contaminate coastal aquifers in a variety of different ways. Direct invasion of coastal aquifers by saltwater is common, as shown in Figure 14.19*b*. Pumping can also mobilize old, saltwater sequestered in deeper units. The source of that water might be from an older, natural episode of seawater intrusion related to interglacials when sea levels were higher than present day (Barlow and Reichard, 2010).

Seawater intrusion into coastal aquifers of the continental United States occurs at a variety of locations along the Atlantic, Pacific, and Gulf coasts. On the Atlantic Coast, one of the most intensively studied problems is associated with the Biscayne

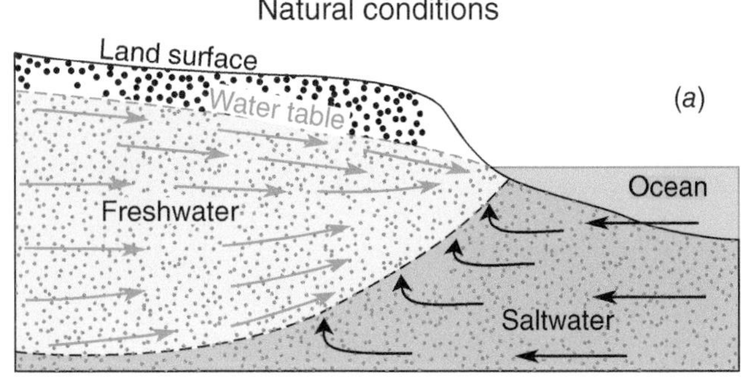

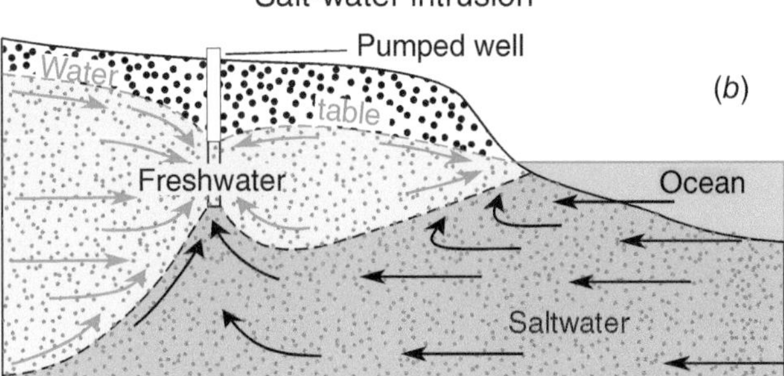

Figure 14.19 Under natural conditions (*a*), freshwater overrides salt water to discharge into oceans. The boundary between them is approximated here as a sharp interface. In real systems, a wide zone of mixing develops. With nearshore pumping (*b*), seawater can invade freshwater aquifers with the possibility for upconing at a well shown here (U.S. Geological Survey USGS, 1999/Public domain).

Aquifer along Florida's southeastern coast (Barlow and Reichard, 2010; Renken et al., 2005). Farther north, coastal areas of the Floridian aquifer have been affected (Barlow and Reichard, 2010), as have several coastal areas of southwestern Long Island (e.g., Lusczynski and Swarzenski, 1966; Stumm et al., 2020). Along the Pacific coast, seawater intrusion is a problem affecting coastal basins in southern California in and around Los Angeles, and the Salinas River Groundwater Basin further north. We explore the problems there in the next section.

14.5.1 Salinas River Groundwater Basin

One of the long-standing problems of seawater intrusion involves groundwater pumping in the Salinas River Basin in California. This section presents an overview on the development of the problem with a focus on the historical progression of seawater intrusion over more than 50 years. In the next chapter, we examine strategies used in controlling the problem.

The Salinas basin is located west of the San Joaquin Valley (Figure 14.20). It is a deep and narrow structural trough, ~150 mi long and oriented approximately southeast–northwest Brown and Caldwell (2015). It is surrounded on three sides by mountain ranges. The Salinas River drains ~5000 mi^2 and discharges into the Pacific Ocean north of Monterey, California. Much of the valley floor (Figure 14.20) is taken up with farming with Salinas as the major urban center. The region is noted vegetables, wine grapes, and other fruits.

Natural rainfall across the valley is only ~14 in so that crop require irrigation. Demands of water for irrigation and urban supplies are met by groundwater with agriculture as the dominant user (Brown and Caldwell, 2015). Groundwater production began in earnest in the 1920s and by 1944 reached 350,000-acre ft/yr mostly around Salinas. Production peaked in 1970 and subsequently declined subsequently to ~500,000 acre ft/yr in 2013.

Seawater intrusion became evident in coastal areas west of Salinas in the 1930s. It progressively worsened with continued pumping. The pattern of seawater intrusion is complicated by complexities in the hydrogeologic setting. In the coastal area west of Salinas, seawater is spreading through two principal aquifers—the Pressure 180-Foot aquifer (P-180) and the Pressure 400-Foot aquifer (P-400). Figure 14.21 shows a simplified view of the hydrogeological setting. Overlying the P-180 aquifer is a locally perched aquifer and confining bed referred to as "Upper units" in Figure 14.21. The sand and gravel, P-180 aquifer is separated from the deeper P-400 aquifer by a "leaky" confining bed that is variable in thickness and possibly absent in places. North and northwest of Salinas is a large cone of depression with water levels during the irrigation season, which can be >30 m below sea level. Much of the early pumping occurred from the P-180 aquifer with later development of the deeper P-400 aquifer.

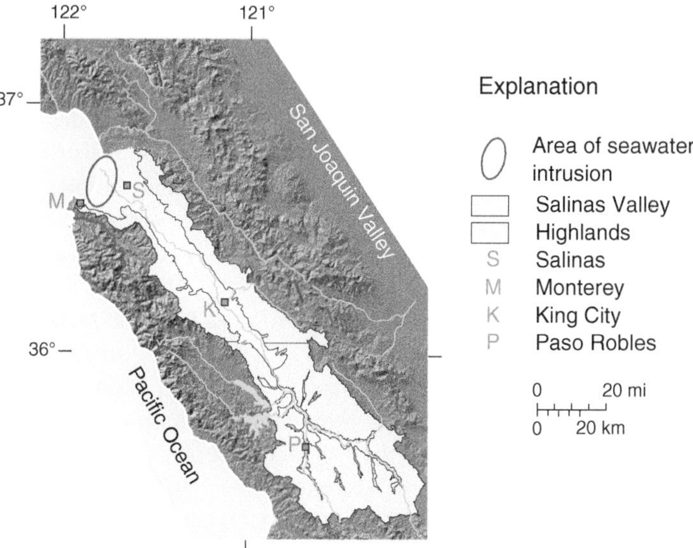

Figure 14.20 Farming in the Salinas River basin depends on groundwater for irrigation. Overpumping in the coastal area around Salinas has caused significant seawater intrusion as shown on the map (U.S. Geological Survey Burton and Wright, (2018)/Public domain).

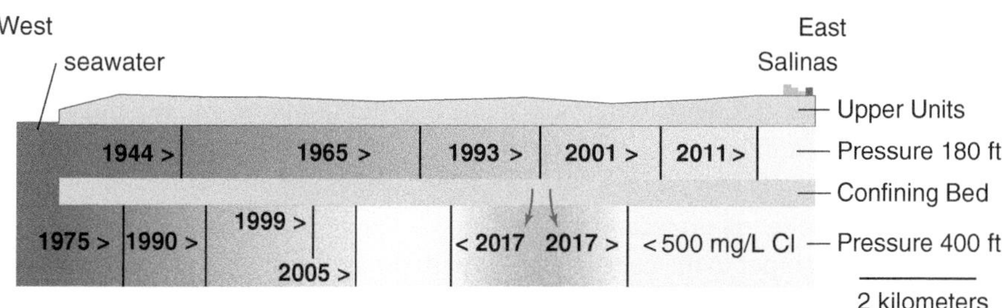

Figure 14.21 Progressive intrusion of seawater through the P-180 and P-140 aquifers. Various strategies have slowed the advance except for downward leakage shown by the red arrows (Original figure (FWS) based on information from Brown and Caldwell, 2015).

Figure 14.21 summarizes the seawater intrusion since 1944 for the P-180 aquifer. The vertical lines with associated dates show the progressive invasion of water with Cl > 500 mg/L By 2013, the front had progressed ~12 km to just west of Salinas. Various strategies have been successful in halting further advances. Seawater intrusion of the P-400 aquifer became evident in 1975 and has progressed through 2005. An isolated portion of the P-400 aquifer, ahead of the main spreading front (see Figure 14.21), was impacted by >500 mg/L-Cl water in about 2015 (Brown and Caldwell, 2015). This zone has continued to grow. Water with Cl >500 mg/L is leaking downward from the P-180 aquifer. The exact cause is not known. The confining bed could be absent, or water could be leaking through damaged wells or wells screened in both aquifers (Brown and Caldwell, 2015).

Preserving the sustainability of coastal aquifers requires measures to control and potentially reverse problems of seawater intrusion. A variety of different strategies have been put in place to control seawater intrusion, particularly in California. We take up these strategies in the following chapter.

Problems of seawater intrusion are difficult to tackle. They are in effect mega-scale problems of point-source contamination. Lobes of invading seawater follow trends in hydraulic conductivity and hydraulic gradients associated with onshore withdrawals of groundwater. Typically, their large areal extents preclude dedicated monitoring systems with water supply-wells are called upon to provide water samples. When monitoring wells are installed, they end up being expensive and few (Prinos et al., 2014). The resolution provided with such sampling is coarse, most useful in defining the extent of seawater invasion. Complexity is also added because seawater is denser than freshwater even a simple measurement of hydraulic head is complicated by the need to know the density of water in the well column. In recent years, helicopter electromagnetic surveying has shown promise in providing improved three-dimensional spatial resolution (Prinos et al., 2014). Downhole, electromagnetic induction logging in PVC-cased wells through time provides excellent spatial (vertical) and temporal resolution of bulk electrical conductivity (Prinos et al., 2014).

14.6 Introduction to Groundwater Modeling

Earlier chapters on well hydraulics explained how analytical solutions could provide future estimates of the drawdowns due to pumping. Yet, these approaches are idealized and mostly designed for the evaluation of a production well or a small well field. They cannot handle complexities such as, spatially distributed or time-varying recharge, complicated groundwater/surface-water interactions, many pumping wells, and complex heterogeneous aquifer systems with irregular boundaries. Numerical models have proven to be powerful and sophisticated tools for solving real problems. This section provides readers with a sense of how groundwater models work, the kinds of data necessary, and an example that illustrates their use in practice.

Hydrogeologists commonly work with codes from the MODFLOW family or functionally equivalent commercial versions, and accessory codes, such as PEST (Doherty, 2018) for parameter estimation or Groundwater Vistas for pre-and post-processing. Another popular code is FEFLOW, a commercially available code that includes capabilities of modeling groundwater flow with coupled heat and mass transport in saturated and unsaturated systems.

The examples presented in the book make use of MODFLOW that has evolved through six major releases to MODFLOW 6 (Langevin et al., 2017), the most recent version. Variants of the code exist, which provide capabilities for contaminant transport, variable-density flow, advanced gridding approaches, and more (Sajil Kumar, 2019).

Classic analytical solutions to problems of well hydraulics provide drawdown at some future time (t) at a reference point that is located some distance (r) from a pumping well. Drawdown is calculated by substituting values of (t) and (r) along with parameter values, such as pumping rate, transmissivity, and storativity, into the analytical solution.

Numerical modeling involves a fundamentally different mathematical approach to solving the flow equation, but the basic approach is similar—organize the basic hydrogeological information, pass that information to the computer code, and interpret calculated hydraulic heads at prescribed time intervals. With numerical modeling, the first step, i.e., organizing the hydrogeological information is typically much more difficult because data needs are more than single transmissivity and storativity values. The focus of numerical modeling is real systems and problems, which are inherently complicated in terms of data needs. Experience with numerical modeling projects suggests that 30–40% of the overall effort is finding the data and organizing them in the prescribed way, as input to the code. The steps with the actual modeling are specialized and difficult, far outside the scope of an introductory book such as this.

What we will emphasize is (1) how information about real systems is used to create conceptual and numerical models, (2) the types of results that come from a model-based analysis of a problem, and (3) the amazing roles that models play in quantitatively analyzing aquifer problems, discussed earlier in this chapter.

14.6.1 Conceptual Model

A conceptual model represents the generalized description of a groundwater system in terms of (1) the hydrogeologic framework, such as aquifers and confining beds, (2) patterns of groundwater flow, recharge, and discharge, (3) rates of withdrawals of water by pumping wells, and (4) boundary conditions. It is also quantitative such that elements of the hydrogeologic framework, such as layer thicknesses, geometry, and hydraulic properties as a function of space, are represented in terms of numbers. Similarly, it should include an assessment of hydraulic head measurements to provide an understanding of present patterns of groundwater flow. The quantification of a real system is uncertain because some parameters like recharge rates or hydraulic conductivities are difficult to measure, and it is impractical to assess all the pumping wells.

Uncertainties cause the initial set of model parameters to be adjusted so that calculated values of hydraulic head, streamflows, etc. match values observed in the field. This adjustment process is known as calibration, the topic of an upcoming section.

The solution of a differential equation requires specification of boundary conditions. Although not always obvious in well hydraulics, boundary conditions are necessary to derive solutions. For example, solution of the Theis equation requires boundary conditions, one at the well to specify the constant pumping rate and another at infinity indicating that the drawdown there is always zero. Boundary conditions are also necessary with numerical models along the top, bottom, and sides of the simulation domain. Two commonly used boundary conditions are specified head boundaries and water-flux boundaries. A no-flow boundary is a special case of a water-flux boundary.

In developing a conceptual model, the boundary conditions receive special consideration. The size of the modeled area needs to be larger than the region of interest so that the somewhat artificial conditions along the boundaries do not inappropriately influence hydraulic head values within the region of interest. In most studies, values for the constant heads or for inflow/outflow rates along boundaries are poorly constrained. One strategy to overcome this problem is to place model boundaries along "natural" hydrogeologic boundaries. An example of this boundary is a large river defined by constant hydraulic-head values equivalent to the stage of the river. Another example is an imaginary no-flow boundary created by a major watershed divide. The boundary is no-flow because the gradient is such that flow moves away from a divide in both directions.

Features of the hydrogeologic setting also help in assigning boundaries. For example, the top of a thick, low-hydraulic conductivity unit at depth can be selected as the bottom of the simulation domain with no flow boundary. At some depth in almost any system, it should be possible to define a no-flow boundary, implying that deeper circulation is minimal.

For a simple problem, a conceptual model can be a schematic figure. More commonly, a conceptual is represented by a report that contains figures, and data tabulations. A report for a real problem would include maps for all key data varying spatially, such as topography, thicknesses for key hydrostratigraphic units, hydraulic conductivities, recharge rates, etc. For a large modeling problem, the conceptual model might reside largely as data layers within a geographic information system in a form that can be directly assigned to the model (Stanton et al., 2010).

Figure 14.22 is a simple example of a conceptual model that includes an unconfined aquifer occurring in association with a stream network (Barlow and Leake, 2012). The light blue shading on Figure 14.22*a* defines the extent of the aquifer and shape of the boundaries. The facing side boundary provides information on the relative thickness of the aquifer. Notice that in the vicinity present-day stream channel, the aquifer is thickest. Water recharged from precipitation across the area flows within the aquifer and discharges at nearby streams or downstream as underflow (Figure 14.22*a*). Typically, information on well locations and pumping rates is critically important. Here, the pumping rate for each well (A and B) would be specified. Aquifer properties, such as hydraulic conductivity are assumed to be a single-valued constant along with the recharge rate.

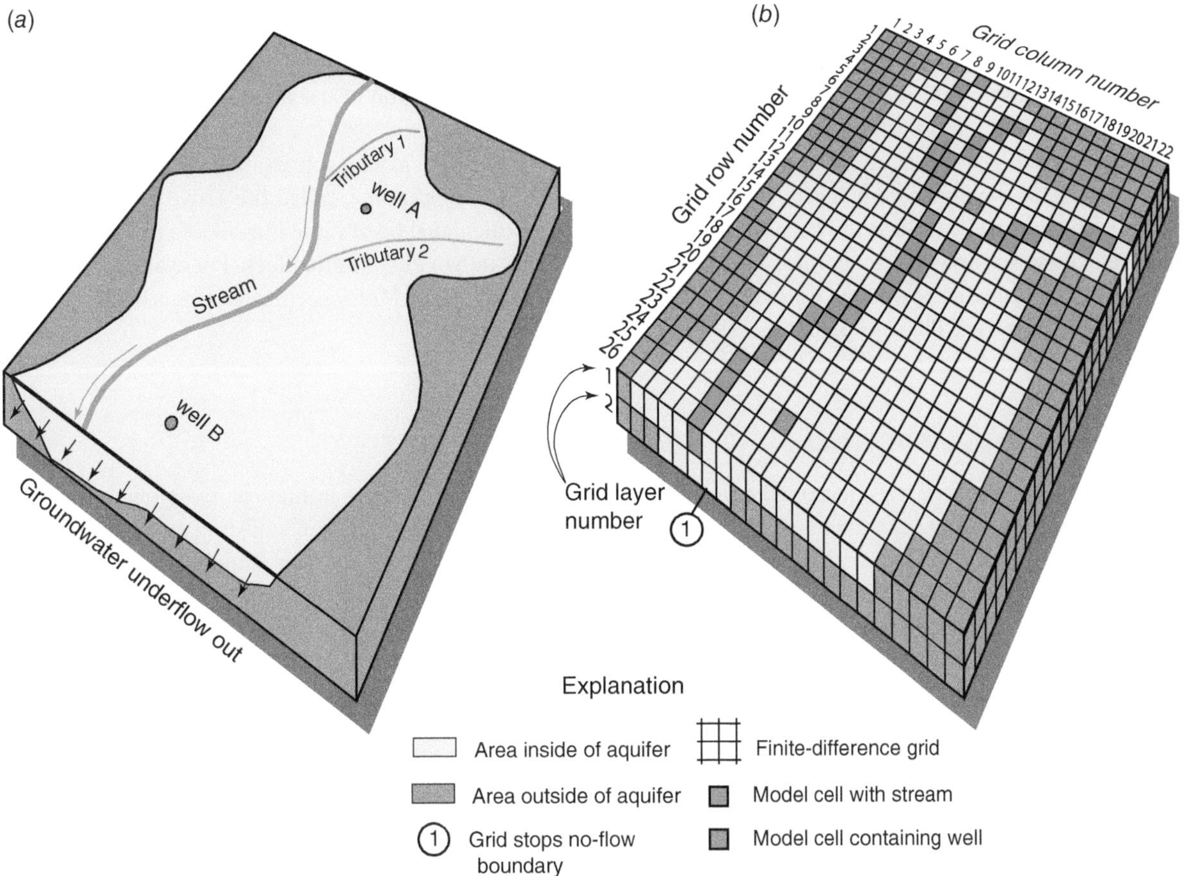

Figure 14.22 The conceptual model for a simple system. (*a*) shows an unconfined aquifer that includes a stream and tributaries. The aquifer, shown in blue, is bounded laterally and along the bottom by a gray-colored unit with a low hydraulic conductivity. The aquifer is discretized with 2 layers, 26 rows, and 22 columns (*b*) (U.S. Geological Survey Barlow and Leake, 2012/Public domain).

Turning to boundary conditions, we assumed the gray units to have an extremely low hydraulic conductivity (Figure 14.22*a*). With this assumption, the logical boundary conditions would be a vertical no-flow boundary around perimeter of the model where the gray unit is present, as well as along the bottom. The boundary condition on the top surface would be a constant flux, the mean recharge. Nodes representing the streams are internal boundaries that are assigned constant heads equivalent to their stages. The end boundary associated with groundwater underflow (Figure 14.22*a*) is designated as a constant-head boundary.

14.6.2 Model Design

The conceptual model naturally to leads a design step that involves creation of the model grid. All numerical modeling approaches require that the study area be subdivided into grid blocks. Sometimes, the term *discretization* is used to describe this step. Simple implementations of MODFLOW utilize a regular discretization, where an aquifer is subdivided into a series of rectangular grid blocks. In a two-dimensional model, each model cell has a length and width, Δx, Δy, and a thickness m. In a three-dimensional model consisting of aquifers and confining beds, individual units are subdivided vertically into grid blocks of a specified thickness, length, and width.

Figure 14.22*b* shows the grid design for the simple example with 2 layers 26 rows and 22 columns and 1141 grid blocks altogether (Barlow and Leake, 2012). Each of the grid blocks are uniquely identified by row/column/layer numbering. The griding provides a basis for taking the hydrostratigraphic and hydraulic data and organizing them in a manner required for modeling. For example, in the top layer (i.e, layer 1), the light blue blocks will receive the known hydraulic conductivity value and gray blocks will be considered as inactive. Similar information is provided for layer 2.

The gridding provides the capabilities for creating the three-dimensional shape of the aquifer, the hydraulic conductivity distribution, and the no-flow boundary conditions on the sides of the aquifer (Figure 14.22*b*). By default, a no-flow boundary is also defined where the grid stops. At point ① (Figure 14.22*b*), for example, the bottom of the blue grid blocks in layer 2 is a no-flow boundary. All other information concerning the problem, such as the location of stream channels, recharge rates, and locations and pumping rates for wells is organized in relation to the model grid.

In the middle of each grid block, there is a node, the point where the model calculates hydraulic head values. In the example (Figure 14.22*b*), everywhere in layers 1 and 2 where the grid block is light blue, called the active grid, the model calculates a value of hydraulic head at the node point. Thus, the array of calculated head value is available to be contoured and to interpret patterns of flow. The model also calculates the water fluxes into or out of a grid block. For example, here one might be interested in knowing how much water is moving into or out of the stream. More specifically, this is the flux across the boundary between a dark blue and light blue cell.

14.6.3 Model Calibration and Verification

The set of model parameters from the conceptual model serves as the starting point for the numerical modeling. Invariably, these starting values will require adjustments. Calibration is the term describing this process of adjusting model parameters to achieve a model that convincingly simulates measured features of the actual system. Thus, as a model steps through time, it reproduces data measured in the field, such as hydraulic heads, stream baseflow hydrographs, or spring flows. In practice, calibration should always include some flux-type data in addition to hydraulic head data to avoid problems of nonuniqueness (Anderson et al., 2015). For example, there is no unique set of parameters (e.g., hydraulic conductivity and recharge values) in certain cases where only head data are used for calibration (Hunt et al., 2020).

These days, calibration is accomplished with powerful, nonlinear estimation codes, such as PEST (Doherty, 2018) or PEST++ (Welter et al., 2015). There is a voluminous literature on the topic and interested readers can refer to Anderson et al. (2015). Once calibration is complete, a verification test is commonly added to check that the model is a valid representation of the hydrogeologic system. Verification involves using the calibrated model to simulate a hydrologic response that is known. For example, one might hold back results from one or more large-scale aquifer tests and examine how well the calibrated model simulates the test. Again, the errors between the observed and simulated hydraulic-head values can be quantified in terms of the error measures. If the model successfully passes this last test, then it can be used to investigate the problems at hand.

14.6.4 Predictions in Modeling

The power of numerical models facilitates the analyses of problems associated with overpumping. Thus, models are useful in understanding problems of groundwater depletion from pumping, and to test potential management strategies for mitigation.

Although models can be run into future time, simulation results should not be thought of as "predictions." Models inherently come along with an "aura of correctness" (Bredehoeft and Konikow, 1993), represented by many calculations, which sometimes cause users to lose sight of the uncertain data on which models are built. Oreskes et al. (1994) posited that "models are representations, useful for guiding further study but not susceptible to proof." Groundwater systems are always poorly characterized, and influencing factors, such as climate, agricultural economics, and government interventions are unpredictable. Thus, the model design depends significantly on the "informed judgment" of its builder rather than on real information. This uncertainty does not disappear once a model is constructed.

If model predictions are suspect, of what use are they? In addressing this issue, Oreskes et al. (1994) pointed out:

> *"Models can corroborate a hypothesis by offering evidence to strengthen what may be already partly established through other means. Models can elucidate discrepancies in other models. Models can also be used for sensitivity analysis—for exploring 'what if questions—thereby illuminating which aspects of the system are most in need of further study, and where more empirical data are needed."*

Computer models are amazing tools that need to be used with full knowledge of their limitations.

14.7 Application of Groundwater Modeling

Section 14.1 provided a discussion of the potential impacts of overpumping on aquifers. The USGS applied MODFLOW to elucidate issues associated with pumping of the Mississippi River Valley Alluvial (MRVA) aquifer (Gillip and Czarnecki, 2009; Reed, 2003). It is located along the Mississippi River mainly in Arkansas, Tennessee, Mississippi, Louisiana, and Tennessee (Figure 14.23). This prolific sand and gravel aquifer has an areal extent of 32,000 mi^2 (82,800 km^2). Groundwater withdrawals in 2000 were third highest in the nation at approximately 35 billion liters per day or 12.8 km^3/yr (Maupin and Barber, 2005 in Yasarer et al., 2020). About 98% of this production is used for the irrigation of row crops (Yasarer et al., 2020).

The focus of the model studies was the MRVA in north-eastern Arkansas, west of the Mississippi River (Figure 14.23). There is a long history of irrigated agriculture that began in 1900. From 1965 to 2005, production of groundwater increased significantly (Gillip and Czarnecki, 2009). During 2005, production was 23.6 billion liters per day or 8.6 km^3/yr, mostly for the irrigation of rice and fish farming (Holland, 2007). The original saturated thicknesses of groundwater in the MRVA aquifer exceeded 61 m (200 ft). However, more than 100 years of pumping created large cones of depression. Saturated thickness in places was less than 6 m (20 ft) thick (Reed, 2003). There was also a reduction of flows in several larger rivers (Czarnecki, 2006).

A model study by Reed (2003) determined that pumping at 1997 rates was unsustainable. A follow-on study (Gillip and Czarnecki, 2009) updated Reed's model and evaluated impacts by 2049 with two scenarios. The alluvial system was conceptualized as two units.

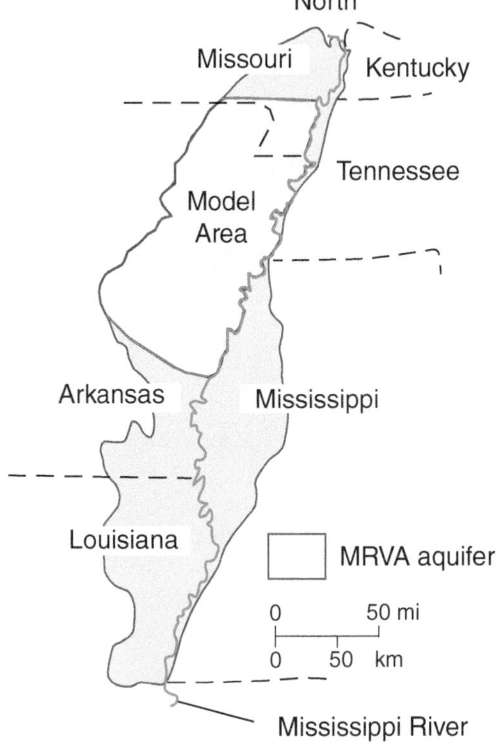

Figure 14.23 The Mississippi River Valley Alluvial aquifer (red outline) stretches along the Mississippi River from the Missouri-Kentucky border southward through eastern parts of Arkansas and Louisiana, as well as parts of western Mississippi. The yellow area has been the focus of groundwater-model investigations (U.S. Geological Survey, adapted from Mahon and Poynter, 1993/Public domain).

The lower unit, the effective aquifer, consists of sands and gravels, fining upward, overlain by a less permeable, upper unit (Gillip and Czarnecki, 2009). The average thickness of the alluvial aquifer is about 30.5 m (100 ft), ranging in thickness from 0 to 50 m (180 ft). The upper confining unit has a variable thickness with a mean of 7.5 m (25 ft). Across the modeled area, 10 zones were defined with recharge values and hydraulic conductivity values for each of the two units (Reed, 2003). The zonation provided a basis for automatic parameter calibration, also supplemented by local, manual adjustments (Reed, 2003).

Recharge arises from several sources. Natural recharge represents an important component of recharge, with 49 in. of precipitation annually. The relatively low hydraulic conductivity of the upper unit, however, constrains infiltration. Another source of recharge is induced infiltration from rivers. Thus, major rivers represent an important feature of the conceptual model (Reed, 2003). Other sources of recharge include smaller water bodies or up-flow from deeper bedrock aquifers.

The model grid had uniform cells, 1610 m by 1610 m (1 mi by 1 mi) with 156 columns, 184 rows (Figure 14.24). The number of cells in the active grid (i.e., 14,104 cells) is much smaller than the total number of cells (i.e., 28,704 cells). The large

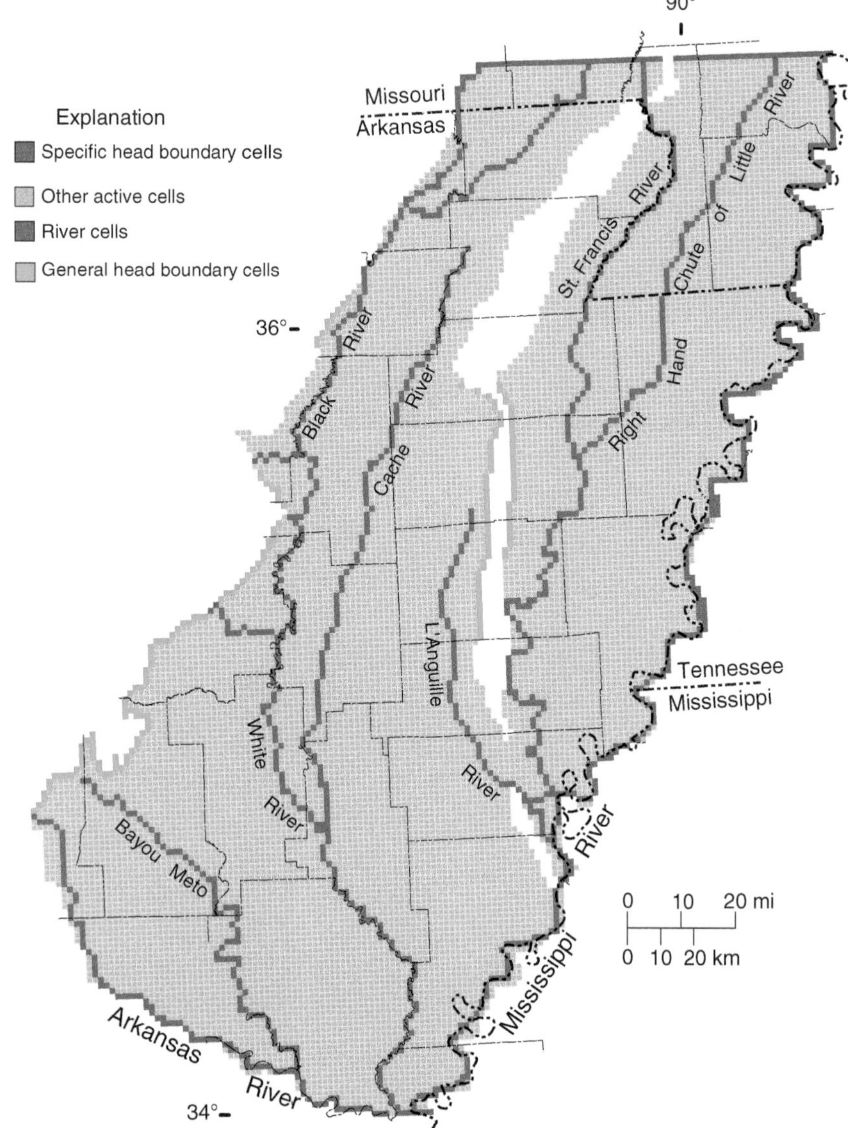

Figure 14.24 The model was constructed with a uniform grid comprised of 156 columns and 184 rows. Major river systems were mapped onto the grid and provided internal boundaries for recharge and discharge. External boundary conditions are specified on the legend (U.S. Geological Survey, adapted from Reed, 2003/Public domain).

inactive zone through the central part of the grid system is Crowleys Ridge. This rock ridge functions as a hydraulic barrier between the east and west zones of the MRVA aquifer. The model grid also shows the location of rivers which represent potential recharge or discharge areas.

The Arkansas and Mississippi Rivers provide natural side boundaries to the simulation domain. Rivers are included in the model using the MODFLOW River Package, which represents rivers as constant head nodes with fluxes constrained by vertical hydraulic conductivity of the riverbed. The northern side boundary (red line—Figure 14.24) is represented as a constant flux boundary. The western side is a mix of no flow and general head segments.

The Gillip and Czarnecki (2009) model beyond the first stress period was transient, stepping forward through time with 24 stress periods or time steps to move from 1918 to 2049. Values of some parameters changed from one period to the next, for example, accounting for increasing withdrawals of groundwater and increasing recharge. For specified years, the calculated value of hydraulic heads for active nodes in layer 2 was contoured to indicate areas of water-level declines.

The purpose of the modeling was to simulate two water-use strategies from 2005 to 2049 to assess the potential impacts on MRVA aquifer. Scenario 1 assumed that the groundwater use in 2005 continued at a constant rate until 2049. Scenario 2 assumed that the groundwater use in 2005 increased by 2% each year until 2049.

Figure 14.25*a* and 14.25*b* depicts calculated hydraulic heads values for 2005, and Scenario 1 for the year 2049, respectively. Dry cells shown in red are places where water levels declined to the very bottom of the aquifer. Functionally, the aquifer was dry. The 2005 results (Figure 14.25*a*) suggested the aquifer was severely depleted in one large, rice-growing area with 135 dry cells within a much larger cone of depression. The presence of these cells was indicative of unsustainable withdrawals from the aquifer (Gillip and Czarnecki, 2009). Pumping with Scenario 1 resulted in further depletion in storage.

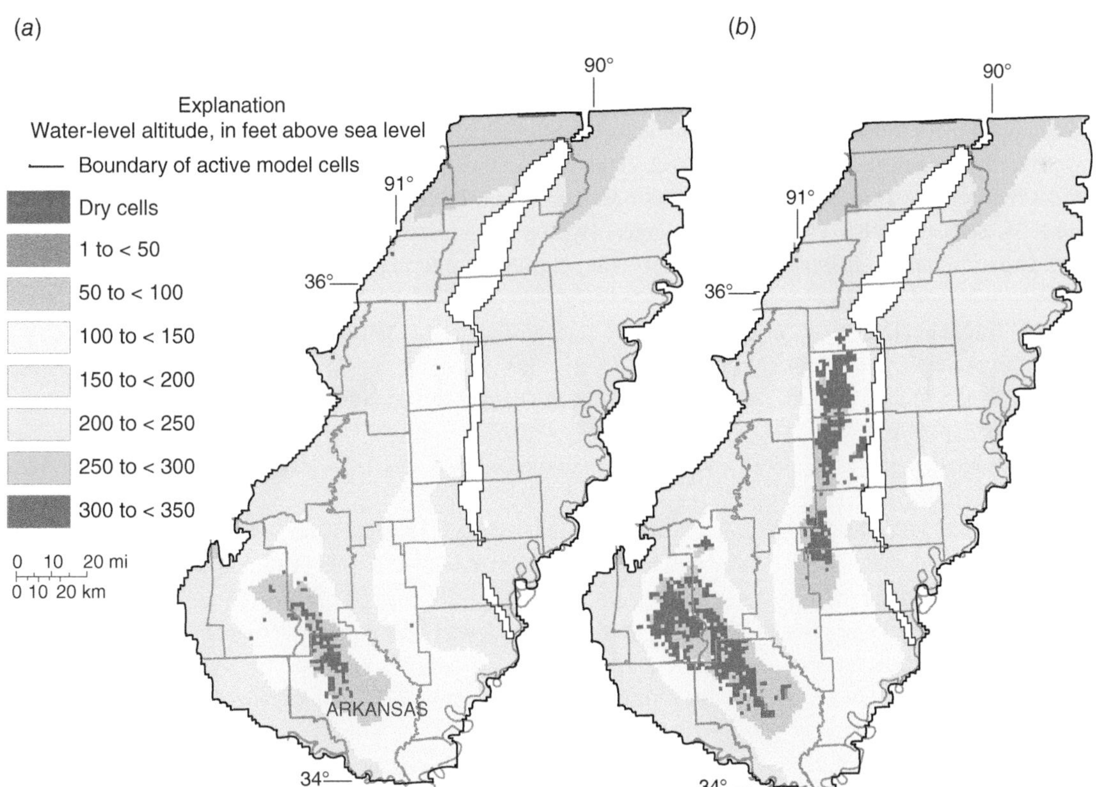

Figure 14.25 Modeled hydraulic heads in 2005 indicated significant drawdown to the south (*a*). By 2049, continuous pumping at 2005 rates (Scenario 1) could lead to 779 dry cells, enlargement of the cone of depression in and around Arkansas County, and the development of a second major cone of depression to the northeast (*b*) (U.S. Geological Survey, adapted from Gillip and Czarnecki, 2009/ Public domain).

References

Alley, W. M., T. E. Reilly, and O. L. Franke. 1999. Sustainability of groundwater resources. U.S. Geological Survey Circular 1186, 79 p.

Anderson, M. P., W. W. Woessner, and R. J. Hunt. 2015. Applied Groundwater Modeling: Simulation of Flow and Advective Transport. Academic Press.

Barlow, P. M., and S. A. Leake. 2012. Streamflow depletion by wells–Understanding and managing the effects of groundwater pumping on streamflow. U.S. Geological Survey Circular 1376, 84 p.

Barlow, P. M., and E. G. Reichard. 2010. Saltwater intrusion in coastal regions of North America. Hydrogeology Journal, v. 18, no. 1, p. 247–260.

Belitz, K. R., and F. J. Heimes. 1990. Character and evolution of the groundwater flow system in the central part of the western San Joaquin Valley, California. U.S. Geological Survey Water Supply Paper 2348, 28 p.

Borchers, J. W., V. Kretsinger Grabert, M. Carpenter, B. Dalgish, and D. Cannon. 2014. Land subsidence from groundwater use in California. Report for the California Water Foundation, Sacramento, California.

Bredehoeft, J. D., and L. F. Konikow. 1993. Editorial—Groundwater models: validate or invalidate. Groundwater, v. 31, no. 2, p. 178–179.

Brown and Caldwell. 2015. State of the Salinas River Groundwater Basin—hydrology report. Monterey County Water Resources Agency Water Reports 21.

Buchanan, R. C., B. B. Wilson, R. R. Buddemeier, and J. J. Butler. 2015. The High Plains aquifer. Kansas Geological Survey Public Information Circular, 18, 6 p.

Bull, W. B. 1964. Alluvial fans and near-surface subsidence in Western Fresno County, California. U.S. Geological Survey Professional Paper 437-A, 71 p.

Burton, C. A., and M. T. Wright. 2018. Status and understanding of groundwater quality in the Monterey-Salinas Shallow Aquifer study unit, 2012–13: California GAMA Priority Basin Project. *U.S.* Geological Survey Scientific Investigations Report 2018-5057, 116 p.

Cooper, D. J., and D. M. Merritt. 2012. Assessing the water needs of riparian and wetland vegetation in the western United States. U.S. Department of Agriculture, Forest Service General Technical Report RMRS-GTR-282, 125 p.

Czarnecki, J. B. 2006. Simulation of various management scenarios of the Mississippi River Valley alluvial aquifer in Arkansas. U.S. Geological Survey Scientific Investigations Report 2006-5052, 21 p.

Doherty, J. 2018. Model-Independent Parameter Estimation User Manual Part I: PEST. SENSAN and Global Optimisers.

Domenico, P. A., and F. W. Schwartz. 1998. Physical and Chemical Hydrogeology. John Wiley & Sons, New York, 506 p.

Faunt, C. C. (ed) 2009. Groundwater Availability of the Central Valley Aquifer, California. U.S. Geological Survey Professional Paper 1766, 225 p.

Faunt, C. C., M. Sneed, J. Traum et al. 2016. Water availability and land subsidence in the Central Valley, California, USA. Hydrogeology Journal, v. 24, no. 3, p. 675–684.

Feng, W., C. K. Shum, M. Zhong, and Y. Pan. 2018. Groundwater storage changes in China from satellite gravity: an overview. Remote Sensing, v. 10, no. 5, p. 674.

Galloway, D., D. R. Jones, and S. E. Ingebritsen. 1999. Land subsidence in the United States. U.S. Geological Survey Circular 1182, 177 p.

Gillip, J. A., and J. B. Czarnecki. 2009. Validation of a groundwater flow model of the Mississippi River Valley alluvial aquifer using water-level and water-use data for 1998–2005 and evaluation of water-use scenarios. U.S. Geological Survey Scientific Investigations Report 2009-5040, 22 p.

Glover, R. E., and G. G. Balmer. 1954. River depletion resulting from pumping a well near a river. Transactions of the American Geophysical Union, v. 35, no. 3, p. 468–470.

Hanak, E., A. Escriva-Bou, B. Gray et al. 2019. Water and the Future of the San Joaquin Valley. *Public Policy Institute of California*, 100 p.

Holland, T. W. 2007 Water use in Arkansas, 2005. U.S. Geological Survey Scientific Investigations Report 2007-5241, 33 p.

Hsieh, P. A., and J. D. Bredehoeft. 1981. A reservoir analysis of the Denver earthquakes: a case of induced seismicity. Journal of Geophysical Research: Solid Earth, v. 86, no. B2, p. 903–920.

Hubbert, M. K., and W. W. Rubey. 1959. Role of fluid pressure in mechanics of overthrust faulting: I. Mechanics of fluid-filled porous solids and its application to overthrust faulting. Geological Society of America Bulletin, v. 70, no. 2, p. 115–166.

Hunt, R. J., M. N. Fienen, and J. T. White. 2020. Revisiting "An exercise in groundwater model calibration and prediction" after 30 years. Insights and new directions. Groundwater, v. 58, no. 2, p. 168–182.

Hurr, R. T., and P. A. Schneider Jr. 1972. Hydrogeologic characteristics of the valley-fill aquifer in the Sterling reach of the South Platte River Valley, Colorado. U.S. Geological Survey Open File Report 73-126, 4 p.

Jenkins, C. T. 1968a. Computation of rate and volume of stream depletion by wells. U.S. Geological Survey Techniques of Water-Resources Investigations Book 4, Chapter D1, 17 p.

Jenkins, C. T. 1968b. Techniques for computing rate and volume of stream depletion by wells. Groundwater, v. 6, no. 2, p. 37–46.

Kingsbury, J. A. 1996. Altitude of the potentiometric surfaces, September 1995, and historical water level changes in the Memphis and Fort Pillow aquifers in the Memphis area, Tennessee. U.S. Geological Survey Water-Resources Investigations Report 96-4278, 1 pl.

Konikow, L. F. 2013. Groundwater depletion in the United States (1900–2008). U.S. Geological Survey Scientific Investigations Report 2013-5079, 63 p.

Kornei, K. 2017. Sinking of Mexico City linked to metro accident, with more to come. Science https://www.science.org/content/article/sinking-mexico-city-linked-metro-accident-more-come (accessed 12 May 2023).

Kustu, M. D., Y. Fan, and A. Robock. 2010. Large-scale water cycle perturbation due to irrigation pumping in the US High Plains: a synthesis of observed streamflow changes. Journal of Hydrology, v. 390, no. 3, 4, p. 222–244.

Langevin, C. D., J. D. Hughes, A. M. Provost et al. 2017. Documentation for the MODFLOW 6 Groundwater Flow (GWF) Model. U.S. Geological Survey Techniques and Methods Book 6, Chapter A55, 197 p.

Leake, S.A., and J. Haney. 2010. Possible effects of groundwater pumping on surface water in the Verde Valley, Arizona. U.S. Geological Survey Fact Sheet 2010-3108, 4 p.

Leake, S.A., A. D. Konieczki, J. A. H. Rees et al. 2000. Desert Basins of the Southwest. U.S. Geological Survey Fact Sheet 086-00, 4 p.

Long, D., X. Chen, B. R. Scanlon et al. 2017. Global Analysis of spatiotemporal variability in merged total water storage changes using multiple GRACE products and global hydrological models. Remote Sensing of Environment, v. 192, p. 198–216.

Lusczynski, N.J., and W.V. Swarzenski. 1966. Salt-water encroachment in southern Nassau and southeastern Queens counties, Long Island, New York. U.S. Geological Survey Water Supply Paper 1613-F, 76 p.

Mahon, G. L., and D. T. Poynter. 1993. Development, calibration, and testing of groundwater flow models for the Mississippi River Valley alluvial aquifer in eastern Arkansas using one-square-mile cells. U.S. Geological Survey Water-Resources Investigations Report 92-4106, 33 p.

Maupin, M. A., and N. L. Barber. 2005. Estimated withdrawals from principal aquifers in the United States, 2000. U.S. Geological Survey Circular 1279, 46 p.

McGuire, V. L. 2017. Water-level and recoverable water in storage changes, High Plains aquifer, predevelopment to 2015 and 2013–15. U.S. Geological Survey Scientific Investigations Report 2017-5040, 14 p.

McGuire, V. L., M. R. Johnson, R. L. Schieffer et al. 2003. Water in storage and approaches to groundwater management, High Plains aquifer, 2000. U.S. Geological Survey Circular 1243, 51 p.

Miller, C. D., D. Durnford, M. R. Halstead, J. Altenhofen, and V. Flory. 2007. Stream depletion in alluvial valleys using the SDF semianalytical model. Groundwater, v. 45, no. 4, p. 506–514.

NASA EO. 2017. San Joaquin Valley is still sinking. NASA Earth Observatory Image, https://earthobservatory.nasa.gov/images/89761/san-joaquin-valley-is-still-sinking (accessed 10 May 2023).

National Research Council. 2002. Riparian Areas: Functions and Strategies for Management. National Academies Press, 444 p.

Oreskes, N., K. Shrader-Frechette, and K. Belitz. 1994. Verification, validation, and confirmation of numerical models in Earth sciences. Science, v. 263, p. 641–646.

Perkin, J. S., K. B. Gido, J. A. Falke et al. 2017. Groundwater declines are linked to changes in Great Plains stream fish assemblages. Proceedings of the National Academy of Sciences of the United States of America, v. 114, no. 28, p. 7373–7378.

Poland, J. F. 1961. The coefficient of storage in a region of major subsidence caused by compaction of an aquifer system. In Short Papers in the Geologic and Hydrologic Sciences, Articles 1–146, U.S. Geological Survey Professional Paper 424-B, 52–54.

Poland, J. F., and G. H. Davis. 1956. Subsidence of the land surface in the Tulare-Wasco (Delano) and Los Banos-Kettleman City area, San Joaquin Valley, California. Eos, Transactions American Geophysical Union, v. 37, no. 3, p. 287–296.

Poland, J. F., B. E. Lofgren, R. L. Ireland et al. 1975. Land subsidence in the San Joaquin Valley, California as of 1972. U.S. Geological Survey Professional Paper 43 7-H, 78 p.

Prinos, S. T., M. A. Wacker, K. J. Cunningham, and D. V. Fitterman. 2014. Origins and delineation of saltwater intrusion in the Biscayne aquifer and changes in the distribution of saltwater in Miami-Dade County, Florida. U.S. Geological Survey Scientific Investigations Report 2014-5025, 101 p.

Reed, T. B. 2003. Recalibration of a groundwater flow model of the Mississippi River Valley alluvial aquifer of northeastern Arkansas, 1918–1998, with simulations of water levels caused by projected groundwater withdrawals through 2049. U.S. Geological Survey Water-Resources Investigations Report 03-4109, 58 p.

Renken, R. A., J. Dixon, J. Koehmstedt et al. 2005. Impact of Anthropogenic Development on Coastal Groundwater Hydrology in Southeastern Florida, 1900–2000. U.S. Geological Survey Circular 1275, 77 p.

Rodell, M., I. Velicogna, and J. S. Famiglietti. 2009. Satellite-based estimates of groundwater depletion in India. Nature, v. 460, p. 999–1002.

Rodell, M., J. S. Famiglietti, D. N. Wiese et al. 2018. Emerging trends in global freshwater availability. Nature, v. 557, p. 651–659.

Sajil Kumar, P. J. 2019. Assessment of corrosion and scaling potential of the groundwater in the Thanjavur district using hydrogeochemical analysis and spatial modeling techniques. SN Applied Sciences, v. 1, p. 1–13.

Sneed, M., J. T. Brandt, and M. Sol. 2018. Land subsidence along the California Aqueduct in west-central San Joaquin Valley, California, 2003–10. U.S. Geological Survey Scientific Investigations Report 2018-5144, 67 p.

Stanton, J. S., S. M. Peterson, and M. N. Fienen. 2010. Simulation of groundwater flow and effects of groundwater irrigation on stream base flow in the Elkhorn and Loup River Basins, Nebraska, 1895–2055—Phase Two. U.S. Geological Survey Scientific Investigations Report 2010-5149, 78 p.

Stromberg, J., M. Briggs, C. Gourley et al. 2004. Human alterations of riparian ecosystems, In P. F. Ffolliott, and L. F. DeBano (eds.), Riparian Areas of the Southwestern United States: Hydrology, Ecology, and Management. CRC Press, Boca Raton, FL, p. 101–126.

Stumm, F., M. D. Como, and M. A. Zuck. 2020. Use of time domain electromagnetic soundings and borehole electromagnetic induction logs to delineate the freshwater/saltwater interface on southwestern Long Island, New York, 2015–17. U.S. Geological Survey Open-File Report 2020-1093, 27 p.

Szilagyi, J. 1999. Streamflow depletion investigations in the Republican River basin: Colorado, Nebraska, and Kansas. Journal of Environmental Systems, v. 27, no. 3, p. 251–263.

Taylor, C.J. and W.M. Alley. 2001. Groundwater-level monitoring and the importance of long-term water-level data. U.S. Geological Survey Circular 1217, 68 p.

USGS. 1999. Groundwater. U.S. Geological Survey General Interest Publication Reston, Virginia, 1999 revision.

Wahl, K. L., and R. L. Tortorelli. 1997. Changes in flow in the Beaver–North Canadian River Basin upstream from Canton Lake, western Oklahoma. U.S. Geological Survey Water-Resources Investigations Report 96-4304, 56 p.

Welter, D. E., J. T. White, R. J. Hunt, and J. E. Doherty. 2015. Approaches in highly parameterized inversion—PEST++ Version 3, a Parameter estimation and uncertainty analysis software suite optimized for large environmental models. U.S. Geological Survey Techniques and Methods, Book 7, Chapter C12, 54 p.

Williamson, A. K., D. E. Prudic, and L. A. Swain. 1989. Groundwater flow in the Central Valley, California. U.S. Geological Survey Professional Paper 1401-D, 127 p.

Winter, T. C., J. W. Harvey, O. L. Franke et al. 1998. Groundwater and surface water, a single resource. U.S. Geological. Survey Circular 1139, 79 p.

Yasarer, L. M., J. M. Taylor, J. R. Rigby, and M. A. Locke. 2020. Trends in land use, irrigation, and streamflow alteration in the Mississippi River Alluvial Plain. Frontiers in Environmental Science, v. 8, no. Article 66, 13 p.

Zaimes, G. 2007. Chapter 7: Human Alterations to Riparian Areas, In G. Zaimes (ed.), Understanding Arizona Riparian Areas. College of Agriculture and Life Sciences, University of Arizona (Tucson, AZ), p. 83–109.

15

Groundwater Management

CHAPTER MENU

Groundwater has been called upon to provide freshwater for thousands of years. A growing concern is the unsustainable rate of utilization of this resource. The main drivers are irrigated agriculture as an essential component in food security for Earth's 7.5 billion people and technologies that gives individual farmers capacity to tap the water under their feet. In wet regions, groundwater is usually a renewable resource, with the water lost to wells and springs replaced by infiltration of rainwater. The story is often different in semi-arid countries where groundwater already stored in the aquifer is removed faster than it is being replenished. Such continuous overproduction of groundwater without replenishment is unsustainable. Another realization is that overpumping is not the only way to kill off a groundwater supply. Contamination related to various agricultural and industrial activities creates sources of contamination that further limit the usability of water supplies.

The concept of groundwater sustainability developed in this chapter is a strategy that sets out "push back" against wanton destruction of groundwater resources. In effect, it provides a pathway to preserving the groundwater resource and its associated functions forever. This effort is unabashedly worthwhile. Groundwater, besides an essential requirement for life as we know it, represents an amazing water resource for reasons we explore in the next section.

What is most troubling is that as a largely "invisible" resource groundwater is being exploited with impunity. Sandra Postel's (1997) book brought the problems of declining groundwater levels and dry rivers and lakes to the world's attention. Scientific work in areas of sustainability became visible about the same time, especially through the World Bank's Groundwater Management Advisory Team (GW-MATE). For a decade, beginning early in the 2000s reports of the Team studied seriously impacted groundwater systems and provided practical strategies aimed towards sustainability.

15.1 The Case for Groundwater Sustainability

Since the beginning of the 20th century, the worldwide rate of withdrawal of groundwater for irrigation has increased substantially to meet the food needs of rapidly growing populations. This growth in groundwater production was made possible with new knowledge and technologies.

The accelerated use of groundwater in agriculture has occurred in waves. The first began in the early 1900s with countries that included Italy, Mexico, Spain, and the United States (van der Gun, 2012). In a second wave in the 1970s, populous countries in Southeast Asia, (i.e., China and India), the Middle East, and in northern Africa began to exploit groundwater. A developing third wave now encompasses parts of Africa, Sri Lanka, and other Asian nations, like Vietnam (van der Gun, 2012). Increasingly, the demands of water can only be met by mining of groundwater from aquifers at rates greater than

Fundamentals of Groundwater, Second Edition. Franklin W. Schwartz and Hubao Zhang.

Companion website: www.wiley.com/go/schwartz/fundamentalsofgroundwater2

natural replenishment. Moreover, pollution is further reducing the availability of groundwater and surface waters. Worldwide the use of groundwater is increasing by 1–2% per year.

Around the world, the history of the groundwater development has followed a consistent pattern. Withdrawals start out small, constrained by small populations or a lack of technical capabilities in pumping the water out of the ground, moving it from one place to another and using it in a practical manner. As needs increase and the benefits of using groundwater become evident, its use begins to grow. This phase of development is slow, growing gradually over decades. For users at this stage of development, the groundwater supply seems reliable and inexhaustible. For these reasons, there is little concern about management of the supply, and development is usually dictated by an individual decision at a farm or factory. This stage of rather uncontentious and peaceful evolution has been called the silent revolution (Llamas and Martínez-Santos, 2005; Llamas et al., 2009). At some point, things change, as problems and limits to growth in pumping become much more obvious as uses begin to reach unsustainable levels. Schwartz and Ibaraki (2011) characterized this end game in development as "a hell-bent assault on groundwater to wring more water from this dwindling and degrading source." Some accumulation of problems or a shock associated with overpumping appears seems to be a necessary pre-requisite for centralized government action.

Since the late 1990s, there has been progress in documenting the severity of problems associated with overpumping of groundwater. Konikow and Kendy (2005) called attention to significant problems of groundwater depletion in North Africa, the Middle East, South and Central Asia, North China, North America, and Australia, and numerous smaller aquifer systems. Konikow's (2011) more recent quantitative estimates of the net depletion of groundwater found that from 1900 to 2000 the loss was about around 3400 km^3 (Figure 15.1). In the first eight years of the new millennium, that number increased to ~4500 km^3. What is particularly thought provoking about these results was the exponential growth exhibited by the global total and some of the other curves, particularly India and the "rest of the world."

A study by van der Gun (2012) examined the present status of some of the world's most important aquifers. His list of aquifers was divided into two groups—those that received natural recharge and were somewhat renewable and those not presently significantly recharged. Of aquifers receiving recharge, four systems stand out—the High Plains aquifer system in the Midwestern U.S. and the Central Valley System of California and alluvial aquifers of northern India and northwest India with approximately 72 km^3 produced each year (van der Gun, 2012). These aquifer systems are discussed in other chapters. Among aquifers with little to no modern recharge the Nubian Sandstone aquifer system and the Saudi Arabia Platform aquifers are most important (van der Gun, 2012).

Until recently, the full extent of the groundwater problems associated with overpumping was largely unappreciated. Most of the large, populous countries of Africa, Asia, and Asia were "black holes" for hydrogeological data. Little way of useful information was able "escape" these countries because it mostly did not exist or was closely held. Thus, it was not surprising that the first large regional geospatial assessments of these countries using GRACE (Gravity Recovery and Climate Experiment) (Famiglietti, 2014; Rodell et al., 2009), were illuminating. It was like switching on the lights in rooms of a dark museum. Some of these estimates of groundwater utilization using GRACE ended up being scaled back somewhat (Long et al. 2017; MacDonald et al., 2016); yet the damage being wrought on the world's aquifers was there for all to see.

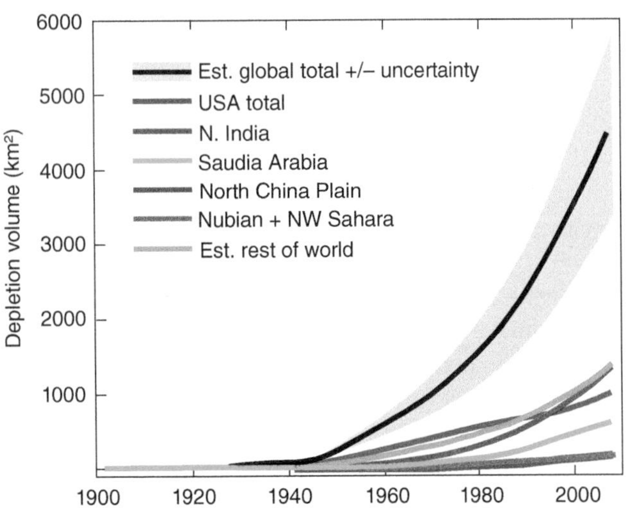

Figure 15.1 History of net global depletion for key countries, aquifers, or regions (U.S. Geological Survey, adapted from Konikow, 2011. Not subject to U.S. Copyright. Published by American Geophysical Union).

15.2 Groundwater Sustainability Defined

The present concepts of sustainable yields have developed over more than century with ideas refined by contributions of historically important hydrogeologists, such as Meinzer, Theis, Lohman, Todd, and many others (Gungle et al. 2016). The starting place was with the concept of safe yield in the management of large aquifers, which is more than 100 years old. Even then, hydrogeologists knew there was a "limit to the quantity of water which can be withdrawn regularly and permanently without dangerous depletion of the storage reserve" (Lee, 1915). Gungle et al. (2016) discussed how concept of safe yield evolved through time with various embellishments toward a more general concept of sustainable yield, which for groundwater arrived in 1982 with foundational work by Bredehoeft et al. (1982). Alley et al. (1999) defined groundwater *sustainability* as the "development and use of groundwater in a manner that can be maintained for an indefinite time without causing unacceptable, environmental, economic, or social consequences."

There is another very different world of sustainability that exists around communities that produce groundwater from shallow aquifers with limited productivity. Such groundwater sources can be unreliable in places, such as India, with a relatively short monsoonal season, punctuated by extended months of dryness. Pumping is often unsustainable even over the period of one or several dry seasons. However, human ingenuity trumped theory hundreds of years ago to create innovative strategies like percolation pits and percolation tanks, which facilitate the storage of rainwater in these shallow aquifers. The important takeaway is that sustainability issues exist at different scales and solutions need to be tailored to these different situations. We will return to this theme in a subsequent section.

The modern concept of groundwater sustainability has evolved to acknowledge inherent complexities of hydrologic systems. For example, groundwater in aquifers is not isolated within the hydrologic cycle but connected to processes on the land surface through recharge and discharge. Thus, chronic overpumping of aquifers not only impacts aquifers but also can contribute to reduced stream flows, drying of wetlands, and destruction of vegetation along streams. Another more recent realization is that contamination of groundwater is a significant risk to sustainability, as is illustrated by problems of salinization in India (Foster et al., 2018).

The list of unacceptable consequences in the definition of groundwater sustainability seemed rather concise—"environmental, economic, or social consequences." Yet, the unstated implications buried within those few general statements are that no prescribed lists of consequences exists to guide sustainability assessments. In other words, sustainable planning for groundwater should consider issues appropriate to key problems of concern.

Historically, concerns about sustainability have always been about the resource itself. Recently, Gungle et al. (2016) provided a rationale for prioritizing environmental needs for water first. In effect, this is a constraint-based based approach where the environmental or "non-extractive" water needs effectively constrain how much water is left over for economic needs (e.g., irrigated agriculture) or social needs (e.g., domestic water supplies).

15.2.1 Sustainability Initiatives

Concerns around sustainability practices are not new. For example, Vrba et al. (2007) coedited a volume concerned with the development of sustainability indicators. This UNESCO-sponsored effort (United Nations Educational, Scientific and Cultural Organization) led to two families, (1) groundwater indicators and (2) indicators of the social and economic relevance of groundwater. These indicators could be useful for global comparisons of the status of countries with respect to groundwater sustainability.

An important step forward was with legislation in California with the Sustainable Groundwater Management Act (SGMA) to address groundwater problems relevant to California. The Act primarily addressed problems associated groundwater overproduction, which in this legislation were termed "*undesirable consequences*." The list of undesirable consequences included (1) chronic declines in water levels, (2) unsustainable reductions in the quantities of stored groundwater, (3) seawater intrusion, (4) groundwater contamination, (5) subsidence of the land surface, and (6) various problems due to pumping-related losses in surface waters. Each of these undesirable consequences was associated with a specific sustainability indicator—a specific metric capable of tracking the potential problem through time in a systematic matter. For example, declining water levels in aquifers (1) and land-surface subsidence (5) can be monitored using wells and extensometers, respectively. Degradation in water quality due to seawater intrusion (3), or industrial or other contamination (4), can be tracked by regular chemical analyses of water samples.

By 2040, users of groundwater in California must have sustainability plans fully implemented and have achieved sustainability. The legislation is also important as a unique policy example with far-reaching implications (Lubell et al., 2020).

The California framework focuses largely on economic and social consequences of substantial diminishment of the resource. The environmental issues around sustainability are mostly bundled with undesirable consequences (6), surface-water depletion. Moreover, the list of undesirable consequences for California are the problems associated with overpumping that we discussed in Chapter 14. Not surprisingly, California's list of undesirable consequences cover key issues associated with other large aquifer systems in the United States and around the world.

For California, surface water impacts, specifically the loses in beneficial uses, stand out among the other consequences in terms of their complexity. Part of the complexity comes from the fact that surface waters have many different beneficial uses. Stanford University released a guidance document to support sustainability work in California (Belin, 2018). It provided a list of the most important beneficial uses for surface waters, which would be relevant to sustainability planning. This list is helpful in illustrating the variety and scope of features to be protected:

- habitats necessary for species established as rare, threatened, or endangered,
- areas like refuges, parks, sanctuaries, etc. where natural resources require special protection,
- cold water ecosystems with respect to vegetation, fish, or wildlife, including invertebrates,
- warm water ecosystems supporting aquatic habitats, vegetation, fish, or wildlife, including invertebrates,
- high-quality aquatic habitats suitable for reproduction and early development of fish,
- habitats necessary for migration or use by anadromous fish (e.g., salmon),
- estuarine ecosystems,
- terrestrial ecosystems, e.g., terrestrial habitats, vegetation, and wildlife (Belin 2018).

Another piece of the complexity puzzle stems from the very different treatment afforded surface water versus groundwater. Although now understood to be a single resource (Winter et al., 1998) from a legal perspective, groundwater and surface water are treated differently in most jurisdictions. California is a useful example in this respect (Lubell et al., 2020).

15.2.2 Sustainability Indicators for the Sierra Vista Subwatershed in Arizona

Sustainable management of groundwater can involve large jurisdictions, such as the State of California, or single sites like the San Pedro Riparian National Conservation Area, Arizona within the larger Sierra Vista Subwatershed (Figure 15.2). We added this discussion here as a counter-example to the broad, statewide sustainability indicators appropriate to California. The list of indicators with the San Pedro are focused on the preservation of the unique riparian ecosystem that exists there. It includes a diverse collection of birds, mammals, reptiles, and amphibians (Healy et al., 2007), associated with a cottonwood-willow riparian forest. Success in the protection of this unique ecosystem involves the restoration and maintenance of sustainable yields within the Sierra Vista Subwatershed to provide stable discharges to the San Pedro River and baseflows (Gungle et al., 2016). Other than sporadic summer storms, perennial flow in the San Pedro River depends on long-term groundwater discharge.

The San Pedro River rises in the Sonoran Desert and flows northward into Arizona and the Gila River. Within the 2560 km^2 (950 mi^2) Sierra Vista Subwatershed (Figure 15.2), the river has a length of ~56 river kilometers, with 35 km of perennial flows (Gungle et al., 2016). The regional surficial aquifer in the subwatershed receives mountain-front recharge, which discharges into the river.

The sustainability concern was that historically high and growing pumping of groundwater had the potential to dry out the river. The population in the Sierra Vista Subwatershed was approximately 68,000 in 2002 and 80,000 in 2012 (Gungle et al., 2016; Healy et al., 2007). Aggressive management of the groundwater began with extensive and comprehensive monitoring of groundwater, surface water, and biological systems. As was the case with California, monitoring here was organized around sustainability indicators (Table 15.1) that were developed through the active participation of stakeholder groups.

The consequence of overpumping, in this case, is reflected in the choice of 14 sustainability indicators that are different from those developed for California (Table 15.1). These evident differences reiterate the point made earlier, namely, to expect that sustainability indicators will vary from place to place because concerns are different. However, certain generalizations are possible. Indicators for special or unique problems, like Sierra Vista, will have an obvious "specificity" because of the environmental focus on the health of the riparian ecosystem (Gungle et al., 2016). The list of sustainability indicators used for managing groundwater in many places across regional and larger scales, such as California, is of necessity much more broadly focused, anticipating a variety of different problems and local situations and settings (Gungle et al., 2016).

It was known for a long time that California's groundwater was being used in an unsustainable manner (Lubell et al., 2020). As is often the case, the relative invisibility of groundwater relative to surface water led to a series of incremental

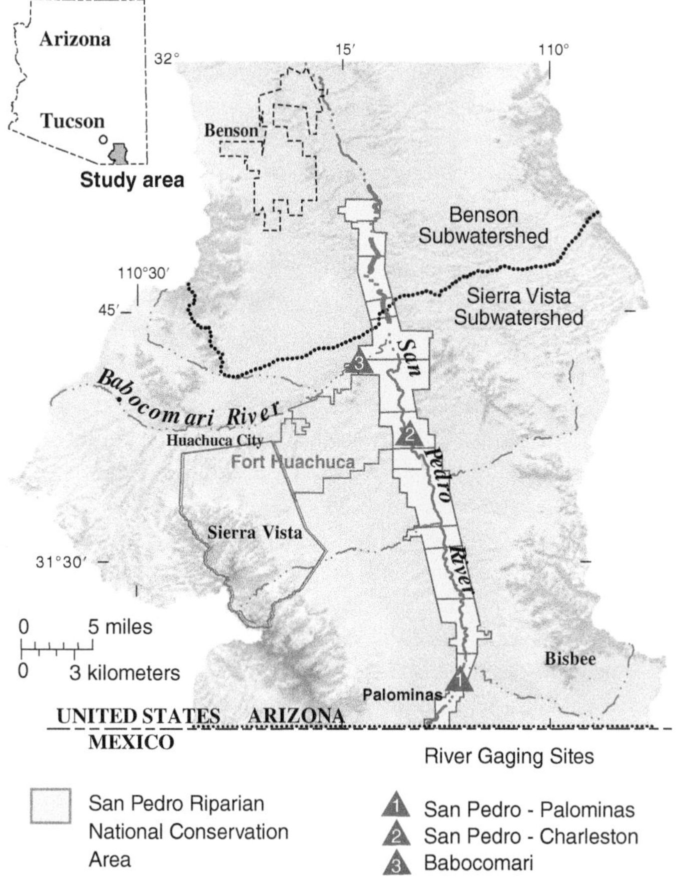

Figure 15.2 Map showing the location of the San Pedro Riparian National Conservation Area southeastern Arizona (insert map). The focus of studies has been the impact groundwater pumping across the Sierra Vista Subwatershed. The San Pedro River rises in Mexico is being gaged, along with the Babocomari River at sites shown on the map (U.S. Geological Survey, adapted from Leenhouts et al., 2006/Public domain).

Table 15.1 Sustainability indicators for management of groundwater in the Sierra Visa Subwatershed in Arizona.

No.	Indicator Group
	Group 1, Subwatershedwide indicators
1	Regional-aquifer water levels
2	Horizontal gradients (regional-aquifer wells)
3	Aquifer-storage change measured with microgravity
4	Annual groundwater-budget balance
	Group 2, Riparian system indicators
5	Near-stream alluvial-aquifer water levels
6	Near-stream vertical gradients
7	Annual fluctuation of near-stream alluvial-aquifer water levels
	Group 3, San Pedro River indicators
8	Streamflow permanence
9	Base-flow on San Pedro and Babocomari Rivers
10	June wet-dry status
11	San Pedro River water quality
12	San Pedro and Babocomari Rivers isotope analysis
	Group 4, Springs indicators
13	Springs discharge
14	Springs water quality

Source: U.S. Geological Survey Gungle et al. (2016)/Public domain.

measures that were ineffective in a management sense. The result was a situation where surface water use was closely watched, yet groundwater was there for the taking, "a literal wild west" (Lubell et al., 2020).

While progress in groundwater management was slow for nearly a century, California's severe drought 2012–2016 led to widespread impacts with declining water levels, subsidence, and risks of lost supplies. These were factors leading to the passage of the Sustainable Groundwater Management Act in California. One unique feature of the legislation in California was its novel strategy for managing a "common-pool resource" (Lubell et al., 2020). Unrestricted production of groundwater from an aquifer, leading to the depletion of the resource, is a textbook example of "the tragedy of the commons" at work. Decisions of individuals to maximize their benefit by aggressive groundwater production are usually in conflict with the sensible management of the resource. Although California's efforts toward sustainable groundwater over the next 20 years is a work in progress, it represents an important effort toward a common societal goal.

In other countries, progress toward sustainability is sometimes lacking even the basic framework for action (Schwartz et al., 2020). This framework is the government's commitment to the management of the resource. It usually involves working with stakeholders to develop a vision for groundwater management. This step leads naturally to the creation of laws and policies (Smith et al., 2016) necessary to provide the authorities and mechanisms needed for sustainability. Specific examples in a sustainability context includes enabling legislation for locating and registering wells and using licenses or "obligations" to regulate production (Smith et al., 2016). The government also needs powers to manage activities, such as the construction of new wells in impacted areas, enforcement of specified withdrawal rates (Garduno and Foster, 2010), and to prosecute undesirable practices, such as pumping water illegally or contaminating groundwater (Smith et al., 2016). Experience in the United States points to the importance of complementary regulations to rigidly control the handling and disposal of hazardous chemicals, and to protect capture zones in and around well fields.

Some countries have laws and regulations in place, yet the groundwater remains effectively unmanaged. Missing in this respect is hands-on, field-level enforcement to implement governments' mandates around groundwater. Countries without an operational framework often lack the institutional capacity for implementation. For example, in large populous countries, such as India, Pakistan, or Indonesia, the commitment to manage coupled groundwater/surface water systems would be massive, given the numbers of wells involved and complimentary needs for hydrologic data and aquifer characterizations.

At a high level (state or national levels), an important role of government is to provide oversight funding, coordination, and overall management. Of necessity, coordination activities need to extend to other governmental agencies with interests in surface water, agriculture, energy, etc. where their policies have implications for groundwater. Other governmental organizations and their employees play a role at the scale of river basins, aquifers, or local communities (Smith et al., 2016).

Meaningful implementation of groundwater policies also needs tacit support and participation from stakeholders. When the views of the water users are represented, especially during the development of policies, there is broader compliance, and willingness to provide groundwater data and to solve problems locally (Garduno and Foster, 2010; Smith et al., 2016).

15.2.3 Socioeconomic Policies and Instruments

Another important feature of a general structure for groundwater is management instruments. These are powerful socioeconomic tools that provide governments with a "carrot and stick" approach that can move a country toward groundwater sustainability in various ways. Properly conceived, they can reduce groundwater utilization and contamination.

Garduno and Foster (2010) listed four socioeconomic tools useful in promoting a reduction in the use of groundwaters. These are summarized from a GW-MATE report (Garduno and Foster, 2010) documenting approaches effective in reducing excessive withdrawals of water for irrigation. First was the enforcement of regulations to reign in excessive uses of groundwater. Success with these regulatory would require renewable rights or licenses to use specified quantities of water in addition to a collection of easy-to-understand management actions, such as drilling bans or production caps for existing wells. Success with regulatory actions requires the existence of well-based data to document impacts and compliance.

Garduno and Foster (2010) also emphasized the need to recognize the importance of water by charging for its use, especially with large commercial users. Implicit in approaches to charge for water is a requirement to monitor quantities of groundwater being pumped. Metering of irrigation wells and charging for the water use are uncommon practices in most countries and from a farmer's perspective would be a major governmental intrusion on historical practices of free groundwater use. Garduno and Foster (2010) suggested possibilities for indirect monitoring using data on electrical power use or crop production, as well as remote sensing.

Third on their list were strategic, governmental investments to provide for the more efficient implementation of irrigation. Examples included efficiencies in the delivery of surface water or more efficient irrigation systems. However, there must be careful to assure that water savings are real, rather than saving water by reducing aquifer recharge (Garduno and Foster, 2010). Similarly, water savings from improvements in irrigation efficiencies needed to translate to reduced pumping (Garduno and Foster, 2010). There was a tendency to used water saved from irrigation efficiencies simply to grow more food. They also noted that in some cases improvements in irrigation efficiency would be at the expense of reduced recharge.

Finally, reductions in groundwater utilization could come about through beneficial macro-policies. However, a necessary first step would be to reverse a historical collection of macro-policies that promotes groundwater over-use (Garduno and Foster, 2010). Thus, activities at the national level must anticipate unintended consequences from actions taken in other parts of the government. For example, government subsidies to farmers through cheap electrical power, fertilizers, and pesticides might lead directly to expansion of groundwater irrigation (Garduno and Foster, 2010). Examples of such policies in India included price support for crops that required lots of water and low-cost rural electric power, fertilizer, and agrochemical to stimulate food production. Unfortunately, people tend to resist actions requiring the payment of market prices for things that historically were virtually free.

15.3 Overview of Approaches for Sustainable Management

This section introduces analytical approaches that contribute in a significant way to the sustainable management of groundwater. They provide the bases for evaluating the status of groundwater/surface water systems within a sustainability context. In other words, there are analyses needed to determine whether present or future groundwater pumping is sustainable, and to plan necessary project interventions. Similar analytical approaches can be used in the future to validate the success of project interventions and to maintain sustainable conditions. In this section, we will introduce three of these competencies, indicator tracking, water budget analyses, and modeling. In a California context, the most serious problems have required the use of all these approaches.

Data needs for sustainability work are large. The fact that the collection of some types of data takes time and that groundwater systems need to be monitored for decades requires forethought to anticipate eventual needs for groundwater-related data (Alley et al., 1999). Absent the necessary data, there are risks of not achieving sustainability goals and wasteful spending.

Data are necessary to provide a clear conceptual understanding of the state of the aquifers and groundwater systems. Examples might include water-level monitoring data, test data that characterize hydrogeological properties, maps, and cross sections showing well locations, distributions of key units and patterns of groundwater flow, information on the location of recharge and discharge areas, and water quality analyses. Beyond groundwater, there are need for data concerned with weather, topography, land use and land cover, irrigation systems, soils, surface water (e.g., stream flows, diversions, and water quality), and sources of contamination, etc.

15.3.1 Indicator Tracking

The techniques available to evaluate groundwater sustainability vary in complexity. The simplest approach relies on tracking sustainability indicators, both historically and into the future. It provides a straightforward approach, which makes it clear whether indicators are moving in an appropriate direction. However, interventions with projects or programs will require more sophisticated analyses.

Indicator tracking has been used in the United States for more than 50 years to assess the response of aquifers to pumping. In areas with large groundwater withdrawals, monitoring of water levels in aquifers has been important in identifying serious depletion in stored water and motivating efforts to conserve water or reduce the quantities of water being pumped. Indicator tracking has usually focused on water-level monitoring but also has extended to monitoring of subsidence and seawater intrusion.

Historically, groundwater in the San Joaquin Valley began to exhibit serious declines in the early 1940s. Accompanying those declines in stored water was extraordinary land subsidence (Poland et al., 1975). The U.S. Geological Survey was active in those years monitoring both groundwater levels and subsidence. Figure 15.3 illustrates the abrupt declines of water levels in wells being pumped and large subsidence southwest of the Town of Mendota (Poland et al. 1975). Pumping levels in wells declined approximately 330 ft with approximately 27 ft of subsidence. During the 1950s and 1960s, there was no thresholds established that would have mandated corrective action to halt declines these unsustainable declines. However, earlier experience in and around San Jose, California would have indicated the seriousness of these problems and the need for

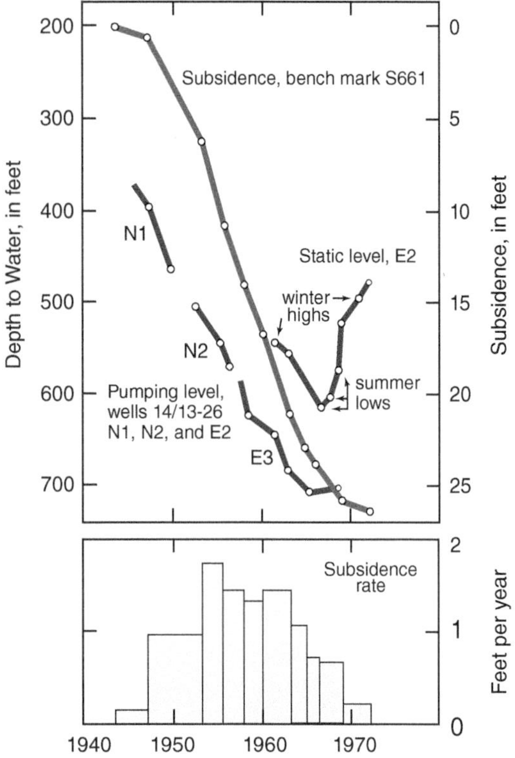

Figure 15.3 In the San Joaquin Valley, large declines in groundwater levels and extraordinary subsidence were addressed by a reduction in groundwater withdrawals through the importation of surface water (U.S. Geological Survey, adapted from Poland et al., 1975/Public domain).

projects to reduce the quantities of water being pumped. As imported surface water for irrigation became available, starting in 1968, water levels began to recover (Figure 15.3) and the rate of subsidence declined markedly (Poland et al., 1975).

The groundwater sustainability program in California makes use of indicatory tacking and provides guidance on its implementation (CDWR, 2017). To provide consistency among the various plans for basins across California, regulations basically prescribe what sustainability metrics to use. These are the "sustainability indicators" that we listed previously. For example, declining water-level elevations in monitoring wells, or elevated levels of chloride in the groundwater samples in a collection of wells indicate a need to address specific problems. What is new with the California legislation is a predetermined threshold for action with each of the sustainability indicators. Violation of that threshold automatically would signal the possibility of an undesirable result (CDWR, 2017) and requires corrective action to fix the problem. There is a detailed process to establish and justify the sustainability thresholds (CDWR, 2017).

In practice, data on water levels from a collection of wells is developed to monitor for potential impacts. A subset of the wells would be designated "as representative monitoring sites" (CDWR, 2017). As an example, consider the undesirable consequence of chronic declines in water levels. Water-level elevation would be tracked with monitoring wells, as the indicator for this undesirable consequence. If declines reached the threshold for action. Appropriate corrective measures would be required to reverse the water-level declines and to the basin back into compliance.

An example in the application of the tracking of sustainability indicators is provided in sustainability assessments of the Sierra Vista groundwater basin. The report on activities (Gungle et al., 2016) presented historical data for each of their 14 sustainability indicators (see Table 15.1) and discussed the implications as far as sustainability was concerned. For example, Figure 15.4 shows partial results for their Indicator 9—Base Flow on San Pedro and Babocomari Rivers (see Figure 15.4). They estimated a comparative baseflow number using the 3-day low flow for the month of January, a period with low evapotranspiration. The values shown in Figure 15.4 represented the smallest mean discharge on any three consecutive days in January (Gungle et al., 2016). Indications were that base flows on the San Pedro River are declining both at the upstream site at Palominas and downstream at Charleston (Figure 15.4). The report integrated the implications of groundwater pumping, as indicated from the 14 indicators, in a useful summary table.

These indicator approaches are using data to tracking trends. As the examples implied, there is a compelling need for data to provide context for the sustainability analyses. Not surprisingly, for many countries such data are unavailable, making even qualitative assessments impractical.

15.3.2 Water Balance Analyses

Water-balance approaches are another integral tool for sustainability analyses. Unlike indicator tracking, the methods provide quantitative information on scope of sustainability issues and form a basis for designing interventions. In both California programs and the Sierra Vista Subwatershed in Arizona, groundwater budgeting provided an important analytical tool.

Water balances are commonly applied to hydrologic features within the hydrologic cycle capable of storing water (i.e., accounting units), such as lakes, reservoirs, watersheds, and aquifers (Healy et al., 2007). Our particular focus is on aquifers as accounting units, but readers should understand that comprehensive sustainability analyses will also involve other units as well. As the term "accounting" implies, water balance calculations track rates of water inflows and outflows to aquifers to calculate the change of storage within the aquifer. In principle, the approach is completely analogous to accounting for

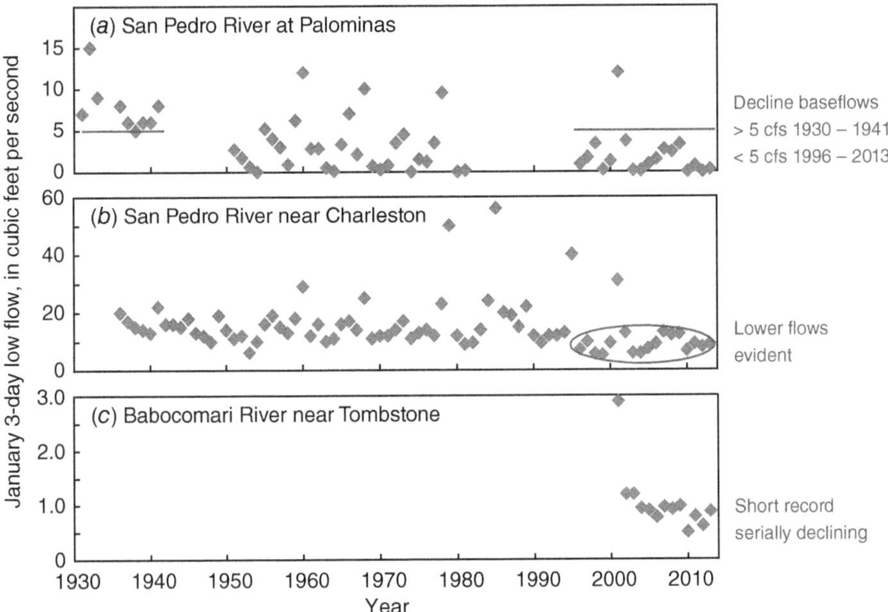

Figure 15.4 Several of the results associated with Indicator 9, 3-day winter baseflow on San Pedro and Babocomari Rivers. The locations of the sites are shown in Figure 15.2. Shown in red are comments to highlight features of the assessment (U.S. Geological Survey Gungle et al., 2016/Public domain).

monthly cash flows in and out of a bank checking account (i.e., the accounting unit) to establish the change in in account balance either plus or minus.

In words, a general water balance equation can be written as

$$\text{inflows}_{(1,2,3...)} - \text{outflows}_{(6,7,8...)} = \text{change in storage} \tag{15.1}$$

where the subscripts 1, 2, 3, etc. identify specific components of inflow and 6, 7, 8, etc. the components of outflow. In water balance applications, the next step is to take Eq. (15.1) and customize it, based on a conceptual model of the aquifer system.

The obvious question is where do the numbers that go into Eq. (15.1) originate. The field-based approaches use monitoring data of all kinds and inventories to make estimates of individual components. Because these approaches are data-based, they are retrospective, looking at historical inflows and outflows. The second common approach to constructing mass balances is modeling based. For example, creation of an aquifer model of the system of interest normally provides numerical values necessary to calculate a mass balance. MODFLOW-type codes normally provide global mass balance information as one measure of the accuracy of the calculation. Users can also output information on fluxes across specific internal and external boundaries of the model.

Importantly, the model-based approaches provide both retrospective and predictive water budgets by starting the model at some time in the past and running it into the future. This capability is necessary to determine the suitability of different corrective actions in correcting some undesirable condition affecting the aquifer system. Similarly, the modeling can provide an assessment of impacts of extreme conditions, such as long-term droughts or global climate change, on sustainability.

To illustrate the use of mass balance calculations, we will discuss two applications. The first is a model-based approach (Luckey and Becker, 1999) that involves the central portion of the High Plains aquifer (Figure 15.5). As discussed in this chapter, the High Plains aquifer system is among America's most important aquifers. Their mass-balance calculations compared two conditions for the aquifer in the study area, a predevelopment state before pumping began, and conditions in 1997. The report describes methods for collecting data necessary for the model construction. For example, initial estimates of recharge came from previous studies and were refined with model calibration. The quantities of water pumped from the aquifer were estimated based on the product of calculated monthly irrigation demand (i.e., the inches (cm) of irrigated water required for four key crops in each county, based on growth curves), the size of the area being irrigated and an assumed irrigation efficiency (Luckey and Becker, 1999).

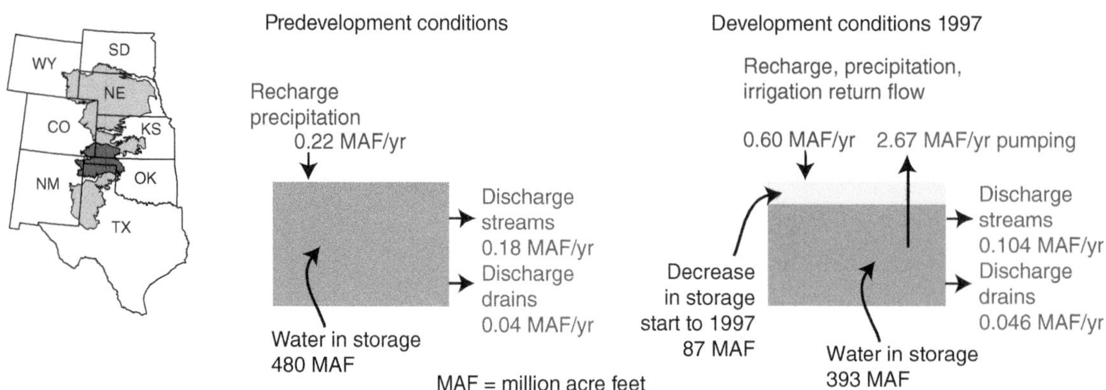

Figure 15.5 Mass balance assessment of groundwater conditions in the central portion of the High Plains aquifer system. Under predevelopment conditions on average, the annual inflow (km^3) equals the annual outflow and there is no change in the large quantity of water stored in the aquifer. Pumping for irrigation inflow creates a large annual depletion of groundwater, which mostly comes from aquifer storage and a reduction in outflows (U.S. Geological Survey, adapted from Luckey and Becker, 1999; Healy et al., 2007/Public domain).

The pre-development water budget calculations are straightforward. Inflow of water to the aquifer was precipitation of 0.22 MAF/yr (million acre-feet per year) or 0.27 km^3/yr. Outflow from the aquifer was 0.18 MAF/yr (0.22 km^3/yr) as natural discharge to streams and 0.04 MAF/yr (0.05 km^3/yr) to drains (Figure 15.5). Before development, the aquifer held 480 MAF (592 km^3) of water in storage. The mass balance equation for this situation has one inflow component, natural recharge (R_N) from precipitation and two outflow components, natural discharge to streams (D_S) and discharge to drains (D_D). The resulting water balance equation then subtracts the outflow components from the inflow component to yield the change in storage (ΔS) for that year:

$$R_N - (D_S + D_D) = \Delta S \tag{15.2}$$

Substituting the known values into this equation computes $\Delta S = 0$ MAF/yr for an average predevelopment year, as shown here:

$$0.22 - (0.18 - 0.04) = 0 \tag{15.3}$$

The zero value for the change in storage was expected because on average for a natural system the recharge equals the discharge.

By 1997, conditions were quite different because of the development of agriculture. One difference was an increase in recharge to 0.60 MAF (0.74 km^3/yr) that includes recharge due to precipitation, and the added recharge coming from dryland cultivation and irrigation. Because lands tend to be commonly over-irrigated, water infiltrates below the root zone and ends up as recharge. Such recharge is termed irrigation return flow. In the water balance equation, this recharge is represented as R_C, where the subscript "C" means combined, irrigation return flow and natural recharge. For the developed case in 1997, the two previous discharge components (D_S and D_D) are still present with D_S reduced, 0.104 and 0.046 MAF/yr (0.128 and 0.056 km^3/yr), respectively. Pumping for irrigation (Q_P) represents the dominant discharge component at 3.29 km^3/yr. The water balance equation for 1997 is written now as:

$$R_C - (D_S + D_D + Q_P) = \Delta S \tag{15.4}$$

Again, substituting values in to (15.4) provides the change in storage

$$0.60 - (0.104 + 0.046 + 2.67) = -2.22\ \text{MAF/yr}. \tag{15.5}$$

The calculation reveals that the groundwater being pumping for irrigation (Q_P, 2.67 MAF/yr) came largely out of storage (ΔS) to the extent of −2.22 MAF/yr. The loss of natural discharge to streams is relatively large with (D_S) declining from 0.18 MAF/yr (predevelopment) to 0.104 MAF/yr in 1997. The case study of the Beaver and North-Canadian River in Chapter 14 (Wahl and Tortorelli, 1997) provided actual observations on the decline in surface water in this same general area.

Luckey and Becker (1999) independently determined that over the time since pumping began to 1997 the decline in aquifer storage for the study area was approximately 87 million acre feet (107 km^3). This is approximately 18% of the water originally stored in the aquifer system. Model predictions in their paper suggested that by 2019 decline of water from storage would be approximately 28%.

The next example comes from the Sierra Vista Subwatershed in Arizona, which illustrates the field-based approach to mass balance calculations (Gungle et al., 2016), as well as a model-based approach (Pool and Dickinson, 2007). Individual estimates of inflow and outflow components derived from field measurements, such as hydraulic head, hydraulic conductivity, and spatial changes in streamflow etc. and annual surveys on the quantities of water being pumped.

One of the sustainability indicators for the Sierra Vista Subwatershed (Figure 15.2) is annual water budget calculations (Gungle et al., 2016). The actual water balance equation was rather complicated, comprised of 14 inflow/outflow components organized around four themes, namely natural aspects of the system (NAS), groundwater pumping (GP), active-management measures, (AMM), and unintentional recharge (UR). Each of the themes can have inflow and/or outflow components as shown in the water balance equation:

$$\text{NAS}\left[\text{inflow}_{(1,2)} - \text{outflow}_{(3,4,5)}\right] + \text{GP}\left[-\text{outflow}_{(6,7,8,9)}\right] + \text{AMM}\left[\text{inflow}_{(10,11,12)}\right] + \text{UR}\left[\text{inflow}_{(13,14)} = \Delta S \qquad (15.6)$$

Table 15.2 provides a brief description of each of the themes and associated components. Clearly, the water-balance equation was uniquely designed for the groundwater situation in the arid Sierra Vista Subwatershed. Organizing water budgets around themes helps identify those inflow/outflow out of human control (NAS, UR) and those that can be modified by human interventions (GP, AMM). The relatively large number of inflow/outflow components for a management problem like this is helpful because it helps to target interventions around the most critical components.

Gungle et al. (2016) calculated annual water budgets from 2002 to 2012. They provided illustrative values for 2002 and 2012 to demonstrate the changes that occurred over the decade, along with estimates of uncertainty. We depicted their results in the form of water-budget equations for each of the two years (Figure 15.6*a* and 15.6*b*), respectively. The actual equations are written in terms of subtotals for each of the four themes. For 2002, values are provided for each of the 14 components with units of cubic meters (per year ×1000) (Figure 15.6*a*). For 2012, only the totals for the four themes are provided (Figure 15.6*b*).

Table 15.2 Listing and description of the four themes and 14 components that comprise the water budgets for the groundwater in the Sierra Vista Subwatershed.

Budget Themes and Components		Description
Natural Aspects of System		
1.	Natural recharge	Recharge from ephemeral channels basin floor, mountain-front recharge
2.	Groundwater inflow	Groundwater inflow to subwatershed across southern boundary
3.	Groundwater outflow	Groundwater outflow from subwatershed across northern boundary
4.	Stream discharge baseflow	Groundwater discharging into San Pedro River from aquifers
5.	Riparian evapotranspiration	Groundwater lost as ET from riparian corridors San Pedro + Babocomari R
Groundwater Pumping		
6.	Municipal + Water Co. pumping	Municipal wells Huachuca City, Fort Huachuca, etc. private water suppliers
7.	Rural/exempt water pumping	Unmetered use of groundwater by rural users with small-capacity wells
8.	Industrial pumping	Includes pumping for turf irrigation, sand/gravel mining, and stock tanks
9.	Irrigation pumping	Pumping to irrigate vineyards, orchards, and pastures
Active Management Measures		
10.	Mesquite + tamarisk treatment	Eradication programs for these nuisance trees that use groundwater
11.	Municipal-effluent recharge	Detention ponds next to wastewater treatment plants recharge groundwater
12.	Detention basin recharge	Stormwater retention basins at Fort Huachuca, Sierra Vissta recharge groundwater
Unintended Recharge		
13.	Total Incidental recharge	Includes recharge from septic tanks, golf courses, and leaking water systems
14.	Urban enhanced recharge	Urban infrastructure, roofs, pavements, etc. locally concentrate runoff

Source: U.S. Geological Survey, Adapted from Gungle et al. (2016)/Public domain.

	Natural aspects of system		Groundwater pumping		Active-management measures		Uninetentional recharge		ΔS
(a) 2002	1	16652.							
	2	3700.	6	−13198.					
	3	−1480.	7	−1480.	10	493.			
	4	−3207.	8	−1974.	11	1604.	13	1110.	
	5	−15048.	9	−3084.	12	247.	14	2344.	
	Subtotals	617. +		−19736. +		2344. +		3454. =	−13321. m³x1000
(b) 2012	Subtotals	617. +		−14987. +		3946. +		4194. =	−6230. m³x1000

Figure 15.6 Annual groundwater budgets for the Sierra Vista Subwatershed for 2002 (*a*) and 2012 (*b*). All numerical values have units of $m^3 \times 1000$ (U.S. Geological Survey, adapted from Gungle et al., 2016/Public domain).

From 2002 to 2010, the annual deficit in storage (ΔS) improved from −13,321 to $-6230\,m^3 \times 1000$. In other words, although water was being lost from groundwater storage, the quantity was substantially reduced. This progress was evident despite a growth in population from ~70,000 people in 2002 to ~81,000 in 2012. Close examination of the numbers for 2012 provides a good understanding of the issues. For example, the five, natural components (1–5) show that without the people, the inflows to the groundwater basin (1 natural recharge and 2 groundwater) slightly exceed the natural outflows (3 groundwater outflows, 4 base-flow discharge, and 5 riparian evapotranspiration). What this means is that the groundwater basin cannot sustainably accommodate withdrawals of groundwater from wells. The decline of storage in the 2002 water balance ($-13{,}321\,m^3 \times 1000$) was due entirely to pumping of groundwater attenuated by active management measures and unintentional recharge (Figure 15.6*a*).

The improvement that was evident in the water balance for 2012 (Figure 15.6*b*) came mostly from a reduction in urban and irrigation pumping with continuing contributions in active management and growth in unintentional recharge. Given the water needs for people in the basin are likely irreducible, sustainable management of the groundwater basin could require treatment and recharge of all urban water used within the basin.

15.3.3 Model-Based Analyses of Sustainability

Field-based water budgets are useful in framing problems of sustainability and helpful in explaining issues in a simple way. However, meaningful analyses of complicated sustainability problems will invariably require modeling. Groundwater models are the most powerful tool in our arsenal of methods for analyzing complicated problems. The proper construction and calibration of a groundwater model leads to a sophisticated understanding of processes and how water moves through the aquifer system(s) (CDWR, 2016). Thus, while computational models are indeed useful for calculating water budgets, they have many more important roles to play in creating sustainable systems and assuring compliance with regulations.

The process of constructing a numerical model itself has several important benefits (CDWR, 2016). Firstly, modeling requires that key data be assembled and organized in an appropriate manner. Gaps in available data will become readily apparent and provide a basis for designing future field programs. Additionally, model runs involving calibration using historical data should make clear (1) what aspects of the physical hydrologic setting and resource utilization are influencing sustainability and (2) the impact of year-to-year variability in key parameters. These types of analyses are useful for identifying and understanding situations, which might pose a threat to sustainability (CDWR, 2016).

Another important application of models is to examine the likely impact of specific strategies or combinations of strategies in achieving specific sustainability goals. In other words, modeling can serve as a tool for screening strategies within a cost and effectiveness framework. As management programs are implemented, the actual progress toward sustainability as determined from monitoring programs can be evaluated in terms expected performance as forecasted by a model analysis. These types of analyses will be useful in "tuning" management strategies for sustainability as experience is gained with time.

Changes in climate, population, land use/land cover expected for California over the next 50 years means that hydrologic responses are likely to change as well (CDWR, 2016). Modeling provides a potential strategy for dealing with such uncertainties.

15.4 Strategies for Groundwater Sustainability

This chapter addresses strategies to address the undesirable consequences of groundwater overproduction. The focus is with modern best practices and robust techniques that are designed to adjust aquifer water-budgets to provide sustainability. According to the water balance equation (Eq. 15.1), a continuing deficit in storage can be eliminated by increasing inflows to the aquifer with managed aquifer recharge and reducing the outflows, mainly the quantities of water being pumped (Figure 15.7).

15.4.1 Increasing Inflows

15.4.1.1 Managed Aquifer Recharge (MAR)

The term, *managed aquifer recharge*, or MAR refers to "the purposeful recharge of water to aquifers for subsequent recovery or environmental benefit" (Dillon et al., 2009). The definition acknowledges other types of human activities that lead to recharge of groundwater. A second activity is unmanaged recharge from storm water or septic tanks. MAR is a key technology at sites around the world.

A "MAR system" encompasses all the elements contributing to a MAR operation. Dillon et al. (2009) listed seven elements common to these systems, "water capture, pretreatment, recharge, recovery, posttreatment, and end use." Figure 15.8 illustrates two examples of these configurations for a confined aquifer recharged by a well and an unconfined aquifer recharged via spreading basins (Alley et al., 2022; Dillon et al., 2009).

The first element in the MAR system is a water capture zone that reflects the reality that a source of water is required for MAR to work. There many different water sources available in this respect, such as water imported from unallocated sources of surface water, diversions from periodic flood flows in rivers, urban wastewater, and brine-filled aquifers. Pretreatment (Figure 15.8) is necessary with poor-quality water to produce water acceptable for drinking or irrigation and to avoid pore-clogging problems affecting inflows to the subsurface. Experience has shown it to be difficult to return water to the ground at high rates because of clogging due to suspended sediments, precipitates, or bacteria. Thus, the common strategies for MAR, spreading basins with wet/dry cycling, and injection wells (Figure 15.8) not only come along with expenses in operations but also costs to restore rates of injection or percolation. As will become clear in later chapters, the chemical challenges associated with the use of novel new sources of groundwater are a common issue to be addressed.

It is beyond the scope of this chapter to do a deep dive into the technical approaches for MAR for unconfined aquifers. Beyond the infiltration basins depicted in (Figure 15.8), various MAR strategies exist for returning water to unconfined aquifers, such ponds and ditches, and river-bank filtration. Figure 15.9 shows a conceptual model of a riverbank filtration system. In addition to storing groundwater an unconfined aquifer, the method is useful in removing particulate pathogens from the water and reducing the concentrations of some organic contaminants (Galloway et al., 2003). Traditional approaches like percolation tanks and recharge pits will be discussed in more detail upcoming sections.

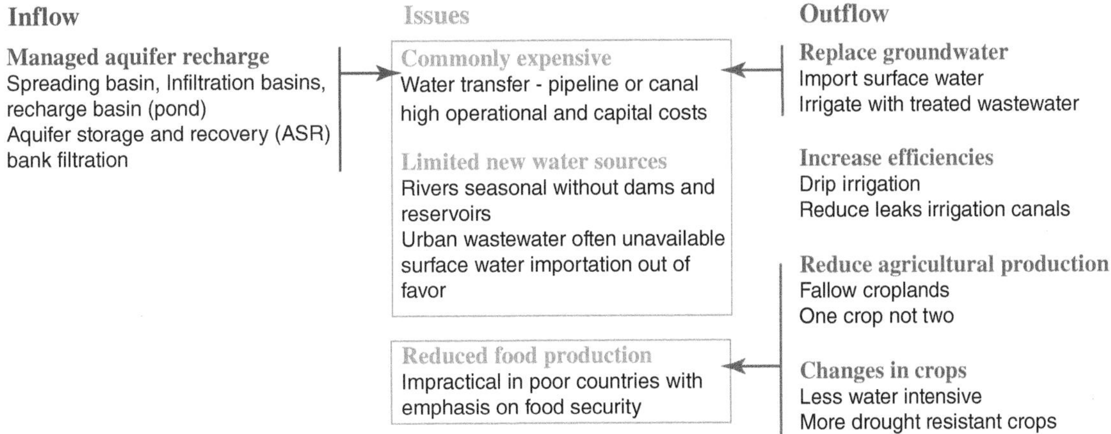

Figure 15.7 Achieving sustainability in an overpumped aquifer requires adjustments to the water balance equation, which involve increasing inflows to the aquifer by MAR and/or decreasing groundwater pumped by finding alternative sources of water or agronomic adjustments. The important strategies are constrained by a variety of issues (FWS).

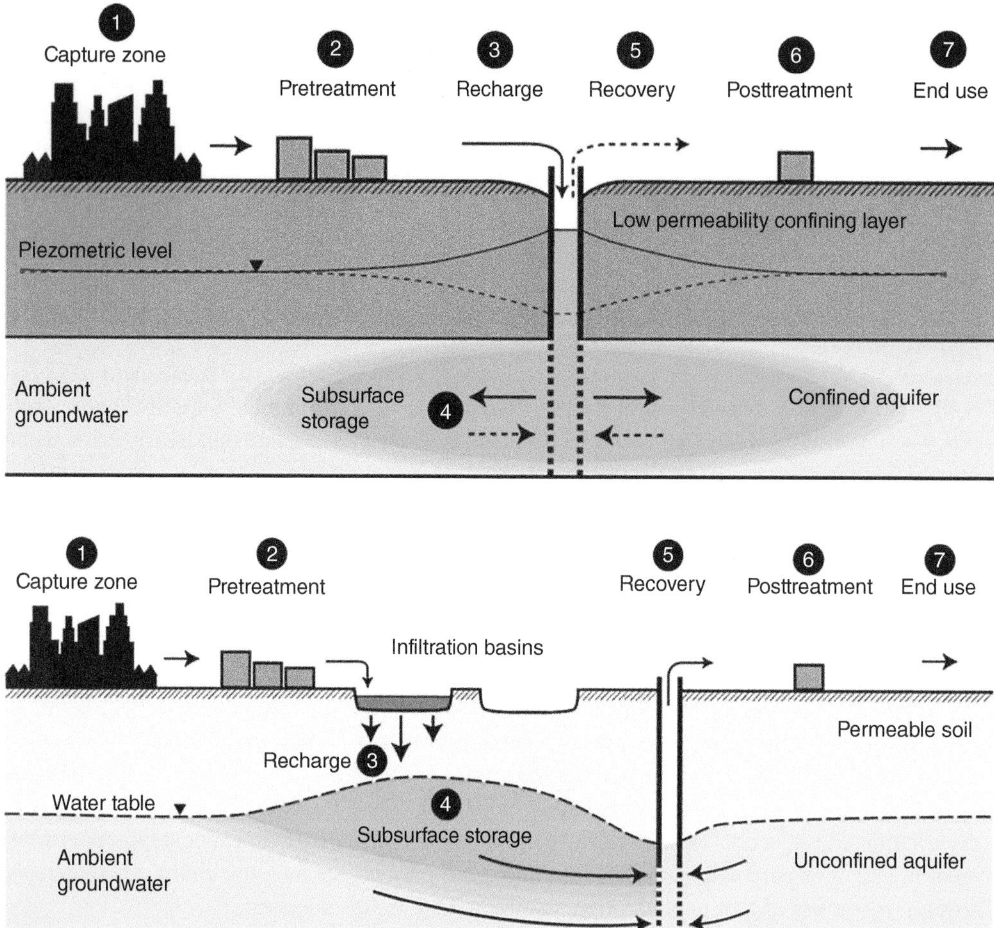

Figure 15.8 MAR system installed in confined aquifers would be designed around the seven basic elements depicted. Where they differ is in the method of by which water is returned to the ground. Injection wells are used with confined aquifers and infiltration basins are used with unconfined aquifers (Dillon et al., 2009. With permission Dept. Agriculture, Water and the Environment, Australia).

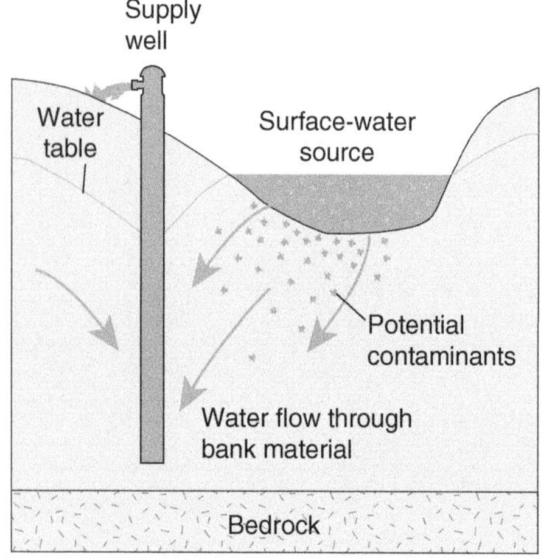

Figure 15.9 Riverbank filtration is a MAR approach that involves storage of surface water in the subsurface through the use well systems emplaced adjacent to streams, which induce flows to the aquifer. It also is effective in reducing the concentrations of some contaminants (U.S. Geological Survey Galloway et al., 2003/ Public domain).

MAR systems also must deal with the potential quality issues with water recovered from subsurface storage. Treatment of recovered water could include metals, such as arsenic, iron and manganese, hydrogen sulfide, and radionuclides (Vanderzalm et al., 2022).

Not surprisingly, modern MAR systems are relatively expensive to install and operate. Systems like these in developed countries might serve several million people, numbers that are relatively small relative to Asian countries such as India or China with large needs for irrigation water.

15.4.1.2 Traditional MAR Approaches

India has a long tradition in using managed aquifer recharge (MAR) to address problems associated with limited availability of water in shallow, bedrock aquifers. Percolation tanks and recharge pits (Figure 15.10) are designed to capture surface water from the summer monsoon for use through dry seasons. A *percolation tank* (Figure 15.10*a*) is a small in-stream or off-stream reservoir placed over high permeability zones, which facilitate the injection and storage of captured runoff. These projects are widely viewed as a useful way supplementing groundwater sources for drinking and for agriculture and are slated for expansion in the future. They are most common in peninsular India in shallow, fractured rock aquifers.

Percolation-tank technologies are scalable through the creation for "tank cascade systems," installed in an organized manner along streams and tributaries. In Shri Lanka, cascade systems there were constructed as early as the 3rd century BC to cope short periods of monsoonal rains in the "dry" areas (Jayasena et al., 2011). The tanks are typically dense with approximately 1 tank per 1.2 km^2. The designs of the various networks of cascades were optimized in terms rainfall and local topography (Jayasena et al., 2011).

Recharge pits are constructed to promote aquifer recharge in situations where shallow deposits have a low permeability (Figure 15.10*b*). Designs are usually flexible, depending upon the physical setting. Commonly, smaller systems are filled with boulders and sand to minimize plugging and to facilitate cleaning. In some implementations, the pit serving to store water and remove fines is completed with an injection well. These days in India there is a booming business in the construction of recharge pits sized for capturing rainfall from one or several houses.

The wide-spread use of these traditional methods to promote groundwater recharge and storage make India world's leading country in terms of the number of installed systems for MAR (Dillon et al., 2019). There are "several million recharge structures" in place with an additional ~11 million more planned. These approaches may be beneficial on a local scale, in hard-rock settings. However, their efficacy in promoting has not been studied extensively using modern approaches (Dashora et al., 2018; Dillon et al., 2019). A recent study concluded that recharge from tanks in crystalline rock areas is relatively unimportant as compared to other sources of recharge and details of the physical, hydrogeological setting (Brauns et al., 2022). In terms of sustainability, these projects may only marginally improve the longer-term viability of small aquifer systems.

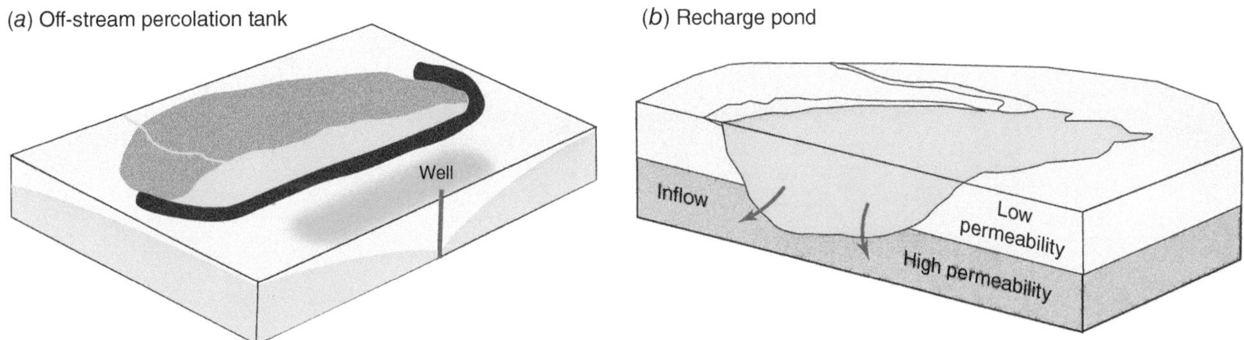

Figure 15.10 Traditional approaches for capturing monsoonal rainfall as recharge. Panel (*a*) shows an off-stream percolation tank. Water collected in a shallow basin recharges a shallow aquifer. Panel (*b*) is a small dug pond to facilitate recharge of a deeper aquifer through a lower permeability unit (FWS).

15.4.1.3 "Sponge City" and Opportunities for Unmanaged Aquifer Recharge

As compared to India, China has fewer of these traditional recharge projects in operation (Dillon et al., 2019). Instead, they are investing in a more modern concept of a "sponge city." The term refers to a collection of "green" technologies for fundamentally redesigning cities in China to better manage the storage and throughput of water. The rapid urbanization over the last two decades has come with serious flooding in cities due to the increasing areas of impervious land areas, and the loss hydrologic features such as "city lakes and ponds, canals, and peri-urban wetlands" (Zevenbergen et al., 2018). The collection of practices aligned with the "sponge-city" idea are low impact approaches practiced in many other countries to manage urban runoff and improve water quality (Zevenbergen et al., 2018). What is different about China is the extent to which these ideas are being embraced by managers.

The sponge city idea is a collection of strategies to restore the function of the urban hydrologic cycle, such as capabilities to store water in a predictable way that facilitates infiltration and evaporation, and the utilization of water by plants. Examples include raingardens, rooftop gardens, permeable pavements, and aesthetic basins for storm runoff.

Enhancing the infiltration of surface waters across a city like Beijing has potential to contribute to unmanaged aquifer recharge of the North China Plain Alluvial aquifer. This effort toward sustainability could have important synergies with diverted surface water from southern China. However, in some settings, the classical designs for sponge-city components may not yet take advantage of the storage potential of the subsurface (Figure 15.11*a*) (Lancia et al., 2020). The alternative would be to route infiltration into the underlying aquifer rather than piping flows through the aquifers (Figure 15.11*b*) to accommodate extreme events (Lancia et al., 2020).

Another constraint on the effective utilization of the subsurface is that sponge-city designs highlighting MAR may not exist everywhere (Lancia et al., 2020). For example, a conceptual model of the hydrologic setting for Shenzhen, China showed upland areas underlain by low permeability bedrock, and lowland coastal plains with a high-water table and at risk for waterlogging, both of which would be inappropriate areas for MAR (Lancia et al., 2020). Useful areas with permeable alluvium and a deeper water table are available but would require additional work to identify (Lancia et al., 2020).

These traditional and modern approaches to unmanaged aquifer recharge are no panacea. The tradition approaches to water harvesting in India are not well suited for hard-rock areas, impact downstream users, and often lead to more pumping (Pahuja et al., 2010). With sponge-city concepts, civil engineers will naturally be risk adverse to depend upon natural properties of groundwater systems. Additional work is needed to understand the full potential and constraints of these approaches.

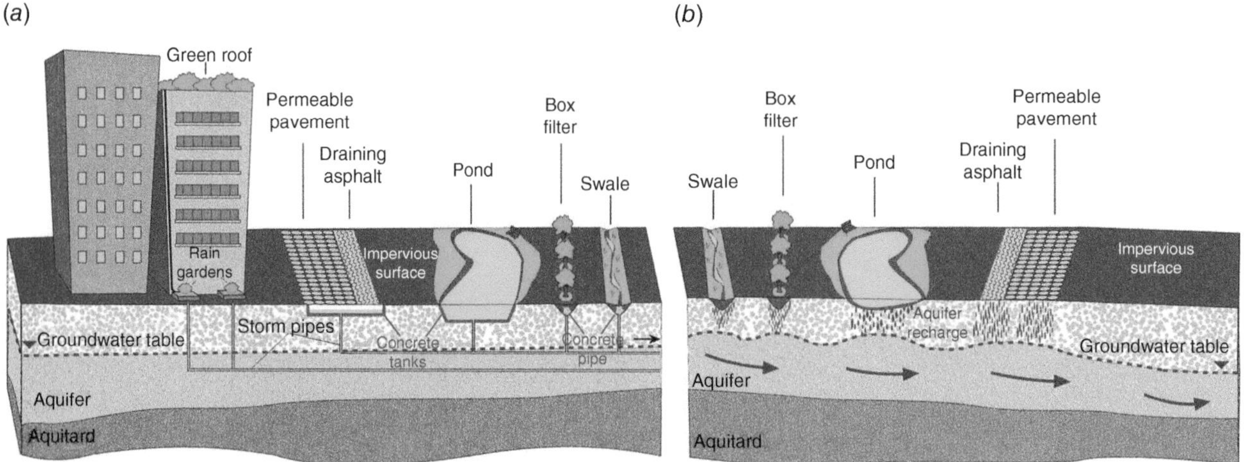

Figure 15.11 Conceptual models for the implementation of "sponge city" projects. Panel (*a*) is classic design where water is captured with engineered storage. Panel (*b*) is a design that takes advantage available storage in groundwater for excess water (Lancia et al. 2020/ Elsevier / CC BY 4.0).

15.4.2 Reducing Outflows

Examples of outflow adjustments include (1) shifting some portion of the irrigation water needs from groundwater to surface water, and (2) using water more efficiently with improved irrigation, and adjustments in cropping.

15.4.2.1 Replacing Groundwater with Surface Water

A favored approach to address excessive groundwater withdrawals is importing surface water to replace groundwater. Such projects can be expensive but are straightforward from an engineering perspective. Water is diverted from a river or storage-water reservoir into aqueducts to the farms or cities where it is needed. For example, in China, diversions of water from the Yangtze River to the North China Plain, including Beijing, has helped to reduce reliance on groundwater. In the USA, importation of surface water helped problems of groundwater overuse in Albuquerque, New Mexico, and Phoenix, Arizona through the diversion of water from the Colorado River. In San Jose, California, and the San Joaquin Valley historical problems of groundwater overuse leading to serious subsidence were recognized and addressed through importation of surface water from northern California. In these cases, groundwater levels rebounded, and subsidence was substantially reduced (Galloway et al., 1999). In Arkansas, a project reduced the serious declines in the Sparta Aquifer. Work to divert excess flood flows of the White River to rice growing areas is designed to help with overpumping of Mississippi River Valley alluvial aquifer.

Planning for new sources of water to augment groundwater these days is complicated because surface waters have already been allocated or simply unavailable in some locations. Coastal cities can utilize desalinated seawater. However, this alternative is expensive both to build and to operate. Another strategy involves ways to use short-duration flood flows more efficiently, increasing the quantity of available surface water. There are also ideas of managing forests in California to increase runoff. The Orange County Water District uses purified urban wastewaters to recharge groundwater.

15.4.2.2 Reduction in Water Used for Irrigation

A simple but controversial strategy is simply to reduce the footprint of groundwater-irrigated agriculture. Approaches include reducing acreage irrigated with groundwater, eliminating double cropping, especially during dry seasons, or replacing low value animal feeds (e.g., alfalfa) by smaller acreages of higher value crops. For many poorer, highly populated countries the cost of reducing agricultural production from irrigated lands in arid areas is reduced food security.

Although irrigation is not a major focus of hydrogeologists, it is a fascinating piece of the sustainability puzzle. Contributions from "increasing efficiencies" are added to Figure 15.7 because conservation of water for irrigation seems an obvious choice to be there. Yet, in many cases, efficiencies in irrigation do not contribute significantly to sustainability. There is no doubt that drip or microspray irrigation that target the application of water to the plants would require less pumped groundwater than a sprinkler irrigation with excess return flows. However, the excess water used in sprinkler irrigation is often not "wasted" (Clemmens et al., 2008). In an ideal situation, excess irrigation water returns to the underlying shallow groundwater for subsequent reuse, or some might runoff to rivers for use by the next irrigator. Thus, irrigation return flows of quality water back to a freshwater aquifer is not a problem requiring more efficient irrigation (Clemmens et al., 2008). The story would change if the excess water, as return flows or runoff to streams, suffered from quality problems making it unusable.

Clemmens et al. (2008) point to savings with irrigation methods capable of eliminating "unnecessary evaporation and transpiration," e.g., non-cropped areas, such as furrows, and non-crop plants, such as "weeds, trees, phreatophytes, etc." True saving can also come from rationing the irrigation water to the plants at the cost of reduced yields or simply not growing crops (Clemmens et al., 2008).

15.4.3 Scaling Issues with Sustainability

The severity and complexity of sustainability problems scale with the magnitude of water volumes and physical areas involved. For example, at the community level in India, dry-season water problems were sometimes solved with a percolation tank. The effort needed for this kind of MAR is modest. In the San Joaquin Valley, towns and communities have been able to manage groundwater by adding a MAR system that banks surface water during wet years and makes that groundwater available in dry years (Scanlon et al., 2016).

Larger aquifers supporting much larger numbers of wells and much larger withdrawals become exponentially more complicated and costly. In California, the aquifer beneath Orange County, near Los Angeles, supplies a significant proportion of water needs for 2.4 million people in an urban setting. There, sustainability has been achieved by a collection of large projects, up to and including the treatment and further purification of urban wastewaters. A critical issue there and with other aquifer systems is where to find the quantities of "new" water for MAR to make aquifers sustainable.

With large groundwater sustainability problems, it becomes difficult to find the water necessary to deal with large groundwater overdrafts and expenses to transport water to areas of need. In populous countries, much of the surface water is already allocated to other uses. Commonly, in S. E. Asia, large and growing cities are underserved with water and have difficulties keeping up with growth. At this stage, purification of urban wastewater is a solution for relatively prosperous places, such as California, Israel, and Singapore (Schwartz et al., 2020). The urban infrastructure necessary to support the collection and treatment of urban wastewater may not exist in large cities, such as Jakarta, Indonesia or parts of Delhi, India.

The reality is that significant money and effort are needed to support several million people with a sustainable groundwater system. Think about the step up needed to support hundreds of millions of people. If MAR was a component of a truly sustainable system at this stage, there would be distributed needs for new water sources across a huge area. Of necessity, there would be many individual projects, spread out over a huge area with no indication that the money and the trained workforce would be there to make it happen.

15.5 Global Warming Vulnerabilities

In some places, global warming creates significant vulnerabilities for the sustainable management of groundwater. Those places, for example, can include watersheds where the water cycle depends on the winter accumulation of a snowpack that melts to provide stream flows to arid settings downstream. In the southwestern United States, the best example is the Colorado River system that is a major source of water for irrigated agriculture and cities of the southwest. The mountainous basin headwaters that comprise 15% of the area contribute 85% of total runoff (Lukas and Harding, 2020). An analogous situation exists in the Central Valley (including the San Joaquin Valley), where streamflow in the Sacramento, San Joaquin Rivers, and other rivers depend on the quantity of snowmelt runoff.

In the Colorado River basin temperatures have increased over the past 40 years. Since 2000, temperatures are approximately ~2 °F higher than the average of the previous century (Lukas and Harding, 2020). Temperatures this high have not occurred over the past 2000 years. During the last 20 years, drought conditions have reduced flows substantially in the Colorado River. The warming story is the same is the same for California with rising temperatures and similar indications from many of the 35 other indicators tracked by California Environmental Protection Agency (OEHHA, 2018). For example, snowpack runoff on the Sacramento River has declined 9% since 1906, and 6% for the San Joaquin River. Snowwater equivalents have commonly been below average since about 2000. Figure 15.12 compares the spatial distribution of snow in the southern Sierra Nevada Mountains from May 10, 2003, considered a normal year, with May 2, 2015, the worst snowpack in 70 years with 5% of average. The absence of snow in early May 2015 in the midst of a severe drought was apparent especially toward the north and Lake Tahoe.

Mann and Gleick (2015) explained that modern droughts in California are more than a loss of precipitation. Increasing temperatures enhance desiccation (drying), magnifying the impact. Considered in this framework the 2013 and 2014 drought years were "very hot and very dry" with 2014 the most extreme year of the entire 110-year instrumental record. Apart from the record snow year of 2017, precipitation and runoff have been below normal in California has been common conditions since 2000 (Mann and Gleick, 2015). The renewed drought (2019–2021) is also very dry and very hot with large reductions in imported surface water for irrigation available to the San Joaquin.

In the California setting, there are important vulnerabilities with using surface water to replace groundwater. There is already evidence that the emergence of very dry and very hot droughts has reduced the availability of surface runoff. Continued global warming will exacerbate drought impacts. Model studies on the Colorado River basin indicate future declines in annual runoff from all parts of the basin in addition to smaller snowpacks and earlier runoff (Lukas et al., 2020).

Spring snowmelt runoff from Sierra Nevada Mountains is moved via canals and aqueducts to support irrigation projects in the San Joaquin Valley and cities in the San Francisco Bay area, Central Valley and Los Angeles. Agriculture in the San Joaquin Valley may be vulnerable to the reallocating of irrigation water away from agriculture to rapidly growing cities.

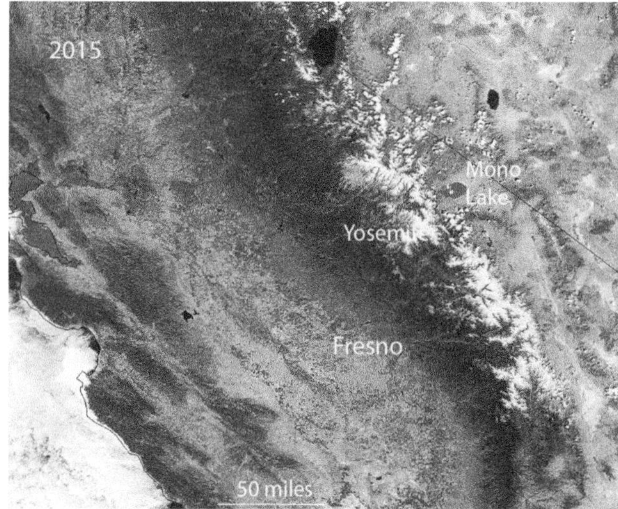

Figure 15.12 Distribution of snow in the Sierra Nevada Mountains in an average year, May 10, 2003, (left) and a drought year, May 2, 2015 (right) (Dan Pisut / NASA).

Moreover, in times of drought, the demands of cities may be "hardened" to the extent that cuts in allocations to cities are smaller than those of irrigators.

Groundwater has always been considered as an important hedge against droughts because the large available supply was usually sufficient to outlast them. Now with 20th century deletion of aquifers, the situation has flipped to where surface water is supplementing unsustainable groundwater utilization. With global warming, the obvious risk in some places is a loss of both surface water and groundwater. The San Joaquin Valley might be one of these places.

The last chapter highlighted the groundwater problems associated with pumping of groundwater for irrigation in the San Joaquin Valley of California. Historical pumping for irrigation, 1940s through to the 1960s, caused water levels to decline more than 105 m (300 ft) (Figure 14.8), which in turn contributed to land subsidence that in places was >8.5 m (28 ft). The unsustainable exploitation of the aquifer was addressed through projects that made surface water available to replace groundwater for irrigation. The Delta-Mendota Canal and California Aqueduct (Figure 14.5) brought water from the Sacramento-San Joaquin Delta, approximately 240 km (150 mi) north of Fresno. The Friant-Kern Canal diverts water from the San Joaquin River north and south along the eastern side of the Valley. The extensive use of surface water led to increases in groundwater-levels and to slowing in rates of subsidence (Galloway et al., 1999).

Experience, however, has been that the quantity of surface water available to irrigators is variable. Water balance calculations for the San Joaquin Valley from 1988 to 2017 found the average annual water use to be 20.6 km^3 (15.7 million acre-feet, MAF) with most water used in irrigation (87%) with the remainder for "natural vegetation and managed wetlands (10%) and cities (3%)" (Hanak et al., 2019). This water came from (1) local sources (70%), such as rain, diversions from rivers from the Sierra Nevada, and groundwater, (2) water imported from the Sacramento-San Joaquin Delta (19%), and (3) reductions groundwater storage (11%) or 2.2 km^3/yr (1.8 MAF/yr) (Hanak et al., 2019).

The long-term sustainable management of groundwater in the San Joaquin Valley is complicated by reoccurring droughts. These have created shortfalls in the water available from Delta imports and local sources (except for groundwater) from year to year (Hanak et al., 2019). Moreover, increasing quantities of Delta water have been passed through to Southern California (Hanak et al., 2019). Not surprisingly, any shortfalls in rain or surface water from local or imported sources is made up by pumping more groundwater. As mentioned in the previous chapter, groundwater for irrigation provides 30% of water requirements in wet years to ~70% in extremely dry years (Faunt et al., 2016).

Figure 15.13 is a histogram showing the net additions and net reductions in groundwater storage on a year-by-year basis (Hanak et al., 2019). Over the thirty years of record, the wet and dry years are close to even, 14 wet years and 16 dry years. The wet years are characterized by groundwater overdrafts that are at or less than the mean. Nine of the wet years were characterized by net additions to groundwater storage (Figure 15.13). However, in every dry year, there was a net depletion of groundwater, often two times or more larger than the average overdrafts (Figure 15.13). Overdrafts on average have approximately doubled from 2003–2017, as compared to 1988–2002. During the severe drought in California,

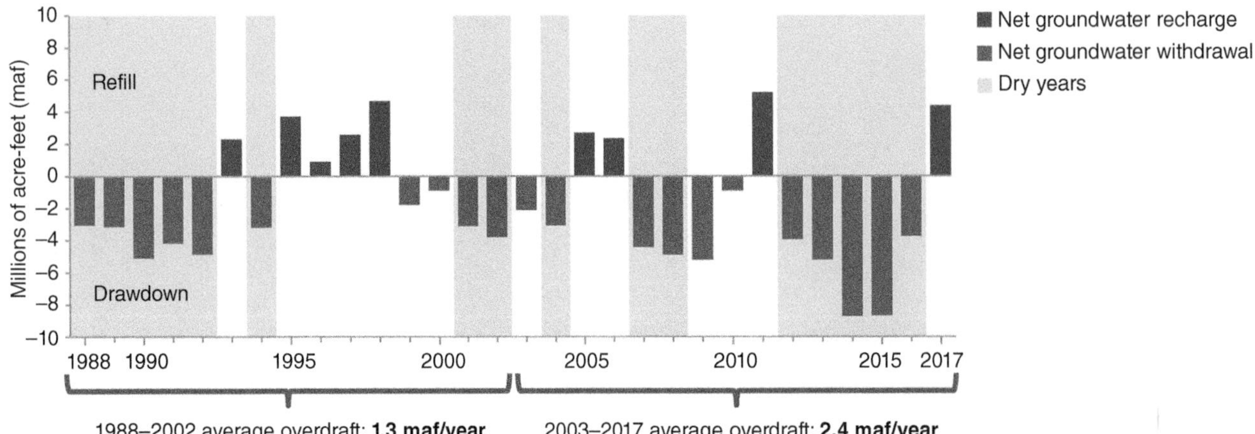

Figure 15.13 Histogram illustrating how groundwater overdrafts increase during dry years and the acceleration of overdrafts more recently (figure excerpted from "Water and the Future of the San Joaquin Valley" by Ellen Hanak et al., 2019. Public Policy Institute of California).

2012–2016, rates of groundwater depletion for two years reached approximately 11.1 km^3 (9 MAF), or 54% of the total water requirements.

Not surprisingly, subsidence was associated with the pumping during the two most recent drought periods (Sneed and Brandt, 2015).

15.6 Chemical Impacts to Sustainability

Discussions of sustainability so far have been cast in terms of depleting groundwater resources by pumping and amplifying problems, like subsidence, streamflow depletion, and destruction of riparian zones. Yet, there are contamination-related impacts that effectively reduce the quantity of available water by making water in aquifers unfit for drinking, irrigation, and other uses. With the chemical impacts, there are effectively no short-term strategies able to reclaim affected groundwater. At best, efforts are aimed at stopping problems from worsening or making water usable through various treatment technologies.

Table 15.3 introduces the key problems known to impact the sustainability of some of the largest and most important aquifers in the world. In some case, these are just "other" problems that affect aquifers such as the San Joaquin portion of the Central Valley aquifer system, or in the case of the Indo-Gangetic aquifer system the most important problems influencing sustainability.

15.6.1 Salinization

Foster et al. (2018) have created a conceptual model describing how groundwaters become salinized in arid areas. In some areas, salts naturally accumulate in soil zones (Figure 15.14). Over-irrigation and rainfall have the potential to transport these salts downward into shallow groundwater. Another possibility is irrigation practices that enable salt accumulation in the soil zone with mobilization from periodic downward flushing.

At depth, paleo-saline groundwater can occur because of hot and arid conditions of the past (Figure 15.14). Irrigation with deeper wells has the potential remobilize this saline water, leading to concentration by evaporation and contamination of shallow groundwater with return flows (Foster et al., 2018). *Waterlogging* is caused by the presence of a shallow water table and leads to evaporation of groundwater and precipitation of salts on or close to the ground surface (Figure 15.14). These processes lead to contamination of shallow groundwater and can develop in arid areas in the vicinity of leaking surface-water canals and over-irrigation of areas that are poorly drained (Foster et al., 2018).

Table 15.3 Key chemical problems affecting the long-term sustainability of aquifers.

Key Problems	Risks	Cause	Examples
1. Salinization	GW unfit for specified uses, Ag impacts	Irrigation practices	IGB, SJV
		Waterlogging	IGB, SJV
		Pumping deep saline gw	IGB
2. Geogenic contamination	Health impacts	Arsenic	IGB
		Uranium	IGB
		Fluoride	Pervasive
3. Ag contamination	Groundwater unfit	Nitrate	IGB, SJV
	Health impacts	Pesticides	emerging
4. Seawater intrusion	GW unusable without treatment	Overpumping coastal aquifers	IGB, SRV, OCWD

gw, groundwater; sw, surface water; ET, evapotranspiration; ag, agriculture; IGB, Indo-Gangetic Basin; SJV, San Joaquin Valley; SRV, Salinas River Basin; OCWD, Orange County Water District.
Source: FWS.

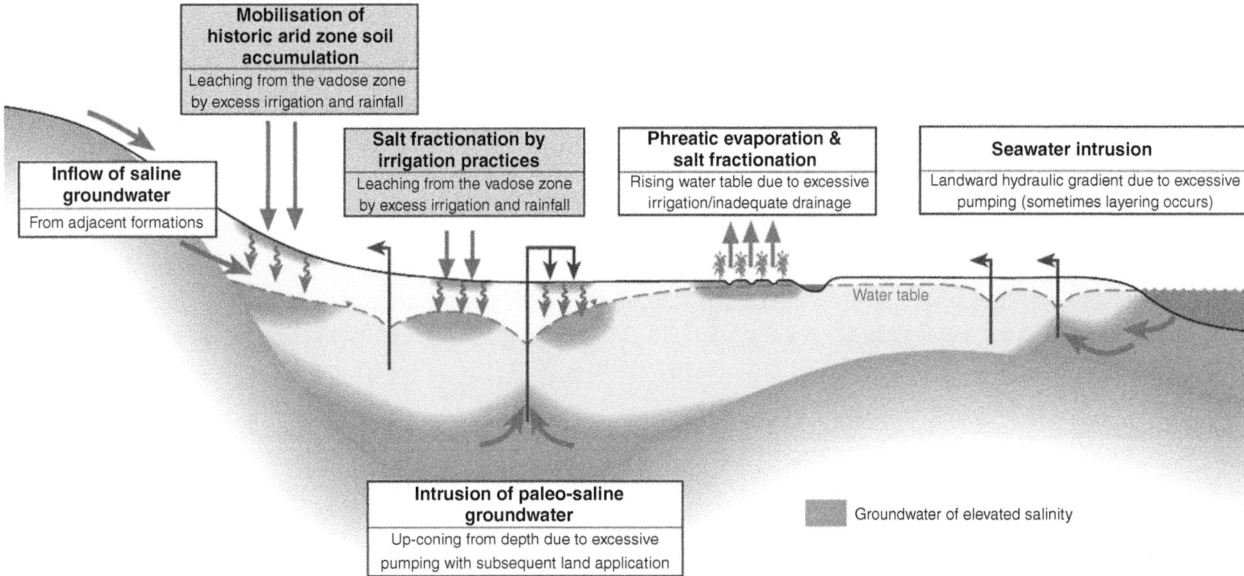

Figure 15.14 Conceptual model illustrating the ways that groundwater can become contaminated by dissolved constituents (Foster et al., 2018 / Springer Nature / CC BY 4.0).

15.6.2 Geogenic and Aenthropogenic Contamination

The presence of geogenic (geologically sourced) contaminants can severely impact aquifer sustainability because their occurrence reduces the quantity of water available for urban and rural water supplies without treatment. Table 15.3 list three examples of these contaminants, namely arsenic, uranium, and fluoride. Even relatively low concentrations of these constituents can cause human-health problems. Serious problems with arsenic and uranium in groundwater occur in alluvial aquifers sourced from the Tibetan Plateau. Unhealthy levels of fluoride are commonly associated with fractured crystalline and rocks in parts of Asia, Africa, and South America (Raj and Shaji, 2017). Subsequent chapters will discuss aspects of the geochemistry and potential health impacts of these contaminants.

A third important driver of sustainability issues is contamination related to agricultural activities. Important contaminants in this respect include nitrate and potentially pesticides. The focus in this section will be with, nitrate.

15.6.3 Salinity and Contamination–Indo-Gangetic Basin (IGB) Alluvial Aquifer

All four problems (Table 15.3) impact the long-term sustainability of the Indo-Gangetic Basin (IGB) alluvial aquifer (Figure 15.15). This aquifer is a source of drinking water for irrigation and drinking water for ~500 million people in Bangladesh, India, Nepal, and Pakistan. It is an alluvial aquifer formed by sediments that were eroded from the Himalayan Mountains and transported by the major rivers, such as the Indus, Ganges, Brahmaputra, and their tributaries (MacDonald et al., 2016). The 2010 production from the IGB aquifer was 205 km^3/yr, with ~122 km^3 in India, 48 km^3 in Pakistan, and ~34 km^3 in Bangladesh and the remainder from Nepal (MacDonald et al., 2015).

Almost all that water is used for irrigation, which was increasing at an estimated 2–5 km^3/yr (MacDonald et al., 2016). GRACE-based estimates of storage depletion were 17.76 ± 4.5 km^3/yr (Rodell et al., 2009) and 14 ± 0.4 km^3/yr (Long et al., 2017). MacDonald et al. (2016) using field measurements determined declines in storage of 8.0 ± 3.0 km^3/yr, with 5.2 ± 1.9 km^3/yr within India. Fortunately, there is prolific recharge from monsoonal rains, irrigation canals, and return flows from irrigation.

Although overpumping of the IGA aquifer system has been evident for some time, the problem is not yet concerning for several reasons. The IGB aquifer system is extensive with generally shallow water tables. MacDonald et al. (2015) estimated that the upper 200 m of the aquifer stores ~30,000 km^3 $\pm$ 10,000 km^3. Another reason is the 100-year legacy of historical irrigation, and canal leakage has served as unintended aquifer recharge system. MacDonald et al. (2015) indicated that about half the surface water diverted from rivers for irrigation leaks into the groundwater with much of it ending up as recharge. The scale of the groundwater recharge provided by irrigation canals is corroborated by the widespread evidence of groundwater table rise and subsequent waterlogging throughout the 20th century.

However, the IGB aquifer system suffers from chemical problems that include salinity, arsenic, and uranium in groundwater (Foster et al., 2018; MacDonald et al., 2016; Young et al., 2019). The problems of salinity in the shallow groundwater of the IGB alluvial aquifer system are complicated by natural processes exacerbated by poorly managed irrigation practices. Figure 15.16 is a salinity map of groundwater in the upper 200 m (MacDonald et al., 2015).

Generally, the most saline groundwaters are evident in Pakistan along the lower and middle reaches of the Indus River where total dissolved solids (TDS) are >2500 mg/L (Figure 15.16). At these levels of salinity, the groundwater is not potable and limited in potential agricultural uses (MacDonald et al., 2015). Near the Indus River and its tributaries, the groundwater tends to be much less saline with TDS values in a range from 500 to 1000 mg/L (Figure 15.16). In India, northwest of Delhi, the salinity of groundwater is commonly >1000 mg/L TDS. In the middle to lower reaches of the Ganges River and into Bangladesh, groundwater has salinities <1000 mg/L TDS (Figure 15.16). Salinities along coastal areas of Bangladesh

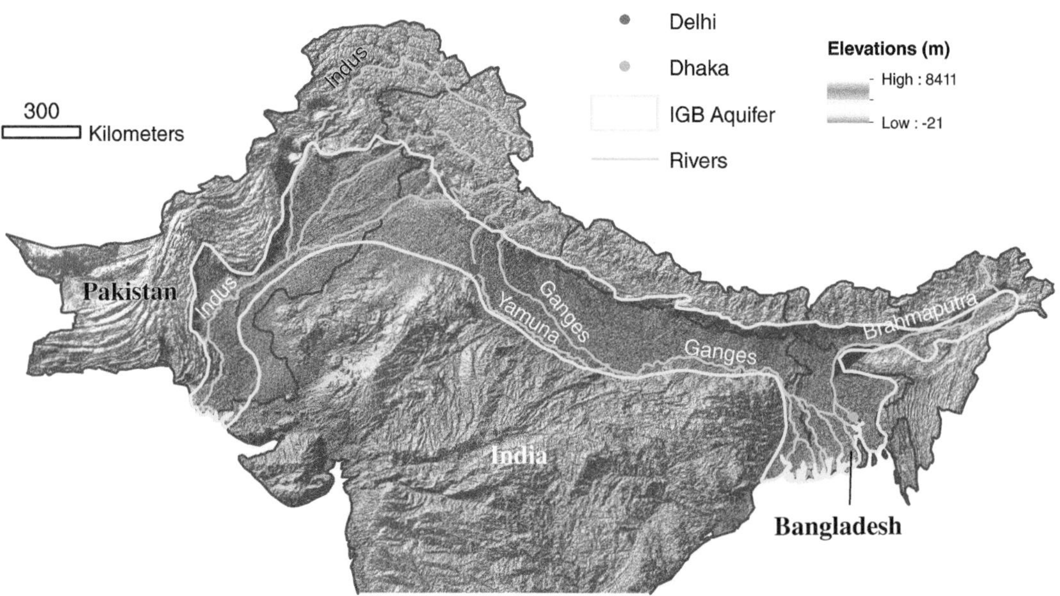

Figure 15.15 The IGB alluvial aquifer is draped across the top of India, extending southward into Pakistan and Bangladesh. The important rivers and their tributaries drain from the Himalayas and receive most of their water locally from summer monsoons (courtesy Ganming Liu).

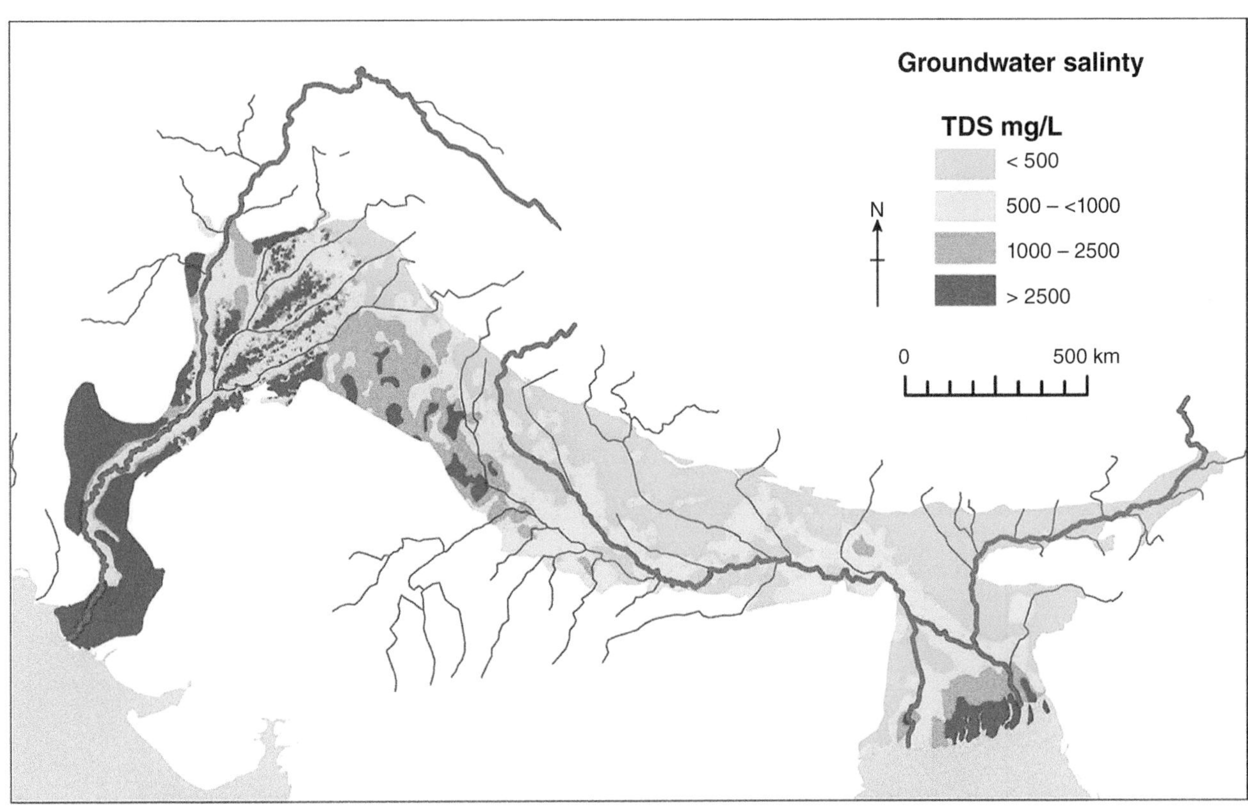

Figure 15.16 Map of showing the total dissolved content of shallow groundwater. Problem areas include places where the climate is relatively arid (Pakistan and northwest India) (Bonsor et al., 2017 / Springer Nature / CC BY 4.0).

(Figure 15.16) are due to seawater intrusion. Overall, approximately 20–25% of groundwater in the IGB alluvial aquifer has salinities more than 1000 mg/L TDS (MacDonald et al., 2015).

The elevated salinity to the north and west is attributed to the historical formation of natural salts during drier periods in the past (MacDonald et al., 2015). The highly salinized groundwater in the lower and middle reaches of the Indus River and northwest India is associated with areas where rainfall is relatively lower as compared to the areas southeast of Delhi with greater rainfall (MacDonald et al., 2015).

Irrigation activities are also contributing to problems of groundwater salinization. For example, the Rechna Doab region of Pakistan exemplifies salinity problems in the western reaches of the IGB aquifer system (Foster et al., 2018). The 3000 km^2 study area within the Indus River Valley (Figure 15.17) is relatively arid with ~400 mm/yr in monsoon rainfall. Deep saline groundwater from earlier times tended to be displaced from beneath the valleys by losing rivers, as reflected by the 1850 configuration of the water table (Figure 15.17). Unintended recharge from leakage of freshwater from irrigation canals subsequently raised groundwater levels between the Chenab and Ravi Rivers through the 1920s into the 1960s (Figure 15.17). The first salinity problems became evident in the 1960s caused by water logging associated with a shallow water table (Foster et al., 2018). More severe salinization was associated with the increasing use of higher salinity groundwater, which began in the 1950s and intensified as deeper wells remobilized deeper, saline groundwater (Figure 15.17). Irrigation return flows also enhanced salinity due to soil-zone processes (Foster et al., 2018).

The IGB aquifer in Bangladesh has also been the focus of worldwide attention concerned with severe health impacts related to arsenic in groundwater. The health effects are many, including skin lesions, various cancers, problems with pregnancies, and cognitive problems in children (Ahmad et al., 2018). Figure 15.18 shows delta areas of Bangladesh and eastern India where shallow groundwater (<100 m) can be contaminated with arsenic, at concentrations 10 to 100 times greater than the common health limits of 10 μg/L (MacDonald et al., 2015). Lower-level contamination extends upstream in the IGB aquifer in areas along the Ganges River and tributaries. As Figure 15.18 shows, arsenic contamination is expected to occur in middle and upper reaches of the Indus River.

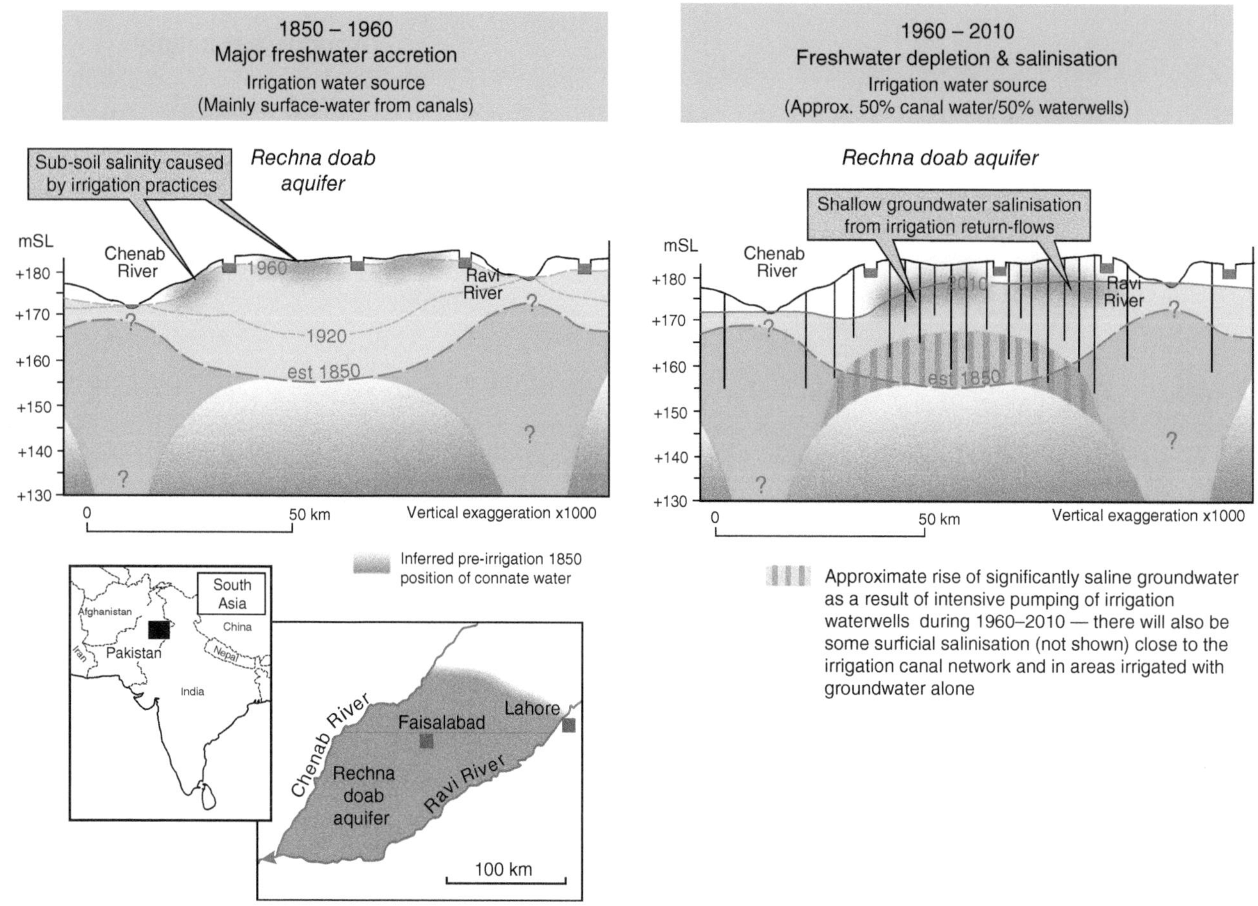

1850 – 1960
Major freshwater accretion
Irrigation water source
(Mainly surface-water from canals)

Sub-soil salinity caused by irrigation practices

Rechna doab aquifer

mSL +180 +170 +160 +150 +140 +130

Chenab River — Ravi River — 1960 — 1920 — est 1850

0 — 50 km — Vertical exaggeration x1000

Inferred pre-irrigation 1850 position of connate water

South Asia — Afghanistan — Pakistan — Iran — China — Nepal — India

Chenab River — Faisalabad — Lahore — Rechna doab aquifer — Ravi River — 100 km

1960 – 2010
Freshwater depletion & salinisation
Irrigation water source
(Approx. 50% canal water/50% waterwells)

Rechna doab aquifer

Shallow groundwater salinisation from irrigation return-flows

mSL +180 +170 +160 +150 +140 +130

Chenab River — Ravi River — 2010 — est 1850

0 — 50 km — Vertical exaggeration x1000

Approximate rise of significantly saline groundwater as a result of intensive pumping of irrigation waterwells during 1960–2010 — there will also be some surficial salinisation (not shown) close to the irrigation canal network and in areas irrigated with groundwater alone

Figure 15.17 Historically, losses of water from rivers created localized zones of freshwater. Irrigation and canal leakage (1850–1960) led to water-table rise and local salinization. Subsequently (1960–2010), the proliferation of wells led to salinization through irrigation return flows and mobilization of paleo-saline water (Foster et al., 2018 / Springer Nature / CC BY 4.0).

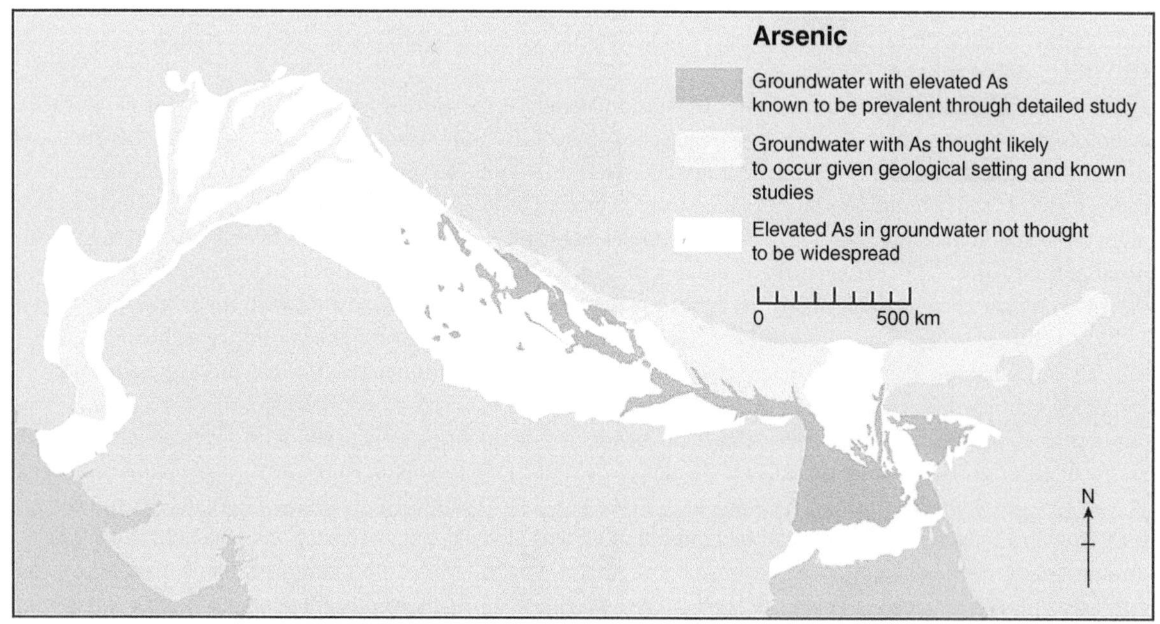

Figure 15.18 Map showing the broader and likely distribution of arsenic in groundwater (Bonsor et al., 2017 / Springer Nature / CC BY 4.0).

The presence of arsenic in the shallow groundwater in places has implications for aquifer sustainability. The usability of the groundwater resource is diminished by the inability to use the water given current levels of knowledge and potential health risks. Water treatment systems are not widely used and in places the risks due to arsenic have yet to be determined.

Arsenic is not the only important contaminant of concern. Lapworth et al. (2017) reported uranium concentrations in groundwaters of the IGB aquifer northwest of Delhi in northern Punjab State. Most samples fell in a concentration range from ~0.9 to 70 μg/L with 10% of values exceeding the health standard of 30 μg/L (Lapworth et al., 2017). Typically, uranium values were higher in deeper groundwater with exceedances associated with urban/peri-urban settings (Lapworth et al., 2017). The source is likely grain coatings or minerals containing uranium. Uranium occurs more broadly in groundwater across India (Coyte et al., 2018). Known health impacts include chronic kidney disease and nephrotoxic effects (Coyte et al., 2018).

This same area in India is also concerning because of the relatively high concentrations of nitrate in shallow groundwater in agricultural areas (Lapworth et al., 2017). Typically, concentrations of N as NO_3 are below the WHO standard of 50 mg/L. However, nitrate contamination is present at depths exceeding 100 m. The study pointed out the vulnerability of deep groundwater to contamination by pesticides and other anthropogenic contaminants (Lapworth et al., 2017).

The next chapter focuses on contaminant concentrations in relation to health standards. Interestingly, the water quality concerns for parts of the San Joaquin Valley are the same as those of the IGB aquifer. Salinity is a serious problem with elevated levels of nitrate and arsenic of concern. Uranium concentrations are anomalously high in groundwater southeastern parts of the San Joaquin Valley (Rosen et al., 2019). Twenty-four percent of 257 water samples collected from wells over an area of 110,000 km^2 contained uranium above the US health standard of 30 μg/L. A number of samples had uranium concentrations >100 μg/L and as high as 449 μg/L (Rosen et al., 2019). Areas of the Sierra Nevada Mountains contain uranium which is transported by physical weathering of bedrock to alluvial fans along the eastern margin of the San Joaquin Valley. Subsequent mobility in groundwater is promoted by elevated concentrations of HCO_3 in irrigation water forming uranium-bicarbonate complexes (Rosen et al., 2019).

15.6.4 Seawater Intrusion

Seawater intrusion is another mechanism reducing the quantity of usable water within an aquifer. The problem is created by overpumping of coastal aquifers and inducing the landward migration of seawater. Like the other problems of chemical contamination in this section, sustainability efforts are usually focused on containing the problem.

A common approach to containing seawater intrusion is utilize a line of injection wells to provide a hydraulic pressure barrier to halt invading seawater. The line of wells create hydraulic heads higher than sea level, which provide fresh groundwater flows from the wells inland towards the developing cone of depression. Thus, pumping of coastal aquifers can continue with protection provided by the hydraulic barrier. Figure 15.19 is a schematic of the system deployed in the Los Angeles groundwater basin (Edwards and Evans, 2013). Pumping of groundwater in coastal areas began in the late 1800s (Panel *a*). By the 1920s increased pumping lowered water-levels below sea level leading to evident seawater intrusion (Panel *b*). Early in 1950, the West Coast Basin Barrier in Los Angeles County was the first of several hydraulic barriers implemented through the injection of freshwater (Figure 15.19, Panel *c*). The Alamitos Barrier was constructed in 1964 followed by Dominguez Gap Barrier in 1964. In 1976, Orange County Water District, near Los Angeles installed the Talbert Barrier in 1976, which was

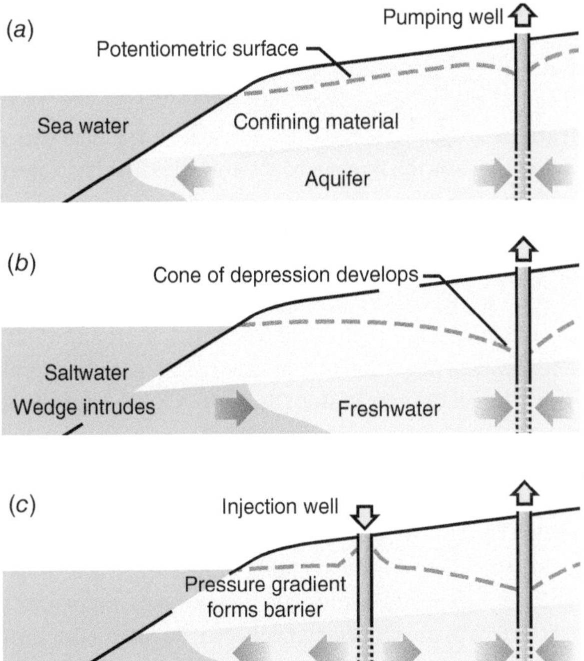

Figure 15.19 Initial lowering of water levels in the 1800s (*a*) initiated sea water intrusion. With time (*b*), water level declines promoted further intrusions. A line of wells installed parallel to the coast created a hydraulic barrier (*c*) (U.S. Geological Survey, adapted from Edwards and Evans, 2013/Public domain).

Table 15.4 Management strategies that have been helpful in controlling seawater intrusion into groundwater in the Salinas River basin in California.

Strategies	Implementations
Increasing Inflows	
Dry season releases surface water	Enhancements of upstream reservoirs enabled dry-season releases for in-stream recharge
Downstream recharge	Seasonal damming Salinas River facilitated dry season recharge of ponded water
Decreasing Outflows	
Replacing groundwater with urban/agricultural wastewater	Tertiary-treated wastewater is piped to farms for irrigation in areas affected by seawater intrusion
Replacing groundwater with surface water	Direct diversion of surface water from downstream seasonal reservoir to augment treated wastewater
Long-term water conservation	Improvements in irrigation practices have reduced groundwater utilization

Source: FWS.

subsequently expanded to include 109 casings at 13 sites. This system presently involves the injection of treated and purified urban wastewater at a rate of ~100 m^3/min (Herndon and Markus, 2014).

Fundamentally, seawater intrusion is a problem of overpumping, which can also be controlled by the sustainability strategies discussed in Section 15.5. These involve projects that increase inflows to an aquifer and decrease outflows. The state-of- the-art is the comprehensive sustainable management of groundwater basins to maximize the use of groundwater and eliminate seawater intrusion.

Two problem areas on the west coast of the United States illustrate principles of sustainable management at work in maintaining a groundwater supply. We discussed the problems of seawater intrusion in the Salinas River basin in Section 14.5. Long-term pumping for irrigation in this predominately agricultural area caused seawater intrusion. The Orange County Water District, just introduced, is interesting not only for hydraulic control of seawater intrusion but also integrated and comprehensive approach to use the coastal aquifer and the subsurface storage.

Table 15.4 lists the management strategies that have been successful in slowing seawater intrusion along the Salinas River. Increasing the storage of upstream reservoirs has provided for dry season releases of water as needed. Some of that water infiltrates through the riverbed as water moves downstream. A small rubber dam near the river mouth provides a seasonal reservoir that facilitates infiltration to areas of low water levels. Tertiary-treated wastewater has replaced groundwater for irrigation in near-shore settings. Additional surface water is also available from the downstream reservoir. Conservation measures have also been effective reducing groundwater utilization. A pressure barrier system is being considered but the freshwater needs for this project continue to be a problem.

The Orange County Water District (OCWD) has successfully dealt with problems of seawater intrusion for more than 50 years. Now, those efforts are part of the District's sustainable water programs to provide water for ~2.4 million people and industries. Two-thirds of their water needs are provided by hundreds of deep, high-capacity water wells across their 900 km^2 service area (Figure 15.20). The aquifer provides ~345 Mm^3/yr of water on a continuing basis (Herndon and Markus, 2014), which is slightly less than five times the natural recharge, ~65 Mm^3/yr. The increased inflows needed to support this pumping sustainably (Table 15.5) comes from aggressive MAR at 23 sites using (1) water from the Santa Ana River, (2) purified municipal wastewater and storm runoff, and (3) imported surface water (OCWD, 2015).

About 65% of the wastewater collected and treated at the Orange County Sanitation District (OCSD) treatment facility and purified at the nearby GWRS facility is piped to the recharge basins at Anaheim (Herndon and Markus, 2014). Purified water is also used to maintain the seawater intrusion barrier (Figure 15.13). Besides controlling seawater intrusion, the barrier also prevents outflows of freshwater to the Pacific Ocean.

Another key to sustainable management is active management with respect to both inflows and outflows (Table 15.5). On the inflow side, managers anticipate the needs for new MAR facilities, and continually maintain the injection capacity of existing systems. Inflows at the infiltration basins are monitored daily. Most basins are designed to be drained and dried to facilitate removal of clogging layers that can reduce recharge rates. The OCWD also imports surface water to augment the other sources of water.

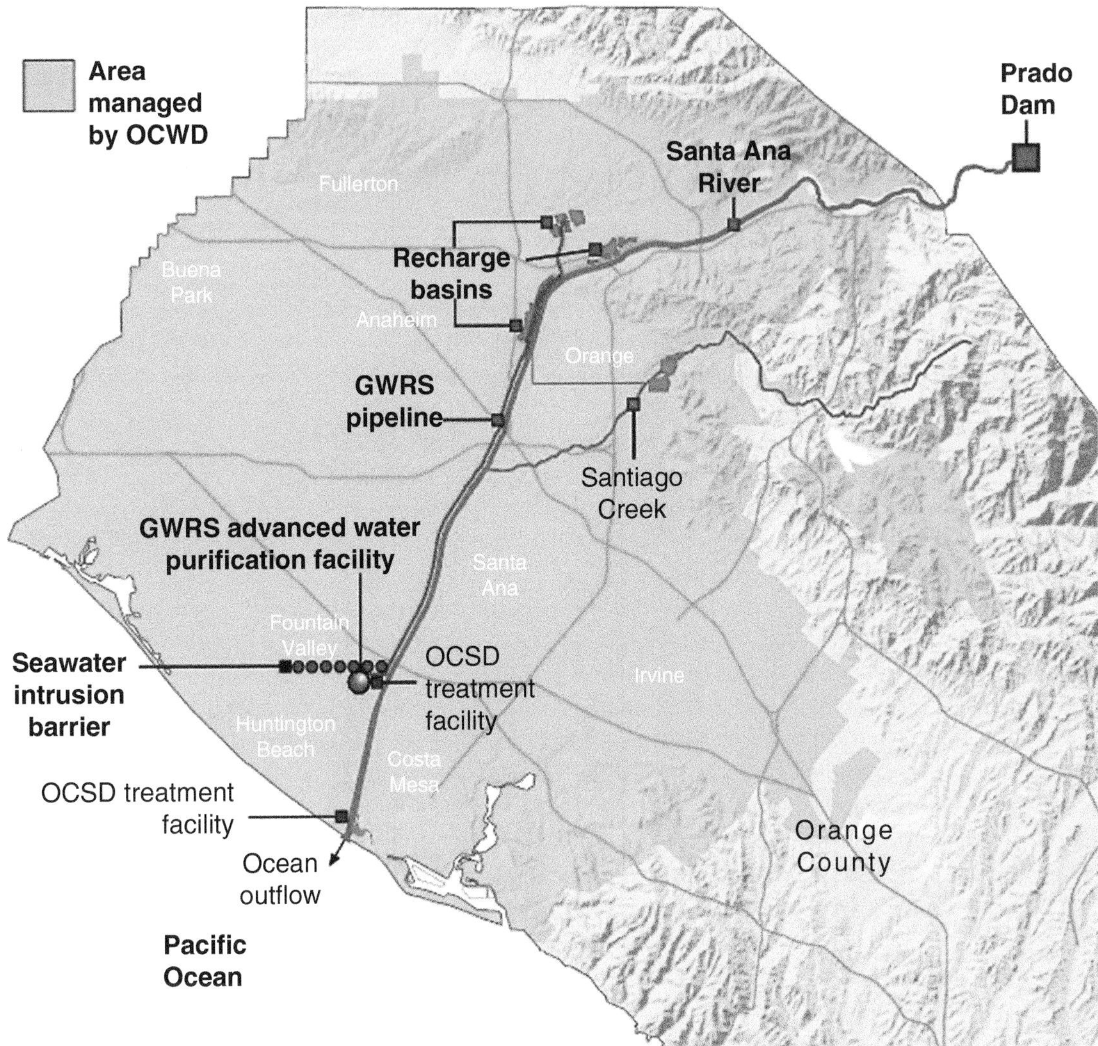

Figure 15.20 Facilities operated by the OCWD (with permission Orange County Water District, OCWD, 2015).

Table 15.5 Strategies for sustainable management used by the OCWD in California.

Strategies	Implementations
Increasing Inflows	
Recharge groundwater from Santa Ana River and other	Seasonal flows in Santa Ana River diverted to nearby spreading basins for MAR
Groundwater recharge with purified urban wastewater	Advance treatment purifies wastewater to drinking water standards
Importation surface water	Surface water imported from sources outside basin
Active maintenance MAR	Maintaining infiltration capacities infiltration basins
Decreasing Outflows	
Hydraulic barrier	Preventing seawater intrusion barrier reduces outflows from the basin
Active management of pumping	OCWD district actively manages and monitors all major groundwater withdrawals

Source: FWS.

On the outflow side, water use, and in-ground reserves are carefully monitored and managed. For example, production data for their high-capacity supply wells are collected monthly. Water from wet years is "banked" in the aquifer systems and becomes available during dry years. Groundwater modeling is used for various planning tasks, such as examining likely outcomes with future MAR projects, tracking water storage, and for providing another method for water budgeting (OCWD, 2015).

References

Ahmad, S. A., M. H. Khan, and M. Haque. 2018. Arsenic contamination in groundwater in Bangladesh: implications and challenges for healthcare policy. Risk Management and Healthcare Policy, v. 11, p. 251–261.

Alley, W.M., T. E. Reilly, and O. L. Franke 1999. Sustainability of groundwater resources. U.S. Geological Survey Circular 1186, 79 p.

Alley, W., P. Dillon, and Y. Zheng. 2022. Basic Concepts of Managed Aquifer Recharge, In P. Dillon, Y. Zheng, W. Alley, and J. Vanderzalm (eds.), Managed Aquifer Recharge: Overview and Governance. IAH Special Publication, p. 1–12.

Belin, A. 2018. Guide to Compliance with California's Sustainable Groundwater Management Act: How to Avoid the "Undesirable Result" of "Significant and Unreasonable Adverse Impacts on Beneficial Uses of Surface Waters". Stanford Digital Repository https://purl.stanford.edu/kx058kk6484 (accessed 29 March, 2023).

Bonsor, H. C., A. M. MacDonald, K. M. Ahmed et al. 2017. Hydrogeological typologies of the Indo-Gangetic basin alluvial aquifer, South Asia. Hydrogeology Journal, v. 25, no. 5, p. 1377–1406.

Brauns, B., S. Chattopadhyay, D. J. Lapworth et al. 2022. Assessing the role of groundwater recharge from tanks in crystalline bedrock aquifers in Karnataka, India, using hydrochemical tracers. Journal of Hydrology X, v. 15, 100121.

Bredehoeft, J. D., S. S. Papadopulos, and H. H. Cooper. 1982. Groundwater: The Water Budget Myth. National Research Council and Geophysics Study Committee, Scientific Basis of Water-resource Management, p. 51–57.

CDWR (California Department of Water Resources). 2016. Best Management Practices for the Sustainable Management of Groundwater: Modeling. California Department of Water Resources, 41 p.

CDWR (California Department of Water Resources). 2017. Best Management Practices for the Sustainable Management of Groundwater: Sustainable Management Criteria. California Department of Water Resources, 35 p.

Clemmens, A. J., R. G. Allen, and C. M. Burt. 2008. Technical concepts related to conservation of irrigation and rainwater in agricultural systems. Water Resources Research, v. 44, no. 7, p. 16 p.

Coyte, R. M., R. C. Jain, S. K. Srivastava et al. 2018. Large-scale uranium contamination of groundwater resources in India. Environmental Science & Technology Letters, v. 5, no. 6, p. 341–347.

Dashora, Y., P. Dillon, B. Maheshwari et al. 2018. A simple method using farmers' measurements applied to estimate check dam recharge in Rajasthan, India. Sustainable Water Resources Management, v. 4, p. 301–316.

Dillon, P., Pavelic, P., Page, D. Beringen, H. and Ward, J. 2009. Managed aquifer recharge: An introduction. Waterlines Report No. 13, 65 p.

Dillon, P., P. Stuyfzand, T. Grischek et al. 2019. Sixty years of global progress in managed aquifer recharge. Hydrogeology Journal, v. 27, no. 1, p. 1–30.

Edwards, B.D., and Evans, K.R. 2013. Saltwater intrusion in Los Angeles area coastal aquifers—The marine connection. U.S. Geological Survey, Fact Sheet 030–02, 2 p.

Famiglietti, J. S. 2014. The global groundwater crisis. Nature Climate Change, v. 4, no. 11, p. 945–948.

Faunt, C. C., M. Sneed, J. Traum, and J. T. Brandt. 2016. Water availability and land subsidence in the Central Valley, California, USA. Hydrogeology Journal, v. 24, no. 3, p. 675–684.

Foster, S., A. Pulido-Bosch, Á. Vallejos et al. 2018. Impact of irrigated agriculture on groundwater-recharge salinity: a major sustainability concern in semi-arid regions. Hydrogeology Journal, v. 26, no. 8, p. 2781–2791.

Galloway, D., Jones, D.R. and Ingebritsen, S.E. 1999. Land subsidence in the United States. U.S. Geological Survey Circular 1182, 177 p.

Galloway, D.L., Alley, W.M., Barlow, P.M., Reilly, T.E., and Tucci, P. 2003. Evolving issues and practices in managing groundwater resources: Case studies on the role of science. U.S. Geological Survey Circular 1247, 73 p.

Garduno, H., and S. Foster. 2010. Sustainable groundwater irrigation: approaches to reconciling demand with resources. World Bank GW MATE Strategic Overview Series, v. 4, 40 p.

van der Gun, J. 2012. Trends, Opportunities and Challenges. United Nations Educational, Scientific and Cultural Organization, World Water Assessment Programme, 38 p.

Gungle, B., Callegary, J.B., Paretti, N.V. et al. (2016). Hydrological conditions and evaluation of sustainable groundwater use in the Sierra Vista Subwatershed, Upper San Pedro Basin, southeastern Arizona (ver. 1.3, April 2019). U.S. Geological Survey Scientific Investigations Report 2016–5114, 90 p.

Hanak, E., Escriva-Bou, A., Gray, B. et al. (2019). Water and the future of the San Joaquin Valley. Public Policy Institute of California, 100.

Healy, R.W., Winter, T.C., LaBaugh, J.W., and Franke, O.L. 2007. Water budgets: Foundations for effective water resources and environmental management. U.S. Geological Survey Circular 1308, 90 p.

Herndon, R., and M. Markus. 2014. Large-scale aquifer replenishment and seawater intrusion control using recycled water in Southern California. Boletín Geológico y Minero, v. 125, no. 2, p. 143–155.

Jayasena, H. A. H., R. Chandrajith, and K. R. Gangadhara. 2011. Water management in ancient Tank Cascade Systems (TCS) in Sri Lanka: Evidence for systematic tank distribution. Journal Geological Society Sri Lanka, v. 14, p. 29–34.

Konikow, L. F. 2011. Contribution of global groundwater depletion since 1900 to sea-level rise. Geophysical Research Letters, v. 38, no. 17, p. L17401.

Konikow, L. F., and E. Kendy. 2005. Groundwater depletion: a global problem. Hydrogeology Journal, v. 13, p. 317–320.

Lancia, M., C. Zheng, X. He et al. 2020. Hydrogeological constraints and opportunities for "Sponge City" development: Shenzhen, southern China. Journal of Hydrology: Regional Studies, v. 28, 100679.

Lapworth, D. J., G. Krishan, A. M. MacDonald et al. 2017. Groundwater quality in the alluvial aquifer system of northwest India: New evidence of the extent of anthropogenic and geogenic contamination. Science of the Total Environment, v. 599, p. 1433–1444.

Lee, C. H. 1915. The determination of safe yield of underground reservoirs of the closed-basin type. Transactions of the American Society of Civil Engineers, v. 40, p. 148–251.

Leenhouts, J.M., Stromberg, J.C., and Scott, R.L. 2006. Hydrologic requirements of and evapotranspiration by Riparian Vegetation along the San Pedro River, Arizona. U.S. Geological Survey Fact Sheet 2006-3027, 4 p.

Llamas, M. R., and P. Martínez-Santos. 2005. Intensive groundwater use: silent revolution and potential source of social conflicts. Journal of Water Resources Planning and Management, v. 131, no. 5, p. 337–341.

Llamas, M. R., L. M. Cortina, and A. Mukherji. 2009. Specific aspects of groundwater use in water ethics, In M. Ramon Llamas, L. Martinez Cortina, and A. Mukherji (eds.), Water Ethics. CRC Press, p. 213–230.

Long, D., Y. Pan, J. Zhou et al. 2017. Global analysis of spatiotemporal variability in merged total water storage changes using multiple GRACE products and global hydrological models. Remote Sensing of Environment, v. 192, no. April, p. 198–216.

Lubell, M., W. Blomquist, and L. Beutler. 2020. Sustainable groundwater management in California: a grand experiment in environmental governance. Society & Natural Resources, v. 33, no. 12, p. 1447–1467.

Luckey, R.R., and Becker, M.F. 1999. Hydrogeology, water use, and simulation of flow in the High Plains aquifer in northwestern Oklahoma, southeastern Colorado, southwestern Kansas, northeastern New Mexico, and northwestern Texas. U.S. Geological Survey Water-Resources Investigations Report 99–4104, 64 p.

Lukas, J. J., and B. Harding. 2020. Chapter 2. Current understanding of Colorado River Basin climate and hydrology, In J. Lukas, and E. Payton (eds.), Colorado River Basin Climate and Hydrology: State of the Science. Western Water Assessment, University of, Colorado, BO, p. 42–81.

Lukas, J. J., E. Gutmann, B. Harding et al. 2020. Chapter 11. Climate change-informed hydrology, In J. Lukas, and E. Payton (eds.), Colorado River Basin Climate and Hydrology: State of the Science. Western Water Assessment, University of Colorado, Boulder, p. 384–449.

MacDonald, A.M., Bonsor, H.C., Taylor, R. et al. 2015. Groundwater resources in the Indo-Gangetic Basin: Resilience to climate change and abstraction. British Geological Survey.

MacDonald, A. M., H. C. Bonsor, K. M. Ahmed et al. 2016. Groundwater quality and depletion in the Indo-Gangetic Basin mapped from in situ observations. Nature Geoscience, v. 9, no. 10, p. 762–766.

Mann, M. E., and P. H. Gleick. 2015. Climate change and California drought in the 21st century. Proceedings of the National Academy of Sciences of the United States of America, v. 112, no. 13, p. 3858–3859.

OCWD (Orange County Water District) 2015. Groundwater Management Plan 2015 Update, https://www.ocwd.com/wp-content/uploads/groundwatermanagementplan2015update_20150624.pdf (accessed 28 March, 2023).

OEHHA (Office of Environmental Health Hazard Assessment). 2018. Indicators of Climate Change in California. California Environmental Protection Agency 326 p.

Pahuja, S., C. Tovey, S. Foster, and H. Garduno. 2010. Deep Wells and Prudence: Towards Pragmatic Action for Addressing Groundwater Overexploitation in India. The World Bank, Washington, DC, p. 97.

Poland, J.F., Lofgren, B.E., Ireland, R.L. et al. (1975). Land subsidence in the San Joaquin Valley, California, as of 1972. *U.S. Geological Survey Professional Paper* 437-H, 78 p.

Pool, D.R. and Dickinson, J.E. 2007. Groundwater flow model of the Sierra Vista Subwatershed and Sonoran portions of the Upper San Pedro Basin, southeastern Arizona, United States, and northern Sonora, Mexico. U.S. Geological Survey Scientific Investigations Report 2006-5228, 48 p.

Postel, S. 1997. Last Oasis: Facing Water Scarcity. WW Norton & Company.

Raj, D., and E. Shaji. 2017. Fluoride contamination in groundwater resources of Alleppey, southern India. Geoscience Frontiers, v. 8, no. 1, p. 117–124.

Rodell, M., I. Velicogna, and J. S. Famiglietti. 2009. Satellite-based estimates of groundwater depletion in India. Nature, v. 460, no. 7258, p. 999–1002.

Rosen, M. R., K. R. Burow, and M. S. Fram. 2019. Anthropogenic and geologic causes of anomalously high uranium concentrations in groundwater used for drinking water supply in the southeastern San Joaquin Valley. CA. Journal of Hydrology, v. 577, 124009.

Scanlon, B. R., R. C. Reedy, C. C. Faunt et al. 2016. Enhancing drought resilience with conjunctive use and managed aquifer recharge in California and Arizona. Environmental Research Letters, v. 11, no. 3, 035013.

Schwartz, F. W., and M. Ibaraki. 2011. Groundwater: A resource in decline. Elements, v. 7, no. 3, p. 175–179.

Schwartz, F. W., G. Liu, and Z. Yu. 2020. HESS Opinions: The myth of groundwater sustainability in Asia. Hydrology and Earth System Sciences, v. 24, no. 1, p. 489–500.

Smith, M., K. Cross, M. Paden et al. 2016. Spring–Managing Groundwater Sustainably. IUCN, Gland, Switzerland, p. 133.

Sneed, M., and J. T. Brandt. 2015. Land subsidence in the San Joaquin Valley, California, USA. 2007-2014. Proceedings of the International Association of Hydrological Sciences, v. 372, p. 23–27.

Vanderzalm, J., D. Page, P. Dillon et al. 2022. *Considerations for Water Quality Management, in Managed Aquifer Recharge:* Overview and Governance. IAH Special Publication, p. 90.

Vrba, J., J. Girman, J. van der Gun et al. 2007, In A. Lipponen (ed.), Groundwater Resources Sustainability Indicators. UNESCO, Paris.

Wahl, K.L., and Tortorelli, R.L. 1997. Changes in flow in the Beaver-North Canadian River basin upstream from Canton Lake, western Oklahoma. U.S. Geological Survey Water-Resources Investigations Report 96-4304, 58 p.

Winter, T.C., Harvey, J.W., Franke, O.L. et al. 1998. Groundwater and surface water-a single resource. U.S. Geological Survey Circular 1139, 79 p.

Young, W. J., A. Anwar, T. Bhatti et al. 2019. Pakistan: Getting More from Water. World Bank, p. 163.

Zevenbergen, C., D. Fu, and A. Pathirana. 2018. Transitioning to sponge cities: Challenges and opportunities to address urban water problems in China. Water, v. 10, no. 9, p. 1230.

16

Water Quality Assessment

CHAPTER MENU

Many modern-day studies of groundwater are concerned with issues of water chemistry. Information on water chemistry is used to make decisions about how water can be used, or whether it contains contaminants that potentially might be hazardous to human or ecological health. This chapter introduces the study of dissolved mass in groundwater. It describes the constituents present in natural and contaminated groundwater, the practical aspects of working with the chemical data, and how wells are used for chemical sampling.

Although the overall goals for this chapter remain unchanged from Schwartz and Zhang (2003) there has been paradigm shift in terms of what groundwater geochemistry is about and the information needed to address the new challenges. The last 60 years have seen a transition from basic geochemical science looking to explain the geochemistry of natural groundwaters to studies necessary to assure the sustainability, safety of groundwater and to mitigate health risks associated with geogenic, anthropogenic contaminants.

The geochemistry of water is complicated. There can be hundreds of dissolved constituents that are mostly present at extremely small concentrations. Outside of academic pursuits, our interests of what constituents to study in groundwater are focused on the inherent suitability of groundwater as a resource. Before the community understood contamination, geochemistry focused on salinity as the issue of concern in determining the value of an aquifer and the water it contained. Thus, analyses focused on concentration of the dominant constituents in groundwater, Ca, Mg, Na, HCO_3, SO_4, and Cl. Measurements also considered constituents like iron that affected water taste and F that came with cosmetic dental and more serious implications. Specific conductance was the key field measurement to provide a direct indication of the total dissolved solids content of the water. All the important graphical approaches, i.e., Collins (1923) bar diagram, the Piper (1944) diagram, Stiff (1951) pattern diagram, and hydrochemical facies diagrams (Back, 1961) were designed to identify patterns in the dominant constituents. Other constituents commonly were neither analyzed nor interpreted.

The important paradigm shift, intensifying over last past 20 years is that contamination of all kinds could impact the usability of water as a resource. From a health perspective, the dominant constituents moved down the list of things we care about in terms of water chemistry. Now many contaminants present in water at low concentrations are more important. It also has taken a long time to realize that groundwater can be naturally contaminated with trace constituents (e.g., arsenic and uranium) that may carry immense health risks. That list of constituents is growing with new contaminants of concern (COCs) that are of interest here and in later chapters.

Fundamentals of Groundwater, Second Edition. Franklin W. Schwartz and Hubao Zhang.

Companion website: www.wiley.com/go/schwartz/fundamentalsofgroundwater2

The reality from intensive chemical characterization of water is that contaminants are prevalent and waiting to be discovered in aquifers. Protection of public health with community groundwater systems requires a much deeper dive into the analyses of water samples. With more constituents under consideration, graphical presentations of results have become more complicated and reduced the importance of the historical methods.

In retrospect, this broadening in scope of groundwater geochemistry is to be expected. There has been an ever-increasing capability to analyze many more constituents in groundwater at increasingly smaller concentrations. Our community has also been shocked into action by the discoveries in the late 1970s of pervasive contamination due to the careless disposal of industrial chemicals, serious health impacts to people in Bangladesh and other countries from arsenic contamination, and most recently by the emergence of per- and polyfluoroalkyl substances (PFAS) that includes PFOA, PFOS, GenX, and 4700 other compounds. Of the PFAS chemicals, perfluorooctanoic acid (PFOA) and perfluorooctane sulfonic acid (PFOS) have been linked to a variety of serious health effects including cancer with PFOA thyroid hormone issues with PFOS (Fenton et al., 2021)

Geochemistry has emerged as an issue of concern Ferguson et al. (2018) wrote about the stress to groundwater quality "from the top down and the bottom up." At the top, shallow groundwaters are being affected by recharge over the last hundred years that has brought with it a variety of novel contaminants. For example, across the Delmarva Peninsula, southeast of Washington D.C., nitrate is widespread in the shallow surficial aquifer occasionally above the Federal standard (Denver et al., 2004). Pesticides are commonly detected at concentrations less than 1 μg/L (Denver et al., 2004). As studies expand globally, these farming-related impacts will be more commonly observed. From below, there are concerns with wastewater from oil and gas activities that are injected in groundwater basins (Ferguson et al. 2018). What is also troubling are indications that the total availability of potable groundwater worldwide is much less than previously thought (Ferguson et al., 2018).

16.1 Dissolved Constituents in Groundwater

Rocks and minerals dissolve in water to form ions. Positively charged species, such as Ca^{2+} or K^+, are *cations*. Negatively charged species, such as HCO_3^- or Cl^- are *anions*. Organic substances can also dissolve to form organic cations or anions. However, most organic compounds (e.g., trichloroethene, TCE) dissolve as *nonionic* (i.e., uncharged) molecules. Most often ions or organic molecules turn up in water from the dissolution of minerals (e.g., halite or calcite), liquids (e.g., trichloroethylene, TCE), or gases,

halite dissolution:	$NaCl = Na^+ + Cl^-$
calcite dissolution:	$CaCO_3 + H^+ = Ca^{2+} + HCO_3^-$
TCE dissolution:	$TCE = TCE_{aq}$

16.1.1 Concentration Scales

Concentration measurements inform us how the concentrations of particular ion or organic molecules dissolved in water. In practice, concentrations are reported using a variety of different scales for different purposes. For example, here are some of the common uses for scales:

- molar concentrations are used in geochemical calculations, such as determination of saturation of groundwater with respect to minerals
- equivalent charge concentrations are commonly needed with plots of geochemical data, and
- mass per unit volume units (i.e., mg/L or μg/L) are used in reporting the results of water analyses and are important.

There are five concentration scales particularly relevant to groundwater practice. *Molar concentration* (M, also mM, μM) represents the number of moles of a species per liter of solution (mol/L). A mole is the formula weight of a substance expressed in grams. For example, a one-liter solution containing 1.42 g of Na_2SO_4 has a (Na_2SO_4) molarity of

1.42/(2 × 22.99 + 32.06 + 4 × 16.00) or 0.010 M. Chemical reactions are written in terms of moles used and moles produced. For example, according to the following reaction, one mole of Na_2SO_4 dissolved in water produces two moles of Na and one mole of SO_4

$$Na_2SO_4 = 2Na^+ + SO_4^{2-}$$

Thus, 1.42 g of Na_2SO_4 dissolved into a liter of water (0.010 M) produces molar concentrations in Na^+ and SO_4^{2-} of 0.02 and 0.01 mol/L, respectively.

Molal concentration represents the number of moles of a species per kilogram of solvent (mol/kg). Typical units are m, mm, or μm. In dilute solutions, this scale for concentrations is almost the same as molar concentrations because a one-liter solution has a mass of approximately 1 kg. For more concentrated solutions, the two scales are increasingly different.

Equivalent charge concentration is the number of equivalent charges of an ion per liter of solution with units such as eq/L or meq/L. The equivalent charge for an ion is equal to the number of moles of an ion multiplied by the absolute value of the charge. For example, with a singly charged species such as Na^+, a 1 M solution of Na^+ (i.e., 1 mol/L) equals 1 eq/L. With a doubly charged species such as Ca^{2+}, 1 mol/L of Ca^{2+} equals 2 eq/L. Equivalent concentrations can also be represented as equivalent charges per unit mass of solution with units such as eq/kg or equivalents per million (epm).

Mass per unit mass concentrations is a scale representing the mass of a species or element per total mass of the system. Many older analyses have been reported using this scale with concentrations in parts per million (ppm) or parts per billion (ppb). More recently, these units of concentration have given way to corresponding concentrations in mg/kg or μg/kg.

Mass per unit volume concentration is the most common scale for reporting concentrations. It defines the mass of a solute dissolved in a unit volume of solution. Concentrations are reported in units such as mg/L or μg/L. Again, there is a close correspondence between these last two scales of concentration. For dilute solutions, 1 ppm = 1 mg/kg = 1 mg/L.

Concentration conversions involve a few simple equations. The most common scale change takes the reported values of chemical analyses in mg/L (also mg/kg or ppm) and converts them to molar concentrations:

$$\text{molarity} = \frac{\text{mg/L} \times 10^{-5}}{\text{formula weight}} \tag{16.1}$$

Conversion from mg/L to meq/L is needed sometimes to plot data or to check chemical analyses using cation–anion balances. This conversion equation is

$$\text{meq} = \frac{\text{mg/L}}{\text{formula wt/charge}} \tag{16.2}$$

The following example illustrates scale conversions.

Example 16.1 The measured concentration of SO_4^{2-} in water is 85.0 mg/L. Express this concentration as molarity, and meq/L. The atomic weights of sulfur and oxygen are 32.06 and 16 gm/mol, respectively

$$\text{mol/L} = \frac{85 \times 10^{-3}}{32.06 + 4 \times 16.0} = 0.89 \times 10^{-3}$$

$$\text{meq/L} = \frac{85}{(32.06 + 4 \times 16.0)/2} = 1.77$$

In a typical water analysis, the report gives the total concentration of the constituent, for example, sulfate (SO_4) without indicating the form of the species, such as sulfate ion (SO_4^{2-}). For dilute solutions, the sulfate ion concentration is approximately equal to the total SO_4 concentration reported from the laboratory. With concentrated solutions, some of the sulfate may be tied up in other constituents besides sulfate ion (i.e., complexes, such as $NaSO_4^-$ and $MgSO_4^0$). In these cases, it takes a computer code to figure out the sulfate ion concentration as well as other potential species containing sulfate.

There also are subtleties in reporting concentrations. For example, nitrate in groundwater occurs as NO_3 or NO_3^- as the ion. However, with analytical reports, the nitrate concentration is often reported as NO_3-N, nitrate-nitrogen. The US standard for nitrate in groundwater is 10 mg/L as NO_3-N. Certain applications might require that you use one or the other concentration conventions.

The atomic weight for nitrogen is about 14.0 and 16.0 for oxygen. Thus, the formula weight for NO_3 is 62. The proportion of N to the formula weight is 14/62 or 0.226. Thus, if the laboratory reports nitrate as NO_3, multiply that concentration in mg/L by 0.226 to convert to NO_3-N. If the laboratory reports as NO_3-N, multiply that concentration by 1/0.226 or 4.43 to convert to NO_3.

Example 16.2 The results of a water analysis indicate a concentration of nitrate-nitrogen (NO_3-N) of 10 mg/L, the same as the U.S. drinking water standard. Convert this concentration to nitrate (NO_3).

$$10\,\text{mg/L}\,(NO_3 - N) \times 4.43 = 44.3\,\text{mg/L}\,(NO_3)$$

16.2 Constituents of Interest in Groundwater

Minerals and organic solids, organic liquids, and gases found in the subsurface dissolve in groundwater to some extent. Thus, elements present in rocks and sediments and organic matter present within the soil zone end up in groundwater through natural processes of weathering and water capability to dissolve materials. Thus, the variety of solutes in groundwater is not surprising. As mentioned, a few dominant cations, Ca, Mg, Na, and anions, HCO_3, SO_4, and Cl, make up most of the dissolved mass in groundwater. Yet, there are many other inorganic and organic constituents present at much lower concentrations. The dissolved organic compounds in groundwater could number in the hundreds. They are typically present in minor or trace quantities. The important dissolved gases in groundwater are oxygen, carbon dioxide, hydrogen sulfide, and methane.

To emphasize these ideas, we present the results of a water analysis from a study area in west central California (Parsons et al., 2014). As expected, the major cations and anions have higher concentrations ranging from 467 to 33.6 mg/L (Figure 16.1). The minor cations and ions, potassium, fluoride, and bromide have concentrations ranging from 3.4 to 0.22 mg/L. The trace elements for this sample mostly occur at concentrations below about 35 μg/L, although strontium, boron, and manganese were higher (Figure 16.1). Total nitrogen and phosphorus in this sample were relatively low.

The plot of data for an illustrative sample shows how widely concentrations can vary. There are eight log cycles represented on the figure ranging, 1×10^4 to 1×10^{-4} mg/L. Yet, modern measurement capabilities provide measurements of constituents, into the range of nanograms and beyond. For example, gold concentrations in water can be measured reliably down to about 5 ng/L (Buskard et al., 2019). The California State Water Control Board established notification levels for PFOA and PFOS at 5.1 and 6.5 ng/L, respectively, and action levels at 10 and 40 ng/L. These levels of analytical are motivated by gold exploration in the first example, and a risk of contamination in the second.

Much of our curiosity about groundwater geochemistry in practice extends to the constituents that are relevant to the usability of the resource. In the years, before contaminants were broadly understood to be a problem (i.e., pre-1975), the list was short. Now, that list has grown to include a myriad of contaminant species, including some with serious health impacts.

16.2.1 Gases and Particles

Water also contains gases dissolved in water or present as a free phase (e.g., gas bubbles). Table 16.1 lists some of the abundant and less abundant gases found in groundwater. Gases most abundant in the atmosphere, nitrogen, oxygen, argon, and carbon dioxide are dissolved in precipitation and enter the groundwater with recharging groundwater. Reactions within the soil zone often increase dissolved carbon dioxide levels as compared to those in precipitation. As will become apparent, both oxygen and carbon dioxide dissolved in water are important in controlling the chemistry of groundwater. Thus, dissolved oxygen concentration is an important variable to be measured in water-quality investigations. Carbon dioxide concentrations are usually calculated from measurements of pH and bicarbonate concentrations. In certain situations, dissolved oxygen reacts with organic matter to produce carbon dioxide.

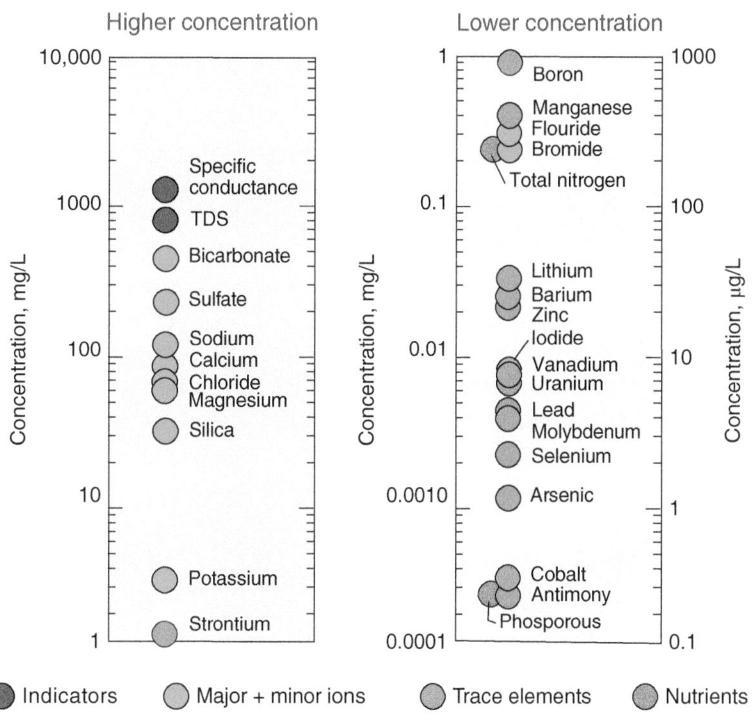

Figure 16.1 An illustrative example of the range in variability that can be expected among four types of constituents, indicators, major and minor ions, trace elements, and nutrients in groundwater. The sample data come from a study area in west central California (U.S. Geological Survey, adapted from Parsons et al., 2014/Public domain).

Table 16.1 A list of gases found in groundwater, along with their common sources.

Gas	Source			Comments
	Atm[a]	S[b]	AC[c]	
Nitrogen N_2				Common in groundwater
Oxygen O_2				Common, control metal chemistry
Carbon dioxide CO_2				Common, important for carbonate chemistry
Hydrogen sulfide H_2S				Locally impairs water quality
Hydrogen H_2				Local from rock-water interaction
Methane CH_4				Local explosive hazard
Neon Ne				Noble gas, useful as tracer
Argon Ar				Noble gas, useful as tracer
Krypton Kr				Noble gas, useful as tracer
Xenon Xe				Noble gas, useful as tracer
Helium He				Noble gas, industrial product, tracer
Radon Rn				Noble gas, radioactive, geogenic, hazard
Chlorofluorocarbons				Synthetic chemicals, useful as tracer
Hydrocarbon gases				Oil and gas contamination

They occur in groundwater through dissolution of atmospheric gases, production by reactions in the subsurface, and from contamination.

[a] Atmospheric.

[b] Subsurface, biogenic, thermogenic, radiogenic.

[c] Anthropogenic contamination.

Source: FWS.

Hydrogen sulfide is generated in the subsurface through biochemical reactions involving organic matter and sulfate in the absence of oxygen. Drinking water with H_2S has a distinct rotten-egg smell. Hydrogen can locally accumulate in the subsurface, probably generated through rock/water interactions.

Methane is an important gas found in groundwater both dissolved and as a separate gas phase. There are cases where methane can form through biochemical reactions that are entirely natural. It can also end up in shallow groundwater as an anthropogenic contaminant, associated with gas and oil practices such as hydraulic fracturing, and several other processes. Accumulations of methane in closed spaces can be the cause of explosions and fires.

The table lists six noble gases. As a family, noble gases are chemically unreactive with respect to other elements. Four of the six (Ne, Ar, Kr, and Xe) are present in the atmosphere and partitioned into water recharging the aquifer. The other two, Rn and He, are generated in the subsurface as products of the decay of solids containing uranium, thorium, and tritium in water (^3He). The noble gases are useful as tracers. Radon is known to cause health effects through inhalation and is commonly monitored in regional assessments of groundwater.

Chlorofluorocarbons (CFCs) are manufactured gases used in various applications, like refrigeration. They have been associated with destruction of the ozone layer. Added to groundwater via recharge they have been used for indirect age dating and flow-path tracing. Last on the list are hydrocarbon gases, such as ethane, propane, etc. that can be present in active gas leaks around producing oil and gas wells.

16.2.2 Routine Water Analyses

For many years, hydrogeologists relied on *routine* analyses that involved measuring the concentration of a standard set of the "major" constituents and a few "minor" constituents. These tests formed the basis for assessing the suitability of water for human consumption or various industrial and agricultural uses and are still important. The major constituents have concentrations >5 mg/L. Minor constituents have concentrations in a range between 0.01 and 10 mg/L. Most other constituents at concentrations <0.01 mg/L were considered "trace" constituents and not analyzed. Laboratories typically reported concentrations results in mg/L.

A routine analysis often includes a few other measured properties in addition to concentrations (Table 16.2), such as pH, total dissolved solids (TDS) reported in mg/L, and specific conductance, reported in microsiemens or an older unit micromhos/cm.

The *TDS content* is the total quantity of solids when a water sample is evaporated to dryness. *Specific conductance* is a measure of the ability of the sample to conduct electricity and provides a proxy measure of the total quantity of ions in

Table 16.2 Example of a routine water analysis.

Parameter	mg/L	(Other)	Parameter	mg/L	(Other)
pH	7.7	(Unitless)	Conductivity	2300	(μS)
Calcium	1[a]		Magnesium	1	
Sodium	50		Potassium	3.5	
Iron	8.7		Nitrate-N	0.1[a]	
Nitrite-N	0.1[a]		Chloride	45	
Sulfate	59		Fluoride	0.25	
Bicarbonate	1315		Hardness, T as $CaCO_3$	8	
Alkalinity, T as $CaCO_3$	1078				
TDS[b]	1321				
Balance	1.01	(unitless)			

[a] concentration "less than."
[b] Total dissolved solids.
Source: Domenico and Schwartz (1998), *Physical and Chemical Hydrogeology*.

Table 16.3 Example in the calculation of the cation/anion balance.

Cation concentration			Anion concentration		
	mg/L	meq/L		mg/L	meq/L
Ca	1	0.05	HCO_3	1315	21.6
Mg	1	0.08	SO_4	59	1.22
Na	550	23.9	Cl	45	1.27
K	3.5	0.09	F	0.25	0.01
Fe	8.7	0.31		Total	24.1
	Total	24.4	Cation/anion ratio = 1.01		

Source: Domenico and Schwartz (1998), *Physical and Chemical Hydrogeology*.

solution. The measurement is approximate because the specific conductance of a fluid with a given TDS content varies depending upon the ions present.

The routine analysis identifies nearly all the mass dissolved in a water sample except if the water is highly contaminated. Unanalyzed ions and organic compounds in natural waters are usually a negligibly small proportion of the total dissolved mass. Thus, a good quality analysis is helpful in illustrating a key property of water, namely that the water is electrically neutral. More specifically, the sum of concentrations of cations in milliequivalent per liter is equal to the sum of the anions, or the ratio of the sums (cation-anion ratio) is one.

Table 16.3 illustrates this calculation for a routine water analysis. The conversion of concentrations from mg/L to meq/L are simple using Eq. (17.2) and a table of atomic weights. While the cation/anion ratio is not exactly one in this example (Table 16.3), it is close. Calculating the *cation/anion ratio* provides one simple check that concentration determinations from the laboratory are not grossly in error. Typically, a ± 0.05 range is acceptable for most analyses.

Large numbers of routine analyses collected over time provide basic data for health and water-resource assessments. These data must be used with care. Errors can occur because of a failure to measure rapidly changing parameters in the field, to preserve the samples against deterioration due to long storage, and to assure the quality of the laboratory determinations. Moreover, it has become abundantly clear that the choice of a constituent to analyze is not necessarily reflected by its absolute concentration but where its concentration falls in relation to standards that reflect the health risk.

16.2.3 Contamination: Expanding the Scope of Chemical Characterization

Groundwater practice now depends on more extensive geochemical characterizations of groundwater. Aquifers can be impacted locally by contamination from industrial chemicals, and more broadly from agricultural chemicals used in crop production, wastes disposed in the subsurface, and geogenic contamination. The investigation of the scope of these problems and potential cleanups requires detailed knowledge of the contaminants involved, their concentration, and associated health risks.

Not surprisingly, the number of constituents of interest in hydrogeochemical investigations has grown to be large with novel contaminants continuing to be discovered. The difficulties in this respect are what to analyze given an appreciation that quality water analyses are expensive, how to collect, analyze, and present results. These are topics we touch on in this chapter.

16.2.3.1 Contaminated Sites

Awareness of the possibilities of contamination has complicated how waters are analyzed. Now, the list of potential contaminants turning up is large, involving literally hundreds of constituents across families such as metals, nutrients, pesticides, organic compounds, and more. Moreover, it is often not clear what contaminants are involved. Consider the pervasive problems of organic contaminants in many different settings. There are literally thousands of different compounds that could be present. At some sites, likely contaminants can be identified by examining purchasing records or knowing the products produced there. With multiple contaminant sources, complicated sites (e.g., landfills), or abandoned facilities, it becomes necessary to screen for contaminants, based on what is likely to be there, compounds of regulatory concern.

A place to start is with the list of 126 (originally 129) priority pollutants prepared by the U.S. Environmental Protection Agency. The list came out in the 1970s in consent decree prescribing a list of chemicals to be regulated. The list contained 13 metals, 26 pesticides, and various organic compounds, including 31 volatile organic compounds. Creation of this relatively short list recognized the great difficulty in analyzing hundreds of potential organic compounds at low concentrations and the need for a manageable list (Keith and Telliard, 1979). The list (https://www.epa.gov/sites/production/files/2015-09/documents/priority-pollutant-list-epa.pdf) in part reflects the most frequently occurring compounds of those times.

Historical practice involved screening sites using a list of common volatile organic contaminants (VOCs). It was effective in discovering the many solvent problems associated with common dissolved contaminants, such as trichloroethene (TCE), tetrachloroethene (PCE), and carbon tetrachloride CTET. A typical laboratory report for VOCs for a sample of water from a site would list the 35 or so compounds, their concentration in the sample, and the reporting limit. The reporting limit is the lowest concentration value that can be expected to be reliable. A report on VOCs would also include the common constituents associated with contamination of groundwater by gasoline and diesel fuel spills, the BTEX compounds (i.e., benzene, toluene, ethylbenzene, and xylene).

16.2.4 Comprehensive Surveys of Water Quality

Establishing the suitability of groundwater as a source of drinking water is complicated by the number of potential contaminants that could be present. Broad surveys, as exemplified by studies by U.S. Geological Survey (USGS) within the National Water-Quality Assessment (NAWQA) program, have addressed this issue by increasing the contaminants that are being screened. Thus, the scope of water characterization within this program has been broad. Each sampled water yielded as many as six basic properties and 215 concentration measurements (Toccalino et al., 2010). Table 16.4 illustrates the comprehensiveness of analyses across the different families of dissolved constituents.

The USGS has followed two strategies in the interpretation of water quality data within NAWQA. The first involves placing water-quality measurements within a study area in the context of results from other study areas (Toccalino and Norman, 2006). NAWQA was designed to facilitate these kinds of comparisons by providing standardized and comprehensive water-quality data from a large and diverse set of study areas. The second involved comparisons of measured concentrations with standards protective of human health (Toccalino and Norman, 2006). These standards are discussed in the next section.

Screening studies for potential health risks are being carried out in other places. For example, the Groundwater Ambient Monitoring and Assessment (GAMA) Program was initiated in California in 2000. The program goals were to improve groundwater quality monitoring across the state and public awareness in areas of groundwater contamination and water quality. This program has a broad focus to elucidate water quality conditions at a basin and larger scales, to identify longer-term trends, and to support efforts in groundwater sustainability. Operationally, the program is like NAWQA in terms of the constituents analyzed (Table 16.4). Thus, results are appropriate for inclusion in national syntheses carried out by the USGS (Fram, 2017).

Table 16.4 Six focus areas and a few examples of the 215 constituents analyzed for public-well samples.

Focus areas (number)	Typical analytes or properties
Basic properties (6)	pH, specific conductance, TDS, temperature, dissolved oxygen, etc.
Inorganic constituents	
Major Ions (9)	Br, Ca, Cl, F, Mg, K, Si, Na, SO_4
Trace elements (23)	As, Ba, Bo, Cd, Fe, Mn, Mo, Se, Sr, U, Zn, etc.
Radionuclides (4)	Radium-226 + Radium 228, Radon-222, etc.
Nutrients + DOC (8)	Ammonia as N, total nitrogen as N, phosphorous dissolved as P, etc.
Organic constituents	
Pesticides (83)	Alachlor, aldicarb, atrazine, cyanazine, diazinon, dieldrin, malathion, etc.
VOCs (85)	Benzene, 1,1-dichloroethane, carbon tetrachloride, trichloroethene, etc.
Fecal indicators (3)	Coliphage, *Escherichia coli*, total coliforms

Source: U.S. Geological Survey, adapted from Toccalino et al. (2010)/Public domain).

16.3 Water Quality Standards

Chemical analyses are commonly interpreted in the context of whether water meets various standards for use by humans or in supporting the health of aquatic ecosystems. These standards change sometimes and vary from country to country or even within countries. This section discusses current standards in various countries and the European Union. Interested readers can examine our cited websites to find links to pages listing those standards.

In the United States, the National Primary Drinking Water Regulations (NPDWRs) are the legally enforceable standards that apply to public water systems. They are designed primarily to protect public health by requiring that contaminants or naturally occurring constituents in water (e.g., arsenic) be less than certain limits. As of 2020, there were 94 regulated contaminants. The last addition of new contaminant was in 2006. On the horizon is the likely regulation of PFOA and PFOS compounds.

Families of contaminants and constituents covered by these regulations include microorganisms, disinfection and disinfection byproducts, inorganic chemicals, organic chemicals, and radionuclides. Disinfection and disinfection byproducts relate to chemicals added (or inadvertently produced) during water treatment to remove microorganisms in water. The list includes a goodly number of the 129 priority pollutants.

Standards are presented in terms of MCLs and MCLGs. The *maximum contaminant level* (MCL) is the highest level of a contaminant that is allowed in drinking water. MCLs are enforceable standards that must be met by public drinking water systems. The *maximum contaminant level goal* (MCLG), also referred to as the public health goal is the level of a contaminant in drinking water below which there is no known or expected health risk. These non-enforceable goals are targets to shoot for in developing water supplies and treatment systems. In certain cases, it is not "economically or technically feasible" to set an MCL, and the regulations require a treatment technique (TT), which must prevent adverse health effects "to the extent feasible." Slightly more than 10% of regulated contaminants are in this category, the most important include pathogens, such as cryptosporidium parasites, *Giardia lamblia* another parasite, *Legionella* bacteria, and viruses.

The drinking water regulation are health based. Commonly, the list of regulated contaminants comes along with a brief description of potential health effects. Also, a few potential sources of contamination are also listed. Figure 16.2 illustrates the information provided for each of the regulated contaminants.

16.3.1 Health-Based Screening Levels—USGS

Placing information from surveys of water in a health-based context depends upon the existence of health-based standards. However, the number of contaminants that potentially could exist in groundwater far exceed those for which MCLs exist (Toccalino and Norman, 2006). Researchers at the USGS have worked to fill this gap with what are called "health-based screening levels" (HBSLs). Noncarcinogenic HBSLs represent the "non-enforceable concentrations of contaminants in water below which adverse noncarcinogenic health effects are not expected over a lifetime of exposure" (Norman et al.,

National Primary
Drinking Water Regulations

Contaminant	MCL or TT (mg/L)	Potential health effects long-term exposure above the MCL	Common sources of contaminant drinking water	Public Health Goal (mg/L)
Arsenic	0.010	Skin damage or problems with circulatory systems, and may have increased risk of getting cancer	Erosion of natural deposits; runoff from orchards; runoff from glass & electronics production wastes	0
Atrazine	0.003	Cardiovascular system or reproductive problems	Runoff from herbicide used on row crops	0.003
TCE	0.005	Liver, kidney or immune system	Discharge from metal degreasing sites and other factories	zero

TT: Treatment technique
TCE: Trichloroethene
Inorganic chemical
Organic chemical

Figure 16.2 The general layout of the Drinking Water Regulations is shown here with three examples of long list of contaminants. Also listed are MCLs and public health goals, along with additional information on health effects and common sources (https://www.epa.gov/sites/production/files/2016-06/documents/npwdr_complete_table.pdf Public domain).

2018). For potential carcinogens, the HBSL values provide a range of values representing estimated lifetime cancer risk of one-in-one million (10^{-6}) to one-in-ten thousand (10^{-4}) (Toccalino and Norman, 2006). What this means for certain contaminants, for example, 1,1,2,2-tetrachloroethane, there is a non-cancer HBSL benchmark value of 100 μg/L and cancer HBSL concentration values ranging from 0.2 to 20 μg/L that correspond with excess deaths of 1 per 10^6 (i.e., 10^{-6}), and 1 per 10^4 (i.e., 10^{-4}), respectively.

These guidelines were established following the same methods as USEPA and toxicology information relevant to human health (Toccalino et al., 2010). Thus, the HBSL information is consistent with the USEPA list of regulated contaminants but non-enforceable in terms of regulatory actions.

These HSBLs have been important for the implementation of health-based assessments of groundwater. For example, Toccalino et al. (2010) screened water quality data from 932 public-supply wells, sampled before treatment, i.e., raw water, from samples collected from 1993–2007 from major aquifers across the United States. Their study made use of 58 MCL values from the USEPA list, and 96 HBSL values. There were no health-based guidelines available for the 67 other measured results.

The searchable database for HBSL values (Norman et al., 2018) early in 2021 contained information for 808 contaminants. The list included 79 MCLs, 174 HBSLs, and 140 HHBPs. This last set of benchmark data refers to Human Health Benchmarks for Pesticides, compiled by the USEPA. These are also non-enforceable guidelines to inform groundwater users whether detection of a pesticide would constitute a health risk.

16.3.2 Secondary Standards for Drinking Water

In addition to the National Primary Drinking Water Regulations, the USEPA also provides guidance in water quality with The National Secondary Drinking Water Regulations (NSDWRs). These nonregulatory standards help in judging the "aesthetics" of drinking water (taste and odor) and cosmetic effects (staining) and in reducing equipment damage. Table 16.5 provides MCLs for 15 contaminants or properties. Copper and fluoride on this list are also regulated constituents on the Primary Drinking Water Regulations with a treatment technique and MCL, respectively.

Table 16.5 Secondary Drinking Water Standards help with the management of drinking water supplies to minimize noticeable effects, such as taste, color, and odor.

Contaminant	Secondary MCL	Noticeable effects above the secondary MCL
Aluminum	0.05–0.2 mg/L	Colored water
Chloride	250 mg/L	Salty taste
Color	15 color units	Visible tint
Copper	1.0 mg/L	Metallic taste; blue-green staining
Corrosivity	Noncorrosive	Metallic taste; corroded pipes/fixtures staining
Fluoride	2.0 mg/L	Tooth discoloration
Foaming agents	0.5 mg/L	Frothy, cloudy; bitter taste; odor
Iron	0.3 mg/L	Rusty color; sediment; metallic taste; reddish or orange staining
Manganese	0.05 mg/L	Black to brown color; black staining; bitter metallic taste
Odor	3 TON (threshold odor number)	"Rotten-egg," musty or chemical smell
pH	6.5–8.5	Low pH: bitter metallic taste; corrosion
		High pH: slippery feel; soda taste; deposits
Silver	0.1 mg/L	Skin discoloration; graying of the white part of the eye
Sulfate	250 mg/L	Salty taste
Total dissolved solids (TDS)	500 mg/L	Hardness; deposits; colored water; staining; salty taste
Zinc	5 mg/L	Metallic taste

Source: U.S. EPA https://www.epa.gov/sdwa/secondary-drinking-water-standards-guidance-nuisance-chemicals#what-are-secondary (accessed 3 May 2023).

A routine water analysis provides a check as to how well some of these secondary regulations are being met. However, people may drink water with higher iron concentrations or total dissolved solids contents. Home treatment can help but many people simply adapt to water they have available.

16.3.3 Standards for Irrigation Water

There are other important uses for water besides drinking. Water is used for irrigation of agricultural crops, or in boilers for making steam. For many of these other uses, there are water quality guidelines that apply. With diligence, one can discover them. The main issues in the application of irrigation water are requirements to match the salinity of the irrigation water to the salt tolerance of selected crops, to avoid salt buildup in the soil, and to avoid a breakdown of the soil structure and a reduction in permeability by using water high in sodium ions low in calcium and magnesium.

The definitive guidelines for water use in irrigation are those presented by (Ayers and Westcot, 1985). These guidelines are used in the United States and outlined in USDA (1993) Chapter 2, Table 2-32. Generally, below a TDS of 450 mg/L there are no salinity problems associated use of groundwater for irrigation. From 450 to 2000 mg/L the risks are slight to moderate. Using water with a TDS >2000 mg/L the risk of salinity problems is severe.

There is a specific problem with irrigating water enriched in sodium. Too much Na in water changes the character of clay minerals, leading to a reduction in infiltration. One can expect problems with SAR values larger than about 8. The parameters of interest in assessing this risk are sodium adsorption ratio (SAR) and TDS together. SAR is calculated as using concentration for Ca, Mg, and Na (as meq/L) in the irrigation water as:

$$\mathrm{SAR} = \frac{\mathrm{Na}}{\sqrt{(\mathrm{Ca}+\mathrm{Mg})/2}} \tag{16.3}$$

The general approach for determining the suitability of water for irrigation begins by using chemical data to calculate the SAR. The value is assigned to one of five ranges, such as 0–3, 3–6, etc. (Figure 16.3). For the given range, use the TDS to determine where in the three TDS categories where within the three risk bands your sample falls. Interestingly, for any of the five SAR ranges, the greatest risk comes with samples with the lowest salinity. In cases where specific conductance is the measure of salinity, use the scale at the bottom of the figure. It is expressed in units of dS/m and can be converted to our more usual scale of μS/cm by multiplying by 1000.

Example 16.3 A water sample has a TDS of 502 mg/L with measured Ca concentration of 28 mg/L, Mg of 8 mg/L, and Na of 150 mg/L. Determine its suitability for irrigation.

The first step is to convert the concentrations to meq/L or 1.39 for Ca, 0.66 for Mg, and 6.53 for Na (see Example 17.1). Next, calculate the SAR using Eq. 16.3 as:

$$\mathrm{SAR} = \frac{6.53}{\sqrt{(1.39+0.66)/2}} = 6.4$$

The data are plotted on Figure 16.3 Notice there is no problem with respect to salinity. However, the combination of SAR and salinity suggest could be problematic.

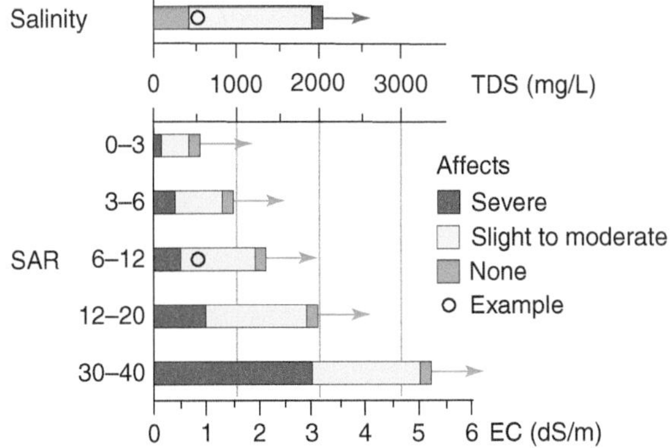

Figure 16.3 Graphical summary of criteria for suitability of irrigation water with respect to salinity and sodium adsorption ratio (SAR). The open circle shows data for Example 17.3 (Graphic by FWS with data from Ayers and Westcot, 1985).

16.4 Working with Chemical Data

This section illustrates some of the basic approaches available for working with chemical data. As will become clear here and later in later chapters, the investigative strategies for local and regional geochemical investigations have changed as knowledge accumulated. In the absence of information, studies of regional aquifers or contaminated sites (landfills, abandoned industrial properties, etc.) of necessity start out broad to identify the issues. In developed countries, preexisting studies usually provide hints of likely problems, but a broad and comprehensive focus is warranted. In less developed countries, most every place is a blank slate in terms of preexisting knowledge.

With screening studies focused on problem identification, every water sample might be analyzed for hundreds of potential contaminants. The reason for this is clearly the large numbers of contaminants associated with important health effects, even at what seem to be low concentrations. In the United States, the NAWQA program has been influential in the development of these comprehensive strategies.

With time, the identification of problems can transition to more focused investigations around a smaller number of contaminants and issues. For example, the contaminants associated with improper disposal of solvents at industrial sites can be narrowed down relatively quickly. However, comprehensive screening studies will need to continue for obvious reasons. First, contamination associated with agrochemicals is likely to continue progressively impacting deeper groundwater. Irrigation leading to problems of salinization will continue as well. Secondly, contaminants of concern like PFOA and PFAS will require screening, given the growing concerns. New contaminant families, such as pharmaceuticals and pathogens, are problematic in surface water with potential to develop to some extent in groundwater.

16.4.1 Relative Concentration and Health-Based Screening

One definition of *potable* is water that is drinkable or fit to drink. This definition is elastic depending upon how thirsty a person is. A second common definition is water that is "safe" to drink. These definitions are not the same because safety as a goal is an expensive add-on to "drinkable." Most people in developing countries drink potable water. Those fortunate to live in developed countries mostly consume water both potable and safe to drink because there are standards and water treatment plants in cities to assure safety. If there are concerns, it is with users taking water from domestic wells, which is never comprehensively treated and nor screened chemically for safety. Thus, regional screening studies of groundwater are helpful in identifying safety issues for rural water users.

Regional screening of groundwater quality is complicated by quantities of data that need to be organized and analyzed. These days chemical data reside in large computer databases. Analytical approaches are usually graphical so that stakeholders can easily understand study results. With a focus on safety, concentrations of constituents need context provided by the health-based drinking water regulations. The concept of relative concentration provides a straightforward way of identifying constituents of potential concern (Toccalino and Norman, 2006; Rowe et al., 2007). Relative concentration is the ratio of the measured concentration of the constituent of interest to that constituent's regulatory or nonregulatory benchmark. Mathematically, this relationship is written as:

$$RC_i = \frac{c_i}{B_i} \tag{16.4}$$

where RC_i is the relative concentration of the ith constituent in the water sample, c_i is the measured concentration of the ith of the constituent, and B_i is the benchmark for that constituent. Note that units of concentration for c_i and B_i need to be the same, for example, mg/L. In applications, (e.g., Fram, 2017), the benchmark values can be drinking water MCLs, HBSLs, secondary drinking water standards, action levels, notification levels, etc. Commonly, benchmarks will change depending upon, the state or country involved. Working in California, one would defer to MCL-CA (i.e., the state benchmark value) as compared to MCL-US (i.e., the USEPA benchmark). As an example, the MCL-CA value for ethylbenzene is 300 μg/L as compared to the MCL-US of 700 μg/L (Fram, 2017).

Example 16.4 In a regional water-quality survey, the concentration of arsenic in a sample of untreated groundwater from a domestic well was 23 μg/L. Calculate the relative concentration for arsenic (RC_{As}) using the USEPA MCL for drinking water.

From Figure 16.2, note that the MCL for arsenic is .010 mg/L or 10 μg/L. The calculation is straightforward:

$$RC_{As} = \frac{c_{As}}{B_{As}} = \frac{23}{10} = 2.3$$

The relative concentration for arsenic is 2.3 times higher than the standard. Terminology developed by the USGS researchers consider relative concentrations >1 to be "high."

Plotting data in terms of relative concentration has significant advantages. The relationships of constituents to health benchmarks can be shown in a simple and consistent manner. The examples (Figure 16.4) came from water-quality screening of two areas in the western San Joaquin Valley of California (Fram, 2017). For simplicity, we only show relative concentration data plotted for three constituents, nitrate, total dissolved solids, and iron. In the study (Fram, 2017), plots include a total of 14 constituents. The circles are relative concentrations for 30 samples from the Delta Mendota study area. The squares represent RC results from sampling 9 wells in the Westside study area.

Color shading on the figure facilitates the subdivision of relative concentrations into low, moderate, and high categories. A *RC* value ≥ 1 (high) indicates that the measured concentration is at or above the health benchmark or secondary indicator. Samples with a *RC* > 4, four times higher than the standard, specifically record the actual wells of concern. As will become clear, this study used both 0.5 and 0.1 values for RC to separate the moderate and low categories (Fram, 2017).

Another way of presenting relative concentration information is to present data for all the classes of contaminants and indicators together. On Figure 16.5, there are two classes of organic compounds, (1) volatile organic compounds (VOVs) and (2) pesticides, fumigants, and constituents of special interest. The inorganic classes include (3) nutrients, (4) trace elements, and (5) secondary, unregulated indicators. The data points on Figure 16.5 represent the highest concentration of the indicated constituent among all the samples. Our figure is simplified from the original (Fram, 2017) to focus on the high and medium categories of relative concentrations.

The plot indicates few organic and special interest contaminants with high *RC*s (Figure 16.5). With inorganic classes, the untreated groundwater is above standards or indicators, including (1) hardness (Figure 16.5), (2) salinity (total dissolved solids and sulfate, and chloride), (3) trace metals (arsenic, molybdenum, and strontium selenium), also boron and

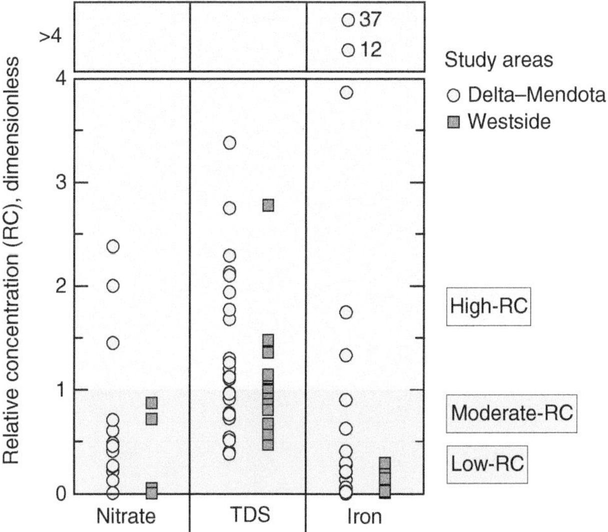

Figure 16.4 Examples of relative concentrations plotted for nitrate, total dissolved solids (TDS) for the 39 samples, and 4 additional samples. Non-detects are shown as zero values. Nitrate has a health-based standard, while TDS and iron are nonregulated indicators (U.S. Geological Survey, adapted from Fram, 2017/Public domain).

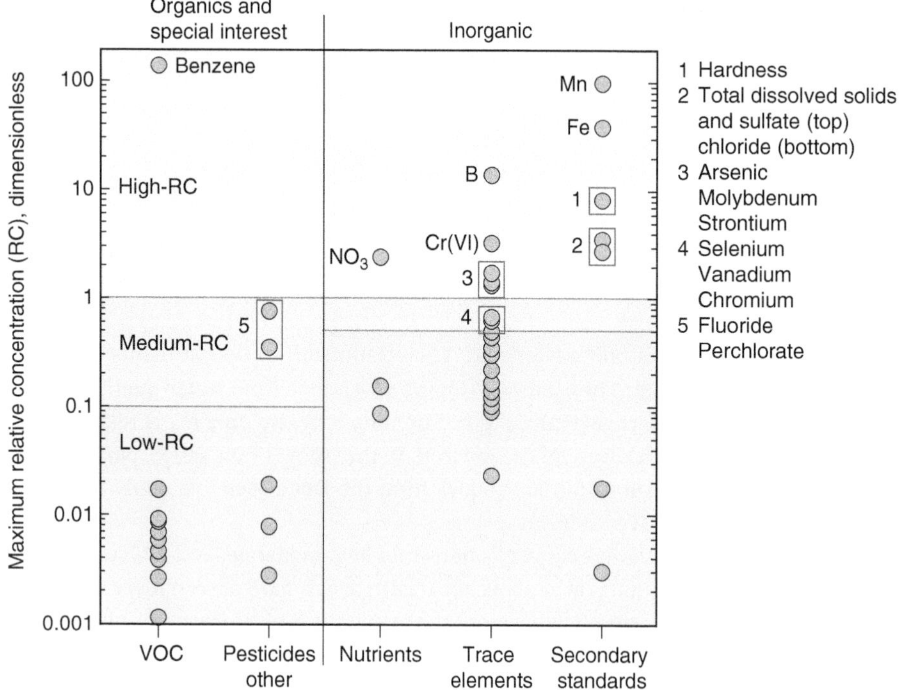

Figure 16.5 Simplified version of a diagram to illustrate maximum relative concentration values of a constituent among all samples. Notice the high-RC values in the western San Joaquin Valley are mainly with respect to secondary standards and mostly with inorganic constituents (U.S. Geological Survey, adapted from Fram, 2017/Public domain).

chromium (VI) and nitrate. Because of consistency in analyses within California's GAMA program, the results for the two study areas of the western San Joaquin could be placed in the context of study areas in California (Fram, 2017).

16.4.2 Scatter Diagrams and Contour Maps

Another important strategy for working with chemical data involves plotting data to understanding spatial/temporal variability in the distribution of constituents or properties, and relationships among variables. One commonly used approach involves scatter diagrams superimposed on various types of maps or cross sections. Figure 16.6 is a map of total dissolved solids superimposed on a digital elevation map, which came from a report by Ging et al. (2020). The dots represent some of the water-supply wells where the samples were collected in the Hueco Bolson study area of Texas and New Mexico. The color of the sample point represents a concentration range. Where appropriate, ranges are designed with benchmarks in

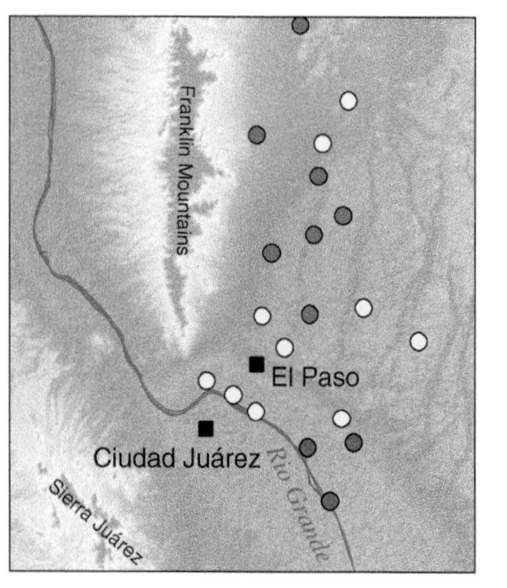

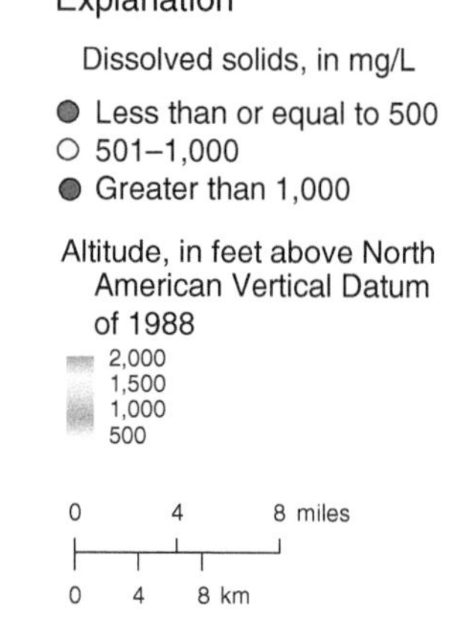

Figure 16.6 Example of a scatter diagram illustrating variability in total dissolved among samples collected from water supply wells in the Hueco Bolson study area in New Mexico and Texas (U.S. Geological Survey, adapted from Ging et al., 2020/Public domain).

mind. Thus, in Figure 16.6, the legend for TDS, includes the 500 mg/L secondary standard for TDS. In this example, there is a tendency for groundwater along the river to exhibit relatively higher salinities. Other things to think about with the legend are keeping the colors to three categories (high, low, and in-between) or four, and choosing ranges so that there are "goodly" number of sample values in each of the categories. If most of the sample values fall in a single category, there may not be a spatial trend worth noting.

Scatter plots also work well for depicting spatial trends in water quality along cross sections. These are designed to capture variability in a constituent as a function of depth. Interested readers can refer to a conceptual diagram showing how a vertical decline in dissolved nitrate results from denitrification ($NO_3 \rightarrow N_2$ gas) deeper in the aquifer after decades of residence under anoxic (i.e., reducing) conditions (Warner and Ayotte, 2014). Cross-sectional scatter plots are not used as commonly as maps because documenting vertical changes in water chemistry along cross sections has greater needs in numbers of data. Data from a large collection of wells completed at different depths are needed, many of which should occur close to the line of the cross section.

As figures become more complicated other design considerations come into play. Shown in Figure 16.7 is a segment of a larger map from Flynn (2003) that shows the concentrations of nitrate plus nitrite (as N) in shallow aquifers along the Front Range Urban Corridor in Colorado from Denver northward. The design of the map illustrates several innovative design features that overcame problems that are common with scatter maps. For example, on maps with many sample locations, data points are often covered up. In Figure 16.7, colors are added to the circumference of the data points, leaving most of the area open to see data points below. Another strategy could involve the use of transparent data points.

A second problem with scatter maps is a background so detailed and multicolored that it competes for attention from the collection of data points. In the example (Figure 16.7), two land-use categories (urban and rural areas) and rivers are shown. These were represented on the map as pastels, colors with white mixed in to soften their look. The color choices somewhat emphasized the rivers, which were an important source of nutrients (Flynn, 2003). The final issue is that on maps with many data points trends can be obscured by unusual combinations of colors associated with data points. This problem emerges in situations where more than three bright colors are used. With the example, colors focus the reader's attention to the higher concentrations in samples from wells located close to the rivers. Colors associated with lower concentration data points were muted as compared to the higher concentration values (Figure 16.7).

16.4.3 Contour Maps

Another approach to visualizing chemical data is with contour maps, especially creating plume maps of contamination. The procedures for contouring data on a map are well-known and will not be discussed here. However, with geochemical data, it

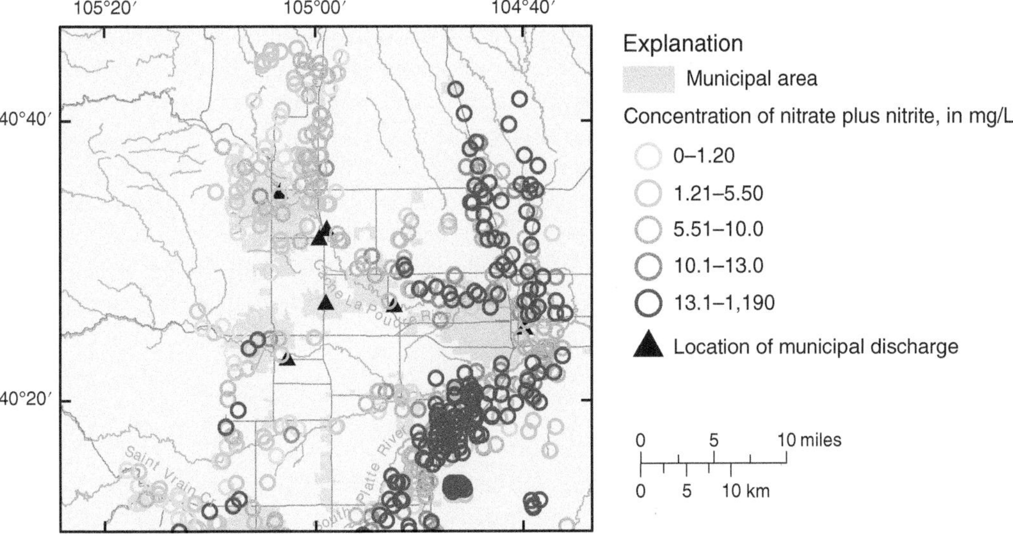

Figure 16.7 Scatter map showing the concentration distribution of nitrate plus nitrite (as N) in shallow groundwater along the Front Range Urban Corridor in Colorado from Denver northward (U.S. Geological Survey, adapted from Flynn, 2003/Public domain).

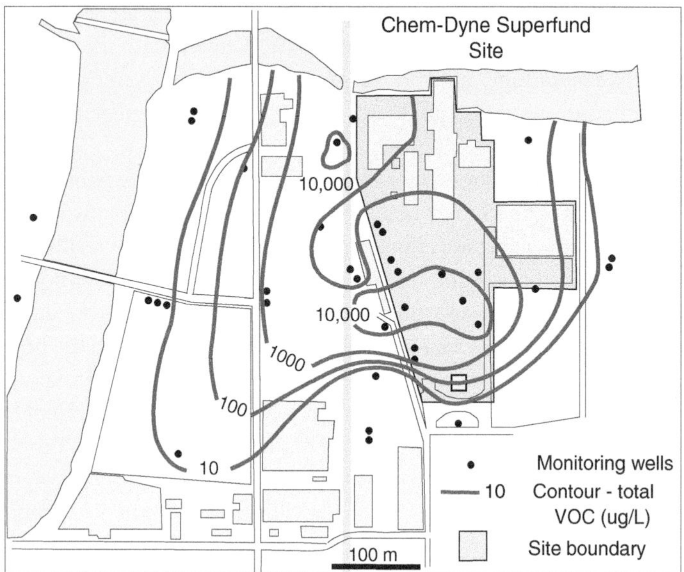

Figure 16.8 Contour map showing the distribution of total volatile organic compounds in the shallow groundwater at the former Chem-Dyne Superfund site in 1989. Concentrations ranged from <10 to >10,000 μg/L (U.S. Environmental Protection Agency, adapted from U.S. EPA, 1989).

is important to think carefully about contour intervals. The range in concentration can be large, commonly falling across several logarithmic cycles (i.e., concentrations from 1 to 1000). Thus, it is useful to use logarithmic contour intervals to make sure that trends in relatively low concentration values are also represented. For example, with concentrations varying between 1 and 1000, choosing to scale arithmetically, i.e., five contour intervals of 200, will cause all low concentrations to fall in a single range from 0 to 200.

Figure 16.8 is a simplified contour map showing the distribution of volatile organic compounds (total VOC as μg/L) in samples collected from shallow monitoring wells at the former Chem-Dyne Superfund site. Operations at the site involved the transfer, storage, and disposal of hazardous organic compounds, which led to contamination of the alluvial sand and gravel aquifer located along the Great Miami River in Ohio. Concentrations of VOCs in the groundwater there varied from <10 to >10,000 μg/L. Choosing logarithmic contour intervals, in this case, were beneficial in defining two hotspots, where concentrations were >10,000 μg/L, and the plume boundary along the 10 μg/L contour line (Fig 16.8). Using only sample data from shallow wells likely reduced the effects of vertical variability in concentrations. An implicit assumption in the creation of a two-dimensional map of a three-dimensional system is that concentrations do not vary in the 3rd dimension, here depth.

Contour maps generally work well with problems of contamination where (1) the contaminant is a manufactured compound not otherwise present in the groundwater, (2) there is an obvious source area with extremely large, dissolved concentrations, and (3) a dense, three-dimensional array of monitoring wells in an around the plume. Scatter maps work well to capture trends in noisy regional data. Noise is variability in concentration due to several factors. Often regional surveys involved collecting samples from preexisting domestic and public water supply wells. Widely spaced wells (tens of kilometers) are unable to capture local trends due to changes in land use (e.g., farmed versus unfarmed) or the presence of surface water sources. Similarly, certain preexisting wells might be shallow, others deep, adding complexity when there are vertical trends. Deep wells can pump old water without agrochemicals, shallow wells might sample young water impacted by irrigation return flows containing pesticides and nutrients. Thus, by chance, a collection of shallow wells surrounded by a collection of deeper wells might suggest localized hotspots.

16.4.4 Piper Diagrams

Describing the concentration or relative abundance of major and minor constituents and the pattern of variability is part of many groundwater investigations. As discussed, scatter plots and contour maps have proven useful for this purpose. Other specialized graphical techniques have been used historically, although most are no longer used. One approach that is still used in practice is Piper diagrams (Piper, 1944).

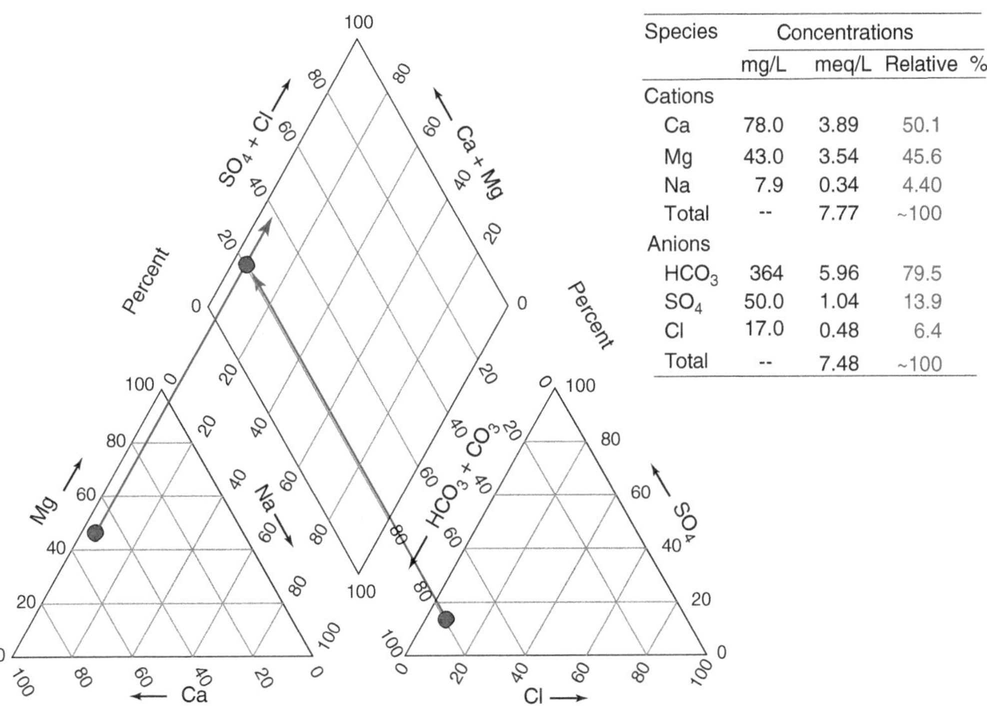

Species	Concentrations		
	mg/L	meq/L	Relative %
Cations			
Ca	78.0	3.89	50.1
Mg	43.0	3.54	45.6
Na	7.9	0.34	4.40
Total	--	7.77	~100
Anions			
HCO_3	364	5.96	79.5
SO_4	50.0	1.04	13.9
Cl	17.0	0.48	6.4
Total	--	7.48	~100

Figure 16.9 A Piper diagram has triangular cation and anion fields and a central quadrilateral. The table illustrates how data are processed from mg/L to meq/L to relative concentration as a percent (FWS).

Piper plots are useful in showing the relative abundances of the major constituents in groundwater and trends associated with these constituents. A Piper diagram has three separate parts. There are two triangular diagrams that display data for the major cations and anions and a quadrilateral where cation and anion data are plotted together (Figure 16.9). At a minimum, construction of Piper diagram requires concentration data for three cations Ca, Mg, and Na, and three anions HCO_3, SO_4, and Cl. Variations can include the addition of K, as Na + K, and/or nitrate (NO_3) as Cl + NO_3.

What is plotted on these diagrams are not actual concentration but the relative abundance of cations and anions. Recalling Table 16.2, we explained that sum of the cation concentrations expressed in meq/L is approximately equal to the sum of the anions. With Piper plots, one extra step is needed to convert units of meq/L to percent. For example, assume that the sum of the concentrations of major cations, i.e., Ca, Mg, and Na is 10 meq/L with the calcium concentration equal 5 meq/L. Of the total cations in the sample, Ca constitutes 50%:

$$\frac{5\ \text{meq/L}}{10\ \text{meq/L}} \times 100 = 50\% \tag{16.5}$$

The same calculation with concentrations (meq/L) of Mg and Na gives their relative concentration as percent. Together, the three relative concentrations add up to 100%. These three relative concentrations provide a data point for the cation triangle.

Similarly, the plotting relative abundance of Cl, SO_4, and $HCO_3 + CO_3$ creates a data point on the anion triangle. Straight lines projected from the two triangles into the quadrilateral field intersect to define the data point for the third field.

Figure 16.9 illustrates the processing of concentration data into data points on the Piper diagram. The inserted table shows the concentrations (as mg/L) for the six major ions, converted to meq/L and next to relative concentrations in percent. The relative concentrations for cations and anions are plotted with lines extending into the quadrilateral to provide the third point. A variety of computer programs are available these days to automate this process.

One disadvantage of a Piper plot is that relative concentrations do not carry information about actual concentrations. For example, major ion data for rain, which has small concentrations, might end plotted on a Piper diagram together with results for a brine. In practice, the shape or color of data points can carry additional information on salinity (TDS) or any other parameters of interest. However, too many colors or data points on the same Piper diagram can sometimes limit its usefulness.

16.5 Groundwater Sampling

Most geochemical investigations rely on wells to provide access to groundwater for sampling. Regional resource assessments involve sampling from public or domestic wells. With a pump and water lines in place, sampling is expedited, as compared to studies needing the installation of monitoring wells. The investigator has little control over the well design and pumping up the sample but needs to be concerned about well locations, completion details, and access to water before it undergoes treatment and storage. Contamination studies or research investigations are more challenging because the investigator has decisions to make on many aspects from the design of the monitoring program to the drilling and logging of the boreholes, to installation and development of the wells, and how to collect samples.

16.5.1 Selecting Water Supply Wells for Sampling

As part of a regional water quality study, it is necessary to select wells to be sampled. Among the important considerations in this respect is a diverse distribution of wells likely to capture major features and trends in water quality and to meet the project objectives. Geographic diversity in sampling is essential to provide appropriate coverage of the study area and different land uses. Hydrogeologic diversity implies appropriate coverage of the units comprising aquifers, logical depth ranges, and other factors necessary to understand what factors explain the water-quality monitoring program. To illustrate these points, we return to the Groundwater Ambient Monitoring and Assessment (GAMA) Program in California that was discussed earlier in this chapter.

The Madera and Chowchilla study unit within the GAMA Program is in the San Joaquin Valley, north of Fresno California (Shelton et al., 2013). It is included here to illustrate a technique for the unbiased selection of wells and for identifying information necessary to address project objectives. The sampling strategy involved subdividing the 860 mi^2 (2225 km^2) study area into 30 equal area (30 mi^2 or km^2) grids blocks (Figure 16.10). One well (a grid well) was selected randomly from a list of potential wells located within each cell. In addition to the grid wells, another 5 wells were selected as understanding wells. They were located along the northern boundary of the study area and included to understand better how water quality changed as a function of depth (Shelton et al., 2013).

The design of the sampling program together with a conceptual model of groundwater setting facilitated two different kinds of assessments. The first was to determine the present status of groundwater quality, using the kinds of detailed water-quality characterizations discussed in Section 16.2. Care was taken within their analysis to demonstrate that the collection of samples represented hydrogeologic variability in terms of the (1) types of aquifer lithologies, for example, alluvial

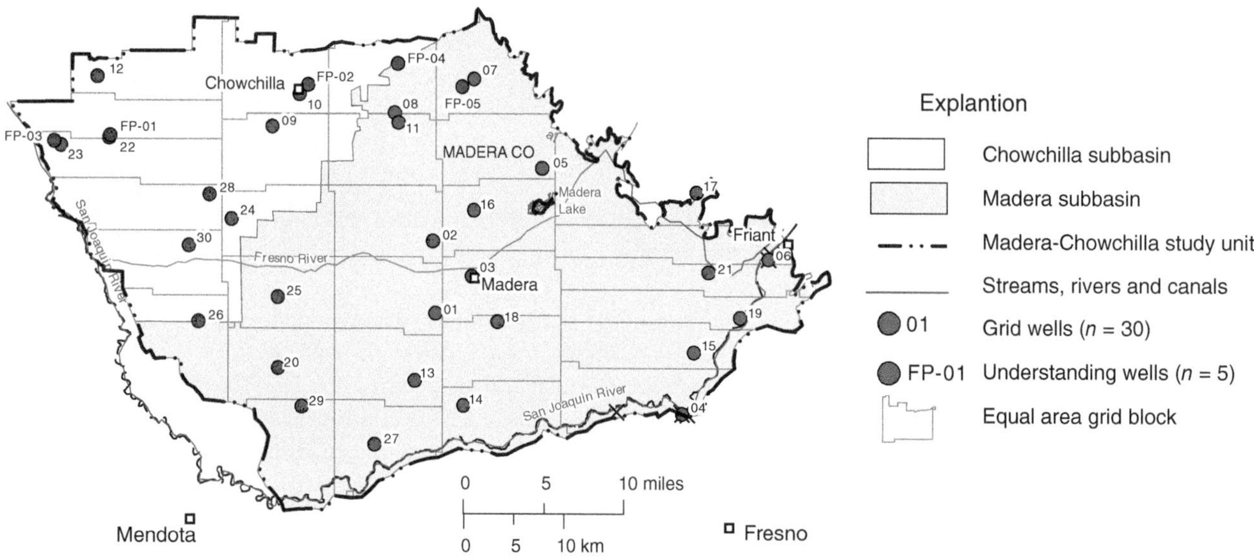

Figure 16.10 In the Madera-Chowchilla study unit, unbiased spatial sampling in water quality involved creation of a sampling grid (30 blocks) and selecting chemical data from one well per grid cell (red dots). The understanding wells (blue dots) provide additional information for elaboration of controlling factors (U.S. Geological Survey, adapted from Shelton et al., 2013/Public domain).

fan deposits, basin deposits, and units above and below the Corcoran Clay, (2) land surface elevations, (3) deep versus shallow samples, (4) old versus young water, etc. The purpose of the second type of assessment was to examine how natural and anthropogenic factors worked to control the concentrations of contaminant families, including nutrients (e.g., nitrate), trace elements (e.g., arsenic and vanadium), unregulated benchmarks (e.g., TDS, manganese), etc. (Shelton et al., 2013). For this purpose, each of the wells was characterized in terms of potential explanatory factors, such as land use, depth, position relative to the Corcoran Clay, and normalized lateral position from the eastern valley wall, and geochemical conditions in terms of oxidation-reduction characteristics, pH, and cation ratios.

16.6 Procedures for Water Sampling

The days are gone when a textbook like ours can provide detailed guidance in procedures for the collection of water samples. Similarly, the *ad hoc* procedures used by many researchers and practitioners are insufficient to assure quality and reliability in low concentration and specialized measurements. Detailed and comprehensive guidance documents are available through in the "National Field Manual for the Collection of Water-Quality Data" (NFM) created and maintained by the USGS (NFM, undated). The NFM manual is a living document available on USGS NFM home page.

The NFM for water-quality data consists of an introduction followed by 10 chapters. It discusses aspects of getting ready to sample, equipment needs, equipment cleaning, how samples should be collected and processed, and many other aspects. Chapter A4 in NFM is particularly informative with helpful checklists for scheduling and activities to get ready for field work. It also includes information with respect to quality control, field blanks, replicates, spiked samples and more. Following here is a general overview of activities done in the field.

16.6.1 Well Inspection and Measurements

Well inspection, the first step, provides (1) a check on the integrity of the well and (2) an initial water-level measurement before any water is removed from the well. The integrity check looks for signs of deterioration, vandalism, or tampering with the well. With the checks out of the way, water levels are measured using an electric tape with the top of the casing as a surveyed measurement point. In most cases, the same tape is sent down different monitoring wells. Thus, care must be taken to decontaminate the electric tape in moving from one hole to another. Most standard operating procedures (SOPs) for water-level measurements provide specific guidance on how to clean the tape before making a measurement. Cross-contamination can destroy wells for water-quality monitoring purposes. The National Field Manual has many different guidelines for maintaining cleanliness in the field with sampling.

16.6.2 Well Purging

Well purging is designed to get rid of water in the well casing and to bring new water from the aquifer (or geologic unit) into the well for sampling. Water that resides in the casing for a long time can lose key dissolved gases or react with the casing and seal materials, causing the chemistry to change. Stimulating this flow of "new" water to the well yields a water sample with a chemical composition representative of *in situ* conditions. If water purged from a well is contaminated, it must be collected at the surface and sent for appropriate disposal.

Purging the well is part of the sampling process and part of its role is "to condition the sampling equipment with well water" (NFM Chapter A4). As a rule, purging requires removing about three well volumes of water. However, real-time monitoring of key parameters, such as pH, temperature, specific conductance, etc. to constant values confirms appropriate purging.

Pumping for development and sampling is much preferred over bailing. For monitoring wells without a pump installed, Barcelona et al. (1985) suggest bladder pumps as the most appropriate for sampling. A *bladder pump* works with a squeezing action like human heart. Gas pressure compresses the sample chamber in the pump and moves the water up the sample line through a one-way valve. When the air pressure is released, water from the well refills the sample chamber, and the pumping cycle is repeated. This method uses positive pressures (not suction) to move the water, and the water moves up the sample line without contacting atmospheric gases. A Bennett pump is also suitable. It is a low-flow pump with a reciprocating piston motor driven by compressed air.

An *electrical submersible* pump is also useful for the purging and the sampling of wells. This pump combines a waterproof electric motor and a turbine pump that are installed down in the well. This kind of pump is great for purging a well but not quite as desirable for collecting the actual sample. This type of pump could impact concentrations of gases or volatile organic compounds in the water (Boulding and Barcelona, 1991). If the purging pump is moved from well to well, then great care must be taken to clean the pump and sampling line after each use. This cleaning can be tedious, and time-consuming. In some cases, it is more cost-effective to dedicate a pump and a sample line to each monitoring well.

When water levels are close to the ground surface, you might consider purging and sampling using a *vacuum pump*. This pumping system works like a straw in a glass of milk—the vacuum sucks the fluid up a rigid sampling line. This method is inexpensive and easy to implement. In practice, however, it can cause gases and volatiles to be lost. For this reason, it is not commonly recommended (Boulding and Barcelona, 1991).

16.6.3 Sample Collection, Filtration, and Preservation

With sampling, it is necessary to provide sample bottles that are appropriate for the constituents being sampled. Moreover, some samples may require filtration and other treatments for preservation. It is beyond the scope of this book to document the various protocols for sample preservation, and container selection. The NFMs provide a useful starting point.

Exercises

16.1 The routine analysis of a water sample provides the following concentrations (as mg/L). Calculate the concentrations of the ions in terms of molarity and milliequivalents per liter (meq/L).

Concentrations in mg/L: Ca—120; Mg—27; Na—53; K—4.4, Fe—0.5; HCO_3—546; SO_4—10; Cl—26; NO_3—0.1; pH 7.6

16.2 Assess in a preliminary way the quality of the analytical results in question (16.1) by determining the cation/anion balance.

16.3 Develop a relatively general Excel spreadsheet that will take ion concentrations in mg/L and convert them to mol/L and meq/L. Plan for each water sample to be processed on a separate Sheet. Add labels to make it clear what you are doing. Save your program on a disk. Show that your spreadsheet works by repeating the calculations in Questions 1 and 2.

16.4 Represent the following chemical data graphically using a Piper diagram (as mg/L): Ca^{2+}—93.9; Mg^{2+}—22.9; Na^+—19.1; HCO_3^-—334.; SO_4^{2-}—85.0; Cl^-—9.0 and a pH of 7.20

References

Ayers, R. S., and D. W. Westcot. 1985. Water quality for agriculture. FAO Irrigation and Drainage, Paper 29, Food and Agriculture Organization, Rome.

Back, W. 1961. Techniques for mapping of hydrochemical facies. U.S. Geological Survey Professional Paper 424-D, p. 380–382.

Barcelona, M. J., J. P. Gibb, J. A. Helfrich, et al. 1985. Practical guide for groundwater sampling. Illinois State Water Survey ISWS Contract Report 374, 94 p.

Boulding, J. R., and M. J. Barcelona 1991. Geochemical sampling of subsurface solids and groundwater. Site Characterization for Subsurface Remediation, U.S. Environmental Protection Agency, EPA/625/4-91/026, p. 123–154.

Buskard, J., N. Reid, and D. Gray. 2019. Parts per trillion (ppt) gold in groundwater: can we believe it, what is anomalous and how do we use it? Geochemistry: Exploration, Environment, Analysis, v. 20, no. 2, p. 189–198.

Collins, W. D. 1923. Graphic representation of water analyses. Industrial & Engineering Chemistry, v. 15, no. 4, p. 394–394.

Denver, J. M., S. W. Ator, L. M. Debrewer, et al. 2004. Water quality in the Delmarva Peninsula, Delaware, Maryland, and Virginia, 1999–2001. U.S. Geological Survey Circular 1228, 36 p.

Domenico, P. A., and F. W. Schwartz. 1998. Physical and Chemical Hydrogeology. John Wiley & Sons, New York, 506 p.

Fenton, S. E., A. Ducatman, A. Boobis et al. 2021. Per-and polyfluoroalkyl substance toxicity and human health review: current state of knowledge and strategies for informing future research. Environmental Toxicology and Chemistry, v. 40, no. 3, p. 606–630.

Ferguson, G., J. C. McIntosh, D. Perrone et al. 2018. Competition for shrinking window of low salinity groundwater. Environmental Research Letters, v. 13, no. 11, 114013.

Flynn, J. L. 2003. Groundwater-quality assessment of shallow aquifers in the Front Range Urban Corridor, Colorado, 1954–98. U.S. Geological Survey Water-Resources Investigations Report 02–4247, 32 p.

Fram, M. S. 2017. Groundwater quality in the Western San Joaquin Valley study unit, 2010: California GAMA Priority Basin Project. U.S. Geological Survey Scientific Investigations Report 2017-5032, 130 p.

Ging, P. B., D. G. Humberson, and S. J. Ikard. 2020. Geochemical assessment of the Hueco Bolson, New Mexico and Texas, 2016-17. U.S. Geological Survey Scientific Investigations Report 2020–5056, 30 p.

Keith, L., and W. Telliard. 1979. ES&T special report: priority pollutants: I-A perspective view. Environmental Science & Technology, v. 13, no. 4, p. 416–423.

NFM. undated National field manual for the collection of water-quality data. U.S. Geological Survey Techniques of Water-Resources Investigations, Book 9. https://pubs.er.usgs.gov/publication/twri09 (accessed on 07 March 2023).

Norman, J. E., P. L. Toccalino, and S. A. Morman. 2018. Health-based screening levels for evaluating water quality data (2nd edition). U.S. Geological Survey National Water-Quality Assessment Program. https://water.usgs.gov/water-resources/hbsl/ (accessed on 10 April 2023).

Parsons, M. C., T. C. Hancock, J. T. Kulongoski, et al. 2014. Status of groundwater quality in the Borrego Valley, Central Desert, and Low-Use Basins of the Mojave and Sonoran Deserts study unit, 2008–2010 – California GAMA Priority Basin Project. U.S. Geological Survey Scientific Investigations Report 2014-5001, 88 p.

Piper, A. M. 1944. A graphic procedure in the geochemical interpretation of water-analyses. Eos, Transactions American Geophysical Union, v. 25, no. 6, p. 914–928.

Rowe, B. L., P. L. Toccalino, M. J. Moran et al. 2007. Occurrence and potential human-health relevance of volatile organic compounds in drinking water from domestic wells in the United States. Environmental Health Perspectives, v. 115, no. 11, p. 1539–1546.

Toccalino, P. L., and J. E. Norman. 2006. Health-based screening levels to evaluate US Geological Survey groundwater quality data. Risk Analysis, v. 26, no. 5, p. 1339–1348.

Toccalino, P. L., J. E. Norman, and K. J. Hitt. 2010. Quality of source water from public-supply wells in the United States, 1993–2007. U.S. Geological Survey Scientific Investigations Report 2010-5024, 209 p.

Schwartz, F. W., and H. Zhang. 2003. Fundamentals of Groundwater. John Wiley & Sons, New York, 583 p.

Shelton, J. L., M. S. Fram, K. Belitz, et al. 2013. Status and understanding of groundwater quality in the Madera-Chowchilla Study Unit, 2008 – California GAMA Priority Basin Project. U.S. Geological Survey Scientific Investigations Report 2012–5094, 86 p.

Stiff, H. A. Jr. 1951. The interpretation of chemical water analysis by means of patterns. Journal of Petroleum Technology, v. 3, no. 10, p. 15–17.

USDA (U.S. Department of Agriculture). 1993. National Engineering Handbook, Part 623, Chapter 2, Irrigation Water Requirements Guide: Soil.

U.S. EPA. 1989. Guidance Document on the Statistical Analysis of Groundwater Monitoring Data at RCRA Facilities. Interim Final Guidance. Office of Solid Waste Management Division, Washington, D.C.

Warner, K.L., and Ayotte, J.D. (2014). The quality of our Nation's waters – water quality in the glacial aquifer system, northern United States, 1993–2009. U.S. Geological Survey Circular 1352, 116 p.

17

Key Chemical Processes

CHAPTER MENU

As dissolved mass is carried along through a groundwater system, a variety of chemical reactions are potentially available to change concentrations. In natural groundwater systems, these can result in (1) a decline in the concentration of constituents initially present in recharge entering the aquifer, such as dissolved oxygen or dissolved organic compounds, (2) the addition of constituents due to the dissolution of minerals that the water encounters along the flow path, and (3) exchanges of ions in solution with those sorbed onto the surfaces of minerals.

In contaminated systems, dissolved mass at sources, such as landfills, dumpsites for industrial wastes, chemical spills, etc. behave in similar ways. With organic contaminants, a variety of microbial processes might breakdown contaminants to other compounds or basic components, such as CO_2 and H_2O. Certain organic constituents may not react biologically at all, such as chlorinated solvents in oxic systems or members of the family of per- and polyfluoroalkyl substances (PFAS).

Figure 17.1 that was modified from Spruill et al. (2005) and Denver et al. (2010) is included here to illustrate a few of these processes at work. A portion of the nitrogen fertilizer applied to a corn crop is dissolved in soil water and enters the groundwater system. As the flow arrows illustrate, moving groundwater physically transports dissolved NO_3 downward and laterally in the surficial aquifer system. The figure also shows the concentration of nitrate-nitrogen measured in samples from a series of nested monitoring wells. Interestingly, water sampled from wells completed in the zone of oxic groundwater, where O_2 concentrations are >0.5 mg/L, have much larger concentrations of NO_3-N than those completed in anoxic groundwater. Anoxic groundwater has oxygen concentrations ≤0.5 mg/L and creates conditions where NO_3 is eliminated in a chemical reaction known as denitrification. In detail, NO_3 reacts to produce nitrogen gas N_2. Chemical and other processes can also influence concentrations. This illustrative example shows how a chemical reaction can substantially change the concentration of a dissolved constituent.

17.1 Overview of Equilibrium and Kinetic Reactions

We begin the exploration of chemical reactions in groundwater with an introduction to the equilibrium and kinetic frameworks that are essential for describing chemical and biological reactions. This section also illustrates how mineral saturation is used to describe the saturation state of groundwater with respect to key minerals.

Fundamentals of Groundwater, Second Edition. Franklin W. Schwartz and Hubao Zhang.

Companion website: www.wiley.com/go/schwartz/fundamentalsofgroundwater2

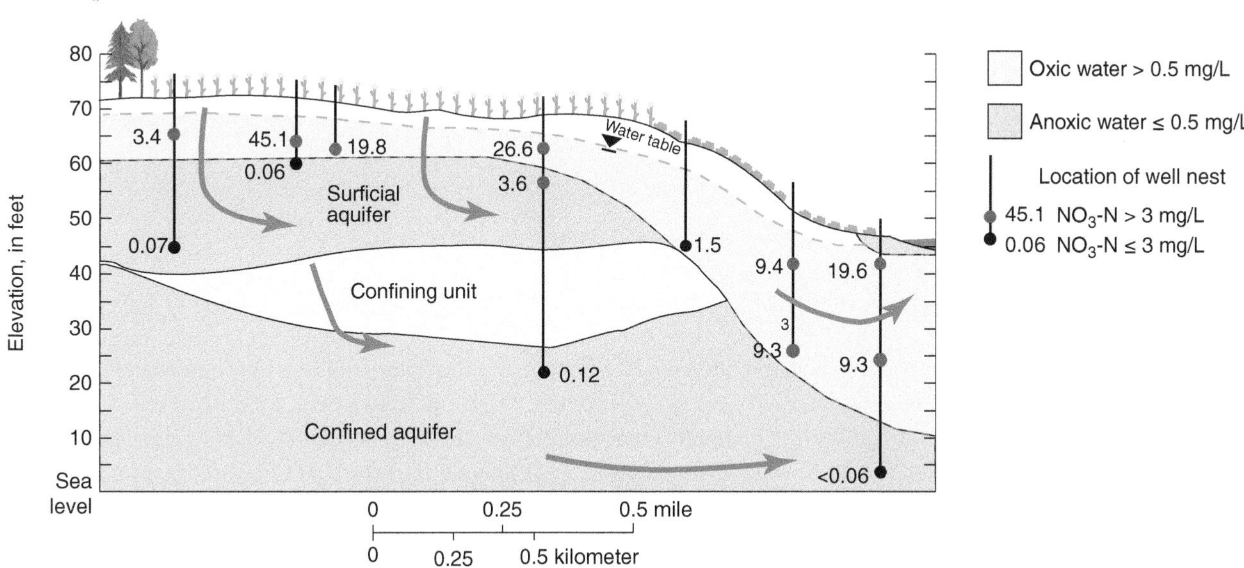

Figure 17.1 Example of physical and chemical processes at work in controlling the concentrations of NO_3-N in a shallow flow system. At this site in North Carolina, USA. Nitrate is advected downward and laterally toward a stream valley to the right side. The shading shows the water in the vicinity of the water table to be oxic and anoxic deeper. This condition facilitates the conversion of NO_3 to nitrogen gas (U.S. Geological Survey, adapted from Spruill et al., 2005 and Denver et al., 2010.Trends and transformation of nutrients and pesticides in a coastal plain aquifer system, United States. *Journal of Environmental Quality* 39(1): 154–167. Copyright © 2010 by the American Society of Agronomy, Crop Science Society of America, and Soil Science Society of America. All rights reserved).

17.1.1 Law of Mass Action and Chemical Equilibrium

Equilibrium and kinetic concepts provide a basis for treating chemical reactions. Let us begin with the following simple example of a chemical reaction, where nitrogen and hydrogen gases react to produce ammonia

$$N_2(g) + 3H_2(g) = 2NH_3(g) \tag{17.1}$$

where the (g) indicates that the compounds are gases, the equal sign indicates that the reaction is an equilibrium reaction (right and left arrows are equivalent) and the numbers in front of the elements or compound (e.g., $3H_2$) are stoichiometric coefficients. When there is no number as is the case for N_2 assume there to be a 1 by convention. Because this is an equilibrium reaction, the ratio of the product (s) found on the right side of the equation is equal to the reactants on the left side of the equation and the reaction can proceed forward or backward to maintain this ratio. The law of mass action expresses this ratio of products and reactions in terms of an equilibrium constant K or:

$$K = \frac{(NH_3)^2}{(N_2)^1(H_2)^3} \tag{17.2}$$

where (NH_3), (N_2), and (H_2), are the molal (or molar) concentrations for reactants and products. The exponents in the equation are the stoichiometric coefficients for the reaction.

The example (Eq. (17.1)) was written in terms of gases to keep things simple. There are conventions to be followed in writing the mass law expression when solids and/or water are involved in the reaction. Equation (17.3) is an example of such a reaction, where gypsum, a solid, dissolves in water

$$CaSO_4 \cdot 2H_2O(s) = Ca^{2+} + SO_4^{2-} + 2H_2O \tag{17.3}$$

By convention the concentration of gypsum is 1, as is (H_2O). Thus, the mass law expression rather than being written as follows

$$K = \frac{(Ca^{2+})(SO_4^{2-})(H_2O)^2}{(CaSO_4 \cdot 2H_2O(s))} \tag{17.4}$$

is written as:

$$K = (Ca^{2+})(SO_4^{2-}) \tag{17.5}$$

Numerical values for the equilibrium constants (K) can be calculated using thermodynamic principles or from tabulations in geochemical textbooks.

Even these simple calculations begin to help in understanding factors that control the salinity of groundwater. The equilibrium constant (K) in Eq. (17.5) is also known as the solubility product, which can be used to calculate how much Ca^{2+} and SO_4^{2-} will be dissolved in water at equilibrium, assuming that there are no other sources. What this means for gypsum dissolution is at equilibrium the concentrations will stop increasing. Knowing that the solubility product for gypsum is $10^{-4.60}$ and that the solution will end up with equal concentrations, i.e., (Ca^{2+}) = (SO_4^{2-}) in solution, we can rewrite Eq. (17.5) as:

$$10^{-4.60} = (Ca^{2+})^2 \text{ giving molar conc } (Ca^{2+}) = 10^{-2.30} = (SO_4^{2-})$$

Conversion of the molar concentration to mg/L provides $Ca^{2+}$and SO_4^{2-} (see Chapter 16)

$$\text{mg/L} = \text{M} \times 1000 \times \text{formula weight}$$

$$\text{Ca(mg/L)} = 10^{-2.30} \times 1000 \times 40.1; \text{SO}_4\text{(mg/L)} = 10^{-2.30} \times 1000 \times 94 = 471.$$

Values of solubility products for simple dissolution reactions provide a sense of how much the dissolution of a specific mineral might contribute to dissolved mass of the groundwater. For example, the solubility product for calcite, $CaCO_3$, is 4.5×10^{-9}—4 orders of magnitude lower than gypsum. Halite, NaCl, has a solubility product of ~38 orders of magnitude higher and when present, can contribute large quantities of dissolved Na and Cl to groundwater. Yet things are not necessarily simple. For example, in groundwater, the presence of dissolved CO_2 makes calcite more soluble.

The concept of equilibrium with respect to groundwater chemistry is important because it constrains how much dissolved mass can end up in groundwater. For example, when water recharges an aquifer with carbonate mineral grains, e.g., calcite, dolomite, equilibrium keeps the major constituents, Ca, Mg, and HCO_3 well within secondary standards for drinking water for TDS. If the aquifer contains gypsum layers, the continued dissolution toward equilibrium with respect gypsum can make the make water too saline to be potable.

Commonly, groundwaters might only be saturated with respect to a few minerals. One reason is that an aquifer may not contain many soluble mineral species. A second reason is that minerals can be there but are reacting slowly. These minerals are on their way toward equilibrium but so slowly they may never get there. Thus, an important aspect of groundwater geochemistry involves examining the relative importance of dissolution kinetics constraints in controlling the chemistry of the groundwater.

17.1.2 Complexities of Actual Groundwater

The calculations in the previous section assumed idealized solutions with one dissolving solid and a dilute solution. Actual groundwater contravenes both assumptions. There are a variety of reacting solids, gases, and liquids. Moreover, the solution is no longer dilute adding significant additional complexity. Instead of using the actual laboratory-measured concentrations, (Ca^{2+}), the calculations require activities [Ca^{2+}]. *Activity* for a species like (Ca^{2+}) is an adjustment to the measured concentration to account for the presence of all the other dissolved species. In practice, one writes the mass action equations in terms of activities instead of molar concentrations. For the case of species Y, the adjustment involves multiplying the molar concentration by a factor:

$$[Y] = \gamma_Y (Y) \tag{17.6}$$

where γ_Y is the *activity coefficient*. Typically, γ is close to one in dilute solutions and decreases as salinity increases. Activities of species with more than one charge usually are smaller than those with a single charge. At relatively high salinities, the activity coefficients may increase and even exceed one.

In practice, one calculates γ values depending upon features of the ions and abundance of other ions, calculated as the "ionic strength." The simplest calculations involve the Debye-Huckel equation

$$\log \gamma_i = -Az_i^2(I)^{0.5} \tag{17.7}$$

where A is a constant that is a function of temperature (0.508 at 25°C), z_i is the ion charge, and I is the ionic strength of the solution. Ionic strength has units of mol/L is given as

$$I = 0.5\sum M_i z_i^2 \tag{17.8}$$

with (M_i) as the molar concentration of species i having a charge z. *Ionic strength* is a measure of the total concentration of ions that emphasizes the increased contribution of species with charges greater than one to solution nonideality.

Use of the Debye-Huckel equation is limited to solutions with ionic strengths less than 0.005 M, which is fresh, potable groundwater. With waters of higher ionic strength, it is necessary to move to the extended Debye-Huckel equations or other more sophisticated approaches.

Example 17.1 Water contains ions in the following molar concentrations Ca^{2+} 1.63×10^{-3}, Na^+ 0.486×10^{-3}, HCO_3^- 2.87×10^{-3}, and SO_4^{2-} 0.44×10^{-3}. Calculate the ionic strength. of the water. The constant "A" in Eq. (17.7) has a value of 0.508 at 25°C

Solution

First, calculate the ionic strength from Eq. (17.7) or

$$\begin{aligned} I &= 0.5\left(1.63 \times 2^2 + 0.48 \times 1^2 + 2.87 \times 1^2 + 0.4 \times 2^2\right) \times 10^{-3} \\ &= 0.0058 \text{ M} \end{aligned}$$

This value of I is at the upper end of the range of applicability of the extended Debye-Huckel equation. Now, determine the γ values from Eq. (17.9)

$$\begin{aligned} \log \gamma_{Na^+} &= -0.508 \times 1^2(0.0058)^{0.5} \\ &= -0.039 \\ \gamma_{Na^+} &= 10^{0.39} = 0.91 \end{aligned}$$

and

$$\begin{aligned} \log \gamma_{Ca^{2+}} &= -0.508 \times 2^2(0.0058)^{0.5} \\ &= -0.154 \\ \gamma_{Ca^{2+}} &= 10^{-1.54} = 0.70 \end{aligned}$$

17.1.3 Deviations from Equilibrium

Equilibrium calculations commonly establish whether groundwater is in equilibrium with respect to one or more minerals. For example, if we added gypsum crystals to distilled water, the solution would be initially undersaturated with respect to gypsum. Gypsum would dissolve until the dissolution reaction reached equilibrium, at which point the dissolution would stop. The initial accumulation of dissolved constituents in the unsaturated zone can be viewed as a reaction of rainwater with the assemblage of minerals comprising the porous medium. Some of these minerals would dissolve rapidly and the water would rapidly achieve equilibrium with respect to these minerals. In other cases, the groundwater may not have reached equilibrium with respect to others. The dissolution of minerals would continue as water moved through the groundwater system. The term, *partial equilibrium system,* describes a complex mineral-water system where reactions may not be at equilibrium (e.g., dissolution or precipitation reactions). Hydrogeologists use water chemistry data to sort out the equilibrium status with respect to various minerals. These calculations in a regional water quality investigation help to explain why the major and minor ion chemistry is changing due to mineral dissolution and precipitation reactions.

The departure of a reaction from equilibrium is determined as a dimensionless ratio of the *ion activity product* to the equilibrium constant (IAP/*K*). The ion activity product is calculated by substituting sample activity values in the mass law expression for a reaction. For example, with measured concentrations converted to activities, $[Ca^{2+}]$ and $[SO_4^{2-}]$, the ion activity product (IAP) for Eq. (17.3) is:

$$IAP = \left[Ca^{2+}\right]\left[SO_4^{2-}\right] \tag{17.9}$$

If the calculated to the known constant K is greater than 1, the reaction (Eq. (17.3)) is progressing from right to left reducing $[Ca^{2+}]$ and $[SO_4^{2-}]$ and increasing gypsum. If IAP $< K$, the reaction is proceeding from left to right. The IAP at equilibrium is equal to the equilibrium constant, K.

This theory provides the saturation state of a groundwater with respect to one or more mineral phases. When $IAP/K < 1$, the groundwater is *undersaturated* with respect to the given mineral. When $IAP/K = 1$, the groundwater is in chemical equilibrium with the mineral, and when $IAP/K > 1$, the groundwater is *supersaturated.* Undersaturation with respect to a mineral results in net dissolution provided the mineral is present. Supersaturation results in the net precipitation of the mineral should suitable nuclei be present.

The saturation state is often expressed in terms of a *saturation index (SI)*, defined as log (IAP/*K*). When a mineral is in equilibrium with respect to a solution, the SI is zero. Undersaturation is indicated by a negative SI and supersaturation by a positive SI. Example 17.2 illustrates how to calculate the state of saturation with respect to a mineral.

Example 17.2 Given a groundwater with the measured molar composition listed below, calculate the saturation state with respect to the minerals calcite, $CaCO_3$, and dolomite, $CaMg(CO_3)_2$. The activity coefficients for Ca^{2+}, Mg^{2+}, and CO_3^{2-} are 0.57, 0.59, and 0.56 respectively. The equilibrium constants defining the solubility of calcite and dolomite are 4.9×10^{-9} and 2.7×10^{-17} respectively, and the measured molar concentrations are $(Ca^{2+}) = 3.74 \times 10^{-4}$, $(Mg^{2+}) = 4.11 \times 10^{-6}$, and $(CO_3^{2-}) = 5.50 \times 10^{-5}$.

Solution

Write the reactions.

$$CaCO_3(s) = Ca^{2+} + CO_3^{2-} \text{ and } CaMg(CO_3)_2(s) = Ca^{2+} + Mg^{2+} + 2CO_3^{2-}$$

Write the mass law equations in terms of activities and replace the term for equilibrium constants (i.e., K values) by IAP (refer to Eq. (17.9)). Recall that the measured concentration, for example, (Ca^{2+}) will need to be converted to $[Ca^{2+}]$ using Eq. (17.6). Thus,

$$\begin{aligned} IAP_{cal} &= \left[Ca^{2+}\right]\left[CO_3^{2-}\right] \\ &= 0.57 \times 3.74 \times 10^{-4} \times 0.56 \times 5.50 \times 10^{-5} \\ &= 6.56 \times 10^{-9} \end{aligned}$$

$$\{IAP/K\}_{cal} = 6.56 \times 10^{-9}/4.90 \times 10^{-9} = 1.34$$

The sample is slightly oversaturated with respect to calcite.

$$IAP_{dol} = \left[Ca^{2+}\right]\left[Mg^{2+}\right]\left[CO_3^{2-}\right]^2 = 4.89 \times 10^{-19}$$

$$\{IAP/K\}_{dol}\} = 4.89 \times 10^{-19}/2.7 \times 10^{-17} = 0.018$$

The sample is strongly undersaturated with respect to dolomite.

Saturation calculations with respect to selected minerals provide a useful way of exploring how the pore fluid has interacted with the minerals making up the porous medium and understanding the pattern of chemical evolution along the flow system.

Example 17.2 showed how these calculations can be approximated in hand calculations. However, the best way to do these equilibrium calculations is to utilize a computer code like PHREEQC (Parkhurst and Appelo, 2013) The advantage an existing code is that the burden of calculations is carried by the computer and the consistency of the data base of equilibrium constants facilitates the comparison of results from similar studies.

17.1.4 Kinetic Reactions

Equilibrium theory yields no information about the time required to reach equilibrium or the reaction pathways involved. Kinetic theory provides this information and is applicable to any reaction. However, it is especially important for reactions for which equilibrium theories do not apply. Examples include irreversible reactions, i.e., reactions that go in one direction or reversible reactions that are slow in relation to mass transport. The reaction that takes a shiny steel nail and turns it to rust is irreversible. There is no way the rusty nail will turn back to shiny steel.

Radioactive decay is an example of an irreversible reaction. For example, carbon-14 decays to nitrogen and an electron with an equation of the following form

$$^{14}\mathrm{C} \rightarrow \, ^{14}\mathrm{N} + \mathrm{e} \tag{17.10}$$

This reaction is written with an arrow and not an equal sign, signifying that it is irreversible, proceeding only in one direction. The kinetic equation for ^{14}C in this reaction has the following form:

$$\frac{d(^{14}\mathrm{C})}{dt} = -k_1 \left(^{14}\mathrm{C}\right)^1 \tag{17.11}$$

where (^{14}C) is the radioactivity (for simplicity concentration) of carbon-14, t is time, and k_1 is the rate constant for the reaction. The negative sign in front of k_1 indicates that the concentration of carbon-14 is being reduced in the reaction. Notice the superscript 1 that is added to the term (^{14}C). In most cases, this superscript is omitted when writing the equation when the value is one.

Nevertheless, we need to keep track of the superscripts on the right-hand side of kinetic equations because these will tell us the *order* of the reaction. As we will learn shortly, Eq. (17.11) is an example of a first-order kinetic reaction with the general form

$$\frac{dC}{dt} = -kC \tag{17.12}$$

Most of the important reactions that we deal with in groundwater investigations are first-order reactions.

Equation (17.12) is an ordinary differential equation that can be solved given an initial condition such as $C(0) = C_0$. In words, this condition states that at time equal to zero there is some initial quantity of a constituent, C_0. The solution to Eq. (17.12) lets us determine the concentration of C for all times greater than zero. This is the wonderful property of kinetic equations that lets us track the decrease or increase of dissolved constituents in aqueous solution with time. Mathematically, the solution has the following form:

$$C = C_0 e^{-kt} \tag{17.13}$$

In words, this equation states that some initial concentration of carbon-14 will be reduced exponentially as a function of time and the rate constant k. The following calculation illustrates these points.

Example 17.3 A chemical constituent in groundwater disappears following a first-order kinetic rate law. Given some initial concentration of 100 mg/L and a rate constant k_1 of 0.00693 days^{-1} calculate how much of the constituent will remain after 100 days.

Solution

Begin by substituting the numbers into Eq. (17.13).

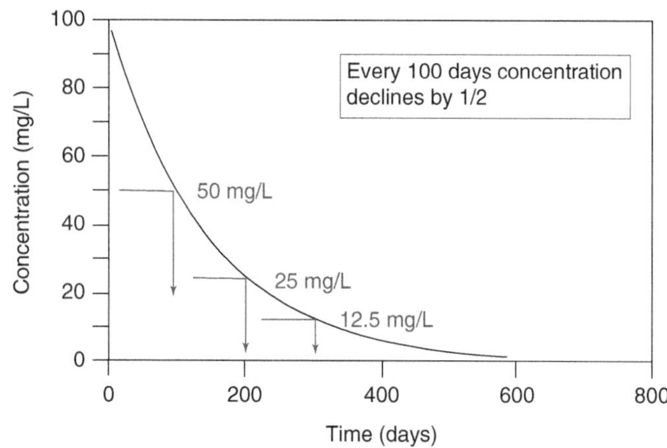

Figure 17.2 The exponential loss of a constituent as a function of time in a first-order kinetic reaction (FWS).

$$C = C_0 e^{-k_1 t}$$

$$C = \left(100 \frac{\text{mg}}{\text{L}}\right) \times e^{-0.00693 \times 100}$$

$$C = \left(\frac{100\ \text{mg}}{\text{L}}\right) \times 0.5 = 50\ \text{mg/L}$$

Thus, at the end of 100 days, there is 50 mg/L of the constituent remaining.

A plot of the function $C = 100e^{-0.00693t}$ using an arithmetic concentration scale shows the typical exponential loss of the constituent as a function of time (Figure 17.2). Plotting the equation as a semi-log plot, however, yields a straight line.

The rate constant in a kinetic equation can also be expressed by an equivalent parameter termed the half-life ($t_{1/2}$). The half-life is the time required to reduce the concentration by one-half. The relationship between $t_{1/2}$ and k can easily be determined starting from:

$$C = C_0 e^{-k_1 t}$$

$$\frac{C}{C_0} = \frac{1}{2} = e^{-kt_{1/2}}$$

$$\ln \frac{1}{2} = -k_1 t_{1/2}$$

$$t_{1/2} = -\frac{\ln 1/2}{k_1} = \frac{0.693}{k_1}$$

For our example problem, k_1 had value of 0.00693, which when substituted into the equation above gives $t_{1/2} = 100$ days.

Not all reactions are the simple, first-order type considered in the previous section. A more general theory exists to analyze more complicated reactions. Interested readers can refer to more specialized geochemical textbooks, such as Appelo and Postma (2004).

17.2 Acid–Base Reactions

Here and in the sections that follow we introduce the families of reactions that are important in controlling the distribution of dissolved species in groundwater. Within those families, there are specific reactions that are essential to know something about. We identify these critical reactions in boxes and tables. Acid–base reactions involve the transfer of hydrogen ions (H^+) among the ions present in the aqueous phase. The concentration of H^+ determines the pH of the solution, which is a key factor controlling many other processes. For example, at low pH water can transport a large load of metals dissolved in solution. With a high pH, the metals tend to precipitate as solids. *pH* is defined formally as the negative logarithm of the hydrogen ion activity or $\text{pH} = -\log[H^+]$. Water is acidic when $\text{pH} < 7$, neutral when $\text{pH} = 7$, or basic when $\text{pH} > 7$.

Table 17.1 Important weak-acid-base reactions important in water systems.

Reaction	Mass law equation	Eq.	−log K (25°C)
$H_2O = H^+ + OH^-$	$K_w = (H^+)(OH^-)$	(17.14)	14.0
$CO_2(g) + H_2O = H_2CO_3^*$	$K_{CO_2} = \frac{(H_2CO_3^*)}{P_{CO_2}(H_2O)}$	(17.15)	1.46
$H_2CO_3^* = HCO_3^- + H^+$	$K_1 = \frac{(HCO_3^-)(H^+)}{(H_2CO_3^*)}$	(17.16)	6.35
$HCO_3^* = CO_3^{2-} + H^+$	$K_2 = \frac{(CO_3^{2-})(H^+)}{(HCO_3^-)}$	(17.17)	10.33

Source: Domenico and Schwartz (1998), Physical and Chemical Hydrogeology.

Listed below are important key acid–base reactions.

- dissociation of water: $H_2O = H^+ + OH^-$
- dissolution of CO_2 in water: $CO_2(g) + H_2O = H_2CO_3^*$
- two stage ionization of carbonic acid: $H_2CO_3^* = HCO_3^- + H^+$

$$HCO_3^- = CO_3^{2-} + H^+$$

With the dissociation of water, an extremely, small quantity of liquid water breaks down into hydrogen (H^+) and hydroxyl (OH^-) ions. The last three reactions are particularly important in helping to regulate pH. They also show that when CO_2 gas is added to water it partitions among three other dissolved constituents—carbonic acid ($H_2CO_3^*$), bicarbonate ion (HCO_3^-), and carbonate ion (CO_3^{2-}). The '*' is added to $H_2CO_3^*$ to indicate that by convention two constituents are represented or:

$$(H_2CO_3^*) = (CO_2\ aq) + (H_2CO_3)$$

Think about CO_2 in this way. If you received a large quantity of cash money, it would quickly change form to food, transportation, and gifts. Some cash might remain but otherwise most of the cash is redistributed.

Table 17.1 shows the mass-law equations for these reactions along with their equilibrium constants. Note that the constants are given as $-\log K$. Thus, a K value of 14 on the table is 10^{-14} for calculations. Calculations with the equations can help understand how mass is distributed among the carbonate species H_2CO_3, HCO_3^-, and CO_3^{2-}, as a function of changing pH. Following is an example illustrating how CO_2 gas is redistributed when added to water.

Example 17.4 Assume that CO_2 is dissolved in water so that the total carbon dioxide or $(CO_2)_T = 10^{-3}$ M. As mentioned, it is distributed three carbonate species $H_2CO_3^*$, HCO_3^-, and CO_3^{2-}. What is the concentration of the three carbonate species at a pH of 6.35?

Solution

A mass balance equation for CO_2 can be written as:

$$(CO_2)_T = 10^{-3} = (H_2CO_3^*) + (HCO_3^-) + (CO_3^{2-}) \tag{17.18}$$

In this form, Eq. (17.18) cannot be solved because there are three unknowns. However, the equations in Table 17.1 provide a way Start by finding the mass law expression in Table 17.1 to substitute:

$$H_2CO_3^* = HCO_3^- + H^+$$

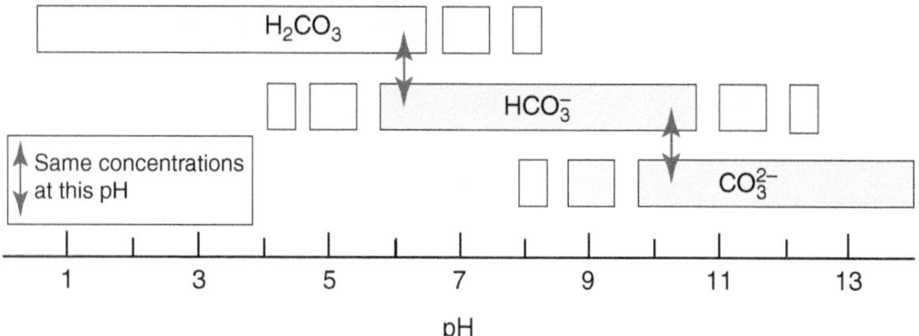

Figure 17.3 The distribution of key carbonate species as a function of pH assuming that $(CO_2)_T = 10^{-4}$ M (FWS).

or

$$K_1 = \frac{(HCO_3^-)(H^+)}{(H_2CO_3)} \tag{17.19}$$

Substitution of the known pH and K_1 values into Eq. (17.21) gives

$$10^{-6.35} = (HCO_3^-)(10^{-6.35})/(H_2CO_3^*) \tag{17.20}$$

Simplification gives $(HCO_3^-) = (H_2CO_3^*)$. Next, we check on (CO_3^{2-}) concentration. Substitution into Eq. (17.17) provides

$$10^{-10.33} = (CO_3^{2-})(10^{-6.35})/(HCO_3^-)$$

and shows that $(HCO_3^-) \gg (CO_3^{2-})$. Assuming for the moment that (CO_3^{2-}) is negligible, a solution to Eq. (17.18) gives $(HCO_3^-) = (H_2CO_3^*) = 10^{-3.31}$ M. Substitution into Eq. (17.17) gives $(CO_3^{2-}) = 10^{-7.29}$ M.

Thus, at the specified pH, (6.35), the concentrations of $H_2CO_3^*$ and HCO_3^- are the same and CO_3^{2-} is much lower.

We could repeat this calculation over a broad range of pH. The important take away here is that across a range of pH from 2 to 13 dominance among the carbonate species changes as well. Below pH 5, $H_2CO_3^*$ is dominant (Figure 17.3). In the range from approximately pH 7 to 9 (typical of many waters), HCO_3^- dominates. The carbonate ion (CO_3^{2-}) is dominant when pH is above 10. Over a pH range of natural groundwater (5–8.5), HCO_3^- is the most abundant species. As the shading implies, even though one species dominates, the others are still present. However, their concentrations will be lower. There are certain pHs where two species are present at equal concentrations (Figure 17.3).

Another important group of acid–base reactions involve the reaction of carbonate minerals with hydrogen ion. These reactions are particularly important because they produce the load of dissolved mass (i.e., ions) found in natural groundwaters and influence pH. Here are examples of these reactions:

- dissolution/precipitation of calcite: $CaCO_3 + H^+ = Ca^{2+} + HCO_3^-$
- dissolution/precipitation of silicate minerals: silicate + H^+ = cations + H_2SiO_3

The spectacular landforms and cave features that are seen around the world are generated as groundwater dissolves calcite ($CaCO_3$) and adds Ca^{2+} and HCO_3^- to the groundwater.

The previous section introduced several acid–base reactions involving minerals. Here, we expand that list and consider additional reactions involving soluble salts, and the dissolution of liquids.

17.3 Mineral Dissolution/Precipitation

CO_2-charged groundwater is effective in dissolving minerals. The most common reactions involve the weak acids of the carbonate and silicate systems and strong bases from the dissolution of carbonate, silicate, and alumino-silicate minerals. This process causes the weak acids to dissociate. In the carbonate system, the relative abundance of

HCO_3^- and CO_3^{2-} increases at the expense of $H_2CO_3^*$. In addition, mineral dissolution cation concentrations increase. Following are generic and specific examples of these reactions.

- carbonate minerals + H^+ = cations + HCO_3^-

  e.g., calcite: $CaCO_3(s) + H^+ = Ca^{2+} + HCO_3^-$
- silicate minerals + H^+ = cations + H_2SiO_3

  e.g., enstatite: $MgSiO_3(s) + 2H^+ = Mg^{2+} + H_2SiO_3$
- alumino-silicate minerals + H^+ = cations + H_2SiO_3 + secondary minerals

  e.g., anorthite: $CaAl_2Si_3O_8(s) + 2H^+ + H_2O$ = kaolinite + Ca^{2+} and

  albite: $2NaAlSi_3O_8(s) + 2H^+ + 5H_2O$ = kaolinite + $4H_2SiO_3 + 2Na^+$

Soluble salts commonly occur disseminated within geologic units (e.g., near-surface deposits in arid areas) or as thick sequences of evaporite deposits in sedimentary basins. Mineral salts are extremely soluble and when present can dissolve to produce saline waters and even brines. The actual composition of the water depends upon the minerals present (e.g., halite, anhydrite, gypsum, carnalite, kieserite, and sylvite). Here are examples of salts and resulting ion species.

$$\text{halite: } NaCl(s) = Na^+ + Cl^-$$
$$\text{anhydrite: } CaSO_4(s) = Ca^{2+} + SO_4^{2-}$$
$$\text{gypsum: } CaSO_4 \cdot 2H_2O(s) = Ca^{2+} + SO_4^{2-} + 2H_2O$$
$$\text{carnalite: } KCl \cdot MgCl_2 \cdot 6H_2O(s) = K^+ + Mg^{2+} + 3Cl^- + 6H_2O$$
$$\text{kieserite: } MgSO_4 \cdot H_2O(s) = Mg^{2+} + SO_4^{2-} + H_2O$$
$$\text{sylvite: } KCl(s) = K^+ + Cl^-$$

17.3.1 Organic Compounds in Water

In the subsurface, other liquids occur. These can include naturally occurring oils and anthropogenic contaminants, such as gasoline or industrial solvents. These liquids can migrate as a separate liquid phase or dissolved in groundwater. Organic compounds differ widely in their overall solubility (Mackay et al., 1985). Some solutes like methanol are extremely soluble, while others such as PCBs or DDT are sparingly soluble. As a rule, the most soluble organic compounds are charged species or those containing oxygen or nitrogen. Examples of this latter group are alcohols or carboxylic acids.

17.4 Surface Reactions

Surface reactions can cause wholesale changes to the geochemistry of both natural and contaminated groundwater. They involve reactions between constituents in solution and solid materials comprising the skeleton of an aquifer. For example, water in clay-rich marine deposits can experience a loss of Ca, and Mg and a gain of Na due to cation exchange processes with clay minerals. These reactions naturally soften the water in much the same way as a commercial softener that people install in their houses. Another example is the tendency for certain organic compounds to sorbed on aquifer materials, which effectively reduces the velocity at which contaminants spread in the groundwater.

This section begins with theories around sorption isotherms, which provided measurement-based approaches to describe sorption. Early work in the assessment of sites for the storage of nuclear work exposed the major weakness of these kinds of empirical approaches. The problem was that one-parameter models were simply inadequate to describe complicated sorption involving radionuclides and metals. Basic theory is presented here because the partitioning of uncharged organic contaminants into solid organic matter in aquifers is well described by simple theory.

17.4.1 Sorption Isotherms

The term *sorption* refers to uptake of a dissolved constituent in water by a solid. This term is general in the sense that sorption might only involve a grain surface (adsorption) or the entire grain volume (absorption). Historically, information on the tendency of a material to sorb a dissolved constituent was gathered from laboratory experiments. Tests involved creating a batch of five or more bottles each containing a constant volume of solution with a unique concentration of the aqueous constituent at a constant temperature. The experiment begins by adding a constant mass of aquifer material and waiting long enough so that equilibrium is achieved with the sorption reaction.

Once mixed with solid material, the initial concentration of the constituent (C_i) is reduced as mass is sorbed onto the solid. The equilibrium concentration (C) is measured for each test solution in the batch. The last step involves the calculation of the mass of constituent sorbed per gram of solid

$$S = \frac{(C_i - C)(\text{solution volume})}{\text{sm}} \tag{17.21}$$

where the starting concentrations, C_i, and the subsequent equilibrium concentrations (C) have units such as mg/L or μg/L, ms is the sediment mass (e.g., g), sv is the volume of the match solution, (e.g., mL) and S is the quantity of mass sorbed on the surface (e.g., mg/g or μg/g). The data from a single bottle provides a single point "a" on an S versus C plot. By repeating, the procedure at the same temperature (hence, isotherm) with different values of C_i, a line is drawn through the points to create the sorption isotherm.

Real isotherms have no prescribed shape. They can be linear, concave, convex, or a complex combination of all these shapes. Sorption is modeled by fitting an experimentally derived isotherm to theoretical equations. The Freundlich isotherm has an equation of the following form where K is the Freundlich sorption equilibrium constant and n is a fitting parameter.

$$S = KC^n \tag{17.22}$$

Langmuir isotherms can be described by equations of the following form:

$$S = \frac{Q^0 KC}{1 + KC} \tag{17.23}$$

where K is the Langmuir equilibrium constant reflecting the extent of sorption, n is a constant usually ranging between 0.7 and 1.2, and Q^0 is the maximum sorptive capacity for the surface.

Figure 17.4 depicts Freundlich isotherms calculated with a K value of 1.5 and values of n of 0.5, 1.0, and 1.5, and a Langmuir isotherm described by a K value of 0.5 and a Q^0 value of 30 mg/g. Notice how the Langmuir isotherm is asymptotically approaching Q^0 the maximum sorptive capacity for the surface. These examples illustrate the range in curve shapes that can be fit with these two equations. If the fit is not satisfactory, there are other equations.

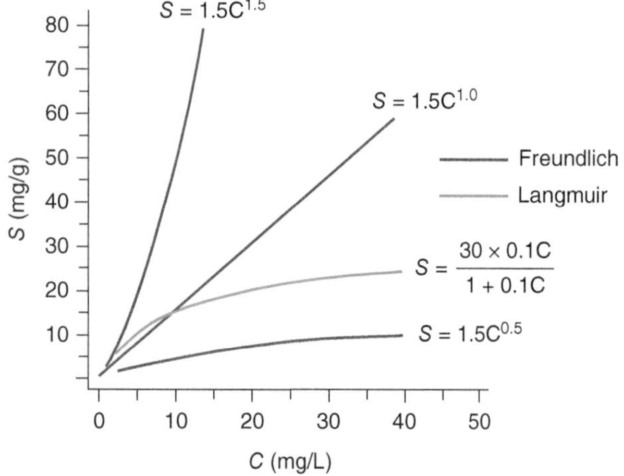

Figure 17.4 Example of Langmuir and Freundlich isotherms (Domenico and Schwartz, 1998, Physical and Chemical Hydrogeology. Copyright © 1990, 1998 John Wiley & Sons, Inc. All Rights Reserved. Reproduced with permission).

A Freundlich isotherm with $n = 1$ is a special case because this linear isotherm is easy to incorporate into mass-transport models. The following equation relates S to C

$$S = K_d C \tag{17.24}$$

where K_d with units like cm^3/g is the *distribution coefficient* (i.e., the slope of the linear sorption isotherm). Large K_d values are indicative of a greater tendency for sorption.

17.4.2 Sorption of Organic Compounds

Hydrophobic (or water-hating) organic molecules, which are nonionic (i.e., they are neither cations nor anions), tend to partition preferentially into solid organic matter (e.g., humic substances and kerogen). This solid matter occurs in small quantities in the porous media, as discrete solids, films on individual grains, or as stringers of organic material in grains. Nonpolar organic molecules, dissolved in water hate water and prefer the organic matter. Overall, the more hydrophobic a compound is, the greater is its tendency to partition into a solid phase. Generally, the larger the size of the molecule, the greater is the tendency to partition out of the water.

The following example illustrates the tendency for an organic compound to be sorbed by aquifer materials. Also, at relatively small concentrations, this sorption can be described by a linear isotherm.

Example 17.5 Figure 17.5 summarizes data collected from batch sorption experiments with an organic constituent using 25 mg of porous media in 50 mL of solution. The batch involved five test solutions. On the figure C_i is the concentration of the constituent before the sediment was added. The concentration after equilibrium was established is C, the lower of the two values on the bottles. Use the given data to calculate the K_d value.

Solution

The first step is to calculate S for teach of the test solutions using Eq. (17.21) as shown on Figure 17.5. Next, the quantity of mass sorbed, S, is plotted versus the equilibrium concentration, C. The K_d value is the slope, which is 5.7 cm^3/g.

Normally, one might expect running batch sorption experiments to determine sorption parameters for organic compounds. Fortunately, there is theory that takes advantage of detailed compilations of sorption of various organic compounds into octanol. In practice, octanol was involved as a fat-like solvent needed for studies of drug and pesticide partitioning.

The distribution coefficient can be expressed as the product of constants describing the contaminant and the porous medium, or

$$K_d = K_{oc} f_{oc} \tag{17.25}$$

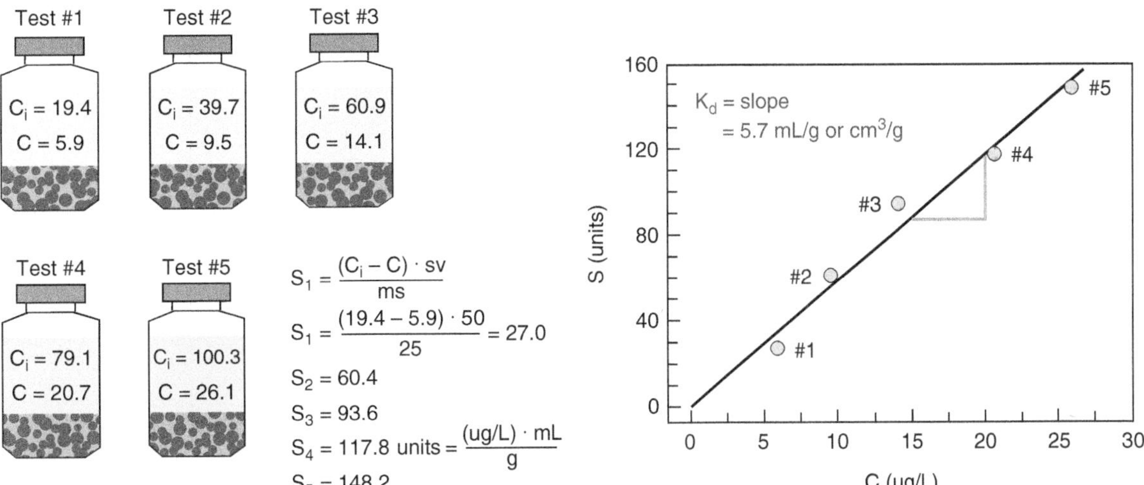

Figure 17.5 The figure illustrates the steps in calculating a K_d value for a batch test involving five test samples (FWS).

Table 17.2 A synthesis of data on the organic carbon content of sediments.

Site name	Type of deposit	Texture	Organic carbon content
Borden, Ontario[a]	Glaciofluvial	Fine-medium sand	0.0002
Gloucester, Ontario[b]	Glaciofluvial	Sands and gravel	0.0006
North Bay, Ontario[c]	Glaciofluvial	Medium sand	0.00017
Woolrich, Ontario[c]	Glaciofluvial	Fine-medium sand	0.00023
Chalk River, Ontario[c]	Glaciofluvial	Fine sand	0.00026
Cambridge, Ontario[c]	Glaciofluvial	Medium sand	0.00065
Rodney, Ontario[c]	Glaciofluvial	Fine sand	0.00102
Wildwood, Ontario[c]	Lacustrine	Silt	0.00108
Palo Alto, Baylands[d]	?	Silty sand	0.01
River Glatt, Switzerland[e]	Glaciofluvial	Sand, gravel	<0.000–0.01
Oconee River, Georgia[f]	River sediment	Sand	0.0057
		Coarse silt	0.029
		Medium silt	0.02
		Fine silt	0.0226

[a] Mackay et al. (1986).
[b] Jackson (personal communication, 1989).
[c] J. Barker (personal communication 1987).
[d] MacKay and Vogel (1985).
[e] Schwarzenbach and Giger (1985).
[f] Karickhoff (1981).
Source: Domenico and Schwartz (1998), Physical and Chemical Hydrogeology.

where K_{oc} is the partition coefficient of a compound between organic carbon and water with typical units such as cm^3/g and f_{oc} is the weight fraction of organic carbon (dimensionless), defined as g_{oc}/g_s or grams of solid organic carbon to grams of total aquifer solids (Karickhoff et al., 1979; Schwarzenbach and Westall, 1981).

Organic carbon contents can be measured in the laboratory on porous-medium samples. However, this parameter is not well characterized, and the range reported for geologic materials is large (Table 17.2). It is the values of K_{oc}, which are determined on the basis of correlations between log K_{oc} and log K_{ow}, the *octanol-water partition coefficient.* This correlation exists because the partitioning of an organic compound between water and organic carbon is not much different than between water and octanol.

Regression equations in practice describe the relationship between K_{oc} and K_{ow} (Griffin and Roy, 1985)

Karickhoff et al. (1979):

$$\log K_{oc} = -0.21 + \log K_{ow} \tag{17.26}$$

Schwarzenbach and Westall (1981):

$$\log K_{oc} = 0.49 + 0.72 \log K_{ow} \tag{17.27}$$

Hassett et al. (1983):

$$\log K_{oc} = 0.088 + 0.909 \log K_{ow} \tag{17.28}$$

where K_{ow} is a tabulated constant (dimensionless), and K_{oc} has units of cm^3/g. Overall, the equations have similar results. The following example illustrates the application of the empirical equations in estimating a distribution coefficient.

Example 17.6 An aquifer has an f_{oc} of 0.01. Estimate the K_d value characterizing the sorption of 1,2-dichloroethane having a log $K_{ow} = 1.48$.

Solution

Starting with the basic equation $K_d = f_{oc}K_{oc}$ and taking logs of both sides gives

$$\log K_d = \log f_{oc} + \log K_{oc}$$

Substitution for log K_{oc} with the Schwarzenbach and Westall (1981) equation yields

$$\log K_d = \log 0.01 + 0.49 + 0.72(1.48) = -0.445$$

The K_d value is $10^{-0.445}$ or 0.36 cm^3/g.

This approach for estimating values of K_d values is useful for rough scoping analyses when other information is lacking. In more comprehensive field studies, K_d values estimated from these equations need to be improved with field or laboratory experiments.

17.4.3 Ion Exchange

Mineral surfaces often bind cations and anions. Because these ions are held "loosely," there is an opportunity for them to exchange with ions in solution. This kind of exchange can result in a dramatic change in the chemistry of the water. A case in point is the natural water-softening reactions that exchange Ca and Mg ions in the water with Na ions sorbed onto clay minerals. Relatively high concentrations of Ca + Mg make the groundwater hard. Hard water tastes better than soft water but does not clean as well because soaps have difficulty in lathering.

Here are the key reactions describing these processes

$$Ca^{2+} + \text{Na-clay} = 2Na^{+} + \text{Ca-clay}$$
$$Mg^{2+} + \text{Na-clay} = 2Na^{+} + \text{Mg-clay}.$$

In contamination problems, ion exchange processes can reduce the mobility of hazardous ions (e.g., radionuclides or trace metals).

Surfaces bind cations and anions in several different ways. For example, ion sorption occurs because of imperfections or elemental substitutions in the crystal lattice of minerals. The result is a surface with a net negative or positive charge, which is balanced by ions (called counter ions) attracted on to the surface. The most common situation is where negative charges in the mineral structure are balanced by cations (Figure 17.6). These cations are exchangeable as are those more mobile cations further from the surface within the diffuse or Gouy layer, where there is an excess of cations as compared to anions (Figure 17.6).

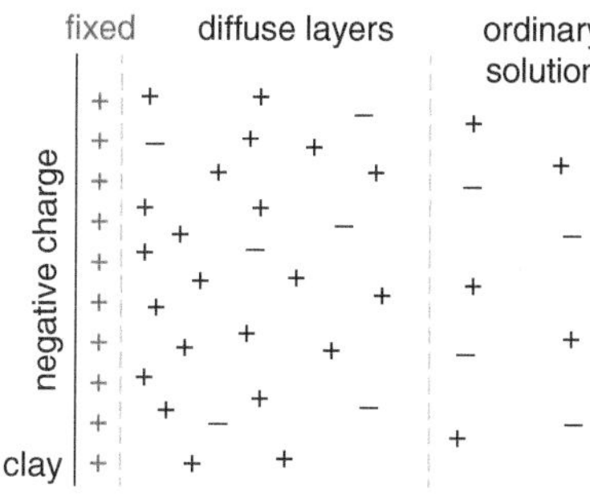

Figure 17.6 Schematic of the surface of a typical clay mineral where the structure negative of clay is balanced by cations in a fixed layer and diffuse layer (Schwartz and Zhang, 2003/John Wiley & Sons).

Clay minerals are the most important group of minerals carrying a significant fixed negative surface charge. Cations bound to the clays are available to exchange with cations in the groundwater. The negative charge in the clays is due to substitutions of cations of a lower valence (e.g., Al^{3+} for Si^{4+} in the lattice), and to a lesser extent because of broken bonds at the edges of the mineral.

The *cation exchange capacity (CEC)* of an exchanger describes how good the exchanger really is. CEC is defined as the quantity of cations bound to the mineral surface, which are available for exchange with ions in solution. Because electrical charges are involved, equivalent units of concentration are the most convenient. CEC is expressed as the number of milliequivalents that can be exchanged at pH 7 in a sample with a dry mass of 100 g. Thus, the greater the CEC the more exchange that is possible.

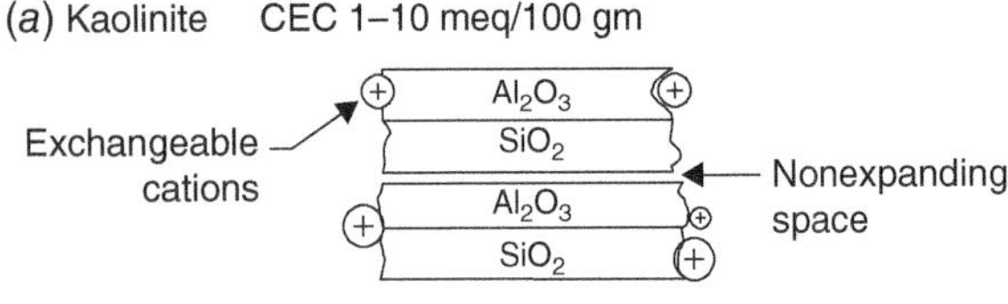

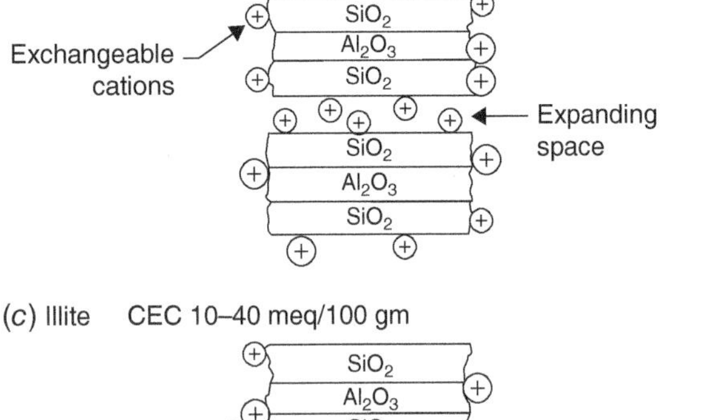

(c) Illite CEC 10–40 meq/100 gm

Exchangeable cations

SiO_2 / Al_2O_3 / SiO_2 / SiO_2 / Al_2O_3 / SiO_2

Nonexchangeable potassium in nonexpansible space

Figure 17.7 Structure of illustrative clay minerals (*a*) kaolinite, (*b*) montmorillonite, and (*c*) illite. Typical values for cation exchange capacity are shown, with the abundance of exchangeable cations shown reflecting the magnitude of CEC (U.S. Department of Agriculture NRCS 2012).

In most cases, the clay minerals have fixed charges or mostly fixed charges. In other words, the surface charge does not change as the chemistry of the pore fluid changes. Figure 17.7 illustrates how the structure of the clay mineral structure influences the CEC.

The exchange sites for kaolinite are due to broken bonds along the edges (Figure 17.7*a*). Because those edges are hydroxylated, the CEC can change as a function of pH. The montmorillonite structure (Figure 17.7*b*), as drawn, shows a larger number of exchangeable cations that mainly reflect the availability of sites due to isomorphous substitution and broken bonds. The result is a much higher range in CEC from 80 to 100. Because isomorphous substitution represents a chemical property of the mineral, the available charge does not change much as a function of groundwater pH. Illite (Figure 17.7*c*) is like montmorillonite sorbs due to fixed charges associated with isomorphous substitution. Vermiculite (not shown) exhibits the largest sorption capacity with CEC values mostly in a range from 100 to 200 meq/100 gm.

Clay minerals exhibit a preference as to which ion occupies an exchange site. For example, clay minerals would rather have Ca and Mg on the exchange sites than Na. Attempts have been made over the years to establish a selectivity sequence, where equivalent amounts of cations are arranged according to their relative affinity for an exchange site. Many variations of this sequence have been published, such as the following one provided by Yong (1985).

$$Li^+ < Na^+ < H^+ < K^+ < NH_4^+ < Mg^{2+} < Ca^{2+} < Al^{3+}$$

In general, cations with larger charges have a greater affinity for an exchange site. The idea of an affinity sequence is a useful concept for addressing exchange preference but is too simplified to be applied rigorously.

17.4.4 Clay Minerals in Geologic Materials

Clay minerals are abundant in fine-grained sediments (e.g., glacial till and lake clays) and sedimentary rocks of all kinds (e.g., shale and mudstone). Often, coarser-grained sediments (e.g., sand and gravel) and sedimentary rocks (sandstone) contain small but important quantities of clay minerals.

An indirect way of quantifying the abundance of clay minerals in the sample is with a grain size (granulometric) analysis. This standard laboratory analysis provides a measure of the weight percent of a sample in sand, silt, and clay-sized fractions.

The sizes are determined according to the diameter of the grains making up the sediment. Following is the nomenclature for a commonly used classification scheme:

$$\text{gravel} \rightarrow 2\,\text{mm} \leftarrow \text{sand} \rightarrow 0.062\,\text{mm} \leftarrow \text{silt} \rightarrow 0.004\,\text{mm} \rightarrow \text{clay}.$$

This wording choice is unfortunate because the "clay" size fraction can include tiny fragments of other minerals (e.g., quartz). Commonly, however, the clay-size fraction is made up mostly of clay minerals. The silt and sand-size fractions commonly include the common-rock forming minerals other than clay minerals.

A variety of factors come into play in determining exactly what clay minerals will be present in a particular fine-grained deposit. Modern clay-rich sediments being transported to oceans consist of weathering products like montmorillonite and kaolinite. With time and burial, there is a progressive conversion of these minerals to illite and chlorite. Thus, old sedimentary rocks (e.g., Paleozoic) contain mostly illite. Glacial sediments on the continents have a diverse clay mineralogy that depends in large measure on the rocks that were eroded to form glacial till, from which other glacial deposits (e.g., outwash sands and glaciolacustrine deposits) developed.

17.4.5 Sorption to Oxide and Oxyhydroxide Surfaces

Another important type of sorption reaction involves solids whose sorption properties change as a function of pH. This property is important because at low pHs the surfaces will sorb anions. At high pHs, cations will sorb. Thus, these types of reactions are relevant to problems of contamination involving metals, and other negatively charged species of interest like NO_3^-, and PO_4^{3-}. Examples of such solids include kaolinite, metal oxides (e.g., SiO_2 and Al_2O_3) and metal oxyhydroxides (e.g., $Fe(OH)_3$, $Si(OH)_4$). The metal oxides and oxyhydroxides typically form a thin coating on sand grains.

This model of sorption assumes the surface is populated by a hydrolyzed species XOH^0 that is available to react with dissolved constituents. The "X" identifies this as constituent on the surface that has a concentration and reacts with ions in solution. However, in water, the "type" surface species reacts to produce associated species XO^-, and XOH_2^+ (Morel and Hering, 1993). The relative abundances of these three surface species change as a function of solution pH, as described by the following reactions

$$XOH = H^+ + XO^-$$

$$XOH + H^+ = XOH_2^+$$

The three surface constituents, XOH_2^+, XOH, and XO^- are always present with a relative abundance determined by pH. At a low pH, XOH_2^+ dominates with other constituents present at many orders of magnitude smaller concentrations. At high pHs, XO^-, is dominant in terms of concentration with the others insignificant. The charged species XO^- and XOH_2^+ invite equilibrium reactions between cation and anions in solution, respectively. Suppose lead (Pb) was present in solutions. As a cation, it would react onto the surface at relatively high pH according to the following reaction

$$XOPb^+ = XO^- + Pb^{2+}$$

If phosphate was present as the constituent, four surface species could be involved making any calculations much more complicated.

As mentioned, the surface has an abundance of XOH_2^+ at low pH's, which favors the sorption of anions. However, at higher pH's, the surface species changes to XO^-, favoring the sorption of cations. There is a pH in between, known as the *zero point of charge* (pH_{zpc}), where the surface has zero net charge. The term *isoelectric point* defines the pH_{zpc} when the binding and dissociation of protons (H^+) are the only reactions affecting surface charge. These surface-charge relationships are unique for given solids and solutions and usually must be determined experimentally. Listed in Table 17.3 are estimates of the isoelectric points for various solids.

Stollenwerk (1991) studied the sorption of molybdate (MoO_4^{2-}) on ferrihydrite, which coats quartz grains of the outwash aquifer on Cape Cod, USA. The $XFeOH^0$ sites behave in the manner just described causing the sorptive properties to change as a function of pH. At pH's below about 7, the ferrihydrite coatings are capable of sorbing molybdate (Figure 17.8). In this pH range, below the pH_{zpc} for ferrihydrite, the surface has a net positive charge. At very lower pH's, almost 100% of the total molybdate ends up sorbed on the surface. Once the pH climbs above the pH_{zpc}, the surface has a net positive charge and loses its capability of sorbing molybdate.

Table 17.3 Values for the isoelectric points for various solids.

Solid	pH
Quartz α-SiO_2	2–3.5
Albite $NaAlSi_3O_8$	2.0
Kaolinite $Al(Si_4O_{10})(OH)_8$	<2–4.6
Montmorillonite	≤2.5
Hematite Fe_2O_3	5–9
Magnetite Fe_3O_4	6.5
Goethite FeOOH	6–7
Corundum Al_2O_3	9.1
Gibbsite $Al(OH)_3$	~9

Source: Domenico and Schwartz (1998), Physical and Chemical Hydrogeology.

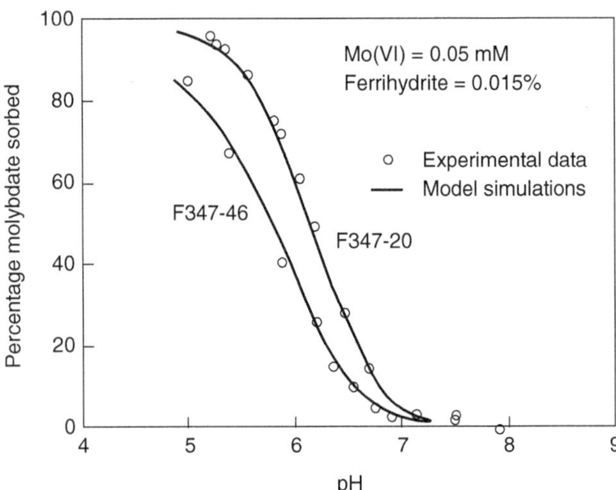

Figure 17.8 Variation in molybdate sorption as a function of pH. The results of model simulation are fitted to data from sorption experiments. The lines differ because of elevated concentrations of SO_4^{2-} and PO_4^{3-} that compete with molybdate for sorption sites (U.S. Geological Survey, Stollenwerk, 1991).

17.5 Oxidation–Reduction Reactions

Oxidation–reduction or redox reactions are enormously important in controlling the geochemistry of groundwaters. They are unlike other reactions so far because they involve the transfer of electrons and are mediated by microorganisms. The microorganisms act as catalysts speeding up what otherwise are extremely sluggish reactions.

Redox reactions involve elements (e.g., C, O, N, Fe, and Mn), which can gain or lose electrons. These transfers of electrons imply that these elements have different oxidation states. The oxidation number refers to charges that are assigned to an element. For example, in the case of CO_3^{2-}, the oxidation number for carbon is (+IV) and for oxygen (−II) giving the molecule a net negative charge of −2. By convention, oxidation numbers are written as roman numerals.

Oxidation is the removal of electrons from an atom forcing a change in the oxidation number of an element. For example, the oxidation of Fe^{2+} can be written as

$$Fe^{2+} = Fe^{3+} + e^- \tag{17.29}$$

where e^- is an electron. The oxidation number for iron changes from (+II) to (+III).

Reduction refers to the addition of an electron to lower the oxidation number

$$Fe^{3+} + e^- = Fe^{2+} \tag{17.30}$$

Table 17.4 Partial list of elements important in natural waters and contaminants having more than one oxidation state, along with examples of associated ions, gases, or solids.

Element		Typical redox states	Examples of ions, gases, and solids		
Natural waters					
Mn	Manganese	II, III, IV	Mn^{2+} II	$MnO_2(s)$ IV	
Fe	Iron	II, III	Fe^{2+} II	Fe^{3+} III	$Fe(OH)_3(s)$ III
S	Sulfur	−II to VI	$H_2S(g)$ −II	HS^- −II	SO_4^{2-} VI
C	Carbon	−IV to IV	$CH_4(g)$ −IV	HCO_3^- IV	CO_3^{2-} IV
Contaminants					
N	Nitrogen	−III, 0, III, IV	NH_4^+ −III	$NH_3(g)$ −III	$N_2(g)$ 0 NO_3^- V
As	Arsenic	III, V	$H_3AsO_3^0$ III	$H_2ASO_4^-$ V	$HASO_4^{2-}$ V
Cr	Chromium	III, VI	Cr^{3+} III	CrO_4^{2-} VI	
Se	Selenium	−II, 0, IV, VI	SeO_3^{2-} IV	SeO_4^{2-} VI	
U	Uranium	IV, VI	$UO_2(s)$ IV	UO_2^{2+} VI	$UO_2(CO_3)$ VI
V	Vanadium	III, IV, V	$V(OH)_3^+$ IV	$H_2VO_4^-$ V	HVO_4^{2-} V

Source: FWS.

All redox reactions transfer electrons and involve elements with more than one oxidation number. Listed in Table 17.4 are some important elements, their typical oxidation states, and a few of the ions and solids that form. A host of trace metals not included in the table also have variable oxidation numbers.

Redox reactions involve a transfer of electrons from a *reductant* (electron donor) to an *oxidant* (electron acceptor). In any reaction, there are both oxidation and reduction reactions occurring (i.e., a pair of redox reactions). Thus, no free electrons result from a redox reaction. Redox reactions have this general form

$$Ox_1 + Red_2 = Red_1 + Ox_2 \tag{17.31}$$

where Ox and Red refer to oxidants and reductants respectively. The more specific example makes this point about electron transfers more obvious. Electrons generated by the oxidation of Fe^{2+} are accepted by the O, which is reduced in the reaction

$$\begin{array}{llll} O_2 + 4Fe^{2+} & + 4H^+ = & 2H_2O + 4Fe^{3+} & \\ O(0) \quad Fe(+II) & & O(-II) & Fe(+III) \\ Ox_1 \quad Red_2 & & Red_1 & Ox_2 \end{array} \tag{17.32}$$

The overall reaction can be represented as two separate *half-redox reactions* Ox_1-Red_1 and Red_2-Ox_2

$$\begin{aligned} &O_2 + 4H^+ + 4e^- = 2H_2O \\ &4Fe^{2+} = 4Fe^{3+} + 4e^- \end{aligned} \tag{17.33}$$

Adding these two reactions together gives the original reaction Eq. (17.34). Half reactions are unique in that electrons are reactants or products. Otherwise, these reactions are like other equilibrium reactions.

Mass law equations can be written in terms of the concentration or activities of reactants and products (including the electrons), and an appropriate equilibrium constant. For example, the mass law expression for the half-reaction

$$Ox + ne^- = Red$$

is

$$K = \frac{[Red]}{[Ox][e^-]^n} \tag{17.34}$$

Rearrangement of Eq. (17.34) gives the electron activity $[e^-]$ for a half-reaction

$$[e^-] = \left\{\frac{[\mathrm{Red}]}{[\mathrm{Ox}]K}\right\}^{1/n} \tag{17.35}$$

In the same way that pH defines $[H^+]$, pe define electron activity, $[e^-]$. Rewriting Eq. (18.35) by taking the logarithm of both sides gives

$$\mathrm{pe} = -\log[e^-] = \frac{1}{n}\left\{\log K - \log\frac{[\mathrm{Red}]}{[\mathrm{Ox}]}\right\}$$

When a half reaction is written in terms of a single electron or $n = 1$, the log K term is written as pe^0 so that

$$\mathrm{pe} = \mathrm{pe}^0 - \log\frac{[\mathrm{Red}]}{[\mathrm{Ox}]} \tag{17.36}$$

Morel and Hering (1993, p. 430) and Pytkowicz (1983, p. 249) provide an extensive tabulation of half-reactions and values of pe^0 or log K. Examples of a few half reactions important for groundwater systems are presented in an upcoming section.

The redox state of soils and water can be expressed in terms of E_H, the redox potential. Values of E_H have units of volts, acknowledging that a redox reaction involves the transfer of electrons. E_H is related to pe by the following equation

$$E_H = \frac{2.3RT}{F}\mathrm{pe} \tag{17.37}$$

where F is the Faraday constant defined as the electrical charge of one mole of electrons (96,500 Coulombs) with $2.3RT/F$ equal to 0.059 V at 25°C. In practice, pe is taken to be the calculated value of electron activity, and Eh is the measured electrode potential for an electrochemical cell. In other words, pe is a calculated quantity and Eh is a measured one. The following example illustrates how pe of a groundwater is calculated assuming redox equilibrium.

Example 17.7 A groundwater has $(Fe^{2+}) = 10^{-3.3}$ M and $(Fe^{3+}) = 10^{-5.9}$ M. Calculate the pe at 25°C assuming that the activities of Fe-species are equal to their concentrations. What should be the measured E_H of this solution? The half-reaction for the reduction of Fe^{3+} to Fe^{2+} is

$$Fe^{3+} + e^- = Fe^{2+} \text{ with } \mathrm{pe}^0 = 13.0$$

Solution

Take the given information and substitute into Eq. (17.36) as follows

$$\mathrm{pe} = \mathrm{pe}_0 - \log\{(Fe^{2+})/(Fe^{3+})\}$$

$$\mathrm{pe} = 13.0 - \log\{10^{-3.3}/10^{-5.9}\}$$

$$\mathrm{pe} = 13.0 - 2.6 = 10.4$$

$$E_H = \frac{2.3RT}{F}\mathrm{pe} = 0.059 \times 10.4 = 0.61\ \mathrm{V}$$

17.5.1 Kinetics and Dominant Couples

So far, we have presented redox phenomena as equilibrium processes, implying that the half-redox reactions are in equilibrium. However, redox reactions are not usually at equilibrium because the number of active microorganisms could be small, or microorganisms do not metabolize some of the reactants easily. Also, redox reactions with extremely large equilibrium constants are essentially irreversible. Take for example a reaction where an excess of dissolved oxygen oxidizes organic matter (as CH_2O)

$$CH_2O + O_2 \rightarrow CO_2 + H_2O$$

The reaction progresses to the right until all the organic matter disappears. This example raises an important question—what good are equilibrium concepts involving pe or E_H if the reactions are not at equilibrium. Fortunately, a constant value of pe or Eh exists when the concentration of one of the couples, O_2/H_2O in this case, is much greater than the others. Because utilization of all the organic matter will require a negligibly small proportion of the oxygen, the redox potential defined by the O_2/H_2O couple does not change as a function of the reaction progress.

At one time, it was thought that the less dominant couples would be controlled by the dominant couples. For example, if the O_2/H_2O couple is dominant, the other couples, like Fe^{2+}/Fe^{3+} would adjust their concentrations according to the pe defined by the dominant couple. However, research has shown that redox equilibrium among all the couples does not occur (Lindberg and Runnells, 1984). Thus, while the concept of some master pe is alluring, each half-reaction has a different pe (Lindberg and Runnells, 1984). The main use for pe calculations is to provide indications of the direction in which a system is evolving (Stumm and Morgan, 1981).

17.5.2 Biotransformation of Organic Compounds

Many half reactions involve the oxidation of organic compounds into simpler inorganic forms such as CO_2 and H_2O and a variety of electron acceptors. These reactions are referred to as *biodegradation or biotransformation reactions* because they are microbially catalyzed. In the subsurface bacteria use these reactions as energy sources. The free energy released by the oxidation of organic matter is different depending upon the electron acceptor. The energetically favorable reactions occur first and so it is common for reactions to follow the preference in electron acceptors $O_2 > NO_3^- > Mn^{4+} > Fe^{3+} > SO_4^{2-} > CO_2$, called the redox ladder. Following here is the reaction describing the oxidation of a generic organic constituent.

$$(1/4)CH_2O + (1/4)O_2(g) = (1/4)CO_2(g) + (1/4)H_2O$$

However, when O_2 is unavailable other species, for example NO_3^-, Mn(+IV), Fe(+III), SO_4^{2-}, and CO_2 accept electrons. The key redox half reactions (Jurgens et al., 2009) include:

Oxygen reduction:	$O_2(g) + 4H^+ + 4e^- = 2H_2O$
Denitrification:	$2NO_3^- + 12\,H^+ + 10e^- = N_2(g) + 6H_2O$
Nitrate reduction:	$NO_3^- + 10\,H^+ + 8e^- = NH_4^+ + 3H_2O$
Mn(IV) reduction:	$MnO_2(s) + 4H^+ + 2e^- = Mn^{2+} + 2H_2O$
Fe(III) reduction:	$Fe(OH)_3(s) + H^+ + e^- = Fe^{2+} + H_2O$
Sulfate reduction:	$SO_4^{2-} + 9H^+ + 8e^- = HS^- + 4H_2O$
CO_2 reduction:	$CO_2(g) + 8H^+ + 8e^- = CH_4(g) + 2H_2O$

Of the electron acceptors listed, all are common constituents at different locations in flow systems. For example, concentrations of O_2, NO_3^-, and CO_2 are elevated in shallow settings in recharge areas, sulfate may accumulate along flow system or units with gypsum or anhydrite. Iron and Mn are important constituents in rocks or grain coatings. As groundwater moves along flow systems, constituents, first O_2, and subsequently NO_3^- are depleted. Once NO_3^- is depleted bacteria utilize Mn and Fe present as solids in the aquifer. Next, comes SO_4^{2-} and CO_2.

For all these reactions, reduced organic carbon serves as the electron donor. Several sources of organic carbon exist including, dissolved carbon generated in the soil zone, particulate matter within aquifers, diffusion of dissolved organic compounds, and recycling of microbial biomass.

Biotransformation reactions are important when the organic compound is a groundwater contaminant. If biotransformation occurs, the reaction will reduce contaminant concentrations, and perhaps alter the major ion chemistry. Later discussions will examine possibilities of reductive dichlorination of common solvents, such as tetrachloroethene, as a mechanism for natural attenuation.

17.5.3 pe-pH and E_H-pH Diagrams

Redox reactions form the basis for understanding the behavior of important constituents in groundwater and trace metals. One example is sulfate/sulfide system (H_2S-SO_4^{2-}). Assume that the total sulfur in a system (S_T) exists as three possible

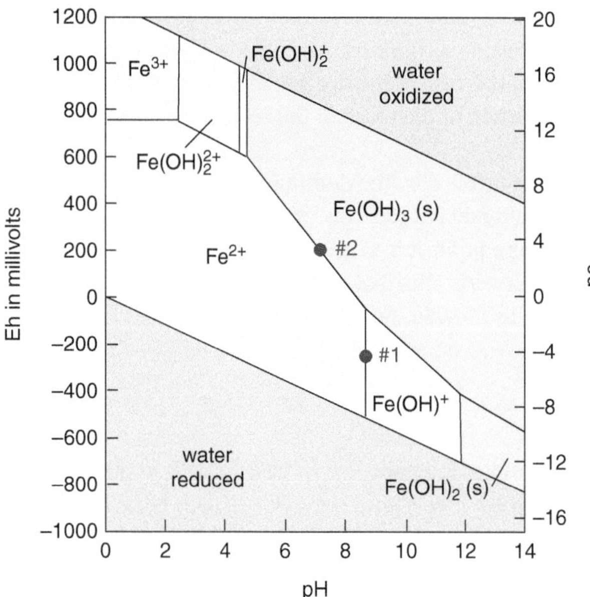

Figure 17.9 Example of an E_H-pH diagram showing the stability relationships of iron hydroxide. In areas of yellow shading the majority of Fe will be dissolved in solution. In areas of blue the majority of iron will occur in a solid. Sample point #1 plots at position where the activities of $Fe^{2+}$and $Fe(OH)^+$ are equal. At sample point #2, Fe^{2+} is in equilibrium with respect to the solid mineral $Fe(OH)_3$ (U.S. Geological Survey adapted from Back and Barnes, 1965 and Hem and Cropper, 1959).

species, SO_4^{2-}, H_2S, and HS^-, where in the first sulfur occurs as S(+VI) and in the last two as S(−II). The relationship among H_2S, HS^-, and S^{2-} is fixed by half-redox reactions. In oxic environments, SO_4^{2-} is the dominant species of the couple with essentially none of the S(−II) species present. In an anoxic environment, the other three species dominate.

Adding a metal ion to the mix, such as Fe^{2+}, increases the complexity because the iron partitions among several species, especially pyrite FeS_2(s). Once the system becomes reducing, almost all the iron and sulfur ends up as pyrite.

Historically, *pe-pH or E_H-pH diagrams* provide a useful way of presenting the essential features of the redox chemistry. Textbooks (e.g., Drever, 1988) discussed pe-pH diagrams and their preparation in detail. Shown in Figure 17.8 is a pe-pH diagram for the ferric-ferrous system (Back and Barnes, 1965; Hem and Cropper, 1959). The upper and lower lines on the field represent the oxidation of water to O_2 and the reduction of water to H_2. These two half-reactions define the upper and lower limits for oxidation or reduction. The other lines are boundaries for what are known as *stability fields*. The boundary lines are calculated based on the key redox half-reactions that were developed earlier in this section. Each stability field is labeled with the one dominant ion or solid for the specified E_H and pH.

A pe-pH diagram is a theoretical construct, with boundaries that change depending upon the activity of Fe^{2+}. Thus, Figure 17.9 has units of pe on the right side reflecting its theoretical origins. However, on the left side axis, E_H is plotted with units of millivolts. It is there because in field settings redox conditions can be approximated by measurements of oxidation–reduction potential on a water sample using a platinum electrode (e.g., Back and Barnes, 1965). Thus, the addition of the E_H scaled permits the plotting of field measurements. An E_H-pH diagram is simply a plot containing field data.

When a data point ends up plotted in the field for an ion (e.g., Fe^{2+}), the element is mobile under the conditions of Eh and pH because a small proportion of the metal occurs as a solid or other species. In the stability fields for solid phases (e.g., $Fe(OH)_3$), nearly all the iron will exist as that solid and is essentially immobile. A data point falling on the boundary between two aqueous species indicates that both species have the same activity (e.g., #1, Figure 17.8). If instead, the boundary line separates the field for an ion and a solid, a point of the boundary implies the ion is in equilibrium with respect to the solid (e.g., #2, Figure 17.9). However, as will be evident in the next section, this approach is problematic and an indicator of conditions in the field.

17.5.4 Quantifying Redox Conditions in Field Settings

Unlike pH measurements that are relatively straightforward, measuring E_H conditions in groundwater is complicated by theoretical and practical problems. The traditional and commonly used approach involves field measurements of oxidation–reduction potential (E_H) with platinum electrodes (Chapelle et al., 1996). Yet, this procedure comes along with several problems noted by Chapelle et al. (1996) that include:

- "unique redox potentials in natural waters do not exist,"
- "electrode-measured E_H values do not agree with E_H values calculated from measured concentrations of redox couples,"
- "groundwater is seldom, if ever, in a state of full redox equilibrium," and
- "platinum electrodes are subject to a variety of interferences."

A common concern has been that electrode-based measurements of E_H in groundwater samples do capture conditions in highly reduced settings that could exist with plumes of organic contaminants. There have been other approaches to characterize redox settings. One approach has been to interpret concentrations of hydrogen (H_2) dissolved in water in redox context (Lovley et al., 1994; Chapelle et al., 1995). In groundwater, anaerobic decomposition of organic matter produces hydrogen as a product. This hydrogen is subsequently utilized in reduction reactions involving nitrate, manganese-iron, sulfate, and carbon dioxide. However, the efficiency of the utilization decreases for strongly reducing systems (Lovley and Goodwin, 1988) such that H_2 is detectable at 1–4 nM with sulfate-reducing bacteria and >5 nM with carbon dioxide reducers (Dinicola, 2006). This approach provides identifies domains that are strongly reducing rather than specific electron acceptors (Dinicola, 2006).

A study by Chapelle et al. (1996) compared differences between E_H measurements and H_2 gas analyses in interpreting redox processes in a contaminated, unconfined aquifer. The plume included dissolved organic compounds, such as benzene, toluene, ethylbenzene, and xylenes (BTEX) known to be biodegradable in aerobic and anaerobic settings. A line of multi-level piezometers was installed across the dissolved plume (Figure 17.10).

E_H measurements provided the zonation depicted in Figure 17.10*a* with the lowest E_H (i.e., slightly less than −150 mV) evident. Other calculations indicated these measured redox potentials were indicative of redox equilibrium with the $Fe(OH)_3$—Fe^{2+} couple. The redox zones derived from H_2 zonation are substantially different. The system is evidently more strongly reducing than Eh measurements would suggest. The hydrogen-based interpretation is also supported by data showing expected patterns in the concentrations of electron acceptors, such as oxygen nitrate, sulfate, and the presence of reaction products, such as Fe(II), hydrogen sulfide, and methane. The use of H_2 gas in conjunction with other constituents provided a much more convincing and nuanced interpretation of the redox zonation at this site.

Comprehensive investigations of the geochemistry of groundwater have formed the basis for a practical approach for interpreting redox conditions (Chapelle et al., 1995; Paschke et al., 2007). The method expanded on the idea of using quantitative information on the concentrations of various reactants and products in the family redox half-reactions involved with biodegradation. A modification (McMahon and Chapelle, 2008) to the earlier interpretive approach used concentration data from O_2, NO_3^- Mn^{2+}, Fe^{2+}, and SO_4^{2-}. A subsequent modification (Chapelle et al., 2009) incorporated the Fe^{2+}/sulfide ratio, where sulfide is the sum of H_2S, HS^-, and S^{2-} to overcome a limitation in the ability to distinguish between Fe(III) and SO_4^{2-} reduction (McMahon et al., 2009). The classification system is presented in Table 17.5.

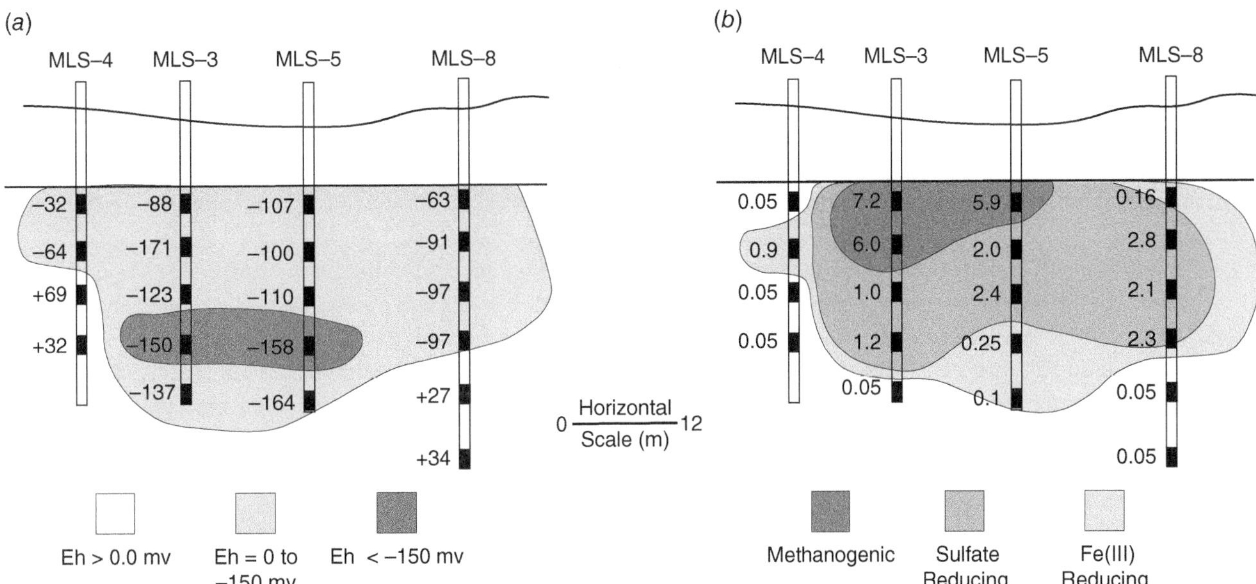

Figure 17.10 Comparison of redox zonation along a cross-section using measured E_H values (*a*) with those determined based on measure values of hydrogen (*b*), which are calibrated as [>5 nM, methanogenic; 1–4 nM sulfate reducing; 0.2–0.8, Fe(III) reducing] (Adapted with permission from Chapelle, F.H., Haack, S.K., Adriaens, P. et al., 1996. Comparison of E_h and H_2 measurements for delineating redox processes in a contaminated aquifer. *Environmental Science & Technology*, *30*(12), 3565–3569. Copyright 1996 American Chemical Society).

Table 17.5 Classification approach for identifying a predominant redox process using water chemistry thresholds.

General redox category	Predominant redox process	Distinguishing Fe(III)–from SO_4^{2-}-reduction	Water-chemistry criteria (mg/L)					Fe^{2+}/H_2S mass ratio
			O_2	NO_3^-	Mn^{2+}	Fe^{2+}	SO_4^{2-}	
Oxic	O_2 reduction	—	≥0.5	—	<0.05	<0.1	—	—
Suboxic[a]	—	—	<0.5	<0.5	<0.05	<0.1	—	—
Anoxic	NO_3^- reduction	—	<0.5	≥0.5	<0.05	<0.1	—	—
	Mn(IV) reduction	—	<0.5	<0.5	≥0.05	<0.1	—	—
	Fe(III)/SO_4^{2-} reduction	—	<0.5	<0.5	—	≥0.1	≥0.5	—
	—	Fe(III) reduction	<0.5	<0.5	—	≥0.1	≥0.5	>10
	—	Mix—Fe(III)/SO_4^{2-} reduction	<0.5	<0.5	—	≥0.1	≥0.5	≥0.3 and ≤10
	—	SO_4^{2-} reduction	<0.5	<0.5	—	≥0.1	≥0.5	<0.3
	Methanogenesis	—	<0.5	<0.5	—	≥0.1	<0.5	—
Mixed[b]	—	—	—	—	—	—	—	—

[a] Further definition of redox processes not feasible.
[b] Criteria for more than one redox process are met.
Source: U.S. Geological Survey (McMahon et al. 2009) as modified from McMahon and Chapelle (2008) *Groundwater* 46: 259–271; and Chapelle et al. (2009) *Groundwater* 47: 300–305. No copyright claims to original US government works.

17.5.5 Redox Zonation

It has been known for a long time that aquifers can exhibit redox patterning. Moving along a flow system beginning in a recharge area, the zones are defined by redox reactions on the redox ladder. For example, studies, such as Champ et al. (1979) and Jackson and Patterson (1982), have shown transitions from redox zones with oxygen, iron–manganese, and sulfate reduction. The probable half reactions controlling pe in these zones are the reduction of oxygen to water, the reduction of iron or manganese oxides/hydroxides, and sulfate reduction to HS^- or H_2S. In a few cases, a methane zone can form from the reduction of CO_2.

With an unconfined, aquifer the redox gradient is often vertical reflecting downward flow conditions (McMahon et al., 2011). In a laterally, extensive confined aquifer zonation tends to occur laterally. Figure 17.1, presented earlier in the chapter showed a generalized pattern of vertical redox zonation. The uppermost zone was an oxic zone where oxygen reduction was occurring. The zone labeled as anoxic on the figure was comprised of a thin nitrogen-reducing zone and a thicker iron-reducing zone.

Reduced sulfide and methane zones tend to develop in confined flow systems containing excess oxidizable DOC and a lack of recharge-containing oxygen downgradient in the system. Thus, redox zones will be observed in an extensive artesian aquifer that receives recharge from a limited area of outcrop or in an aquifer with confining units.

In the United States, comprehensive geochemical screening together with classification-based interpretations of redox zonation have advanced understanding of redox processes. There has been evident progress in defining spatially variable redox zonation, but also in discovering factors influencing that zonation, and finally, an ability to better understand problems of geogenic contamination. A case in point is the assessment of redox conditions in the Western San Joaquin study area (Fram, 2017). Figure 17.11 displays results from the collection of wells from the Central area as a function of depth with a west-to-east orientation (Fram, 2017). The wells include the grid wells and several additional wells. Figure 17.11 and associated text explain this terminology. Notice first that the largest number of wells are suboxic or anoxic but only moderately reducing, having progressed to manganese and iron reducing (Figure 17.11). Second, waters to the east are more reduced in general than those to the west. This pattern in redox status has been explained in terms of aquifer lithology (Fram, 2017). Aquifer sediments on the right half of the figure were derived from the Sierra Nevada Mountains to the east. The western portion of the aquifer received sediments from the Coast Range (Fram, 2017).

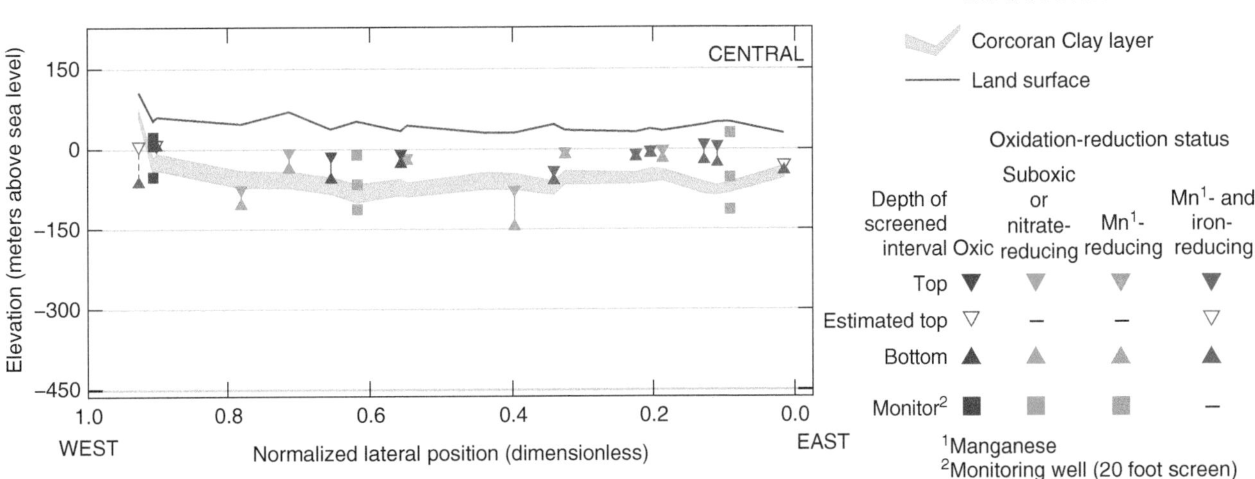

Figure 17.11 The oxidation–reduction status of groundwaters within the Central region of the Western San Joaquin Valley, which was sampled by the USGS. Waters to the east are somewhat more reduced as compared to those to the west because of differences in the sources of aquifer sediments (U.S. Geological Survey, adapted from Fram, 2017).

17.6 Microorganisms in Groundwater

Microorganisms occur ubiquitously in the subsurface. Although bacteria predominate, protozoa and fungi are also common. *Bacteria* are small single-celled organisms. To live, they metabolize dissolved organic matter present naturally or due to contamination. Protozoa are somewhat larger, single-celled organisms that can feed on bacteria. Fungi live by degrading other organic matter and effectively recycle plant and animal debris (Chapelle, 1993).

Bacteria occur in the subsurface in different ways. Some are adapted to move freely in the water with flagella that propel them. The direction of motion can be either random or directed in response to concentrations of organic compounds (*chemotactic*). This property of motility can be important for the local redistribution of bacteria, and possibly influence larger-scale transport. Most commonly, bacteria are *immotile*, bound to the surfaces of the aquifer solids. The attached population forms what is called a *biofilm*, which consists of bacteria held together and to the particles by extracellular polymers (McCarty et al., 1984).

In pristine groundwater systems, the bulk of the bacterial population is attached to the particle surface. Theoretical work suggests that when the concentration of metabolizable organic compounds (i.e., the substrate for growth) is low, immotile bacteria have a competitive advantage over motile forms (Kelly et al., 1988). The proportion of motile bacteria may be larger in contaminated systems containing metabolizable organic compounds (Ghiorse and Wilson, 1988). Thus, with increasing substrate concentrations, motile bacteria may become more competitive.

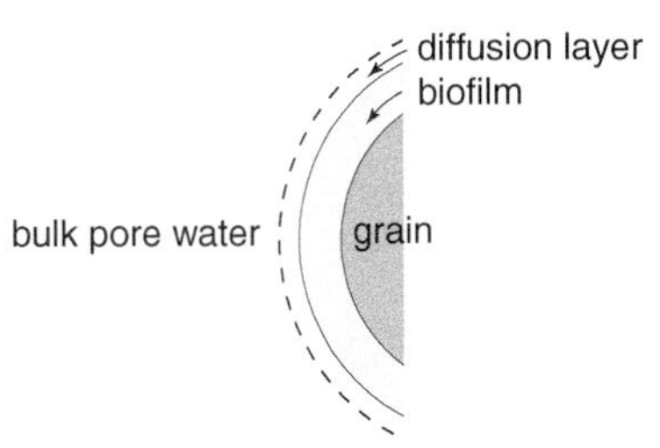

Figure 17.12 Conceptualization of a biofilm at a pore scale. The diffusion layer serves as a transition to the bulk pore water (Adapted from Domenico and Schwartz, 1998).

Biofilms are the site for transformation reactions that are important in reducing the concentrations of some organic contaminants. They are commonly conceptualized in terms of a simple layer with molecular diffusion operating to bring metabolizable organic compounds from the bulk pore fluid to clusters of bacteria cells in the biofilm (Figure 17.12).

Real biofilms are more complex. For example, in cases where organic substrate concentrations are relatively low, patchy or discontinuous biofilms form (Rittmann, 1993). Furthermore, detailed microscopic studies of model biofilms demonstrated fluid flow within biofilm channels (Stoodley et al., 1994). These channels provide a delivery system for nutrients into the biofilm that is more efficient than diffusion.

17.6.1 Quantifying Microbial Abundances

A variety of techniques have been developed to enumerate the abundance of microorganisms in aquatic systems. Depending upon the purpose of the study, one might collect either water samples or samples of soil material. Water samples can be collected from conventional monitoring wells, using sterile techniques (e.g., Harvey et al., 1984; McNabb and Mallard, 1984). However, most often soil cores are collected because most of the microbial population is attached on grain surfaces. Again, care must be exercised in keeping the sampling equipment and core tubes sterile. Readers interested in the details of these sampling approaches can refer to Armstrong et al. (1988) and McNabb and Mallard (1984).

Once the sample of water or solid is collected, techniques are available in the laboratory to characterize the size of microbial population. In the case of a soil sample, however, an additional step is required to remove bacteria from the solid materials. The most common procedure involves shaking sediment-distilled water slurry. Actual microbial numbers are determined via a plate count or direct counting procedures.

A *plate count* involves smearing a sample (water or sediment/water slurry) containing the bacteria on a growth medium, incubating the plate for some time, and counting the bacterial colonies. The number of bacteria can be expressed, for example, as colony forming units per milliliter (CFU/mL) in the case of water samples, or CFU/ [g dry weight of sediment]. A slightly different plating procedure, the most-probable-number (MPN) approach, determines how much a sediment slurry or water sample can be diluted before no growth takes place. The dilution information provides a basis for estimating the number of cells per volume (or mass) of the original sample. Direct counting procedures use a microscope to count stained bacteria. This procedure provides an estimate of the number of cells per gram of material.

17.6.2 Microbial Ecology of the Subsurface

The subsurface is home to a diverse set of microorganisms. In pristine groundwater systems, population densities range between 10^5 and 10^7 cells per gram dry weight (gdw) (Ghiorse and Wilson, 1988). Generally, these population densities are small relative to those found in the soil zone and the unsaturated zone. This difference in the size of microbial populations is caused by a decrease in the concentration of organic matter as recharging groundwater moves through the soil.

The distribution of microorganisms depends in a complex way on the residence time of water in the system, and the geochemistry of organic matter transported along the flow system. As Ghiorse and Wilson (1988) suggest, the quantity of organic matter and nutrients passed along a pristine groundwater system is relatively meager. Thus, microbial populations appear adapted for growth and survival in nutrient-poor (*oligotrophic*) conditions (Balkwill et al., 1989). Theoretically, in a system closed to organic carbon, the concentration of organic matter should decline until it is no longer possible to support the microbial population (Ghiorse and Wilson, 1988). However, this endpoint has not been discovered in large flow systems.

Results from several studies show that even down to several hundred meters, microbial abundance does not diminish with depth. However, there can be considerable spatial variability both vertically and laterally. For example, in holes drilled as part of the Deep Probe Project of the U.S. Department of Energy, Levine and Ghiorse (1990) found a highly significant correlation between bacterial abundance and hydraulic conductivity. Bacteria were more abundant in sandy sediments and much less abundant in clayey sediments. Similar distributions are observed in shallow environments as well. This pattern may reflect the much greater difficulty in originally colonizing the fine-grained units because of straining filtration of migrating organisms (Zachara, 1990). In addition, the larger fluxes of water through the more permeable units would likely make significantly more dissolved organic matter available. Fredrickson et al. (1990) found significant diversity in the ability of bacteria to utilize various types of organic compounds even when nearby samples were compared. Their data indicate that a great diversity in bacteria at the population and organism levels even within the same geologic unit.

Exercises

17.1 Often (CO_3^{2-}) is not reported in water-quality analyses because the concentration is too low to measure directly. Develop an expression from the appropriate mass laws to calculate the concentration of (CO_3^{2-}) given pH and (HCO_3^-).

17.2 Following are the results of a batch sorption experiment in which two grams of porous media were mixed with 10 mL of a solution containing various concentrations of Cd.

Initial (Cd) concentration (mg/mL)	Equilibrium (Cd) concentration (mg/mL) × 10^{-4}
0.0005	0.040
0.001	0.093
0.010	1.77
0.050	39.3
0.096	205.2

A Find the equation for the Freundlich isotherm that best fits these data. Hint: Transform the basic equation and data in terms of the logarithm of concentration.

B Assuming that only the lowest concentrations are of interest, fit the same set of data with a linear Freundlich isotherm (i.e., $n = 1$) to obtain a distribution coefficient (K_d). Be sure to indicate the units for K_d.

17.3 At a site, the following contaminants are found (i) parathion (log K_{ow} 3.80), (ii) chlorobenzene (log K_{ow} 2.71), and (iii) DDT (log K_{ow} 6.19).

A Using the equation of Hassett et al. (1983) in Eq. (17.28) estimate the partition coefficient between solid organic carbon and water (K_{oc}).

B With an f_{oc} of 0.01, estimate the distribution coefficients (K_d) for these compounds.

C On the basis of sorption alone, which of these compounds is the most mobile and which is the least mobile?

17.4 Compare the range in variability of the distribution coefficients for carbon tetrachloride (log K_{ow} 2.83) estimated using the three different regression equations (17.26, 17.27, and 17.28) and an f_{oc} of 0.01.

17.5 The properties of six organic compounds are listed below.

Compound	log K_{ow}
#1	2.04
#2	0.89
#3	0.46
#4	−0.27
#5	2.83
#6	1.48

List the organics in order of decreasing mobility.

17.6 Consider the following half-redox reaction involving NO_3^- with a concentration of 10^{-5} M and ammonium with a concentration of 10^{-3} in groundwater with a pH of 8.

$$1/8NO_3^- + 5/4H^+ + e^- = 1/8\,NH_4^+ + 3/8\,H_2O$$

What is the pe and E_H of the system at 25°C assuming redox equilibrium and log K = 14.9.

17.7 The couple O_2/H_2O is the found to be the dominant couple producing a pe of 12.0. By assuming redox equilibrium, determine how $(Fe)_T = 10^{-5}$ M would be distributed as Fe^{2+} and Fe^{3+} given pe^0 of 13.

References

Armstrong, J. M., L. E. Leach, R. M. Powell et al. 1988. Bioremediation of a fuel spill: Evaluation of techniques for preliminary site characterization, In NWWA/API Conference, Petroleum Hydrocarbons and Organic Chemicals in Groundwater: Prevention, Detection and Restoration. National Water Well Association, Dublin, Ohio, p. 931–944.

Appelo, C. A. J., and D. Postma. 2004. Geochemistry, Groundwater and Pollution. CRC Press, 647 p.

Back, W., and I. Barnes. 1965. Relation of electrochemical potentials and iron content to groundwater flow patterns. U.S. Geological Survey Professional Paper 498-C. 16 p.

Balkwill, D. L., J. K. Fredrickson, and J. M. Thomas. 1989. Vertical and horizontal variations in the physiological diversity of the aerobic chemoheterotrophic bacterial microflora in deep southeast coastal plain subsurface sediments. Applied and Environmental Microbiology, v. 55, no. 5, p. 1058–1065.

Champ, D. R., J. Gulens, and R. E. Jackson. 1979. Oxidation–reduction sequences in groundwater flow systems. Canadian Journal of Earth Sciences, v. 16, no. 1, p. 12–23.

Chapelle, F. H. 1993. Groundwater Microbiology and Geochemistry. John Wiley & Sons, Inc., New York, 424 p.

Chapelle, F. H., P. B. McMahon, N. M. Dubrovsky et al. 1995. Deducing the distribution of terminal electron-accepting processes in hydrologically diverse groundwater systems. Water Resources Research, v. 31, no. 2, p. 359–371.

Chapelle, F. H., S. K. Haack, P. Adriaens et al. 1996. Comparison of E_h and H_2 measurements for delineating redox processes in a contaminated aquifer. Environmental Science & Technology, v. 30, no. 12, p. 3565–3569.

Chapelle, F. H., P. M. Bradley, M. A. Thomas et al. 2009. Distinguishing iron-reducing from sulfate-reducing conditions. Groundwater, v. 47, no. 2, p. 300–305.

Drever, J. I. 1988. The Geochemistry of Natural Waters. 2nd Ed, Prentice Hall, Englewood Cliffs, NJ, 437 p.

Denver, J. M., A. J. Tesoriero, and J. R. Barbaro. 2010. Trends and transformation of nutrients and pesticides in a coastal plain aquifer system, United States. Journal of Environmental Quality, v. 39, no. 1, p. 154–167.

Dinicola, R. S. 2006. Evidence for chloroethene biodegradation in groundwater at former Building 957 drum storage area, Area 2, Operable Unit 2, Naval Undersea Warfare Center, Division Keyport, Washington. U.S. Geological Survey Scientific Investigations Report 2006-5030, 12 p.

Domenico, P. A., and F. W. Schwartz. 1998. Physical and Chemical Hydrogeology. John Wiley and Sons, Inc., New York, 506 p.

Fram, M. S. 2017. Groundwater quality in the Western San Joaquin Valley study unit, 2010: California GAMA Priority Basin Project. U.S. Geological Survey Scientific Investigations Report 2017–5032, 130 p.

Fredrickson, J.K., Brockman, F.J., Hicks, R.J. et al. (1990). Biodegradation of nitrogen-containing aromatic compounds in deep subsurface sediments: Proceedings of the First International Symposium on Microbiology of the Deep Subsurface, Eds. C.B. Fliermans and T.C. Hazen, Westinghouse Savannah River Company, p. 6-27 to 6-44.

Ghiorse, W. C., and J. T. Wilson. 1988. Microbial ecology of the terrestrial subsurface. Advances in Applied Microbiology, v. 33, p. 107–172.

Griffin, R. A., and W. R. Roy. 1985. Interaction of organic solvents with saturated soil-water systems. Environmental Institute for Waste Management Studies, The Univ. of Alabama, Open File Report 3, 86 p.

Harvey, R. W., R. L. Smith, and L. George. 1984. Effect of organic contamination upon microbial distributions and heterotrophic uptake in a Cape Cod, Mass., aquifer. Applied and Environmental Microbiology, v. 48, no. 6, p. 1197–1202.

Hassett, J. J., W. L. Banwart, and R. A. Griffin. 1983. Correlation of compound properties with sorption characteristics of nonpolar compounds by soils and sediments: concepts and limitations, In C. W. Francis, and S. I. Auerbach (eds.), Environment and Solid Wastes: Characterization, Treatment and Disposal (Chapter 15). Butterworth Publishers, London, p. 161–178.

Hem, J. D., and W. H. Cropper. 1959. Survey of ferrous-ferric equilibria and redox potentials. U.S. Geological Survey Water-Supply Paper 1459-A, 31 p.

Jackson, R. E., and R. J. Patterson. 1982. Interpretation of pH and Eh trends in a fluvial-sand aquifer system. Water Resources Research, v. 18, no. 4, p. 1255–1268.

Jurgens, B. C., P. B. McMahon, F. H. Chapelle, et al. 2009. An Excel® workbook for identifying redox processes in groundwater. U.S. Geological Survey Open-File Report 2009–1004, 8 p.

Karickhoff, S. W. 1981. Semi-empirical estimation of sorption of hydrophobic pollutants on natural sediments and soils. Chemosphere, v. 10, no. 8, p. 833–846.

Karickhoff, S. W., D. S. Brown, and T. A. Scott. 1979. Sorption of hydrophobic pollutants on natural sediments. Water Research, v. 13, no. 3, p. 241–248.

Kelly, F. X., K. J. Dapsis, and D. A. Lauffenburger. 1988. Effect of bacterial chemotaxis on dynamics of microbial competition. Microbial Ecology, v. 16, p. 115–131.

Levine, S.N. and Ghiorse, W.C. (1990). Analysis of environmental factors affecting abundance and distribution of bacteria, fungi and protozoa in subsurface sediments of the Upper Atlantic Coastal Plain, USA. In Proceedings of the First International Symposium on Microbiology of the Deep Subsurface, ed. C.B. Fliermans, and T.C. Hazen: 15–19. Westinghouse Savannah River Company, Aiken, S.C

Lindberg, R. D., and D. D. Runnells. 1984. Groundwater redox reactions: an analysis of equilibrium state applied to Eh measurements and geochemical modeling. Science, v. 225, no. 4665, p. 925–927.

Lovley, D. R., F. H. Chapelle, and J. C. Woodward. 1994. Use of dissolved H_2 concentrations to determine distribution of microbially catalyzed redox reactions in anoxic groundwater. Environmental Science & Technology, v. 28, no. 7, p. 1205–1210.

Lovley, D. R., and S. Goodwin. 1988. Hydrogen concentrations as an indicator of the predominant terminal electron accepting reactions in aquatic sediments. Geochimica et Cosmochimica Acta, v. 52, no. 12, p. 2993–3003.

Mackay, D. M., D. L. Freyberg, P. V. Roberts et al. 1986. A natural gradient experiment on solute transport in a sand aquifer: 1. Approach and overview of plume movement. Water Resources Research, v. 22, no. 13, p. 2017–2029.

Mackay, D. M., P. V. Roberts, and J. A. Cherry. 1985. Transport of organic contaminants in groundwater. Environmental Science & Technology, v. 19, no. 5, p. 384–392.

Mackay, D. M., and T. M. Vogel. 1985. Groundwater contamination by organic chemicals: uncertainties in assessing impact, In B. Hitchon, and M. R. Trudell (eds.), Second Canadian/American Conference on Hydrogeology. National Water Well Association, Dublin, OH, p. 50–59.

McCarty, P. L., B. E. Rittman, and E. J. Bouwer. 1984. Micro-biological processes affecting chemical transformations in groundwater, In G. Bitton, and C. P. Gerba (eds.), Groundwater Pollution Microbiology. John Wiley & Sons, New York, p. 89–115.

McNabb, J. F., and G. Mallard. 1984. Microbial sampling in the assessment of groundwater pollution, In G. Bitton, and C. P. Gerba (eds.), Groundwater Pollution Microbiology. John Wiley & Sons, New York, p. 235–260.

McMahon, P. B., and F. H. Chapelle. 2008. Redox processes and water quality of selected principal aquifer systems. Groundwater, v. 46, no. 2, p. 259–271.

McMahon, P. B., T. K. Cowdery, F. H. Chapelle, et al. 2009. Redox conditions in selected principal aquifers of the United States. U.S. Geological Survey Fact Sheet 2009-3041, 6 p.

McMahon, P. B., F. H. Chapelle, and P. M. Bradley. 2011. Evolution of redox processes in groundwater, In P. Tratnyek, T. Grundl, and S. Haderlein (eds.), Aquatic Redox Chemistry. American Chemical Society, Washington, p. 581–597.

Morel, F. M., and J. G. Hering. 1993. Principles and Applications of Aquatic Chemistry. John Wiley & Sons, New York, 608 p.

NRCS. 2012. National Engineering Handbook. Part 631 Chapter 3. Engineering Classification of Earth Materials. USDA online, https://directives.sc.egov.usda.gov/OpenNonWebContent.aspx?content=31847.wba.

Paschke, S. S., L. J. Kauffman, S. M. Eberts, et al. 2007. Overview of regional studies of the transport of anthropogenic and natural contaminants to public-supply wells, section 1 of Paschke, S.S., ed., Hydrogeologic settings and groundwater flow simulations for regional studies of the transport of anthropogenic and natural contaminants to public-supply wells—studies begun in 2001. U.S. Geological Survey Professional Paper 1737–A, p. 1-1—1-18.

Parkhurst, D. L., and C. A. J. Appelo. 2013. Description of input and examples for PHREEQC version 3—a computer program for speciation, batch-reaction, one-dimensional transport, and inverse geochemical calculations. U.S. Geological Survey Techniques and Methods 6(A43), 497 p.

Pytkowicz, R. P. 1983. Equilibria, Nonequilibria & Natural Waters, v. II. John Wiley, New York, 353 p.

Rittmann, B. E. 1993. The significance of biofilms in porous media. Water Resources Research, v. 29, no. 7, p. 2195–2202.

Schwarzenbach, R. P., and W. Giger. 1985. Behavior and fate of halogenated hydrocarbons in groundwater, In C. H. Ward, W. Giger, and P. L. McCarty (eds.), Groundwater Quality. John Wiley and Sons, New York NY, p. 446–471.

Schwarzenbach, R. P., and J. Westall. 1981. Transport of nonpolar organic compounds from surface water to groundwater. Laboratory sorption studies. Environmental Science & Technology, v. 15, no. 11, p. 1360–1367.

Schwartz, F. W., and H. Zhang. 2003. Fundamentals of Groundwater. John Wiley & Sons, New York, 583 p.

Spruill, T. B., A. J. Tesoriero, H. E. Mew, Jr., et al. 2005. Geochemistry and characteristics of nitrogen transport at a confined animal feeding operation in a Coastal Plain agricultural watershed, and implications for nutrient loading in the Neuse River Basin, North Carolina, 1999–2002. U.S. Geological Survey Scientific Investigations Report 2004–5283, 57 p.

Stollenwerk, K. G. 1991. Simulation of molybdate sorption with diffuse layer surface-complexation model. U.S. Geological Survey Water-Resource Investigations Report 91-4034, p. 47–52.

Stoodley, P., D. DeBeer, and Z. Lewandowski. 1994. Liquid flow in biofilm systems. Applied and Environmental Microbiology, v. 60, no. 8, p. 2711–2716.

Stumm, W., and J. J. Morgan. 1981. Aquatic Chemistry, 2nd edition. John Wiley & Sons, New York, 780 p.

Yong, R. N. 1985. Interaction of clay and industrial waste: a summary review, In B. Hitchon, and M. Trudell (eds.), Second Canadian/American Conference on Hydrogeology. National Water Well Association, Dublin, OH, p. 13–25.

Zachara, J. (1990) Hydrogeology in relation to microorganisms: Proceedings of the First International Symposium on Microbiology of the Deep Subsurface, Eds. C.B. Fliermans and T.C. Hazen, Westinghouse Savannah River Company: Aiken, S.C. 5-3 to 5-13.

18

Isotopes and Applications

CHAPTER MENU

Isotopes are widely used in research and practice in groundwater hydrology. They can be divided broadly into stable and radiogenic types. The stable isotopes are used mainly for flow system tracing and climate reconstruction; the radiogenic isotopes are used as groundwater tracers for applications such as estimation of recharge rates. This chapter provides a basic understanding of how isotopes are used and the basic interpretive techniques that are involved. The first part of this chapter presents basic definitions and units of measurements. Next, we look in detail at the stable isotopes, oxygen-18 ($\delta^{18}O$) and deuterium (D), which are used most. The last part deals with approaches in dating groundwater, which have applications to problems of geologic waste disposal.

18.1 Stable and Radiogenic Isotopes

Isotopes are atoms of the same element that differ in terms of their mass. For example, hydrogen with an atomic number of 1 has three isotopes, ${}^{1}_{1}H$, ${}^{2}_{1}H$, and ${}^{3}_{1}H$, with mass numbers (superscripts) of 1, 2, and 3, respectively. The first of these isotopes is stable, whereas the last ${}^{3}_{1}H$ (usually written ^{3}H) decays radioactively to ${}^{2}_{3}He$. Thus, radioactive decay is one of the important processes that involve isotopes. In radioactive decay, atoms of a particular isotope change spontaneously to a new, more stable isotope. Isotopic concentrations also change due to processes like evaporation, condensation, or water/rock interactions. Typically, these processes favor one of the isotopes of a given element over others, producing fractionation.

The stable isotopes of hydrogen, oxygen have found vast applications in groundwater-related investigations. Usually, groundwater studies involve deuterium (^{2}H or D), and oxygen-18 (^{18}O). One important reason is that these isotopes are measured for water (H_2O) and serve as relatively non-reactive tracers of precipitation sources and flow systems.

The average abundances of D and ^{18}O in water are small and cannot be measured accurately. The solution is to report isotopic abundances as positive or negative deviations of isotope ratios away from a standard. This convention is represented in the following general equation, derived from Fritz and Fontes (1980):

$$\delta = \frac{R_{\text{sample}} - R_{\text{standard}}}{R_{\text{standard}}} \times 1000 \tag{18.1}$$

Fundamentals of Groundwater, Second Edition. Franklin W. Schwartz and Hubao Zhang.

Companion website: www.wiley.com/go/schwartz/fundamentalsofgroundwater2

where δ, reported as permil (), represents the deviation from the standard and R is the particular isotopic ratio (e.g., $^{18}O/^{16}O$) for the sample and the standard. For example, we would express ^{18}O of a sample

$$\delta^{18}O_{sample} = \frac{\left(^{18}O/^{16}O\right)_{sample} - \left(^{18}O/^{16}O\right)_{standard}}{\left(^{18}O/^{16}O\right)_{standard}} \times 1000$$

A $\delta^{18}O$ value of −20‰ means that the sample is depleted in ^{18}O by 20‰, or 2% relative to the standard. This way of expressing isotopic compositions takes advantage of the ability of mass spectrometers to measure isotopic ratios accurately. The errors involved in determining δ values are typically a small proportion of the possible range of values.

Deuterium and oxygen-18 compositions of water are usually measured with respect to the VSMOW (Vienna Standard Mean Ocean Water) standard. This choice of standard is appropriate because precipitation that recharges groundwater originates from the evaporation of ocean water. However, the isotopic composition of rain or snow in most areas is not the same as ocean water. Evaporation and subsequent cycles of condensation significantly change the isotopic composition of water vapor in the atmosphere.

Certain isotopes decay radioactively, such as tritium (^{3}H) and carbon-14 (^{14}C). These two isotopes have been important in age dating and tracing of groundwaters. Radioactive decay occurs mainly by the emission of an α particle ($^{4}_{2}He$) or a β particle (electron $^{0}_{-1}e$). Often accompanying the emission of these particles is γ radiation, which is electromagnetic energy of short wavelength. This radiation forms when nuclides produced in an excited state (noted by $*$) revert to a so-called ground state. As the following examples illustrate, α-decay changes both the mass number and the atomic number, β-decay changes the atomic number only, and γ-emission changes neither.

$$\alpha - \text{decay: } ^{232}_{90}Th \rightarrow ^{232}_{88}Ra + ^{4}_{2}He$$

$$\beta - \text{decay: } ^{228}_{88}Ra \rightarrow ^{228}_{89}Ac + ^{0}_{-1}e$$

$$\gamma - \text{emission: } ^{236}_{92}U^{*} \rightarrow ^{236}_{92}U + \gamma$$

The arrows in these equations indicate that radioactive decay is irreversible. The quantity of the reacting or parent isotope continually decreases, while the product or daughter isotope increases. Reactions become more complex when the daughter itself decays through a series of other products until a stable form is finally created. We refer to these reactions as decay chains or disintegration series (e.g., $^{232}_{90}Th$, $^{235}_{92}U$, and $^{238}_{92}U$ series).

The decay of radioactive isotopes is independent of temperature and follows a first-order rate law. The rate of decrease in the activity of a radioactive substance is commonly expressed in terms of a radioactive half-life ($t_{1/2}$), which is the time required to reduce the number of parent atoms by one-half. Our particular interest in this chapter is with tritium ($t_{1/2}$ 12.2 years), carbon-14 ($t_{1/2}$ 5730 years), and others, less commonly employed.

Radioactive decay is important for two reasons. When radioactive isotopes occur as contaminants, decay through several half-lives can reduce the hazard when the residence time in a flow system is much greater than the half-life for decay. This capacity to attenuate radioactivity provides the rationale to support the subsurface disposal of some radioactive contaminants. Radioactive decay also forms the basis for various techniques used to determine groundwater "ages." However, this term and concept is highly misleading because of associated complexities (Suckow, 2014).

The concentration of radiogenic isotopes is measured in slightly different ways. Tritium [T] is measured in terms of its relative abundance in the water molecule as compared to hydrogen of mass 1 [H]. *One tritium unit* (TU) corresponds to one atom of ^{3}H in 10^{18} atoms of ^{1}H (Fontes, 1980). Measurements of ^{14}C are reported as *percent modern carbon* (pmc). This measure is the ratio of the ^{14}C activity in the sample to that of an international standard, expressed as a percentage. Using appropriate models, pmc has been interpreted as a water age, although most historical estimates are erroneous. Measurements of ^{36}Cl are reported as the ratio of $^{36}Cl/Cl$. Typical values for meteoric water fall in a range of 100 to 500×10^{-15}.

18.2 ^{18}O and Deuterium in the Hydrologic Cycle

When water changes from a gas to a liquid or from a solid to a liquid, isotopic fractionation occurs. Fractionation causes the relative abundance of the isotopes to change because the heavier isotopes (^{18}O and D) tend to be more abundant in the condensed phase. For example, as water evaporates from the ocean, the heavier molecules $H_2{}^{18}O$ and HDO are more abundant in the water phase than in the vapor phase.

The isotopic composition of rain or snow in most areas is not the same as ocean water, which by definition has δD and $\delta^{18}O$ values of 0‰. Evaporation and subsequent cycles of condensation (rain) significantly change the isotopic composition of water vapor in the atmosphere. Observation shows that water vapor (i.e., evaporated ocean water) in equilibrium with ocean water has $\delta D = -80‰$ and $\delta^{18}O = -10‰$. In the phase change from ocean water ($\delta D = 0‰$ and $\delta^{18}O = 0‰$) evaporating to vapor, the resulting change isotopic composition, for example $^{18}O/^{16}O_{vapor}$, is determined by the fractionation factor (α). Let us dig deeper here using a laboratory measured factor fractionation factor. For the case of the equilibrium phase change from water to water-vapor, the fraction factor is:

$$\alpha = \frac{\left(^{18}O/^{16}O\right)_{water}}{\left(^{18}O/^{16}O\right)_{vapor}} \tag{18.2}$$

or in terms of del notation:

$$\alpha = \frac{1000 + \delta^{18}O_{water}}{1000 + \delta^{18}O_{water}} \tag{18.3}$$

The fractionation factors for liquid-vapor transformations are 1.0098 for ^{18}O and 1.084 for D at 20°C. Knowing that $\delta^{18}O_{water}$ for ocean water is 0‰, by definition, simply rearranging Eq. (18.3) provides a simple way to calculate $\delta^{18}O_{vapor}$

$$\delta^{18}O_{vapor} = \frac{1000 + \delta^{18}O_{water}}{\alpha} - 1000$$

Substituting the known values gives

$$\delta^{18}O_{vapor} = \frac{1000 + 0}{1.0098} - 1000 = -9.7‰$$

A similar calculation for δD gives a value of −77.4‰.

Several additional points need to be made about this calculation. First, the difference in fractionation factors for δD and $\delta^{18}O$ causes an 8× greater fractionation in D as compared to ^{18}O. Second, the fractionation factors are temperature dependent. At 0°C, the fractionation factors for water-vapor transformations are 1.0117 for ^{18}O and 1.111 for D. Because these numbers are bigger, the vapor will be even more depleted in the two isotopes. The following calculation illustrates this point.

Example 18.1 Calculate the isotopic composition of a vapor in equilibrium with seawater at 0°C. The fractionation factors for ^{18}O and D are 1.0117 and 1.111, respectively.

Solution

$$\delta^{18}O_{vapor} = \frac{1000 + 0}{1.0117} - 1000 = -11.6‰$$

$$\delta D_{vapor} = \frac{1000 + 0}{1.111} - 1000 = -99.9‰$$

As expected with the lower temperature, the vapor is more depleted than the ocean water with respect to ^{18}O and D. The lower temperature has enhanced the fractionation. The temperature control on fractionation also explains why winter precipitation is more depleted in D and ^{18}O and why moving north or south away from the equator there is increasing fractionation of the vapor.

These simple equilibrium models help to explain what happens as fractionation occurs. However, natural processes are more complex, and isotopic ratios of water vapor in air masses associated with oceans are not in equilibrium with the water. In general, vapor is more depleted than expected. Rainfall sampled in coastal areas near the equator has an isotopic composition most like seawater, but it is still slightly depleted.

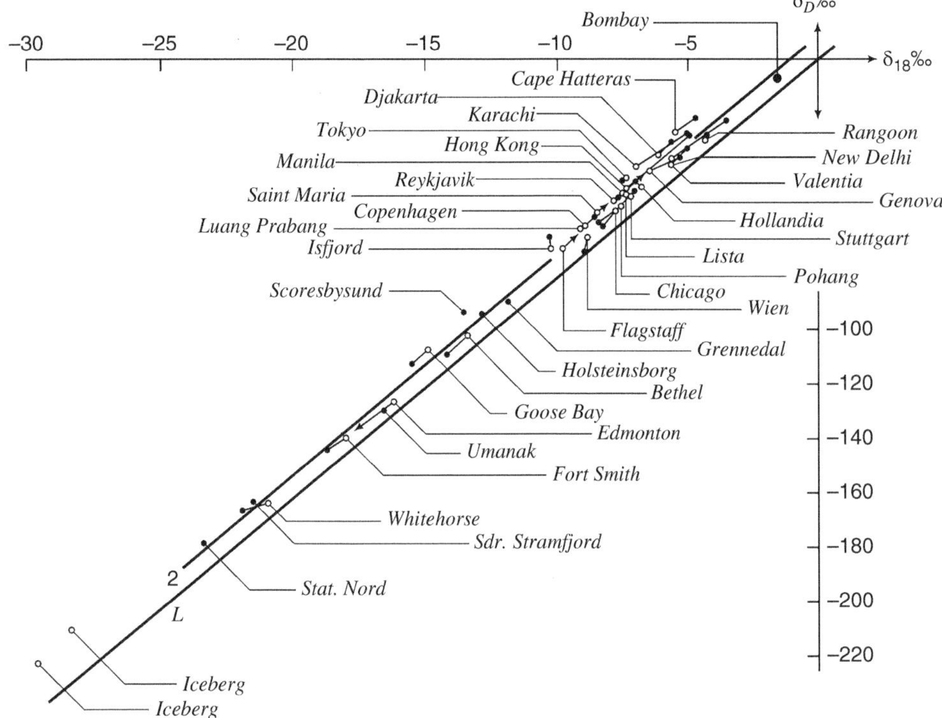

Figure 18.1 Weighted (open circles) and unweighted (dots) means for continental stations in the Northern Hemisphere, except for Africa and the Near East. (Dansgaard, 1964 published in Tellus XVI (1964), 4 and distributed under the terms of the Creative Commons Attribution 4.0 license).

With evaporation and condensation processes, both ^{18}O and D are fractionated in a consistent way, although the extent of fractionation is different in both cases. Thus, water circulating from oceans into the atmosphere and falling as rain has δ^{18}O and δD values that are correlated. In other words, if you know one of the values, you can predict the other. Isotope chemists discovered this fact in the early 1950s when they started analyzing samples of precipitation from many different sites all around the world. When this diverse collection of δD and δ^{18}O values for samples of meteoric water are plotted together, they lie along a straight line known as the *global meteoric water line* (GMWL) (Figure 18.1) (Craig, 1961). The equation for this line is approximately $\delta D = 8\delta^{18}O + 10‰$. The slope of this line, 8, reflects the difference in the fractionation behavior between ^{18}O and D, which we discussed using Eq. (18.1). Where results from a specific site end up plotting on the meteoric line depends upon different factors like temperature, and travel paths of storms to sampling points, etc. Samples of precipitation from hot coastal areas such as Karachi Pakistan or Hong Kong plot towards the upper or "hot" end of the GMWL (Figure 18.1). Other cities in warm, continental settings, such as Flagstaff, Arizona in the USA, are depleted in ^{18}O and D and plot further down the GMWL. Samples from much colder cities such as those in northern Canada, such as Edmonton, Fort Smith are even more depleted (e.g., ~16‰) and plot further yet down the GMWL (Figure 18.1).

A large collection of ^{18}O and D measurements on precipitation samples from a small region may not necessarily fall along the GMWL. However, they still end plotting along a line called the "local meteoric water line" (LMWL). Figure 18.2 shows an example of a LMWL constructed for the Yucca Mountain area of southern Nevada, USA (Moscati and Scofield, 2011). Notice that the slope of the LMWL is 6.4 and lower than the GMWL. At this site, the cause is secondary evaporation of rain, particularly with hot and dry weather during summer months. Other factors contributing to a unique LMWL are related to the source of the water vapor in the ocean and storm tracks. Also shown on Figure 18.2 is an example of an evaporation line. When meteoric water evaporates, its isotopic composition deviates from the GMWL along a line with a shallow slope. This is a topic we will return to in later sections.

Where sample results plot on the GMWL has been framed so far in terms samples from very different geographic locations. However, the discussion glossed over the simple question as to why water samples collected in the same region might exhibit variability in isotopic compositions. One reason is that precipitation falling during summer storm events or winter storm events can vary substantially due mostly to the specific temperature conditions associated with individual storms.

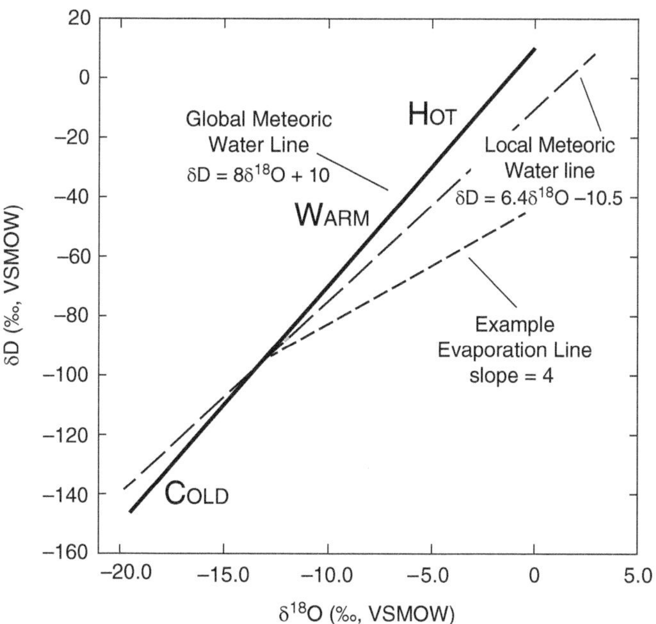

Figure 18.2 In local regions, sample results may align somewhat differently along a so-called local meteoric line, for example here having a slope of 6.4. When influenced by evaporation, samples plot along an evaporation line, which for this example is 4 (U.S. Geological Survey, adapted from Moscati and Scofield, 2011/Public domain).

Data collected at a site in Nevada by Moscati and Scofield (2011) are useful in illustrating this concept. Because on average the temperature of rain in winter was colder than in summer, the winter data (Nov.–April) plot down the LMWL, while data from summer storms (May–Oct.) plot further up the LMWL (Figure 18.3). Thus, the variability among storms within a season and differences between seasons work together to creates a range in δ^{18}O from ~−30‰ to 3‰. Another factor at work in causing variability in summer and winter data is a seasonal difference in storm tracks (Moscati and Scofield, 2011). Winter storms come from the Pacific Ocean and experience depletion due a collection of processes (coming up next) as they cross mountain chains. The summer storms track from the Gulf of California or Gulf of Mexico and do not experience these

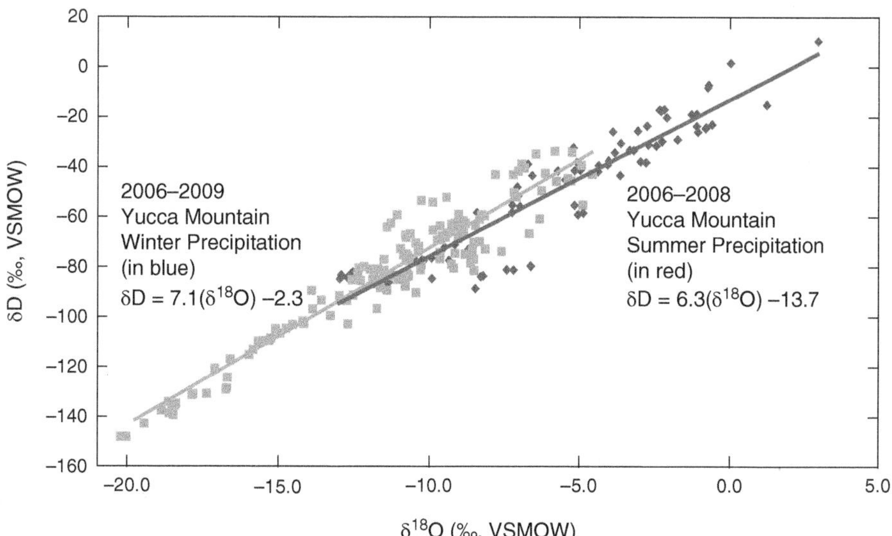

Figure 18.3 δD and δ^{18}O data for precipitation samples from the Yucca Mountain exhibit significant variability both seasonally and between seasons. The seasonal LMWL lines differ because evaporation affects rain differently than snow (U.S. Geological Survey, adapted from Moscati and Scofield, 2011/Public domain).

processes. As a result, precipitation is relatively enriched (Moscati and Scofield, 2011). Another interesting feature of isotopes in precipitation at Yucca Mountain is two similar but different seasonal LMWLs (Figure 18.3). The slope of the summer line (i.e., 6.3) is less like GMWL (i.e., 8) than the winter line (i.e., 7.1). The flatter slope in summer is caused by fractionation of rain due to evaporation in warm temperatures and low humidity. Winter precipitation is snow, which "does not fractionate due to evaporation" (Moscati and Scofield, 2011).

The usual way of interpreting the results of ^{18}O and D analyses in hydrogeological applications is to take the measured δD and δ^{18}O and make a plot of δD versus δ^{18}O. Where the data fall in relation to the global meteoric water line provides clues as to what processes have occurred. Water with an isotopic composition falling on the global meteoric water line is assumed to have originated from condensation of water vapor and to be unaffected by other isotopic processes. The actual location on the global meteoric water can vary because of temperature and other effects, which we will examine in more detail. Deviations from the meteoric water line result from other isotopic processes. In most cases, these processes affect the relationship between δD and δ^{18}O in such a unique way that the position of the data points can help to identify a process.

18.2.1 Behavior of D and ^{18}O in Rain

In the previous section, we discussed global and local meteoric water lines generally in terms geography, seasonality, and features of storms. This section looks more specifically at how fractionation processes work on vapor masses being transported from oceanic sources and the associated precipitation. Figure 18.4 illustrates conceptually how the δ^{18}O composition of water vapor and associated precipitation evolves as a storm moves inland from a coastal source. Water vapor in storm clouds in equilibrium with ocean water (0‰ δ^{18}O) is depleted in δ^{18}O, for example, −10‰ δ^{18}O. As the finite volume of moisture in the storm moves onshore, it will evolve isotopically because rain preferentially enriched in ^{18}O depletes the water vapor (Figure 18.4). Also, the rain reduces the quantity of water being carried by the storm, as reflected by the shrinking size of the clouds. Across the coastal plain there is a modest decline in the δ^{18}O composition of the vapor with a 10‰ enrichment in the associated precipitation. The abrupt change in elevation as the storm moves into the "mountains" (Figure 18.4) has some important effects. First, cooling of vapor masses at increasing elevations causes greater rainfall. This increasing rain depletes the reservoir of water vapor in ^{18}O more rapidly. This decline is magnified by increasing fractionation of ^{18}O between the vapor and precipitation. All these effects are reflected in Figure 18.4 by (1) the rapid decrease in the size of the cloud, (2) the rapid depletion of ^{18}O in water vapor at the mountain front (−14‰ to −21‰), and by (3) increased fractionation between the vapor and precipitation (12‰).

These ideas are embodied in the concepts of a *continental effect*, an *elevation effect*, and a *latitude* effect. The continental effect describes the progressive depletion in the isotopic composition of water vapor moving from oceans onto continents. Because the rain is preferentially enriched in the heavier isotopes (i.e., D and ^{18}O), the reservoir of heavier isotopes in the water vapor is depleted. Each new rain is isotopically lighter than the previous one. The flow of vapor across the coastal plain in Figure 18.4 represents the beginning of a continental effect.

The elevation effect, also depicted in Figure 18.4, describes the acceleration in the depletion of δD and δ^{18}O in water vapor as it moves over mountains. As water vapor is carried to higher elevations, there is a tendency for more rain, which increases the rate of depletion. Also, at higher elevations, the temperature of condensation is lower, which increases the fractionation and intensifies the rate of depletion.

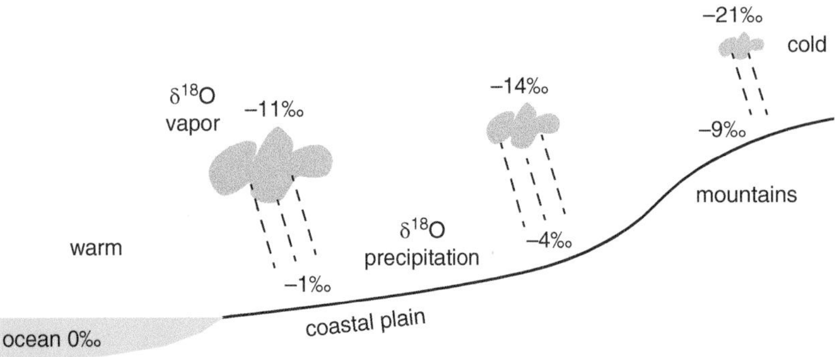

Figure 18.4 Water vapor moving landward from an ocean becomes increasingly depleted in ^{18}O from rainout. As the vapor cools with increasing elevation, rainfall increases with increasing fractionation between the vapor and the rain (FWS).

Observations from around the world have shown that precipitation at higher latitudes is relatively depleted in D and ^{18}O (i.e., more negative) as compared to samples from lower latitudes. This latitude effect is explained by higher precipitation rates and colder temperatures at higher latitudes. For example, moving northward along the western coast of the United States to Canada, values of δD decrease from approximately −22 to −101‰ and values of δ^{18}O decrease from −4 to −14‰ (Gat, 1980).

18.3 Variability in ^{18}O and Deuterium in Groundwater

This section describes some of the processes at work to produce variability in ^{18}O and D in groundwater. Plotting δ^{18}O and δD data for water samples with reference to the global meteoric water line or a local meteoric line provides the framework for identifying these isotopic processes.

18.3.1 Spatial and/or Temporal Variability of δ^{18}O and δD Compositions in Aquifers

Sampling of regional aquifer systems usually results in situations where the δ^{18}O and δD content of groundwater varies from place. Here, we examine two of the more common situations. Figure 18.5*a* represents a simple unconfined aquifer in a warm area with recharge from rainfall [sample 1]. Before development of the aquifer, water [sample 2] recharged to the aquifer flowed laterally to the stream (right side) where it discharged. With addition of wells, pumping induced flow from the nearby river to the well [sample 3]. The source of the river flow was snowmelt runoff from nearby mountains. In the setting of storage depletion by pumping, precipitation and upgradient recharge [samples 1, 2] would remain isotopically similar, both relatively enriched with respect to δ^{18}O and δD and plot on the GMWL (Figure 18.5*a*). In contrast, river water [sample 5] and water induced into the aquifer by pumping [sample 4] would be depleted isotopically, given the cold, high elevation source. Water being pumped from the well [3] would be a mixture of enriched and depleted water and plotting between the end members on the GMWL (Figure 18.5), reflecting the mixture of both sources.

Figure 18.5*b* depicts the case of a confined aquifer in region that had experienced the effects of significant climate change. In this scenario, climate change was caused by the gradual loss of the continental ice with an associated increase in temperature. The rainy, colder climate during glacial times (i.e., >12,000 years ago) gave way to a dry, warmer present-day climate.

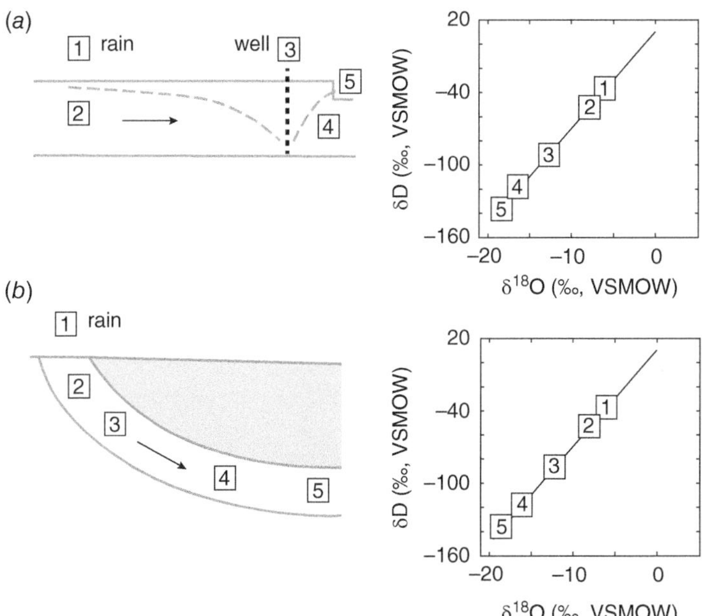

Figure 18.5 Panel (*a*) illustrates how pumping in a well can create an isotopic composition for water that is a mixture of two end members. Panel (*b*) shows the gradual enrichment in δ^{18}O and δD contents associated with a warming climate (FWS).

The aquifer was recharged by rainfall at a single outcrop. The active recharge and flow during wetter, colder times essentially charges aquifer with water that is isotopically depleted with respect to present-day rainfall. The depletion in $\delta^{18}O$ and δD would be caused by colder temperatures and larger rainfall as vapor moved across continents. Thus, samples from most wells in the aquifer (e.g., samples 1 to 4) reflect the slow warming of temperatures with the oldest water most depleted. The least depletion with respect to $\delta^{18}O$ and δD is associated with warmer, present-day rainfall, sampled from shallow wells in the recharge area (e.g., sample 5).

This gradual warming in climate following the most recent glaciation would be common in temperate-zone aquifers in countries around the world. For example, in the Northern Great Plains of the United States and Canada, the greatest availability of water for groundwater recharge is April and May when snowmelt runoff combines with spring rainstorms.

18.3.2 Connate Water in Units with Low Hydraulic Conductivity

There are places in North America where the shallow groundwater is very different isotopically than the precipitation. Old, connate water was trapped in geological materials with an ultra-low hydraulic conductivity. *Connate water* is water that was incorporated in a sediment at the time of its deposition with an isotopic character reflecting the prevailing climate at that time. Despite the passage of time, that water would have been retained because the unit had a low hydraulic conductivity and the groundwater system was being flushed extremely slowly.

Sites like this can exist in glaciated terrains, where thick, clay rich deposits of glacial till or lacustrine clays still contain connate water reflecting the intense cold of glacial conditions (Remenda et al., 1994; Hendry and Wassenaar, 1999) or perhaps water from the last warmer interglacial (Hendry and Wassenaar, 1999). The hydraulic conductivity of these deposits is so low, for example, less than 3×10^{-11} m/sec in laboratory tests (Remenda et al., 1994), that 10,000+ years of recharge from above or upward flow from below was not able to flush the connate water.

Figure 18.6*a* is an idealization of the changing isotopic composition of pore water as a function of depth at sites in Saskatchewan, in Western Canada (Hendry and Wassenaar, 1999). The upper few meters of sediments were commonly fractured and more permeable, with a $\delta^{18}O$ composition like present-day recharge, ranging from −14‰ to −18‰, depending upon location. Otherwise, samples appeared depleted relative to modern precipitation. $\delta^{18}O$ and δD values for samples fell along the meteoric water for this region (Figure 18.6*a*). Figure 18.6*b* illustrates the pattern in variability in $\delta^{18}O$ and

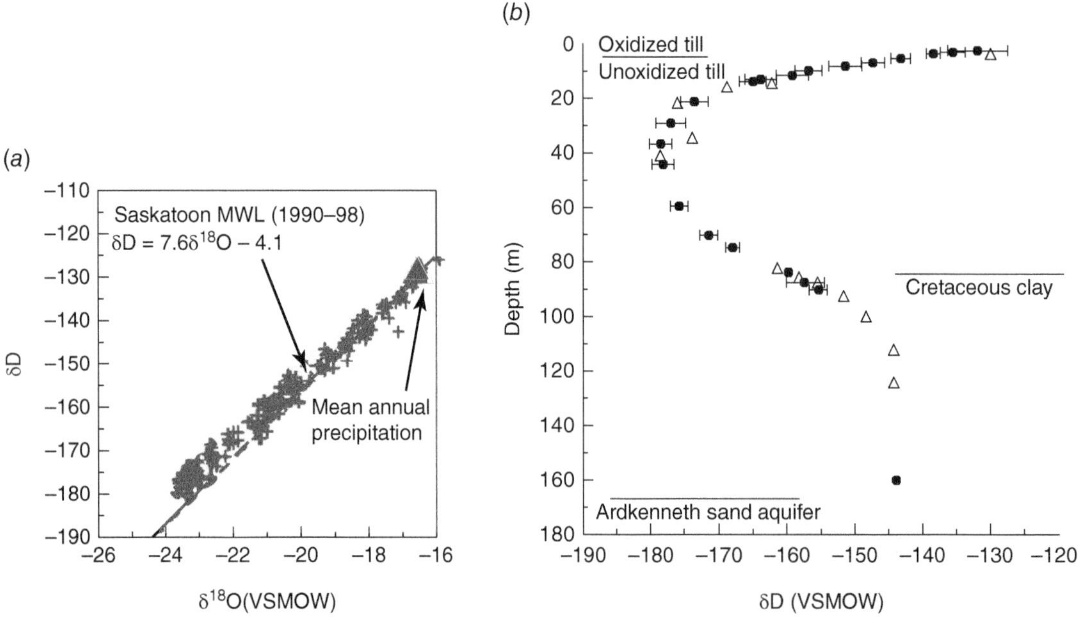

Figure 18.6 In low permeability settings in Western Canada, $\delta^{18}O$ and δD values exhibit significant variability along the meteoric water line due to climate change (*a*). The most depleted samples in unoxidized till (*b*) represent connate water trapped as the till was deposited during glaciation (Hendry and Wassenaar, 1999, Water Resources Research 35(6): 1751–1760. Copyright 1999 by the American Geophysical Union).

δD with respect to depth. At a depth of approximately 40 m, δ^{18}O values for groundwater were significantly depleted, i.e. approximately −24‰, in δ^{18}O and −178‰ in δD reflecting connate, glacial waters as sampled from pore water in a glacial till. Remenda et al. (1994) reported similar values for a site in Manitoba. Given the ultra-low hydraulic conductivity of these units, diffusion remains as an important creating the gradients in isotopic profiles. Similar patterns of isotopic variability have been observed in aquifers in Nevada and in glacial till in Wisconsin and southwestern Ontario, Canada.

Studies like these provide important information on past climatic conditions. They also contribute an important methodology in providing a large-scale measure of permeability. Confirming the presence of connate water in a unit points to an ultra-low hydraulic conductivity unit that could be suitable for siting an engineered facility like a landfill.

18.4 Evaporation and the Meteoric Water Line

The isotopic composition of water can be altered by processes along the hydrologic cycle. Here and in the following section, we emphasize a process commonly encountered in hydrologic applications, namely evaporation. As water evaporates, it gradually becomes enriched in heavier isotopes δ^{18}O and δD. On a plot of δD versus δ^{18}O, data for samples of that that water, which initially plotted on a meteoric water line, will fall away from the meteoric water following a so-called *evaporation line* (Figure 18.2). An evaporation line has a slope around ~5 as compared to ~8 for the meteoric water line. The slope tends to vary from > 5 under humid conditions to < 5 under arid conditions. The slope also can be affected by seasonality in the precipitation (Benettin et al., 2018).

The presence of an evaporation trend in isotopic data naturally lends itself to various applications. For example, a study in Florida (Sacks, 2002) illustrated how mass balance calculations with ^{18}O and deuterium can be used to estimate the groundwater contribution to inflows to small lakes. Figure 18.7 illustrates how the isotopic composition of a lake depends upon the relative quantities of groundwater throughput. Lakes with relatively small inflows and outflows of groundwater would be characterized by enrichment in δD and δ^{18}O in the water indicative of a longer residence time and greater evaporation. With a more robust groundwater system, there would be less residence time and indications of lakewater evaporation as indicated by δD and δ^{18}O compositions.

Using this approach, Sacks (2002) studied 81 lakes in central Florida from the vicinity of Tampa Bay eastward. The lakes were broadly subdivided into three categories of groundwater inflows, low <25% of total inflows to the lake, medium, 25–50% of total inflows, and high, > 50% inflows. Along the coastal lowlands most lakes were in the category. In the central highlands, inflows were in the medium to high category (Sacks, 2002).

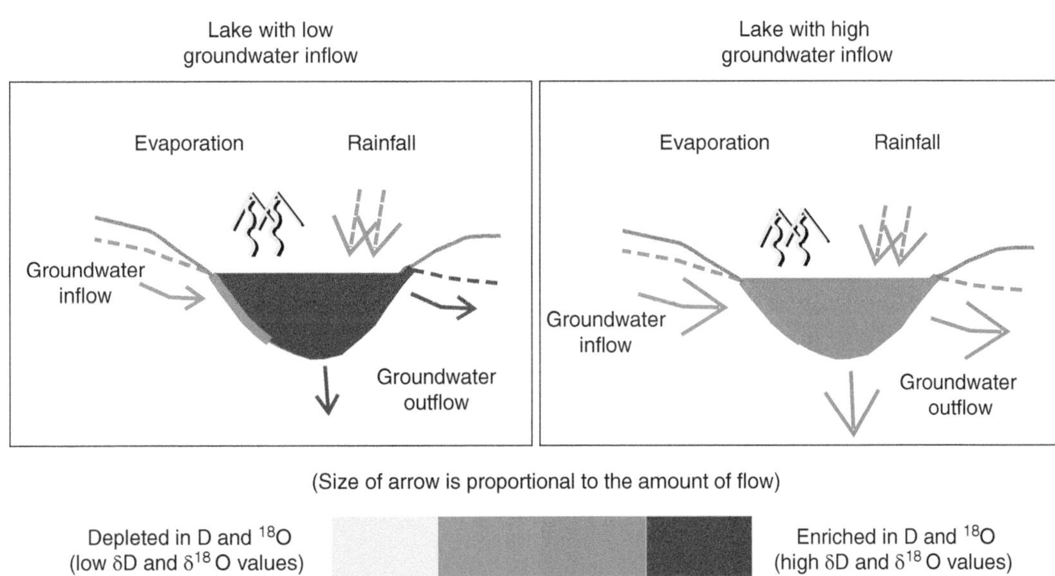

Figure 18.7 With less active through flows of groundwater, the isotopic composition of the lake would reflect more intense evaporation (i.e., enrichment less negative values of δD and δ^{18}O) and greater deviations along an evaporation line. Greater through flows of groundwater would be reflected by more negative values and smaller deviations (U.S. Geological Survey Sacks, 2002/Public domain).

18.4.1 Other Deviations from GMWL

Evaporation is not the only process that modifies the isotopic composition of groundwater. For example, isotopic exchange between minerals and groundwater is important in deep, basinal flow systems or in geothermal systems. The relatively high temperatures enable oxygen and hydrogen to exchange between phases and to achieve an equilibrium distribution. This equilibrium is described by the fractionation factor for the mineral. Savin (1980) summarized fractionation factors for a variety of different minerals and described how they change as a function of temperature.

When only δ^{18}O is involved in mineral exchange, the samples typically fall along a horizontal line, reflecting an *oxygen shift* away from the global meteoric water line in the direction of less negative δ^{18}O. The size of the oxygen shift is proportional to the difference in original δ^{18}O between the water and rock, the temperature, and the time of contact, and inversely proportional to the water/rock ratio (Truesdell and Hulston, 1980). Yellowstone Park, Steamboat Springs, Wairaki, and Salton Sea are well-known examples. With exchange with CO_2, the shift is also horizontal but in the direction of decreasing δ^{18}O.

18.4.2 Illustrative Applications with Deuterium and Oxygen-18

This section examines some of the ways in which deuterium and oxygen-18 are used in practice. As will be evident with the second example, these isotopes often just one component of a much comprehensive investigative strategy.

18.4.2.1 Role of Wetland in Streamflow

The Northern Great Plains in Canada and the United States contains millions of small "pothole" *lakes* and wetlands. They are associated with the unique hummocky topography related to features of the most recent continental deglaciation. The pond water comes from snowmelt runoff, summer storms and groundwater to a lesser extent. Because of the hot, dry summer weather, these waterbodies experience significant evaporation and become enriched isotopically with respect to δ^{18}O and δD. The study by Brooks et al. (2018) took advantage of this evaporative signal in water stored in the wetlands to evaluate their contribution to the associated surface-water system.

The study involved the upper part of Pipestem Creek watershed. It is located in south-central North Dakota, USA and has a surface area of approximately 1672 km^2. However, because of the hummocky topography, the watershed contains numerous closed pothole basins, so the actual contributing area is "likely much smaller." Measurements of δ^{18}O and δD in samples collected from the creek, adjacent wetlands, and groundwater in 2014 and 2015 provided the bases for determining the contributions of prairie-pothole wetlands to creek flow. Shown in Figure 18.8 is plot of D versus ^{18}O for samples collected in the study.

Here are key points for the reader to take away from Figure 18.8.

1) Samples of precipitation collected at the site fell generally along the Global Meteoric Water Line (GMWL; $\delta D = 8 \cdot \delta^{18}O + 10$). The open circle was an estimate of an average annual value. Values of snowmelt runoff (not sampled in 2014) would also have plotted on GMWL but would have been substantially depleted in δD and δ^{18}O because of cold winter conditions and substantially different pathways followed by winter and summer storms.
2) Groundwater (black dot) also plotted along the GMWL at $\sim\delta D = -110$‰ and $\sim\delta^{18}O = -15$‰. This result and other sampling in 2015 indicated that groundwater was recharged by winter precipitation that had not experienced evaporation. Also, for these locations the wetlands did not contribute to groundwater recharge.
3) The wetlands and stream samples all plotted away from the GMWL, along a local evaporation line (LEL; $\delta D = 5.8 \cdot \delta^{18}O - 23.1$). The LEL with a characteristic slope of 5.8 is a clear indication that waters in the stream and the wetlands had been evaporated.
4) The point of intersection of the GMWL and the LEL was an estimate of the composition of the water before evaporation. This point of intersection coincided with the isotopic value of a single groundwater sample in 2014 and others from 2015 not shown here.
5) The isotopic character of the stream during this period of measurement in 2014 was explained as simple mixture of two end-members—depleted groundwater at the lower end of the LEL and evaporated wetland water at the enriched end of the LEL.
6) These results illustrated the importance of wetlands together with groundwater in maintaining the flow in Pipestone Creek during warm summer months, even under low flow conditions.

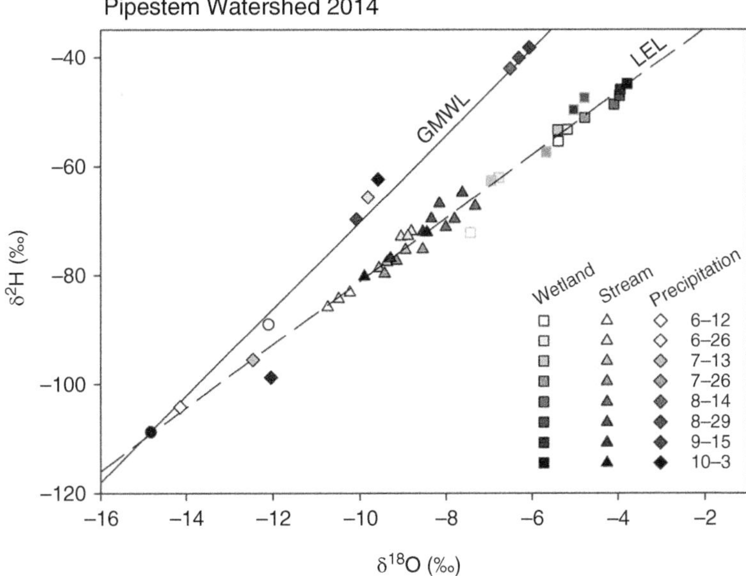

Figure 18.8 Isotopic data were measured for water samples from precipitation, wetlands, and streams in Pipestem Creek watershed, in east-central North Dakota, USA. Data come from a permanent wetland (P1, in black outline) and a temporary wetland (T8, grey outline) and three stream-sampling sites. Time in 2014 is indicated as month-day and coded by shading (U.S. Geological Survey Brooks et al., 2018. U.S. Government work, published in 2018 by the American Geophysical Union).

18.4.2.2 Integrated Study of Recharge Dynamics in a Desert Setting

Heilweil et al. (2007) used a variety of environmental tracers to study infiltration and recharge to groundwater in a desert setting. The 50-km^2 study area in southwestern Utah is part of the eastern Mojave Desert. It overlies the Navajo Sandstone, a prolific and important regional aquifer in parts of Utah, Arizona, Colorado, and New Mexico.

The annual precipitation averaged about 210 millimeters per year. The ground surface slopes gently from an upland area in the southwest towards northeast, and the area is drained by several ephemeral streams. The Navajo Sandstone there is commonly exposed at or near the ground surface and was covered by less than 1.5 m of soil. The coarsest and most permeable soils are prevalent in the uplands along with bedrock outcrops. At lower elevations to the northeast, finer and less permeable soils are present. A thick unsaturated zone 15–45 m thick has developed within the bedrock. Unsaturated bedrock was cored and water samples were extracted from cores by cryodistillation.

Samples of water from precipitation, the vadose zone and groundwater were analyzed for $\delta^{18}O$ and δD (Figure 18.9). The plot of δD and δD includes both the global and local meteoric water lines. Here are points to notice with respect to Figure 18.9.

1) Unevaporated samples did not fall along the global meteoric water line GMWL (dashed line; Figure 18.9). The best-fit line through precipitation data provided a local water line (LMWL; $\delta D = 7.61 \cdot \delta^{18}O - 0.03$). Here, the local meteoric water line fell below the global meteoric water line with a slope of 7.61 as compared to 8 for the GMWL.
2) The isotopic data for precipitation samples generally fell along or close to the local meteoric water line. They exhibited a large range in both $\delta^{18}O$ (−15.5‰ to +3.3‰) and δD (−126.4‰ to +4.2‰). This range was due to large range in temperatures associated with the precipitation. Precipitation in late fall and winter was isotopically lighter.
3) The data for groundwater samples plotted as a relatively tight cluster, −9.7‰ to −11.9‰ in $\delta^{18}O$ (mean −11.1‰) and −79‰ to −94‰ for δD (mean −86‰). Like the precipitation data, the results for the groundwater samples fell along the local meteoric water line.
4) Samples of water from the unsaturated zone are divided into two families—those from sites where recharge was high and those from sites where the recharge was low. Higher infiltration and recharge occurred at sites where surface-water runoff accumulated, or the thin soil cover was coarse grained, or bedrock was exposed. Lower recharge was evident at sites without surface runoff or where the soil was fine grained.
5) Notice that data points for water samples in areas low-recharge samples (red symbols) were located away from the local meteoric line along a LEL. It had a slope of about 3.2, appropriate for the arid setting.

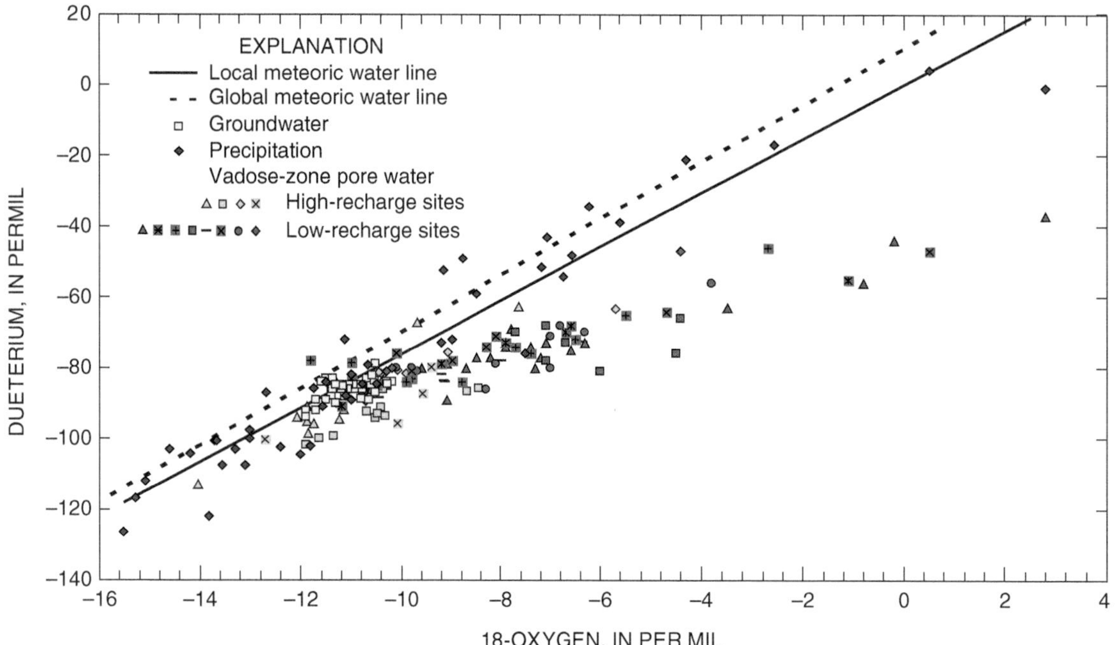

Figure 18.9 Plot of δD versus $\delta^{18}O$ data for a study area in southwestern Utah (U.S. Geological Survey Heilweil et al., 2007/Public domain).

6) The points farther along the evaporation line (mostly the samples in sites with low recharge) were much more evaporated than values from the high-recharge sites. In the higher recharge sites, infiltration into bedrock or coarse soils was sufficiently rapid to move water deeper, out of the influence of evaporation and evapotranspiration. Approximately, 85% of groundwater was recharged through the 50% of the area associated with those deposits.
7) $\delta^{18}O$ and δD values for groundwater were isotopically lighter than weighted mean of values for local precipitation during the cooler winter months. These finding indicated that the bulk of recharge occurred during colder times of the year.
8) Using the age dates associated with the study, researchers found older water, from 50 to a few thousand years of age, was more depleted isotopically than water recharged more recently. One possible interpretation was that the proportion of winter precipitation was higher in the past than it is presently. At this site, winter rainfall came from Pacific winter storms, and summer rainfall from summer monsoonal rainfall.

This section focuses on their application of deuterium and oxygen-18. However, the study also utilized other tracers including bromide, chloride, and tritium. Detailed measurements of tritium concentrations (Heilweil et al., 2007) provided estimates of relatively recent infiltration rates that varied from 1 to >57 mm/yr.

18.5 Radiogenic Age Dating of Groundwater

Groundwater age dating is generally considered as an important tool for the assessment of water resources. For example, these methods are thought to be helpful in estimating the rate at which groundwater supplies are being renewed from recharge, or the pace at which surface sources of contamination is likely to cause problems for aquifers. All this information has potential applications for assessments with respect to sustainability. Also, information on tracer concentrations in the subsurface can contribute to groundwater model development, particularly with conceptual modeling and calibration.

The science around age dating of groundwater, particularly with radiogenic tracers, like carbon-14 and tritium, stretches back more than 50 years with voluminous collections of papers on theory and practical applications. It had always been the hope that age dating would be simple in the sense that a sample of groundwater could be analyzed by itself to provide a straightforward estimate of age, effectively the residence time of water in the groundwater system from the place where it was recharged to where it was sampled. Yet, hopes have been dashed by the increasing sophisticated understanding of the complexity of mass transport processes that operate on age-defining tracers.

It took the development of sophisticated mass transport models to fully appreciate that other physical transport processes, for example, diffusion and dispersion, and advection could also be enormously important in controlling of age-defining tracers. In 2008, Bethke and Johnson proposed a "new way of thinking about groundwater age." They envisioned a water sample not as some closed volume of water tracking along a flow tube from a recharge area to the point of sampling but as "a mixture of waters that have resided in the subsurface for varying lengths of time." This new way of thinking then requires replacing the simple definition of groundwater age (above) with a new definition of groundwater namely—"the average over all the water molecules of the length of time each molecule has spent in the subsurface."

Essentially, their idea is an open system view of how some volume of water moving down a flow line might also mix with nearby groundwater that could be much older or much younger. The water molecules comprising the sample will have travelled a variety of different pathways, fast or slow, before they come together in that sample. Thus, for that sample, there exists what Torgersen et al. (2013) refer to as a "frequency distribution of age." In other words, even though a water sample is a collected from a point, the water molecules there have travelled along a collection of pathways each with a unique age (Suckow, 2014). The problem, as we will explain shortly, is the character of distribution is never known and the measurement of some tracer concentration or activity is usually not the mean of this frequency distribution.

The implications of this "new way of thinking" are enormous. Despite the title for this section, some researchers think that "age," in the way humans think about it, "cannot be determined for groundwater" (Suckow, 2014). The problem is that people think of age in a groundwater sense through common experiences of age, for example, in terms of people, or dogs, or cars. Think about the following analogy. Suppose someone showed you a vintage BMW automobile. With a little work, one could figure out that perhaps it is 25 years old. Now, suppose you looked at a car put together using a miscellaneous collection of parts (1940–2000) from a wrecking yard. How old is this second car? The simplest answer is we do not know for sure.

This idea and others led naturally to definitions of groundwater age that recognizes the inherent complexities. Here are three of the more commonly used definitions (McCallum et al., 2015).

- *Advective age* represents the time required for a particle of water to move from the place of recharge to down the flow system where the groundwater is sampled (McCallum et al., 2015). This time is theoretical in the sense that no processes other than advection are operative. In a modeling framework, these ages/times are those associated with particle tracking. The advective age forms the basis for various interpretive analyses.
- *Tracer age* is the apparent age associated with an environmental tracer assuming that it has a single age. That apparent age is determined as the outcome of a chemical or nuclear process, for example, the radioactive decay of a tracer after it entered a flow system and moved to a point where it was sampled and analyzed. The assumption of a single tracer age is required by absence of modeling tools to estimate the distribution of different ages represented within a water sample (McCallum et al., 2015).
- *Mean Age* is "the average age of water molecules at a location in the aquifer" (McCallum et al., 2015). This accounts for the fact that a water sample is represented by a distribution of ages. The distribution comes from adding macroscopic dispersion in addition to advective dispersion. Think of the dispersion as caused by the water locally followed fast and slow pathways with diffusion added as well. In particle-tracking terms within a highly heterogeneous medium, the travel times for a collection of particles reaching a model cell (sample volume) will be distributed around some mean value. This mean age is not necessarily equal to the advective age.

The use of radioactive tracers commonly developed around the concept of tracer age. Early on, it was recognized that this concept was over-simplified because it did not adequately account for the inherent complexity of geochemical processes involved with a highly reactive tracer like ^{14}C. For example, researchers discovered how certain geochemical processes like mineral dissolution or precipitation, ion exchange, or redox processes involving organic carbon could directly alter the activity of ^{14}C being transported in a flow system. In other words, ^{14}C transport was complicated by other processes besides radioactive decay that could modify the activity of ^{14}C in ways that were difficult to quantify. Simple mathematical approaches for adjusting values to account for these other processes were developed in the 1960s and 1970s. These methods gave way to much more comprehensive geochemical models like NETPATH (Plummer et al., 1991, 1994), which were designed to sort out the influence of some complex collection of reactions influencing carbon species dissolved in water. As capabilities for mass transport modeling developed with analytic solutions and more powerful numerical approach (Bethke and Johnson, 2008; McCallum et al., 2015), it became obvious that transport-related processes of mixing, dispersion, and diffusion also added even more complexity to sorting out age dates.

The evident complexity of age dating has important practical implications for hydrogeological practice. The idea that age dating of tracers like ^{14}C was simply a geochemical effort that involves sampling and activity measurements, correcting for a few processes, and interpreting dates is long gone. Current practice (e.g., Bethke and Johnson, 2008) suggests that beyond a complete characterization of carbon geochemistry, one also needs a comprehensive, model-based description of the physical hydrogeological system. Characterization of necessity would include hard-to-measure features related to fractures, heterogeneity, and characteristics of confining beds, often not well characterized in most studies. As Bethke and Johnson (2008) imply, by the time one has done all this work, you should be able to answer the questions that motivated age dating in the first place.

The age dating literature will need to be examined in a new light—considering that many older case studies provided dates determined using abbreviated or less complete interpretive methods. Retroactively sorting out the likely extent of mixing in the context of an old paper could be impossible.

So, it is a brave new world when 50 plus years of work on age dating is diminished by inadequacies in conceptual models and a growing list of issues affecting processes (Cartwright et al., 2017; Schwartz et al., 2017; Cartwright et al., 2020). Bethke and Johnson (2008) also touch on obstacles created by the challenge of culture in making progress. It used to be that age dating was a largely a geochemical problem of sampling and making laboratory measurements, with certain challenges with correcting for various reactions. Now, "age dating" has morphed into a reactive, mass transport framework that is among the most challenging modeling in hydrogeology.

18.5.1 Exploring Old and New Concepts of Age for Groundwater

Tracer-based dating techniques are based on theory describing the radioactive decay of a nuclide with time. This nuclear, decay process follows a first-order kinetic rate law for decay. At first thinking was that the tracer age of isotope transported in groundwater system could be determined simply by the following equation

$$t = -\frac{t_{1/2}}{\ln 2}\ln\left(\frac{A_0}{A_{obs}}\right) \tag{18.4}$$

where $t_{1/2}$ is the half-life for decay, A_0 is the activity assuming no decay occurs, and A_{obs} is the observed or measured activity of the sample. This simple dating model assumed that some reference volume of groundwater containing a dissolved, radioactive tracer entered a confined aquifer recharge and was carried by flow in that stream tube without dispersion or diffusion (i.e., plug flow) to a point where it is sampled (Figure 18.10). In this case, samples collected at time 1, time 2, time 3 would only reflect the declining activity of the radionuclide due to radioactive decay.

All the assumptions work to create a closed system where neither the water nor the dissolved tracer inside the reference volume interact with water or constituent outside the reference volume. For this idealized situation, the tracer age would be the same as the advective age. So, one can take the ages and make inferences on the velocity of groundwater migration, recharge rates, and rates of contaminant migration. This case, however, is the exception and not the rule. Most often, the advective age is quite different than tracer age. Moreover, without a calibrated transport model, the advective age will be unknown. The temptation might be to assume the advective age is equal to the tracer age and proceed from there. The following example will provide a more example of difficulties in interpretation.

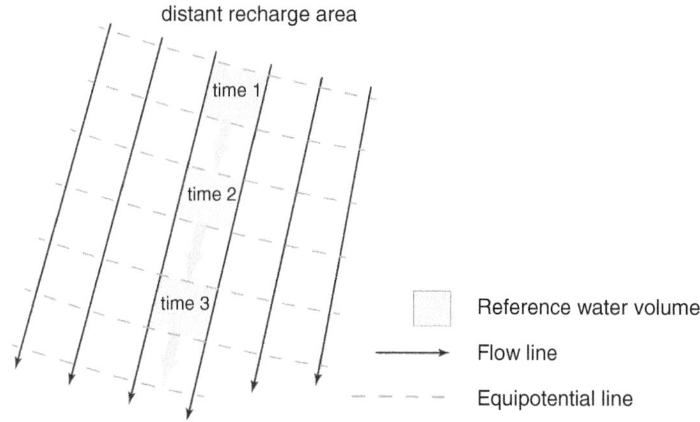

Figure 18.10 Map view of a segment of flow system within a confined aquifer away from a distant recharge area. A reference volume of water is being carried along the flow system as reflected by the time snapshots. Radioactive decay is the only process affecting the tracer activity (FWS).

The example comes from a theoretical, model study by Sudicky and Frind (1981), which examined the importance of diffusion in controlling the activity in ^{14}C in water moving along a hypothetical confined aquifer (Figure 18.11) and more notably what error might exist by assuming the advective age is equal to the tracer age. The flow problem set up here is that of a classical artesian aquifer sandwiched between confining beds. ^{14}C entering the aquifer as recharge is carried along the aquifer by groundwater flow. The only difference with the conceptual model in Figure 18.1 is the possibility for diffusion ^{14}C out of the aquifer into the confining beds (Sudicky and Frind, 1981). This diffusion occurs because the activity of ^{14}C in water in the aquifer is higher than in the confining bed (Figure 18.11). The significance of this diffusion process is determined mainly by the diffusion coefficient D'. A relatively large value, for example, 10^{-6} cm^2/sec makes diffusion in groundwater an important as a process. With values smaller than 10^{-8} cm^2/sec, diffusion is negligible.

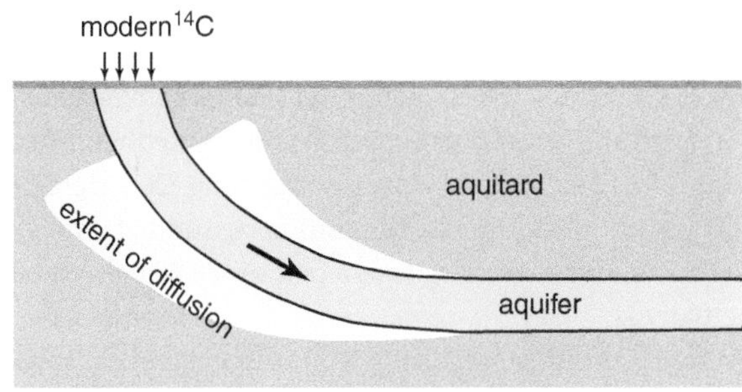

Figure 18.11 Carbon-14 being transported with flow in an extensive confined aquifer is not only lost by radioactive decay but diffusion out of the aquifer into aquitard units (Sudicky and Frind, 1981. Water Resources Research 17(4): 1060–1064). Copyright 1981 by the American Geophysical Union.

By reflecting on the conceptual model (Figure 18.10), one can anticipate what is happening. When diffusion is shut off, this system is essentially closed, and the tracer age is only determined by the decay in the activity of ^{14}C. This is the simple case just discussed with reference to Figure 18.11 and Eq. (18.4) when the advective age equals the tracer age. When diffusion becomes important, ^{14}C is lost from the aquifer not only by radioactive decay but also diffusion of ^{14}C into the aquitards, above and below. Thus, at points along the aquifer the activity of ^{14}C would be less than expected for the closed system, making the water older than expected. In other words, the tracer age at some sample point can be much greater than advective age. The greater the diffusion coefficient the greater the difference in the two ages (Figure 18.12). Using a diffusion coefficient of 10^{-6} cm^2/sec, the age estimates for water samples from this hypothetical aquifer at distance of 40 km are older by a factor of at least 2. In the next section, we describe more of physical processes that add complexity in interpreting ^{14}C.

18.5.2 Carbon-14

This section looks in greater detail at ^{14}C to better understand the complexities inherent in developing useful information from a collection of measurements. This tracer is appealing as an age-dating tracer because of the relative ease of measurement, worldwide distribution in recharge, long half-life that facilitates dating 30,000-year-old water (Cartwright et al., 2020). ^{14}C originates naturally in the upper atmosphere through a reaction involving nitrogen and neutrons. As is the case with tritium, weapons testing in the atmosphere has affected its concentration in recent years. However, except for young waters, this increase does not affect the interpretation.

The source of ^{14}C in groundwater is dissolved inorganic carbon (DIC) in recharge, i.e., the carbonate system in water HCO_3^-, dissolved CO_2, etc. Various sources contributing this carbon is CO_2 in the atmosphere, DIC in surface waters, and most importantly high concentrations of CO_2 in soils, which dissolves into infiltrating waters. Another important source is the dissolution of carbonate rocks as acidic waters move downward through the unsaturated zone (Clark and Fritz, 1997). The activity of ^{14}C in CO_2 gas is approximately usually approximated as a constant at 100 pmc (Fritz and Fontes,

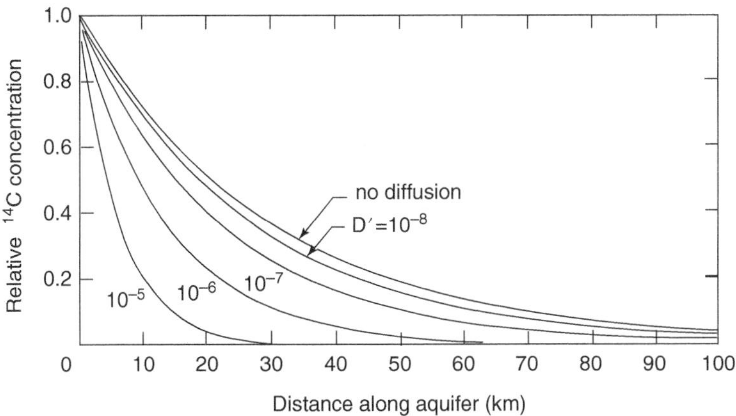

Figure 18.12 As the diffusion coefficient increases more and more ^{14}C is lost from the aquifer. As a result, relative concentrations of ^{14}C at sampling points along the aquifer would be lower with enhanced diffusion into the aquitards (Sudicky and Frind, 1981. Water Resources Research 17(4): 1060–1064. Copyright by the American Geophysical Union).

1980). However, studies show activities in water older than 2000 years BP there is a substantial linear increase in activity to approximately 160 pmc at 30,000 years (Cartwright et al., 2020). Neglecting this loading function underestimates the tracer age. Moreover, the 100 pmc value is not a universal constant. For example, studies in arid or semi-arid regions with lower recharge and deeper water tables find markedly lower values (e.g., 20–50 pmc) deeper in the unsaturated zone (Cartwright et al., 2020). At depths below the root zone, CO_2, depletion in ^{14}C, can be generated from the decay of old organic matter and the loss of CO_2 from root respiration. Assuming a pmc value of 100 in a tracer-age calculation when the actual water entering an aquifer as recharge has a pmc of say 20 will lead to the conclusion that the tracer age of the water is much older than it actually is.

The accumulated evidence points to evident complexities in the use of ^{14}C as an age-dating tracer. Early on, researchers began to realize that reactions existed that modified the ^{14}C signal through additions of "dead carbon" (i.e., no ^{14}C) or removal of ^{14}C. For example, the dissolution of old calcite adds dead carbon and reduces the ^{14}C activity in the water. A fix for this problem came from adjustments to correct for this addition of dead carbon. In terms of Eq. (18.4), theory provided an adjustment with an A_0 below 100 pmc (the ^{14}C activity assuming no decay) to account for the fact that another process besides radioactive decay influenced the ^{14}C activity of the sample. Thinking was that calculations of tracer ages were meaningful as long as A_0 and A_{obs} differed only because of the effects of radioactive decay.

It turned out that a variety of different reactions could alter the ^{14}C activity of groundwater (Mook, 1980; Reardon and Fritz, 1978; Wigley et al., 1978). Examples include:

1) The congruent dissolution of carbonate minerals, which we just discussed. This reaction adds dead carbon which lowers the ^{14}C activity measured for the sample, making the water appear older than it actually is.
2) The incongruent dissolution of carbonate or other Ca-containing minerals, accompanied by the precipitation of calcite. This process will remove ^{14}C as calcite precipitates, and if dolomite is the mineral dissolving, additional dead carbon comes through (1) above. This process could occur in the zone of saturation following the rapid solution of calcite to equilibrium, with subsequent precipitation as dolomite slowly dissolved.
3) The addition of dead carbon from other sources such as the oxidation of old organic matter, sulfate reduction, and methanogenesis. Again. these reduce the ^{14}C activity of the sample.

Approaches of varying degrees of sophistication were developed to estimate groundwater age, taking account of these kinds of interactions. The simplest approaches used ion and isotopic data for a single sample without information from other samples. These approaches are not really used these days because they are restricted to simple or idealized situations. Readers interested in these approaches should refer to an excellent overview by Torgersen et al. (2013).

The more sophisticated approach interprets ^{14}C data using a computer code that accounts for the variety of processes that influence carbon mass balances. The best example is the NETPATH model (Plummer et al., 1991, 1994), which is still commonly used for interpreting ^{14}C data when both isotopic and water-chemistry data are available.

Aravena et al. (1995) used ^{14}C to date water from the Alliston aquifer in Ontario, Canada. Their NETPATH analysis showed that the key reactions influencing the chemical evolution of groundwater in the aquifer included incongruent dissolution of dolomite, ion exchange, methanogenesis, and oxidation of sedimentary organic matter. This study illustrated the complexity of geochemical processes that work at a real site to influence the ^{14}C abundance in groundwater. Yet, one would still need to be cautious in terms of how these kinds of dates are being used, because the effects of dispersion and diffusion were not considered in the analysis.

The next major step forward came with the recent evolution of powerful reactive transport models that could handle the complexities of reactions in addition to physical mass transport, including advection, dispersion, and diffusion. We already have discussed issues around diffusion into confining beds in creating mismatches in the advective ages and tracer ages (Sudicky and Frind, 1981). Other processes, like matrix diffusion with fractured media, or diffusion into lower permeability layers or lenses within aquifers all have the potential to reduce the activity of ^{14}C within an active flow system. Moreover, all these processes could work together. The addition of dispersion creates a distribution of ages, which can be characterized in terms of a mean age. In multi-aquifer systems, there is the potential for cross formational flow producing mixing at an even larger scale.

Modeling studies (Bethke and Johnson, 2008; McCallum et al., 2015) found that tracer ages were invariably different than advective ages in complicated systems. More importantly, even knowing this problem existed corrections are not feasible. Recommendation from both these studies are to move away from the tracer age-correction paradigm to full-blown modeling approaches that simulate ages directly or use the concentration of a reactive tracer (e.g., activity of ^{14}C) or several tracers to calibrate a two or three-dimensional flow and transport model (Bethke and Johnson, 2008; Suckow, 2014; McCallum et al., 2015). This latter approach essentially does away with the notion of groundwater age. This is really no loss because the modeling itself provides estimates of recharge rates, groundwater velocities, etc.; the very information that tracer-age analyses were meant to provide. This model-based approach will be illustrated with a case study in the following section.

18.5.3 Chlorine-36 and Helium-4: Very Old Groundwater

In practice, ^{14}C measurements are useful in systems of moderate size with maximum advective ages for water of ~40,000 years old. However, studies of geological basins or extremely large aquifers require tracers with longer half-lives for use in model calibration.

One useful isotope in this respect is ^{36}Cl ($t_{1/2} = 3.01 \times 10^5$ years) (Phillips, 2000). Given its relatively long half-life, ^{36}Cl can be applied in systems with advective ages from 50,000 to about 1.5 million years. Again, the most important source of this isotope is fallout from the atmosphere. However, ^{36}Cl can be problematic as a tracer because it can be generated in the subsurface. Key processes include the irradiation of elements like ^{35}Cl, ^{40}Ca, or ^{39}K or the decay of uranium or thorium (Cartwright et al., 2017). The quantity of *in situ* production depends upon features of the mineralogy and water chemistry. The problem comes from the fact that these *in situ* sources are difficult to identify and to characterize. The use of ^{36}Cl is complicated by variability in the ^{36}Cl activity of recharge and by groundwater sources of chloride Cl, which is a component part of the measure of abundance, that is $^{36}Cl/Cl$.

There are issues with ^{36}Cl used when used as species to calibrate basin scale models. The precision of measurements is such that there is only a relatively small window for measurements, between one and "several" half-lives (e.g., 300,000 to say 1.5 million years) (Bethke and Johnson, 2008). Thus, there is little calibration power within the younger, more active parts of basinal systems or much older waters in confining beds or more sluggish parts of flow systems. For basinal systems, ^4He seems to be a better choice, as we discuss shortly.

Levels of ^{36}Cl in the atmosphere, as was the case with tritium, were elevated up to 2 or 3 orders of magnitude because of global fallout from high-yield nuclear weapons tests in the 1950s. The presence of so-called bomb-pulse ^{36}Cl in a water sample provides a clear indication of the relatively young age of the sample. This application has turned out to be useful been useful in sorting out the patterns of complex unsaturated zone recharge (Fabryka-Martin et al., 1993). Other previous applications of ^{36}Cl have been in dating groundwater in large regional aquifers—the Great Artesian Basin of Australia (Bentley et al., 1986) and the Milk River aquifer in Alberta, Canada (Phillips et al., 1986).

Although ^4He itself is a stable isotope, its usefulness in studies of large basinal flow systems is a reason for a brief overview. ^4He is a product of the radioactive decay of nuclides like uranium and thorium. Because the half-lives of these nuclides are extremely long, the rate of ^4He production is relatively constant (Bethke and Johnson, 2008). Unlike radionuclides that decay as a function of residence times in the subsurface, ^4He accumulates linearly. Recall Figure 18.11 that shows a reference volume of water moving downgradient in a confined aquifer. In this case, the isotope is ^4He with some initial concentration (c_0) that represents its concentration in the atmosphere. As the water volume moves down the aquifer, ^4He accumulates not only due to the decay uranium and thorium contained in the aquifer solids, and finer-grained lenses or interbeds but also the confining beds above and below the aquifer. The finer-grained units may in fact produce the largest concentrations of ^4He (Bethke and Johnson, 2008). The following equation (Bethke and Johnson, 2008) describes the ^4He concentration (c) at some time (t) in an aquifer due to linear accumulation:

$$c = c_0 + \frac{R_\propto t}{\phi} \tag{18.5}$$

where c_0, is the ^4He concentration (mol/m^3) in recharge, R_α is the rate of production (mol/m^3 yr), ϕ is porosity (dimensionless). Rearranging Eq. (18.5) provides the tracer age for some collected sample:

$$t = \frac{\phi(c - c_0)}{R_\propto} \tag{18.6}$$

Earlier discussions pointed out some of the inherent problems in working with equations (e.g., Eq. (18.6)) based on plug flow assumptions. A paper by Bethke and Johnson (2008) illustrated more rigorous, model-based approaches for working with tracers, which has promise of achieving the original idea of age dating. The original study was presented in a more detailed paper (Bethke et al., 1999), which was summarized in Bethke and Johnson (2008). The basic approach involved two types of numerical models. The first was a sophisticated digital model, beyond the scope of this book, that directly simulates water age. The second is a reactive, chemical transport model, which simulated the steady-state pattern of groundwater flow across a model grid along with concentrations of ^4He.

Briefly, the latter type of model requires a detailed numerical characterization of the hydrogeologic setting, which is based on grid cells in a two-dimensional grid or grid blocks in a three-dimensional grid. The spatial distribution of key hydraulic parameters, like hydraulic conductivity, porosity, etc. reproduces the pattern of layering in terms of aquifers and confining beds. Boundary conditions determine the location of inflows and outflows of water and mass, as does mass loading rates for designated nodes at specified, internal grid points.

The model is calibrated by adjusting the model parameters to produce a spatial distribution in concentration of one or several species that compares favorably with measured tracer concentrations. A satisfactorily calibrated model provides a detailed quantitative description of all aspects of the basin hydrology using the geochemical tracer, but without necessarily worrying about the age date (Bethke and Johnson, 2008).

Modeling reactive transport within a sedimentary basin is a difficult undertaking (Bethke and Johnson, 2008). However, codes are becoming increasingly available. For example, Salmon et al. (2015) describe a flow and transport modeling package that can integrate processes related to groundwater flow, and reactions bearing on the chemical evolution of constituents in the water, as well as $\delta^{13}C$ and the activity of ^{14}C. This approach uses the general reactive transport code PHT3D developed by Prommer et al. (2003). Versions of HydroGeoSphere (Therrien et al., 2006) can do these calculations as well as other codes. The greater problem moving forward is the huge quantity of hydrostratigraphic and hydrogeological data needed to apply these models to real systems. There is uncertainty in key thermodynamic parameters, rate laws, and coefficients for kinetic reactions. Mineralogical data are commonly in poor supply along with values of dispersivity, effective diffusion coefficients, fracture characteristics, recharge rates and more. Unfortunately, these data problems may not be solvable.

18.5.4 Tritium

Tritium is another radiogenic isotope that has important hydrogeological applications. Tritium concentrations are reported in terms of tritium units (TU), with 1 TU corresponding to one atom of ^3H in 10^{18} atoms of ^1H (Fontes, 1980). Tritium occurs naturally in the atmosphere, with concentrations in modern precipitation less than 10 TU. However, tritium, generated by thermonuclear testing in the atmosphere between about 1952 and 1963, swamped the natural production of tritium.

Given its relatively small half-life ($t_{1/2}$ 12.2 years) and its nonreactive character, tritium has found a variety of different uses in water-cycle problems on decadal timescales (Michel et al., 2018). Concentrations of tritium are controlled by the magnitude of loading from rainfall, radioactive decay and mixing of waters within the hydrologic cycle, such as precipitation and groundwater.

Jurgens et al. (2012) presented an example of a problem that illustrates the loading and mixing components, which was developed in an earlier study by Michel (2004). The problem was to explain processes that contributed to measured tritium concentrations in the upper Missouri River at Nebraska City, Nebraska. The conceptual model assumed that tritium concentrations in the river were a binary mixture of (1) rainfall and runoff and (2) groundwater that was recharged in the past with decay due to residence in the subsurface. Figure 18.13 shows the tritium concentrations due to precipitation and measured concentrations in the river (Jurgens et al., 2012). The modeling involved an optimization strategy to find values of mean groundwater age (4.3 years) and mixing ratio (84% groundwater) that yielded the best fit between calculated and observed concentrations of tritium in the river. These calculations were facilitated using TracerLPM, an interactive Excel® (2007 or later) workbook program (Jurgens et al., 2012) for evaluating groundwater age distributions. As mentioned earlier in this chapter, several different mathematical models are available in the package to evaluate prototypical settings and to explain observed measurements of environmental tracers.

The tritium concentrations in precipitation (Figure 18.13) are notable for their amazing variability. For example, at Nebraska City, nuclear weapons testing in 1962–1963 produced tritium levels approaching 5000 TU. By the late 1960s, tritium levels in precipitation had declined by an order of magnitude (~450 TU) with the reduction in atmospheric weapons testing. Concentration steadily declined to present-day levels of ~ 9 TU. The tritium spike for precipitation in Nebraska was relatively high for the continental United States. Comparably high numbers were measured in Ottawa, Canada. Higher concentrations ~10,000 TU were evident in northern Canada. The higher historical concentrations of tritium at higher latitudes in North America were related to the where the nuclear testing occurred. The largest tests conducted in the Union of Soviet Socialist Republics (USSR) were to the far north at Novaya Zemlya.

Figure 18.14 is a map of the total tritium deposited from 1953–1983 (Michel et al., 2018). This time period reflects the tritium generated from weapons testing. For each quadrangle on the map, reconstructed values in monthly tritium

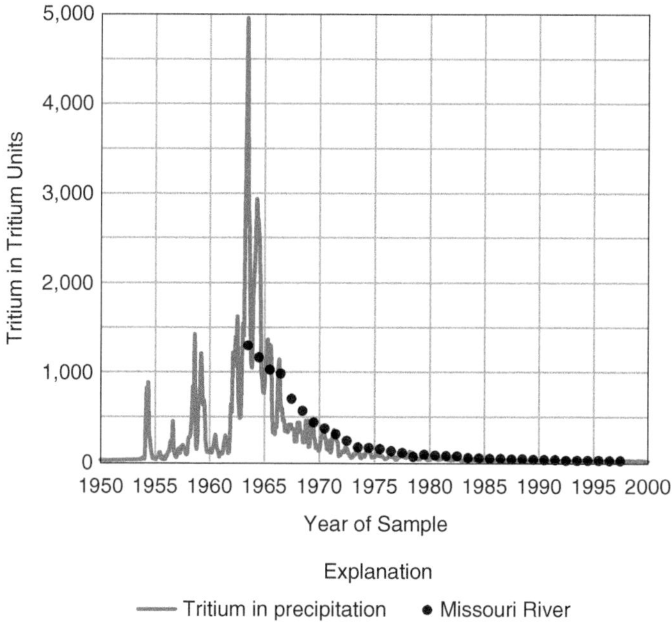

Figure 18.13 Shown in blue is the reconstructed loading of tritium in precipitation at Nebraska City from 1950 to 2000. The black circles are measured tritium concentrations in the Missouri River (U.S. Geological Survey Jurgens et al., 2012, U.S. Government work, adapted from Michel, 2004 published by Wiley InterScience).

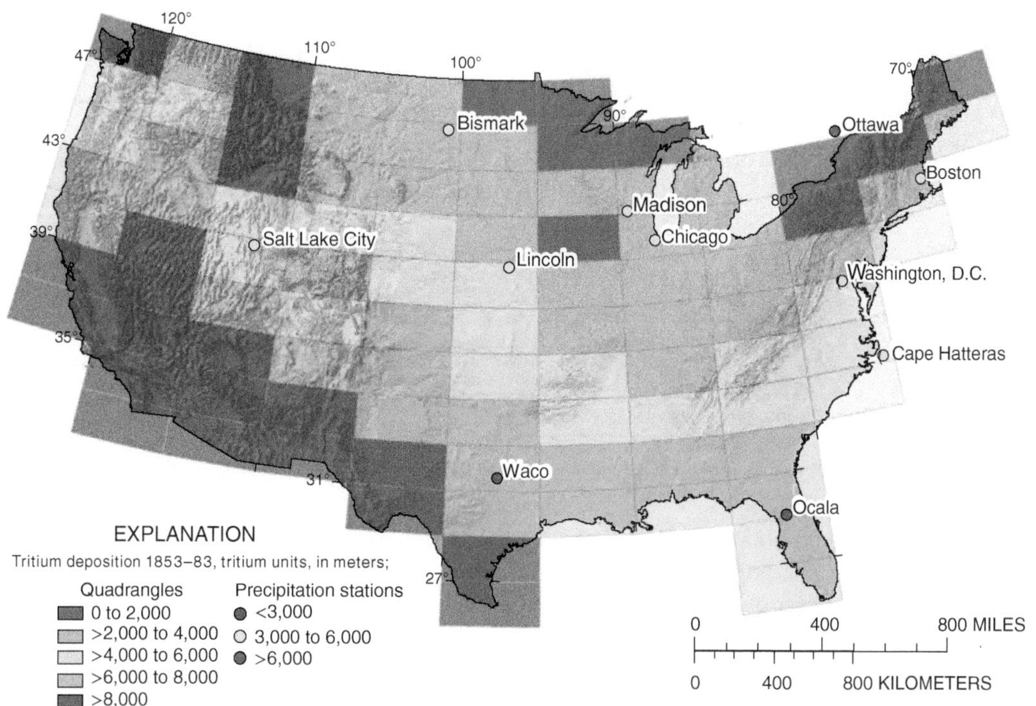

Figure 18.14 Map showing tritium deposition from 1953 to 1983, which reflects the impact of weapons testing (U.S. Geological Survey Michel et al., 2018/Public domain).

concentrations were multiplied by the average precipitation values (in meters [m]) for that month. The sum of the monthly values over the 30 years on the map represents tritium deposition (Michel et al., 2018).

The map clearly shows latitudinal aspects of tritium deposition with northerly stations, for example, Bismarck, North Dakota greater than more southerly stations, for example, Lincoln, Nebraska (Figure 18.14). Michel et al. (2018) also noted a "continental" effect where coastal quadrangles experienced less deposition than interior quadrangles at the same latitude. Coastal precipitation is modified by interactions of atmospheric water vapor with ocean surface water, low in tritium. Cape Hatteras is a coastal station that receives precipitation influenced by ocean interactions. Precipitation also exhibits seasonal variability associated with weather patterns and tropospheric/stratospheric interactions.

Ideally, by knowing the concentration of ^3H in precipitation (the source) and its distribution in groundwater, one should be able to roughly estimate the tracer age of the water. Historically, it had been difficult to use tritium in such a quantitative way. The main problems stemmed from the uncertainty and complexity of atmospheric loading. However, the studies just described have significantly reduced uncertainty with tritium applications. However, with the passage of time, the usefulness of tritium has faded as radioactive decay has largely eliminated the bomb-pulse signal.

This limitation of tritium dating can be overcome by adding measurements of helium-3 (^3He), the stable daughter of tritium decay. The isotopic decay reaction is written as

$$^3H \rightarrow \ ^3He^* + \beta^-$$

where the * indicates a tritiogenic source of ^3He (Solomon et al., 1995). Other sources of ^3He exist that must be appropriately accounted for. When both ^3H and $^3He^*$ measurements are available, summing ^3H (TU) and $^3He^*$ (as TU) eliminates the decay of ^3H (Solomon et al., 1995). Thus, one could determine, for example, the peak concentration coinciding with 1960s bomb testing. In addition, one could determine the $^3H/^3He^*$ age of the water. The $^3H/^3He^*$ age of a groundwater sample can be calculated as

$$t = \frac{t_{1/2}}{\ln 2} \ln\left(\frac{^3He^*}{^3H} + 1\right) \tag{18.7}$$

where $t_{1/2}$ is the half-life for decay of ^3H, and ^3H and $^3He^*$ are tritium concentrations expressed in tritium units.

Example 18.2 A groundwater was found to contain 14.8 TU of tritium and 31.6 TU of tritiogenic helium ($^3He^*$). Calculate the age of the groundwater. The half-life for tritium decay is 12.3 years.

Solution

The age of the water can be determined by substituting values into Eq. (18.7):

$$t = \frac{t_{1/2}}{\ln 2} \ln\left(\frac{^3He^*}{^3H} + 1\right) = \frac{12.3}{0.693}\left(\frac{31.6}{14.8} + 1\right) = 18.2 \text{ years}$$

18.5.5 Categorial Assessments Using Tritium Ages

Historically, one of the important applications of tritium in groundwater investigations is for the categorial assessments based on tritium concentrations. The simplest approach involved sorting groundwaters with tritium measurements into two categories, (1) *modern* that identified groundwater recharged after the start of nuclear weapons testing in the 1950s, and (2) *premodern* that identified water recharged before the 1950s (Lindsey et al., 2019).

This simple type of classification scheme is useful in establishing the risk of potential contamination. For example, aquifers with pore-water categorized as modern, are at risk for modern contaminants, such as industrial chemicals, agrochemicals (herbicides, pesticides, and nutrients), petrochemicals, and landfill leachates. Premodern water would be at risk for geogenic contaminants, such arsenic, and other heavy metals that accumulate over long times (Lindsey et al., 2019). Categorial analyses are also helpful in assuring the quality of obviously old water samples. The presence of tritium where it is not expected could suggest faulty seals on deep piezometers, inadequate development before sampling, or leakage from abandoned wells. Another possibility is a flawed conceptual model, such as overlooking the presence of localized vertical fractures. In other words, the "obviously old water" may not be so obvious.

The initial approach to this categorization was straightforward. A single threshold value, such as 0.5 TU or 1 TU, was used assign a water sample to a category based on a tritium analysis—sample tritium level > threshold was modern, and sample tritium level ≤ threshold was premodern (Lindsey et al., 2019). As information on the spatial and temporal variability in tritium loading was developed for the continental United States, a more sophisticated categorization approach involved unique thresholds for tritium samples colocated in a given quadrangle. In other words, the thresholds take account of when the samples were collected and where it was collected (Lindsey et al., 2019). In water resource assessments, one might encounter measurement on tritium samples collected decades apart from each other. Thus, older tritium analyses would require a higher modern threshold than a recently collected sample. In terms of location, the modern threshold value for samples from Nebraska should be higher than values for samples from Florida collected at the same time. This strategy recognizes that tritium values for rainfall in more northerly locations might be considerably higher than for coastal settings.

Lindsey et al. (2019) also added a mixed category between modern and premodern. This mixed category represents a mixture of modern and premodern water that could be created, for example, by pumping and induced infiltration. Figure 18.15 illustrates how the two cutoff values are established. Briefly, for a given quadrangle in the continental United States, for example, it will be possible to create a tritium loading curve for precipitation from 1953 to present day using two estimates per year. That loading curve would apply to all samples from that quadrangle (dark blue line, Figure 18.15). A second curve (light blue line, Figure 18.15) is created to account for the year the sample was collected by decaying the precipitation curve to 2015. This second line provides modern/mixed threshold for 2015, as the "minimum post-peak concentration of decayed ^{3}H in precipitation" (Lindsey et al., 2019). This value is 2.89 as shown on Figure 18.15.

Next the mixed/premodern threshold is determined first by averaging values of stabilized tritium in precipitation. In this example, the average of tritium concentrations determined for precipitation from 2008 to 2012 is 9.53 TU. The assumption is that present-day samples are no longer influenced by nuclear testing. They in effect approximate tritium concentrations in precipitation before weapons testing (i.e., 1952). A final step is needed to decay 9.53 TU over about 63 years from 1952 to

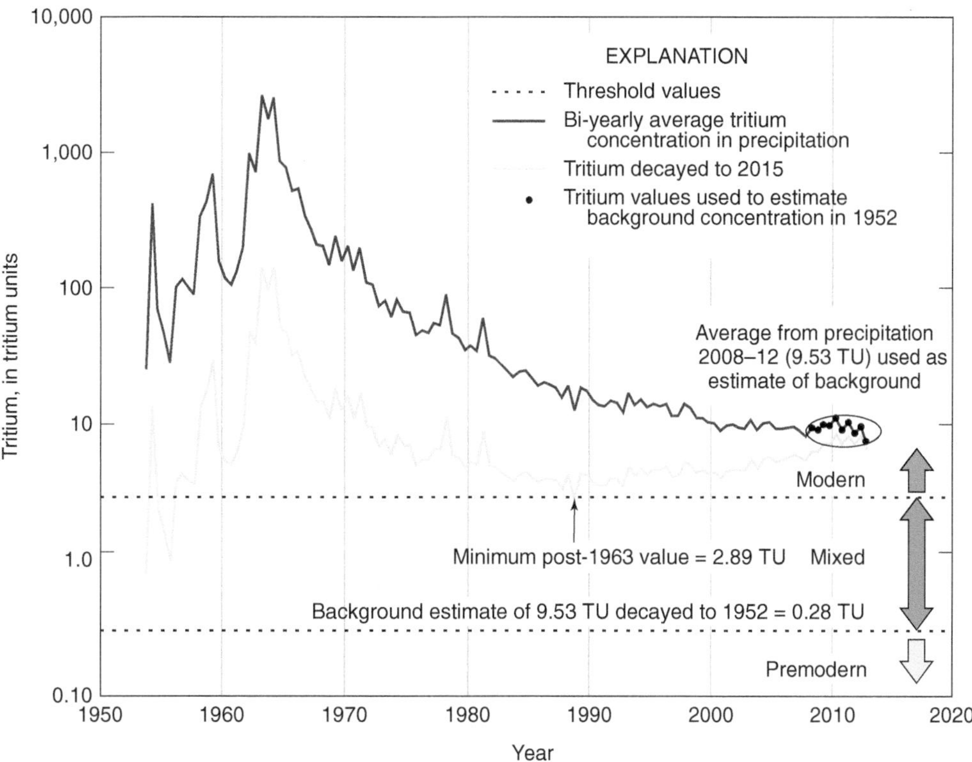

Figure 18.15 Example of a categorial approach based on tritium measurements for separating groundwater into modern, mixed and premodern classes. The threshold values apply to samples collected in 2015 within 41–43°N. latitude and 90–95°W. longitude (U.S. Geological Survey Lindsey et al., 2019/Public domain).

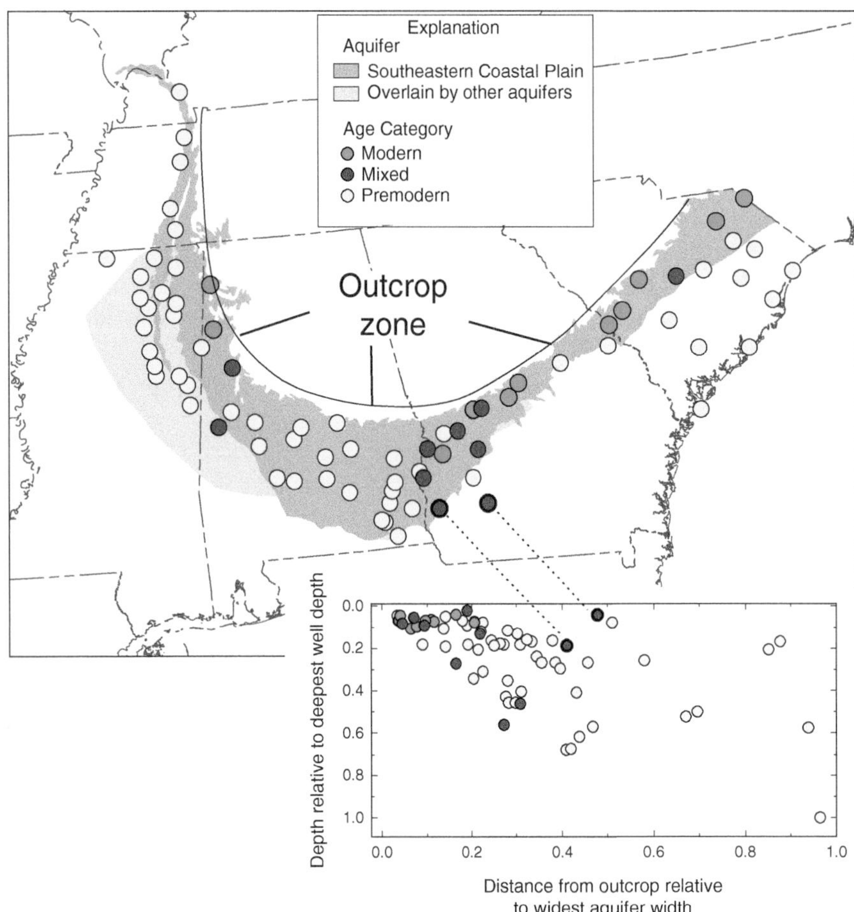

Figure 18.16 Example of a categorial analysis of relative groundwater age based on tritium for the Southeastern Coastal Plain aquifer system in the U.S. The aquifer extends from Mississippi through to South Carolina (U.S. Geological Survey Lindsey et al., 2019/Public domain).

2015. This time period corresponds to decay over about five half-lives for tritium ($t_{1/2} = 12.2$ years), which yields the mixed/premodern threshold of 0.28 TU.

Lindsey et al. (2019) confirmed the efficacy of this classification-based approach at a continental scale using samples from 1,788 water supply wells emplaced in 19 principal aquifers. They also evaluated the approach on a regional scale using samples collected and measured from the Southeaster Coastal Plain aquifer system in Mississippi, Alabama, Georgia, and South Carolina (Figure 18.16). As expected, samples classed as modern were mostly collected from areas where modern recharge is presently occurring, which is at shallow depth in recharge areas. Similarly, mixed samples were found at shallow depths as well (Lindsey et al. 2019).

18.6 Indirect Approaches to Age Dating

Indirect methods depend on interpreting systematic changes in the chemical composition of indicator species or isotopes along groundwater flow paths in relation events or processes that provide an unambiguous time stamp. The indirect approaches work because the isotopic composition or dissolved concentration of some tracer changes with time in recharge. When the time variation is known, one can directly infer the apparent age of a sample from its measured abundance in groundwater. Of course, the simple approaches to interpret these ages provide apparent ages because of errors in not considering dispersion and diffusion. This section discusses the use of two practical approaches that involve environmental isotopes ($\delta^{18}O$ and δD) and atmospheric contaminants with a known history of use (chlorofluorocarbons and sulfur hexafluoride).

18.6.1 Isotopically Light Glacial Recharge

Examples presented earlier in this chapter illustrate how a time-varying signal in the $\delta^{18}O$ and δD levels in precipitation are preserved in large aquifer systems or in units having extremely low hydraulic conductivity. In regions like western Canada or northern Wisconsin, the isotopic composition of water associated with glaciation is reasonably well known. Thus, systematically documenting the isotopic character of water samples from low-permeability sites can provide an indirect age. Water that is isotopically much lighter than present-day meteoric water recharging shallow aquifers is likely relatively old, originating during colder glacial climates.

18.6.2 Chlorofluorocarbons and Sulfur Hexafluoride

Beginning in the 1930s, a variety of gaseous organic substances was developed and began to be released into the atmosphere. Once dissolved into precipitation, they infiltrated the subsurface and eventually made their way into shallow groundwater. An age-dating method developed in the 1990s focused initially on the chlorofluorocarbons (CFCs; i.e., CFC-11, CFC-12, and CFC-113) (Plummer and Friedman, 1999). The residence time in the atmosphere is 44 years for CFC-11 (trichlorofluoromethane), 90 years for CFC-113 (trichlorofluoromethane), and 180 years for CFC-12 (dichlorofluoromethane) (Cartwright et al., 2017). This family of organic compounds was widely used as propellants in aerosol cans and refrigerants (Freon). As environmental tracers, they were particularly useful because their concentration in the atmosphere could be well characterized with concentrations increasing through time into the 1990s (Figure 18.17). Beginning in 1990s, the use of CFCs was curtailed because of concerns related to the destruction of ozone in the Earth's upper atmosphere. Figure 18.17 shows the resulting reduction in atmospheric abundances of CFCs.

This decline in atmospheric concentrations of CFCs has reduced the usefulness of CFCs as an age-dating tracer. In essence, the loading function has become less unique; for example, a relatively high CFC-11 concentration in water could be from 2010 or the early 1990s (Figure 18.17). Nevertheless, the presence of CFCs in groundwater at relatively high concentrations suggests very young groundwater.

SF6 (sulfur hexafluoride) has been developed as an alternative to CFCs in providing apparent ages for very young groundwater (Busenberg and Plummer, 2000). Since the early 1950s, SF6 as a gas has been used in industrial applications (Cartwright et al., 2017). The reconstruction for the northern hemisphere points to its rapid and continuing accumulation in the Earth's atmosphere (Figure 18.17). Unlike CFCs, there are natural sources of SF6 in the atmosphere, but these appear to be small and need not be considered for groundwater applications (Busenberg and Plummer, 2000).

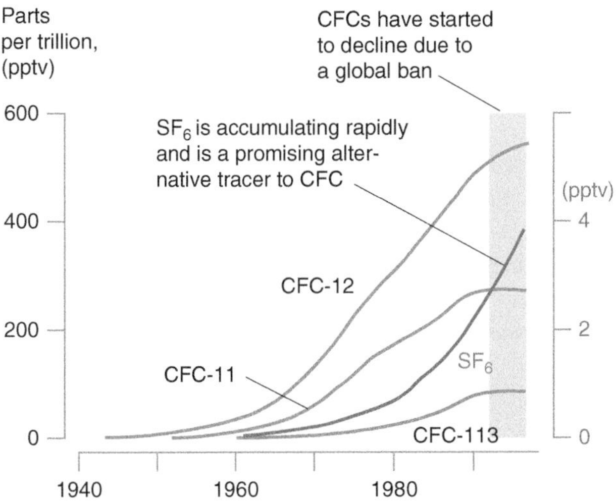

Figure 18.17 Reconstruction of atmospheric concentrations of CFC-11, CFC-12, CFC-113 and sulfur hexafluoride (SF6) since 1940 over North America (U.S. Geological Survey Plummer and Friedman, 1999/Public domain).

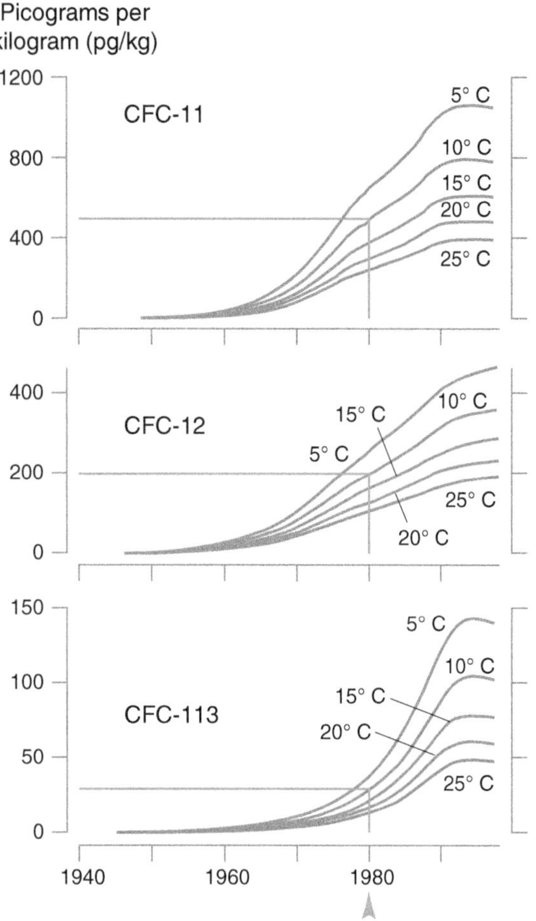

Figure 18.18 The equilibrium concentrations of CFC-11, CFC-12, and CFC-113 dissolved in water depend upon temperature. Knowing the groundwater temperature, the families of curves from 5 to 25°C facilitate the temperature adjustment. For example, analytical results of a water sample with a 10° C recharge temperature; 493 pg/kg with CFC-11, 203 pg/kg with CFC-12 or 28 pg/kg with CFC-113 all provide a time of recharge of 1980 (U.S. Geological Survey Plummer and Friedman, 1999/Public domain).

CFC and SF6 dating methods for groundwater applications owe much of their success to the efforts of scientists at the U.S. Geological Survey (USGS) in overcoming the technical difficulties in simply using these tracers and verifying their efficacy in different settings (Busenberg and Plummer, 1992; Dunkle et al., 1993; Busenberg and Plummer, 2000). The potential use of CFCs in dating young groundwater had been recognized since the mid-1970s (Thompson, 1976; Thompson and Hayes, 1979). These compounds have also been used in tracer experiments (Randall and Schultz, 1976).

The dating method assumes that the concentration of dissolved tracer concentrations in precipitation is in equilibrium with the atmospheric gas-phase concentrations according to Henry's Law. However, the equilibrium constant describing the partitioning of gases into water is temperature dependent. This means that for each of the CFCs there is family of historical loading curves, 0, 5, 10, 15°C, etc. (Figure 18.18).

A simple approach to using CFCs or SF6 in age dating involves collecting a sample of groundwater, measuring the concentrations of the tracer and utilizing those concentrations together with the loading function at the proper temperature to obtain a date of recharge. While simple in concept, the method is technically challenging. To ensure that samples are completely isolated from the air, Busenberg and Plummer (1992) sealed samples in glass ampules by heat fusing immediately following their collection. The laboratory measurements are also difficult to perform because it is imperative that samples (especially old samples) not be contaminated with modern air, which contains high tracer concentrations. Readers interested in a detailed discussion of the analytical techniques can refer to guidance documents.

In using these atmospheric tracers, one also needs to be aware of the problem of "excess air." As Busenberg and Plummer point out, almost all groundwater is supersaturated with air. Excess air is produced by the dissolution of air bubbles entrapped in the water as it infiltrated. With this additional air, concentrations of CFCs or SF6 in the water end higher than the equilibrium concentration and if uncorrected, make the water sample appear younger than it actually is. This problem is less severe for CFCs than SF6, because the Henry's Law Constant for SF6 is smaller than for the CFCs. Interested readers can refer to Busenberg and Plummer (2000) for additional guidance.

An example of the application of CFC dating comes from a 1993 study of groundwater of the Delmarva Peninsula, which is located along the eastern side of Chesapeake Bay (Dunkle et al., 1993). CFC-11 and CFC-12 data were collected for nests of wells along a cross section (F1–F1′) completed in a shallow surficial aquifer near the town of Fairmont. Figure 18.19 indicates the CFC-11 or CFC-12 (pg/kg) ages. Usually, the oldest value from a pair of available dates (CFC-11 or CFC-12) for each sample was selected as the best estimate. This shallow flow system was relatively active with apparent water ages ranging from 1986 to 1963. The tracer ages generally agreed with relative ages based on hydrologic arguments (Dunkle et al., 1993). The concentration of nitrate was also plotted on the cross section. Notice in a farmed area in the vicinity of wells 68 and 69, nitrate concentrations in shallow exceeded the 10 mg/L standard. Beneath the forested area in the vicinity of wells 70 and 71, the nitrate concentration was 0.9 mg/L, much lower. However, nitrate contaminated water appears to be discharging to surface water (Figure 18.19).

Like many other age dating techniques, there are limitations. For example, CFC concentrations in anaerobic environments may be reduced through microbial degradation. Processes like dispersion and sorption can also influence CFC concentrations (Dunkle et al., 1993). In spite of limitations, the approach has potential for dating relatively young waters.

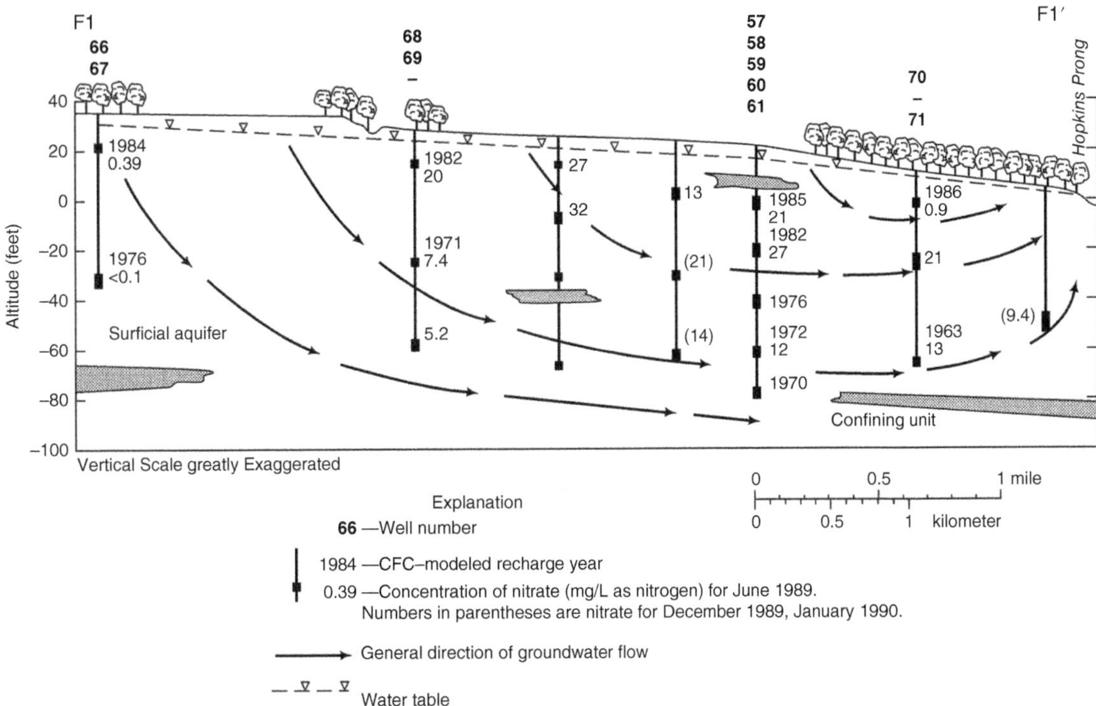

Figure 18.19 Cross section of Fairmount network along F1–F′ showing CFC-modeled recharge years, nitrate concentrations (in milligrams per liter as N) and generalized groundwater flow paths (U.S. Geological Survey Dunkle et al., 1993/U.S Government work published in 1993 by American Geophysical Union).

Exercises

18.1 A large lake has $\delta^{18}O$ and δD values of −14‰ and −104‰, respectively. Calculate the isotopic composition of vapor in equilibrium with this water at 10°C.

18.2 Explain why the $\delta^{18}O$ and δD composition of rainfall changes as air masses move inland on the continents.

18.3 Shown in Figure 18.20 are isotopic data collected from the confined aquifer at Alliston, Ontario, Canada (Aravena et al., 1995). The rectangle shows the isotopic composition of present-day meteoric water at the site.

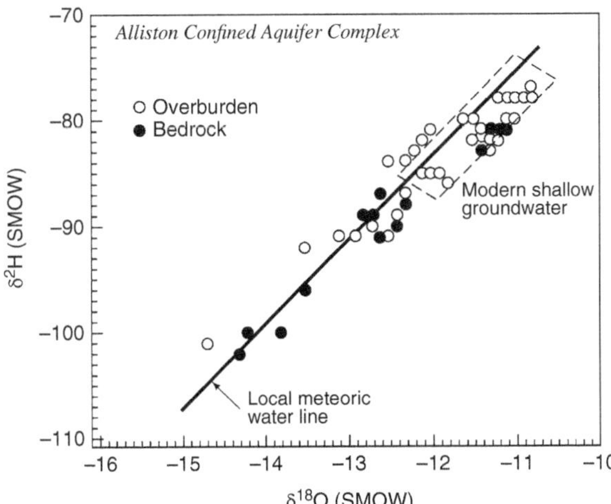

Figure 18.20 Plot of $\delta^{18}O$ versus δD contents for groundwater samples from the Alliston Aquifer (Aravena et al., 1995. Water Resources Research 31(9): 2107–2317. Copyright 1995 by the American Geophysical Union).

A Is there evidence that groundwater has been evaporated in the past? Explain your answer.
B Given that the study area is located in the southern part of Canada, suggest one reason why some groundwater samples are isotopically different from others.

18.4 Tritium can be applied in groundwater studies as both a direct and an indirect way of dating groundwaters. Explain the difference in these approaches.

18.5 An analysis of groundwater found ^{3}H and $^{3}He^{*}$ contents of 27.3 TU and 39.1 TU, respectively. Calculate the age of the water sample.

18.6 The concentration of CFC-11 in four groundwater samples was measured as 1.9, 76.9, 176.1, and 430.4 pg/kg (data from Dunkle et al., 1993). Estimate the year in which recharge occurred, assuming a recharge temperature of 10°C.

References

Aravena, R., L. I. Wassenaar, and L. N. Plummer. 1995. Estimating ^{14}C groundwater ages in a methanogenic aquifer. Water Resources Research, v. 31, no. 9, p. 2307–2317.

Benettin, P., T. H. Volkmann, J. von Freyberg et al. 2018. Effects of climatic seasonality on the isotopic composition of evaporating soil waters. Hydrology and Earth System Sciences, v. 22, no. 5, p. 2881–2890.

Bentley, H. W., F. M. Phillips, S. N. Davis et al. 1986. Chlorine 36 dating of very old groundwater: 1. The Great Artesian Basin, Australia. Water Resources Research, v. 22, no. 13, p. 1991–2001.

Bethke, C. M., and T. M. Johnson. 2008. Groundwater age and groundwater age dating. Annual Review of Earth and Planetary Science, v. 36, no. 1, p. 121–152.

Bethke, C. M., X. Zhao, and T. Torgersen. 1999. Groundwater flow and the ^{4}He distribution in the Great Artesian Basin of Australia. Journal of Geophysical Research, v. 104, no. B6, p. 12999–13011.

Brooks, J. R., D. M. Mushet, M. K. Vanderhoof et al. 2018. Estimating wetland connectivity to streams in the Prairie Pothole Region: an isotopic and remote sensing approach. Water Resources Research, v. 54, no. 2, p. 955–977.

Busenberg, E., and L. N. Plummer. 1992. Use of chlorofluorocarbons (CCl_3F and CCl_2F_2) as hydrologic tracers and age-dating tools: the alluvium and terrace system of central Oklahoma. Water Resources Research, v. 28, no. 9, p. 2257–2283.

Busenberg, E., and L. N. Plummer. 2000. Dating young groundwater with sulfur hexafluoride: natural and anthropogenic sources of sulfur hexafluoride. Water Resources Research, v. 36, no. 10, p. 3011–3030.

Cartwright, I., D. Cendón, M. Currell et al. 2017. A review of radioactive isotopes and other residence time tracers in understanding groundwater recharge: possibilities, challenges, and limitations. Journal of Hydrology., v. 555, p. 797–811.

Cartwright, I., M. J. Currell, D. I. Cendón et al. 2020. A review of the use of radiocarbon to estimate groundwater residence times in semi-arid and arid areas. Journal of Hydrology, v. 580, 124247.

Clark, I. D., and P. Fritz. 1997. Environmental isotopes in hydrogeology. CRC Press, 328 p.

Craig, H. 1961. Isotopic variations in meteoric water. Science, v. 133, no. 3465, p. 1702–1703.

Dansgaard, W. 1964. Stable isotopes in precipitation. Tellus, v. 16, no. 4, p. 436–438.

Dunkle, S. A., L. N. Plummer, E. Busenberg et al. 1993. Chlorofluorocarbons (CCl_3F and CCl_2F_2) as dating tools and hydrologic tracers in shallow groundwater of the Delmarva Peninsula, Atlantic Coastal Plain, United States. Water Resources Research, v. 29, no. 12, p. 3837–3860.

Fabryka-Martin, J., S. J. Wightman, W. J. Murphy et al. 1993. Distribution of chlorine-36 in the unsaturated zone at Yucca Mountain: an indicator of fast transport paths, In Proceedings *Focus '93: Site Characterization and Model Validation*. American Nuclear Society, La Grange Park, Illinois, p. 58–68.

Fritz, P., and J. C. Fontes. 1980. Introduction, In P. Fritz, and J. C. Fontes (eds.), Handbook of Environmental Isotope Geochemistry, v. 1. Elsevier, Amsterdam, p. 1–19.

Fontes, J. C. 1980. Chapter 3. Environmental isotopes in groundwater hydrology, In P. Fritz, and J. C. Fontes (eds.), Handbook of Environmental Isotope Geochemistry, v. 1. Elsevier, Amsterdam, p. 75–140.

Gat, J. R. 1980. Chapter 1. The isotopes of hydrogen and oxygen in precipitation, In P. Fritz, and J. C. Fontes (eds.), Handbook of Environmental Isotope Geochemistry, v. 1. Elsevier, Amsterdam, p. 21–47.

Heilweil, V. M., D. K. Solomon, and P. M. Gardner. 2007. Infiltration and recharge at Sand Hollow, an Upland Bedrock Basin in Southwestern Utah. U.S. Geological Survey Professional Paper 1703-I, In D. A. Stonestrom, J. Constantz, T. P. A. Ferré, and S. A. Leake (eds.), Groundwater Recharge in the Arid and Semiarid Southwestern United States. U.S. Geological Survey, Menlo Park, California, p. 221–251.

Hendry, M. J., and L. I. Wassenaar. 1999. Implication of the transport of δD in pore waters for groundwater flow and the timing of geologic events in a thick aquitard system. Water Resources Research, v. 35, no. 6, p. 1751–1760.

Jurgens, B. C., J. K. Böhlke, and S. M. Eberts. 2012. Tracer LPM (Version 1): an Excel® workbook for interpreting groundwater age distributions from environmental tracer data. U.S. Geological Survey Techniques and Methods Report 4-F3, 60 p.

Lindsey, B. D., B. C. Jurgens, and K. Belitz. 2019. Tritium as an indicator of modern, mixed, and premodern groundwater age. U.S. Geological Survey Scientific Investigations Report 2019–5090, 18 p.

McCallum, J. L., P. G. Cook, and C. T. Simmons. 2015. Limitations of the use of environmental tracers to infer groundwater age. Groundwater, v. 53, no. S1, p. 56–70.

Michel, R. L. 2004. Tritium hydrology of the Mississippi River basin. Hydrological Processes, v. 18, no. 7, p. 1255–1269.

Michel, R. L., B. C. Jurgens, and M. B. Young. 2018. Tritium deposition in precipitation in the United States, 1953–2012. U.S. Geological Survey Scientific Investigations Report 2018–5086, 11 p.

Mook, W. G. 1980. Carbon-14 in hydrogeological studies, In P. Fritz, and J. C. Fontes (eds.), Handbook of Environmental Isotope Geochemistry (Chapter 2), v. 1. Elsevier, Amsterdam, p. 49–74.

Moscati, R. J., and K. M. Scofield. 2011. Meteoric precipitation at Yucca Mountain, Nevada: Chemical and stable isotope analyses, 2006–09. U.S. Geological Survey Scientific Investigations Report 2011–5140, 16 p.

Phillips, F. M. 2000. Chlorine-36, In P. G. Cook, and A. L. Herczeg (eds.), Environmental Tracers in Subsurface Hydrology. Springer, Boston, MA, p. 299–348.

Phillips, F. M., H. W. Bentley, S. N. Davis et al. 1986. Chlorine 36 dating of very old groundwater: 2. Milk River aquifer, Alberta, Canada. Water Resources Research, v. 22, no. 13, p. 2003–2016.

Plummer, L. N., and L. C. Friedman. 1999. Tracing and dating young groundwater. U.S. Geological Survey, Fact Sheet 134-99, 4 p.

Plummer, L. N., E. C. Prestemon, and D. L. Parkhurst. 1991. An interactive code (NETPATH) for modeling NET geochemical reactions along a flow PATH. U.S. Geological Survey Water Resources Investigations Report 91-4078, 227 p.

Plummer, L. N., E. C. Prestemon, and D. L. Parkhurst. 1994. An Interactive Code (NETPATH) for Modeling net Geochemical Reactions along a Flow Path Version 2.0. U.S. Geological Survey Water-Resources Investigations Report 94–4169, 130 p.

Prommer, H., D. A. Barry, and C. Zheng. 2003. MODFLOW/MT3DMS-based reactive multicomponent transport modeling. Groundwater, v. 41, no. 2, p. 247–257.

Randall, J. H., and T. R. Schultz. 1976. Chlorofluorocarbons as hydrologic tracers, a new technology. Arizona-Nevada Academy of Science, Hydrology and Water Re sources in Arizona and the Southwest, v. 6, p. 189–195.

Reardon, E. J., and P. Fritz. 1978. Computer modelling of groundwater ^{13}C and ^{14}C isotope compositions. Journal of Hydrology, v. 36, no. 3–4, p. 201–224.

Remenda, V. H., J. A. Cherry, and T. W. D. Edwards. 1994. Isotopic composition of old groundwater from Lake Agassiz: Implications for late Pleistocene climate. Science, v. 266, no. 5193, p. 1975–1978.

Sacks, L. A. 2002. Estimating groundwater inflow to lakes in Central Florida using the isotope mass-balance approach. U. S. Geological Survey Water-Resources Investigations Report 02-4192, 59 p.

Salmon, S. U., H. Prommer, J. Park et al. 2015. A general reactive transport modeling framework for simulating and interpreting groundwater ^{14}C age and $\delta^{13}C$. Water Resources Research, v. 51, no. 1, p. 359–376.

Savin, S. M. 1980. Oxygen and hydrogen isotope effects in low-temperature mineral-water interactions, In P. Fritz, and J. C. Fontes (eds.), Handbook of Environmental Isotope Geochemistry (Chapter 8), v. 1. Elsevier, Amsterdam, p. 283–327.

Schwartz, F. W., G. Liu, P. Aggarwal et al. 2017. Naïve simplicity: the overlooked piece of the complexity-simplicity paradigm. Groundwater, v. 55, no. 5, p. 703–711.

Solomon, D. K., R. J. Poreda, P. G. Cook et al. 1995. Site characterization using $^3H/^3He$ groundwater ages, Cape Cod, MA. Groundwater, v. 33, no. 6, p. 988–996.

Suckow, A. 2014. The age of groundwater–definitions, models and why we do not need this term. Applied Geochemistry, v. 50, p. 222–230.

Sudicky, E. A., and E. O. Frind. 1981. Carbon 14 dating of groundwater in confined aquifers: implications of aquitard diffusion. Water Resources Research, v. 17, no. 4, p. 1060–1064.

Therrien, R., R. G. McLaren, E. A. Sudicky, et al. 2006. HydroGeoSphere—A Three-dimensional Numerical Model Describing Fully-intergrated Subsurface and Surface Flow and Solute Transport. June 26, 2006, Draft, Groundwater Simulations Group.

Thompson, G. M. 1976. Trichloromethane: a new hydrologic tool for tracing and dating groundwater. Ph.D. dissertation, Department of Geology, Indiana University, Bloomington, 93 p.

Thompson, G. M., and J. M. Hayes. 1979. Trichlorofluoromethane in groundwater—a possible tracer and indicator of groundwater age. Water Resources Research, v. 15, no. 3, p. 546–554.

Torgersen, T., R. Purtschert, F. M. Phillips et al. 2013. Defining groundwater age, In Isotope Methods for Dating Old Groundwater (Chapter 3). International Atomic Energy Agency, p. 21–32.

Truesdell, A. H., and J. R. Hulston. 1980. Isotopic evidence on environments of geothermal systems, In P. Fritz, and J. C. Fontes (eds.), Handbook of Environmental Isotope Geochemistry (Chapter 5), v. 1. Elsevier, Amsterdam, p. 179–226.

Wigley, T. M. L., L. N. Plummer, and F. J. Pearson Jr. 1978. Mass transfer and carbon isotope evolution in natural water systems. Geochimica et Cosmochimica Acta, v. 42, no. 8, p. 1117–1139.

19

Mass Transport: Principles and Examples

CHAPTER MENU

This chapter introduces basic concepts of mass transport. It provides a conceptual framework on how constituents that are dissolved in groundwater are moved or redistributed, and otherwise interact chemically with other dissolved constituents or the porous medium. In other words, *mass transport* describes the delivery/distribution system for dissolved constituents in groundwater. A key feature concerned with transport is the pathways that mass follows as it moves in the subsurface. In most cases, a variety of pathways can exist which differ in terms of their starting and ending locations and the routes they follow in between. Details concerning the geometry of the pathways are important because the chemical reactions that influence dissolved constituents being transported may differ because of variability in mineralogy or geochemical settings (such as redox conditions).

This theory of mass transport is sufficiently general that it applies to both contaminants and natural constituents. Contaminants can be different in terms of where they originate and other unique properties, but like the natural constituents, they are also transported. Because contaminants due to human activities are relatively young (e.g., less than 100 years old), they might have been transported only a relatively short distance along some pathway. Natural constituents and geogenic contaminants like arsenic are much more likely to be transported further distances with much longer times available for transport.

It is difficult to explain the transport processes in a simple way because so much can be happening at the same time along multiple pathways. The chemical and biological processes discussed in Chapter 17 are available as well as nuclear reactions like radioactive decay. Moreover, these same processes might affect the chemistry of water in the unsaturated zone, which is on its way to becoming recharged.

19.1 Subsurface Pathways

As will become clear, advection is the primary process responsible for moving constituents along a pathway from one place to another. *Advection* is mass transport due to the flow of water in which the mass is dissolved. The direction and rate of transport coincides with the flow of groundwater. Thus, the pattern of the groundwater flow broadly determines the paths followed by the mass.

Fundamentals of Groundwater, Second Edition. Franklin W. Schwartz and Hubao Zhang.

Companion website: www.wiley.com/go/schwartz/fundamentalsofgroundwater2

Figure 19.1 provides examples that illustrate the variety and spectrum in complexity for the different pathways. It also illustrates how nitrate (NO_3-N), a pervasive agricultural and urban contaminant, is useful as a tracer for defining pathways in various kinds of aquifer systems.

The first example from Jagucki et al. (2008) (Figure 19.1*a*) illustrates a series of pathways associated with NO_3 transport within the High Plains aquifer in east-central Nebraska (Jagucki et al., 2008). In this example, complexity in pathways comes from pumping a sequence of aquifers consisting of an unconfined aquifer overlying two deeper confined aquifers. Interest in understanding flow paths, in this case, was motivated because nitrate (NO_3-N) in the Unconfined aquifer at relatively high concentrations (i.e., median 16 mg/L) turned up unexpectedly in several public supply wells in the Upper confined aquifer (Figure 19.1*a*). The presence of the thick and low permeability Confining unit logically should have protected the confined aquifers.

Pumping from the confined aquifers for both irrigation and public supplies have lowered hydraulic heads in those units creating Flow Paths (1) (Figure 19.1) that are associated with downward leakage across deeper confining units. These pathways exist because of the rapid adjustments in hydraulic head due to the withdrawals of groundwater from the deeper, confined aquifers. However, the NO_3 should not have moved along these pathways given the low flow

(*a*)

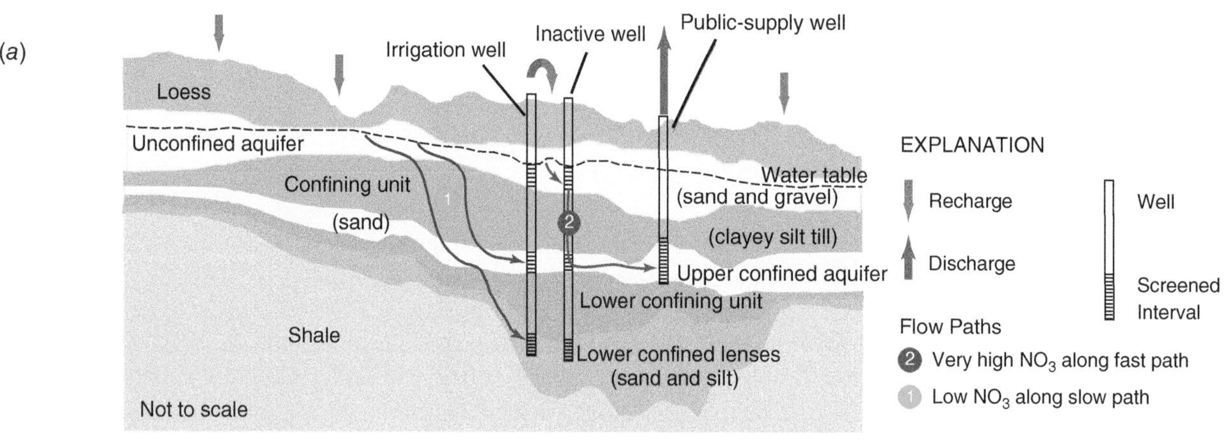

(*b*)

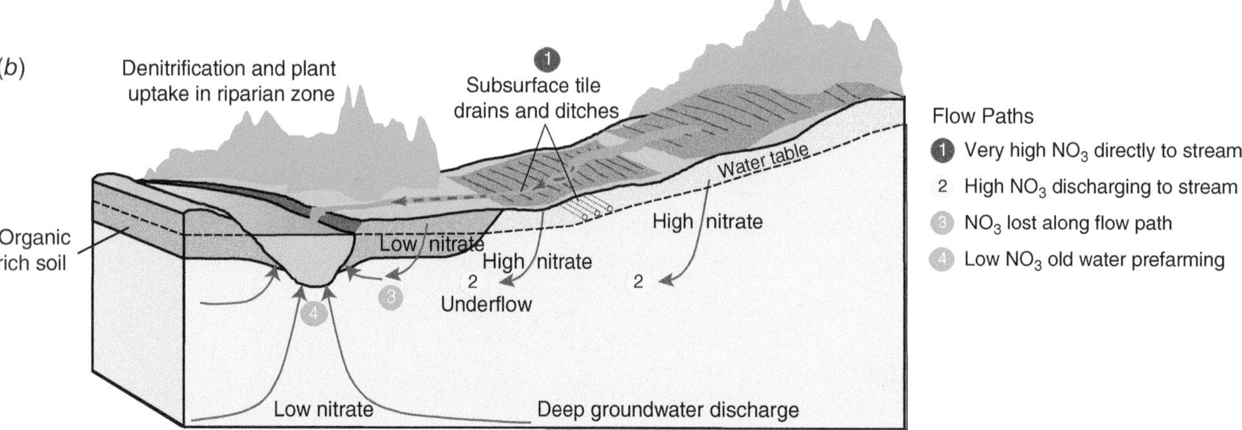

Figure 19.1 It is unlikely in (*a*) that Flow Paths (1) caused NO_3 caused contamination in deeper aquifers. More likely, is fast flow (2) down wells screened in both units. In (*b*) Flow Paths (1) and (2) contribute NO_3 to the stream, whereas water discharging from Flow Paths (3) and (4) are low in NO_3 (U.S. Geological Survey (*a*) adapted from Jagucki et al., 2008, and (*b*) adapted from Dubrovsky et al., 2010/Public domain).

velocities due to the low hydraulic conductivity of the confining beds. Moreover, reducing conditions in the Confining unit should drive a denitrification reaction, $NO_3 \rightarrow N_2(g)$, which would lower concentrations in nitrate along this pathway (Jagucki et al., 2008). Reaction-driven declines in the concentration of a constituent along a flow path is a good example of how the chemical and biological reactions, discussed in Chapter 17, are potentially available to change concentrations.

Because NO_3 transport along Flow Paths (1) is not likely the cause of contamination in the public supply wells, it is necessary to consider the existence of other fast-flow paths. As shown in Figure 19.1*a*, abandoned irrigation wells can be screened in multiple aquifers above and below confining beds. Thus, contamination was possibly due to fast downward flow through abandoned wells screened in different aquifers. In effect, large quantities of contaminated water could be moved downward along Flow Path (2) connecting the Unconfined aquifer to the Upper confined aquifer (Jagucki et al., 2008).

The second example (Dubrovsky et al., 2010) illustrates a more general application of flow path concepts to examine the origin of nitrate (NO_3-N) contamination of a stream within a watershed in an agricultural watershed. The contamination of the stream during baseflow conditions (Figure 19.1*b*) comes from groundwater, possibly contaminated, contributed along four different flow paths. Two flow paths (i.e., 1 and 2) discharge groundwater with elevated concentrations of NO_3, whereas two others (i.e., 3 and 4) discharge water with relatively low concentrations. The NO_3 chemistry of the stream would reflect the volume-weighted average of concentrations of water already in the stream and that contributed along the four flow paths. The very high NO_3 concentration in water moving along the ditch (Flow Path 1) is due to the presence of drainage tiles beneath the agricultural field, which drains groundwater at the water table immediately below fields being fertilized. Shallow groundwater (Flow Paths 2) can contain high concentrations of NO_3 which eventually discharges into the stream after flowing under the riparian zone shown on the figure.

Flow Path 3 (Figure 19.1*b*) is like 1 and 2, except that groundwater recharges within the riparian zone. Any NO_3 in water moving along this path is substantially reduced by denitrification reactions promoted by significant quantities of organic matter in the sediments and uptake by plants there (Dubrovsky et al., 2010). These are examples of reactions or processes that can substantially impact mass being transported by advection through the groundwater system. The contributions of NO_3 to the stream from Flow Paths 4 are small because the water is old, recharged before farming began, and the likelihood of redox reactions removing NO_3 via denitrification reactions.

19.2 Advection

This section looks in greater detail at advection—the process responsible for transporting dissolved mass along a flow system. Because this process is directly associated with groundwater flow, factors important in determining the direction and velocity of groundwater flow apply to mass transport as well. For example, with topographically-driven flow systems, groundwater flow patterns depend on water-table configuration, style of geologic layering, size of the groundwater basin, and pumping or injection. Thus, the background from the earlier treatment of regional groundwater flow allows us to consider advection in just a few pages.

Figure 19.2 illustrates these ideas with the help of an idealized hydrogeological cross section. Water recharges the flow system in upland areas and flows toward a discharge area in the middle of the cross section. The presence of a more permeable unit (shown in yellow, relative $K = 10$) causes deeper and subvertical flow. With a natural constituent or some local source of contamination within the recharge area, advection will result in the transport of the dissolved constituent down a stream tube as illustrated by the red shading. With advection, mass transported along the groundwater flow system will remain in the same stream tube. It takes other processes to move dissolved mass between stream tubes. In a steady-state flow system, the complimentary system of path lines and stream tubes define the direction of mass spreading (Figure 19.2).

For most practical problems, groundwater and dissolved mass will move at the same rate (in the absence of other processes) and in the same direction. Accordingly, the Darcy equation developed in Chapter 3 describes the velocity of advective transport

$$v = -\frac{K}{n}\frac{\Delta h}{\Delta l} \tag{19.1}$$

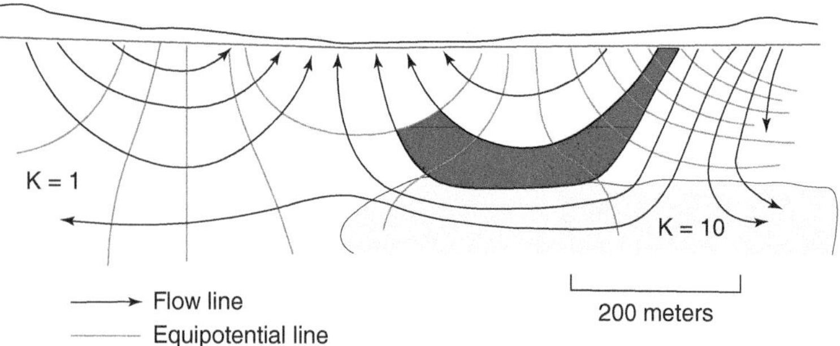

Figure 19.2 Hydrogeological cross section showing a local flow system with a deeper more permeable zone to the right (yellow shading). The red shading illustrates advective transport along a flow tube from a hypothetical source (Adapted from Domenico and Schwartz, 1998. Physical and Chemical Hydrogeology. Copyright © 1990, 1998 John Wiley & Sons, Inc. All Rights Reserved. Reproduced with permission).

where v is the linear groundwater velocity, K is hydraulic conductivity, n is the effective porosity, and $\Delta h/\Delta l$ is the hydraulic gradient. Recall from Eq. (19.1) that the linear groundwater velocity and the velocity of advective transport increases with decreasing effective porosity. This relationship is particularly important in fractured rocks, where the effective porosity can be as low as 1×10^{-4} or 1×10^{-5}, much lower than the total porosity (e.g., 0.20).

Example 19.1 A small volume of tracer is added to an unconfined aquifer that has a hydraulic conductivity of 1 m/day and a porosity of 0.35. The hydraulic gradient is 0.07. Calculate how far the center of mass of the tracer will move in one year.

Solution

Assuming that transport is due mainly to advection, calculate the advective velocity using Eq. (19.1).

$$v = -\frac{K}{n}\frac{\Delta h}{\Delta l}$$

$$v = -\frac{1\ \text{m/day}}{0.35} \times 0.07 = 0.2\ \text{m/day}$$

Knowing the velocity (v) and the travel time (t), calculate the distance traveled as (d):

$$\begin{aligned} d &= v \times t \\ &= 0.2\ \text{m/day} \times 365\ \text{days} = 73\ \text{m} \end{aligned}$$

Thus, the tracer will move 73 m down the flow system in one year.

The term *mass flux* is used to describe the quantity of mass being carried by groundwater. For systems where advection is the main transport mechanism, the advective mass transport flux (J) is written as

$$J_{\text{adv}} = Cq \tag{19.2}$$

where C is the concentration of the constituent, and q is the specific discharge or Darcy velocity. The advective mass transport flux is defined as mass transported through a unit area per time. For example, if the concentrations are expressed in grams, length in meters, and time in days, the mass flux has units of g/m^2/day.

Example 19.2 The specific discharge of groundwater (q) is 0.02 m/day. The concentration of a contaminant in a control volume is 100 μg/L. Calculate the contaminant mass flux out of the control volume in grams.

Solution

The mass flux is calculated using equation (19.2)

Convert the concentration from μg/L to g/m^3 or 100 μg/L = 0.1 g/m^3

$$J_{\text{adv}} = (0.1\ \text{g/m}^3)(0.02\ \text{m/day}) = 2 \times 10^{-3}\ \text{g/m}^2/\text{day}$$

19.3 Dispersion

Dispersion is another physical transport process that influences the way mass is transported in a groundwater system. Dispersion causes a zone of mixing to develop between a fluid of one composition that is adjacent to or being displaced by a fluid with a different composition. Thinking again about flow tubes in a steady-state groundwater system is a good place to begin examining the concept of dispersion. Dispersion spreads mass beyond the region it normally would occupy due to advection alone. In other words, in the direction of flow, some mass spreads farther down a stream tube than expected. Similarly, mass spreads at right angles (i.e., vertically and laterally) to the direction of flow. This process is particularly noticeable with plumes of dissolved contaminants and important because with longer travel distances dispersion plumes become larger.

Figure 19.3*a* illustrates a plume of dissolved and unreactive contaminant at some large concentration c_0 emanating from a cube-shaped source in an unconfined aquifer. In this illustration, advection is the only transport process and the plume is confined to its stream tube with a concentration equal to the source concentration, c_0. In Figure 19.3*b*, both advection and dispersion are operative. With the addition of dispersion, some mass spreads farther down the steam tube than before and sideways, outside the stream tube. Because the plume has the same quantity of mass as shown in Figure 19.3*a*, Concentrations in the plume (Figure 19.3*b*) are lower.

19.3.1 Tracer Tests

Much of what is known about mass transport in porous media has come from tracer tests. First came laboratory experiments using small columns more than 60 years ago. More recently, natural gradient tracer tests have proven to be enormously valuable in validating basic theories at a field scale and in developing new ideas. This test involves pulse loading of small volume of a tracer in an aquifer in a manner that avoided perturbing the flow system. That small volume was intensively monitored using multilevel piezometers as it moved down the flow system. Analyses of the concentration distributions with time provided information on advective velocity, the extent of dispersion in terms of characteristic parameters, and in some cases geochemical parameters. These tests were run to learn more about the manifestations of physical and chemical transport processes at large scale in shallow unconfined aquifers and to validate geostatistical concepts of dispersion.

Tracer tests at Borden, Cape Cod, and Columbus, Mississippi (Table 19.1) were unique because between 5 to 10,000 monitoring points provided precise three-dimensional characterizations of the evolving tracer plumes over travel times of several years.

The U.S. Geological Survey conducted a large-scale natural gradient tracer test in a shallow, stratified sand and gravel aquifer on Cape Cod, Massachusetts (LeBlanc et al., 1991). The test involved several reactive tracers (i.e., lithium and

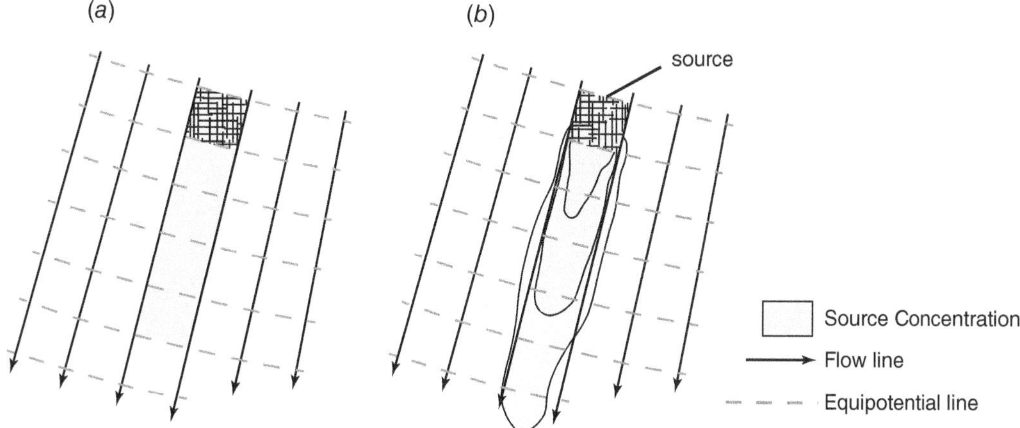

Figure 19.3 Physical transport of a dissolved tracer in identical flow systems for the same time. In (*a*) only advection is operating. In (*b*), both advection and dispersion are at work (Adapted from Domenico and Schwartz, 1998. Physical and Chemical Hydrogeology. Copyright © 1990, 1998 John Wiley & Sons, Inc. All Rights Reserved. Reproduced with permission).

Table 19.1 Information on three important massively instrumented tracer tests.

Site	Aquifer material	Test scale (m)	References
Canadian Forces Base Borden, Ontario	Glaciofluvial sand	90	Mackay et al. (1986) Freyberg (1986) Sudicky (1986)
Cape Cod Massachusetts	Sand and gravel	250	LeBlanc et al. (1991) Garabedian et al. (1991) Hess et al. (1992)
Columbus, Mississippi	Sandy gravel Gravelly sand	280	Boggs et al. (1992) Adams and Gelhar (1992) Rehfeldt et al. (1992)

Source: Domenico and Schwartz (1998). Physical and Chemical Hydrogeology.

molybdate). Initially, we will focus on Br^-, which is assumed to be nonreactive. The term nonreactive means that the tracer does not sorb or otherwise interact with the medium and is transported with the same velocity as the groundwater.

The test began with the "instantaneous" injection 7.6 m^3 of tracer solution into the aquifer. The transport of Br^- along the shallow flow system was monitored in three dimensions with a dense network of thousands of monitoring wells. Figure 19.4 shows snapshots of the Br^- footprint at three times, 33, 237, and 461 days (LeBlanc et al., 1991). The footprint shows those areas of the plumes where Br^- concentrations exceeded 1 mg/L.

Looking closely at the areas of the three Br^- footprints in Figure 19.4, it is evident that their size increased as a function of travel from the source. With transport, the maximum concentrations of Br^- declined from 429 mg/L to 65.2 mg/L to 39.0 mg/L (Figure 19.4). The increasing plume volume, accompanied by a reduction in concentration was the work of dispersion. Another feature evident in Br^- footprints is the tendency for greater elongation in the direction of flow as compared to the direction transverse (i.e., 90°) to the direction of flow (Figure 19.4). The next section provides an in-depth discussion of dispersion.

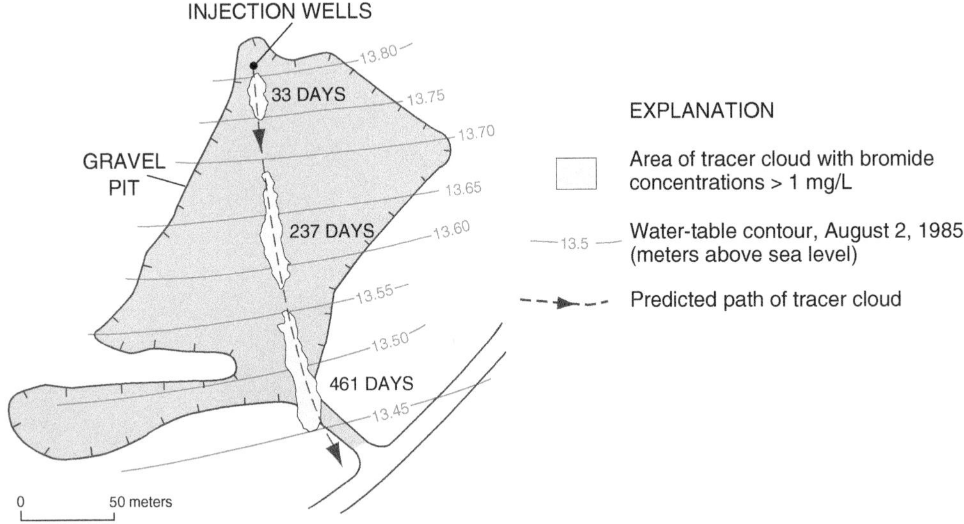

Figure 19.4 Example of a natural gradient tracer test at Cape Cod, Massachusetts illustrating how rapid pulse loading affects the plume geometry. The increasing size of the plume footprint with time illustrates the effects of dispersion (Adapted from LeBlanc et al., 1991. Water Resources Research, 27(5): 2017–2029. Copyright 1991 by the American Geophysical Union).

19.3.2 Dispersion at Small and Large Scales

Dispersion is a process that causes a zone of mixing to develop between a fluid of one composition that is adjacent to or being displaced by a fluid with a different composition. Although relevant to the evolution of natural groundwater and large-scale problems of contamination, the large scale of problems and low resolution of typical monitoring strategies make it difficult to recognize this process in the context of regional studies. Thus, dispersion is a process mostly studied in relation to smaller-scale problems of industrial contamination featuring high-resolution monitoring systems. It is an important process in the context of contamination problems because mass mixing can bring about reduction in contaminant concentrations.

Dispersion operates to spread dissolved constituents, often contaminants, beyond the volume of porous medium they normally would occupy due to advection alone. It requires that mixing water bodies have differences in chemical or isotopic compositions. For example, dispersive mixing was evident with the test at Cape Cod (Figure 19.4), where the concentration of Br^- was substantially higher than background levels.

Dispersion will also be manifest in a large system context when there is some local variability in the chemistry of the recharging groundwater. For example, on a regional scale, changes in water chemistry might depend on patterns of land use, for example, cropped areas as compared to forests and cities. However, with large regional problems, sampling never has the resolution to define concentration differences associated with dispersion.

19.4 Processes Creating Dispersion

Dispersion occurs in groundwater because of two processes, diffusion, and mechanical dispersion. *Diffusion* is a process of mass transport in response to a concentration gradient. With local variability in water chemistry, diffusion transports mass from zones of high concentration to adjacent waters with a low concentration. Solute molecules would move from the body of water with a high concentration into the body with low concentration. Thus, diffusion is a transport process because the dissolved mass moves in a manner that reduces concentrations. This process is important for few kinds of problems with deep systems but is sufficiently complicated to be beyond the scope of an introductory book.

Mechanical dispersion is a process of mixing that occurs because of local variability around some mean velocity of flow. We will illustrate this concept with a "thought" experiment. What would happen if a cluster of "rubber duckies" was released in some river, as shown in Figure 19.5. Would the duckies stay in a tight cluster or spread out? Obviously, they will spread out. Even though the river had some mean velocity, there would be spatial variability in the local velocities about the mean. Thus, some duckies were fortunate and generally found the high velocity pathways to take them ahead of the flock. Others would fall behind, as they moved slower or sideways. This same idea applies to flows within a groundwater system subject due to various sorts of non-idealities.

Mechanical dispersion, unlike diffusion, is an advective process and not a chemical one. With time, dissolved mass occupying some volume becomes gradually more dispersed as different fractions of mass are transported in locally variable velocity regimes. At a pore scale, variability in the direction and rate of transport is caused by nonidealities in the porous medium. At the scale of an aquifer, i.e., tens of meters to thousands of meters, the most important driver in this respect is variability in hydraulic conductivity. With a laboratory column filled with sand at a smaller scale, centimeters to meters, variability in velocity is related to the pore geometry.

Nonidealities at scale of a collection of pores are referred to as microscopic. For example, Figure 19.6 illustrates variability in velocity at microscopic scale across individual pore throats because of drag close to the individual grains. Also, variability in the tortuosity of the flow channels (Figure 19.6) makes for slight differences in the length and travel times along different pathways. Megascopic nonidealities are larger scale effects that create dispersion at an "aquifer" scale. For example, the variability in hydraulic conductivity shown in Figure 19.7 creates dispersion by creating variability in flow at a macroscopic scale. Sudicky (1986) measured hydraulic conductivity in samples from core-holes one meter apart along a cross section of the

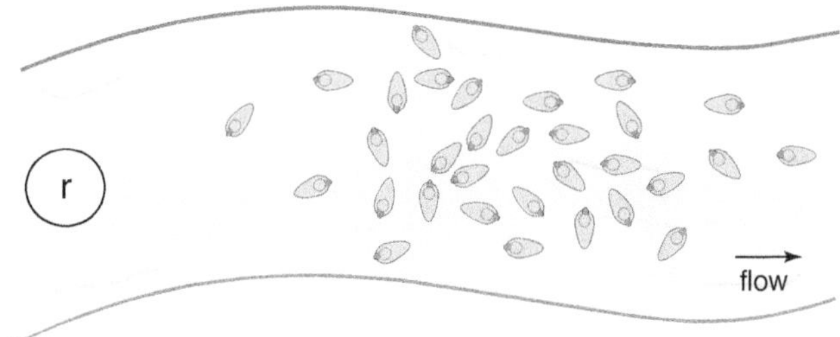

Figure 19.5 "Rubber duckies" released in a river from the circle at "r" end up dispersed due to local variability in the velocity around the mean (Adapted from Schwartz and Zhang, 2003/©John Wiley & Sons).

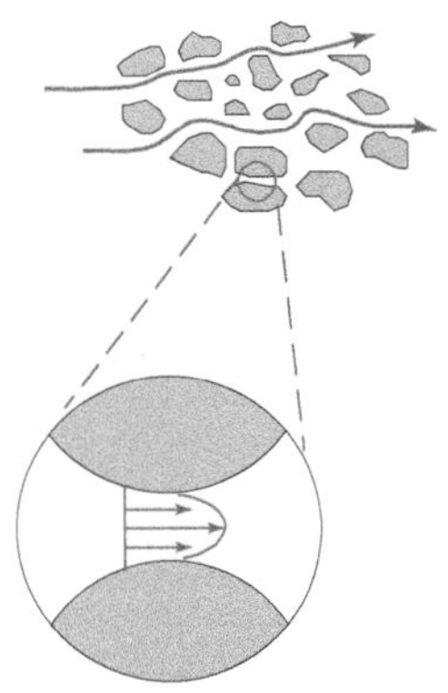

Figure 19.6 Examples of nonidealities at the pore scale, which include variability in velocity across pore throats and local variability in flowpath through the porous medium (Adapted from Domenico and Schwartz, 1998. Physical and Chemical Hydrogeology. Copyright © 1990, 1998 John Wiley & Sons Inc. All Rights Reserved. Reproduced with permission).

shallow unconfined aquifer at Canadian Forces Base Borden. Because hydraulic conductivity values vary over many orders of magnitude, they were plotted as negative natural logarithms or $x = -\ln(K)$, meaning that 3 on Figure 19.7 corresponds to a K value of 0.0497 cm/sec and 5 to $K = 0.0067$ cm/sec. Thus, although samples of the Borden sand appear similar in hand specimens, there are permeable, horizonal, and subhorizontal pathways of varying length.

Di Dato et al. (2016) with the help of different shaped inclusions examined the structure of the hydraulic conductivity field. They described the tendency for inclusions to be longer in the x-direction as compared to the y-direction. This feature causes dispersion in the x-direction to be significantly higher in x-direction as compared to the y-direction. Megascopic nonidealities exist with systems of aquifers and confining beds, although rarely relevant to practical problems.

Two- and three-dimensional experiments in the laboratory and field scale tracer tests (e.g., Cape Cod) showed that mass spreading could occur transverse (i.e., right angles) to the direction of mean groundwater flow. In other words, if groundwater is flowing in the x-direction, dissolved constituents could spread in the y and z directions as well. One component of transverse spreading occurs in the horizontal plane, and the second occur in the vertical plane. With the plumes in Figure 19.8a along the midline plane, longitudinal and horizontal transverse spreading are enlarging the plume with time (like Figure 18.4). The plume grows away from a source. At first (t_1), the footprint of the plume is relatively small but the maximum contaminant concentrations are an order of magnitude higher than at t_2.

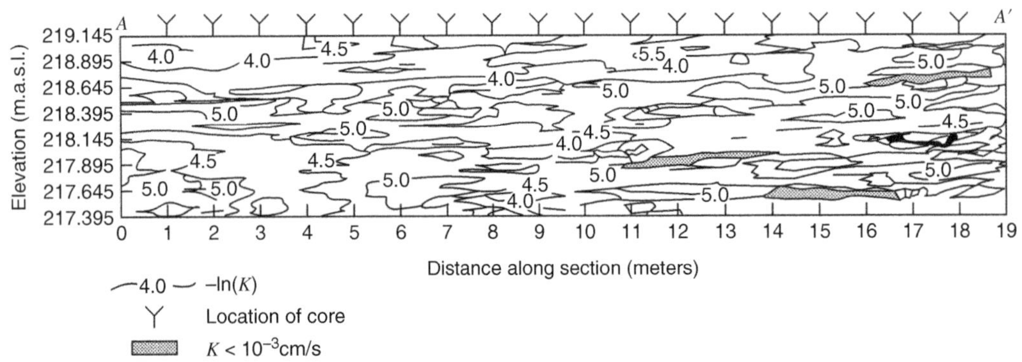

Figure 19.7 An example illustrating the variability in hydraulic conductivity (−ln K) hydraulic conductivity. This inherent variability is responsible for creating mechanical dispersion. (Sudicky, 1986. Water Resources Research 22(13): 2069–2082. Copyright 1986 by the American Geophysical Union).

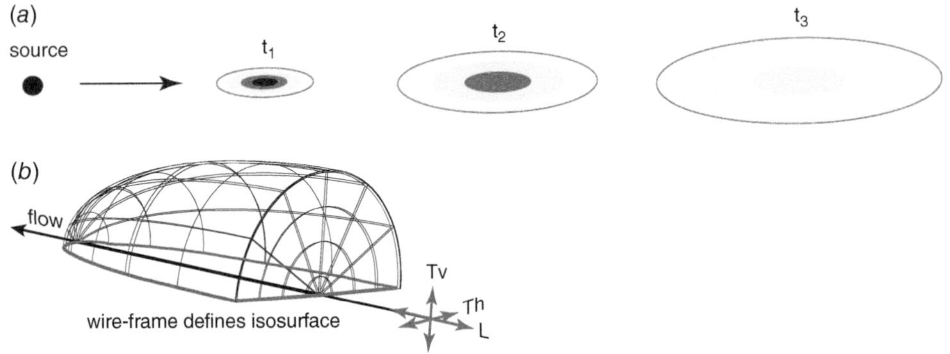

Figure 19.8 Spreading away from the source (a) creates a footprint of the plume, which grows with time, while the concentration declines. The elliptical distribution in concentration (a) comes from greater dispersion in the direction of flow (L) as compared to the transverse horizontal direction (Th). The wireframe (b) represents 1/4 of the three-dimensional shape (i.e., top front). The red lines illustrate the relationship between the 2D and 3D shapes ((a) Adapted from Domenico and Schwartz, 1998. Physical and Chemical Hydrogeology. Copyright © 1990, 1998. John Wiley & Sons Inc. All Rights Reserved. Reproduced with permission; (b) FWS).

With subsequent time steps, the footprint becomes bigger with a marked decline in maximum concentrations. The vertical transverse spreading is not shown in Figure 19.8*a*. In most cases, spreading in the vertical transverse direction is less than the horizontal transverse direction. The inherent horizontal or subhorizontal bedding in most sand and gravel aquifers tends to inhibit transverse vertical dispersion.

The two-dimensional representations of plumes, i.e., footprints (Figure 19.8*a*) are horizonal slices through the 3D cloud of tracer constituents moving down the flow system. Representing the concentration distributions in three dimensions is difficult because it involves uses isosurfaces instead of lines. In terms of concentration, for example, a 5 mg/L isosurface wraps some volume of the plume, separating porewaters with tracer concentrations > 5 mg/L from those <5 mg/L. The isosurface associated with an appropriately low concentration value will define the three-dimensional shape of the entire plume, while that associated with a relatively high concentration value will define the shape plume core, usually just a small proportion of the total volume of contaminated water.

Figure 19.8*b* illustrates a segment of an isosurface for an idealized plume shape as a one-time snapshot for a pulse release of tracer affected by dispersion as it moves downgradient in a flow system. To show the shape of one concentration isosurface, only the top-front volume (i.e., 1/4 of the tracer volume) is shown. The entire 3D plume (Figure 19.8*b*) is a prolate spheroid (shaped somewhat like an American football) with a longer axis in the direction of flow and shorter axes transverse to the direction of flow (horizontal and vertical) of equal length, although often the vertical transverse axis is shorter than the horizontal transverse axis. A close look at Figure 19.8*b* shows the plume to be symmetrical around a vertical plane along the midline of the plume. It also is symmetrical along the horizontal plane, although not obvious with just 1/4 of the plume. The red ellipse on the isosurface is the intersection of the isosurface with the horizontal plane and that line for an appropriately low value of concentration is the outermost contour line for the footprint of the plume (Figure 19.8*b*).

Figure 19.8*b* also features a local x–y–z axis that shows the three spreading directions. That axis is assumed to be moving at the average velocity of the plume. Thus, upgradient spreading is not actually due to mass moving upgradient, but that the back end of the plume is moving slower than the mean. Spreading in the three axial directions is referred to as *longitudinal* (L), *transverse horizontal* (Th) *and transverse vertical* (Tv).

19.5 Statistical Patterns of Mass Spreading

There are also conceptual models that describe the concentration distribution inside of a plume. Laboratory and field experiments show that concentration distributions along the spreading axes tend to have a Gaussian or normal shape. Before proceeding, let us review the characteristics of the "bell-shaped" normal distribution. It is defined by two parameters, a mean and standard deviation (Figure 19.9). The standard deviation (σ) or the variance (σ^2) are measures of the spread in the data about the mean, which here are distances in relation to the axes of spreading. The peak height of the distribution is the maximum concentration. One standard deviation contains about 68.3% of the total tracer mass under the curve. Four standard deviations account for 99.9% of the mass. A simple graphical technique is useful in estimating the standard deviation in concentration distributions (σ). It involves measuring half-width of a normally distributed concentration distribution (Γ) (Figure 19.9).

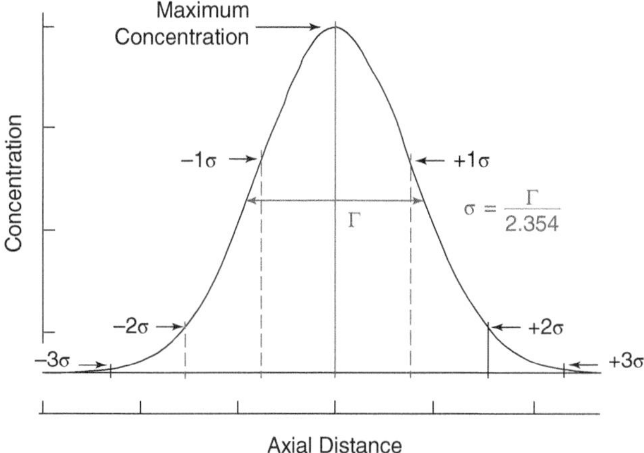

Figure 19.9 In dispersion applications, a normal distribution describes concentrations distributions as a function of axial distance. The standard deviation (σ) represents the spread of mass in axial directions around the maximum concentration. A measurement of distance (Γ) at the half-maximum concentration provides the basis for a graphical estimate of (σ) (FWS).

The tendency for concentrations distributions to be Gaussian facilitates determination of concentration distributions along the spreading axes x, y, and z. Figure 19.10 shows normal concentration distributions in two directions (longitudinal and transverse horizontal) for the two-dimensional transport of a pulse plume. The normalized mean concentration (i.e., $C = 1$) is the concentration

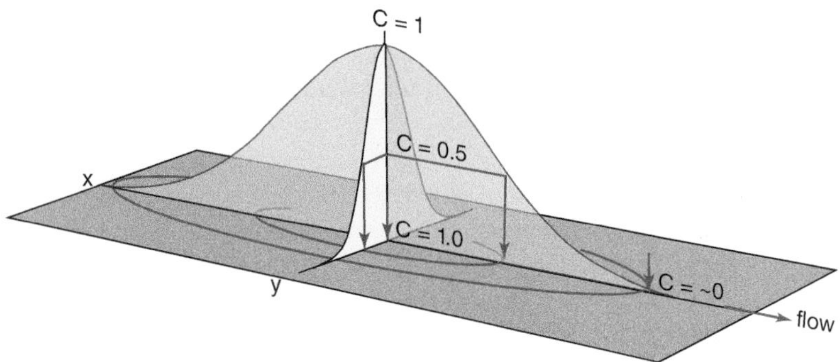

Figure 19.10 Normal distributions describe concentrations in axial directions (*x* and *y*) (*z* not shown). Knowledge of various concentrations values (e.g., here 0.5 and 0) along the axes enables construction of the elliptical concentration distributions along the midline plane (brown) (FWS).

maximum coinciding with the midpoint of the plume. Two constructions on the plot show how relative concentration values, i.e., $C = 0.5$, are defined along the two axes shown and provide the elliptical concentration contours (Figure 19.10).

The theoretical model holds that the variance of the concentration distribution in the longitudinal direction (${\sigma_L}^2$) is proportional to the dispersion coefficient for the system such that

$$D_L = {\sigma_L}^2/2t \tag{19.3}$$

where D_L is the longitudinal dispersion coefficient with units of $[L^2/T]$ and t is time. When mass spreads in two or three dimensions, the distributions of mass sampled normal to the direction of flow transverse horizontal (Th) and transverse vertical (Tv) are also normally distributed with variances that increase in proportion to $2t$. Thus, transverse-horizontal dispersion coefficient (Figure 19.10) is defined as

$$D_{\text{Th}} = {\sigma_{\text{Th}}}^2/2t \tag{19.4}$$

The two-dimensional spread of a tracer in a unidirectional flow field results in an elliptically shaped concentration distribution (Figure 19.10) that is normally distributed in both the longitudinal and transverse directions. Typically, longitudinal dispersion is greater than transverse dispersion (Figure 19.10).

Various kinds of tracer experiments have been helpful in expressing the dispersion coefficients of Eqs. (19.3) and (19.4) (e.g., D_L) in terms of more fundamental parameters, dispersivity and mean groundwater velocity as follows,

$$D_L = \alpha_L v \quad \text{and} \quad D_T = \alpha_T v \tag{19.5}$$

where α_L is the longitudinal dispersivity and α_T is the transverse dispersivity. Dispersivity values have units of length and are a characteristic parameter of the porous medium, just like hydraulic conductivity. The values are easy to interpret, the bigger the number the more mechanical dispersion there is in the system. In essence, dispersivity values reflect spreading in the system described by the variances in the normal distributions. Because there are two directions of spreading in the transverse direction, there are two transverse dispersivities, α_{Th} and α_{Tv}.

Substituting Eq. (19.5) into Eqs. (19.3) and (19.4), respectively provide the mathematical definitions of dispersivity or

$$\alpha_L = \sigma_L^2/2vt \quad \text{and} \quad \alpha_{\text{Th}} = \sigma_{\text{Th}}^2/2vt \tag{19.6}$$

with α_{Tv} similar in form to α_{Th}. Because the travel distance from the source (x) is equal to vt, Eq. 19.6 can be written as

$$\alpha_L = \sigma_L^2/2x \quad \text{and} \quad \alpha_{\text{Th}} = \sigma_{\text{Th}}^2/2x \tag{19.7}$$

The next section will illustrate how dispersivity values are determined and used.

19.6 Measuring, Estimating, and Using Dispersivity Values

The most important use of dispersivity numbers is associated with modeling the behavior of contaminant migration. Modeling is useful in providing guidance as to what a contamination problem might look like in the future without intervention and the efficacy of potential cleanup strategies in meeting cleanup goals. There are examples in the next chapter that illustrate these approaches. This chapter is designed to illustrate how dispersivity values are calculated in practice. The simplest approaches are based on ideas from the last section. For the purposes of continuity, we will start with interpretation of data from natural gradient tracer tests. However, it should be obvious that these approaches are useful in research, but not in practice. The following example is based on data from a natural gradient tracer test at the Borden site located in Southwestern Ontario, Canada (Mackay et al., 1986; Freyberg, 1986).

Example 19.3 Analysis of the Borden Data

Water, containing Cl^- at 892 mg/L, was injected into a shallow, unconfined sand aquifer. What started out as a roughly lenticular plume of tracer, approximately 6 m in diameter and 1 m thick was monitored for almost three years as it moved 60 m down the flow system.

As expected, dispersion worked to spread this lens. Figure 19.11 shows this original Cl^- plume after transport for 462 days and 42 m from the source. The plume had become elliptical in shape, some 30 m long and 10 m wide with a maximum Cl^- concentration of ~65 mg/L. Dispersion was especially effective in elongating the plume in the longitudinal direction. Given the concentration distribution calculate the longitudinal dispersion coefficient and eventually the longitudinal dispersivity.

Solution

The first step is to analyze the Cl^- distribution assuming that the concentration along the longitudinal plume axis (Figure 19.11) is normally distributed. The maximum Cl^- concentration is about 65 mg/L. One might plot these data to create a normal distribution and measure the Γ_L to estimate the dispersivity α_L. But it simple just to measure distance along the curving longitudinal axis from the two intersection points of 32.5 mg/L (half maximum concentration) where an inferred concentration contour of 32.5 crosses the longitudinal axis (i.e., point a to point b; Figure 19.11). Γ_L is about 10 m, which provides

$$\sigma_L = \frac{12}{2.354} = 4.73 \text{ m}$$

The dispersion coefficient (D_L) is given as

$$D_L = \frac{\sigma_L^2}{2t} = \frac{4.73 \times 4.73}{2 \times 462} = 0.024 \text{ m}^2/\text{day}$$

We also know that $D_L = \alpha_L v$, with v given as the distance of travel (x) over the travel time (t). Given that $d = 42$ m and $t = 462$ days, the longitudinal dispersivity is calculated as

$$\alpha_L = \frac{D_L}{v} = \frac{D_L \times t}{x} = \frac{0.024 \times 462}{42} = 0.27 \text{ m}$$

The dispersivity is calculated here is approximately the same as was calculated by Freyberg (1986). It is a relatively small value because the Borden sand aquifer is quite homogeneous.

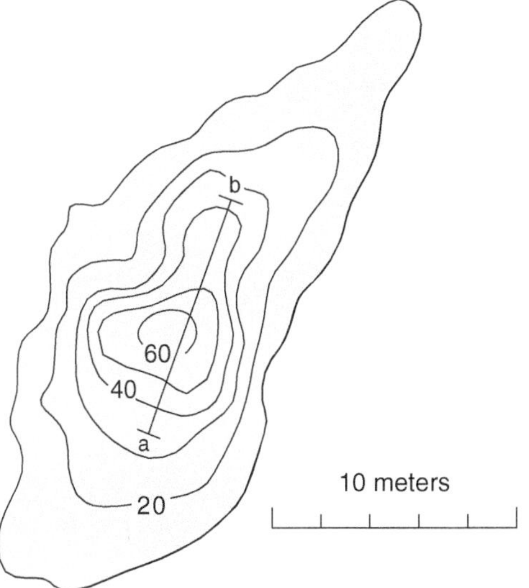

Figure 19.11 Map view of Cl^- concentrations in the natural gradient tracer test at Borden at 462 days (Adapted from Mackay et al., 1986. Water Resources Research 22(13): 2017–2029. Copyright 1986 by the American Geophysical Union).

19.6.1 Sources with a Continuous Release

Plumes with a continuous release of contaminants at a source tend to be much more common than the pulse releases seen with natural gradient tracer tests. However, these statistical approaches with some modifications have applications to these kinds of problems (Robbins, 1983).

(*a*) Spreading of a tracer from a point source

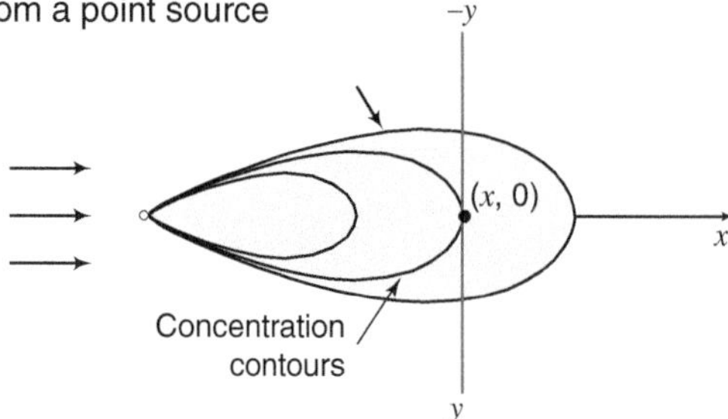

(*b*) Breakthrough curve at (x, 0)

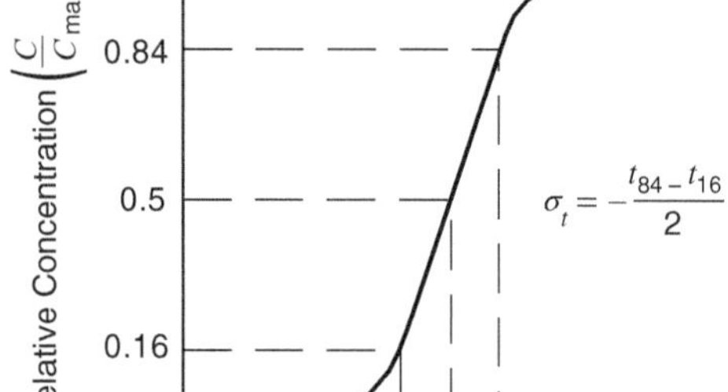

(*c*) Variation in relative concentration along (y, −y)

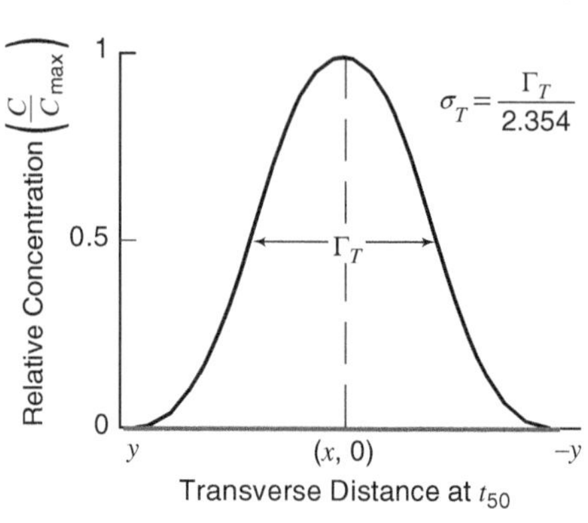

Figure 19.12 Graphical procedure for estimating dispersivities from a continuous point source. The calculation of longitudinal dispersivity is calculated using an estimate of the temporal standard deviation (Panel *b*) with concentration data from (*x*,0) (*a*). Calculation of σ_T (Panel *c*) makes use of concentration data normal to the direction of flow along −*y* to *y* (Robbins, 1983 with permission).

Figure 19.12*a* illustrates continuous spreading from a point source and the associated concentration contours. However, there are not sufficient data to calculate standard deviation in the pattern of longitudinal direction. A *breakthrough curve* (i.e., contaminant concentrations versus time) is needed at the monitoring well (*x*,0) (Figure 19.12*b*). Chemical monitoring at a downstream well through time can establish the initial arrival of plume at this well and the subsequent growth in concentration to some maximum value. With the breakthrough curve established, a simple graphical approach yields the temporal standard deviation and variance (σ_t and σ_t^2 respectively). The time variance (α_t^2), however, needs to be expressed as a variance in space using the following equation

$$\sigma_L^2 = v^2\sigma_t^2 \tag{19.8}$$

where *v* is the linear groundwater velocity. The previous equations are modified accordingly to provide the longitudinal dispersion coefficient (D_L)

$$D_L = \frac{v^2\sigma_t^2}{2t} \tag{19.9}$$

As before, knowing D_L and *v*, permits calculation of the longitudinal dispersivity.

In the transverse direction (*y*, −*y*; Figure 19.12*a*), concentrations ideally should yield the expected normal distribution (Figure 19.12*c*). In this direction, one simply uses the half-width of the normal distribution to estimate σ_{Th}.

19.6.2 Available Dispersivity Values

For many applications, common practice has been to use estimates of dispersivity based on compilations of values. Gelhar et al. (1992) undertook a critical review of field experiments at 59 sites around the world. Test data yielded some 106 values longitudinal dispersivity ranging from 0.01 to 5500 m at scales of 0.75 m to 100 km. Although it appeared that longitudinal dispersivities increased indefinitely with scale (Figure 19.13), this is not the case. Not all the values in Figure 19.13 were reliable and created the impression of increasing longitudinal dispersivity as a function of distance (Gelhar et al., 1992). The most reliable dispersivity values were at the low end of the range and for somewhat smaller scale tests.

From such tests, a consistent view about macroscopic dispersion has emerged. Heterogeneity at the macroscopic scale contributes significantly to dispersion because it creates local-scale variability in velocity. Values of macroscopic dispersivity are in general two or more orders of magnitude larger than those from column experiments. In field experiments, values range from approximately 0.1 to 2 m over relatively short transport distances. Although no reliable, large-scale studies have been carried out, it is probable that longitudinal dispersivity values more than 10 m exist. Values, however, are not likely as large as estimates determined using contaminant plumes and environmental isotopes, which are less reliable (Gelhar et al., 1992).

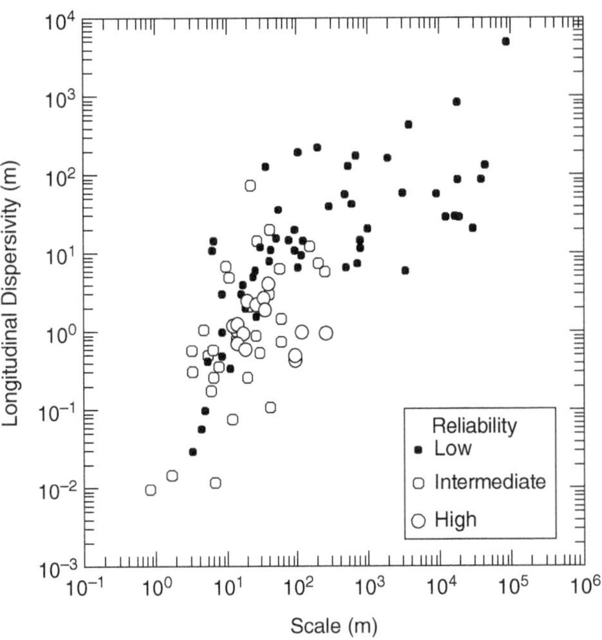

Figure 19.13 Longitudinal dispersivity values versus scale of data and classified by reliability (Gelhar et al., 1992. Water Resources Research 28(7): 1955–1974. Copyright 1992 by the American Geophysical Union).

Experiments in the field commonly show that dispersivity values increase as a tracer moves away from a source. Eventually, however, the dispersivity values become constant. Gelhar et al. (1979) refer to this constant macroscale dispersivity as the *asymptotic dispersivity*. For aquifers, a tracer may have to spread 10's or 100's of meters before the asymptotic dispersivity is obtained. The reason why this behavior is observed goes back to aquifer heterogeneity, which is the main cause of dispersivity at the field scale. When a tracer spreads a few meters, it encounters very little of the heterogeneity present in the aquifer. Thus, at small displacements from the source, there is relatively limited dispersion—in most cases resembling dispersion in a column. When a tracer spreads farther, it will encounter the large-scale heterogeneity (see Figure 19.13), which is much more important in causing dispersion. In effect, there is a transition zones between the microscale and macroscsale heterogeneity in the medium.

Values of transverse dispersivity are also estimated from longitudinal values. In practice, transverse horizontal dispersivity values are 3 to 4 times smaller, and transverse vertical dispersivities are 10 times smaller. These ratios come from natural gradient tracer tests that show dispersion in the longitudinal direction > transverse horizontal ≫ transverse vertical direction.

It is usually not practical in practice to run tracer tests. Common test methods (Domenico and Schwartz, 1998), such as single well pulse tests, two-well tracer tests, and various types of tracer tests with a well, can be expensive, time consuming and subject to subtle errors. Moreover, dispersivity estimates that come tracer tests run in a laboratory are not relevant to contamination problems developing at much greater scales. The dispersion created by microscale processes (Figure 19.6) operating in columns is insignificant as compared to mechanical dispersion created at a macroscale due to heterogeneities in the hydraulic conductivity field.

19.7 Dispersion in Fractured Media

Many basic concepts of dispersion apply to fractured media. Nevertheless, the added complexity is sufficient to warrant treating these media separately. Our focus here is with media where fractures are the only permeable pathways. The hydraulic conductivity of the unfractured volume of the medium is small and advective transport through the matrix is negligible. However, mass can diffuse into the unfractured rock matrix in response to concentration gradients.

Mixing in a fracture system is the result of both mechanical mixing and diffusion. Concentrations in the fractures will be affected by diffusion into the matrix, a chemical process (Figure 19.14*a*); variability within individual fractures caused by asperities, an advective dispersive process (Figure 19.14*b*); fluid mixing at the fracture intersections, a diluting or possibly

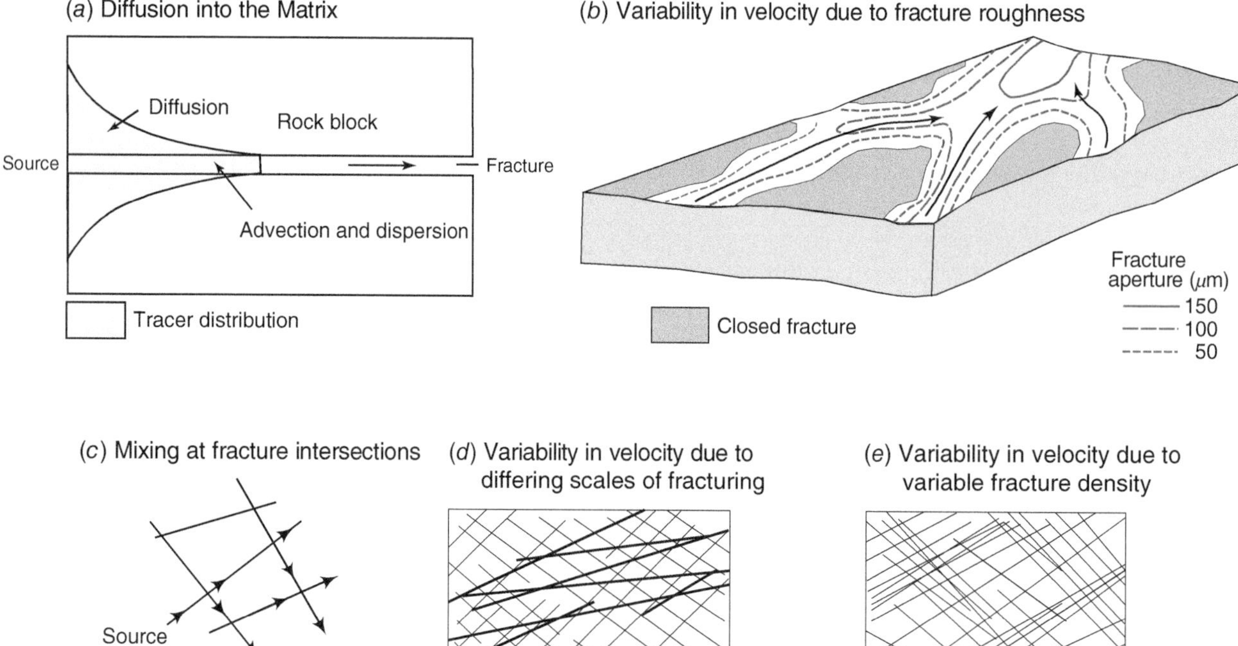

Figure 19.14 In a porous and fractured medium matrix diffusion (*a*) can reduce the apparent velocity of the contaminant through the medium. Dispersion is created because of variability in velocity caused by channeling along a fracture plane creating variability in velocity (*b*). Mixing at fracture intersections (*c*) and various geometrical features (*d, e*) can create dispersion within a network of fractures. (Domenico and Schwartz, 1998, Physical and Chemical Hydrogeology. Copyright © 1990, 1998 John Wiley & Sons, Inc. All Rights Reserved. Reproduced with permission).

diffusive process (Figure 19.14*c*); and variability in velocity caused by differing scales of fracturing (Figure 19.14*d*) or by variations in fracture density (Figure 19.14*e*).

Diffusion into the pores of matrix provides an important process to attenuate the transport of contaminants by an apparent reduction in advective velocity. This result contradicts experience with porous media, where the contribution of diffusion to dispersion is generally swamped by mechanical dispersion. Diffusion is more important in fractured media because localizing mass in fractures provides the opportunity for large concentration gradients to develop. Theoretical studies by Grisak and Pickens (1980) and Tang et al. (1981) explored this process using numerical and analytical models of a single fracture bounded by an infinite porous matrix. Dispersion in the plane of single fracture is caused by the variability in fracture aperture. This variability develops due the roughness of the fracture walls and the precipitation of secondary minerals. At many locations, the fracture may be closed to flow and transport. Such aperture topology (see Figure 19.14*b*) gives rise to *channeling* (Neretnieks, 1985), where mass moves predominantly along networks of irregularly shaped pathways in the plane of a fracture.

This channelization model has developed from tracer tests carried out in fractured rocks. For example, studies at the Stripa mine (Neretnieks, 1985) showed how the inflow of water to tunnels was extremely localized. Approximately one third of the flow entered from approximately 2% of the fractured rocks. These results are strongly indicative of localized channel flow within individual fractures.

Channels can be so poorly interconnected that they may not interact with one another over appreciable distances (Neretnieks, 1985). This behavior is termed *pure channeling*. From this discussion, it is easy to understand why the smooth, parallel plate model used by some investigators is really a simplified representation of a fracture.

Dispersion at the next larger scale occurs when the geometry of the three-dimensional network begins to influence mass transport. A tracer moving along a fracture to an intersection (Figure 19.14*c*) partitions into two or more fractures. Because the water is also partitioned and because concentration is the mass per unit volume of solution, partitioning by itself does not affect the concentrations. However, concentrations will be affected if there is mixing with water that does not contain the tracer. This dilution depends directly the type of mixing process in the intersection and quantity of water flowing along the fractures.

The dispersion in Figures 19.14*d* and *e* is analogous to that caused by heterogeneities in porous media. Differing scales of fracturing or spatial variability in fracture density create variability in local velocity. Even larger scale dispersion could develop if individual discontinuities exist on a regional scale or the variability in fracture density includes several units.

19.8 Chemical Processes and Their Impact on Water Chemistry

This section examines the role of geochemical processes in mass transport. Including reactions with physical transport processes facilitates the analysis of traditional problems like the diagenesis of carbonate rocks, karst formation, and the chemical evolution of groundwater. Examples presented in this chapter are not designed to comprehensively describe all the different types of water found in natural settings. Instead, they provide illustrative examples of a process-oriented approach to understand the controls on the geochemical evolution of groundwater.

Chemical reactions in the shallow subsurface generally play the dominant role in controlling the chemistry of groundwaters. Rainfall or snowmelt that ultimately ends up recharging most groundwater systems has little dissolved mineral matter. Yet, once this water disappears into the ground, it quickly accumulates a relatively large load of dissolved mineral matter. Armed with knowledge of how various chemical processes impact water chemistry, and features of the geologic setting, it is possible to understand why the chemistry has evolved as it has or to predict what the water chemistry is likely to be for some new area. The important chemical reactions begin in the unsaturated zone as water percolates through the soil zone and moves deeper. Key reactions in this respect include

i) the dissolution of soil gases, like O_2 and CO_2,
ii) the dissolution/precipitation of minerals, like calcite, dolomite, or feldpsar,
iii) cation exchange reactions, and
iv) the dissolution/utilization of organic compounds.

The dissolution reactions are particularly important because water is so efficient at dissolving gases and solids found in the vadose zone. Thus, by the time water gets to the water table, it has acquired a significant load of dissolved mineral matter.

The chemistry of groundwater not only depends upon the processes in the vadose zone but also the reactions operating along the saturated flow system. Most of the same processes affecting ion concentrations in the unsaturated zone are also operative in the saturated zone including the dissolution and precipitation of various minerals, and cation exchange. Redox reactions that occur along flow paths become increasingly important in transferring mass among the various aqueous species.

Generally, the most rapidly dissolving minerals have the greatest impact on the chemistry of water. Thus, as water moves through the vadose zone, minerals like calcite {$CaCO_3$}, dolomite, {$CaMg(CO_3)_2$} and gypsum {$CaSO_4{\cdot}2H_2O$} dissolve to saturation if they are present. Minerals like quartz and feldspar dissolve more slowly at low temperatures and will only reach equilibrium in water having a long residence time in the subsurface. Thus, think of water-rock system being a partial-equilibrium chemical system, in equilibrium with some minerals and out of equilibrium with respect to others that are slow to dissolve or possibly not even present.

19.8.1 Gas Dissolution and Redistribution

The dissolution and redistribution of $CO_2(g)$ are important soil-zone processes that have a profound influence on the chemistry of groundwater. Rainwater or melted snow contain relatively small quantities of mass, are somewhat acidic and have a P_{CO_2} of about $10^{-3.5}$ atm. As this water moves downward, it rapidly dissolves CO_2 that occurs in soil at partial pressures larger than the atmospheric value. Elevated CO_2 pressures are due primarily to root and microbial respiration and to a lesser extent the oxidation of organic matter (Palmer and Cherry, 1984). Values for soils range generally from $10^{-3.5}$ atm to more than 500 times larger (Palmer and Cherry, 1984). CO_2 dissolved in water is further redistributed among the weak acids of the carbonate system according to the following set of reactions

$$CO_2(g) + H_2O = H_2CO_3^*$$

$$H_2CO_3^* = HCO_3^- + H^+$$

$$HCO_3^- = CO_3^{2-} + H^+$$

A direct result of dissolving $CO_2(g)$ in water is a rapid increase in the total carbonate content of the water and a decrease in pH. For a pH range of 4.5–5.5, $H_2CO_3^*$ is the dominant carbonate species with HCO_3^- and H^+ the most dominant anion and cation, respectively. In general, because the partial pressures of CO_2 gas are high in the soil zone infiltrating water becomes quite acidic and rich in dissolved carbonate species.

Another important soil zone/atmospheric process is the dissolution of $O_2(g)$. The resulting levels of dissolved oxygen are often sufficiently large at least initially to control the redox chemistry in shallow groundwater.

19.8.2 Mineral Dissolution/Precipitation

CO_2-charged water is effective in dissolving minerals. The most common reactions involve the weak acids of the carbonate and silicate systems and strong bases from the dissolution of carbonate, silicate, and alumino-silicate minerals. As we showed in Section 17.2, this process causes weak acids to dissociate. In the carbonate system, the relative abundance of HCO_3^- and CO_3^{2-} increases at the expense of $H_2CO_3^*$. Overall, both alkalinity and cation concentrations increase. Following are both generic and specific examples of weak acid strong base reactions

carbonate minerals + H^+ = cations + HCO_3^-

for example, calcite: $CaCO_3(s) + H^+ = Ca^{2+} + HCO_3^-$

silicate minerals + H^+ = cations + H_2SiO_3

for example, enstatite: $MgSiO_3(s) + 2H^+ = Mg^{2+} + H_2SiO_3$

alumino-silicate minerals + H^+ = cations + H_2SiO_3 + secondary minerals

for example anorthite: $CaAl_2Si_3O_8(s) + 2H^+ + H_2O$ = kaolinite + Ca^{2+}

albite: $2NaAlSi_3O_8(s) + 2H^+ + 5H_2O$ = kaolinite + $4H_2SiO_3 + 2Na^+$

Soluble salts commonly occur within certain geologic units, for example, soils or near-surface deposits in arid areas or as thick evaporite deposits in sedimentary basins. Mineral salts are extremely soluble and when present can dissolve to produce saline waters and even brines. The resulting composition of the water depends upon the particular minerals present (e.g., halite, anhydrite, gypsum, carnalite, kieserite, and sylvite). Here are examples of salts and ion species found in water.

halite: $NaCl(s) = Na^+ + Cl^-$
anhydrite: $CaSO_4(s) = Ca^{2+} + SO_4^{2-}$
gypsum: $CaSO_4{\cdot}2H_2O(s) = Ca^{2+} + SO_4^{2-} + 2H_2O$
carnalite: $KCl{\cdot}MgCl_2{\cdot}6H_2O(s) = K^+ + Mg^{2+} + 3Cl^- + 6H_2O$
kieserite: $MgSO_4{\cdot}H_2O(s) = Mg^{2+} + SO_4^{2-} + H_2O$
sylvite: $KCl(s) = K^+ + Cl^-$

Sulfide oxidation is one of the important redox reactions within the unsaturated zone. Minerals like pyrite or marcasite are dissolved through oxidation reactions to produce $Fe(OH)_3(s)$, SO_4^{2-} and H^+. Indeed, pyrite oxidation is one of the most important acid producing reactions in geological systems (Moran et al., 1978). In coal mining areas like the Appalachians, this reaction can be the cause of serious problems of acid-mine drainage (Moran et al., 1978). Following is an example of sulfide mineral oxidation,

$$4FeS_2 + 15O_2 + 14H_2O = 4Fe(OH)_3 + 16H^+ + 8SO_4^{2-}$$

Precipitation reactions operate to remove mass from solution within the soil zone and along the groundwater flow system. While such mass losses are important, they are difficult to notice in chemical data because they tend to be swamped by mass additions. It is these precipitation processes that give rise to the formation of cements and authigenic pore-filling minerals.

19.8.3 Cation Exchange Reactions

The most common exchange reactions are the water softening reactions where Ca^{2+} and Mg^{2+} in the water exchange with sorbed Na^+ as groundwater moves through clayey material, for example

$$\begin{matrix} Ca^{2+} & & Ca^{2+} \\ Mg^{2+} + Na\text{-clay} & = & 2Na^+ + Mg\text{-clay} \\ Fe^{2+} & & Fe^{2+} \end{matrix}$$

where Na-clay is Na adsorbed onto a clay mineral.

Exchange reactions can involve clay minerals within the unsaturated zone. In large flow systems, cation exchange is usually related to clay minerals deposited in a marine environment, which contain a large reservoir of exchangeable Na. One does not have to go far in the United States or Canada to find clays or shales capable of ion exchange.

19.8.4 Dissolution/Utilization of Organic Compounds

Work by Wallis et al. (1981), Thurman (1985), and Hendry et al. (1986) have identified a variety of important organic reactions such as dissolution of organic litter at the ground surface, complexation of Fe and Al, sorption of organic compounds, and oxidation of organic compounds.

The dissolution of organic litter at or close to the ground surface is the major source of dissolved organic carbon (DOC) in soil water and shallow groundwater. DOC concentrations typically fall in a range from 10 to 50 mg/L in the upper soil horizons and less than 5 mg/L deeper in the unsaturated zone (Thurman, 1985). Concentrations are highest at their source and decline with depth through sorption and oxidation. The most important fraction of the DOC is a group of humic substances, consisting mainly of humic and fulvic acids. Tannins and lignins, amino acids, and phenolic compounds are often present in smaller concentrations (Wallis et al., 1981).

The complexation of Fe and Al with organic matter is an important process facilitating the transport of these poorly soluble metals from the A horizon of the soil to the B horizon (Thurman, 1985). This is one key feature of the soil-forming process called podzolization (Thurman, 1985). In terms of groundwater systems, this reaction is not particularly important because most of these complexed metals sorb in the B horizon.

Dissolved organic compounds originating in the upper part of the soil horizon are not particularly mobile due to sorption. Many of the sorption models we discussed in Chapter 17 operate within the soil zone. Hydrophobic sorption occurs because of the relatively large quantities of solid organic matter present in the upper part of many soil horizons (Thurman, 1985). Similarly, the abundance of metal oxides, hydroxides, and clay minerals leads to surface complexation reactions and electrostatic interactions. Again in terms of the chemistry of shallow groundwater, this is one of the processes that keeps the quantity of DOC in recharge at relatively low concentrations.

Oxidation reactions involving organic matter can influence the chemistry of shallow groundwater. For example, the oxidation of dissolved organic matter (represented as CH_2O) provides a source of CO_2 gas within the unsaturated zone, which is readily dissolved in soil water. A second reaction, involving the oxidation of a sulfur-containing compound (represented by the amino acid cysteine), is thought to play a major role in the accumulation of gypsum in shallow soils (Hendry et al., 1986). This reaction is the organic counterpart to the pyrite oxidation reaction. In arid areas, this process can also produce recharge with large SO_4^{2-} concentrations (Hendry et al., 1986).

19.8.5 Redox Reactions

Redox processes operate along groundwater flow paths and play an important role in controlling (1) the chemistry of metal ions and solids (e.g., Fe^{2+}, Mn^{2+}, and Fe_2O_3), (2) species or solids containing sulfur (e.g., SO_4^{2-}, H_2S, and FeS_2), (3) NO_3^- an important contaminant in groundwater, and (4) dissolved gases containing carbon (e.g., CO_2 and CH_4). These reactions are mediated by bacteria that gain energy from the oxidation of organic carbon and reduction of O_2, NO_3^-, etc. as indicated in the following half redox reactions important in aquifers.

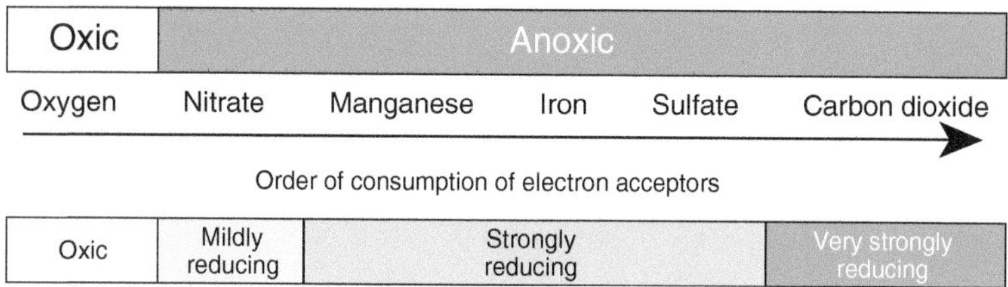

Figure 19.15 The electron acceptors form a redox ladder that explain the redox zonation that can develop along a flow system with transport away from a recharge area (U.S. Geological Survey, adapted from Kingsbury et al., 2014/Public domain).

Oxygen reduction:	$O_2(g) + 4H^+ + 4e^- = 2H_2O$
Denitrification:	$2NO_3^- + 12\ H^+ + 10e^- = N_2(g) + 6H_2O$
Mn(IV) reduction:	$MnO_2(s) + 4H^+ + 2e^- = Mn^{2+} + 2H_2O$
Fe(III) reduction:	$Fe(OH)_3(s) + H^+ + e^- = Fe^{2+} + H_2O$
Sulfate reduction:	$SO_4^{2-} + 9H^+ + 8e^- = HS^- + 4H_2O$
CO_2 reduction:	$CO_2(g) + 8H^+ + 8e^- = CH_4(g) + 2H_2O$

The oxidation reactions typically involve organic matter (e.g., as CH_2O). The complete reaction for denitrification, for example, shows CH_2O as the electron donor and NO_3^- as the electron acceptor or

$$5CH_2O + 4NO_3^- + 4H^+ = 5CO_2(g) + 2N_2(g) + 7H_2O$$

We have not yet emphasized the ordering of the half-redox reduction reactions within a transport context. Taken together, this collection of reactions (See Textbox) forms a redox ladder with conditions becoming increasingly reducing moving from the equation involving the reduction of O_2 to CO_2.

It is usually possible along flow paths to define redox zones. These zones are parts of an aquifer in which pe is controlled by one of the redox half reactions shown in the textbox. As is indicated with Figure 19.15, there is a definite order in which the various electron acceptors are utilized along a transport pathway. For example, when dissolved oxygen is present close to the water table in a recharge area, the O_2/H_2O half reaction is dominant because oxygen reduction yields the most energy to bacteria present in the system. With continuing advective transport along the flow path, oxygen is consumed and microorganism favor nitrate, the next electron donor in the ladder, through the $NO_3/N_2(g)$ half reaction. Although a common agricultural contaminant, nitrate may not be present. In this case, $MnO_2(s)$ or $Fe(OH)_3(s)$ reduction occurs.

Reduced sulfide and methane zones tend to develop in confined flow systems containing excess oxidizable DOC and a lack of recharge-containing oxygen downgradient in the system. Thus, redox zones will be observed in an extensive artesian aquifer that receives recharge from a limited area of outcrop or in an aquifer with confining units.

Redox reactions along flow systems assist in attenuating certain contaminants like nitrate but can create problems of geogenic contamination with trace metals like arsenic. Nitrate fertilizers applied to farm crops often end up as NO_3^-, which contaminates shallow unconfined aquifers. However, as shown by the redox ladder, once oxygen is used up, denitrification might substantially reduce nitrate concentrations to levels below regulatory concern. This essentially a process of natural attenuation where a microbial community converts nitrate to nitrogen gas.

In the case of arsenic, harmful to human health, reducing conditions promote the release of arsenic from solid phases into groundwater, where it may be consumed by drinking contaminated groundwater. This problem is most serious in alluvial aquifers, such as the Indo-Gangetic Alluvial aquifer in South Asia, the Central Valley aquifer in California, and the Mississippi River Alluvial aquifer. In these aquifers, quartz grains often have an iron mineral coating, $Fe(OH)_3$, that can contain arsenic if a nearby source exists. Under strongly reducing conditions (Figure 19.15), reductive dissolution of Fe oxyhydroxides also releases As into solution. This process has created severe health problems in Bangladesh and has been an important contaminant of concern in the United States.

19.9 Examples of Reactions Affecting Water Chemistry

This section provides examples of how the chemical processes presented in the previous section work to control the chemistry of groundwater. For each example, we highlight the most important reaction and describe the water evolves chemically. For most of these examples, the mineralogy of the hydrogeologic unit and climate play major roles in determining how the water evolves. Thus, with information about the geologic setting and the climate, one can begin to anticipate how water might evolve chemically.

19.9.1 Chemical Evolution of Groundwater in Carbonate Terrains

- dissolution of CO_2 gas
- dissolution of one or two carbonate minerals, e.g.,

$$CaCO_3(s) + H^+ = Ca^{2+} + HCO_3^-$$

This simple example shows how water evolves as a consequence of two simple chemical processes, the dissolution of CO_2 gas in the shallow soil zone and the dissolution of calcite and/or dolomite. Rainwater enters the subsurface and rapidly dissolves CO_2 gas, creating an aggressive, acidic water. At the same time, it begins to dissolve carbonate that it encounters. The aggressiveness of the water (reflected by pH) depends upon the partial pressure of CO_2 gas initially present in the water. As shown above, calcite dissolves to increase the concentrations of both Ca^{2+} and HCO_3^- in groundwater. HCO_3^- is also generated through the rapid dissolution of CO_2 into water. When calcite is common in soils, Ca^{2+} and HCO_3^- and pH values increase quickly along a flow path. Thus, shallow groundwater in such a setting is likely at equilibrium with respect to calcite with typical Ca^{2+} concentrations of 55 mg/L and HCO_3^- of 260 mg/L.

A carbonate terrain provides a field setting where the dissolution of CO_2 gas and calcite and/or dolomite influences the chemistry of soil water and groundwater. Langmuir (1971) compared the chemistry of groundwater from surface springs and wells in a carbonate region in central Pennsylvania. The spring water reflected relatively short travel distances, and probably approximates the composition of water leaving the unsaturated zone. The most abundant ions were Ca^{2+} and HCO_3^- with relatively small concentrations of Na^+ and Cl^-. The calculated P_{CO_2} for a single spring sample was $10^{-2.08}$ atm. Generally, spring waters were undersaturated with respect to calcite and dolomite. Presumably, transport through fractures had been so rapid that calcite equilibrium was not achieved. Farther along the flow system, calcite equilibrium finally constrains increases in Ca^{2+} and HCO_3^- concentrations. Groundwater approached saturation with respect to both calcite and dolomite with longer residence times. Typically, water from a predominantly limestone source rock more closely approaches saturation with respect to calcite than dolomite.

The example of evolving composition in a predominantly carbonate terrain (Langmuir, 1971) illustrated how the composition of the starting rainwater simply proceeds toward equilibrium with respect to those minerals available for dissolution (e.g., calcite and dolomite).

19.9.2 Shallow Brines in Western Oklahoma

- dissolution of halite and gypsum/anhydrite

This example is like the previous one in that the presence of a few rapidly dissolving minerals controlled the chemical evolution of the groundwater. In the shallow groundwater of western Oklahoma and the southeastern part of the Texas Panhandle, the minerals in shallow units are so soluble that brines are generated. Freshwater recharged through permeable units moves downward until it encounters salts at depths ranging from 10 to 250 m (Figure 19.16). The dissolving salt produces cavities at the updip limit or the top of the salt (Johnson, 1981). Periodically, the rocks overlying the cavities collapse. The process is apparently self-perpetuating because the collapse and fracturing of overlying units provides improved access to the salt for fresh water.

The evaporites mainly halite and gypsum/anhydrite are interbedded with a thick sequence of red-beds. Given the salts involved, high concentrations of Na^+ and Cl^- in groundwater were not surprising. Clearly when groundwater encounters large quantities of soluble salts in the subsurface, the impact on the chemistry can be considerable.

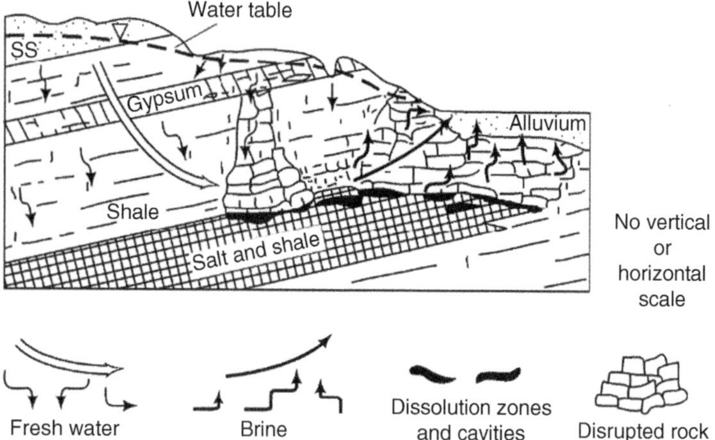

Figure 19.16 Continued dissolution of soluble minerals along the top of the "salt and shale" unit due to circulation of shallow groundwater leads to collapse of overlying units and the generation of brines (Johnson, 1981. Reproduced from Journal of Hydrology 54(1–3): 75–93. Copyright 1981, with permission from Elsevier Science).

19.9.3 Chemistry of Groundwater in an Igneous Terrain

- CO_2 dissolution and redistribution
- dissolution of alumino-silicate minerals

Igneous and metamorphic rocks contain silicates and alumino-silicate minerals that are slow to react even under attack by acidic waters. Thus, the mass dissolved in porewater will be relatively small when only these minerals are present. Garrels and MacKenzie (1967) looked at a case where waters rich in CO_2 reacted with plagioclase, K-feldspar, and biotite in an igneous rock. The study documented the chemical evolution of water from rain (low concentrations cation and anions) through groundwater with a short residence time (ephemeral springs) to groundwater with a longer residence time (perennial springs). In ephemeral water, the concentration of all ions is low (e.g., Ca^{2+}—3.1 mg/L; Na^{+}—3.0 mg/L and HCO_3^-—20 mg/L) because only relatively slow-reacting minerals, such as plagioclase feldspar, biotite and K-feldspar dissolve. In fact, the concentration of most ions is so low that the initial composition of precipitation had to be considered when Garrels and MacKenzie worked with the data. Dissolution apparently took place under closed-system conditions with about half of the CO_2 consumed.

With longer residence time, the quantity of mass dissolved in the groundwater increased (e.g., Ca^{2+}—10.1 mg/L; Na^{+}—6.0 mg/L and HCO_3^-—55 mg/L). The concentrations of cations increased due to the continued reaction of biotite and plagioclase, as does the alkalinity (reflected in the HCO_3^- concentration and pH). An important source of Ca^{2+} is the dissolution of small quantities of carbonate minerals. In the deeper parts of the system, montmorillonite occurs as a weathering product of plagioclase in addition to kaolinite. There are then settings with poorly soluble mineral that are slow to dissolve where the water is acidic with relatively low concentrations of dissolved constituents.

Fractured crystalline rocks are an important source of water worldwide. However, the geochemistry in regional aquifer systems is complicated to the extent that the conceptual model just discussed (i.e., Garrels and Mackenzie, 1967) needs to be applied with caution. The essential problem is that simple natural systems contain so little dissolved mass that other processes or sources can significantly alter the water chemistry. This feature is magnified because shallow fracturing in the rocks effectively connects local activities on the ground surface to the groundwater.

Crystalline rocks are also inherently complicated because they include so many different rock types—such as igneous rocks (granite, basalt, etc.), and metamorphic rocks. Their mineralogy can vary significantly and for most investigations is poorly studied.

A study of the water quality of the New England crystalline rock (NECR) aquifers (Flanagan et al., 2018) occurring in New England, (i.e., Connecticut, Massachusetts, Maine, New Hampshire, Rhode Island, and Vermont), and small areas of New York, and New Jersey illustrates this complexity and variability. This collection of fractured rock aquifers is a major source of water across that extends over 186,000 km^2 (72,000 mi^2) in the U.S. (Figure 19.17) and yielded 74 Mgal/d for various uses.

As is the case with most fractured rock systems, productivity and geochemistry depend upon the character of fracture development and rock lithology that are usually variable and poorly known. For the NECR study area is complicated with more than 1000 "formations," reclassified into 37 lithogeochemical groups and finally, ultimately three major lithology categories—felsic igneous rocks, mafic igneous and metamorphic rocks, and metasedimentary rocks (Flanagan et al., 2018). Additional geologic complexity also has arisen because of the occurrence of glacial till in upland areas and outwash deposits along valley bottoms.

Studies of domestic wells, which were age dated, indicated that younger waters typically had larger concentrations of dissolved oxygen and CO_2 (lower pH) than older waters which were more evolved chemically. For example, weathering of plagioclase along the flow system consumed carbon dioxide to increase pH (Flanagan et al., 2018). With younger groundwater, Ca and HCO_3 tended to be most dominant with absolute concentrations dependent upon whether reactive minerals (e.g., carbonates) were available in the soil zone. Thus, Ca concentrations could vary from much less than 20 mg/L to greater than 50 mg/L. HCO_3 concentrations, similarly, could be much less than 75 mg/L or greater than 125 mg/L. Waters with lower Ca-HCO_3 concentrations typically had lower pH values, often less than 6.5. With longer residence times, groundwater could evolve towards Na-HCO_3 associated with the dissolution of Na-feldspars or cation exchange. Again, the variation in Na concentrations were typically large, some water less than 6 mg/L with others greater than 20 mg/L Thus, in comparison to other major aquifer systems in the U.S., NECR aquifers, the total dissolved content and specific conductance of groundwaters were low, given the crystalline rock setting (Flanagan et al., 2018).

Figure 19.17 A large area of New England states and small pieces of New York and New Jersey are underlain by crystalline rock aquifers, collectively referred to as the New England crystalline rock (NECR) aquifers (U.S. Geological Survey, adapted from Flanagan et al., 2018/Public domain).

Variability in the geochemistry system came from unexpected sources or processes. For example, elevated concentrations of Na and (or) Cl in some groundwaters were associated with relict seawater or connate water from the time of formation. Another important source included road salt, applied for road deicing in populated areas (Flanagan et al., 2018). Groundwater chemistry is also influenced by anthropogenic contaminants from farming and wastewater disposal in septic tanks.

19.9.4 Evolution of Shallow Groundwater in an Arid Prairie Setting

- CO_2 dissolution and redistribution
- O_2 dissolution
- dissolution/precipitation of calcite, pyrite and gypsum
- cation exchange

A study of shallow groundwater from a study area in North Dakota (Moran et al., 1978) illustrates how many of the processes that we have considered work together. The system response to CO_2 dissolution and calcite distribution is the same the previous examples. What is most important in this system is the aridity that results in abundant gypsum in the soil and ultimately large solute loads in the groundwater.

Over much of the plains region of the United States and Canada, annual potential evaporation exceeds annual precipitation by a considerable amount. Thus, water that infiltrates in normal precipitation years evaporates and deposits a small quantity of gypsum. With repeated rain or snowmelt, gypsum accumulates in the upper part of the soil horizon (Moran et al., 1978). Exceptional recharge can dissolve some of this soluble material and move it down and into the groundwater system. In some arid areas, recharge water could have SO_4^{2-} concentrations >5000 mg/L (Hendry et al., 1986).

Data from 39 water samples collected from various near surface glacial deposits in North Dakota found mean Na^+ and Ca^{2+} concentrations of 469 mg/L and 62 mg/L, respectively (Moran et al., 1978). The dominant anions were HCO_3^- and SO_4^{2-} with concentrations of 748 mg/L and 577 mg/L, respectively. Cation exchange processes replaced Ca^{2+} that resulted from gypsum

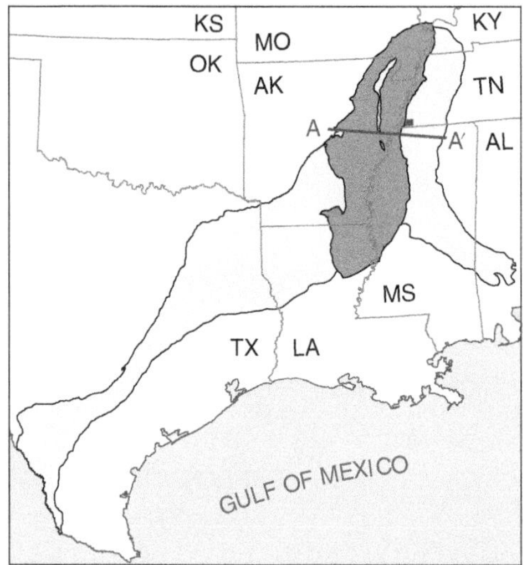

Figure 19.18 Map showing location of the Mississippi River Valley alluvial aquifer (grey blue) and the underlying embayment-uplands aquifers (yellow). A-A′ is the approximate location the hydrostratigraphic cross section. Memphis Tennessee is shown in the southwestern corner of Tennessee (TN) (U.S. Geological Survey, adapted from Kingsbury et al., 2014/Public domain).

($CaSO_4{\cdot}2H_2O$) dissolution with Na^+. Thus, the abundance of Na^+ and SO_4^{2-} in the shallow groundwater were attributed mainly to gypsum dissolution and cation exchange. Reaction of pyrite with oxygen also adds SO_4^{2-} to the groundwater.

19.10 A Case Study Highlighting Redox Processes

Redox processes have an amazing potential to influence the geochemical character of groundwater. This case study discusses the redox chemistry of an interesting assemblage of aquifers along the Gulf Coast of the United States (Kingsbury et al., 2014). These aquifers are part of the larger Mississippi embayment–Texas coastal uplands aquifer system and Mississippi River Valley alluvial (MRVA) aquifer that extend from southwest Texas eastward to Tennessee and Alabama. Besides the Mississippi River Valley alluvial aquifer system, the work of Kingsbury et al. (2014) included five associated aquifers—four older embayment-uplands sand aquifers, which include the (1) middle Claiborne, (2) lower Claiborne-upper Wilcox, and (3) middle Wilcox, (4) lower Wilcox and (5) the local terrace deposits aquifer found in and around Memphis (Figure 19.18). Readers interested in a more detailed discussion of these aquifers can refer to Chapter 4.

Figure 19.18 describes the stratigraphic arrangement of these aquifers along the approximate cross section A-A′ (Kingsbury et al., 2014). For simplicity, the aquifers below the middle Claiborne aquifer—i.e., the lower Claiborne-upper Wilcox, middle Wilcox and lower Wilcox aquifers are lumped together. In areas associated with the MRVA aquifer, most of the groundwater production from older units comes from the middle Claiborne (shallow) aquifer and to a lesser extent lower Claiborne-upper Wilcox aquifers (deeper). In Arkansas, the middle Claiborne Group is often known as the Sparta sand.

The pie charts contain information on the redox status of groundwater as interpreted from water-quality analyses of samples mainly from domestic wells. Because of relatively abundant organic matter in the Mississippi River Valley alluvial aquifer and Fe and Mn on grain coatings, groundwater is typically low in dissolved oxygen (DO) and strongly reducing (i.e., anoxic) (Figure 19.19). Waters from the terrace deposits aquifer, are mostly oxic because of lower contents of organic matter and clean quartz as the aquifer matrix. Water in the shallow embayment-uplands aquifers tend to be oxic because recharge for the same reason. However, with continued migration to deeper, confined aquifers, waters have become anoxic with the loss of oxygen (Figure 19.19).

19.10.1 Iron and Manganese

The impact of redox reactions on groundwater chemistry are most evident in the MRVA aquifer. Strongly reducing conditions resulted in the dissolution of mineral coatings on quartz grains leading to the release of iron (Fe) and manganese (Mn) into groundwater. Iron dissolved in groundwater severely impacts the taste of water to the extent of requiring treatment before drinking, and severe staining of bathtubs, toilets, sinks, and laundry. There is no health-based standard for Fe, although a nonmandatory limit of 0.3 mg/L is assigned with the USEPA National Secondary Drinking Water Regulations. Evaluation of 749 water analyses from the MRVA aquifer found a median Fe concentration of 3.29 mg/L, more than 10 times higher than the secondary standard with 86% of samples >0.1 mg/L (Knierim et al., 2021). Typically, Fe concentrations ≥0.1 mg/L are indicative of strongly reducing conditions, beyond Mn(IV) reduction (McMahon and Chapelle, 2008).

By comparison, the median concentration of Fe was 0.16 mg/L for the middle Claiborne aquifer (shallow Embayment-uplands aquifer) in this area, and 0.12 mg/L for the lower Claiborne aquifer (deep Embayment-upper aquifer) (Knierim et al., 2021). As indicated in Figure 19.19, oxic conditions were more prevalent in these aquifers.

Mn is present in oxide coatings on quartz grains and has also found its way into groundwater given the strong reducing conditions in the MRVA aquifer. Although present at relatively low concentrations, Mn is known to cause neural problems (Kingsbury et al., 2014). Mn is not regulated by the USEPA. However, there is a Health-Based Screening Level (HBSL) of

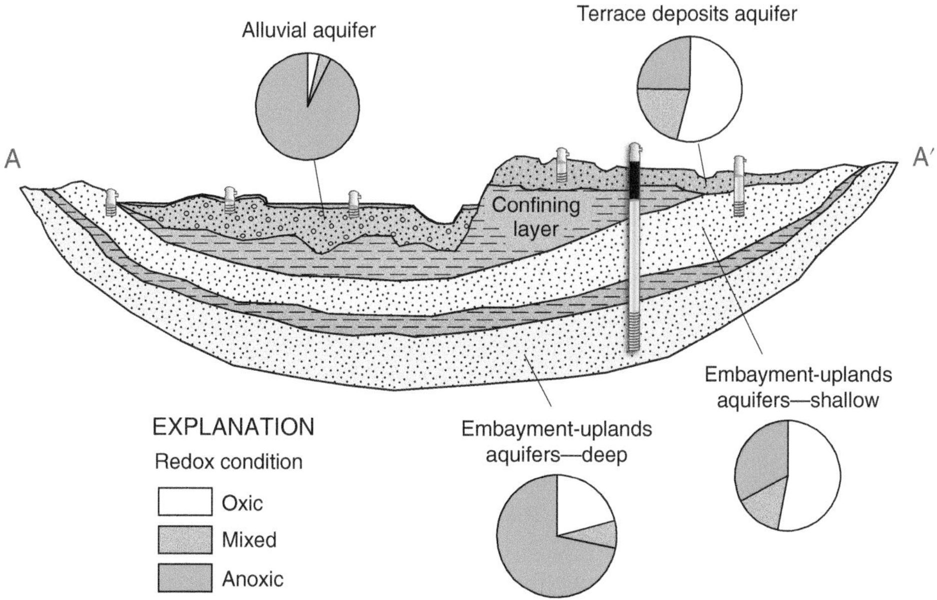

Figure 19.19 Idealized cross section showing deeper and shallower embayment-uplands aquifers, Mississippi River Valley alluvial aquifer, and terrace deposits aquifer in the vicinity of Memphis. The pie charts indicate the redox conditions for various units as oxic, mixed, and anoxic. Mixed waters have characteristics of both oxic and anoxic waters (U.S. Geological Survey, adapted from Kingsbury et al., 2014/Public domain).

300 μg/L (0.3 mg/L). This non-enforceable, water-quality benchmark was developed by the USGS following methods comparable to the USEPA procedures with priority pollutants (Norman et al., 2018).

The potential seriousness of the health problem with Mn in groundwater in the MRVA aquifer is evident with Figure 19.20. Approximately 80% of water sampled exceeded the benchmark standard of 300 μg/L (Kingsbury et al., 2014). With the embayment-uplands aquifers and the terrace aquifer, concentrations of Mn in groundwater were mostly below the benchmark standard (Figure 19.20). These aquifers tended to contain less organic matter and Fe and Mn solids, and consequently were less reduced with comparatively large concentrations of dissolved oxygen (Figure 19.20).

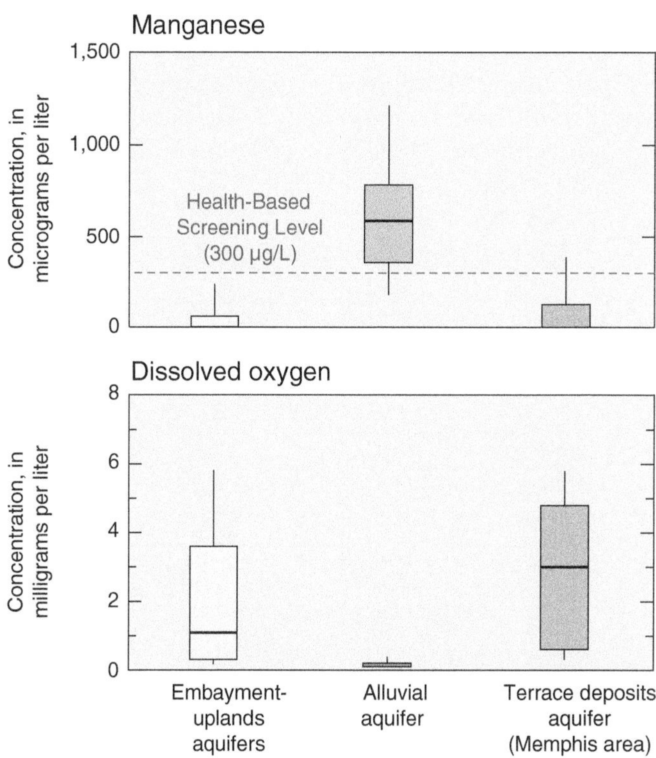

Figure 19.20 The problem of manganese in groundwater is mostly associated with the alluvial aquifer (MRVA aquifer) where many samples indicated concentrations above the benchmark standard of 300 μg/L. As evidenced by the results for dissolved oxygen, the MRVA aquifer is much more reducing than the embayment-uplands aquifers and terrace aquifer (U.S. Geological Survey, adapted from Kingsbury et al., 2014/Public domain).

19.10.2 Arsenic

Over the last 20 years, the immense health effects associated with relatively low concentrations of arsenic in groundwater have become apparent, especially in India and other Asian countries. Long-term exposures to arsenic through drinking of untreated groundwater can lead to skin lesions, keratosis, leading to skin and other cancers. Other problems of arsenic exposure include learning deficits in children, high blood pressure, heart disease and more. In the U.S., As is a priority pollutant with an MCL (maximum contaminant level) of 10 μg/L.

Arsenic can be associated with oxide coatings on quartz grains in aquifers. The necessary conditions are that naturally occurring sources of As(V) need to be present and able to accumulate via sorption on the oxide coatings. Under oxic conditions, these oxide coatings are stable solids, and As, although potentially present, is sequestered with Fe and Mn. Once the redox condition becomes anoxic to the extent where Fe oxides serve as electron acceptors, these coatings dissolve and Fe, Mn and As are released into the groundwater (Kingsbury et al., 2014). Thus, aquifers with water quality impaired by high concentrations of redox-related, dissolved Fe can be at risk health impacts from As.

The strongly reducing conditions associated with the MRVA aquifer and the relatively high iron concentrations in the water are concerning because of the possibility for As-related problems. Of the water sampled from the MRVA aquifer, approximately 10% had As concentration that exceeded the 10 μg/L MCL. Kingsbury et al. (2014) noted a correlation between As and Fe with waters from the MRVA aquifer, as has been the case in studies elsewhere. Values of As approached and exceeded the MCL of 10 μg/L with a few of the samples with higher Fe concentrations (i.e., >4 mg/L). As expected, arsenic was not a problem with the nearby shallow embayment-uplands aquifers in Arkansas and Tennessee (Kingsbury et al., 2014).

19.10.3 Nitrate

The area associated with the MRVA aquifer is farmed extensively. Chapter 14 discussed problems of storage depletion caused by overpumping of groundwater.

Another impact of farming associated with an unconfined aquifer is the possibility of nitrate (NO_3) contamination of shallow groundwater. Interestingly, as nitrate contaminated water recharges this aquifer, oxygen is depleted a few feet below the water table and NO_3 becomes the next available electron acceptor. Figure 19.21 is a conceptual model of the transport processes and the zonation in terms of electron acceptors. (Kingsbury et al., 2014).

Along a downward pathway, denitrification over a depth range from approximately 5 to 14 ft below the water table transforms NO_3 to nitrogen gas. Consequently, NO_3 concentrations, which otherwise might have approached the MCL of 10 mg/L (NO_3-N) declined to concentrations approaching the laboratory reporting level of 0.05 mg/L. As nitrogen disappears, iron/manganese coatings on quartz grains take over as the next electron acceptor leading to strongly reducing conditions and the associated increases in Fe, Mn, and As concentrations (Kingsbury et al., 2014).

Nitrate loading through applications of fertilizer to areas underlain by embayment-uplands aquifers and terrace deposits aquifer is about half as much as the MRVA aquifer (Figure 19.22). Nevertheless, NO_3 concentrations in those aquifers were noticeably higher than the MRVA aquifer, especially with the terrace deposits aquifer (Kingsbury et al., 2014). Because these

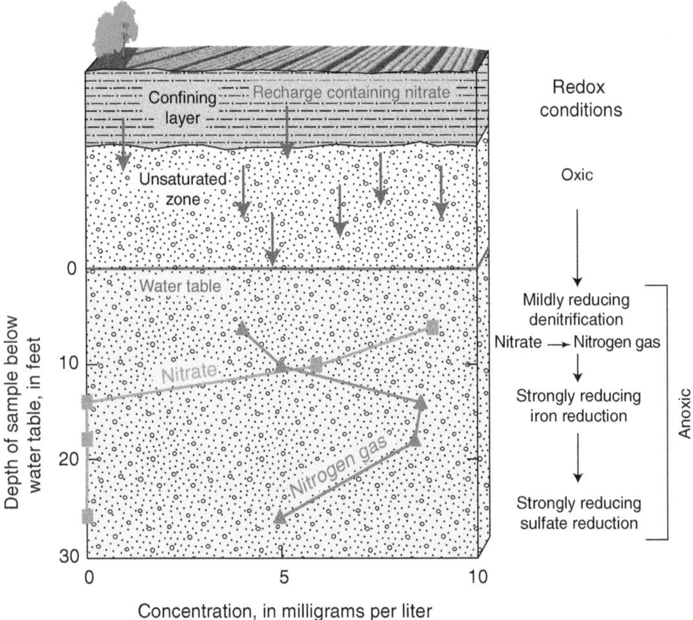

Figure 19.21 Water infiltrating the unsaturated zone is commonly oxic and can contain relatively high concentrations of NO_3. Over short travel distances, DO is utilized and denitrifications reduces NO_3 to nitrogen gas, leading to the removal of NO_3. As strong redox conditions develop, iron coatings are reduced, substantially increasing concentrations of Fe concentrations in the groundwater, as well as Mn and As (U.S. Geological Survey Kingsbury et al., 2014/Public domain).

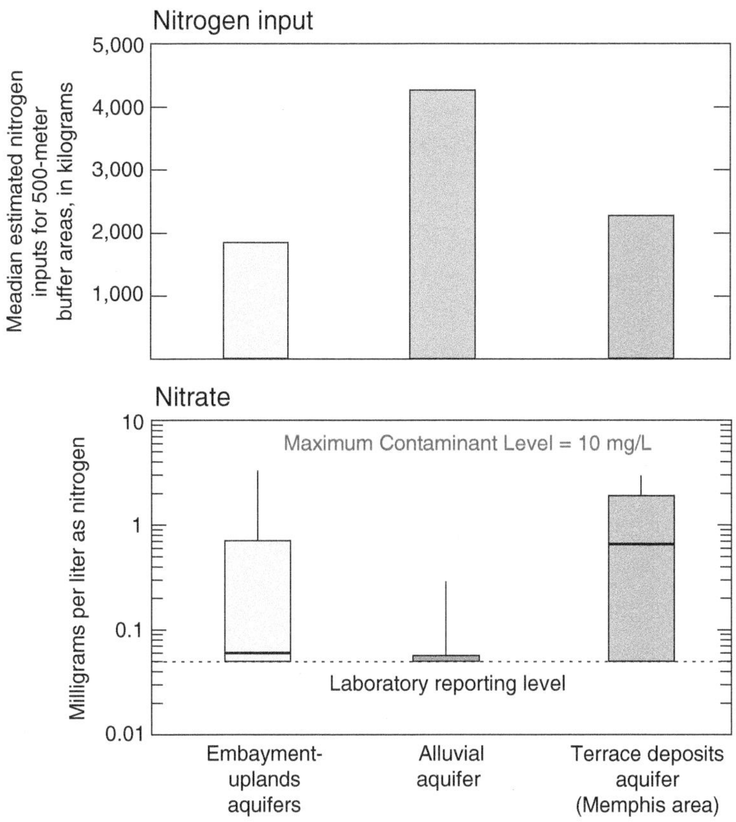

Figure 19.22 Histogram of median nitrogen loading to circular areas around wells (500 m radius) shows relatively high fertilization of crops, especially for areas associated with the MRVA aquifer. Nitrate concentrations are relatively low in all aquifers, due to relatively low infiltration rates and attenuation through denitrification, particularly in the MRVA aquifer (U.S. Geological Survey, adapted from Kingsbury et al., 2014/Public domain).

units are cleaner sands with less organic matter, the oxic or milder redox conditions were less effective in reducing NO_3 concentrations through denitrification.

19.10.4 Machine Learning for Mapping Redox Conditions

An emerging tool for the study of redox conditions is machine learning. This modeling technology encompasses a variety of computational approaches designed to recognize patterns in large data sets that might include many variables (Knierim et al., 2021). Recently, it has been used to predict and visualize patterns in groundwater chemistry. This section discusses an application to predict threshold probabilities for dissolved oxygen (DO) and Fe concentrations (Knierim et al., 2021). For DO, this probability might involve "the probability of exceeding a DO concentration of 1 mg/L was greater than 80%" (Knierim et al., 2021). The combination of data on DO probabilities and Fe provided the bases for mapping redox zonation according to Table 19.2.

Table 19.2 Criteria for the categorization redox zone using the predicted probability of dissolved oxygen (DO) concentration exceeding a threshold of 1 mg/L and iron concentrations.

Redox category	DO (probability)	Iron (μg/L)
Oxic	>0.8	<1,000
Mixed oxic	>0.8	>1,000
	≤0.8 and ≥0.1	<500
Mixed anoxic	≤0.8 and ≥0.1	≥500
Anoxic	<0.1	NA

Source: Knierim et al. (2021)/Public domain.

A boosted regression tree (BRT) modeling approach was applied to three key aquifers of the Mississippi embayment physiographic region (Knierim et al., 2021). The study area is located in the Mississippi River embayment which included the MRVA and adjacent units of the middle and lower Claiborne aquifers.

The model depended upon measured data on DO and Fe from groundwater samples from wells for training and validation, as well large numbers of explanatory variables. *Explanatory variables* in this application represented parameters with some power to influence the predicted variables (DO and Fe concentrations). Examples of explanatory variables for this study included information about wells, such as location, depth, screen elevation, land use, soil types, and information extracted from a groundwater flow model (Knierim et al., 2021). For the Fe model, output from the DO model (i.e., predicted probability of a DO concentration > 1 mg/L) was used as one of the explanatory variables.

The numbers of different explanatory variables were large—130 with DO and 132 for Fe (Knierim et al., 2021). A much smaller number these variables were influential in the prediction—33 for DO and 37 for Fe. With DO, the most significant of the 33 variables for prediction tended to be proxies for distance along the groundwater flow paths. Their usefulness implied the existence of evolving redox conditions from upland recharge areas along some flow path. Similarly, with the prediction of Fe concentrations, the most important of the reduced list of 37 explanatory variables included geochemical parameters, like predicted values of specific conductance and predicted DO, which evolved along flow paths (Knierim et al., 2021).

The map showing the concentrations of Fe across the MRVA aquifer (Figure 19.23) is illustrative of the results coming from the modeling with machine learning (Knierim et al., 2021). Notice the patterns that exist in iron concentrations. The small areas shown in blue with concentrations <500 μg/L are likely topographically higher and obvious recharge areas.

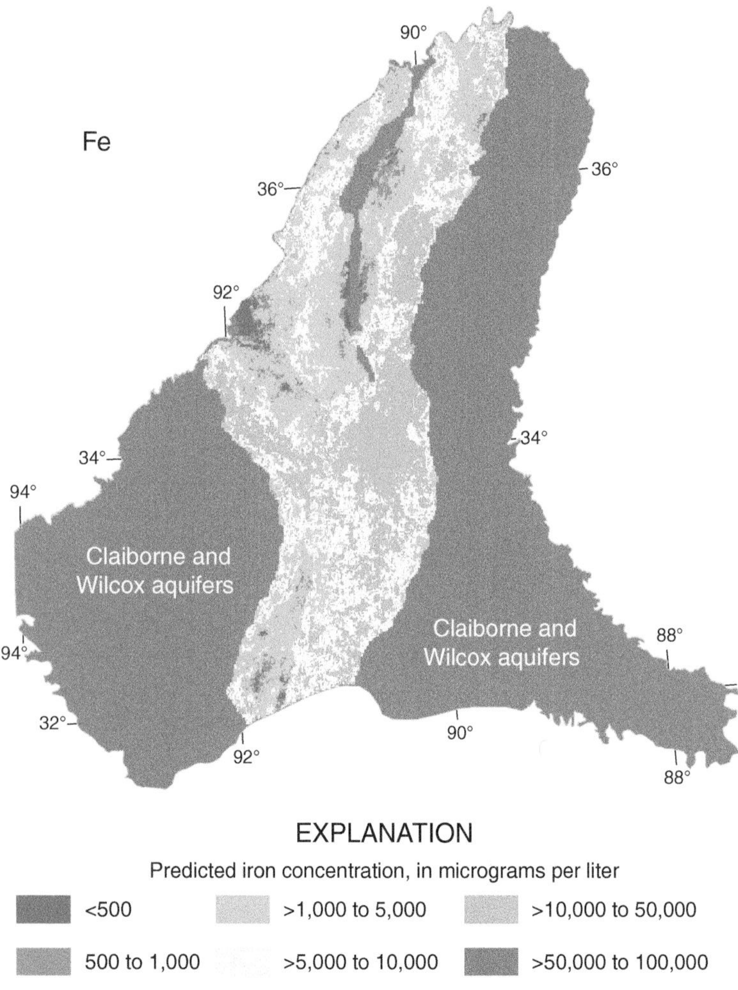

Figure 19.23 Map of iron concentrations in μg/L constructed for the MRVA aquifer using machine learning approaches. Higher values evolve downgradient in flow systems where residence times are longer and DO values has been consumed (U.S. Geological Survey, adapted from Knierim et al., 2021/Public domain).

The relatively systematic transitions from these lower concentrations to much higher values, such as 10–50 mg/L (shown in orange) suggests increasing concentrations in Fe along flowpaths. Maps like those shown in Figure 19.23 would be helpful in identifying areas where drinking untreated groundwater could give rise to health-related problems associated with Fe and Mn

Similarly, modeling with DO provided the basis for categorizing the redox conditions and displaying them in map form (see Plate 3, Knierim et al., 2021). The predicted Fe concentrations were used to confirm the categorization. Similar analyses involved the upper and lower Claiborne aquifers, and altogether supported and refined the general descriptions of redox conditions of key aquifers of the Mississippi Embayment presented earlier in this section.

Exercises

19.1 The hydrogeologic cross section, Figure 19.24 illustrates the pattern of groundwater flow along a local flow system. Assume that of source of contamination develops with advection as the only operative transport process. Describe the pathway for contaminant migration and estimate at what time in the future the plume will reach the stream.

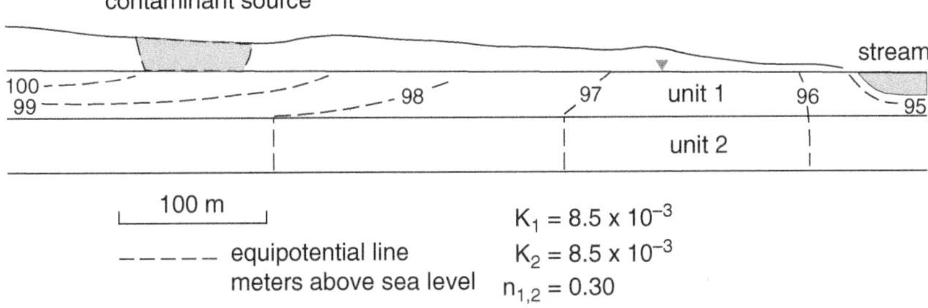

Figure 19.24 Hydrogeological cross section. (Domenico and Schwartz, 1998, *Physical and Chemical Hydrogeology*. Copyright © 1990, 1998 John Wiley & Sons, Inc. All Rights Reserved. Reproduced with permission).

19.2 A contaminant is added as a point source to groundwater flowing with a constant velocity of 4×10^{-6} m/sec. Assuming longitudinal and two transverse dispersivities (*y* and *z* directions) of 1.0, 0.1, and 0.01 m, determine the spatial standard deviations ($\sigma_{x,y,z}$) in the plume size after 400 m of transport.

19.3 Explain why the nitrate composition of groundwater found in an alluvial aquifer in parts of the United States may become depleted in agricultural nitrate after moving relative short distances away from a recharge area.

19.4 Over the plains region of the Northern United States and Canada, carbonate-rich till overlies marine shale and sandstone. Recharge from snowmelt has the following chemical composition as it moves downward through till and into shale bedrock:

Concentration (milligrams per liter)							
Unit	Ca^{2+}	Mg^{2+}	Na^+	HCO_3^-	SO_4^{2-}	Cl^-	pH
Till	79.0	50.0	210.0	436.0	61.0	14.0	7.80
Shale	5.0	0.5	450.0	1044.0	6.0	53.0	8.10

Source: Domenico and Schwartz, 1998, *Physical and Chemical Hydrogeology*. Copyright © 1990, 1998 John Wiley & Sons, Inc. All Rights Reserved. Reprinted with permission.

Interpret the chemical evolution of the water in terms of the most likely mass transport processes.

19.5 In carbonate rocks subject to recharge, the most dramatic changes in major ion chemistry occur over a relatively short distance as infiltration first enters the unit. Later changes are often almost insignificant by comparison. Explain why using arguments related to kinetics and mineral equilibrium.

19.6 Using the concept of redox zones, explain why H_2S gas is rarely found in groundwater close to the water table in recharge areas.

References

Adams, E. E., and L. W. Gelhar. 1992. Field study of dispersion in a heterogeneous aquifer: 2. Spatial moments analysis. Water Resources Research, v. 28, no. 12, p. 3293–3307.

Boggs, J. M., S. C. Young, L. M. Beard et al. 1992. Field study of dispersion in a heterogeneous aquifer: 1. Overview and site description. Water Resources Research, v. 28, no. 12, p. 3281–3291.

Di Dato, M., F. P. de Barros, A. Fiori et al. 2016. Effects of the hydraulic conductivity microstructure on macrodispersivity. Water Resources Research, v. 52, no. 9, p. 6818–6832.

Domenico, P. A., and F. W. Schwartz. 1998. Physical and Chemical Hydrogeology. John Wiley and Sons, New York, 506 p.

Dubrovsky, N. M., K. R. Burow, G. M. Clark, et al. 2010. The quality of our Nation's waters—Nutrients in the Nation's streams and groundwater, 1992–2004. U.S. Geological Survey Circular 1350, 174 p.

Flanagan, S. M., J. D. Ayotte, and G. R. Robinson, Jr. 2018. Quality of water from crystalline rock aquifers in New England, New Jersey, and New York, 1995–2007 (ver. 1.1, April 2018). U.S. Geological Survey Scientific Investigations Report 2011–5220, 104 p.

Freyberg, D. L. 1986. A natural gradient experiment on solute transport in a sand aquifer: 2. Spatial moments and the advection and dispersion of nonreactive tracers. Water Resources Research, v. 22, no. 13, p. 2031–2046.

Garabedian, S. P., D. R. LeBlanc, L. W. Gelhar et al. 1991. Large-scale natural gradient tracer test in sand and gravel, Cape Cod, Massachusetts: 2. Analysis of spatial moments for a nonreactive tracer. Water Resources Research, v. 27, no. 5, p. 911–924.

Garrels, R. M., and F. T. MacKenzie. 1967. Origin of the chemical compositions of some springs and lakes, In W. Stumm (ed.), Equilibrium Concepts in Natural Water Systems (Chapter 10). American Chemical Society, Washington DC, p. 222–242.

Gelhar, L. W., A. L. Gutjahr, and R. L. Naff. 1979. Stochastic analysis of macrodispersion in a stratified aquifer. Water Resources Research, v. 15, no. 6, p. 1387–1397.

Gelhar, L. W., C. Welty, and K. R. Rehfeldt. 1992. A critical review of data on field-scale dispersion in aquifers. Water Resources Research, v. 28, no. 7, p. 1955–1974.

Grisak, G. E., and J. Pickens. 1980. Solute transport through fractured media: 1. The effect of matrix diffusion. Water Resources Research, v. 16, no. 4, p. 719–730.

Hendry, M. J., J. A. Cherry, and E. I. Wallick. 1986. Origin and distribution of sulfate in a fractured till in southern Alberta, Canada. Water Resources Research, v. 22, no. 1, p. 45–61.

Hess, K. M., S. H. Wolf, and M. A. Celia. 1992. Large-scale natural gradient tracer test in sand and gravel, Cape Cod, Massachusetts: 3. Hydraulic conductivity variability and calculated macrodispersivities. Water Resources Research, v. 28, no. 8, p. 2011–2027.

Jagucki, M. L., M. K. Landon, B. R. Clark, et al. 2008. Assessing the vulnerability of public-supply wells to contamination—High Plains Aquifer near York, Nebraska. U.S. Geological Survey Fact Sheet 2008-3025, 6 p.

Johnson, K. S. 1981. Dissolution of salt on the east flank of the Permian Basin in the southwestern USA. Journal of Hydrology, v. 54, no. 1–3, p. 75–93.

Kingsbury, J. A., J. R. B. Barlow, B. G. Katz, et al. 2014. The quality of our Nation's waters—Water quality in the Mississippi embayment–Texas coastal uplands aquifer system and Mississippi River Valley alluvial aquifer, south-central United States, 1994–2008. U.S. Geological Survey Circular 1356, 72 p.

Knierim, K. J., J. A. Kingsbury, and C. J. Haugh, 2021. Machine-learning predictions of redox conditions in groundwater in the Mississippi River Valley alluvial and Claiborne aquifers, south-central, United States. U.S. Geological Survey Scientific Investigations Map 3468, pamphlet 16 p., 3 sheets, scale 1:3,400,000.

Langmuir, D. 1971. The geochemistry of some carbonate groundwaters in central Pennsylvania. Geochimica et Cosmochimica Acta, v. 35, no. 10, p. 1023–1045.

LeBlanc, D. R., S. P. Garabedian, K. M. Hess et al. 1991. Large-scale natural gradient tracer test in sand and gravel, Cape Cod, Massachusetts: 1. Experimental design and observed tracer movement. Water Resources Research, v. 27, no. 5, p. 895–910.

Mackay, D. M., D. L. Freyberg, P. V. Roberts et al. 1986. A natural gradient experiment on solute transport in a sand aquifer: 1. Approach and overview of plume movement. Water Resources Research, v. 22, no. 13, p. 2017–2029.

McMahon, P. B., and F. H. Chapelle. 2008. Redox processes and water quality of selected principal aquifer systems. Groundwater, v. 46, no. 2, p. 259–271.

Moran, D. R., G. H. Groenwold, and J. A. Cherry, 1978. Geologic, hydrologic, and geo-chemical concepts and techniques in overburden characterization for mineral-land reclamation. North Dakota Geological Survey Report of Investigation No. 63, 152 p.

Neretnieks, I. 1985. Diffusivities of some constituents in compacted wet bentonite clay and the impact on radionuclide migration in the buffer. Nuclear Technology, v. 71, no. 2, p. 458–470.

Norman, J. E., P. L. Toccalino, and S. A. Morman. 2018. Health-based screening levels for evaluating water-quality data (2nd ed.), U.S. Geological Survey web page, accessible at https://water.usgs.gov/water-resources/hbsl/ (accessed on 8 April 2023).

Palmer, C. D., and J. A. Cherry. 1984. Geochemical evolution of groundwater in sequences of sedimentary rocks. Journal of Hydrology, v. 75, no. 1–4, p. 27–65.

Rehfeldt, K. R., J. M. Boggs, and L. W. Gelhar. 1992. Field study of dispersion in a heterogeneous aquifer: 3. Geostatistical analysis of hydraulic conductivity. Water Resources Research, v. 28, no. 12, p. 3309–3324.

Robbins, G. A. 1983. Determining dispersion parameters to predict groundwater contamination. Ph.D. Dissertation. Texas A&M University.

Schwartz, F. W., and H. Zhang. 2003. Fundamentals of Groundwater. John Wiley & Sons, Inc., 583 p.

Sudicky, E. A. 1986. A natural gradient experiment on solute transport in a sand aquifer: spatial variability of hydraulic conductivity and its role in the dispersion process. Water Resources Research, v. 22, no. 13, p. 2069–2082.

Tang, D. H., E. O. Frind, and E. A. Sudicky. 1981. Contaminant transport in fractured porous media: analytical solution for a single fracture. Water Resources Research, v. 17, no. 3, p. 555–564.

Thurman, E. M. 1985. Organic Geochemistry of Natural Waters. Kluwer Academic Publishers Group, Dordrecht, 497 p.

Wallis, P. M., H. Hynes, and S. A. Telang. 1981. The importance of groundwater in the transportation of allochthonous dissolved organic matter to the streams draining a small mountain basin. Hydrobiologia, v. 79, no. 1, p. 77–90.

20

Introduction to Contaminant Hydrogeology

CHAPTER MENU

Contamination problems are organized around basic concepts of flow and mass transport. The physical processes advection and dispersion, as well chemical, biological, and nuclear processes, are at the heart of every contamination problem. However, there are a few other processes yet to be discussed, for example, the flow of contaminants that are themselves liquids or nonaqueous phase liquids (NAPLs). Another fundamental concept is that of a source zone, which encompasses those places where the contaminant is available to be influenced by those processes discussed above. As will become evident, a plume is a manifestation of a source zone but may not always be present.

The earlier chapters have developed a fundamental understanding of different transport processes. The study of groundwater contamination requires the integration of these processes, considering the inherent variability in the extent to which the processes operate. Additional complication comes from the spectrum of different investigative techniques needed to address different kinds of contamination problems. Fortunately, there are basic concepts that help organize our thinking about contamination problems with respect to the origin of the problem, strategies for investigation, and remediation (Figure 20.1).

Three basic principles are helpful in organizing and describing contamination problems—point and nonpoint-source problems, families of contaminants, and NAPLs. These are the main topics then for this chapter.

20.1 Point and Nonpoint Contamination Problems

The terms "point" or "nonpoint" describe the degree of localization of the contaminant source. Figure 20.1 illustrates an example of a *point source problem*. The source is identifiable and small in scale (meters to kilometers), such as a leaking storage tank, one or more disposal ponds, or a sanitary landfill. Associated plumes of contaminants could be large, kilometers in length. They are often well-defined because concentrations of dissolved contaminants can be significantly higher than background levels. Large quantities of wastes localized in the source can keep a source active for decades to centuries.

A *nonpoint* problem refers to larger-scale, relatively diffuse contamination originating within a large source area, many sources, widely distributed and poorly defined (Figure 20.1). As the figure illustrates, an entire study area is a logical source area. Examples of nonpoint contaminants could include geogenic contaminants like arsenic and uranium, nitrates that

Fundamentals of Groundwater, Second Edition. Franklin W. Schwartz and Hubao Zhang.

Companion website: www.wiley.com/go/schwartz/fundamentalsofgroundwater2

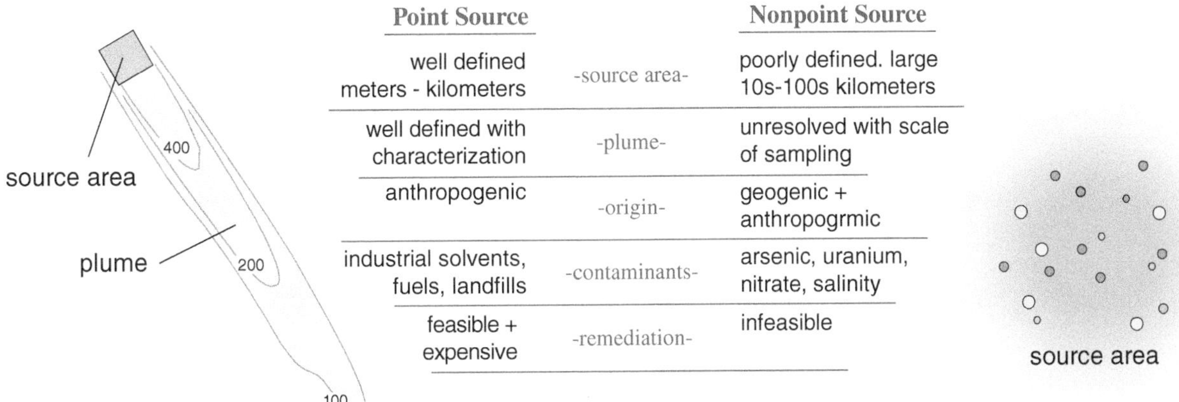

Figure 20.1 Point and nonpoint source contamination as a conceptual model for organizing contamination problems (FWS).

originate from fertilizer, or salinity associated with both geogenic and anthropogenic sources, such as salt used to de-ice highways in winter. Typically, there is a large enclave of contamination with variable concentrations of contaminants at the scale that sampling occurs.

Because of the localization of point-source problems, it is usually feasible to remediate them by removing or isolating the source and quickly minimizing the potential health risks by in-home water treatment or replacing supplies. A variety of specialized investigative techniques are available to identify patterns of plume migration and areas of risk. Nonpoint problems are beyond remediation because contamination is so variable and widespread. Over a long term, certain problems, such as nitrate contamination, might improve with nutrient management approaches in agriculture. Investigative approaches for non-point problems are designed around the sampling of existing wells and can identify potential hotspots and health-related risks. However, in places like Bangladesh, drinking water from individual wells would need to be tested individually to mitigate arsenic risks or perhaps installing deeper wells with potentially less risk. Examples illustrating point and non-point sources are given as follows.

In 1970s, people recognized the importance of point-source contamination due to the careless disposal of industrial chemicals, solid wastes, and so on. In North America, Europe, and other industrialized places, the historical characteristic of the waste disposal, sometimes from the early 1900s, has meant that sources might be associated with large plumes of dissolved contamination. Old, unregulated landfills might contain a variety of urban refuse and hazardous industrial wastes placed together in a relatively small source area. Commonly, waste materials were used to fill in holes in the ground created from the extraction of sand and gravels.

An example of one of these historical point-source problems was the Babylon landfill, located on Long Island in the United States (Figure 20.2) (Kimmel and Braids, 1980). Landfilling at the Babylon site began in 1947 with urban refuse, incinerated garbage, cesspool waste, and industrial refuse. Disposal of wastes in an unengineered landfill in shallow, sandy materials facilitated their dissolution and the creation of large contaminant plumes. Numerous plumes developed, including unregulated constituents, Ca^{2+}, Mg^{2+}, Na^{+}, K^{+}, HCO_3^-, SO_4^{2-}, and Cl^-, at concentrations elevated above background; nitrogen species such as NH_4^+ and NO_3^-; metals, particularly iron and manganese; and organic compounds. Plumes spread within the permeable surficial sand with the regional flow to the south–southeast.

Nonpoint contamination problems develop across large spatial scales, due to anthropogenic activities like farming or mining, and/or geogenic processes with constituents such as arsenic, uranium, selenium, nitrate, and fluoride. These contaminants are important because of their prevalence and potential health impacts. Chapter 16 introduced the Madera and Chowchilla study unit in the San Joaquin Valley, north of Fresno California (Shelton et al., 2013). This study area is notable for interesting examples of nonpoint contamination with respect to selected trace elements (i.e., arsenic, barium uranium, and vanadium) (Shelton et al., 2013). Except for barium, the relative concentrations of the three others exceeded a relative concentration (RC) of 1. Recall that the relative concentration of a constituent such as arsenic is the measured concentration of arsenic in the sample divided by the accepted health standard, which for arsenic is 0.010 μg/L. In this setting, the occurrence of these four constituents is an example of nonpoint contamination.

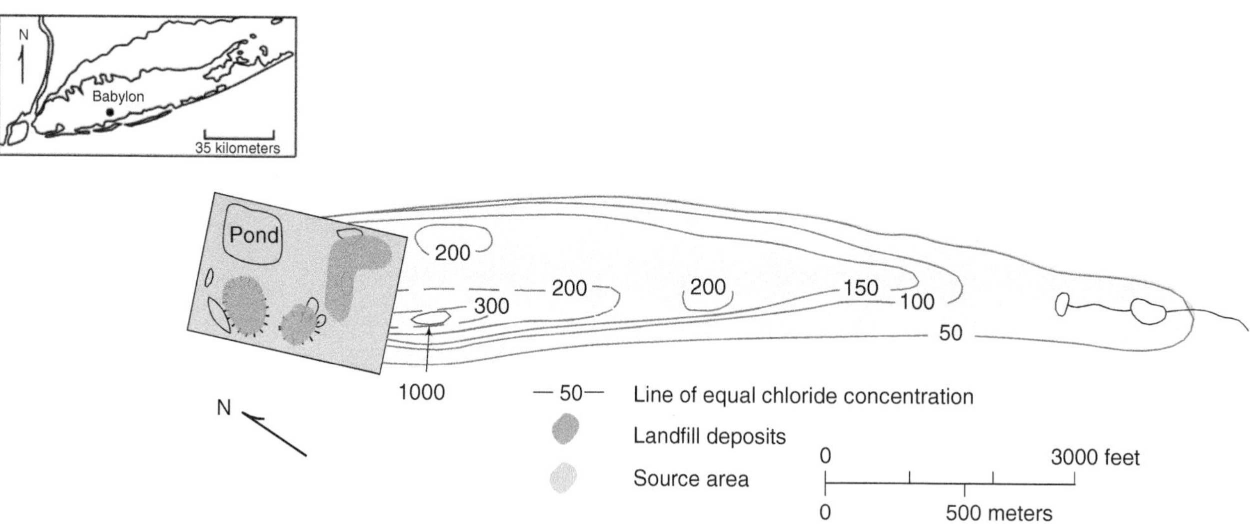

Figure 20.2 The Babylon landfill was located on Long Island, New York. A slice map between 9.1 and 12.1 m below the water table showed the concentration of Cl^- within a long plume that moved in a southeasterly direction (U.S. Geological Survey, adapted from Kimmel and Braids, 1980/Public domain).

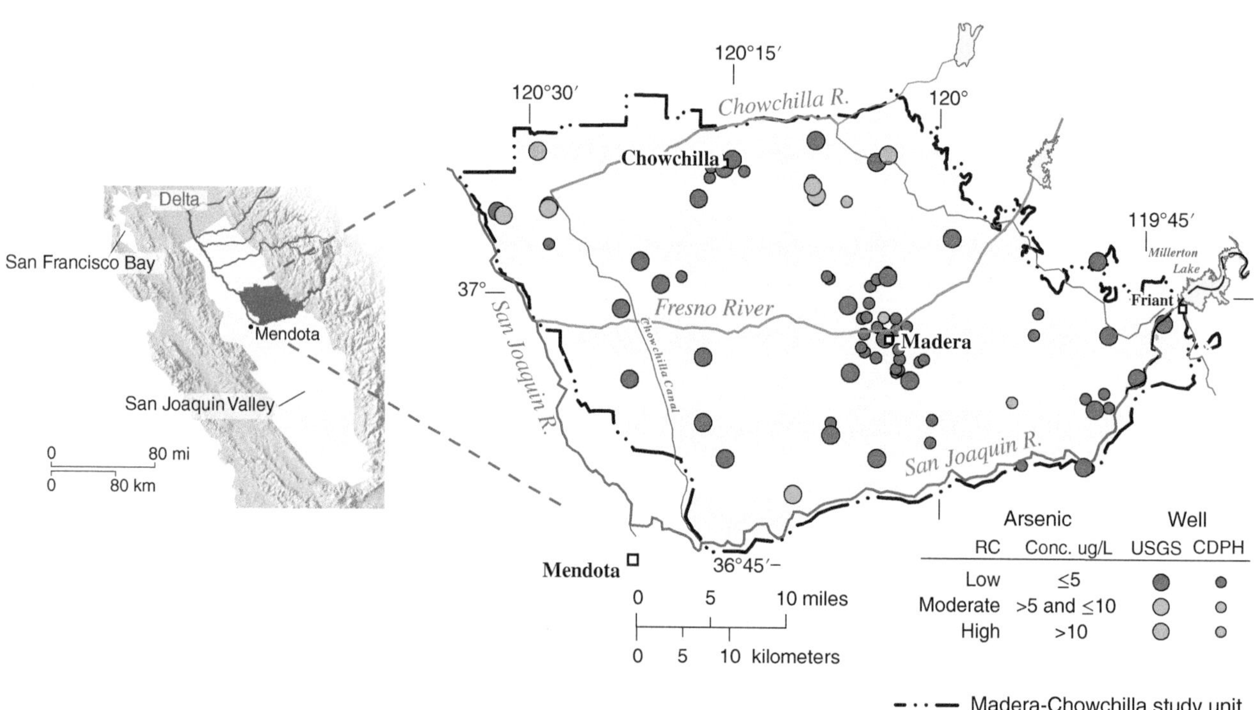

Figure 20.3 Scatter map for arsenic in groundwater in the central eastern portion of the San Joaquin Valley in California. Besides indicating arsenic concentrations, the scatter map differentiates samples collected by the USGS and California Department of Public Health (CDPH). Locally, the relative concentration of arsenic was moderate to high (i.e., >10 μg/L) in several locations. (U.S. Geological Survey, location map from Galloway and Riley, 1999 and arsenic map adapted from Shelton et al., 2013/Public domain).

The scatter map for arsenic (Figure 20.3) illustrates an essential feature of a nonpoint problem (Shelton et al., 2013). The entire study area was a potential source area for arsenic with some samples exceeding a RC value of 1, indicating concentrations greater than the regulatory standard. Because sampling involved existing wells, differences in well construction, i.e., depth and screen lengths, associated land uses, and in local geology, made it difficult to describe spatial patterns. In general, arsenic concentrations tended to be higher in deep wells where manganese-reducing conditions prevailed (Shelton et al., 2013). In other areas of the San Joaquin Valley, the occurrence of this redox condition along the axial trough of the valley was somewhat more predictable in a spatial sense (Belitz et al., 2003).

20.2 Families of Contaminants

A second conceptual way to think about contamination problems involves families of contaminants. Groundwater can contain many different dissolved constituents, sometimes at concentrations that can cause serious health impacts or substantially impair the uses of water without treatment. In industrialized or populous countries, a variety of different industrial, agricultural, and domestic activities contribute to contamination. In other places, subtle problems of geogenic contamination can be problematic as well. The extraordinarily large numbers of potential contaminants and variety of potential sources add significant complexity with respect to the investigation of problems. Additional complexity arises because new contaminants are constantly emerging.

The concept of contaminant families involves organizing contaminants around element type and mode of occurrence. This approach ends up grouping together dissolved constituents that are affected similarly by chemical or biological processes. Rather being comprehensive, our list includes those contaminants that are common in practice. The major families include (1) trace elements, (2) nutrients, (3) unregulated inorganic compounds on the secondary list, (4) organic contaminants, (5) pathogens, and (6) radionuclides. These contaminants have the potential to produce health problems or affect the usability of water. In moving through the groups, we will point out the common issues. As a broad generalization, too much of anything in water can produce health problems in humans.

20.2.1 Minor/Trace Elements

The first of the families of contaminants on the list is that of minor/trace elements. As a group, they contain the largest proportion of elements found on the periodic table. Constituents of particular concern include arsenic, uranium, and fluoride. They most commonly originate from nonpoint sources, are typically geogenic in origin, and known to cause serious health problems.

Exposure to arsenic (As) in drinking water is among the most well-studied public health problems associated with groundwater. Human exposure to arsenic affects "almost every organ system in the body including the brain" (Tyler and Allan, 2014). The most evident impact is skin lesions (i.e., arsenical keratoses) that can lead to skin cancer. There are also significant risks of internal cancers of the bladder, liver, and lungs (Morales et al., 2000). Studies have also confirmed problems in cardiac function at even low arsenic concentrations (Pichler et al., 2019). Arsenic also causes neural dysfunction in children, which are manifested as deficits in memory and learning, as well as mood disorders (Tyler and Allan, 2014). Uranium might also be problematic in drinking water not from its radiological properties but its chemical impacts, particularly chronic kidney disease (Coyte et al., 2018).

Fluoride is probably the best example of a trace nonmetal occurring as a contaminant. It is controversial because the concentration of fluoride that is commonly added to water to prevent tooth decay is somewhat close to the MCL of 4 mg/L. Above the MCL, fluoride in groundwater is associated with health problems. Examples of these health impacts have included enamel fluorosis of the teeth, which comes along with yellow/brown staining and pitting of the enamel, and musculoskeletal effects such as skeletal fluorosis and weakened bones, leading to fractures (NRC, 2006). At concentrations above the MCL, fluoride might accumulate within bones, increasing their density and promoting the growth of osteophytes. The associated impairments range from joint stiffness and pain to mobility problems associated with chronic pain, arthritis-like symptoms, and calcification of ligaments (NRC, 2006). Fluoride in drinking water can lead to problems of neurotoxicity in a developing fetus and infants (Grandjean and Landrigan, 2014) with impacts such as reduced scores on IQ tests.

20.2.2 Nutrients

This group of potential contaminants includes those ions or organic compounds containing nitrogen or phosphorus. By far, the most concerning nutrient species in groundwater are nitrates (NO_3^-), and to a lesser extent, ammonium ion (NH_4^+). These are anthropogenic compounds related to the use of synthetic fertilizers containing nitrogen, manure from animal feeding operations, and wastewater effluents from septic tanks and induced infiltration from contaminated rivers. The nutrients are typically nonpoint contaminants. But large, septic tank systems can cross-over to become point source problems.

Historically, the health standard for nitrate was set in consideration of methemoglobinemia. This is a rare blood disorder which impairs oxygen transport in infants (Ward et al., 2018). More recently, other adverse health effects of nitrate in drinking water have extended to "colorectal cancer, thyroid disease, and neural tube defects" such as spina bifida (Ward et al., 2018). There are indications of risks at concentrations below the regulatory limit. There are also lingering concerns of endogenous nitrosation that involves the formation of cancer-causing compounds (for example, nitrosamines) in the body with consumption of nitrate-contaminated water (Ward et al., 2018).

20.2.3 Other Inorganic Species

This miscellaneous group includes constituents present in nontrace quantities such as Ca, Mg, Na, Fe, and Mn and nonmetals such as ions containing carbon and sulfur (for example, HCO_3^-, HS^-, CO_3^{2-}, SO_4^{2-}, and H_2CO_3) or other species such as Cl^-. Many of these ions are major contributors to the overall salinity of groundwater. Extremely high concentrations of some combination of species make water unfit for human consumption and for many industrial uses. The health-related problems are not serious, as compared to those caused by regulated constituents in the other families. However, high concentrations of even relatively nontoxic salts, particularly Na^+, can disrupt cell or blood chemistry with serious consequences. At lower concentrations, an excessive intake of Na^+ may cause less serious health effects such as hypertension. The potential sources of major ion salinity include (1) groundwater in contact with geogenic deposits of mineral salts, (2) leachate from mine tailings, mine spoil, or sanitary landfills, and (3) industrial wastewater that often has large concentrations of common ions in addition to heavy metals or organic compounds.

Certain constituents in water impair the usability of water without treatment. For example, an abundance of and Ca and Mg ions contributes to problems of water hardness, such as precipitates leading to the replacement of equipment, soap residues in clothes, and itchy skin. Fe concentration in water above a few mg/L can impact the taste of water and the food/beverages prepared using that water. Both Fe and Mn can lead to staining of plumbing fixtures and clothes. The presence of iron in wells under reducing conditions can lead to well-screen encrustation and aquifer biofouling (Walter, 1997). This problem is associated mostly with wells that are shut down for long periods. In places like Ohio with pressure relief wells installed to depressurize zones below dams, constant maintenance is required to maintain their functionality.

20.2.4 Organic Contaminants

Contamination of groundwater by organic compounds is a logical consequence of the large quantities of unrefined petroleum products and man-made organic compounds in use. These compounds come from a variety of anthropogenic sources, which can include inappropriate disposal of hazardous wastes, mixed-waste landfills (i.e., domestic and hazardous wastes), leaks of all kinds (i.e., tanks, pipes, and pipelines), applied chemicals such as herbicides and pesticides, and organic compounds infiltrating from rivers. Table 20.1 and associated discussions introduce the most important families of organic contaminants. Other sections will examine the processes and problems associated with these different families of contaminants.

20.2.4.1 Petroleum Hydrocarbons

Petroleum hydrocarbons and derivatives are made up of carbon and hydrogen that are derived from crude oil, natural gas, and coal. There are many organic compounds in crude oil, and commonly we worry about just a few. Those compounds listed on Table 20.1, as fuels are aromatic hydrocarbons, contain at least one *benzene ring* (that is, C_6H_6). Examples of these compounds include benzene (C_6H_6), toluene ($C_6H_5CH_3$), ethylbenzene (C_8H_{11}), and xylene ($C_6H_4(CH_3)_2$). These so-called BTEX compounds are both extremely soluble in water and toxic and are common index species for contamination associated with gasoline.

The *polynuclear aromatic hydrocarbons* (PAHs) are also components of concern in petroleum hydrocarbons. These compounds form from a series of benzene rings. Examples include anthracene and phenanthrene. These compounds are poorly soluble and readily sorb, so they are most commonly of concern in surface water. PAHs are found in mineral

Table 20.1 Important families of organic contaminants in groundwater.

Chemical family	Examples of compounds
Hydrocarbons and derivatives	
Fuels	Benzene, toluene, *o*-xylene, butane, and phenol
PAHs	Anthracene and phenanthrene
Alcohols and ethers	Ethanol and methyl tertiary butyl ether (MTBE)
Creosote	*m*-Cresol and *o*-cresol
Ketones	Acetone
Halogenated aliphatics	Tetrachloroethene, trichloroethene, trichloroethane, dichloromethane, and PFAS
Halogenated aromatics	Chlorobenzene and dichlorobenzene
Polychlorinated biphenyls	2,4-PCB, 4,4-PCB

Source: Domenico and Schwartz (1998), Physical and Chemical Hydrogeology.

oil products such as coke, bitumen, coal tar (and creosote), and vehicle fuels and can be released due to incomplete combustion.

Alcohols and ethers are blended with gasoline as octane enhancers, which provide for cleaner burning and reducing "smog-forming" toxic pollutants. Worldwide ethanol and methyl tertiary butyl ether (MTBE) are commonly used, although the use of MTBE was phased out in the United States in 2006. Ethanol as an alcohol is totally miscible in water and biodegrades rapidly. Ethanol is a strong electron donor creating highly reduced groundwater that inhibits the biodegradation of BTEX compounds that favor oxidizing conditions (Rasa et al., 2013). MTBE is moderately soluble in water. A 10% blend of MTBE in gasoline has a solubility of ~5000 mg/L in water at 25°C and resists biodegradation (Squillace et al., 1996).

Coal–tar creosote is a combination of many different organic compounds, such as phenols, PAHs, and various heterocyclic compounds. Coal–tar creosote has been used as a wood preservative for lumber and railway ties, but their use has been phased out. The common source of problems were facilities that treated lumber.

Overall, the compositional differences among the fuels have important implications for monitoring and cleanup. For example, the low boiling fractions like gasoline contain highly volatile components and constituents like benzene, ethylbenzene, toluene, and xylene that dissolve to form plumes where gasoline and oils are spilled. The high-molecular weight compounds are much less volatile and tend to be present in water at low concentrations. The common oxygenates ethanol and MTBE designed to reduce emissions both exacerbate problems of fuel contamination but in different ways.

20.2.4.2 Halogenated Aliphatic Compounds

These compounds are formed from chains of carbon and hydrogen atoms where some of the hydrogens may be replaced by chlorine, fluorine, or bromine atoms. Examples of these compounds include industrial solvents such as tetrachloroethene (PCE), trichloroethene (TCE), and carbon tetrachloride (CT) and PFAS, per-and polyfluorinated substances that are used in the manufacture of coatings and products to resist heat, grease, and stains, and to provide waterproofing. Historically, the use of solvents or their improper disposal has given rise to many of the most serious problems of dense nonaqueous phase contamination (DNAPLs) encountered in hydrogeological practice. The problems of PFAS have only gained visibility in the last decade or so, and the full scope of groundwater issues remains to be determined.

20.2.4.3 Halogenated Aromatic Compounds

These compounds are formed from benzene rings with substituted halogens. Examples include chlorobenzene and dichlorobenzene (DCB), which are used in various industrial and agricultural applications. Again, by virtue of their relatively large specific gravity, these compounds occur as DNAPLs in groundwater. While these compounds will turn up as contamination problems, they are not nearly as prevalent to chlorinated solvents used in degreasing of metals (PCE, TCE, etc.)

20.2.4.4 Polychlorinated Biphenyls

Polychlorinated biphenyls were widely used in the 1960s and 1970s in transformers and capacitors. Their environmental persistence toxicity has made them an important contaminant, even though their production has been curtailed. Chemically they consist of chlorine-substituted benzene rings joined together.

20.2.4.5 Health Effects

There are important health effects related to drinking water contaminated by organic compounds. However, as Craun (1985) points out, it is difficult to establish which compounds are most toxic because not all have been tested, and health risks are inferred from studies of laboratory animals, poisoning or accidental ingestion, and occupational exposures. Furthermore, there is a serious lack of information on the health effects related to the combined effect of several compounds, and on the epidemiology of populations consuming contaminated water. Organic contamination of drinking water is a cause of cancer in humans and animals and a host of other problems including liver damage, impairment of cardiovascular function, depression of the nervous system, brain disorders, and various kinds of lesions. More detailed information on health effects related to organic materials can be obtained from Craun (1985).

20.2.5 Biological Contaminants

The important biological contaminants include *pathogenic bacteria, viruses, or parasites*. It does not take a degree in medicine to be aware of the serious health problems from typhoid fever, cholera, polio, and hepatitis. Other less serious abdominal disorders are often too well-known by travelers to countries with poor sanitation. As a group, these health effects are some of the most significant related to the contamination of groundwater.

The main source of biological contamination is from human and animal sewage, or wastewater. Groundwaters become contaminated due to (1) land-disposal of sewage from centralized treatment facilities or septic tank systems, (2) leachates from sanitary landfills, and (3) various agricultural practices such as the improper disposal of wastes from feedlots.

Problems of biological contamination are prevalent in domestic wells. For example, in the lower Susquehanna River basin, found mostly in southeastern Pennsylvania, USA, 70% of groundwater samples had indications of contamination with the total coliform bacteria indicator (Lindsey et al., 1997). Analyses also found specific indications of contamination by fecal coliform (25% of samples) and *Escherichia coli* (30% of samples). A major contributing factor appeared to be well construction that omitted sanitary caps on the casing and grouting of the casing (Lindsey et al., 1997). However, associations with fractured limestone terrains point to permeable geologic pathways from septic tanks, manure applications etc. The study noted that the "presence of bacteria in water from rural wells is one of the most important water-quality issues related to human health" (Lindsey et al., 1997).

Samples of groundwater, collected before treatment from municipal wells in a shallow sand and gravel aquifer in La Crosse, Wisconsin, detected enterovirus, rotavirus, hepatitis A virus, and noroviruses in 50% of 48 samples (Borchardt et al., 2004). The likely source was the nearby Mississippi River, which contained infectious levels of the same viruses, as well as other sources of sewage. At this site, bacterial contaminants were attenuated, pointing to greater mobility for viruses in this coarse sand and gravel aquifer.

20.2.6 Radionuclides

Ra-222 or radon is the most concerning contaminant in this category. Human exposure to radon gas through inhalation of indoor air is known to cause lung cancer (Gross, 2017). Radon occurs naturally in water from the decay of radium-226, which is a decay product of uranium-238. Uranium itself is most abundant in shale, "metamorphic rocks derived from shales," and granitic rocks (Gross, 2017). There are several pathways for radon entry to buildings. Most commonly, radon gas generated in the unsaturated zone directly leaks into buildings. Remedial action is warranted if indoor air has activities >4.0 pCi/L or greater. Action levels for indoor air might also be expressed as becquerels per cubic meter (Bq/m^3), where 1 pCi/L = 37 Bq/m^3.

However, radon gas can be dissolved in groundwater with activities of hundreds to thousands of pico-curies per liter (pCi/L). Presently, there is no USEPA maximum concentration level (MCL) for radon dissolved in water, although there was a proposed MCL of 300 pCi/L. With groundwater, human exposure can come as radon is released from common household uses of water with showering and in washing machines and dishwashers and add to other sources of radon. Drinking water containing radon gas is less concerning with weak associations to cancer.

20.3 Presence or Absence of Nonaqueous Phase Liquids (NAPLs)

Another way to think about contamination problems is in terms of whether there are nonaqueous phase liquids (NAPLs) present. In effect, the presence of an NAPL means there is one or more fluids present, besides water. For example, a leaky underground tank at a filling station means that pure phase liquid, the gasoline, can leak into the groundwater. NAPL problems are complicated for several reasons. The organic fluids commonly have a density different than that of water, which is 1.00 g/mL. *Dense nonaqueous phase liquids (DNAPLs)*, also known as "sinkers", have a density greater than 1, (e.g., 1.46 g/mL for trichloroethylene) and flow downward in an aquifer. *Light nonaqueous phase liquids (LNAPLs) "floaters"* have a density less than 1 (e.g., 0.87 g/L for benzene) and flow "downhill" on the water table.

Another source of complexity occurs when flowing LNAPL or DNAPL moving through a porous medium leaves behind a trail of pores that contain some of the NAPLs. This residual fluid although unable to move remains in pores and is available for slow dissolution into groundwater to form plumes of aqueous constituents. The complication comes from the fact that the growing source created by flowing NAPL is difficult to identify in the subsurface. Finally, if the fluid is volatile, vapors are free to migrate within the unsaturated zone, adding yet another pathway for contaminants to move along and cause problems.

Figure 20.4 illustrates these ideas in more detail with a simple case of a point-source as compared to DNAPL spill. Leakage of water containing a dissolved contaminant from a lagoon can create a plume of contamination down-gradient of the source (Figure 20.4*a*). In this case, water is the only fluid present in the discrete well-defined source. The DNAPL case represents a spill on the ground surface. The groundwater in this case is flowing from left to right, whereas the DNAPL, the second liquid, is moving downward through the aquifer by virtue of a density greater than that of water (Figure 20.4*b*). The only mobile liquid is DNAPL present in pools of contamination ponded on the clay lens with some indication of deeper flow. As the mobile DNAPL moved deeper, small quantities are left behind as residual saturation in pores. Both the ponded and residual DNAPL contributed to the formation of a plume of dissolved DNAPL migrating to the right with the flowing groundwater. Such DNAPL sources in the subsurface are diffuse and difficult to define explicitly. Volatile DNAPL above the

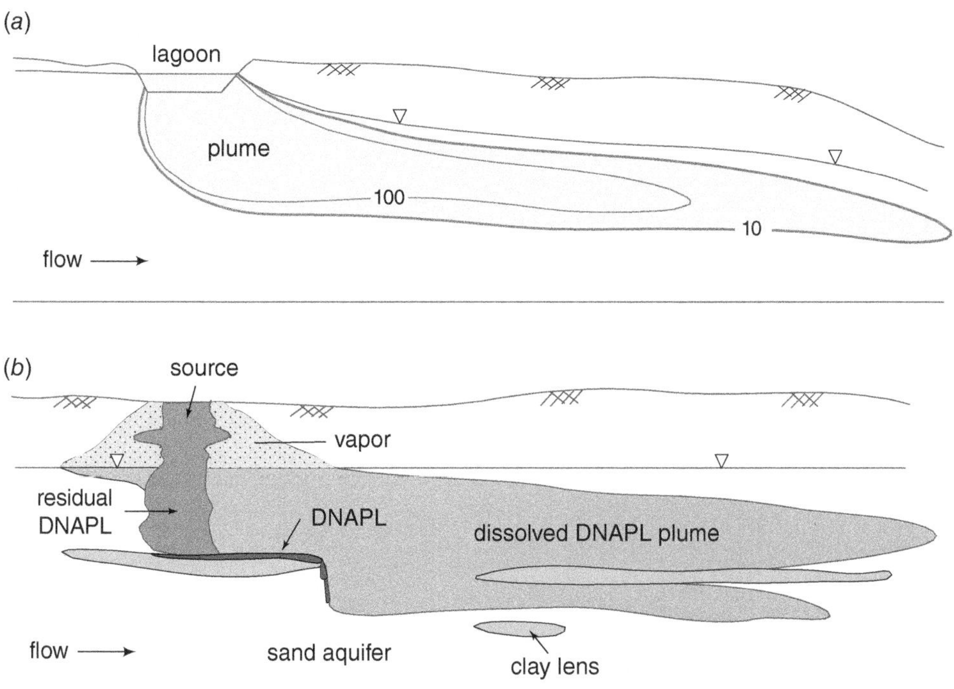

Figure 20.4 Example of a simple plume of point-source contamination developed from a discrete source (*a*). With a DNAPL spill (*b*), a second liquid is added that tends to flow generally downward. The ponded and residual DNAPL dissolves to create a plume of dissolved DNAP that moves laterally (Domenico and Schwartz, 1998, Physical and Chemical Hydrogeology. Copyright © 1990, 1998 John Wiley & Sons, Inc. All Rights Reserved. Reproduced with permission).

water creates a gaseous vapor that can migrate laterally and to contaminate subsequent infiltration moving downward through the vapor. Volatilization of residual DNAPL leads to contamination of the unsaturated zone.

20.4 Roles of Source Loading and Dispersion in Shaping Plumes

Chapters 17 and 19 have introduced transport and other factors that determine the distribution of contaminants and other constituents in the subsurface. This section discusses these concepts in relation to the development of plumes of dissolved contaminants associated with point sources. Plumes develop as dissolved contaminants are mobilized from a source and are transported down the flow system. If the source is old (i.e., 50+ years) and the contaminant is nonreactive, plumes can be kilometers in length. A separate plume develops for each dissolved contaminant because the behavior of the different constituents in reactions, such as sorption and biodegradation, can be highly variable. For example, in an aerobic aquifer, TCE does not biodegrade. However, benzene in solution is biodegraded and much less mobile. At complex sites, ten to twenty different plumes can co-exist and may need to be examined in detail.

20.4.1 Source Loading

Loading history describes how the concentration of a contaminant or its rate of production varies as a function of time at the source. A spill is an example of pulse loading, where the source produces contaminants at a fixed concentration for a relatively short time (Figure 20.5*a*). This loading could occur from a one-time release of contaminants from a storage tank or storage pond. In the simplest cases, plumes are isolated bodies moving away from the source down the flow system. Long-term leakage from a source is termed continuous source loading. Figure 20.5*b* illustrates one type of continuous loading where the concentration remains constant with time. This loading might occur, for example, when small quantities of contaminants are leached from a volumetrically large source over a long time. LNAPLs or DNAPLs present at a source can dissolve at a slow rate over many decades to provide this kind of source loading. In this case, plumes remain connected to the source and grow longer with time (e.g., Figure 20.4*a*).

Most sources of long-term contamination cannot be described in terms of a constant loading function. For example, the concentration of chemical wastes added to a storage pond at an industrial site can vary with time due to changes in a manufacturing process, seasonal or economic factors, or the addition of other reactive wastes (Figure 20.5*c*). Leaching rates for solid wastes at a sanitary landfill site could be controlled by seasonal factors related to recharge, or by a decline in source strength as components of the waste, such as the biodegradation of organic components. This latter source behavior could result in the loading history shown in Figure 20.5*d*. Typically, plumes remain connected to the source and grow. However, complexity in the source loading function leads to increasingly variable concentration distributions.

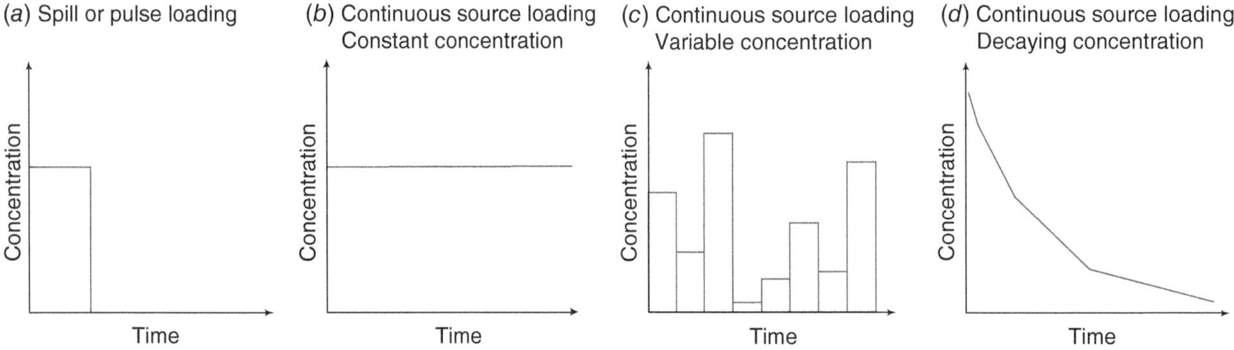

Figure 20.5 Examples of functions used to characterize contaminant loading from a spill (*a*) or long-term leakage (*b*, *c*, and *d*) (Domenico and Schwartz, 1998, Physical and Chemical Hydrogeology. Copyright © 1990, 1998 John Wiley & Sons, Inc. All Rights Reserved. Reproduced with permission).

20.5 How Chemical Reactions Influence Plumes

The largest plumes with a point-source problem of contamination are due to chemically nonreactive species that behave like Cl^-. They are mobile and are advected at the linear groundwater velocity. Dispersion is the only other process that influences the concentration distribution of a non-reactive species. When contaminants react, they are generally less mobile and end up moving at a velocity that is less than linear groundwater velocity. Contaminants that are strongly sorbed or biodegrade with short half-lives only spread a short distance from the source. Moreover, a plume of a biodegrading constituent at some point will stop spreading because the concentration eventually falls below detection within the plume.

This section examines two important kinds of reactions—kinetic degradation reactions, such as radioactive decay or biodegradation, and sorption reactions that move constituents out of the solution onto the aquifer solid—and their influence on contaminant migration. Figure 20.6 illustrates how plume spreading is controlled by the rate of degradation (Panels *a*, *b*) and strength of sorption (Panels *c* and *d*), in addition to modest longitudinal and transverse dispersion. The upper two cross-sections (Figures 20.6*a*,*b*) compare the relative concentration distributions for a contaminant degrading via a simple first-order kinetic reaction with half-lives, 30 and 3.0 years. The simple, one-layer aquifer system is recharged at the left end of the cross-section with discharge at the right end.

Clearly, the faster the constituent is consumed by the degradation reaction, (i.e., shorter half-life), the less mobile the constituent is, for example, as compared to a nonreactive species. With smaller half- lives, the plume may reach steady state and effectively stop spreading.

Figures 20.6*c*,*d* illustrate the importance of sorption in controlling plume spread. The controlling parameter in this case is the selectivity coefficient that controls the extent of sorption. A relatively small value (such as 0.1) indicates weak sorption and results in a longer plume (Panel *c*), while a value of 10 indicates strong sorption. Selectivity coefficients behave just like the K_D values discussed with respect to sorption of organic contaminants. With weak sorption, the contaminant is transported with a velocity close to the advective velocity. In other words, sorption is not effective in controlling contaminant migration. With strong sorption, the contaminant moves at a velocity much less than the advective velocity with a smaller plume (Panel *d*). In practice, strongly sorbed contaminants are not mobile.

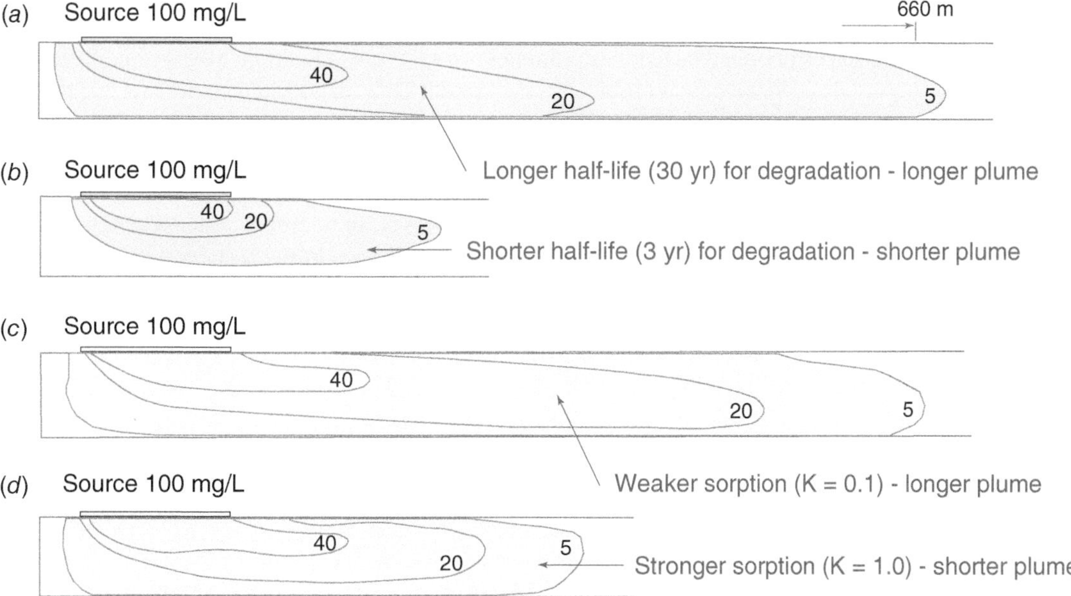

Figure 20.6 Chemical reactions, such as degradation (*a*, *b*) and sorption (*c*, *d*), apparently slow down the velocity of plume migration and in the case of degradation halt the spreading of the plume. Faster degradation and stronger sorption result in relatively smaller plumes (*b*, *d*) (FWS).

20.5.1 Biodegradation of Organic Contaminants

With *primary utilization*, a dissolved organic compound provides the main source of energy and carbon for the microorganisms. The microorganisms produce energy by the electron transfer involved with redox reactions. A variety of organic compounds can serve as the primary substrate for microbial metabolism (NRC, 1993). Following are examples of half-reactions involving the primary utilization of organic contaminants

benzene oxidation: $12H_2O + C_6H_6 = 6CO_2 + 3OH^- + 30e^-$
toluene oxidation: $14H_2O + C_6H_5CH_3 = 7CO_2 + 36H^+ + 36e^-$

Certain chlorinated solvents dissolved in groundwater, like tetrachloroethene (PCE), trichloroethene (TCE), trichloroethane (TCA), dichloroethene (DCE), and dichloroethane (DCA), are relatively oxidized and resist biotransformation in aerobic settings because they do not provide much energy. For this reason, in aerobic settings, plumes of TCE and TCA can easily be several kilometers long. However, in anerobic settings, these compounds can be biotically transformed in reductive dehalogenation reactions. Here are a few examples of these redox reactions

PCE reductive dehalogenation: $C_2Cl_4 + H^+ + 2e^- = C_2HCl_3 + Cl^-$
TCE reductive dehalogenation: $C_2HCl_3 + H^+ + 2e^- = C_2H_2Cl_2 + Cl^-$

20.5.2 Degradation of Common Contaminants

Biotransformation reactions are important because they influence the transport of the most important families of contaminants (Table 20.1)—hydrocarbons and halogenated aliphatics. Commonly, biotransformation processes are examined in relation to two prototypical settings. In *aerobic settings*, oxygen is the major electron acceptor for the various biotransformation reactions. In *anerobic settings*, oxygen is absent, and other compounds, like nitrate, sulfate, and carbon dioxide, function as electron acceptors.

Hydrocarbons in various families constitute one of the most common contaminants in groundwater. In aerobic groundwater systems, aromatic hydrocarbons with up to two benzene rings are mineralized relatively rapidly with minimal lag times. *Lag time* refers to the time before a microbial population begins to degrade the organic compound. For example, Nielsen and Christensen (1994a) found that benzene, toluene, and *o*-xylene degraded with short lag times (about four days). Initial concentrations of approximately 150 μg/L of benzene and toluene decreased to <2 μg/L in less than one month. The degradation of *o*-xylene required about three months and was substantially less complete. These laboratory findings of relatively rapid degradation rates were also corroborated by field tests. Studies of an oil spill at Bemidji, Minnesota, (Cozzarelli et al., 1989) showed that hydrocarbons also degraded in anerobic systems as well. Thus, dissolved constituents due to spills of gasoline (i.e., BTEX compounds) tend to form small plumes that are attenuated by biodegradation.

Other components of fuels (or coal–tar derivatives) are polyaromatic hydrocarbons (PAHs) such as naphthalene, anthracene, or pyrene. These compounds comprised a series of benzene rings. In general, PAHs are biodegradable in aerobic settings with the rate decreasing as the molecular weight of the compound increases. In anerobic systems, PAHs may be somewhat more resistant to degradation. However, Lyngkilde and Christensen (1992) report the reduction of naphthalene concentrations under ferrogenic conditions.

Phenolic compounds are also found as components in fuels or as industrial contaminants. Phenols are characterized by an aromatic ring with an attached hydroxyl group. Examples of these compounds include phenol, nitrobenzene, *o*-cresol, and *o*-nitrophenol. Microcosm studies by Nielsen and Christensen (1994b) show that these compounds are mineralized in aerobic systems. Indications are that they biodegrade relatively rapidly with broad variation in the rates. *p*-, *m*-, and *o*-cresol also degrade under anerobic conditions (Smolenski and Suflita, 1987). Other hydrocarbons, such as alcohols and ketones, are readily degradable in groundwater.

Halogenated aliphatic compounds include the common industrial solvents like carbon tetrachloride (CT), tetrachloroethene (PCE), trichloroethene (TCE), trichloroethane (TCA), vinyl chloride (VC), and dichloroethene (DCE). These compounds are formed from carbon atoms joined together in chains with attached atoms. In aerobic systems, they resist degradation and are persistent. There are no known bacteria that can oxidize these compounds as a primary substrate because the reaction yields so little energy (Chapelle, 1993). Under anerobic conditions (i.e., reducing), halogenated compounds are commonly biotransformed. The reactions involve a sequential loss of chlorine atoms in a process called

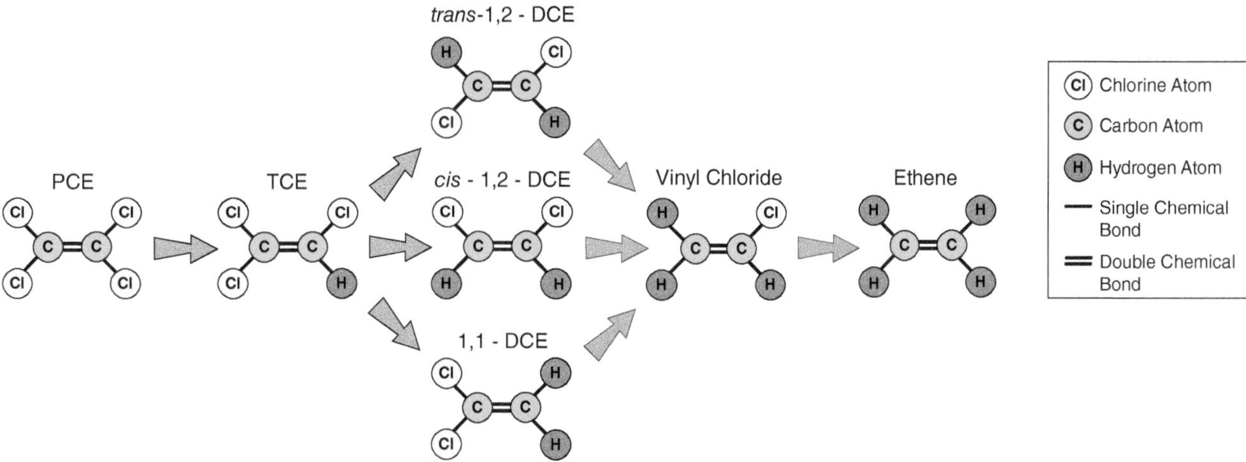

Figure 20.7 Reductive dehalogenation involves the successive removal of chlorine atoms and substitution by hydrogen atoms. There are alternative pathways with the most common being straight across from PCE to ethene (U.S. Environmental Protection Agency, adapted from Wiedemeier et al., 1998).

reductive dehalogenation. Figure 20.7 illustrates reductive dehalogenation reactions beginning with PCE. The last step from vinyl chloride is quite slow, relative to the other dehalogenation reactions, so vinyl chloride often tends to accumulate.

20.5.3 Reactions Influencing Plume Development

Historically, there has not been much interest in understanding the details of chemical processes and their impact on contaminant mobility. The state of the practice usually involved careful mapping of the distribution of a few key contaminant species and inferring what processes were operative. In some cases, it is relatively easy to identify transport processes and their impact on transport. A case in point is sorption of contaminants onto aquifer solids. For example, organic compounds with a large octanol/water partition coefficient tend to be strongly sorbed and form a small plume, whereas a compound with a small K_{ow} should form a large plume.

Studies have shown how sorption can influence the mobility of organic contaminants in groundwater. Jackson et al. (1985) studied the Gloucester site near Ottawa, Canada, where a variety of organic compounds were disposed in a shallow trench. These contaminants included diethyl ether, tetrahydrofuran, 1.4-dioxane, carbon tetrachloride, benzene, and 1,2-dichloroethane. Plume maps for three of these compounds showed variable rates of spreading. However, the extent of spreading was inversely related to the hydrophobicity or "water hating" characteristic of the various compounds as measured by the octanol/water partition coefficient. Thus, 1,2-dichloroethane with a high log K_{ow} (1.48) migrates at a much slower velocity than that of 1,4-dioxane, whose relatively small log K_{ow} (−0.27) produced a much longer plume. This behavior suggests that sorption of the organic compounds is the most important process controlling the spread of contaminants.

With the advent of a remedial action scheme called monitored natural attenuation, there has been much more interest in systematically documenting the variety of reactions that occur. *Natural attenuation* relies on the capacity of natural reactions attenuating contaminants in a timely manner before they impact downstream receptors. Demonstrating the efficacy of natural attenuation as a potential remedy for contamination problems requires that details of the chemical processes be absolutely and unambiguously defined through rigorous field evaluations. Thus, in recent years, there has been a significant leap forward in understanding complex chemical and biological processes in groundwater and the pathways for chemical transformation (Ford et al., 2007).

Rigorous evaluations require a comprehensive evaluation of key processes and especially of reactions, which not only involves contaminants present at the source but also any in-growing contaminants. For example, in some settings, contaminants added at the source (e.g., PCE and TCE) biodegrade to produce hazardous products like DCE and VC. These species "grow in" as transport occurs and can reach their maximum concentrations away from the source. In addition, when contaminant concentrations are high, these reactions can impact the concentration of other "natural" constituents in the groundwater that are caught up in the redox reactions. For example, redox processes can change the concentrations of common electron acceptors like iron, sulfate, and oxygen.

In the United States, there is a relatively high bar to gain approval of monitored natural attenuation as the optimal remedy for cleanup as compared to more active methods. Studies need to be organized to address four key questions

- are contaminants being removed and has this removal arrested the migration of plumes;
- what actual mechanisms (reactions and processes) are at work and at what rate are they attenuating contamination;
- will the reactions and processes keep working over the long term and will contaminants be permanently immobilized and;
- is the remedy still working well into the future (Ford et al., 2007)?

Readers interested in a deeper dive into the world of monitored natural attenuation and the associated should begin with a review of the report by Ford et al. (2007).

20.6 Nonaqueous Phase Liquids in the Subsurface

One of the first considerations in dealing with light nonaqueous phase liquids (LNAPLs) or dense nonaqueous phase liquids (DNAPLs) is defining some measure of abundance. *Saturation* represents the abundance of NAPL in a porous medium as the volume of the ith fluid per unit void volume. For a representative elemental volume,

$$S_i = \frac{V_i}{V_{\text{voids}}} \tag{20.1}$$

where V_i is the volume of the ith fluid. In a multicomponent system (e.g., water, air, and NAPL), the sum of all the saturations is equal to 1.

In most NAPL/water systems, it is difficult for an NAPL to fill the entire pore or to drain out completely. For example, an NAPL entering a porous medium cannot displace all the water; some remaining water saturation is likely. Similarly, when the NAPL drains from a porous medium, some NAPL is left behind. When saturations of one of the fluid components (e.g., water or NAPL) are small, the fluid phase becomes disconnected and essentially cannot flow. The term *residual saturation* defines the saturation at which a fluid component becomes unable to flow. The following section explores this idea in more detail.

20.6.1 Features of NAPL Spreading

Consider a spill of LNAPLs or DNAPLs on the ground surface. With time, free product percolates downward through the unsaturated zone toward the water table. The most important process influencing downward movement of the free product is gravity-driven flow. The NAPL does not displace the water as it goes but moves around it from pore to pore once saturation exceeds the residual saturation.

Several important factors control the flow of NAPLs. In the case of a noncontinuous source or spill (Figure 20.8), the volume of the free product gradually decreases because some of the downward moving NAPL is trapped in each pore at residual saturation. Thus, if the spill is relatively small, downward percolation in the unsaturated zone will stop once the total spill volume is at residual saturation. Another pulse of NAPLs is necessary to move the product downward.

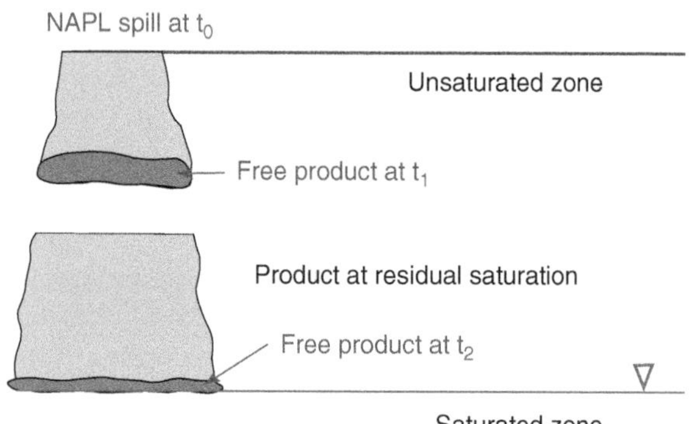

Figure 20.8 Downward percolation of NAPL in the unsaturated zone. As the contaminant moves downward, the quantity of mobile fluid decreases and an increasing volume of the NAPL is trapped as residual saturation (Domenico and Schwartz, 1998. Physical and Chemical Hydrogeology. Copyright © 1980, 1998 John Wiley & Sons Inc. All Rights Reserved. Reproduced with permission).

The main threat to groundwater from such small spills is the opportunity for continuing dissolution of the NAPL by infiltration or vapor phase migration in the vadose zone.

Heterogeneity in hydraulic conductivity is another factor that controls NAPL distributions in the subsurface. Consider the downward spread of LNAPL from a leaking tank (Figure 20.9). As it moves downward, in some case, it can spread laterally in the unsaturated zone. This spreading is due to capillary forces that operate together with gravity forces to control migration. The presence of layers of varying hydraulic conductivity promotes the lateral flow of LNAPL in more permeable units (Figure 20.9). Even a relatively thin, low-permeability unit will inhibit downward percolation and force the free product to move laterally. If such a layer is continuous, the LNAPL will spread only within the unsaturated zone. If the layer is discontinuous, it will flow in "downhill" on that unit until it eventually spills over and continues downward toward the water table (Figure 20.9).

The behavior of LNAPLs in the near surface also may change with the pattern of release, for example, illustrating a problem of older, discontinuous leaks from tanks in a refinery. In this example, the free product from these older spills will have had time to flow away from the sources, leaving behind zones of residual saturation (Figure 20.10). Mobile LNAPL will be present with relatively large spills. The pathway for the flow of mobile LNAPL is downward to the capillary fringe/water table, which acts as a barrier to flow because LNAPL is less dense than water. This barrier creates LNAP ponding that sometimes results in the depression of the water table (Figure 20.10). Mobile LNAPL will flow laterally in the capillary zone in the direction of declining water table elevation. Eventually, the flow will stop once the total spill volume is accounted for with residual saturation. Clearly, other pathways for LNAP flow might exist, both as a vapor and as a dissolved constituent, in flowing groundwater. We will discuss these processes in upcoming sections.

Another feature of a groundwater system that influences LNAPL distributions is water table fluctuations. The complexities in LNAPL saturation outlined above and the lack of mobility may conspire to trap LNAPLs below a rapidly rising water table. Continuing water table fluctuations over a long time will smear the free product above and below the water table.

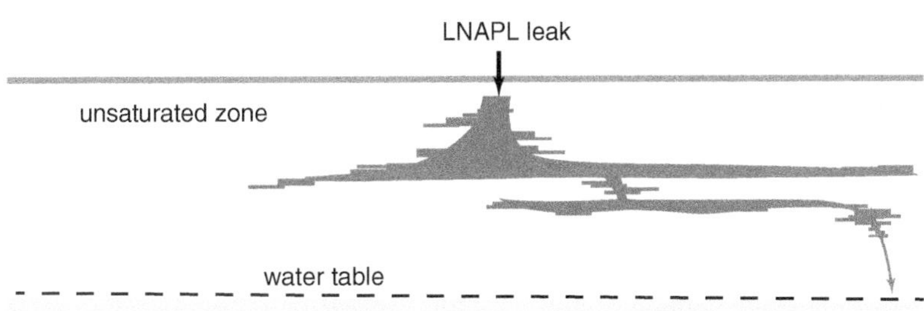

Figure 20.9 Presence of layers of lower hydraulic conductivity within the unsaturated zone can cause LNAPLs to mound and to spread laterally. In this simple example with a continuing release, pathways exist for flow to move deeper in the unsaturated zone (FWS).

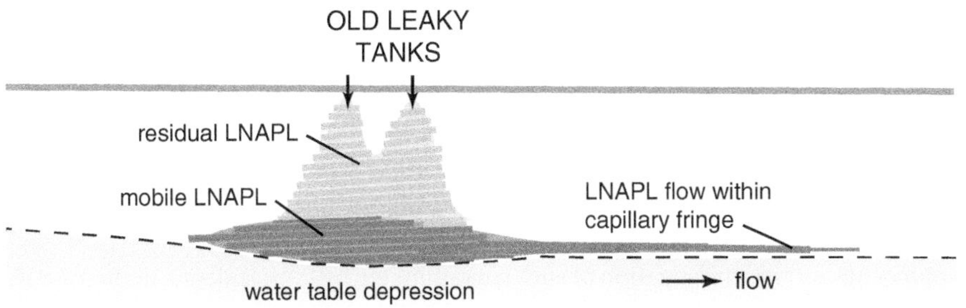

Figure 20.10 With an old, depleted source, mobile LNAPL might end up ponded at the water table, while moving laterally within the capillary fringe. The direction of lateral flow is in the direction of the declining topographic gradient on the water table (FWS).

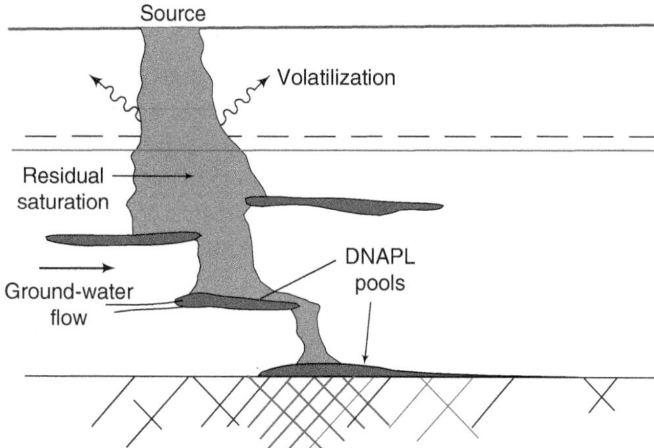

Figure 20.11 Complex distribution of DNAPLs in the vicinity of a subsurface source. Heterogeneities in hydraulic conductivity cause pooling. At depth, DNAPL flows into the bedrock along fractures (Adapted from Domenico and Schwartz, 1998. Physical and Chemical Hydrogeology. Copyright © 1980, 1998 John Wiley & Sons Inc. All Rights Reserved. Reproduced with Permission).

20.6.2 Occurrence of DNAPLs in the Saturated Zone

Unlike LNAPLs that move laterally along the capillary fringe, DNAPLs can flow downward below the water table and even to the base of the aquifer. Downward-moving DNAPLS displace water in the saturated zone because they have a specific gravity much greater than that of water. DNAPL accumulating on low-permeability units will move downhill following the topography of the boundary (Figure 20.11). This flow can be in a direction that is different than that of the groundwater. Spreading continues until the DNAPL is at residual saturation or trapped in *pools*, zones where DNAPL saturation is close to 1. Within both the saturated and the unsaturated flow systems, zones of residual saturation and pools are a source of secondary DNAPL contamination. However, this simple conceptual model disguises the complex patterns of *free product* migration that can occur below the water table. As we have learned in earlier chapters, porous media can have complicated spatial structures at macroscopic and microscopic scales. The presence of fractures might add yet another major source of complexity, contributing to DNAPL flow paths (Figure 20.11).

20.6.3 Secondary Contamination Due to NAPLs

DNAPLs and LNAPLs in the subsurface serve as important sources of *secondary contamination*. Problems develop when organic contaminants present as the free or residually saturated products partition into the soil gas through volatilization or into the groundwater through dissolution. Thus, even a small volume spill of a volatile organic liquid in the unsaturated zone can produce a plume of dissolved contaminants in the groundwater because the broader distribution of vapor in the unsaturated zone contaminates vapor moving into the aquifer (Mendoza and McAlary, 1990). In addition, both LNAPLs and DNAPLs in the subsurface also will dissolve to create plumes of dissolved contaminants that undergo groundwater transport.

20.7 Approaches for the Investigation of Contaminated Sites

In the world's developed countries, there was a 30-year period from about 1980 to 2010, when the focus of hydrogeological practice involved the investigation and remediation of contaminated sites. Academic and industrial efforts were massive, leading to revolutionary new approaches in all aspects of hydrogeology concerned with problems of contamination. To a large extent, work in research and in industrial site clean-ups has dwindled in developed countries. However, work is active in developing countries.

This section provides an overview of approaches for the investigation of sites contaminated by tank leakage, inappropriate disposal of industrial wastes, or newly discovered, legacy problems. Earlier chapters have discussed health-related problems with contamination of agricultural lands. As mentioned, these kinds of problems typically will not make use of the specific and high-resolution approaches used by problems of industrial contamination.

20.7.1 Preliminary Studies

There is much that can be learned about potentially contaminated sites, sometimes without leaving the office. Pre-existing geological and hydrogeological reports and maps might exist and be available on the web. There can also be nearby studies of similar problems. Topographic data that formerly were captured on maps are now available in digital archives. LIDAR data products are sometimes available to provide high-resolution topographic information as needed. Satellite geospatial data products of many different types are available to provide more regional data describing land-use land cover and details concerning streams, rivers, and wetlands. Data on domestic water wells are also available and somewhat helpful.

Information as to where wastes might have been disposed in ponds, or where facilities with buried tanks might have existed, is not available for some sites. One possible approach to help with this problem involves the analysis of historical aerial photographs. Thus, even though a former disposal area may now be buried, it might have been captured in aerial photographs taken say 50 years ago. Often the historical photos capture the layout of old landfills, trenches, drum disposal areas, and disposal ponds. They also might contain information on the location of old facilities (for example, buildings, and storage tanks), which can provide useful clues as to where contamination could be likely.

20.7.2 Reconnaissance Geophysics

Early in an investigation of large, contaminated sites, it was often economical to use rapid investigative approaches to inform more detailed and expensive hydrogeological investigations. Geophysical methods shine in investigating old and abandoned sites, where former landfill sites, waste-storage ponds, or drum-disposal areas could be hidden by cosmetic changes to the ground surface. Of geophysical approaches discussed in Chapter 7, electrical and radar-based approaches are most useful. Typically, equipment like a Geonics EM31 terrain conductivity meter can be used for this purpose (Chapter 7). For example, Jordan and Costantini (1995) describe a site where high-resolution electromagnetic surveying helped pin-point buried drum-disposal sites, buried bulk wastes, and slag in just a few days. Advances in geophysical technologies (Chapter 7) make even larger-scale studies feasible. Ground-penetrating radar (GPR) has also proven to be somewhat useful in site investigations but can be limited by adverse features of the geological setting. It is capable of finding underground targets like storage tanks or drum-storage areas.

20.7.3 Soil Gas Characterization

Characterizing the composition of soil gases is an industry-standard technique for using measured concentrations of shallow samples of gas from the unsaturated zone to infer the presence of volatile organic compounds from deeper in the subsurface. With time, volatiles present as NAPLs or volatile organic compounds (VOCs) dissolved in groundwater at the capillary fringe partition into the soil gas and gradually diffuse upward to the ground surface (Marrin and Kerfoot, 1988). Data collected from systematic surveys are useful in locating source areas or plumes and sites for groundwater monitoring wells. In some studies, the information provided by a soil-gas survey may be all that is required to establish whether or not a contamination problem exists or to identify a source.

The presence of volatiles is established commonly by collecting soil gas from some fixed depth and analyzing the sample with a gas chromatograph. Direct push technologies for soil-gas sampling make it easy and relatively inexpensive to collect samples at a fixed depth across areas of interest. One of the benefits of push technologies is an absence of cuttings that occur with other drilling methods. However, this approach does not detect all organic contaminants because not all are volatile. Also, organic compounds that are too soluble can be problematic because vapors moving through the unsaturated zone will dissolve into any water present.

Volatile organic compounds are common organic contaminants encountered in field studies. Gasoline, for example, contains a number of different VOCs. Many of the most frequently encountered contaminants at former Superfund sites in the United States were VOCs including the BTEX compound and DNAPLs used as industrial solvents (e.g., trichloroethene and tetrachloroethene). Among the top 17 most frequently encountered contaminants at 546 Superfund Sites, ten were VOCs that can are amenable to measurements in soil gases (Table 20.2).

Sampling occurs from devices driven from 2 to 4 m into the unsaturated zone from which a small sample of soil gas is extracted by pumping. Gas samples are usually analyzed on-site with a gas chromatograph (GC). The GC measurement provides a quantitative estimate of the mass of a particular volatile compound per volume of soil gas (for example, μg/L). When plotted on a map, these data can be used to infer zones of contamination in the unsaturated and saturated

Table 20.2 VOCs included on the list most commonly detected organic and inorganic contaminants at 546 Superfund sites in the U.S.

Common constituents ≥15% of 546 sites	Less common constituents <15% of 546 sites
Trichloroethene	1,1,1-Trichloroethane
Toluene	Ethylbenzene
Benzene	Xylenes
Chloroform	Methylene chloride
Tetrachloroethene	*trans*-1,2-Dichloroethene
Phenol	

Source: U.S. Environmental Protection Agency, adapted from Kerfoot and Barrows (1986).

zones. By relating contaminant concentrations in the soil gas to measured concentrations in the groundwater, the soil gas data can be transformed to provide a quantitative estimate of concentrations in groundwater.

There are certain circumstances when these approaches may not work well. A low-permeability layer in the unsaturated zone can inhibit the upward diffusion of vapor and promote extensive horizontal spreading. This situation could produce an estimated distribution of contaminants that is much larger than the actual one. Other problems relate to how the contaminant occurs. With LNAPLs, vapor concentrations in the unsaturated zone will be larger and easier to detect than if the contaminant is dissolved in the groundwater (Reisinger et al., 1987). Furthermore, when contaminants are present in localized fracture zones (for example, in sandstone or limestone), the rates of diffusion away from the fractures may be so slow that vapor phase transport is limited in extent (Reisinger et al., 1987). Thus, care must be exercised in interpreting the results, and where necessary, conclusions should be confirmed using an independent approach.

Soil gas sampling provides a rapid and economical way of surveying large sites for contamination. Thus, it is attractive for reconnaissance studies aimed at discovering what volatile contaminants are present and where they are located. This information often assists in designing a conventional sampling program.

Wittman et al. (1985) presented a case study that demonstrated the potential of soil-gas sampling for delineating contaminated groundwater. The case considered the origin of contamination in the Verona Well Field at Battle Creek, Michigan. Conventional groundwater monitoring pointed to the presence of an additional but unidentified source. Soil gas surveying was utilized to explore a large rail yard. The sampling approach involved driving a metal probe approximately one meter into the soil and extracting the soil gas with a hand pump. Sample analyses were performed on-site using a gas chromatograph. Figure 20.12 shows PCE results for 43 samples. The survey detected three areas of elevated soil gas level in the rail yard. The most concentrated of the three was a small solvent disposal area that had gone undetected in previous investigations. Subsequently, two wells confirmed the PCE source. Overall, the approach worked well, except that it did not reveal the full extent of the PCE plume. Migrating away from the source, the plume moved deeper and lost touch with the water table. With this condition, volatiles were unable to partition into the soil gas.

20.7.4 Distribution of Dissolved Contaminants

The most definitive assessment of contaminant distributions requires the installation of monitoring wells and the collection and analysis of water samples. The associated drilling and piezometer installation also provide a wealth of other samples and data, concerned with stratigraphy, patterns of flow, etc. Because of uncertainties involved with plume definition, a phased approach is usually involved with ongoing interpretations and evaluations. Usually, monitoring continues over a period of at least several years and requires a number of sampling rounds. As discussed previously, a water analysis would typically include the major cation and anions (routine analysis) and packages of organic/and or metal analyses depending upon the type of contamination involved. Detailed information on the installation of piezometers and water table wells is presented in Chapter 7.

The information collected from the chemical monitoring provides the basis for understanding the distribution of contaminants in the subsurface. Knowledge concerning the behavior of plumes in the context of the physical hydrogeological setting is essential for creating strategies for remediation. In Chapter 7, we introduced the concept of contour plots for depicting

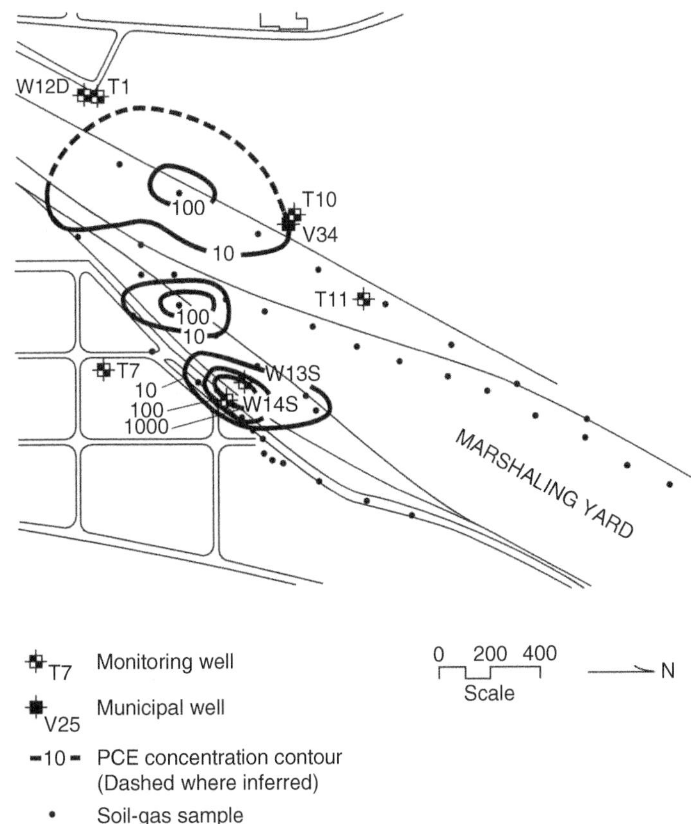

Figure 20.12 Results of a soil-gas survey (July 1984) for PCE in a rail yard in Battle Creek, Michigan (Adapted from Wittmann et al., 1985. Reproduced by permission of the National Groundwater Association. Copyright © 1985. All rights reserved).

patterns in the concentration for some constituent of interest. This earlier discussion, however, glossed over the reality that distributions of NAPLs in the source zone and/or plumes of dissolved contaminants are inherently three-dimensional features. Thus, comprehensive characterization requires three-dimensional monitoring which comes from the installation of *nests* of monitoring wells. In other words, at each location across the site selected for monitoring, there could a nest of three or four piezometers, each completed to a different depth with a relatively short screen.

Guidance documents (e.g., Lovelace, 2007) explain how elements of the design of the monitoring system function to answer key questions. A clear definition of contamination associated with any *source zones* is essential, especially in areas where LNAPLs and DNAPLs are present to serve as sources of dissolved contamination. Source zones likely contain the highest concentrations of dissolved contaminants and are commonly in remediation with removal or isolation. Another critical requirement with monitoring is to carefully identify the flow path down-gradient in the groundwater flow system This pathway is usually the midline of the plume, where local concentrations are highest, and which is tracking through most transmissive parts of the aquifer (Figure 20.13).

Another goal of monitoring is to establish the plume boundaries in the direction of flow and lateral to the direction of flow (Figure 20.13). This information is needed for regulators to determine whether other users of the groundwater are being impacted, other properties are affected, or whether groundwater is discharging to surface waters. Lovelace (2007) also suggested monitoring fringe areas of the plume (Figure 20.13) because concentrations in these places are sensitive indicators of continued expansion of the plume (concentrations ↑) or success with specific remedial strategies (concentrations ↓).

Upgradient wells or in this example a transect of multilevel wells are installed to characterize the background chemistry of uncontaminated groundwater at the site (Figure 20.13). On occasion, these up-gradient wells can be contaminated by other undetected sources of contamination. As shown in the figure, other piezometers may be installed away from the source area and plume to provide hydraulic head data to inform flow understanding over a larger area of the site (Figure 20.13).

Even with a relatively dense array of monitoring wells, visualizing and interpreting geochemical data can be problematic. Source areas can be extremely heterogeneous, leading to local variability in presence/absence of contaminants and their

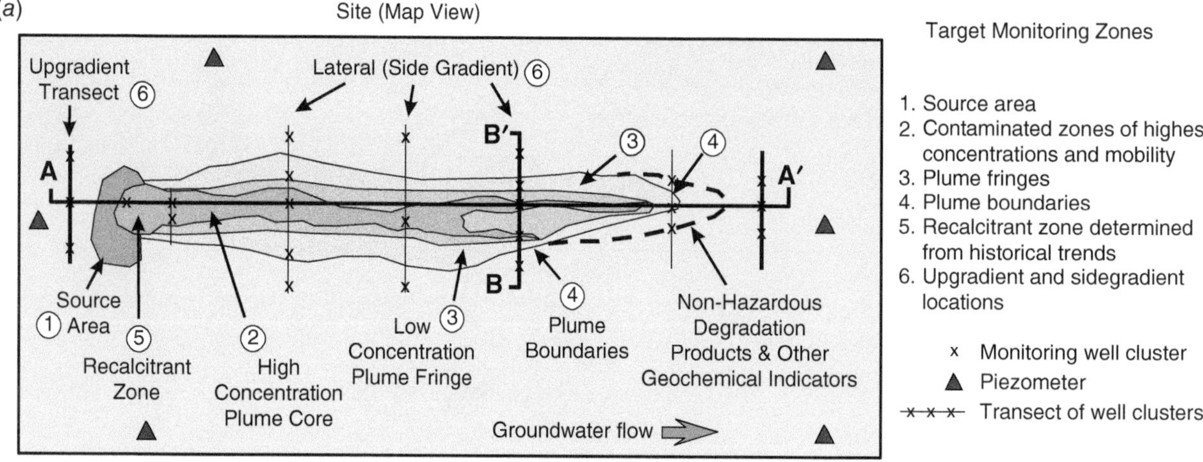

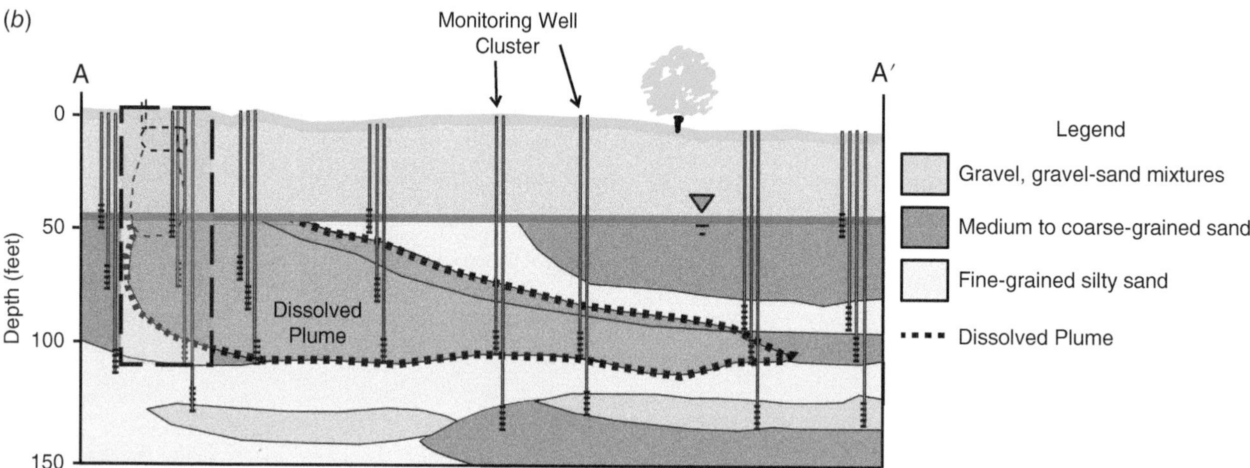

Figure 20.13 A comprehensive approach to monitoring subsurface contamination. Panel (*a*) provides a map of well locations to identify targets of critical interest, such as the source area, plume boundaries, and high-concentration plume core. Transects of cluster wells are installed so as to provide transects to illustrate the plume geometry with depth, such as the longitudinal transect A-A′ (Panel *b*) (U.S. Geological Survey Lovelace, 2007/Public domain).

concentrations. For example, a historical landfill may have received drums of different waste solvents or other contaminants from different companies across many years, which are sprinkled among other refuse. Similarly, redox conditions may change spatially depending upon the specific mix of urban refuse and construction wastes and temporally due to gradual degradation of organic matter in the landfill. Variability also comes along with a plume being transported down-gradient in an aquifer. Such variability can be associated with changes in the mineralogy of the porous medium, fluctuation fraction organic carbon, and locally driven patterns in redox zonation.

20.7.5 Plume Maps

The simplest approach to constructing a map is to produce a simple two-dimensional map-view, often called a *plume map*. It is prepared by plotting the location of monitoring wells on a site map along with the measured concentration for the contaminant of interest. When sampling wells are nested at the same location (that is piezometer nests), one would select the largest concentration value at the point of interest, irrespective of depth. Thus, a plume maps depict the worst case of contamination but provides no information on depths (Figure 20.14). There can be deeper and/or shallower zones with no contamination (Figure 20.14).

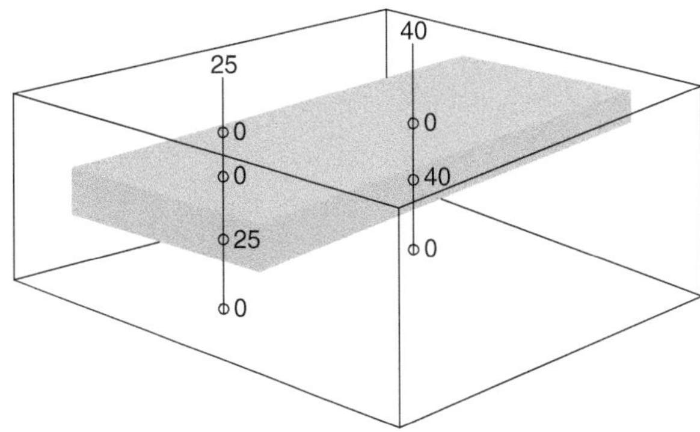

Figure 20.14 A plume map is constructed by plotting the maximum concentration observed at a monitoring site. In this example, values of 25 and 40 would be plotted on a location map showing where the nest of piezometers is located (Schwartz and Zhang, 2003. Fundamentals of Groundwater. Copyright 2003/John Wiley & Sons).

Given the inherent variability in the data, only four or five contour lines would be used to avoid obscuring the basic pattern with by too many, small, closed contours. This guideline implies no requirement for equal contour intervals. Maximum and minimum isoconcentration lines are designed to show the most contaminated zone on the site and the maximum extent of the plume, respectively.

If an aquifer is thick, an alternative is to use slice maps which show the contaminant plume within different depth zones. This approach provides some capability of visualizing the distribution of contaminants as a function of depth but without the complexity of fully three-dimensional renderings. Plume maps constructed for the Islip landfill, Long Island New York, (Kimmel and Braids, 1980) illustrate the application of slice maps to visualize a specific conductance plume. The Islip landfill began operation in 1933 in an old sand pit. Through the years, it extended to a maximum size of 6.9 hectares (17 acres). Because the landfill had no liner or surface seal, water infiltrated and reacted with the various industrial and urban wastes to contaminate the shallow unconfined aquifer at the site. Because the landfill leachate was slightly denser than the ambient groundwater, it sank downward, i.e., variable density flow, to contaminate the entire 170 ft of the aquifer. Slice maps for specific conductance turned out to be useful in visualizing how the leachate (represented by specific conductance) spread both laterally and vertically (Figure 20.15). In this case, the plume is "boot"-shaped in three dimensions. Most spreading is evident in Figure 20.15*d* at a depth of 94–114 ft below the water table. The largest specific conductance values (that is, greatest concentrations of contaminants) are found near the base of the aquifer (see Figure 20.15*d*).

Another way of presenting contamination data is to plot concentration data along cross-sections. Commonly, such sections are oriented along the mid-line of the plume or transverse to the midline. An example of this approach is shown without contours in Figure 20.13.

20.7.6 Mapping the Distribution of NAPLs

It is difficult to measure how much LNAPL is present near the water table. Standard practice is to install a water table observation well (screened for some distance above and below the water table) and to measure the thickness of the free product in the well (Figure 20.16) using an *interface probe*. When the probe is lowered down a well, it beeps at the oil–air interface to provide the depth to the top of the oil in the well and beeps again at the oil–water interface. The thickness, as the difference in these two measured depths, provides the *apparent thickness* of the free product. The measurement is crude because it really cannot establish the complexities of LNAPL saturation near the water table due to variability in saturation (Figure 20.16). Experience shows that problems are probably not as bad as indicated by measurements of free product measurements in observation wells.

When an NAPL is present at a site, usually its abundance is described using an isopach map that shows the thickness and lateral extent. These kinds of maps are most presented for LNAPLs because the distribution of LNAPLs is much more "regular" in comparison to that of DNAPLs that can be spread extensively through an aquifer. Making a map of NAPLs is straightforward. The measured thickness of LNAPL is plotted on a map of the site next to the location of each well. Lines of equal NAPL thickness are contoured to provide the desired map. Recall that the thickness of LNAPL shown on such a map is an *apparent thickness*, which is greater than the actual NAPL thickness. Figure 20.17*a* is an example of an isopach map showing the distribution of mobile LNAPL on-site (Wiedemeier et al., 1996). It is spreading in the southwesterly direction away from the source area. Soluble BTEX compounds have created an associated plume (Figure 20.17*b*).

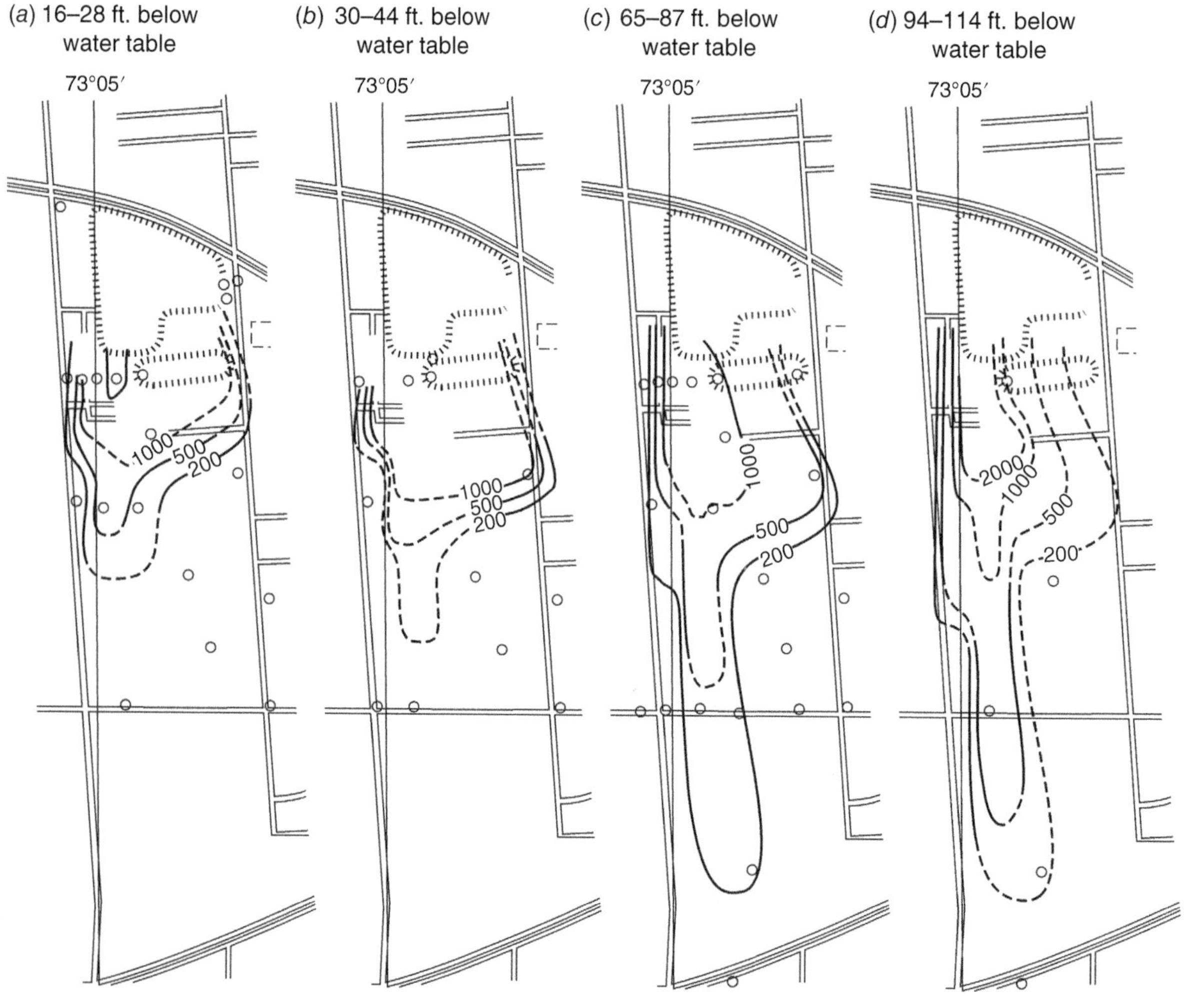

(*a*) 16–28 ft. below water table

(*b*) 30–44 ft. below water table

(*c*) 65–87 ft. below water table

(*d*) 94–114 ft. below water table

73°05′

Figure 20.15 Series of slice maps through the plume at Islip, New York, beginning close to the water table, (*a*) (16–28 ft. below the water table) to (*d*) (94–114 ft.). Specific conductance values were highest at depth (*d*) suggesting density driving forces also contributed to vertical spreading. (U.S. Geological Survey Kimmel and Braids, 1980/Public Domain).

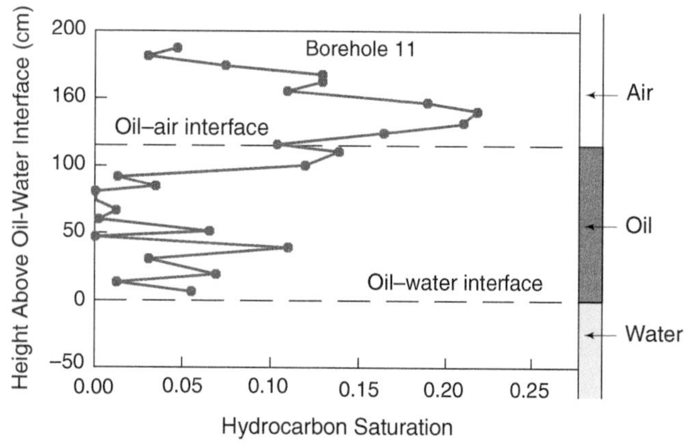

Figure 20.16 Measured hydrocarbon saturation profile compared to the LNAP distribution obtained from an observation well (Huntley et al., 1992. Reproduced by permission of the National Groundwater Association. Copyright © 1992. All rights reserved).

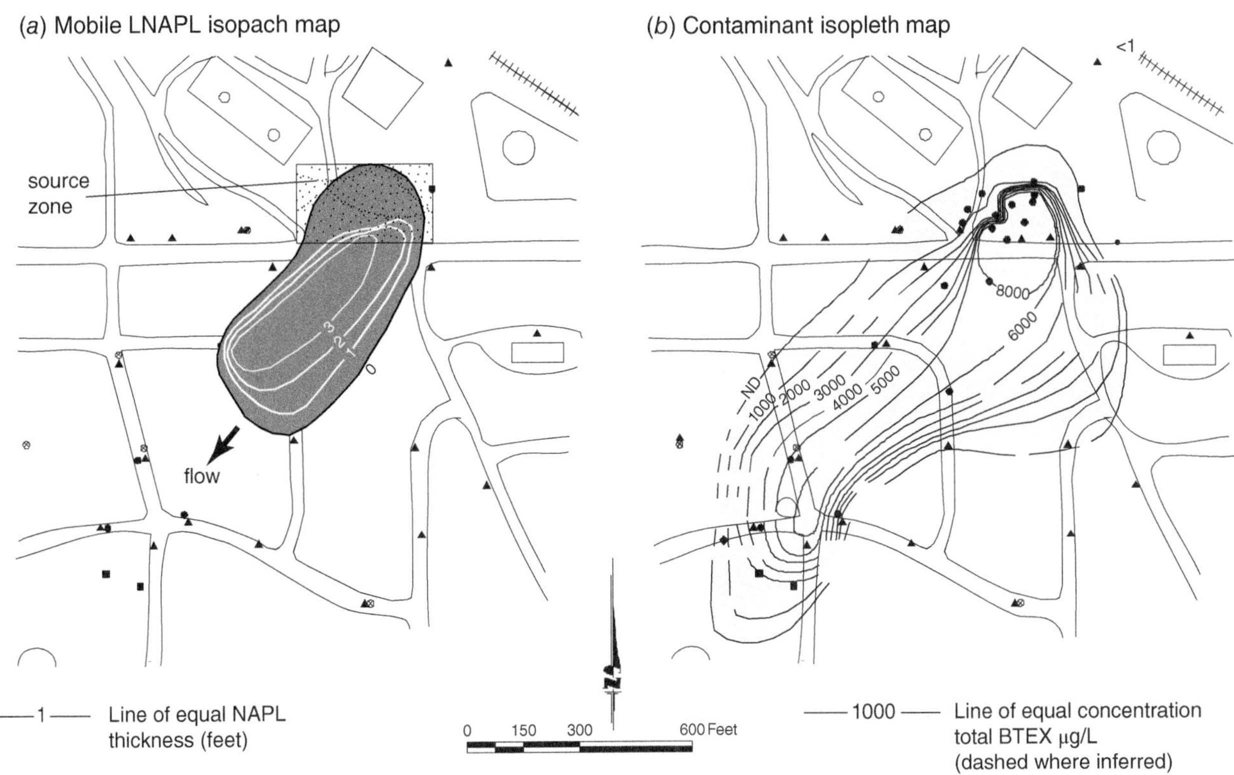

Figure 20.17 Information on apparent NAPL thickness is presented here as a contour map. At the source, the maximum thickness was 4 ft with mobile NAPL flowing to the southwest (*a*). An associated plume of total BETX is shown spreading southwest in the direction of groundwater flow (*b*) (U.S. Environmental Protection Agency, adapted from Wiedemeier et al., 1998).

20.8 Field Example of an LNAPL Problem

It is a straightforward exercise to create a plume map of contaminants, such as *BTEX compounds* (i.e., benzene, toluene, ethylbenzene, and xylene) or a solvent such as TCE. However, such maps by themselves provide little information as to whether biodegradation reactions are important in reducing contaminant concentrations. In some settings, natural degradation processes may be all that is needed to affect site remediation as compared to other remedial strategies. Developing arguments favoring so-called natural attenuation requires specific evidence that redox reactions are operative. For example, the loss of organic contaminants (such as the BTEX compounds) via oxidation would need to be accompanied by a decline in concentration of electron acceptors, such as oxygen, $Fe(OH)_3(s)$, nitrate, and sulfate. (Table 20.3). However, it is sometimes more convenient to measure the increasing concentration of expected reactants. By carefully monitoring indicator products

Table 20.3 Behavior in the concentration in indicator reactants and products during biodegradation.

Redox reaction	Indicator reactants ↓	Indicator products ↑
Oxygen reduction	BETX, $O_2(aq)$	—
Fe(III) reduction—$Fe(OH)_3(s)$	BETX	Fe^{2+}
Denitrification	BETX, NO_3^-	—
Sulfate reduction	BETX, SO_4^{2-}	H_2S
Methanogenesis	BETX	CH_4
Reductive dehalogenation—TCE	TCE	DCE, Cl^-
Reductive dehalogenation—DCE	DCE	Vinyl chloride, Cl^-

Source: Adapted from Wiedemeier et al. (1998).

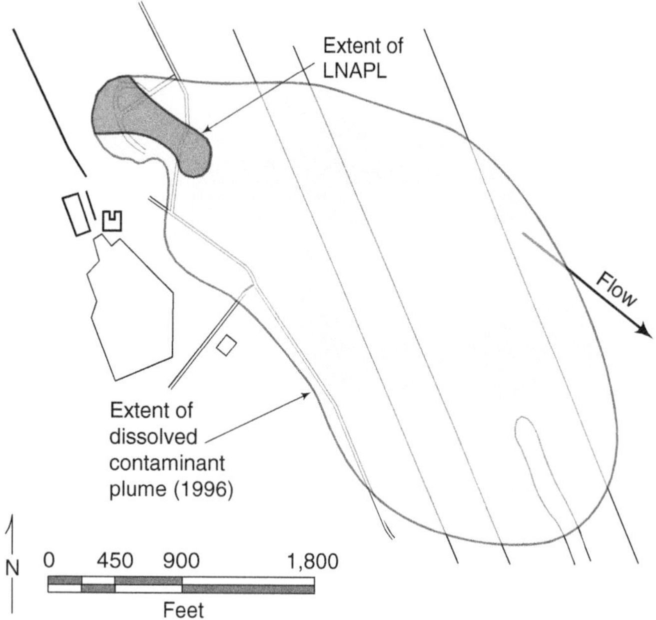

Figure 20.18 Map of the former site at Plattsburgh Air Force Base, New York. There is a small zone where LNAPL is present. Plumes of contaminants and associated species occur downstream away from the source (Air Force Center for Environmental Excellence, adapted from Wiedemeier et al., 1996).

and reactants, in addition to the actual contaminants (e.g., BTEX or TCE), specific biodegradation processes can be identified (Table 20.3).

Wiedemeier et al. (1996) illustrated this approach in defining the complex pattern of chemical changes that accompany the migration of contaminants away a from a training area (Site FT-002) at Plattsburgh Air Force Base (AFB). The base is located 167 miles north of Albany, New York. The contaminants at this site were a unique mixture of fuel (i.e., BTEX compounds) and chlorinated aliphatics (i.e., TCE, DCE, and vinyl chloride). Contamination was caused as jet fuel containing PCE and TCE was spilled during the training exercises. The source area (Figure 20.18) had LNAPL floating on the water table (Wiedemeier et al., 1996).

Plumes associated with the source were a unique collection of dissolved constituents. As will become clear, the BTEX plume was of limited extent, with chlorinated aliphatics distributed over a much larger area. This detailed characterization sought to confirm that biodegradation was controlling plume spreading.

The aquifer is shallow, unconfined, and approximately 50 ft thick. It comprises well-sorted fine-to-medium sand with a mean hydraulic conductivity of 11.6 ft/day. The linear groundwater velocity in this unit is about 0.39 ft/day.

Contaminants present in the LNAPL have dissolved over time to create various plumes. There is a BETX plume that extends for about 2000 ft down-gradient from the source (Figure 20.19*a*). In the source area, BTEX compounds had concentrations of approximately 17 mg/L (Wiedemeier et al., 1996). If biodegradation reactions involving BETX compounds were operative, there should have been evident depletion in specific indicator reactants or increases in associated indicator products (Table 20.3). Concentrations of indicator reactants, oxygen, nitrate, and sulfate associated with the BETX plume were below background values (Figure 20.19*b*,*d*,*e*). Indicator products, Fe^{2+}, and methane (Table 20.3) were present at concentrations above background (Figure 20.19*c*,*f*). Taken together, these results indicated biodegradation reactions occurred and involved all key redox reactions—oxygen reduction to methanogenesis (Table 20.3) The BTEX plume was shown to be at steady state, no longer spreading at that time.

The TCE plume (not shown) was generally larger than the BTEX plume and was also at steady state. Wiedemeier et al. (1996) used a similar approach to establish that reductive dichlorination reactions were also operating at the site. The expected metabolic products with reductive dichlorination, such as dichloroethene, vinyl chloride, and Cl^-, were mapped in association with the other plumes.

Interestingly, nearly all the organic compounds present in the system are being biodegraded. TCE is biotransformed through the process of reductive dehalogenation with the formation of DCE, vinyl chloride (VC), and ethene as Cl atoms are replaced with H. Thus, even though DCE, VC, ethene, and Cl^- are not present at the source as original contaminants, plumes develop as these compounds form from the biotransformation of TCE. Once the BTEX compounds are effectively removed via biodegradation (at about 1500–2000 ft down-gradient from the source), the reductive dehalogenation reactions cease as well.

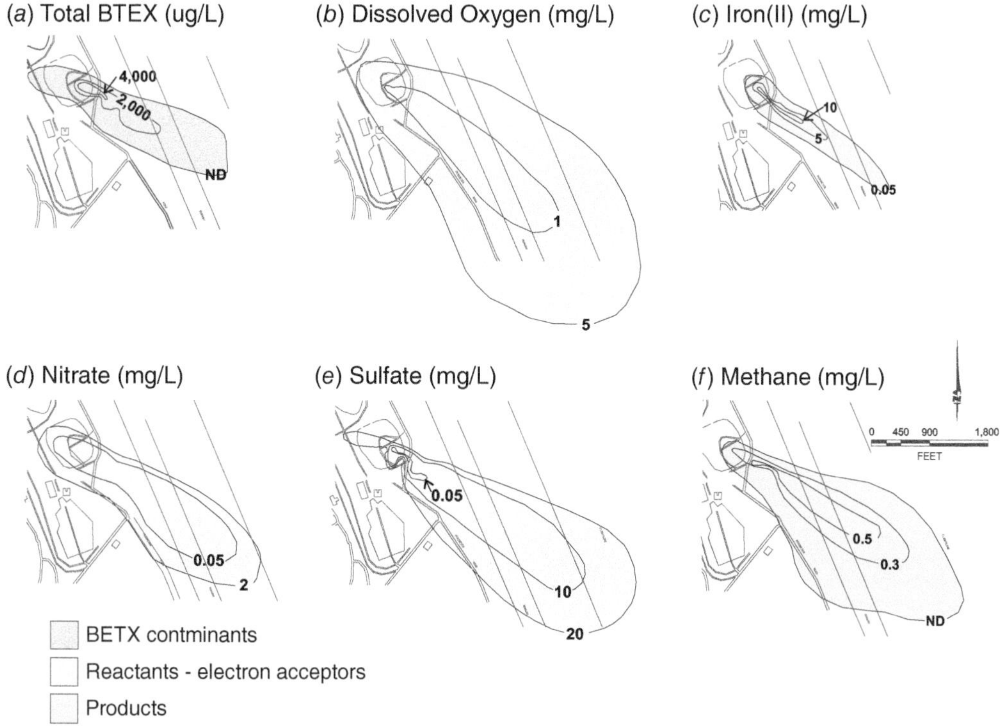

Figure 20.19 Plume maps showing the distribution of dissolved BTEX compounds (*a*), electron acceptors (reactants) involved in the biodegradation reactions (*b*, *d*, *e*), and products (*c*, *f*) (Adapted from U.S. Environmental Protection Agency Wiedemeier et al., 1998).

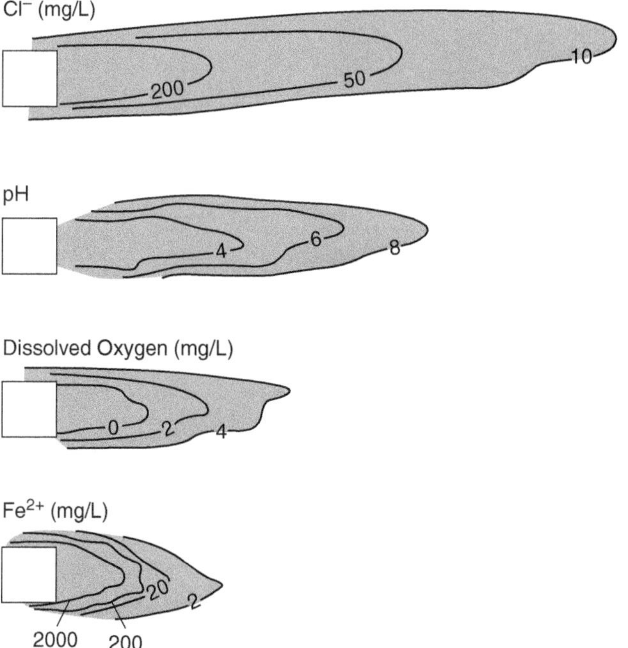

Figure 20.20 Plume maps for several groundwater constituents associated with a hypothetical leaking landfill (Domenico and Schwartz, 1998, Physical and Chemical Hydrogeology. Copyright © 1990, 1998 John Wiley & Sons, Inc. All Rights Reserved. Reproduced with permission).

The mineralization of BTEX compounds requires an electron acceptor. The concentration distributions of oxygen, nitrate, and sulfate (Figure 20.20) as compared to that of BTEX indicates that they function as electron acceptors. This case study shows that by careful study of contaminants, their breakdown products, and natural constituents, one can interpret likely patterns of chemical interaction.

Exercises

20.1 Shown on Figure 20.20 is a series of plumes from a sanitary landfill. Examine these plumes in detail and answer the following questions.

A Qualitatively evaluate the extent to which advection and dispersion are important in controlling contaminant spread at the site.
B Given the type of source and the resulting plume shapes, what can you say about the type of source loading?
C Suggest what processes could be operating to cause the increasing pH away from the source.
D Metals, for example, Fe, tend to be relatively abundant in landfill leachates. However, Fe^{2+} is strongly attenuated relative to mobile species like Cl^-. Explain why iron behaves in this way.

20.2 On cross-sections (a) and (b) shown in Figure 20.21, the contaminant distribution for the organic compounds with the specified chemical properties noted below is illustrated. Consider all important spreading mechanisms.

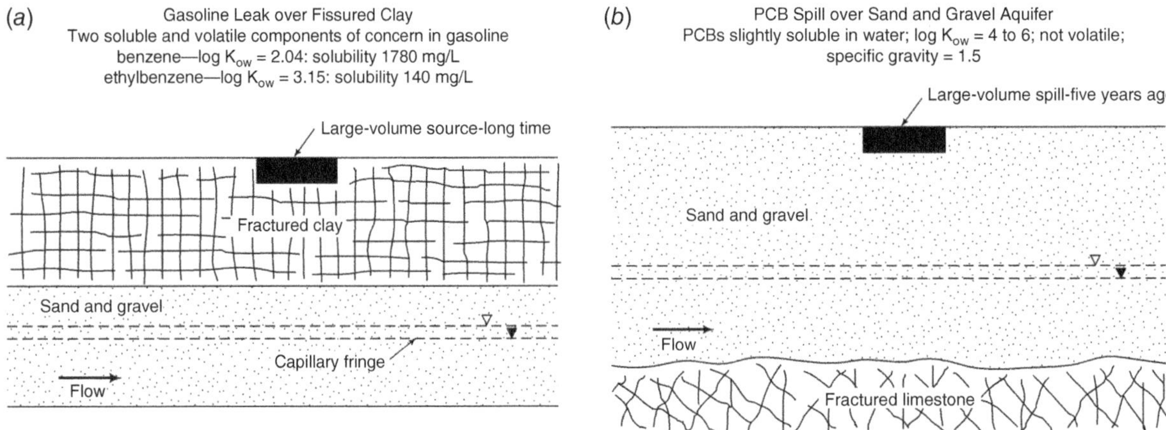

Figure 20.21 Two typical contamination scenarios; (*a*) a gasoline spill over fractured clay, and (*s*) a PCB spill into a sand and gravel aquifer (Domenico and Schwartz, 1998, Physical and Chemical Hydrogeology. Copyright © 1990, 1998 John Wiley & Sons, Inc. All Rights Reserved. Reproduced with permission).

20.3 This problem involves groundwater contamination associated with the sewage disposal system that formerly operated at Otis Air Base on Cape Cod, Massachusetts. The location of sampling wells is depicted in Figure 20.22. Some of the chemical information relating to these wells (1983) is tabulated in Table 20.4.

A At Otis, there are many different contaminants present. Creating a plume map of elevated specific conductance helps define generally where the groundwater contamination is located. Given the information on Table 20.4, construct the specific conductance plume map.
B For some of the wells, organic contamination represented as the total concentration of volatile compounds has been analyzed. Construct a plume map illustrating the distribution of VOCs.
C Explain some of the reasons why the VOC plume is different than that for specific conductance.

20.4 At some of the monitoring sites in Qu. 3, wells are completed to various elevations. Use these data to construct a vertical cross-section of the specific conductance plume. With the help of the cross-section, describe how thick the plume is vertically, and whether it changes with depth moving away from the sand beds.

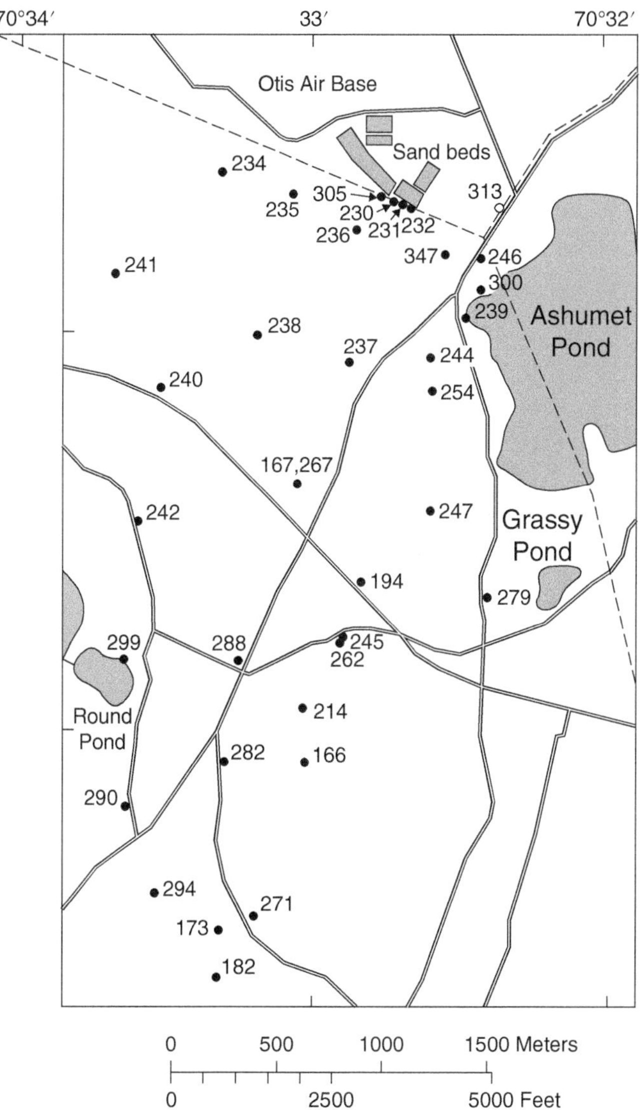

Figure 20.22 At the former Otis Air Base on Cape Cod, Massachusetts sewage tainted with chlorinated solvents and other organic contaminants was disposed of in a surface infiltration system, noted as "sand beds" on the map. The site and downgradient areas were instrumented with various groundwater sampling wells at the numbered locations on the map (U.S. Geological Survey Thurman et al., 1984/Public Domain).

Table 20.4 Summary of some chemical data for groundwater at the former Otis Air Base.

Well no.	Depth (ft)	Specific cond.[a]	Total VOCs[b]	Well no.	Depth (ft)	Specific cond.[a]	Total VOCs[b]
166	67	100	0.5	254	54	220	8.1
167	55	48	na	254	26	70	nd
173	69	122	na	262	159	125	na
182	69	80	nd	262	85	255	46.7
194	57	145	1.1	262	69	200	3.7
214	60	83	na	262	41	90	na
230	48	115	na	267	155	122	na
231	57	145	na	267	136	95	na
232	58	153	0.8	267	111	120	na
234	99	122	na	267	88	195	9.6
235	94	81	na	271	165	125	nd

(Continued)

Table 20.4 (Continued)

Well no.	Depth (ft)	Specific cond.[a]	Total VOCs[b]	Well no.	Depth (ft)	Specific cond.[a]	Total VOCs[b]
236	106	126	1	271	141	132	nd
237	88	128	0.9	271	85	150	nd
238	106	95	na	271	41	55	nd
239	64	190	268.4	279	86	76	na
240	95	57	na	279	61	73	na
241	98	62	na	282	123	143	nd
242	77	51	nd	282	94	208	5.9
244	90	230	390.6	282	70	215	nd
245	25	200	nd	282	49	100	na
246	35	152	na	288	97	142	30.8
247	70	120	na	290	91	90	na
254	246	59	1	294	89	139	0.5
254	168	115	1.3	299	20	76	nd
254	140	175	4.5	300	30	410	3.9
254	107	235	93.7	300	10	38	na

[a] Specific conductance in μmhos/cm.
[b] Total VOVs μg/L.
na—not analyzed; nd—not detected.
Source: U.S. Geological Survey Thurman et al. (1984)/Public Domain.

References

Belitz, K., N. M. Dubrovsky, K. Burow, et al. 2003. Framework for a groundwater quality monitoring and assessment program for California. U.S. Geological Survey Water-Resources Investigations Report 03-4166, 78 p.

Borchardt, M. A., N. L. Haas, and R. J. Hunt. 2004. Vulnerability of drinking-water wells in La Crosse, Wisconsin, to enteric-virus contamination from surface water contributions. Applied and Environmental Microbiology, v. 70, no. 10, p. 5937–5946.

Chapelle, F. H. 1993. Groundwater Microbiology and Geochemistry. John Wiley & Sons, Inc., New York, 424 p.

Coyte, R. M., R. C. Jain, S. K. Srivastava et al. 2018. Large-scale uranium contamination of groundwater resources in India. Environmental Science and Technology Letters, v. 5, no. 6, p. 341–347.

Cozzarelli, I. M., R. P. Eganhouse, and M. J. Baedecker. 1989. The fate and effects of crude oil in a shallow aquifer II. Evidence of anaerobic degradation of monoaromatic hydrocarbons. U.S. Geological Survey Water-Resources Investigations Report 88-4220, p. 21–33.

Craun, G. F. 1985. Epidemiologic studies of organic micropollutants in drinking water. Science of The Total Environment, v. 47, no. December, p. 461–472.

Domenico, P. A., and F. W. Schwartz. 1998. Physical and Chemical Hydrogeology. John Wiley and Sons, New York, 506 p.

Ford, R. G., R. T. Wilkin, and R. W. Puls. 2007. Monitored Natural Attenuation of Inorganic Contaminants in Groundwater. National Risk Management Research Laboratory, U.S. Environmental Protection Agency.

Galloway, D., and F. S. Riley. 1999. San Joaquin Valley, California, In D. Galloway, D. R. Jones, and S. E. Ingebritsen (eds.), Land subsidence in the United States, U.S. Geological Survey Circular 1182. U.S. Geological Survey, Reston, p. 23–34.

Grandjean, P., and P. J. Landrigan. 2014. Neurobehavioural effects of developmental toxicity. The Lancet Neurology, v. 13, no. 3, p. 330–338.

Gross, E. L. 2017. Evaluation of radon occurrence in groundwater from 16 geologic units in Pennsylvania, 1986–2015, with application to potential radon exposure from groundwater and indoor air. U.S. Geological Survey Scientific Investigations Report 2017-5018, 24 p.

Huntley, D., R. N. Hawk, and H. P. Corley. 1992. Non-aqueous phase hydrocarbon saturations and mobility in a fine-grained, poorly consolidated sandstone, In Proceedings of the 1992 Petroleum Hydrocarbons and Organic Chemicals in Groundwater: Prevention, Detection, and Restoration. National Groundwater Association, p. 223–237.

Jackson, R. E., R. J. Patterson, B. W. Graham, et al. 1985. Contaminant Hydrogeology of Toxic Organic Chemicals at a Disposal Site, Gloucester, Ontario: 1. Chemical Concepts and Site Assessment. Environment Canada. National Hydrology Research Institute Paper No. 23, Ottawa, 114 p.

Jordan, T. E., and D. J. Costantini. 1995. The use of non-invasive electromagnetic (EM) techniques for focusing environmental investigations. The Professional Geologist (June 1995), p. 4–9.

Kerfoot, H. B., and L. J. Barrows. 1986. Soil Gas Measurement for Detection of Subsurface Organic Contamination. U.S. Environmental Protection Agency, Las Vegas, Nev., 2 p.

Kimmel, G. E., and O. C. Braids. 1980. Leachate plumes in groundwater from Babylon and Islip landfills, Long Island, New York. U.S. Geological Survey Professional Paper 1085, 38 p.

Lindsey, B. D., C. A. Loper, and R. A. Hainly. 1997. Nitrate in groundwater and stream base flow in the lower Susquehanna River basin, Pennsylvania and Maryland. U.S. Geological Survey Water-Resources Investigations Report 97-4146, 56 p.

Lovelace, J. K. 2007. Chloride concentrations in groundwater in East and West Baton Rouge Parishes, Louisiana, 2004–05. U.S. Geological Survey Scientific Investigations Report 2007-5069, 27 p.

Lyngkilde, J., and T. H. Christensen. 1992. Fate of organic contaminants in the redox zones of a landfill leachate pollution plume (Vejen, Denmark). Journal of Contaminant Hydrology, v. 10, no. 4, p. 291–307.

Marrin, D. L., and H. B. Kerfoot. 1988. Soil-gas surveying techniques. Environmental Science & Technology, v. 22, no. 7, p. 740–745.

Mendoza, C. A., and T. A. McAlary. 1990. Modeling of groundwater contamination caused by organic solvent vapors. Groundwater, v. 28, no. 2, p. 199–206.

Morales, K. H., L. Ryan, T. Kuo et al. 2000. Risk of internal cancers from arsenic in drinking water. Environmental Health Perspectives, v. 108, no. 7, p. 655–661.

NRC (National Research Council). 1993. In Situ Bioremediation, When Does It Work? The National Academies Press, Washington, D.C., 207 p.

NRC (National Research Council). 2006. Fluoride in Drinking Water: A Scientific Review of EPA's Standards. The National Academies Press, Washington, DC, 508 p.

Nielsen, P. H., and T. H. Christensen. 1994a. Variability of biological degradation of aromatic hydrocarbons in an aerobic aquifer determined by laboratory batch experiments. Journal of Contaminant Hydrology, v. 15, no. 4, p. 305–320.

Nielsen, P. H., and T. H. Christensen. 1994b. Variability of biological degradation of phenolic hydrocarbons in an aerobic aquifer determined by laboratory batch experiments. Journal of Contaminant Hydrology, v. 17, no. 1, p. 55–67.

Pichler, G., M. Grau-Perez, and M. Tellez-Plaza. 2019. Association of arsenic exposure with cardiac geometry and left ventricular function in young adults: evidence from the Strong Heart Family Study. Circulation: Cardiovascular Imaging, v. 12, no. 5, 5 p.

Rasa, E., B. A. Bekins, D. M. Mackay et al. 2013. Impacts of an ethanol-blended fuel release on groundwater and fate of produced methane: simulation of field observations. Water Resources Research, v. 49, no. 8, p. 4907–4926.

Reisinger, H. J., D. R. Burris, L. R. Cessar et al. 1987. Factors affecting the utility of soil vapor assessment data, In Proceedings of the First Outdoor Action Conference on Aquifer Restoration, Groundwater Monitoring and Geophysical Methods. National Water Well Association, p. 425–435.

Schwartz, F. W., and H. Zhang. 2003. Fundamentals of Groundwater. John Wiley & Sons, 583 p.

Shelton, J.L., Fram, M.S., Belitz, K. et al. (2013). Status and understanding of groundwater quality in the Madera-Chowchilla study unit, 2008: California GAMA Priority Basin Project. U.S. Geological Survey Scientific Investigations Report 2012–5094, 86 p.

Smolenski, W. J., and J. M. Suflita. 1987. Biodegradation of cresol isomers in anoxic aquifers. Applied and Environmental Microbiology, v. 53, no. 4, p. 710–716.

Squillace, P. J., J. S. Zogorski, W. G. Wilber et al. 1996. Preliminary assessment of the occurrence and possible sources of MTBE in groundwater in the United States, 1993–1994. Environmental Science and Technology, v. 30, no. 5, p. 1721–1730.

Thurman, E. M., L. B. Barber Jr., M. L. Cezan et al. 1984. Sewage contaminants in groundwater, In D. LeBlanc (ed.), Movement and Fate of solutes in a Plume of Sewage-contaminated Groundwater, Cape Cod, Massachusetts, U.S. Geological Survey Open File Report 84-475143-161.

Tyler, C. R., and A. M. Allan. 2014. The effects of arsenic exposure on neurological and cognitive dysfunction in human and rodent studies: a review. Current Environmental Health Reports, v. 1, no. 2, p. 132–147.

Walter, D. A. 1997. Geochemistry and microbiology of iron-related well-screen encrustation and aquifer biofouling in Suffolk County, Long Island, New York. U.S Geological Survey, Water-Resources Investigations Report 97-4032, 37 p.

Ward, M. H., R. R. Jones, J. D. Brender et al. 2018. Drinking water nitrate and human health: an updated review. International Journal of Environmental Research and Public Health, v. 15, no. 7, p. 1557.

Wiedemeier, T. H., M. A. Swanson, D. E. Moutoux, et al. 1998. Technical protocol for evaluating natural attenuation of chlorinated solvents in groundwater. U.S. Environmental Protection Agency EPA/600/R-98/128.

Wiedemeier, T. H., M. A. Swanson, D. E. Moutoux, et al. 1996. Technical protocol for evaluating natural attenuation of chlorinated solvents in groundwater. Air Force Center for Environmental Excellence, San Antonio Texas, 396 p.

Wittmann, S. G., K. J. Quinn, and R. D. Lee. 1985. Use of soil gas sampling techniques for assessment of groundwater contamination., Proceedings of the NWWA/API Second Conference (Houston, USA) on Petroleum Hydrocarbons and Organic Chemicals in Groundwater – Prevention, Detection and Restoration. National Water Well Association, p. 291–310.

Index

Fundamentals of Groundwater, Second Edition. Franklin W. Schwartz and Hubao Zhang.

Companion website: www.wiley.com/go/schwartz/fundamentalsofgroundwater2

d

e

f